introduction to PARASITOLOGY

10th edition

the late

ASA C. CHANDLER, M.S., Ph.D.,

Professor of Biology, Rice Institute, Houston, Texas
Former Officer-in-Charge, Hookworm Research Laboratory,
School of Tropical Medicine and Hygiene, Calcutta, India

CLARK P. READ, M.A., Ph.D.,

Professor of Biology, Rice Institute, Houston, Texas
Former Associate Professor of Parasitology,
Johns Hopkins University School of Hygiene and
Public Health, Baltimore, Maryland

JOHN WILEY & SONS, INC., NEW YORK · LONDON · SYDNEY

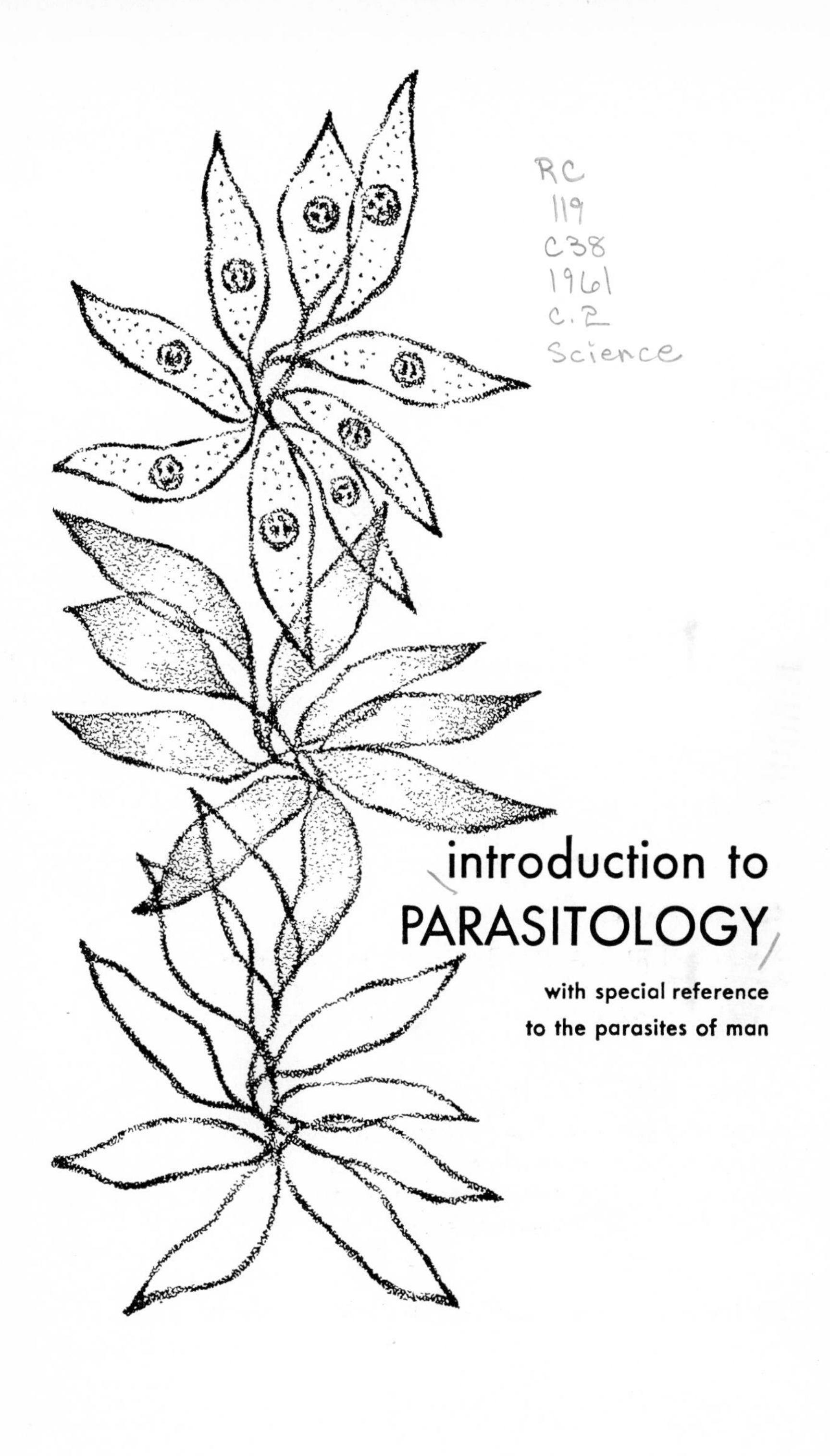

introduction to PARASITOLOGY

with special reference
to the parasites of man

LIBRARY OF CONGRESS CATALOG CARD NUMBER: 61–5670

PRINTED IN THE UNITED STATES OF AMERICA

TENTH EDITION

15 14 13 12 11 10

ISBN 0 471 14487 8

PREFACE

In 1918 the senior author prepared a book on *Animal Parasites and Human Disease* designed to set forth interesting and important facts of human parasitology in a readable form that would make them available to a wide range of intellectually curious readers. Although not so widely taken up by the general public as anticipated, this book was at once accepted as an introductory textbook in parasitology, and year by year was adopted by more and more teachers, colleges, universities, and medical schools throughout the country. With the fourth edition, in 1930, the book was entirely rewritten, rearranged to serve its function as a textbook more efficiently, and presented under a new title, *Introduction to Human Parasitology.* The book was, however, widely used as a *general* introductory textbook, so with succeeding editions its scope has been broadened to include more and more references to, or discussion of, parasites of lower animals, particularly those of importance in veterinary medicine. To reflect this extension the title was again changed in the sixth (1940) edition to its present form. The parasites of man are still most fully considered and are used as examples of their respective systematic groups, but all the parasites of veterinary importance are at least mentioned, and many of them are discussed. It would obviously be impossible to make detailed reference to parasites of other animals in an introductory textbook, but general statements are made concerning the occurrence of representatives of groups of parasites in various types of hosts. For completeness, such groups as the monogenetic flukes, strigeids, Cestodaria, etc., are discussed in this edition.

When *Animal Parasites and Human Disease* was first published, parasitology was taught in only a few universities, but there was a steady, gradual increase in the attention given to the subject up to about the beginning of World War II. It is hoped that this book may have played some part in the development of this gradually increasing popularity by stimulating the interest of students and by making easier the task of the teacher. During World War II and for several years thereafter, there was a very sharp upturn in interest in parasitology, due to a belated realization of the importance of the subject as a factor in world health and in the welfare of military expeditions. With the advent of World War II, parasitology took its rightful place of prominence in the community of sciences and came of age in America. During the war parasitological problems all over the world presented themselves for immediate solution, and the neglect with which parasitology had been treated in the past became painfully apparent. There were distressingly few individuals who had had experience with even such common parasitic diseases as malaria or amebiasis, not to mention schistosomiasis, leishmaniasis, scrub typhus, etc., which few had ever even heard of. Our military forces performed a veritable miracle in correcting the situation. Not only were thousands of people trained for the efficient application of what was already known, but research in parasitology flourished as never before.

Unfortunately, in the last few years interest in parasitology in the United States, particularly unfortunately in medical schools, has tended to fall back to its inadequate prewar status. We have tended to demobilize in our fight against parasites, just as we prematurely demobilized militarily immediately after the war. We have realized our error in the latter instance, but have not yet realized it in the former; however, unless we do something about our neglect of research and education in parasitology we shall inevitably regret it, and perhaps much sooner than we think. The reasons for this are outlined in Chapter 1 (Introduction) and need not be repeated here. It is enough to say that even though parasitic diseases at present are of relatively minor importance within the boundaries of the United States, they are still of vast importance to us, and we are very shortsighted in neglecting them as we are now tending to do.

The rapid advances in knowledge in the field of parasitology, which have made it necessary to revise and largely rewrite this book every four to six years since it first appeared in 1918, have continued. In the years that have elapsed since the ninth edition was written, continual advances have been made in the knowledge of the epidemiology and control of parasitic diseases; thus the book has been extensively revised again. A large number of sections have been largely rewritten. As far as possible, new and up-to-date systems of classification have been adopted throughout. Many of the illustrations which were not new in the preceding edition have been improved or are entirely new in this edition.

The chapter on spirochetes, found in earlier editions, has been deleted. These organisms are briefly treated with the other bacteria, the rickettsias, and the viruses in a chapter dealing with arthropod-borne nonprotozoans.

Only enough classification and taxonomy are included to give the student an understanding of the general relationships of the parasites considered. Outlines of classification of major groups and a number of simple keys to important groups of genera and species of arthropods have been set in small type so that they do not interfere with the readability of the text and can be omitted if not considered necessary. Most students, however, will benefit from a little experience in the use of keys for identification.

Discussions of correct scientific names and synonymy have been mostly omitted as inappropriate in an introductory textbook. An effort has been made to use scientfic names that are most generally accepted as correct. Some of the names used, e.g., *Dibothriocephalus latus* and *Schizotrypanum,* have not yet been accepted by the majority of North American authors, although the authors feel that eventually they will be. In such cases the instructor can, of course, have his students employ the more widely used names if he wishes; no harm will have been done by calling attention to the fact that there *are* differences of opinion. Names that have long been in common use, although not now accepted as correct under rules of zoological nomenclature, are given in parentheses.

Throughout the book special emphasis has been laid on the biological aspects of the subject. Considerable space is devoted to life cycles, epidemiological factors, interrelations of parasite and host, and underlying principles of treatment and prevention, rather than to such phases as classification, nomenclature, and morphology. This book, as an introductory one, is more concerned with fundamental principles than with the details that would interest a specialist. Clinical features of the diseases caused by the parasites are not dealt with sufficiently to satisfy medical students; these are left for the professor to fill in to the extent he desires, but the underlying reasons for the pathologic effects are adequately discussed. Some therapeutic details are also omitted, although the availability of effective drugs, their mechanism of action, reasons for failure, effects on the host, etc., are considered.

Parasitology has grown so rapidly in recent years and covers such a wide field that it is difficult to go very far into the subject within the limits of one book. Nevertheless it is believed that a comprehensive, integrated account of the entire field is much the most desirable method of approaching the subject at the start. Protozoology, helminthology, and medical entomology have many interrelations, and no one of them can be satisfactorily pursued very far without some knowledge of the others. For more advanced work a comprehensive textbook is too cumbersome; the subject naturally splits into its three component parts.

A brief list of references is provided at the end of each chapter for the student who wishes to pursue the subject further. Included are books or papers that give extensive reviews or summarizations of the subjects with which they deal or which contain good bibliographies; also included are a few of the more recent contributions of importance which would not be found in bibliographies of the other works cited, and which contain information beyond that cited in the present book. In the text, references that are included in the bibliographies have the date cited in parentheses; other references are usually made in the form "Smith in 1948 . . . ." It should not be too difficult for a student to trace down most of these references, if he wishes, through such journals as *Biological Abstracts, Helminthological Abstracts, Tropical Diseases Bulletin, Review of Applied Entomology, Index Medicus, Veterinary Bulletin,* etc.

In "Sources of Information" at the end of the book is a list of the leading journals where important articles on parasitology frequently appear. Particular attention is called to the periodicals mentioned in the preceding paragraph. The *Tropical Diseases Bulletin* reviews practically all current work in the field of human parasitology, especially protozoology and helminthology. The *Review of Applied Entomology,* Series B, contains abstracts of all important contributions in the field of medical and veterinary entomology. The *Veterinary Bulletin* reviews important work on diseases of domestic animals. *Biological Abstracts* contains abstracts of interest in parasitology in its sections on parasitology, sanitary entomology, and in appropriate subsections under systematic zoology. The *Index Medicus* and *Quarterly Cumulative Index Medicus* list references to nearly all writings of medical interest, and the *Journal of the American Medical Association* lists references in all the leading medical journals of the world and reviews many of the more important articles. These valuable bibliographic and abstracting journals are necessary for anyone who attempts to keep pace with the progress of parasitology; without them this book could not have been kept up-to-date.

There are few if any of the journals listed under "Sources of Information" or of books or articles listed under chapter references that have not been drawn upon for help in the preparation of this book. All of them, collectively, have made the book possible, and to their authors or contributors are due therefore the thanks both of the authors and of everyone who may profit in any way from the present book.

The senior author passed away during the initial stages of the preparation of this edition. The junior author bears the total responsibility for the chapters dealing with protozoa and helminths and has rewritten the arthropod chapters largely from notes gathered by the senior author.

As Professor Chandler would have done, I wish to express my appreciation of the kindness of many friends and colleagues who have helped in weeding out errors and in suggesting improvements. I hope that those who use the book will continue to offer criticisms or suggestions; they will be given careful consideration in future editions.

Clark P. Read

December, 1960

CONTENTS

Chapter 1

INTRODUCTION

One of the most appalling realizations with which every student of nature is brought face to face is the universal and unceasing struggle for existence which goes on during the life of every living organism, from the time of its conception until death. We like to think of nature's beauties, to admire her outward appearance of peacefulness, to set her up as an example for human emulation. Yet under her seeming calm there is going on everywhere—in every pool, in every meadow, in every forest—murder, pillage, starvation, and suffering.

Man often considers himself exempt from this interminable struggle for existence. His superior intelligence has given him an insuperable advantage over the wild beasts which might otherwise prey upon him; his inventive genius defies the attacks of climate and the elements; his altruism, which is perhaps his greatest attribute, protects to a great extent the weak and poorly endowed individuals from the quick extinction which is the inevitable lot of the unfit in every other species of animal on the earth. Exempt as we are to a certain extent from these phases of the struggle for existence, we have not yet freed ourselves from two other phases of it, namely, competition among ourselves, resulting in war, and our fight with parasites which cause disease.

We have made far more progress toward the latter phase than toward the former. The very inventive genius that has freed us from the great epidemics of infectious disease—cholera, plague, smallpox, yellow fever—that even in the nineteenth century spread terror in the world, has made our struggles with each other constantly more devastating and perilous, until today they threaten complete destruction of our civilization. No epidemics of disease ever threatened anything like that, even in the early days of our civilization when changing ways of life gave epidemics opportunities they never had had before in the history of the world, and when we had not yet developed countermeasures against them.

But our concern in this book is with the more auspicious struggle with parasites. Here progress has been largely one-sided, for the slow process of evolution on which our parasitic enemies must depend is no match for the swift development of advantages afforded by human ingenuity; we purposely refrain from saying "intelligence," since the application of our ingenuity to destruction of each other can hardly be construed as intelligence. With few exceptions as far as man and his domestic animals are concerned, the enemy has been discovered, his resources and limitations known, his tactics understood, and weapons of offense and defense developed. Progress is not *always* one-sided, however. When we developed powerful chemotherapeutic drugs and deadly residual insecticides, we thought we had achieved insuperable advantages, but bacteria and trypanosomes countered with drug resistance, and flies with chemical tolerance. So we invent new chemical weapons and the parasites and vectors new defenses, but the latter seem able to work faster than our chemists.

There is another disquieting aspect of the matter. Most of our progress has been medical or chemical—development of therapeutic and prophylactic drugs, vaccines, and insecticides. These successes, as Vaucel pointed out, along with good environmental and social conditions, have been adequate to protect the privileged Europeans and Americans even when living in undeveloped and underprivileged countries, and also the infinitesimal fraction of natives in these countries (usually called the tropics, but not confined to that area) who live European lives. Our medical successes have had much less effect on the millions of people who are living under practically the same conditions as they lived under several thousand years ago. Some of our great medical victories *have* affected backward populations, but only when they have not involved important changes in the age-old ways of life. The conquest of malaria by residual sprays is an outstanding example; houses are sprayed, but not changed, and everything else remains the same. Sleeping sickness has been greatly reduced in large areas in Africa, but all the native population had to do was to present itself for treatment. Yellow fever in South America, plague in Madagascar, kala-azar in India are other examples of the same thing. In Latin America, however, more progress has been made, concomitant with improvement in economic conditions. Houses are not merely sprayed to kill the bugs that transmit Chagas' disease; efforts are made, with spectacular success in some places in Brazil, to make the houses more suitable for human beings and less suitable for the bugs.

Because of his way of life, the white man never suffered seriously from what he considers the minor plagues of the topics, caused by filariae, *Onchocerca,* guinea worm, hookworms, *Ascaris,* schistosomes, *Fasciolopsis,* trypanosomes, *Leishmania,* and such diseases as relapsing fever, yaws, and tropical ulcer, to mention only a few. But these are all part and parcel of the native's daily life; he cannot avoid them, yet they incapacitate him for work, blind him, mutilate him, and make his life miserable. So little have most of these diseases affected the white man that most of them are probably totally unfamiliar to students starting to study this book.

So, in spite of some spectacular successes, the human race still has far to go in the process of emancipation from parasitic disease. It will require improvement in social and economic conditions of great masses of people—the provision of wells, latrines, decent housing, refuse disposal, proper food, shoes, and elimination of insect vectors—and also education. Of all these items, probably two stand out in importance: proper food, since malnutrition not only causes disease per se, but is a very large factor in ability to fight other diseases (see p. 30); and education, because only by knowing what is dangerous, and why, can mankind hope to win in the struggle with disease.

In spite of the fact that most of the parasites dealt with in this book are now relatively scarce, localized, or entirely eliminated in the United States, it does not follow that they are of little importance to us. In these days, with international travel as common as interstate travel was a generation ago, many a home-town physician has to deal with patients suffering from diseases which previously had been only names to him. Also the opportunity for dissemination of parasites or vectors entering as stowaways in airplanes, or on or in the bodies of passengers, is greater than ever before. Even when it took weeks or months to go from continent to continent, dispersal of parasites was common. Traders brought filariasis from the South Seas to Egypt, slaves brought hookworms and schistosomes from Africa to America, and trading vessels carried yellow fever from the American tropics to New York and Philadelphia. What can be expected when we can have breakfast in Colombia and supper in Florida?

But this is not all. Isolationism is gone, whether we like it or not. The world is fast becoming an economic unit, or at least two competing economic units, and a disease that affects the production of rice in Burma or meat in Argentina or coffee in Brazil inevitably affects us economically, and our stake in the welfare of undeveloped countries, large already, will inevitably increase. We have less than 10% of the world's population and 8% of its area, but we use 50% of the produce of the Free World. We depend on foreign sources for over 40% of our minerals, and 10% of our other raw materials—soon it will be 20%. Undeveloped areas of the world—the areas principally affected by parasitic diseases—supply 60% of our imports and 40% of our exports. Obviously, then, the diseases that profoundly affect the health and productivity of these areas are of very real concern to us. We also import certain parasitic diseases which may spread no further but must be reckoned with in public health planning. For example, approximately 70,000 persons in New York, 2000 in Chicago, and 1500 in Philadelphia are infected with the blood fluke, *Schistosoma mansoni.* These are persons who have migrated to the continental United States in the last few years. The diseases from which underprivileged people suffer are chronic ones, as Wright (1951) pointed out, and sick or incapacitated people are a greater drain on productivity than dead ones.

We cannot credit all our relative freedom from parasitic diseases to our own purposeful efforts. With the progress of civilization, many human parasites have gradually been falling by the wayside, but the less civilization has advanced in an area the fewer have fallen. As M. C. Hall said, the louse had its welfare imperiled when the Saturday night bath supplanted occasional immersion from falling into water; it had a struggle for survival when modern plumbing and laundering facilities laid the foundation for a daily bath even in winter, and clean clothes once a week. The housefly got a severe setback when the automobile replaced the horse, and when modern sewage systems were developed. Mosquitoes suffered with the advent of agricultural drainage and reclamation schemes. With the reduction of these vectors went reduction in the protozoan and bacterial diseases they disseminate—malaria, epidemic typhus, dysentery, etc. Of course, the advent of DDT greatly speeded up the process in some cases, but epidemic typhus had disappeared and malaria had become quite limited geographically in this country *before* that magic chemical and allied substances were discovered. Substitution of sanitary toilets for the rush-covered floors of the Middle Ages and the shaded soil of unsanitated areas spells extinction for hookworms and *Ascaris.* Cooking and refrigeration make life more precarious for *Trichinella* and *Taenia.* Improved water supplies and good sewage disposal are dangerous to most intestinal infections. To the extent that these concomitants of civilization have become part of the way of life of a people, parasitic infections have decreased even without new insecticides, new chemotherapeutics, or new vaccines. These specifically developed weapons have practically completed the white man's freedom from most of the infectious diseases that he once justifiably feared; radical changes have been wrought even since the last edition of this book was published.

For our domestic animals, on the other hand, domestication and increasing concentration have meant increasing parasitization, for they soil their table with their feces, they eat uncooked food, they drink contaminated waters from ponds and streams, they bathe only by accident, and they have hairy bodies that provide ideal playgrounds for ectoparasites. The parasite egg that had to pursue a deer or antelope to a new bedground five miles away was out of luck, said Hall, whereas when millions of eggs are sowed on limited pastures, the parasites have all the advantage. For human parasites, increased concentration had an opposite effect owing to better opportunity for improved water, control of foods, and sanitary sewage disposal. But the parasites of the roaming deer and antelope are in a less vulnerable position than those of cattle or sheep. Some years ago the U. S. Department of Agriculture exterminated Texas fever in the United States, and eliminated *Boophilus annulatus*, but it took years of hard and expensive work. Today warbles, hornflies, screwworms, sheep bots, and cattle lice could probably be exterminated in a fraction of the time and at much less cost.

In spite of spectacular advances in our struggle with parasites, it is obvious, then, that the battle is far from won. In 1947 Stoll made the startling estimate that there are in the world today 2200 million helminthic infections—enough for one for every inhabitant if they were evenly distributed. We have sufficient knowledge to be able to control most, though certainly not all (schistosomiasis is a conspicuous exception) of the infectious diseases of ourselves and of our domestic animals, but they are still mostly unsubdued in vast areas of the world. There is not only need for additions to our knowledge of the causes and control of diseases, but also, and perhaps even more pressing, a need for the efficient application of what we already know. Apathy to parasitic diseases is largely the result of ignorance concerning them. Although this ignorance is most abysmal in still-primitive peoples of undeveloped countries, it is by no means absent in our own country. Some of our neighbors still think that malaria results from damp night air, that vaccination should be done away with, that animal experimentation is unjustified. After all, it is only 200 years since our Pilgrim Fathers boiled witches instead of water to control cholera!

History

Early views. Up to the middle of the seventeenth century knowledge of parasitology was limited to recognition of the existence of a few self-asserting external parasites such as lice and fleas, and a few kinds of internal parasites which were too obvious to be overlooked, such as tapeworms, *Ascaris*, pinworms, and guinea worms. These parasites were, however, thought to be natural products of human bodies, comparable to warts or boils. Even such immortal figures in parasitology as Rudolphi and Bremser at the beginning of the nineteenth century supported this idea. In Linnaeus' time this view gradually gave way to another, that internal parasites originated from accidentally swallowed free-living organisms. Flukes, for instance, were thought to be "landlocked" leeches or "fish"; in fact, the name fluke is said to come from the Anglo-Saxon *floc,* meaning flounder. Until the middle of the seventeenth century the necessity for parents was regarded as a handicap placed upon the higher vertebrates alone. Biology students struggling with required insect collections sometimes wonder how Noah ever succeeded in collecting all the species which must have been known even in his day for rescue in the Ark, but that was no worry of Noah's; he anticipated that insects, worms, snakes, and mice would be spontaneously generated after the flood as well as before.

Redi. The grandfather of parasitology was Francesco Redi, who was born in 1626. In the latter half of the seventeenth century he demonstrated to an unbelieving world that maggots developed from the eggs of flies, and that even *Ascaris* had males and females and produced eggs. He extended the idea of parenthood so far that it is really remarkable that its universal application, even to bacteria, had to wait for Pasteur's ingenious experiments two centuries later. Although Redi's recognition of obligatory parenthood in lower animals was his outstanding achievement, he was the first genuine parasite hunter; he searched for and found them not only in human bowels but in other human organs, in the intestines of lower animals, in the air sacs of birds, and in the swim bladders of fish.

Leeuwenhoek. This same half-century marked the origin of protozoology, for it was then that the Dutch lens grinder, Leeuwenhoek, perfected microscopes which enabled him to discover and describe various kinds of animalculae, many recognizable as Protozoa, in rain water, saliva, feces, etc.; among the organisms in feces he discovered what was probably a *Giardia,* although the first protozoan definitely recognized as a human parasite was *Balantidium coli,* discovered by Malmsten in Sweden in 1856, nearly two centuries later.

Rudolphi. In spite of the work of these pioneers, parasitology made little progress until about a century later, when Rudolphi came upon the scene. He was born in Stockholm in 1771, but did most of his work in Germany. He did for parasitology what Linnaeus did for zoologists in general; he collected and classified all the parasites known up to his time. Zeder, in 1800, recognized five classes of worms which Rudolphi named Nematoidea, Acanthocephala, Nematoda, Cestoda, and Cystica; the last had to be discarded about 50 years later when bladderworms were found to be the larval stages of the Cestoda.

Developments to 1850. During the first half of the nineteenth century numerous new species of parasites were discovered and described by Dujardin, Diesing, Cobbold, Leidy, and others. Meanwhile, observations on the life cycles of flukes and cestodes were being made. O. F. Muller discovered cercariae in 1773 but thought they were Protozoa; Nitzsch, in 1817, recognized the resemblance of the cercarial body to a fluke and regarded the creature as a combination of a *Fasciola* and a *Vibrio;* Bojanus, in 1818, saw the cercariae emerge from "royal yellow worms" in snails, and Oken, the editor of *Isis,* in which the work was published, felt willing to wager that these cercariae were the embryos of flukes; contributions by Creplin, von Baer, Mehlis, von Siebold, von Nordman, and Steenstrup finally added enough pieces to the puzzle so that by 1842 the general pattern of the picture could readily be seen.

Meanwhile, light was also shed on the true nature of bladderworms and hydatids. As the result of observations by Redi, Tyson, Goeze, Steenstrup, von Siebold, and van Beneden, their relationships with tapeworms gradually became apparent, but up to 1850 they were generally regarded as "hydropically degenerated" as the result of development in an abnormal host into which they had accidentally strayed. It was during this period also that *Trichinella* was discovered in human flesh by Peacock (1828), and in pigs by Leidy (1846); that Dubini discovered human hookworms (1842); that Hake discovered the oöcysts of Coccidia in rabbits; that Gluge and Gruby discovered trypanosomes in frog blood (1842); and that Gros found the first human ameba, *Entamoeba gingivalis* (1849).

Introduction of experimental methods. The next important milestone in parasitology was the introduction of experimental methods. Although Abildgaard had observed as far back as 1790 that sexless tapeworms (*Ligula*) from sticklebacks would become mature when fed to birds, experimental work in parasitology really began in the middle of the nineteenth century, when Herbst (1850) experimentally infected animals with *Trichinella,* and Kuchenmeister in 1851, having the right idea about the nature of cysticerci, proceeded to prove it by feeding species of *Taenia* from rabbits to dogs and obtaining adult tapeworms. Two years later Kuchenmeister proved that bladderworms in pigs gave rise to tapeworms in man, as he had suspected because of the similarity of their heads.

These results gave a tremendous impetus to work in parasitology which has persisted to the present day, although it was temporarily eclipsed by the spectacular advances in bacteriology from about 1880 to the end of the century. The name of Leuckart stands out with especial brilliance in the early days of experimental parasitology; other shining lights in helminthology, who began their work before the beginning of the twentieth century, were Braun, Hamann, von Linstow, Looss, Lühe, and Schneider in Germany; Blanchard, Brumpt, Moniez, and Railliet in France; Cobbold and Nuttall in England; van Beneden in Belgium; Odhner in Sweden; Fuhrmann and Zschokke in Switzerland; Galli-Valerio, Grassi, and Stossich in Italy; and Cobb, Curtice, Leidy, Theobald Smith, Stiles, and Ward in America. In protozoology there were Bütschli, Doflein, Koch, von Prowazek, Schaudinn, and von Siebold in Germany; Davaine, Mégnin, Laveran, Leger, Nicolle, Sergent, and Aimé Schneider in France; Bruce, James, and Ross in England; and Leidy, Calkins, and Craig in America.

Insects as intermediate hosts and vectors. Following work on life cycles of helminths came the demonstration of the role of insects as intermediate hosts and vectors of parasites. Leuckart was the pioneer here when in 1867 he observed the development of *Mastophorus* (*Protospirura*) *muris* of mice, a spiruroid, in mealworms. Two years later Leuckart's pupil, Melnikov, showed that *Dipylidium* developed in dog lice, and in the same year Fedschenko observed the development of the guinea worm in *Cyclops.* The pioneer work on the role of bloodsucking arthropods was by Manson in 1878, when he observed the development of *Wuchereria bancrofti* in mosquitoes. This suggested to him the probability of mosquitoes having a comparable role in connection with malaria, and it was his advice and encouragement that led to Ross's proof of it in 1898. Meanwhile, however, two American workers, Theobald Smith and Kilbourne (1893), ingeniously worked out the transmission of Texas fever by ticks; this was the first demonstration of an arthropod as an intermediate host and vector of a protozoan parasite. Two years later Bruce showed that *Trypanosoma brucei* was transmitted by tsetse flies, and this paved the way for proof of the role of tsetse flies in sleeping sickness, though the proof of a developmental cycle in the fly was not made until 1909 by Kleine. The year 1898 brought not only the epoch-making demonstration of the role of mosquitoes in the transmission of malaria made by Ross in India and by Grassi in Italy, but also the discovery of penetration of the skin by hookworm larvae, made by Looss in Egypt. In 1900 the important discovery of the transmission of yellow fever by mosquitoes was made by the American Yellow Fever Commission in Havana. From this time on, discoveries in the life cycles and modes of transmission of parasites came thick and fast.

Chemotherapy. Important progress has also been made in the chemotherapy of parasitic infections. One of the earliest specific remedies known was quinine for malaria, introduced into Europe in the seventeenth century; with the other alkaloids of cinchona it held the field for almost 300 years. It was not until the early part of the twentieth century that other important specific drugs were discovered—Salvarsan for syphilis by Ehrlich in 1910; emetin for amebic dysentery by Rogers in 1912; tartar emetic for leishmaniasis by Vianna in 1914; Tryparsamide for sleeping sickness by Brown and Pearce in 1920–1921. During and after World War II came the discoveries of the amazing action of antibiotics against syphilis, rickettsial diseases, and amebiasis, as well as many bacterial diseases. During this period also came the discovery of Chloroquine to replace quinine and atebrin for treatment and prophylaxis of the blood forms of malaria, and discovery of Primaquine and Daraprim for radical cure by destruction of the tissue stages of malaria parasites.

In the field of anthelmintics a few remedies—male fern, Cusso, and areca nut for tapeworms, and Santonin for nematodes—have long been known. The first great advance was made when some Italian workers established the value of thymol for hookworms in 1880. This held the field for over 30 years but was succeeded by oil of chenopodium in 1913, carbon tetrachloride in 1921, and tetrachlorethylene in 1925. Chenopodium was also very useful for *Ascaris* but was supplanted by hexylresorcinol about 1930, and since World War II has been threatened by Hetrazan and piperazine. The value of antimony compounds for schistosomiasis was discovered by McDonagh and Christopherson. Gentian violet was introduced as an anthelmintic in 1927 but has largely been replaced by chloroquine for *Clonorchis*, by piperazine for *Enterobius*, and will probably be made obsolete for *Strongyloides* by the new drug, dithiazanine. Also, in 1938, Harwood set a landmark when he showed the value of Phenothiazine as a veterinary anthelmintic. Hexachlorethane for *Fasciola* was introduced in Europe in 1926 but was not fully appreciated until 1941. During and after

World War II atebrin was found to be effective against tapeworms, and the value of antimony and arsenic compounds for filariasis was established by Brown, Culbertson, and others. After the war Hetrazan was introduced for filariasis and has been a useful drug against other helminths.

In the field of insecticides the outlook for control of nearly all arthropod parasites and arthropod-borne diseases was revolutionized by the advent of DDT and other chlorinated hydrocarbons for use as residual sprays, beginning about 1943. Development of aerosols and effective repellents has added to the troubles of insect parasites and vectors. A dramatic testimonial to the effectiveness of these developments was the voluntary dissolution of the National Malaria Society in 1951 because of the attainment of its goal—the elimination of malaria as a major public health problem (since then as an endemic disease) in the United States.

Immunity. Study of the nature and mechanism of immunity to parasitic infections is fairly recent, having been developed mainly by American workers. The work of W. H. and L. B. Taliaferro in 1925 on the mechanism of immunity in trypanosome and malaria infections was the beginning; W. H. Taliaferro, with Cannon, Huff, Sarles, and other collaborators, has been prominent in further work in connection with immunity both to malaria and to nematode infections. A pioneer piece of work in acquired metazoan immunity was done by Blacklock and Gordon on the skin maggot (*Cordylobia*) in 1927, and another by Miller (1931) on larval tapeworms in rats. Since then, many important contributions to metazoan immunity have been made by nearly a score of American workers.

Development of parasitology in America. In concluding this historical section a brief résumé of parasitology in America is in order. The only early naturalist in America who took an interest in this subject was Joseph Leidy; during the last half of the nineteenth century he made many and valuable contributions. He is said to have become so absorbed in the study of a worm that he entirely forgot an obstetrical case he had engaged to attend. If Joseph Leidy can be called the grandfather of American parasitology, H. B. Ward may be considered the father of it. He not only made numerous contributions of his own over a period of 50 years, but he also stimulated interest in a host of others. His position in American parasitology can best be appreciated when it is recalled that among the students who started their scientific careers under him at the University of Illinois were Ackert, Cort, Faust, Hunter, LaRue, Manter, Miller, Stunkard, Thomas, and Van Cleave. The only other university which even approaches such an output of senior modern parasitologists is Harvard, among whose sons are Kofoid, Pearse, Sawyer, Smillie, Tyzzer, Wenrich, and Ward himself. A large proportion of the ever-increasing number of the younger generation of parasitologists in America today are the scientific grandchildren of H. B. Ward.

Importance of minor contributions. The discoveries mentioned in this brief résumé of the history of parasitic diseases are but a few of the more conspicuous milestones on the path of progress of modern medicine as related to animal parasites. But not one of the great outstanding discoveries in the field of parasitology and preventive medicine could have been made without the aid of numerous less heralded accomplishments of hundreds of other investigators who, often without any semblance of the honor and recognition which they deserve, work for the joy of the working and feel amply repaid if they add a few pickets to the fence of scientific progress.

The formation of the American Society of Parasitologists in 1926 marked the weaning of parasitology as a science in the United States, but it was not until World War II that it really came of age and took its rightful place in the community of sciences in America. As the senior author pointed out in 1946, parasitology touches upon or overlaps so many other sciences that a parasitologist probably has to stick his nose into more different fields of knowledge than any other kind of biologist. A parasitologist, like an orchid, requires long and careful nurturing, and develops slowly (about 85% of parasitologists have a Ph.D. degree). But when he comes to flower he is a rare and beautiful object, scientifically speaking, and is usually slow in going to seed.

REFERENCES

The following is a list of references of a general nature and general books on parasitology in which students who are interested may find additional information or different viewpoints. Books and references dealing with more limited subjects are listed at the end of the appropriate chapters. These references are not intended to be complete; they include only a few important or comprehensive treatises, mostly recent, to help the student who desires to do so to pursue the subject beyond the hallway to which this book may lead him. Many of the references contain bibliographies of their own which should give an entrée to the literature of the subject.

Ackert, J. E. 1937. *Laboratory Manual of Parasitology.* Burgess, Minneapolis.
Anonymous. 1951. Future of parasitology. *Nature,* 168: 527–529.

Baer, J. G. 1951. *Ecology of Animal Parasites.* University of Illinois Press, Urbana, Ill.

Belding, D. L. 1953. *Textbook of Clinical Parasitology.* 2nd ed. Appleton-Century-Crofts, New York.

Benbrook, E. A. 1952. *List of Parasites of Domesticated Animals in North America.* 2nd ed. Burgess, Minneapolis.

Benbrook, E. A., and Sloss, M. W. 1948. *Veterinary Clinical Parasitology.* Iowa State College Press, Ames, Iowa.

Birch, C. L., and Anast, B. P. 1957. The changing distribution of helminthic diseases in the United States. *J. Am. Med. Assoc.,* 164: 121–126.

Blacklock, D. C., and Southwell, T. 1953. *A Guide to Human Parasitology for Medical Practitioners.* 5th ed. H. K. Lewis, London.

Cable, R. M. 1947. *An Illustrated Laboratory Manual of Parasitology.* Revised. Burgess, Minneapolis.

Cameron, T. W. M. 1934. *The Internal Parasites of Domestic Animals.* A. C. Black, London.

1946. *The Parasites of Man in Temperate Climates.* 2nd ed. University of Toronto Press, Toronto.

Cameron, T. W. M. 1956. *Parasites and Parasitism.* John Wiley, New York.

Chandler, A. C. 1946. The making of a parasitologist. *J. Parasitol.,* 32: 213–221.

Culbertson, J. T., and Cowan, C. 1952. *Living Agents of Disease.* Putnam, New York.

Dubois, A., and Van den Berghe, L. 1948. *Diseases of the Warm Climates.* Grune and Stratton, New York.

Faust, E. C. 1949. Reflections of a medical parasitologist. *J. Parasitol.,* 35: 1–7.

Faust, E. C., and Russell, P. F. 1957. *Clinical Parasitology.* 6th ed. Lea and Febiger, Philadelphia.

Gradwohl, R. B. H. (Editor). 1951. *Clinical Tropical Medicine.* Mosby, St. Louis.

Haberman, R. T., et al. 1954. Identification of some internal parasites of laboratory animals. *Public Health Service Pub.* 343.

Hall, M. C. 1931. The glorification of parasitism. *Sci. Monthly,* 33: 45–52.

1932. Parasitology and its relation to other sciences. *Puerto Rico J. Public Health and Trop. Med.,* 7: 405–416.

Hartmann, F. W., Horsfall, F. L., and Kidd, J. G. 1954. *The Dynamics of Virus and Rickettsial Infections.* Blakiston, Philadelphia.

Hassal, A., et al. 1932–1953. *Index-Catalogue of Medical and Veterinary Zoology. Authors,* Pts. 1–18 and Supplements. U. S. Department of Agriculture, Washington.

Hegner, R. W., Root, F. M., Augustine, D. L., and Huff, C. G. 1938. *Parasitology.* Appleton-Century-Crofts, New York.

Hirst, L. F. 1953. *The Conquest of Plague. A Study of the Evolution of Epidemiology.* Oxford University Press, London.

Hoeppli, R. 1956. The knowledge of parasites and parasitic infections from ancient times to the 17th century. *Exp. Parasitol.,* 5: 398–420.

Hull, T. G. 1947. *Diseases Transmitted from Animals to Man.* 3rd ed. Thomas, Springfield, Ill.

Lapage, G. 1956. *Mönnig's Veterinary Helminthology and Entomology.* 4th ed. Bailliere, Tindall, and Cox, London.

Lapage, G. 1956. *Veterinary Parasitology.* Oliver and Boyd, Edinburgh and London.

Leuckart, R. 1879–1886. *Die Parasiten des Menschen und die von ihnen herruhrenden Krankheiten.* C. I. Winter, Leipzig.

Lwoff, A. (Editor). 1951–1955. *Biochemistry and Physiology of Protozoa.* Vols. I and II. Academic Press, New York.

Mackie, T. T., Hunter, G. W. III, and Worth, C. B. 1954. *A Manual of Tropical Medicine.* 2nd ed. Saunders, Philadelphia.

Manson-Bahr, P. H. 1954. *Manson's Tropical Diseases.* 14th ed. Williams and Wilkins, Baltimore.

Manter, H. 1950. *A Laboratory Manual in Animal Parasitology, with Special Reference to the Animal Parasites of Man.* Revised. Burgess, Minneapolis.

May, S. 1954. Economic interest in tropical medicine. *Am. J. Trop. Med. and Hyg.,* 3: 412–421.

Most, H. (Editor). 1951. *Parasitic Infections in Man.* Columbia University Press, New York.

Nauss, R. W. 1944. *Medical Parasitology and Zoology.* Hoeber, New York.

Otto, G. F. 1958. Some reflections on the ecology of parasitism. *J. Parasitol.,* 44: 1–27.

Pearse, A. S. 1942. *Introduction to Parasitology.* Thomas, Springfield, Ill.

Piekarski, G. 1954. *Lehrbuch der Parasitologie.* Springer-Verlag, Berlin.

Riley, W. A. 1952. *Introduction to the Study of Animal Parasites and Parasitism.* 6th ed. Burgess, Minneapolis.

Shattuck, G. C. 1951. *Diseases of the Tropics.* Appleton-Century-Crofts, New York.

Stoll, N. R. 1947. This wormy world. *J. Parasitol.,* 33: 1–18.

Strong, R. P. 1944. *Stitt's Diagnosis, Prevention and Treatment of Tropical Diseases.* 7th ed., 2 vols. Blakiston, Philadelphia.

Symposium on intraspecific variation in parasitic animals. 1957. *Systematic Zool.,* 6: 2–28.

U. S. Department of Agriculture. 1942. *Keeping Livestock Healthy,* Yearbook. Government Printing Office, Washington.

Volumen jubilare pro Professore Sadao Yoshida. 1939. Vol. 2, Osaka.

Whitlock, J. H. 1938. *Practical Identification of Endoparasites for Veterinarians.* Burgess, Minneapolis.

Wright, W. H. 1951. Medical parasitology in a changing world. What of the future? *J. Parasitol.,* 37: 1–12.

1950. Tropical diseases of veterinary public health importance. *Am. Vet. Med. Assoc. Proc.,* 87th Meeting: 106–116.

Zeliff, C. C. 1948. *Manual of Medical Parasitology, with Techniques for Laboratory Diagnosis and Notes on Related Animal Parasites.* 2nd ed. State College, Pa.

Chapter 2

PARASITES IN GENERAL

Nature of parasitism. The world of animal life consists of communities of organisms which live by eating each other. In a broad sense all animals are parasites, in that they are helpless without other organisms to produce food for them. Plants alone are able to build up their body substance out of sunlight and chemicals. Herbivorous animals, when they feed on vegetation, exploit the energy of the plants for their own use. Carnivorous animals, in turn, exploit the energies of the herbivorous ones, larger carnivores exploit the smaller ones, etc., the whole series thus constituting what ecologists call a food chain; many such chains can be traced in any animal community.

But animals and plants are not preyed upon alone by successively larger forms which overpower and eat them; they are also preyed upon by successively smaller forms which destroy only small, more or less replaceable portions, or even more subtly exploit the energies of the host by subsisting on the food which the host has collected with expenditure of time and energy. Elton (1935) said, "The difference between a carnivore and a parasite is simply the difference between living upon capital and income, between the burglar and the blackmailer. The general result is the same although the methods employed are different." A man's relation to his beef cattle is essentially that of a tiger to its prey; his relation to his milk cattle and hens is essentially that of tapeworms or hookworms to their hosts. There is every gradation between parasites and carnivores, e.g., hookworms, leeches, horseflies, bloodsucking bats, and tigers; there are also all gradations between parasites and saprophytes, or organisms which live on the wastes or leftovers, e.g., rickettsias, incapable of living outside host cells; *Entamoeba histolytica,* feeding on the tissue of the host; *Trichomonas hominis,* feeding, in part, at least, on digested foods which would otherwise be converted into tissues; *Entamoeba coli,* feeding on still undigested particles and bacteria; and the coprozoic amebas, feeding on the waste fecal matter of the host.

General relations to hosts. The popular notion that parasites are morally more oblique in their habits than other animals, as if they were taking some unfair and mean advantage of their hosts, is, as Elton remarks, unjustified. Carnivores and herbivores have no interest in the welfare of their prey and ruthlessly destroy them; parasites, of necessity, cannot be so inconsiderate, for their welfare is intimately bound up with the welfare of the host. "A parasite's existence," says Elton, "is usually an elaborate compromise between extracting sufficient nourishment to maintain and propagate itself, and not impairing too much the vitality or reducing the numbers of its host, which is providing it with a home and a free ride." A dead host is seldom of any use to a parasite in its adult state, though it may capitalize the death of an intermediate host as a means of attaining its destination in the definitive host. Food of the right kind and in sufficient quantity is the burning question in all animal society; for parasites this resolves itself into the question of what to do when the host dies; most internal parasites, as adults, are so specialized for a protected life in the body of the host that they are unable to take any steps to deal with this situation. The result is that they make no attempt to do so and resign themselves to dying with their hosts, leaving it to their offspring to find their way to another host in order to continue the race, and, since the offspring have to run enormous risks in order to succeed, they have to be produced in correspondingly enormous numbers, running into millions.

Parasites and food habits. Since food is the hub of the wheel of animal life, it is natural to find that many parasites have taken advantage of the food habits of their hosts in order to propagate themselves from host to host. Intestinal Protozoa usually solve the problem by entering into a resistant cystic stage in which they can survive outside the body until they can re-enter a host with its food or water. Blood Protozoa, such as trypanosomes and malaria parasites, are adapted to live temporarily in bloodsucking insects which feed on the host and subsequently reinject them into another host. Most flukes and tapeworms lay eggs which develop into larvae in the body of an animal which the host habitually eats, or which is eaten by a third animal, which is then eaten by the definitive host. Some intestinal nematodes, such as the spiruroids, do likewise; others, such as *Ascaris*, follow the tactics of intestinal Protozoa; and still others, such as hookworms, produce self-reliant embryos which actively burrow into the skin of their hosts. Most parasitic arthropods are able to migrate from host to host when these come in contact with each other, directly or indirectly; but the bloodsucking flies have no worries about this matter, and can go at will from host to host.

The result of the dependence of parasites to such a large extent on the food habits of animals is that the food habits largely determine the nature of the parasites harbored. *Ascaris, Trichuris,* intestinal Protozoa, etc., are abundant where unsanitary conditions favor fecal contamination of food or water; many fluke infections of man are abundant in localities in the Far East where fish is habitually eaten raw; species of *Taenia* are abundant where pork or beef is eaten raw or partly cooked; guinea worms are common where infected *Cyclops* is ingested with drinking water; and spiruroid infections occur only accidentally in man because the human animal is nowhere habitually insectivorous in habit.

Origin of parasitism. Parasitism, in the restricted sense of a small organism living on or in, and at the expense of, a larger one, probably arose soon after life began to differentiate in the world. It would be difficult, if not impossible, to explain step by step the details of the process of evolution by which some of the highly specialized parasites reached their present condition. Parasitism at times has probably grown out of a harmless association of different kinds of organisms, one of the members of the association, by virtue, perhaps, of characteristics already possessed, developing the power of living at the expense of the other, and ultimately becoming more and more dependent upon it.

It is easy to understand the general mechanism by which parasites of the alimentary canal were evolved from free-living organisms which were accidentally or purposely swallowed, and which were able to survive in the environment in which they found themselves, and to adapt themselves to it. It is also easy to see how some of these parasites might eventually have developed further territorial ambitions and have extended their operations beyond the confines of the alimentary canal. The development of some of the blood Protozoa of vertebrates, on the other hand, seems clearly to have taken place in two steps: first, adaptation to life in the gut of insects and, second, adaptation to life in vertebrates' blood or tissues when inoculated by hosts with skin-piercing and bloodsucking habits.

Kinds of parasites. Parasitism is of all kinds and degrees. There are facultative parasites which may be parasitic or free-living at will, and obligatory parasites which must live on or in some other organism during all or part of their lives, and which perish if prevented from doing so. There are intermittent parasites which visit and leave their hosts at intervals. Some, as mosquitoes, visit their hosts only long enough to get a meal; others, as certain lice, leave their hosts only for the purpose of molting and laying eggs; some ixodid ticks never leave their host except for the final egg-laying venture from which there is no return; *Ascaris* and intestinal Protozoa live in one host from the time they hatch from an egg or a cyst until they die, but produce eggs or cysts which escape to be transferred to a new host. Some parasites are such during only part of their life cycles; botflies, for instance, are parasitic only as larvae, hookworms only as adults. The final degree of parasitism is reached in those parasites which live generation after generation on a single host, becoming transferred from host to host only by direct contact. Such are the scab mites and many species of lice. Every gradation is found among all the types of parasites mentioned above.

It is sometimes convenient to classify parasites according to whether they are external or internal. External parasites, or ectoparasites, living on the surface of the body of their hosts, suck blood or feed upon hair, feathers, skin, or secretions of the skin. Internal parasites, living inside the body, occupy the digestive tract or other cavities of the body, or live in various organs, blood, tissues, or even within cells. No sharp line of demarcation can be drawn between external and internal parasites since inhabitants of the mouth and nasal cavities, and such worms and mites as burrow just under the surface of the skin, might be placed in either category.

Definitive and intermediate hosts. Some parasites pass different phases of their life cycle in two or more different hosts; in a few kinds of flukes four may be involved. According to a dictionary definition, the host in which the parasite reaches sexual maturity is the definitive host and those in which it undergoes preliminary development are the intermediate hosts. Strict adherence to these definitions, however, leads to some peculiar situations among the Protozoa. For example, since the malaria parasites undergo sexual reproduction in mosquitoes, the mosquito, according to the definition, is the definitive host and man is the intermediate host. But in the case of *Leishmania,* in which no sexual reproduction occurs, shall we consider man or sandfly the intermediate host? Because of these difficulties, the authors prefer to use the terms definitive and intermediate for vertebrates and arthropods, respectively, in relation to protozoan parasites which alternate in their life cycles between these two types of hosts, irrespective of where the sexual reproduction, if any, occurs. An alternative is to avoid these terms altogether for Protozoa, rickettsias, etc., which multiply in both hosts, and simply to speak of vertebrate and invertebrate hosts.

Effects of parasitism on parasites. Aside from the toning down of their effects on the host, parasites are often very highly modified in structure to meet the demands of their particular environment. As a group, parasites have little need for sense organs and seldom have them as highly developed as do related free-living animals. Fixed parasites do not need, and do not have, well-developed organs of locomotion, if, indeed, they possess any. Intestinal parasites do not need highly organized digestive tracts, and tapeworms and spiny-headed worms get along very well without any digestive tract at all. On the other hand, parasites must be specialized, often to a very high degree, to adhere to or to make their way about in their particular host, or the particular part of the host in which they find suitable conditions for existence. Examples of specializations of external parasites are the compressed bodies and backward-projecting spines of fleas, which enable them to glide readily between hairs without backsliding; the clasping talons on the claws of lice; the barbed proboscides of ticks; and the tactile hairs of mites. In these same parasites can be observed marked degenerations in the loss of eyes and other sense organs, absence of wings, and sometimes reduction of legs. Internal parasites are even more peculiar combinations of degeneration and specialization. They possess all sorts of hooks, barbs, suckers, and boring apparatus, yet they have practically no sense organs or special organs of locomotion, a very simple nervous system, and sometimes, as said before, a complete absence of the digestive tube.

Most remarkable are the elaborate specializations of parasites in their reproduction and life histories to insure, as far as possible, a safe transfer to new hosts for succeeding generations. Every structure, every function, every instinct of many of these parasites is modified, to a certain extent, for the sole purpose of reproduction. A fluke does not eat to live, it eats only to reproduce. The inevitable death of the host is the parasite's doomsday, against which it must prepare by producing all the offspring possible, in the hope that enough will survive to keep the race from extinction. The complexity to which the development of the reproductive systems may go is almost incredible. In some adult tapeworms not only does every segment bear complete male and female reproductive systems, but it may bear *two* sets of each. The number of eggs produced by many parasitic worms may run well into the millions. The complexity of the life history is no less remarkable. Not only are free-living stages interposed and intermediate hosts made to serve as transmitting agents, but also often asexual multiplications, sometimes to the extent of several generations, are passed through during the course of these remarkable experiences.

Mutual tolerance of host and parasite. In the course of time a mutual adjustment or tolerance frequently develops between a host and parasite which permits the two to live together as a sort of compound organism without very serious damage to either. It may not be in the best interest of the parasite to destroy its host, for in so doing it would destroy itself. Excessive pain or irritation caused by ectoparasites is likely to lead to their own destruction at the hands or teeth of their irritated hosts. In well-established host-parasite relations the host protects itself against the injurious effects of parasites by placing its blood-forming and tissue-repairing mechanisms on a plane of higher activity and by developing antibodies that interfere in various ways with the activities or welfare of the parasites (see p. 26). There may also be an increase in the number of phagocytes with accompanying enlargement of spleen and liver, and in helminthic infections, an increase in the special type of white blood corpuscles, the eosinophiles, which are probably concerned with neutralization of injurious products of the worms. Failure to respond to parenteral invasions of worms by an increase in eosinophiles (eosinophilia) may indicate that the defensive mechanism of the host is not functioning adequately. Enhanced defensive mechanisms develop in individual hosts, and eventually by natural selection a whole species or race often becomes more tolerant. When a parasite is introduced into a new species of host, the delicate adjustment may be missing and either the parasite may fail to survive or the host may be severely injured or destroyed; this sometimes results when a parasite or parasite strain is introduced into a new locality, or it may result when some environmental or physiological change upsets the balance. A high degree of pathogenicity of a parasite is often evidence of a relatively recent and unperfected host-parasite relation or of a lowered resistance of the host due to some physiological stress. An organism and the parasites that are particularly adapted to live with it may, in a way, be looked upon as a sort of compound organism. When an intermediate host is involved, there is a third party added to the association, and the relationships of intermediate hosts to definitive hosts, as well as those of the parasite to each, may be important.

Host susceptibility and specificity. For a parasite to live habitually in a host there must be (1) suitable conditions for access to the host, involving a dependable means of transmission from one host to another; (2) ability to establish itself in a host when it reaches one; and (3) satisfactory conditions for growth and reproduction after it establishes itself. It is the interplay of these factors that determines in what hosts a parasite lives. Every parasite, of course, has at least one species of host, and sometimes several, in which these conditions are satisfactorily met; otherwise it would cease to exist. Usually there are other hosts in which one or both conditions are only occasionally met, in which case "accidental" parasitism results. Man's failure to utilize insects as food, except accidentally, relieves him from *common* infection with parasites such as spiruroids, *Acanthocephala,* and most tapeworms, which encyst in insects as intermediate hosts; he is, however, susceptible to a great many of these parasites when they *do* get access to him. On the other hand, man must commonly be exposed to infection with such parasites as bird malaria, animal schistosomes, dog and cat hookworms, and bird filariae; yet infection rarely or never occurs because the parasites do not find suitable conditions for normal development in the human body. Such animals as rats, dogs, cats, and various domestic animals must very often be exposed to infection with human parasites, yet they habitually harbor very few of them.

Some parasites are spread by direct or indirect contact with infected parts, e.g., the mouth amebas, itch mites, trichomonads, and free-moving ectoparasites. The parasites of the digestive system may gain entrance by the larvae boring through the skin and then migrating to their final destination, (e.g., hookworms) but more commonly they enter the mouth as cysts, eggs, or encapsulated larvae.

Successful access to a host does not necessarily mean successful establishment in it. Various parasitologists have shown that the escape of protozoa from cysts, of helminths (e.g., *Ascaris*) from eggs, of tapeworms or flukes from their capsules or cysts, and of nematodes from sheaths all depend on complicated and neatly timed interaction of factors involving host and parasite physiology. These interactions may lead to successful escape and consequent establishment of a particular parasite in one host but not in another, and of one parasite and not another, even if closely related, in a particular species of host (Read, 1958).

Most helminths must then invade the skin or mucous membranes of their hosts. Some helminths have mechanical devices to aid them (e.g., stylet cercariae and tapeworm embryos), and some have portals of entry made for them by their vectors (e.g., the filariae), but many depend largely on glandular secretions. These may contain various enzymes to aid them, including a hyaluronidase-like substance or "spreading factor" as Esslinger (1958) demonstrated for screwworms, but particularly enzymes that aid the parasite's progress by liquefying glycoproteins of basement membranes and intercellular cements (see Lewert, 1958). Such enzymes are present in secretions from schistosome cercariae and eggs, *Strongyloides* embryos, tapeworm embryos and developing larvae, *Hypoderma* larvae, and probably others. The basement membranes and ground substance are thicker and denser in older animals, which may be an important factor in age immunity (see p. 26) and in certain hormone deficiencies, whereas the glycoproteins are reduced by pregnancy and by vitamin C deficiency. Invasion may also be facilitated by other organisms. Nematode larvae may help viruses enter the central nervous system; *Histomonas* may be safely carried through the stomach in the eggs of *Heterakis;* and amebas completely fail to invade the mucous membranes of axenic (germ-free) hosts.

After penetration many helminths have to perform extensive migrations through the hosts' tissues to reach their ultimate destination. Human hookworms after penetrating the skin enter the circulatory system, go via the heart to the lungs, escape into the lungs, migrate to the throat and are swallowed. *Ascaris lumbricoides* does the same except that its point of departure is the intestinal mucosa. In normal hosts a series of reactions or stimuli guide them in the right direction, but in strange hosts the physiological road signs are missing and the parasites get lost and wander aimlessly in abnormal locations, failing to mature. *Ancylostoma braziliense* migrates normally in its natural cat and dog hosts, but in man it merely rambles under the skin, causing creeping eruption. *Toxocara canis* of dogs tends to go to the muscles and brain of mice, but primarily to the liver in man (see p. 459). Horse bots (*Gasterophilus*), the lung fluke (*Paragonimus*), and gnathostomes are other parasites that creep aimlessly in the skin or eyes of man. The larvae of the pork tapeworm encyst in the muscles of the normal pig host but, in man, often blunder into the eye or brain. Filariae of the genus *Setaria* live in the mesenteries of their normal hosts but often invade the eyes or brain of "foreign" hosts.

After arriving at a proper destination, parasites must still find conditions suitable for their continued growth, development, and reproduction. Read and his colleagues (1959) showed that the abundance of urea in the spiral valve of elasmobranchs is osmotically essential for certain tapeworms but may inhibit other parasites. There is evidence that the bile salts in correlation with the *p*H and other physical characteristics of the intestine may have similar effects. Ackert in 1938 found a substance in duodenal mucus, abundant in older chickens, but not in baby chicks, that inhibits *Ascaridia.* High temperature may keep mammalian parasites from developing in birds, etc.

There are many ways in which diet may affect the kinds of parasites harbored by a host and how much it suffers from their effects (see Chandler, 1959). Schiller, Read, and Flyger found that when squirrels eat dog biscuits instead of acorns they become congenial homes for the dwarf tapeworm, *Hymenolepis nana,* which is never harbored by them under natural conditions. Oguri and Chu in 1955 found that ducks fed on poultry feed are entirely refractory to a trematode of the cloaca, whereas if fed on squid they become as susceptible as the natural sea-bird hosts. Elsdon-Dew reported that when Africans are transplanted from their native bush villages to a city, with revolutionary changes in what they eat, the *Entamoeba histolytica* that they harbor changes from a relatively harmless inhabitant of the intestinal lumen to a highly pathogenic tissue invader. De Witt in 1956 found that when mice are infected with schistosomes and fed a certain deficient diet, injury to the host enhances susceptibility so that many more worms become established, but the diet also harms the worms so they do not grow or develop normally.

The amount of damage done by hookworms is very largely dependent on the extent to which the diet replaces the iron and protein that is lost in the blood they waste. High protein diets tend to eliminate flagellate infections, which are conspicuously few in strict carnivores. A pure milk diet, lacking in p-aminobenzoic acid, is reported to suppress the malaria parasite, *Plasmodium berghei,* in rats and mice.

Obviously, many factors are concerned in host susceptibility and specificity, and in most instances they are still almost entirely unknown.

Geographic distribution. The distribution of parasites over the surface of the earth is dependent (1) on the presence of suitable hosts, and (2) on habits and environmental conditions that make possible the transfer from host to host. A human parasite that does not utilize an intermediate host is likely to be found in every inhabited region of the world, provided that its particular requirements with respect to habits and environmental conditions are met; and if it can also live as a parasite in other animals it may occur even beyond the limits of human habitation. Parasites such as intestinal Protozoa and itch mites, which require only slight carelessness in habits for their transfer and are largely independent of external conditions, are practically cosmopolitan, but vary in abundance with the extent of the carelessness on which their propagation depends. Helminths that have to live for some time outside the body of the definitive host while the larvae develop to the infective stage, either in the eggs (e.g., *Ascaris* and *Trichuris*), as free-living organisms (e.g., hookworms and the numerous intestinal nematodes of domestic animals) or in intermediate hosts, are more limited since they, or their intermediate hosts, may be affected by such environmental conditions as temperature, humidity, nature of the soil, etc.

The general climate of an area may be of much less importance than the microclimate existing in the immediate locality where the parasite lives—in burrows, under the soil surface, in mines, etc. Of course when intermediate hosts are involved suitable ones must be present, and climatic conditions must be satisfactory for development of the larvae in them. Under favorable climatic conditions, habits which facilitate the transfer of parasites from water, food, soil, intermediate hosts, or, in the case of many parasites of man and domestic animals, reservoir hosts, are often the determining factors in the occurrence and frequency of parasitic infections. Guinea worms thrive where *Cyclops* are swallowed with drinking water, hookworms where feet are bare, schistosomes where there are irrigation ditches, *Clonorchis* where fish are raised in polluted reservoirs and considered a delicacy uncooked, sylvatic plague where there is liaison between reservoir hosts and rats or man, *Fasciola* and *Fasciolopsis* where certain water vegetation is relished for food, *Hymenolepis nana* where food is exposed to mouse droppings, etc.

The Russian parasitologist, Pavlovsky (see Pavlovsky et al., 1955), has called attention to what he calls "natural nidality" of diseases that are maintained in wild animals in ecological nidi (nests) where suitable vectors are present and ecological conditions are favorable for continued transmission, and which are transmissible to man or domestic animals when these intrude upon the nidi. Examples are cited of such nidi, in the Soviet Union, of brucellosis, tularemia, and rickettsial and virus diseases, all involving rodents and their ectoparasites. These are all zoonoses, i.e., diseases transmissible from animals to man, but the idea can be extended to strictly human infections also. In North America many virus, rickettsial, spirochete, and helminthic infections show this "nidarial" character. The ecology of zoonoses has recently been discussed by Audy (1958).

With modern transportation facilities, as remarked in the previous chapter, the possibilities of extension of the range of parasites are increased. With more frequent experimentation, parasites may find new suitable intermediate hosts, and the required environmental conditions, in new places. Yellow fever may have failed, during all the past centuries, to gain access to the Far East only because the long sea journey exceeds the incubation period of the disease and makes it possible to discover cases of yellow fever and prevent them, or mosquitoes which might have fed on them, from entering. Today the danger is greater.

Spread to new hosts. All animals tend gradually to extend their range by adapting themselves to slightly different conditions. Parasites, however, are at a disadvantage as compared with free-living animals, for while a song sparrow can find an infinite number of intergrading conditions between the damp, cool forests of the northwest and the dry, hot deserts of the southwest, an *Ascaris* can find no intergrading conditions between the conditions in the intestine of a pig and those in the intestine of a human being. The change must be made in a single jump or not at all.

The closeness of the bond between parasites and their hosts varies greatly; some parasites, e.g., *Schizotrypanum cruzi,* are very indiscriminate; others, like *Taenia saginata,* are limited to a single species. Some genera and even species have maintained their allegiance to particular hosts or their descendants through vast periods of time. One species of tapeworm, for instance, is parasitic in ostriches in Africa and rheas in South America, but in no other birds. Parasites have been thought to give true clues to phylogenetic relationships (e.g., Cameron, 1952; Rothschild and Clay, 1952), but there is considerable doubt that host-parasite associations can be so used in any general fashion. (See Baer et al., 1957.)

Since the parasites have a less changeable environment than their hosts, they tend to undergo evolutionary changes more slowly, so that while a host is differentiating into new species, genera, families, or even orders, the parasites may change relatively little. In some cases, however, inherent host characteristics are of less importance than similar environmental conditions resulting from similarity of habits or diet and a parasite may adapt to phylogenetically unrelated hosts. These parasites tend to become physiologically adapted to the new hosts and thus are not as readily transferred from one kind of host to another. Examples are *Ascaris* and *Trichuris* in pigs and man, dwarf tapeworms of man, rats, and mice, and itch mites and *Trichomonas* of a variety of hosts. Some parasitologists are inclined to consider very closely related parasites in different hosts as distinct species until proved otherwise, whereas others tend to lump them all together. It seems to the authors preferable to regard them as hostal varieties or races which can be referred to as *Sarcoptes scabiei* of horses, man, etc., for instance, rather than giving them definite species or subspecies names about which troublesome questions of priority, identity, etc., are sure to arise. Much the same situation exists for geographical races or subspecies of free-living organisms. When a parasite becomes adapted to a new host it is still uncertain whether this is due to selection of the fittest of randomly occurring genetic types, to mutations induced by a changed environment, or to somatic adaptation.

Resistance and Immunity

Knowledge of the means by which animals resist infectious disease, and the mechanisms by which this resistance is increased against specific organisms or their products by prior exposure to them, either naturally or by artificial means, began with Pasteur in the last half of the nineteenth century. Since then, this young science of immunology has enjoyed rapid growth and today provides us with tools for preventing, ameliorating, or curing many diseases, as well as with means of diagnosing them by immunological tests.

In the early days of its development this science dealt almost exclusively with bacteria or their products, or with nonliving antigens, and was largely concerned with the demonstration of various antigen-antibody reactions, such as toxin neutralization, agglutination, precipitation, lysis, complement fixation, increased phagocytosis (opsonification), and allergic sensitization. Most of the observations were made *in vitro,* and little attention was paid to *functional* immunity, i.e., the actual protection afforded, except in the case of toxins and antitoxins, and later of viruses and "neutralizing" antibodies.

It was well into the twentieth century before it became clear that the fundamental principles of immunity are the same for protozoans, helminths, and arthropods as for bacteria and toxins, though there are differences in degree or in details. Study of the development of resistance to metazoan parasites has been particularly fruitful in explaining functional as compared with *in vitro* demonstrations of immunity, as will be seen.

Natural immunity. The natural immunity to particular parasites that is the birthright of species, races, or even individuals is usually due to the various factors affecting susceptibility and specificity which are considered in a preceding section. It is not infrequently broken down by physiological derangements or handicaps, e.g., removal of the spleen or thyroid, dietary deficiencies, injury by concurrent infections, or other debilitating factors. Natural immunity may be due to the presence of "natural" antibodies which sometimes, at least, are really antibodies developed against microbial inhabitants of the skin, intestine, or respiratory tract, or their products, or possibly against constituents of food that have to have antigens that are shared by the parasites. Such "natural antibodies" are present in human sera against many species of trypanosomes and trichomonads. They are not present in new-born infants or germ-free animals.

Age resistance. Age resistance is often in reality acquired immunity, or it may be due to increased speed of development of acquired immunity. The ability to cope with infectious disease develops with age, just as does ability to digest beef steak or to solve mathematical problems. Babies cannot mobilize phagocytes or produce antibodies as efficiently as older individuals. However, there is a true age immunity of some animals to some infections. Sandground in 1928 wrote that he thought that age resistance is usually associated with abnormal or imperfectly adapted hosts; any incompatibility between host and parasite appears to become intensified with age. It is significant that most cases of human infections with "foreign" worms, belonging in other animals, are recorded in children. Ackert in 1938 found a tangible basis for age resistance of chickens to *Ascaridia* in the increase with age of intestinal goblet cells, the mucus of which he showed to have an inhibitory effect on the worms. Dietary changes and consequent changes in the bacterial flora of the intestine and its *p*H may also be a factor in age resistance.

Acquired immunity. Recovery from disease confers immunity to that particular disease, sometimes for life, sometimes for only a short period. Some diseases are held in check by the defenses of the host without being completely eliminated, so that they go into a relatively quiescent chronic state. In infections where the organisms multiply in the body, e.g., tuberculosis and syphilis and such protozoan diseases as Chagas' disease, malaria, etc., as long as the parasites remain in the body the host is protected against reinfection; this condition is called premunition. The host and the parasite exist together in a more or less delicately balanced state. When the resistance begins to fall off, the parasites multiply sufficiently to renew it, sometimes causing a temporary relapse in the process; however, the rise in resistance again ceases before the infection is entirely eliminated. In helminth infections such as schistosomiasis, hookworm, etc., there is a comparable phenomenon, but here, when the pendulum swings in favor of the parasites, they cannot, of course, multiply, but there is renewed acquisition of them when there is opportunity, and those already present renew metabolic activity and reproduction. Immunity to bites of particular arthropods, which most individuals can develop (see p. 529), is also kept up by continual exposure, but falls off in its absence.

The basis for all specific acquired immunity is presumed to be due to the development of new kinds of globulin molecules (one of the blood proteins) which are chemically modified by the presence of an antigen in such a way that they have an affinity for the particular antigen and combine with it. According to the nature of the antigen and environmental circumstances, this results in detoxification of toxins, neutralization of viruses, agglutination, precipitation, lysis, complement fixation, increased phagocytosis (opsonification), and allergic sensitization. In the early days of the development of immunology as a science, most work was done with bacteria or their toxins or with nonliving antigens such as egg albumin. It was well into the twentieth century before it became clear that the fundamental principles of immunity are the same for protozoans, helminths, and arthropods as for bacteria or toxins.

It may now be assumed that acquired immunity develops against (1) all parenterally located parasites, (2) all parasites with a parenteral phase, even if it is only temporary invasion of the mucosa (e.g., esophagostomes and cysticercoids of *Hymenolepis nana*), and (3) parasites that break the mucosa sufficiently to inject antigens from the mouth while feeding, as trematodes and most nematodes do. The tapeworms and acanthocephalans are exceptional in that some of them at least fail to get any antigen inoculated into their hosts and thus fail to stimulate any specific immunity. Refractoriness of hosts carrying these parasites to reinfection was shown by Chandler (1939) and Burlingame and Chandler (1941) to be due entirely to a crowding effect. A unique situation exists in the case of the tapeworm, *Hymenolepis nana,* which may infect its host either as eggs, in which case there is a parenteral cysticercoid phase in the villi, or as cysticercoids developed in beetles, in which case there is no parenteral phase and, as might be expected, no specific immunity. It is now clear that antibody reaction to parasites may (1) directly or indirectly destroy them by combining with their body substance; (2) interfere with their nutrition, and consequently with their growth and reproduction, by combining with secreted products, probably by inhibition of necessary enzymes; or (3) protect the host by neutralizing toxic products. A fourth type occurs in virus infections—an antibody that combines with an integral portion of a virus, the portion which enables the virus to attach to host cells and enter them.

Antibody action of the first type, combination with body substances of the parasites, including their capsular substances, results in their immobilization, agglutination, enhanced susceptibility to phagocytosis, and lysis. Amebas and trypanosomes, as well as bacteria, exhibit such phenomena. To be affected by these antibodies, parasites must be directly exposed to them. Intracellularly located parasites, such as *Schizotrypanum cruzi, Leishmania,* and *Toxoplasma,* escape, but are vulnerable when made homeless by the disintegration of a host cell until they can invade another. Also, some extracellularly located parasites, such as relapsing fever spirochetes in the brain or leptospiras in kidney tubules, may be more or less protected from antibodies.

Antibodies of the second type, that interfere with nutrition and consequently with growth and reproduction, have been largely neglected in bacterial immunology, although they have been demonstrated in anthrax infections. Inhibition of reproduction of *Trypanosoma lewisi* in rats was demonstrated by Taliaferro and Taliaferro in 1922 and was later shown to be due to an antibody which was named "ablastin," but it was not until 1957 that Thillet and Chandler demonstrated that this was due to host reaction to metabolic products of the trypanosomes and could be induced by immunization with these products entirely free of trypanosomes. In 1958, Taliaferro and his colleagues found that ablastin inhibited synthesis of protein and nucleic acid. This could be either a cause or more likely, in the senior author's opinion, a result of inhibition of reproduction which is attributable to interference with processes of nutrition and secondarily with respiration. Chandler (1932–1937) and others found that a similar phenomenon occurred in infections in rats with the nematode, *Nippostrongylus,* manifested by (1) slower development to the adult stage, (2) stunted growth, and (3) inhibition of reproduction. It was suggested (Chandler, 1935) that these manifestations were likewise due to antibodies of the ablastin type, directed against secretions of the parasites, presumably involving enzymes utilized in nutrition. It was suggested that this might be a very widespread phenomenon and might be of great importance in functional immunity; it would account for the far greater effectiveness of migrating living larvae than dead ones in producing immunity. In bacterial infections, also, it is recognized that living organisms are more effective as vaccines than dead ones. Strong support for the idea of antibody reaction against metabolic products, now recognized as a common if not universal reaction in helminth infections, was provided by the demonstration of precipitates in the intestine and at the external openings of metazoan parasites placed in immune serum. This was first demonstrated by Gordon, Blacklock, and Fine in 1930 for the skin maggot, *Cordylobia,* later for *Nippostrongylus* larvae by Sarles in 1938, and still later for numerous other helminths (adult and larval nematodes, cercariae, miracidia, and eggs of schistosomes, and tapeworm embryos). Thorson in 1953 was the first actually to demonstrate increased resistance to helminth infections by immunization with metabolic products of *Nippostrongylus;* this was quickly followed by similar demonstrations by him for hookworms, by Campbell and others for *Trichinella,* and by Kagan for schistosomes.

It is interesting to note that such worms as *Trichinella* and *Nippostrongylus,* dependent for food on the intestinal mucosa, which would be expected to become saturated with antibody before there was a high titer of antibodies in the blood, produce immunity much more quickly than bloodsuckers such as hookworms or *Haemonchus.* A similar local concentration and consequent local manifestation of immunity was demonstrated by Blacklock and Gordon in *Cordylobia* infections in the skin of guinea pigs (see p. 780). Even when antigens are liberated into the general circulation they may stimulate more or less local immunity since they may have special affinity for, and be taken up by, particular tissues. This might explain results obtained in parabiotic twins (rats) with *Nippostrongylus* infections by Chandler (1935) and with *Trichinella* infections by Zaiman (1953). It is reasonable to suppose that antibodies might also be formed against parasite penetration enzymes (see p. 20), but the evidence thus far is meager.

Antibodies of the third type, which protect the host by combining with toxic products of the parasites, comparable with antitoxins against excreted bacterial toxins, are not conspicuous in parasitic infections since few if any protozoa or helminths seem to produce comparable toxic agents. However, antibody reaction against arthropod bites falls into this category and is considered further on pp. 529–530. It also was demonstrated by Esslinger in 1958 that screwworms secrete toxic substances against which the host can produce antibodies.

In this connection we must also consider the importance of allergic reactions. These are due to heightened sensitivity of tissues to "foreign" proteins, including secretions, waste products, and substances liberated from dead parasites, which may not be intrinsically toxic. Actually the symptoms associated with parasitic infections are in large part allergic in nature, e.g., excretions or secretions of such worms as *Trichinella,* hookworm larvae, and filariae; feces, shed cuticles, etc., of itch mites and *Hypoderma;* and salivary secretions of lice, fleas, red bugs, etc. The heightened sensitivity (allergy) is due to antibody production in cells directly exposed to the parasite antigens and is therefore often local. Combination of antigen with antibodies situated in or on the cells is injurious to the cells. Eventually, enough free circulating antibodies may be produced to neutralize antigens before they come in contact with antibody-loaded cells, and then we have immunity. Inflammatory tissue reactions due to allergy not only irritate the host, however, but may also interfere with the welfare of the parasites by impeding their migrations, or by preventing ingestion of food (e.g., by ticks, red bugs, and lice). On p. 28 we mentioned three phenomena associated with development of ablastic immunity to helminths. There are two others: (1) resistance to reinfection and (2) expulsion of worms already harbored upon continued reinfection, first demonstrated by Stoll in 1929 in sheep infected with *Haemonchus* (see p. 444) and termed "self cure." Both these phenomena are probably in large part allergic in character; the resistance to reinfection is due partly to quickened cellular reaction and encapsulation of migrating parasites, and the "self cure" apparently to irritation of sensitized tissues by the metabolic products associated with the exsheathment of fresh incoming worms (Stewart, 1955). Cortisone has an inhibitory effect on antibody production and decreases inflammatory tissue reactions of allergic nature. In infections such as trichinosis and filariasis, where allergic symptoms are of outstanding significance, cortisone may be of great benefit to the host (see pp. 411 and 483), whereas in infections such as *Trypanosoma lewisi* or *Haemonchus*, where antibody production is primarily inimical to the parasites, it may be harmful.

Diet seems to play an important role in the ability of an animal to develop resistance against parasitism. This was first strikingly demonstrated by Foster and Cort (1932, 1935) working with hookworms in dogs, and by Whitlock and others working with intestinal nematodes of domestic animals. The level of hookworm disease in a community appears to be more a measure of nutritional adequacy than of exposure to infection. Antibody production, nevertheless, seems to have a high priority for available protein in the body, so there may be more devious ways in which diet affects resistance. Read (1958) called attention to one; malnutrition may increase cortisone production which depresses resistance responses.

For most adult tapeworms, which do not stimulate antibody production, the diet, as long as it contains adequate carbohydrate of usable type, does not affect the worms but *does* help the host, permitting him to feed the worms as well as himself without injurious loss of proteins and vitamins.

Immunological diagnostic tests. There are numerous effects of antibody reaction against parasites or their products that are demonstrable *in vitro* or by skin tests. In schistosome infections, for instance, Kagan (1958) summarized the serological reactions as follows: Complement fixation with antigens extracted from cercariae or adult worms; precipitation of extracts of various life cycle stages; circumoval precipitates formed around living eggs by antibodies reacting with metabolic products of the enclosed miracidia; "cercarien hüllen reaction" (CHR)—formation of a transparent membrane around living cercariae; flocculation of inert particles on which antigens were adsorbed; agglutination of cercariae; agglutination of blood cells sensitized with tannic acid and coated with antigens extracted from eggs, cercariae, or adults; and immobilization of miracidia. In addition, skin reactions to injection of cercarial antigen are positive during active infection and for years after cure; whereas reaction of egg antigens are negative during active infection but become positive 6 months after cure. Another skin test is the Prausnitz-Küstner (PK) reaction; when immune serum is injected into the skin of an uninfected person, and followed by an injection of an antigen at the same site 24 hours later, a wheal and erythema develop. All these tests appear to be specific for schistosomes, but only the circumoval precipitin test is species-specific. In filarial infections there is little specificity as far as different kinds of filariae are concerned; hence in human filarial infections easily obtained antigens from *Dirofilaria* are used for tests. Possibly occasional false positive tests are due to accidental immunization with larvae of nonhuman filariae inoculated by mosquitoes. Where direct demonstration of parasites or their eggs or larvae is difficult or unreliable, serological or skin tests may be valuable in diagnosis, e.g., in chronic protozoan infections (extra-intestinal amebiasis, kala-azar, Chagas' disease, and toxoplasmosis) or such helminthic infections as schistosomiasis, hydatid disease, cysticercosis, and filariasis. Such a test is badly needed for visceral larva migrans and research is under way to find one. Serological tests will be mentioned later in appropriate chapters. Sometimes special serological tests have been developed, e.g., the "milky gel" tests for kala-azar and the dye test for toxoplasmosis. Precipitin tests can also be used to identify the source of meals in blood-sucking arthropods.

The Names of Parasites

In all branches of natural history it has been found not only expedient but also necessary to employ scientific names, for there are estimated to be more than 10 million species of animals. Common names, like nicknames, vary from place to place, and often the same name is applied to quite different organisms in different places. Linnaeus, in the eighteenth century, devised a system of "binomial names" which consisted of the genus name, beginning with a capital letter, followed by a species name, in zoology beginning with a small letter, and both Latinized in form, since Latin came nearer to being a universal language than any other. Strictly, the genus and species names are followed by the name of the man who first gave the species name, in parentheses if the genus name is not the one he originally used, but in ordinary references to species this is omitted. The genus name may be likened to a surname and the species name to a given name, e.g., *Ascaris lumbricoides* is comparable to Smith, John.

Family names in zoology always have the ending "idae" attached to the root of the type genus, e.g., Muscidae from *Musca,* Ascarididae from *Ascaris* (root ascarid); superfamily names end in "oidea," but there is no standard ending for orders or classes. In botany the family ending is "aceae," e.g., Spirochaetaceae.

In order to avoid confusion there were adopted (in 1904) rules of nomenclature, known as the International Code of Zoological Nomenclature, which makes it impossible for any two animals to have the same name. A genus name can apply to only one genus in the entire animal kingdom, and a species name to only one species within a genus. The tenth edition of Linnaeus' *Systema Naturae* (1758) is accepted as the starting point for the names, no name proposed prior to that time having any standing. The first valid name given an animal is considered the correct one. Of course, if an animal is put in the wrong genus, it must be transferred to the right one. If a genus is split up, the animal may have to be placed in a new genus; for example, the old genus *Oxyuris* has been split into a number of genera. *Oxyuris equi* of the horse was the earliest one placed in the genus, therefore the *restricted* genus *Oxyuris* must contain this species and any others which fall into its subdivision of the old genus; since the human oxyurid falls into a different subdivision it comes out with the next available genus name, *Enterobius.* For the same reasons *Filaria bancrofti* is now *Wuchereria bancrofti,* etc. If two genera are combined, the older genus name applies to all the members of the merged genera. If the same animal is given different species names by different workers, the earliest name applies.

Although this system was established to prevent confusion, in many instances strict application of the rules has resulted in just the opposite. The number of possible errors and misinterpretations are disheartening, in consequence of which names, long recognized and accepted, have to be discarded for others, because someone shows that the established name was really first applied to another species, or an earlier name was overlooked, or for some other reason. Unfortunately, the commoner animals are the ones which suffer most, for they are the most likely to have been redescribed by various workers and to have been shifted about from genus to genus. Unfortunate as this situation is, it is better than having no rules at all, and steps are now being taken to make names which have been in common usage for many years inviolable. The synonymy, or list of aliases, of some of our common parasites is already deplorably long. In some instances there is a difference of opinion as to what the correct name should be.

Although the scientific names are sometimes barbarously long and at first may be very annoying and even terrifying, every student of parasitology, as of every other branch of biology, must overcome any childish aversion he may have for them, and become used to accepting and using them. They are not obstacles to be avoided, but valuable tools without which there would be hopeless confusion.

REFERENCES

Audy, J. R. 1958. The localization of disease with special reference to the zoonoses. *Trans. Roy. Soc. Trop. Med. Hyg.*, 52: 308–328.

Baer, J. G., et. al. 1957. *Symposium Spéc.ficité Parasit.* Univ. Neuchatel, Switzerland.

Ball, G. H. 1943. Parasitism and evolution. *Am. Naturalist,* 78: 345–364.

Becker, E. R. 1933. Host specificity and specificity of animal parasites. *Am. J. Trop. Med.*, 13: 505–523.

1953. How parasites tolerate their hosts. *J. Parasitol.*, 39: 467–480.

Bozicevich, J. 1951. Immunological diagnosis of parasitic diseases; in Most, *Parasite Infect'ons in Man,* Chapter 4, 37–55.

Brand, T. von. 1952. *Chemical Physiology of Endoparasitic Animals.* Academic Press, New York.

Cameron, T. W. M. 1952. Parasitism, evolution and phylogeny. *Endeavor,* 9: 44.

Caullery, M. 1950. *Le Parasitism et la Symb'ose.* 2nd ed. G. Doin, Paris.

Chandler, A. C. 1923. Speciation and host relationships of parasites. *Parasitology,* 15: 326–339.

1932. Experiments on resistance of rats to superinfection with the nematode, *Nippostrongylus muris. Am. J. Hyg.*, 16: 750–782.

1935–1937. Studies on the nature of immunity to intestinal helminths. (Ref. at end of Chapter 11.)

1948. Factors modifying host resistance to helminthic infections. *Proc. 4th Intern. Congr. on Trop. Med. Malaria,* 2: Sect. 6, 975–983.

1959. The role of malnutrition in helminthic infections. *Proc. 6th Intern. Congr. on Trop. Med. Malaria,* Lisbon.

Culbertson, J. T. 1941. *Immunity against Animal Parasites.* Columbia University Press, New York.

Elton, C. 1935. *Animal Ecology,* Chap. 6. Macmillan, New York.

Faust, E. C. 1931. The nosogeography of parasites and their hosts. *Puerto Rico J. Public Health Trop. Med.*, 6: 373–380.

Hall, M. C. 1930. The wide field of veterinary parasitology. *J. Parasitol.*, 16: 175–184.

International Rules of Zoological Nomenclature. 1926. *Proc. Biol. Soc. Wash.*, 39: 75–103.

Kagan, I. G. 1958. Contributions to the immunology and serology of schistosomiasis. *Rice Inst. Pamphlet*, 45: 151–183.

Lewert, R. M. 1958. Invasiveness of helminth larvae. *Rice Inst. Pamphlet*, 45: 97–113.

May, J. M. 1958. *The Ecology of Human Disease. MD Publ.*, New York.

Metcalf, M. M. 1929. Parasites and the aid they give in problems of taxonomy, geographical distribution, and palaeogeography. *Smithsonian Misc. Collections*, 81: No. 8.

Oliver-Gonzalez, J. 1944. Blood agglutinins in blackwater fever. *Proc. Soc. Exptl. Biol. Med.*, 57: 25–26.

1946. Functional antigens in helminths. *J. Infectious Diseases*, 78: 232–237.

Pavlovsky, E. N., et al. 1955. Natural nidi of human disease and regional epidemiology. (In Russian.) Summaries in *Rev. Appl. Entomol. B*, 1956. 44: 145.

Read, C. P. 1958. Status of behavioral and physiological "resistance." *Rice Inst. Pamphlet*, 45: 36–54.

Smith, Theobald. 1934. *Parasitism and Disease.* Princeton University Press, Princeton, N. J.

Stewart, D. F. 1955. "Self-cure" in nematode infestations of sheep. *Nature*, 176: 1273–1274.

Strong, R. P. 1935. The importance of ecology in relation to disease. *Science*, 82: 307–317.

Stunkard, H. W. 1929. Parasitism as a biological phenomenon. *Sci. Monthly*, 28: 349–362.

Taliaferro, W. H. 1929. *Immunology of Parasitic Infections.* Century, New York.

1940. The mechanism of acquired immunity in infections with parasitic worms. *Physiol. Revs.*, 20: 469–492.

Part I

PROTOZOA

Chapter 3

INTRODUCTION TO PROTOZOA

Place of Protozoa in the animal kingdom. It is usual for zoologists to divide the entire animal kingdom into two great subkingdoms, the Protozoa and the Metazoa. These groups are very unequal in number of species. The Metazoa include all the animals with which the majority of people are familiar, from the simple sponges and jellyfishes, through the worms, mollusks, and the vast hordes of insects and their allies, to the highly organized vertebrate animals, including man himself. The Protozoa, on the other hand, are with few exceptions microscopic or almost microscopic animals, whose very existence is unknown to the average lay person. There is no question but that in point of numbers of individuals the Protozoa exceed the other animals, millions to one; a pint jar of stagnant water may contain many millions of these minute animals. Over 15,000 species of Protozoa have been described, but it is probable that there are thousands more which are not yet known to science.

Although Protozoa are usually considered to be fundamentally different from Metazoa by being unicellular instead of multicellular, the distinction is not as sharp as it would at first appear; some Protozoa form multinucleated plasmodial masses suggestive of syncytial tissues in Metazoa, and some colonial forms not only have somatic and reproductive cells differentiated from each other, but the colonies can also move and respond as units, and exhibit some degree of differentiation of anterior and posterior ends. The difference between such Protozoa and the simpler Metazoa is merely one of degree. Besides, although some complex Protozoa are not divided into cells, they have a greater variety of structurally different parts than some of the Metazoa (Fig. 1). For this reason some biologists prefer to think of the Protozoa as noncellular organisms rather than as single-celled ones, since the latter designation suggests that they are to be compared with individual cells of a metazoan body.

The distinctions between Protozoa and other primitive organisms that exist as single cells or simple colonies is even more difficult, for there are transitional forms which link them to bacteria, fungi, and algae. In general they differ from bacteria in having distinct membrane-bound nuclei and in exhibiting sexual phenomena and often complicated life cycles, but there are a few Protozoa which have the

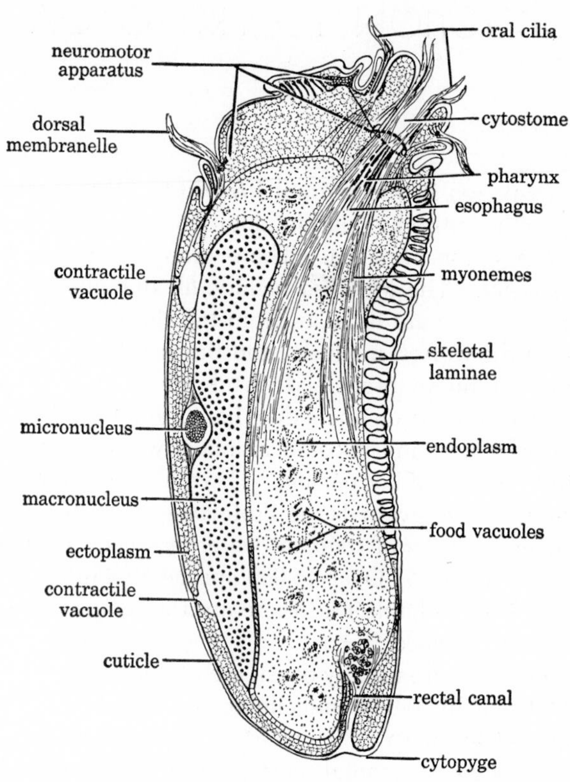

Fig. 1. A complex ciliate, *Diplodinium ecaudatum,* showing highly developed organelles. (After Sharp, *Univ. Calif. Publ. Zool.,* 13, 1914.)

nucleus broken up into many parts that are little more than granules, and some in which no sexual processes have been observed. Besides, even bacteria have been suspected of having sex, although on somewhat shaky grounds. Chemical reactions, staining properties, and the like are sometimes resorted to as distinguishing characters, but on these bases the spirochetes should be aligned with the Protozoa. Nevertheless, these organisms have now quite generally been abandoned to a botanical fate. The slime molds, which the protozoologists call Mycetozoa and include with the Sarcodina, are also claimed by the botanists, who put them in with the fungi and call them Myxomycetes.

Even more confusing is the case of the green flagellates, which protozoologists put in a subclass of the Mastigophora while botanists include them among the green algae. As Hall (1953) remarked, this suggests that protozoologists are unable to distinguish between animals and plants, which is disconcerting to those who like their taxonomy simple and consistent. To remedy this situation Calkins (1933) ejected the entire group of chlorophyll-bearing flagellates (Phytomastigophorea) from their relatives among the Protozoa. Some of the nonchlorophyll-bearing forms, obviously close cousins which had secondarily lost their chlorophyll, were arbitrarily transferred to Zoomastigophorea and thus retained in the Protozoa, but this kin-splitting did not make anybody very happy. It became even more difficult when it was shown that green flagellates could be "cured" of their chlorophyll by treatment with streptomycin.

The principal difference between plants and animals, when we get down to these primitive forms, is in their manner of nutrition. Plants synthesize their organic compounds from simple inorganic substances like CO_2, H_2O, and nitrates, with the aid of chlorophyll, whereas animals utilize ready-made organic compounds or break down more complex ones and reassemble the parts to suit their needs. But according to this criterion a green flagellate (or alga!) like *Euglena* can be, and is, a plant by day and an animal by night. It is clear that between the higher plants and higher animals there is a broad no-man's land of single-celled organisms which might be segregated into a buffer state, for which the name Protista was suggested by Ernst Haeckel many years ago. The boundary between Protista and Metazoa is fairly sharply defined, but that between the Protista and the lowest forms of Metaphyta (algae and fungi) is much more arbitrary.

Structure. A protozoan, in its simplest form, conforms to the usual definition of a cell: a bit of cytoplasm containing one or more nuclei. In most Protozoa, even though in some cases there may be a number of nuclei present, these are all of one kind, but in the ciliates, except the primitive opalinids that inhabit the rectum of Amphibia, there are two quite distinct types of nuclei, a macronucleus filled with densely staining granules, and a micronucleus which is vesicular in structure, more like the nuclei of other Protozoa (Fig. 1); sometimes there may be a number of one or both kinds. In many Protozoa, e.g., the intestinal amebas, there is an endosome near the center, and in some of these there are deep-staining granules encrusted on the inner surface of the nuclear membrane (Figs. 3, 4). The endosome may or may not contain chromatin; the chromosomes may form out of a zone of minute granules between the endosome and the nuclear membrane.

In most, but not all, cases division of the nucleus is accomplished by some form of mitosis or a process at least hinting at it; there is, however, no uniformity in the process as there is in Metazoa. Nature seems to have been experimenting with nuclear division in the Protozoa. Typical chromosomes are formed in some Protozoa, e.g., many amebas, but often there is no clear evidence of them. In *Entamoeba* mitosis takes place entirely within the nuclear membrane; a characteristic feature is the division of a *centriole* in the endosome into two, which migrate to opposite ends of the intranuclear spindle, but remain connected by a deep-staining strand called an *intradesmose* until division of the six or eight chromosomes is completed. In the ciliates the macro- and micronuclei are formed after sexual reproduction from a single micronucleus.

The body of some of the simpler flagellates and amebas has no true cortex or pellicle, although the outer layer of the cytoplasm may be denser and less granular, forming an *ectoplasm,* in contrast to the *endoplasm.* Most other Protozoa have some sort of pellicle or cortex, giving them more or less definite shape. In ciliates the cortex is thick and contains a variety of structures (Fig. 1). Many Protozoa, particularly Sarcodina, produce shells of cellulose, chitin, cemented sand grains, silica, lime, or other substances, and some flagellates and ciliates have transparent chitinous loricas or tests, sometimes with collars (Fig. 2*E*).

Organelles. The term *organelle* is used in place of organ for structures that are only parts of a single cell. The organelles contained in a protozoan's body may be many and varied. Those connected with movement or locomotion differ in different groups. The simplest type of movement is by means of simple outflowings of the body cytoplasm known as *pseudopodia* (Fig. 2*A*). These are used both for locomotion and for the engulfing of food. In some species, e.g., the amebas, they are blunt, lobelike projections of the body, but in others they are very slender and tapering, and in the Foraminifera they branch and anastomose into complex food-trapping networks. Some pseudopodia are permanently supported by axial rods, and these are called *axopodia* (Fig. 2*B*). Pseudopodia are the characteristic organs of locomotion of the entire class Sarcodina, to which the amebas belong, but many flagellates and Sporozoa, e.g., the malaria parasites, also have the power of ameboid movement by means of pseudopodia.

Flagella and *cilia* are usually constant in arrangement and form.

Flagella (Fig. 2*C*) are characteristic of the class Mastigophora, but they also occur in some stages in the life cycle of certain amebas and in the spermlike microgametes of Sporozoa. They are long, whiplike outgrowths, capable of violent lashing or of rippling movements, and are composed of a fine filament, the *axoneme,* surrounded by a thin film of cytoplasm. The majority of species has only one or two flagella,

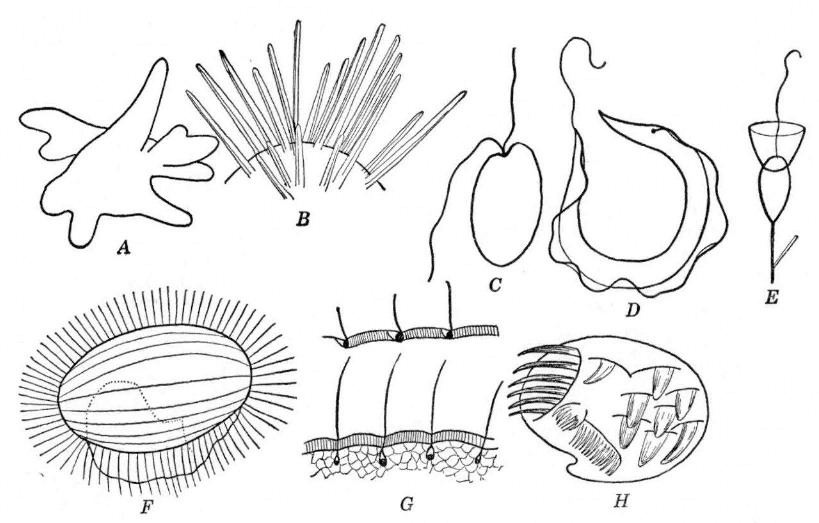

Fig. 2. Types of organs of locomotion in Protozoa: *A*, *Amoeba* with pseudopodia; *B*, a heliozan with axopodia; *C*, *Bodo*, with two free flagella; *D*, *Trypanosoma* with flagellum attached to undulating membrane; *E*, choanoflagellate with flagellum and collar; *F*, *Pleuronema* with cilia and membranelle formed of fused cilia; *G*, modes of insertion of cilia; *H*, *Aspidisca* with cirri. (Figures *F* to *H* adapted from Calkins, *Biology of the Protozoa.*)

although many parasitic forms may have up to eight, but in some of the parasites (or symbionts) of termites and wood-roaches there may be hundreds. The flagella may be directed forward or trail behind, or may be attached to the side of the body by a delicate *undulating membrane* (Fig. 2*D*); if more than one is present they all may be alike, and perform similar functions, or they may be widely different.

A flagellum always arises from a minute deep-staining body called a *basal granule* or *blepharoplast* (Fig. 15). In many parasitic flagellates there is another deep-staining body called the *parabasal.* In the trypanosomes and their allies a similar body is called a *kinetoplast* (Fig. 17); it differs in being Feulgen-positive, therefore containing desoxyribonucleic acid, an essential ingredient of chromatin. In some

Protozoa the blepharoplast is connected with the nucleus by a fiber called a *rhizoplast.* The kinetoplast may also be connected with the blepharoplast by a fibril or even by a cone of fibrils (Fig. 24). Some flagellates have other specialized organelles connected with the blepharoplast, e.g., the *costa,* extending along the base of the undulating membrane in trichomonads, and one or more *axostyles,* varying from delicate filaments to a stout rod, also found in trichomonads as well as in some other flagellates, and acting as supporting rods (Fig. 10).

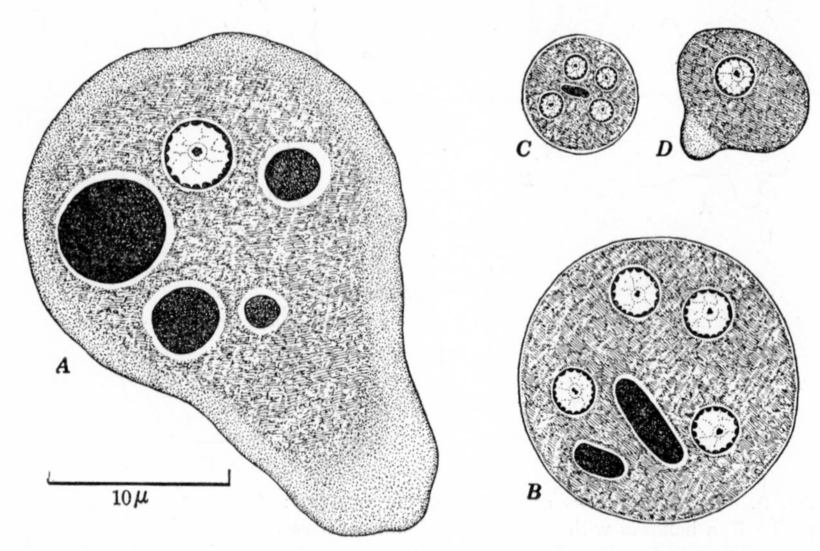

Fig. 3. *Entamoeba histolytica:* A and *B*, trophozoite and cyst of invasive, large-race form; *C* and *D*, cyst and trophozoite of lumen-dwelling, small-race form (*E. hartmanni* of European authors). ×2500.

Cilia (Fig. 2*F*), which are characteristic only of the subphylum Ciliophora, have a structure similar to flagella, and like them arise from individual basal granules (Fig. 2*G*), but they are much shorter, more numerous, and beat rhythmically by a bending to one side. Cilia have much more coordination of movement than flagella, and regular waves of beats of the cilia can be seen passing over the body of a ciliate. The cilia may be fairly evenly distributed over the body in rows or may be in patches, and there are usually enlarged or specialized cilia, or membranelles formed of fused cilia, in the peristomal or other regions (Fig. 2*F*). In some creeping forms there are tufts of cilia fused together into stout organs called *cirri* (Fig. 2*H*).

Many Protozoa possess delicate contractile fibrils called *myonemes* (Fig. 1) which run in various directions in the ectoplasm or pellicle of the animal. In some flagellates and ciliates fibrils and minute deep-staining bodies have been described and have been interpreted as a more or less highly organized *neuromotor apparatus,* i.e., a definitely arranged and organized substance having a nervous control over the myonemes and cilia or flagella.

Organelles for food-taking occur chiefly in the flagellates and ciliates. Such Protozoa may have a *cytostome* or cell mouth for the ingestion of food (Fig. 1) and a *cytopyge* or cell anus for the elimination of waste matter. They also have a delicate membranous *cytopharynx* for leading the food material into the endoplasm, and *food vacuoles* into which the food is accumulated and where it is circulated inside the body. In some protozoans, namely the Suctoria, a much modified group of ciliates, there are developed sucking tentacles for the absorption of food. In others there are tiny capsules in the ectoplasm, the *trichocysts,* containing minute threads which can be shot forth when stimulated and used either for overpowering prey or for protection from enemies.

Protozoa, particularly those living in fresh water or soil, usually have one or more contractile vacuoles (Fig. 1). These are little cavities in the cytoplasm in which water collects, and which periodically contract, forcing their contents to the outside of the cell, sometimes through definite excretory pores on the pellicle. Their primary function seems to be as hydrostatic regulators, to get rid of water which enters the denser cytoplasm of the organisms by osmosis. To a minor degree these act as excretory organs for elimination of metabolic wastes, but the many parasitic and marine Protozoa which lack contractile vacuoles, not needing them as osmotic regulators, seem to suffer no inconvenience; they get rid of their metabolic wastes through the body wall quite satisfactorily. The presence of a contractile vacuole is one feature by which free-living amebas in feces can be distinguished from true inhabitants of the intestine.

Sense organs in the form of pigment spots sensitive to light and processes sensitive to chemical substances, giving, perhaps, a sensation comparable to taste, are present in some free-living species.

Although no protozoan possesses all these organelles, many possess a considerable number of them and exhibit a degree of complexity and organization almost incredible in a single-celled animal which is barely, if at all, visible to the naked eye.

Physiology and reproduction. In their physiology and manner of life the Protozoa differ among themselves almost as much as do the Metazoa. Some ingest solid food through a cytostome or wrap themselves around the food; others possess chlorophyll and are nourished in a typical plant manner, and still others absorb nutriment by osmosis from the fluids or tissues in which they live. Ingested food particles are surrounded by fluid, forming *food vacuoles* (Fig. 1), which circulate in the endoplasm. In ciliates they follow a regular course. Digestive fluids appear to be secreted into the vacuoles, and the vacuoles develop an acid reaction during digestion, later becoming neutral again. Undoubtedly the substances that can be digested vary widely with different Protozoa. Some species, e.g., *Entamoeba histolytica,* excrete substances which dissolve blood corpuscles and tissue cells outside the body, the soluble product being then absorbed through the body wall. Indigestible residue from solid food is extruded through the body wall; in forms having a pellicle this takes place through a cytopyge. Reserve food material is stored as glycogen, fats, oils, and other substances. Many parasitic forms store food in the form of *volutin* or *metachromatic granules,* which stain like chromatin. As cysts form, some amebas form *chromatoid bodies* (Figs. 4 and 5) which may contain reserve protein material.

Most free-living Protozoa are aerobic, using free oxygen in their respiration, but some, like certain bacteria, are anaerobic. Among parasites the respiration is sometimes aerobic and sometimes anaerobic, or it may be aerobic *or* anaerobic, according to the availability of oxygen.

The malaria parasites, leishmanias, and some trypanosomes have an aerobic respiration, with or without the participation of iron-containing systems as in higher organisms, whereas such parasites as the trichomonads and the intestinal amebas are characterized by an anaerobic metabolism.

The multiplication or reproduction of Protozoa is of two quite distinct types, an asexual multiplication, more or less comparable with the multiplication of cells in a metazoan body, and sexual reproduction, comparable with a similar phenomenon in the higher animals. Several common asexual methods of multiplication occur amongst protozoans, namely, *simple fission,* or division into two more or less equal parts; *budding,* or separation of one or more small parts from the parent cell; and multiple fission or *schizogony,* which results from multiple or repeated division of the nucleus before the cytoplasm divides, thus producing a whole brood of offspring. In the flagellates simple fission is longitudinal, usually beginning with the blepharoplast, whereas in ciliates it is transverse. In flagellates the old flagella may be retained by one daughter, and new ones grow out from the blepharoplasts for the other, or the old ones may disappear and new ones form. Multiplication occurs in encysted forms in some species but not in others.

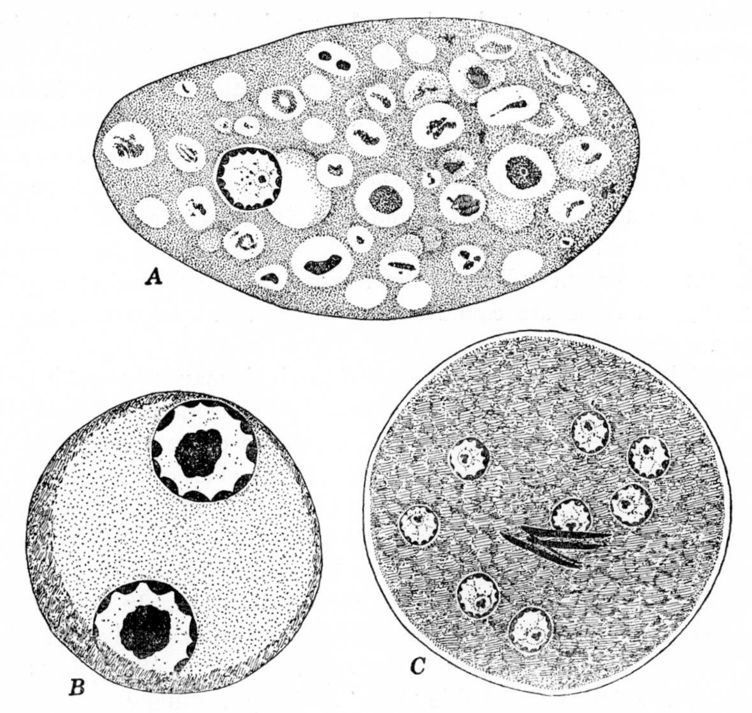

Fig. 4. *Entamoeba coli: A,* trophozoite; *B,* 2-nucleated cyst with large glycogen vacuole; *C,* young, mature cyst with splinterlike chromatoid bodies. ×2500.

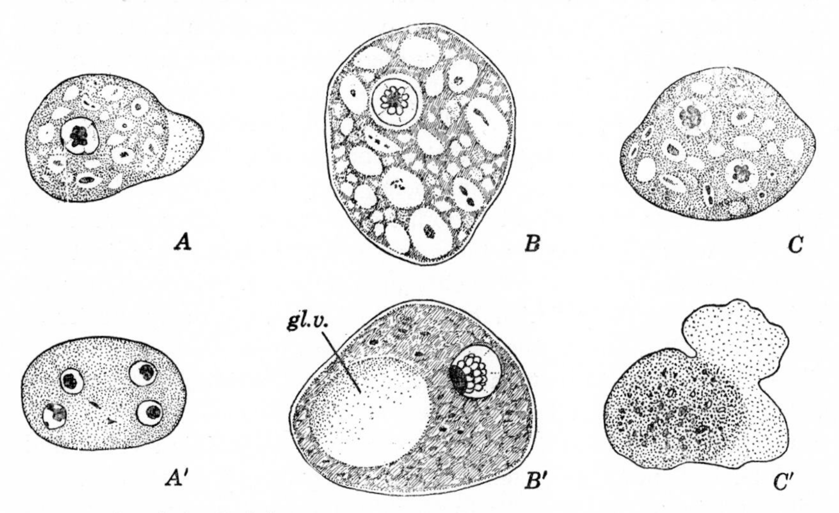

Fig. 5. *A* and *A′*, *Endolimax nana,* trophozoite and cyst; *B* and *B′*, *Iodamoeba bütschlii,* trophozoite and cyst, latter showing large glycogen vacuole (*gl.v.*); *C* and *C′*, *Dientamoeba fragilis,* stained and living specimens, the latter with flat leaflike pseudopodia. ×2500.

Protozoa which are in the phase of asexual multiplication are called *trophozoites,* in contrast to *gametocytes* which give rise to sex cells, and to *cysts,* which do not grow or multiply, although, early in their formation, the nuclei may multiply. The trophozoites of the Sporozoa, which multiply by schizogony, are called *schizonts,* and the daughters resulting from multiple division are called *merozoites.* After the sexual process a different form of multiple fission occurs called *sporogony,* ending in the formation of *sporozoites.* Cells intermediate between the parent cells and the merozoites or sporozoites may be formed, and these are called agametoblasts and sporoblasts, respectively.

Multiplication by one of the asexual methods may go on with great vigor for a long time, but sooner or later some modification of the process occurs. In many Protozoa a process comparable to sexual reproduction in higher animals occurs. In the ciliates this takes place by *conjugation,* i.e., a temporary union of two individuals during which time a daughter nucleus of one enters the other and fuses with a daughter nucleus, and vice versa; at the end of the process the two individuals separate, each being now a fertilized cell. In many other Protozoa two individuals, the *gametes,* unite permanently and their nuclei fuse, a process which is known as syngamy. Sometimes the gametes are indistinguishable from ordinary asexually multiplying individuals, whereas in other instances the gametes are cells produced by a special process of multiplication; the parent cell is then called a *gametocyte.* When there is no visible difference between the gametes, the process of fusion is called *isogamy,* whereas when the gametes differ in size, form, motility, etc., the process is called *anisogamy.* There are, however, all gradations between isogamy and a condition of anisogamy in which one gamete, the *macrogamete,* corresponds closely to an ovum, being large, immobile, and with a relatively large amount of cytoplasm charged with reserve food material, whereas the other, the *microgamete,* is relatively minute, is actively motile by means of flagella, and contains very little cytoplasm, being thus essentially similar to a spermatozoon. In many species of parasitic Protozoa, e.g., the malaria parasites, the sexual cycle takes place in an alternate host; in others, e.g., the Coccidia, it takes place outside the body of a host.

In many parasitic Protozoa, for example, the parasitic amebas and the intestinal and blood flagellates, no sexual process has been observed with certainty. In those Protozoa which do not show true sexual reproduction, i.e., exchange of nuclear material between different individuals, resulting in mixing of hereditary characters and also in rejuvenation of the cells, it is likely that some process occurs which brings about the rejuvenation. Protozoan cells tend to grow old after continued asexual multiplication and lose their youthful vitality and reproductive power, just as do the cells of a metazoan animal. The recent demonstration by Honigberg and Read of an induced transformation in a trichomonad protozoan suggests that recombination of genetic characters may occur without the usual forms of sexual reproduction. Such transformations should also be sought in the trypanosomes.

Encystment. A great many Protozoa, at some time in their life cycle, are able to form more or less impervious protective capsules around their bodies, enabling them to survive unfavorable environmental conditions such as desiccation, unfavorable temperatures, injurious chemicals, or lack of oxygen. This process is called *encystment.* It is by this means that many parasitic Protozoa are able to survive conditions outside the body and to pass through the inhospitable environment of the stomach to reach the intestine or other organs of new hosts. Most parasitic Protozoa that are not transmitted by intermediate hosts resort to cyst formation to gain access to new hosts, though a few, e.g., *Trichomonas* and *Dientamoeba,* manage without this.

In many Protozoa of water and soil, encystment occurs as a reaction to desiccation, but in the parasitic amebas and flagellates it is a normal phase and the life cycle, cysts being formed even when conditions are entirely satisfactory for continued multiplication of trophozoites. Encystment and excystment may both occur in the same culture medium, but each is favored by certain chemical and physical characteristics of the environment.

In the Sporozoa cyst formation is associated with sexual reproduction, the zygotes (fertilized gametes) being enclosed in *oöcysts* in which the sporogonic multiplication occurs, ending in the formation of a few to thousands of *sporozoites.* The sporozoites may be free in the oöcysts, as in the Haemosporidia, to which the malaria parasites belong; they may be enclosed, singly or in groups, in capsules of their own called spores, as in the gregarines; or they may be enclosed in sporocysts inside the oöcysts, as in the Coccidia. Trophozoites of amebas, flagellates, or ciliates prepare to encyst by ceasing ingestion of food and extruding food residue, but they frequently store up considerable amounts of reserve food in the form of glycogen, volutin granules, or chromatoid masses. This is the precystic stage. Then the cyst wall forms, and in many species a multiplication of the nucleus ensues. Mature cysts of *Entamoeba coli,* for instance, have eight nuclei and those of *E. histolytica* have four, but in *Iodamoeba* cysts there is only one.

When an encysted organism arrives in a favorable environment it excysts and begins to multiply in the trophozoite stage. In some amebas there is a complicated series of nuclear divisions in the multinucleate individual that escapes from the cyst, before any unicellular amebas are set free.

Classification. It is little wonder that the varied assemblage of single-celled animals constituting the group Protozoa should be difficult to classify. Many undergo profound modifications in the course of their life cycles, and the entire life cycle must be considered in any scheme of classification.

For a long time it was customary to divide Protozoa into four classes: the Rhizopoda or amebalike forms, the Mastigophora or flagellates, the Ciliata or ciliates, and the Sporozoa or spore-forming parasitic forms. Doflein, however, modified this by first splitting the entire phylum Protozoa into subphyla, the Plasmodroma and the Ciliophora, an arrangement that has been quite generally followed by protozoologists. For our purposes we shall adopt the classification used by Jahn and Jahn (1949) and by Hall (1953), using the uniform endings for zoological names which Pearse (1936) suggested: for phyla and subphyla, *a;* class, *ea;* subclass, *ia;* order, *ida;* suborder, *ina;* and as everywhere used, for family, *idae* (*aceae* in botany). These have not been uniformly adopted by zoologists, but the idea is a good one. The classification presented below is a tentative one. Radical revisions in classification of Protozoa are now being worked out, but general agreement among specialists has not yet been attained. Further classification of the subphyla Mastigophora and Sporozoa is given on pp. 86–87 and 161–164, respectively.

Subphylum I. **Mastigophora.** With flagella throughout most of the life cycle, and a definite pellicle usually covering body. Sexual reproduction, where known, by syngamy. Flagellates.

Class I. **Phytomastigophorea.** Majority with chromatophores containing chlorophyll. Plantlike flagellates, arranged in six orders, and including such well-known free-living forms as *Euglena* and dinoflagellates.

Class II. **Zoomastigophorea.** No chromatophores; store lipids and glycogen, but no starch or paramylum, and with no cellulose membranes or tests. Arranged in five orders, including all the parasitic flagellates.

Subphylum II. **Sarcodina.** Possess pseudopodia throughout most of life cycle, but may have flagella at some stage. Body with very delicate pellicle or none, but many free-living forms with tests or shells.

Class I. **Actinopodea.** Floating or sessile free-living organisms with axopodia (see p. 40). Peripheral cytoplasm foamy in character. Includes Radiolaria and Heliozoa.

Class II. **Rhizopodea.** Pseudopodia of lobose, filar, or anastomosing network types, but never axopodia or foamy peripheral cytoplasm. Arranged in five orders, including slime molds, amebas, and Foraminifera.

Subphylum III. **Sporozoa.** Parasitic forms with complicated sexual and asexual phases of the life cycle, and without locomotor organs in the adult stage.

Class I. **Telosporidea.** Zygote undergoes multiple fission (sporogony), producing sporozoites which after entering a host become trophozoites (schizonts), which multiply by schizogony and eventually produce gametocytes, except in the Eugregarinida, in which the trophozoite eventually becomes a gametocyte. Arranged in three subclasses and six orders. Includes gregarines, Coccidia, Haemosporidia, etc.

Class II. **Cnidosporidea.** Zygotes give rise to one or more trophozoites, without sporogony. Trophozoites produce characteristic spores composed of several cells, including one or more polar capsules containing coiled filaments that can be shot out, and one or more *sporoplasms,* analogous to sporozoites of the Telosporidea. Arranged in four orders, containing many important parasites of fish (Myxosporida) and of invertebrates (Microsporida).

Class III. **Acnidosporidea.** Life cycle somewhat like Cnidosporidea, but spores without polar capsules. Includes Haplosporidea, parasites of fishes and invertebrates.

Subphylum IV. **Ciliophora.** Possess cilia in some stage of the life cycle.

Class I. **Ciliatea.** Cilia or structures made from them (cirri, membranelles) present throughout life cycle. Arranged in two subclasses: Protociliatia, with nuclei all alike and sexual reproduction by syngamy; and Euciliatia, with nuclei of two types, macronucleus and micronucleus, and sexual reproduction by conjugation.

Class II. **Suctorea.** Adults without cilia, but with suctorial tentacles. Larval forms produced by budding, have cilia for short period. Sexual reproduction by conjugation in most forms.

The subphylum Sarcodina includes mainly free-living forms inhabiting ocean, fresh water, and soils. Some marine forms, e.g., the Foraminifera, are instrumental in building up chalk deposits out of their shells, and are useful in distinguishing geological strata. Others, the Radiolaria, form vast deposits of "radiolarian ooze." Only a few are parasitic, and these are all typical amebas which produce pseudopodia from any part of the naked body.

The subphylum Mastigophora includes a vast assemblage of organisms called flagellates, many of which bridge the gap between plants and animals. Here again the majority of the included forms are free-living; many of them possess chlorophyll and live like typical plants. Others have cytostomes through which they ingest solid food as do animals, and still others absorb dissolved substances by osmosis through their cell walls. Some, like *Euglena,* physiologically may be plants in the daytime and animals by night. All the parasitic species are of animal nature, feeding either by ingestion or by osmosis. Formerly the spirochetes were associated with the flagellates because of a supposed relationship with the trypanosomes, but this idea has long since been exploded, and spirochetes are now placed in a group by themselves, associated with bacteria rather than Protozoa, though in some respects they show affinity to the latter.

The subphylum Sporozoa includes a varied assemblage of parasitic forms, the relationships of which are discussed at the beginning of Chapter 9. They include numerous important agents of disease not only for all sorts of vertebrates, but also for invertebrates. Man is seriously afflicted only by the malaria parasites, but domestic animals are attacked by a number of different types.

The subphylum Ciliophora includes the most highly organized Protozoa. In one subclass, Protociliatia, are placed the ciliates which have from two to several hundred nuclei all of one kind, and which reproduce by fusion of gametes, thus being intermediate between the other Protozoa and the higher ciliates with functionally distinct nuclei and sexual reproduction by conjugation. Some zoologists consider the Protociliata to belong to the Mastigophora. Nearly all the protociliates are inhabitants of the large intestine of frogs and toads. Members of the subclass Euciliatia have the most complicated organization of any Protozoa. The majority are free-living forms found in abundance in foul water, hay infusions, etc.; whence the name "Infusoria" sometimes applied to them. Many inhabit the rumen or intestines of herbivorous animals, but, since they prey on bacteria and debris and do not attack the host itself, they may be regarded as commensals rather than parasites. Some are even symbionts, since they digest cellulose and in other ways contribute to the nutrition of the host. A few are at least potentially true pathogenic parasites, e.g., *Balantidium* (see p. 106), which parasitizes man, monkeys, and pigs.

The class Suctorea, which lose their cilia and acquire suctorial tentacles as adults, are for the most part free-living organisms attached to various objects in water, but a few are parasitic on ciliates, and one, *Allantosoma intestinalis,* is of interest as a parasite of ciliates in the cecum of horses.

Parasitism and host specificity. It is very likely that parasitism among the Protozoa arose in the beginning by the ingestion by animals of free-living forms. Some of these may be conceived of as having found conditions of life satisfactory in the digestive tracts of animals which devoured them; in the course of time such forms would become more and more perfectly adapted to the new environment, and eventually lose their power to live and reproduce in the outside world. Such parasitism would be expected to occur first in cold-blooded aquatic animals and subsequently to extend to warm-blooded land animals. It is significant that most of the common genera of intestinal Protozoa of man, e.g., *Entamoeba, Chilomastix, Trichomonas,* and *Giardia,* have representatives in the Amphibia, in some cases so closely similar to the human species as to have cast doubts on their specific distinctness.

Many of the blood Protozoa have undoubtedly arisen by a process only slightly more complicated. They first adapted themselves to the digestive tracts of invertebrates; in bloodsuckers they would then become adapted to living in the presence of the blood on which the invertebrates fed; having survived this probationary treatment, such parasites might then be capable, if inoculated into the blood stream or tissues of the vertebrates on which their invertebrate hosts habitually fed, of adapting themselves to life in this new environment, which had thus been approached in an indirect manner. There is little room for doubt that the leishmanias and trypanosomes of vertebrates arose in this manner.

The specificity of protozoan parasites for particular hosts is a much disputed question. The striking similarity between such parasites as the various amebas, *Trichomonas,* and *Chilomastix* in different species of mammals, together with the fact that nearly all the species from man are transferable to rats and other animals, throws grave doubt on the idea of fairly strict specificity which has been advanced by some protozoologists. Some intestinal parasites seem to have progressed in evolution to the point where they can inhabit only one or a few closely related hosts, but many have not evolved beyond the stage of hostal varieties, i.e., mere races of a single species, for the time being especially adapted to a particular host species by virtue of having lived in that host for a long time, but capable of transfer to a different host under favorable circumstances. Blood and tissue parasites, in general, show more specificity than intestinal parasites, but even among these there is much variation. The human malaria parasites are strictly confined to primates, but the malaria parasites of birds show much less specificity. *Schizotrypanum cruzi* is an example of a blood and tissue parasite that can infect such widely different mammals as opossums, armadillos, bats, rodents, dogs, and man. The question of host specificity is important from an epidemiological standpoint, since it involves the question of the extent to which other animals may act as reservoirs for human parasites.

Protozoan vs. bacterial disease. The general course of the diseases caused by Protozoa is different in some respects from that of the majority of bacterial diseases. Most bacteria attract leucocytes and are attacked by them; when they invade the body there is an immediate sharp attack by the leucocytes, followed by mobilization of the larger phagocytic cells of the body. The battle usually continues unabated until either the host succumbs or the bacteria are completely destroyed, with not a survivor left. The waxy-coated acid-fast bacteria of tuberculosis and leprosy constitute an exception. After a preliminary struggle, a sort of truce is struck and the disease settles down to a comparatively mild, chronic state in which there is a balance of power between invader and host, each one, however, ready to take advantage of the slightest circumstance which tips the balance in its favor. This is essentially the course taken by most protozoan infections also. Often, after an initial flare-up, there may be no symptoms whatever for a time, but the parasites are still present, suppressed but not destroyed, and ready to stage an insurrection the moment the resistance of the host is weakened by other invasions, or by exhaustion, malnutrition, etc.

A good example of the difference between a protozoan and a typical bacterial infection can be seen in the nature and course of amebic as compared with bacillary dysentery, the former with no pus and of long duration, the latter with abundant pus and of short duration. The survival of protozoan infections in a chronic state seems to be due to immunity from attack by leucocytes, and the tendency of the other phagocytic cells of the body to relax their activity before their job is completed.

It is interesting, and perhaps suggestive, that spirochete diseases tend to be chronic like protozoan diseases, whereas the diseases caused by the rickettsias and by the insect-borne filtrable viruses (yellow fever, dengue, and sandfly fever) end in complete elimination of the parasites. In most bacterial diseases stimulation of the natural defenses of the host by means of vaccines or serums is more effective than it is in protozoan diseases. Chemotherapy by means of more or less specific drugs, e.g., iodine and arsenic compounds for amebiasis, antimony compounds for leishmaniasis, quinine and Chloroquine for blood stages of malaria, and 8-aminoquinolines for exoerythrocytic stages of malaria, is effective in most protozoan infections, whereas against bacteria no good chemotherapeutic drugs were known for a long time. Now, however, the sulfonamides and antibiotics have wide ranges of effectiveness against many types of bacteria. The effectiveness of antibiotics thus far known is far greater against bacteria than against Protozoa. It is interesting to note that even in their reaction to drugs, spirochetes are more or less in an intermediate position. Like Protozoa, they were long known to be specifically affected by a particular group of drugs (arsphenamines), yet, like bacteria, they are highly susceptible to antibiotics.

REFERENCES

Calkins, G. N. 1933. *Biology of the Protozoa.* 2nd ed. Lea and Febiger, Philadelphia.

Calkins, G. N., and Sumners, F. M. (Editors). 1941. *Protozoa in Biological Research.* Columbia University Press, New York.

Craig, C. F. 1948. *Laboratory Diagnosis of Protozoan Diseases.* 2nd ed. Lea and Febiger, Philadelphia.

Doflein, F. 1949. *Lehrbuch der Protozoenkunde.* 6th ed., revised by E. Reichenow. Teil I. Fischer, Jena.

Faust, E. C., Sawitz, W., Tobie, J., Odom, V., Peres, C., and Lincicome, D. R. 1939. Comparative efficiency of various technics for the diagnosis of Protozoa and helminths in feces. *J. Parasitol.* 25: 241–262.

Grassé, P. 1953. *Traité de Zoologie.* Vol. I, Parts 1–2. Masson et Cie, Paris.

Grell, K. G. 1956. *Protozoologie.* Heidelberg.

Hall, R. P. 1953. *Protozoology.* Prentice-Hall, New York.

Hegner, R. W. 1927. *Host-Parasite Relations between Man and His Intestinal Protozoa.* Century, New York.

1928. The evolutionary significance of the protozoan parasites of monkeys and man. *Quart. Rev. Biol.,* 3: 225–244.

Hoare, C. A. 1949. *Handbook of Medical Protozoology for Medical Men, Parasitologists and Zoologists.* Bailliere, Tindall and Cox, London.

Hyman, L. H. 1940. *The Invertebrates: I. Protozoa through Ctenophora;* and 1959. V. *Smaller Coelomate Groups,* pp. 698–713. McGraw-Hill, New York.

Jahn, T. L., and Jahn, F. F. 1949. *How to Know the Protozoa.* Brown, Dubuque, Iowa.

Kudo, R. R. 1954. *Protozoology.* 4th ed. Thomas, Springfield, Ill.

Lwoff, A. (Editor). 1951–1955. *Biochemistry and Physiology of Protozoa.* Vols. I and II. Academic Press, New York.

Morgan, B. B., and Hawkins, P. 1948. *Veterinary Protozoology.* Burgess, Minneapolis.

Pearse, A. S. 1936. *Zoological Names, a List of Phyla, Classes and Orders.* Duke University Press, Durham, N. C.

Taliaferro, W. H. 1926. Host resistance and types of infection in trypanosomiasis and malaria. *Quart. Rev. Biol.,* 1: 246–269.

Wenrich, D. H. 1952. Protozoa as material for biological research. *Bios.,* 23: 126–145.

Wenyon, C. M. 1926. *Protozoology.* 2 vols. Bailliere, Tindall and Cox, London.

Wenyon, C. M., and O'Connor, F. W. 1917. *Human Intestinal Protozoa in the near East.* Bale, Sons and Danielsson, London.

Chapter 4

AMEBAS

Amebas are animated bits of naked protoplasm, familiar to every freshman biology student who peers through a microscope with appropriate marvelings at the simplicity of animal life in its most primitive state. The vast majority are free-living animals inhabiting soil, water, and decaying organic matter everywhere, and play an important role in the control of bacteria in some of these situations. In view of their wide adaptability and the frequent contamination of food or drinking water by their cysts, it is not surprising that some species of them have adapted themselves to living out the active phase of their lives in the intestines of animals. The majority even of these are harmless commensals, content to use the intestine as a haven of refuge where food is abundant and enemies scarce, but a few have developed a taste for live meat, and have taken to feeding upon the wall of the intestine that shelters them.

The amebas belong to the subphylum Sarcodina, class Rhizopodea, and order Amoebida (see p. 48). The members of this order are characterized by having lobelike pseudopodia and no tests or shells. The species which live as commensals or parasites in the intestines of various animals, from termites and roaches to man, are placed in the family Endamoebidae; they are distinguished by having no flagellated stage and no contractile vacuoles. Recently a species morphologically indistinguishable from *Entamoeba** *histolytica* was found living in sewage.

Because of their small size, simple life cycles, and scarcity of good distinguishing characters, classification of Endamoebidae into genera and species requires a great deal of care and patience. Separation of Endamoebidae into genera is based on the character of the cysts and minute structural differences in the nuclei, and separation of species on still finer details of these characters.

Habits of trophozoites. All the commensal and parasitic amebas,

* For use of "Entamoeba" vs. "Endamoeba," see p. 59.

as far as known, inhabit the large intestine of their hosts, with the exception of *Entamoeba gingivalis,* which makes itself at home in the mouth. *E. histolytica* and perhaps some of the others occasionally invade the lower part of the small intestine just above the ileocecal valve, and they can frequently be found in the appendix. They all multiply in the active or trophozoite phase by simple fission; in most species this is initiated by a division of the nucleus, but in one genus, *Dientamoeba,* the nuclear division usually occurs shortly after cell division, resulting in a high proportion of individuals with two nuclei. Most of the species are mere scavengers, feeding on bacteria, cysts, and various debris in the contents of the large intestine, but *E. histolytica,* and to some extent *E. gingivalis,* are more fastidious. Both have histolytic and cytolytic power, and undoubtedly feed in part on the juice of tissue cells which they dissolve. *E. gingivalis* occasionally picks up a bacterium or other types of food, but in its natural habitat in the mouth it lives mainly on leucocytes or their nuclei.

Encystment. Most of the parasitic amebas form cysts, which are better able to withstand conditions outside the body than are the trophozoites, but *Entamoeba gingivalis* and *Dientamoeba fragilis* manage to survive without them. When preparing to encyst, the amebas eliminate all food vacuoles, round up, and shrink somewhat, probably by a condensation of the cytoplasm, so that the nucleus becomes relatively large. This is the precystic stage. A delicate cyst wall develops to protect the organism during its hazardous existence outside the body while waiting for an opportunity to infect a new host. As long as the amebas are alive the cyst walls are relatively impervious to many substances, including dyes and weak disinfectants, but they have only feeble resistance to desiccation; *E. coli* cysts are more resistant than those of *E. histolytica.* The cysts are unaffected by either chlorine or dilute silver ions in the proportions used for killing bacteria in drinking water, but they are killed in a few minutes by 5% acetic acid. The fact that living cysts are not ordinarily stained by dilute eosin, whereas dead cysts are, has been used extensively as a test of the viability of cysts in experimental work, but there is some question of its reliability. In some species glycogen is stored during encystment in more or less well-defined vacuoles, and there may be deep-staining "chromatoid bodies." These gradually disappear as the cysts grow older.

In most of the species of amebas some multiplication of the nucleus takes place during the formation of the cysts, but in *Iodamoeba* the nucleus remains single. In *Entamoeba histolytica* and *Endolimax nana* four nuclei are normally produced, and in *Entamoeba coli* eight.

The exact conditions which induce encystment are unknown. In most cultures no encystment occurs in the absence of rice starch. It has been suggested that in cultures, at least, encystment is related to population growth, but whether to crowding or accumulation of waste products is unknown. The presence of various organic compounds, with a low oxygen tension, seems to be necessary for encystation in cultures. Cysts are not found in dysenteric or liquid stools and are never formed in tissues or liver abscesses. Encystment should be considered a naturally recurring phenomenon in the life cycle; its sole purpose is the safe transfer of the parasite from one host to another. A single cyst, at least of *E. coli,* may be sufficient to establish an infection (Rendtorff, 1954).

Excystment. The conditions which lead to excystment are also little understood, but most cysts "hatch" in the small intestine above where the trophozoites ultimately settle down. The amebas escape from their cysts through a perforation in the cyst wall. The process was first described by Dobell in 1928 for *Entamoeba histolytica.* The four-nucleated ameba draws itself in and out of the cyst several times before escaping. It then undergoes a complicated series of nuclear and cell divisions, resulting ultimately in eight little amebulas with a single nucleus each.

Species found in man. Before the appearance, in 1919, of Dobell's book, *Amoebae Living in Man,* the amebas found in man were in a terrible muddle, and most of the earlier literature cannot be relied upon as far as species are concerned. Since the publication of this valuable work there have been a number of suggested modifications or additions, but none of them has stood the test of time; today six species of amebas living in man are recognized and are separated into four genera, just as Dobell arranged them. Most protozoologists recognize the following genera and species: *Entamoeba gingivalis,* inhabiting the mouth; *E. histolytica,* a pathogenic intestinal form; *E. coli, Endolimax nana,* and *Iodamoeba bütschlii,* harmless intestinal forms except for one amazing case (see p. 79); and *Dientamoeba fragilis,* an intestinal form which is at least sometimes pathogenic.

There are two races of *E. histolytica* which differ in the size of their cysts, and also in their pathogenicity, the small-cyst race being almost if not entirely nonpathogenic to man, whereas the large-cyst race is a potentially pathogenic tissue invader. The small-cyst, nonpathogenic form has been considered worthy of recognition as a distinct species, *E. hartmanni,* by European workers, although most American workers refer to it as "small-race" *E. histolytica.* These races appear to breed true, although there is one record of change from small to large cysts after five years in culture, but this observation needs confirmation.

Differentiation of genera. The outstanding characteristics of the genera of amebas which occur in human beings are as follows:

Entamoeba. Nucleus vesicular with chromatin arranged in a peripheral layer of beadlike granules of fairly uniform size, and a small compact endosome; a capsulelike structure, can usually be seen surrounding the endosome. Cysts, if produced with 4 or 8 nuclei similar in structure to those of the free forms, and including also glycogen masses and refractile chromatoid bodies, though these masses and bodies commonly disappear before or soon after the cysts become mature. (See Figs. 3 and 4.)

Endolimax. Nucleus vesicular without a distinct peripheral layer of chromatin. A fairly large compact endosome in the interior, usually more or less eccentric and connected by threads or processes with one or more smaller masses. Mature cysts oval, with 4 nuclei in the known species, similar in structure to those of the free forms. The cysts contain, in addition to the nuclei, a number of small refractile volutin granules. The young cysts also contain masses of glycogen. (See Fig. 5*A* and *A'*.)

Iodamoeba. Nucleus vesicular with moderate-sized central endosome and well-developed membrane without a distinct peripheral zone of chromatin, but with a single layer of rather large granules between the endosome and the outer membrane; cysts very characteristic, formerly known as iodine cysts or I cysts, often of irregular shape, containing, besides a single nucleus, a number of brightly refractile granules and a relatively large, sharply defined solid mass of glycogen which stains deeply in iodine. The cyst nucleus is peculiar in that the endosome comes to lie peripherally in contact with the nuclear membrane. (See Fig. 5*B* and *B'*.)

Dientamoeba. Mature individuals with 2 similar nuclei; these are vesicular with the endosome represented by a cluster of small granules near the center; nuclear membrane very delicate without distinct peripheral chromatin; trophozoites burst in water; cysts not found. (See Fig. 5*C* and *C'*.)

Host specificity. There has been much discussion about the identity of the human species of amebas and morphologically identical ones found in other animals. Strict host specificity on the part of intestinal amebas can no longer be accepted, although some protozoologists have grimly adhered to belief in it in spite of growing evidence against it. The genus *Entamoeba* is an excellent one to illustrate the situation. In man there are two universally recognized intestinal species, *E. coli* and *E. histolytica.* The latter species can be successfully transferred to a variety of animals such as monkeys, rodents, rabbits, cats, dogs, and pigs. *E. coli* has been transferred to monkeys, cats, and rats. In various species of lower primates, from spider monkeys (*Ateles*) to apes indistinguishable forms of one or both of these types of amebas have been described. Five of the species of human amebas (and four human flagellates) have been found by Kessel in *Macaca* monkeys, differing in no morphological or physiological respect from the corresponding Protozoa in man. Although these might have been acquired from contact with human beings, this is probably not true of forty-four wild Philippine monkeys, obtained where they probably had had no chance to be contaminated by their Protozoa-infested human compatriots, and which Hegner found to harbor eleven different species of human intestinal, oral, and vaginal protozoa. One was a veritable zoological garden for human Protozoa, harboring eight different species, and none had less than two.

Natural infections with *E. histolytica* occur in most of the animals in which experimental infections have been produced. Carnivorous animals, e.g., cats and dogs, rarely harbor amebas of any kind, whereas nearly all species of omnivorous and herbivorous animals harbor one or more species of their own as well as being susceptible to the human species. One species, *Entamoeba polecki,* common in pigs and various ruminants, occasionally infects man; it produces one-nucleated cysts. Reptiles harbor a number of species, including one, *E. invadens,* which is strikingly similar to *E. histolytica* and may cause fatal infections in lizards and snakes. Amebas belonging to the same genera as those found in man occur also in insects and other invertebrates. A species found in roaches, *Endamoeba blattae,* is thought by some to belong to the same genus as the species of *Entamoeba* in mammals; hence the name "*Endamoeba*" is often used, especially by American writers, but the correctness of "*Entamoeba*" is being more and more widely recognized. For a detailed discussion of this question see Kirby (*J. Parasitol.* 31: 177–188, 1945).

Cultivation. All the intestinal amebas, as well as the flagellates and ciliates (except *Giardia*), can be grown in artificial cultures, but most of them only in the presence of living "associates," of which bacteria, including many single species, *Schizotrypanum cruzi,* and cultured tissue cells will suffice. A vast amount of effort has been put into a study of the growth requirements of amebas, especially *E. histolytica,* in an effort to develop an axenic (free of other organisms) culture medium. Shaffer and Frye (see Shaffer et al., 1949) came near it with a preconditioned clear medium containing few bacteria. There is some evidence that bacteria may be important in the initiation of amebic disease (see p. 71). The growth of *E. histolytica* under axenic conditions was finally accomplished by using media containing a particle-containing juice prepared from chick embryos (Reeves, Meleney, and Frye, 1957).

Many media have been developed; some are entirely liquid (e.g.,

St. John, *Am. J. Trop. Med.*, 12, 1932; Balamuth, *Am. J. Clin. Pathol.*, 16, 1946), and some are coagulated slants of whole egg, liver infusion agar, or serum, overlaid with various fluids, usually containing albumin or serum, and with rice starch added, on which amebas feed gluttonously. Some widely used media of this type are those of Boeck and Drbohlav (*Am. J. Hyg.*, 5, 1925); Dobell and Laidlaw (*Parasitology*, 18, 1926); Cleveland and Sanders (*Science*, 72, 1930), now available as Difco's desiccated "Endamoeba medium"; E. C. Nelson (*Am. J. Trop. Med.*, 27, 1947), and various modifications of these.

Coprozoic amebas. Cysts of free-living amebas and flagellates, and sometimes even ciliates, may enter the body with food and pass through the alimentary canal unhatched and undigested. Some of these find conditions satisfactory for rapid multiplication in the feces after passage, and may confuse an unwary laboratory technician. All such amebas, however, are distinguishable by the presence of one or more contractile vacuoles. There is no evidence that these coprozoic forms can ever establish themselves and multiply in the intestines; they become progressively more abundant in stale feces, whereas the trophozoites of the true intestinal species die out very rapidly, usually within a few hours after leaving the intestines. The discovery of an ameba in sewage, *Entamoeba moshkovskii* (see Neal, 1953), which can pass for *E. histolytica* even on close inspection, complicates diagnosis, for there is no reason to doubt that cysts of this organism may be swallowed under unsanitary conditions. That would make it necessary for a long-suffering technician, in order to make a positive identification of *E. histolytica* in asymptomatic cases, to culture the organism at 37°C. but not at 20° or below, since the parasitic species will not grow at the low temperatures but *E. moshkovskii* will.

Diagnosis. Diagnosis depends upon (1) finding the organisms in the feces and (2) making a correct identification of them. Cysts are rarely found in liquid or dysenteric stools, whereas trophozoites, except those of *Dientamoeba*, are less frequently found in formed stools. Cysts of intestinal amebas, and of other intestinal Protozoa also, are voided intermittently, so a single examination cannot be relied upon to bring to light all infections. When a series of examinations of a single group of people is made over a period of weeks, some parasites in the stools may show up regularly, whereas others appear and reappear without rhyme or reason, although there is no reason to doubt that the infection has been there throughout. The percentage of existing infections found at a single examination may vary from 20 to 65% depending in part on the technique employed, time used, and skill and experience of the technician, but probably also on degrees of exposure and resistance of the people examined. In Egypt a single examination for 10 minutes by MIF reveals 50 to 65% of existing infections with the three commonest species of amebas, when the incidences on six examinations were shown to be 97 to 98%, whereas Sawitz and Faust (1942) found much lower percentages of actual infections revealed by a single examination in the United States. Better results are obtained by making examinations at intervals of several days than on successive days, and there is little advantage in examining several slides from the same stool.

Formed stools to be examined for cysts can be kept in the icebox for 2 or 3 days before examination, but trophozoites in liquid stools degenerate very rapidly unless preserved and should be searched for as soon as possible after being passed, while the stool is still warm. If the stools are kept at body temperature the trophozoites are usually still identifiable for about 30 minutes. A higher percentage of infections can be found in a single purged or diarrheic stool examined for trophozoites than in a single formed stool examined for cysts. Oil-purged stools are useless for examination.

Direct smear examinations are made by comminuting a small amount of feces in saline, to which may be added 1:1000 aqueous eosin to stain the debris pink, leaving the living trophozoites and cysts unstained. After spreading the smear over the width of two cover glasses, apply a cover to one side for examination of living organisms; to the other side, before covering, add a small drop of D'Antoni's iodine (1.5 grams iodine in 100 cc. standardized potassium iodide, filtered after standing 4 days). The iodine will stain nuclei, chromatoid bodies, etc., well enough for identification. Additional aqueous instead of saline smears facilitate diagnosis by destroying fungus and *Blastocystis* and by showing certain characteristics of *Dientamoeba.*

An improved iodine-staining technique was described by Sapero et al. (1951). Brooke and Goldman (1949) showed that addition of fixative to 5% elvanol, a polyvinyl alcohol, fixes and preserves trophozoites and small cysts very well, but large cysts are distorted. Smears so fixed can be dried on slides and subsequently stained without harm to the trophozoites. Sapero and Lawless (1953) described the MIF technique by which specimens can be collected in the field, home, etc., and preserved in vials by unskilled workers. The specimens can be examined months later with the trophozoites and cysts excellently preserved and showing easily recognizable diagnostic characters.

Permanent, stained preparations showing minute details of structure can be made by fixing wet films with Schaudinn's fluid (two parts of saturated $HgCl_2$, one of 95% alcohol) with or without polyvinyl alcohol; if the latter is used the films may subsequently be dried, otherwise not. These films are then stained with iron hematoxylin, in the use of which there are many variations (see Craig, 1944), all of them requiring skill, patience, and time. Lawless (1953) described a permanent-staining technique that requires no special training in its use and takes less than 3 minutes, which is very satisfactory for routine diagnosis. It should be noted that trophozoites are much more frequently found in MIF-stained preparations than in permanently stained slides, in which many are evidently lost.

Faust et al. (1939) worked out a method of concentrating both the cysts of Protozoa and eggs of worms by a zinc sulfate flotation technique. Saturated NaCl cannot be used for flotation, as it can for worm eggs, since it shrinks the cysts and makes them unrecognizable. The method is good only for cysts, but concentrates these to a considerable degree, after which they can still be stained (see pp. 253–254).

Other diagnostic methods are culturing, which some workers (Michael and Cooray, 1952) consider more accurate than fecal examination for intestinal infections; and complement fixation, which is particularly useful for amebiasis of the liver or other organs outside the intestine, but which is often not positive in intestinal cases, especially if tissue invasion is slight (Hussey and Brown, 1950; Kenney, 1952; McDearman and Dunham, 1952). Brown and Whitby (1955) described an ameba-immobilization test with sera from persons with amebiasis but found it to be no better than complement fixation and considerably more trouble. Norman and Brooke reported in 1954 that cultivation was an effective diagnostic technique for fresh purged stools, but not for stale normally passed specimens.

Entamoeba histolytica

Distribution and incidence. Because of its capacity for causing disease, its wide geographical distribution, and its discomforting frequence as a resident of the human colon, *Entamoeba histolytica* must be ranked as one of the most important human parasites. Although it once had the reputation of being mainly a tropical parasite, it is by no means so limited. It has world-wide distribution, and is almost as frequently present, though fortunately not so frequently pathogenic, in the temperate zones as in the warmer areas of the world. The only reason that this ameba often inhabits more people in tropical than in temperate localities is that the people in the tropics take less pains to avoid devouring its cysts with contaminated food or water. In the

Kola Peninsula of Russia, lying entirely within the Arctic Circle, a 60% incidence of *E. histolytica* was recorded.

When routine examinations are made by competent microscopists, seldom less than 5 to 10% of the entire population, even in northern Europe and the United States, are found to be infected. In a statewide survey of Tennessee, Meleney et al. in 1932 found more than 11% of the rural population infected, and in one group of counties above 22%; these findings on one examination indicate probably twice as great actual incidence. Available information indicates that this prevalence has probably undergone little change. In one group of 27 individuals in 5 backward families, 23 were carriers of *E. histolytica.* The Chicago outbreaks of 1933 and 1934 show how well this parasite can prosper far from the haunts of dark skins and palm trees. In the United States about 150 deaths annually between 1946 and 1953 were attributed to amebiasis. In many parts of the tropics incidences of 30 to 60% have been reported; in some Egyptian villages near Cairo where careful repeated examinations were made by personnel of the U.S. Naval Medical Unit No. 3 and by the senior writer, the incidence of what is morphologically *E. histolytica* approaches, if it does not actually reach, 100% in people over one year of age.

Infants under a year old are rarely infected with this or other amebas, although they frequently harbor flagellates. The incidence gradually increases during childhood and usually reaches its highest incidence in young adults. There seems to be no racial discrimination against the organism. In areas where moderate incidences of 5 to 20% occur, there are usually about twice as many *E. coli* infections, probably because the thicker and more resistant cyst wall of that species makes its transmission easier.

Morphology. The trophozoites of *E. histolytica* (Fig. 3) are large, usually 20 to 30 μ in diameter in the tissue-invading, erythrocyte-eating forms, but smaller, usually 12 to 15 μ in diameter, in the lumen-dwelling *minuta* form and in the small-cyst race. The amebas have a thick outer layer of clear, refractile ectoplasm enclosing the more fluid granular endoplasm. In the fresh state, when warm, the amebas are very active, and travel along in a straight line in a manner which Dobell describes as suggesting a slug moving at express speed. In this condition the rapidly advancing end of the body consists of a single clear pseudopodium, while ingested red corpuscles, if present, flow and roll about as though in a mobile liquid. Other amebas have more tendency to stay in one place, where they extend and retract their pseudopodia without making much headway.

After stools have been passed and allowed to cool, the amebas become abnormal and die quickly. The amebas under these circumstances remain in one place and throw out large, dome-shaped, clear pseudopodia from different parts of the body; the endoplasm becomes full of vacuoles, and bacteria invade the dying body. The nucleus also disintegrates and presents abnormal appearances in both fresh and stained preparations. Even in this condition the large amount of clear ectoplasm serves as a means of differentiation from *E. coli.*

The nucleus is so delicate in structure that it is practically invisible in fresh active forms except under a phase contrast microscope. After being fixed and stained the nucleus has a characteristic structure. The nuclear membrane is encrusted with uniform fine granules of chromatin, and a small dotlike central endosome is surrounded by an indefinite, clear halo. Between the endosome and the nuclear membrane is a clear area devoid of granules, marked by a linin network which often has a spokelike radial arrangement. *E. coli,* on the other hand, has coarser and more irregular peripheral granules and a larger endosome, eccentric in position, with a more definite halo and with a few chromatin granules strung on the linin network surrounding the halo; the nucleus of this species is visible as a bright refractile ring in fresh, living organisms. When stained with iodine the nuclear membrane and endosome of entamoebas show as refractile bodies, and the cytoplasm stains a greenish yellow.

E. histolytica multiplies by simple fission and a modified form of mitosis in which, according to Kofoid and Swezy, six chromosomes are formed. When preparing to encyst, the amebas become smaller and rounded, lose their food vacuoles, and then lay down the delicate cyst wall. The relatively large nucleus then divides into two and then four progressively smaller ones, but with the same morphology as the nucleus of the trophozoite. Rarely, *E. histolytica* overshoots the mark and produces eight nuclei in a cyst.

Most precystic or young cystic amebas (Fig. 3*B* and *D*) lay down in the cytoplasm one or two bar-shaped chromatoid bodies which are refractile in living or iodine-stained cysts and which stain deep black with iron hematoxylin; these chromatoid bodies are quite different from the less massive splinterlike ones found in *E. coli.* In most young cysts there is some stored glycogen, usually in less well-defined vacuoles than in young cysts of *E. coli.* Both the chromatoid bodies and the glycogen vacuoles disappear as the cysts grow older. In fresh preparations the cysts have a faintly greenish tint and are refractile; if a preparation containing numerous cysts is viewed with a low-power objective slightly out of focus the cysts appear as little shining spheres. The size of the cysts varies from about 5 to 20 μ

in diameter. As noted on p. 57, there is a small-cyst and a large-cyst race. The former has cysts less than 10 μ (mean 7 to 8 μ) in diameter, the latter has cysts over 10 μ (mean 12 μ) in diameter; the large-cyst race, even when the trophozoites are in the lumen-dwelling *minuta* phase, produces large cysts.

The mature four-nucleated cysts of *E. histolytica* are characteristic enough for any trained technician to be able to identify them. Their differentiation from those of *E. coli* and *Endolimax nana,* with which they are most likely to be confused, is indicated by the table below:

	Entamoeba histolytica	*Entamoeba coli*	*Endolimax nana*
Size	5–20 μ	10–33 μ	4 × 5 μ – 10 × 14 μ
Shape	Round	Round	Usually oval
Nuclei, number	Usually 4	Usually 8	Usually 4
Nuclear structure	Membrane encrusted with fine chromatin granules; small central endosome	Membrane encrusted with coarser granules; larger, usually eccentric endosome; a few scattered chromatin granules	Chromatin in a single or lobed mass, large relative to size of nucleus
Chromatoid bodies (when present)	Barlike	Splinterlike	Absent or dotlike
Glycogen vacuoles (when present)	Usually diffuse, ill-defined	May be fairly well defined	None

Habits and biology. Like most other parasitic amebas, *Entamoeba histolytica* is normally an inhabitant of the large intestine, frequently invading the appendix and occasionally venturing into the lower part of the small intestine. Although amebic ulcers may be found anywhere along the 6 ft. of the large intestine from ileocecal valve to anus, they are most frequent in the cecum and ascending colon, and next most frequent at the opposite end, in the sigmoid flexure and rectum. These are the regions where the contents of the intestine are usually allowed a temporary halt in their otherwise rough and restless journey through the alimentary canal.

The motile forms, or trophozoites, live for only a few hours after leaving the body even if the feces are kept warm, and they are killed immediately by drying, acids, or other unfavorable conditions.

Cysts have never been found in the tissues except in the liver of dogs fed with raw liver or liver extract. They apparently form in the lumen of the large intestine, and are often passed before they have fully matured, but they are capable of completing their development outside the body if the cyst wall has been formed. The cysts are rarely found in liquid stools, where the trophozoites frequently abound, but they are often the only forms present in normal formed stools.

The cysts, if kept moist and cool, will live for a number of weeks outside the body. They may remain viable for a week or two in feces if kept cool, and for as long as 10 days in water at room temperature. In a refrigerator they can be kept alive in water for 6 or 7 weeks. They will not stand desiccation, however, and have been found to die in 5 to 10 minutes when dried on the hands. They will live 24 to 48 hours in the intestines of either flies or roaches. Although cold is favorable for their survival (this in itself is enough to throw suspicion on their limitation to the tropics), they are susceptible to moderately high temperatures, even as low as 115° to 120°F. They are therefore killed by pasteurization of milk and by heating of water; heating and filtration are the only practicable methods yet known for destroying them in drinking water. Their specific gravity is only about 1.06, so they settle very slowly in contaminated water; it is estimated that it would take them 4 days to settle 10 ft. in perfectly quiet water.

Food habits, races, and tissue invasion. As already noted, when in the lumen of the intestine in symptomless cases, the amebas commonly do not contain food vacuoles, although ingested bacteria or other debris is sometimes seen, but tissue invaders freely ingest red blood cells. In fact, this is the sole ingested food; the number engulfed usually ranges from one to ten, but may reach forty. *E. coli,* which eats without discrimination, may occasionally ingest a red cell if available, but any ameba found in a dysenteric stool and containing only red cells may safely be regarded as *E. histolytica.* In cultures the amebas feed voraciously on bacteria and starch grains. (For culture methods, see p. 59.) Hoare (1952), who recognizes the large-race and small-race forms as distinct species, and considers the small-race form as harmless as *E. coli,* summarized evidence that even the large-race form has a commensal, lumen-dwelling phase in which it ingests bacteria and in which the trophozoites are very small, whence the name "*minuta* form." However, it seems probable that these amebas would apply themselves to the surface of the mucous membrane, as they do to solid surfaces in cultures, and then by their lytic excretions dissolve the superficial cells and nourish themselves on the cell juices. The relative infrequence of solid food in the food vacuoles supports this idea.

Under favorable conditions, e.g., low host resistance, and with help in the form of a mucous membrane irritated by chemicals, injured by bacteria, viruses, or worms, or unhealthy state due to malnutrition, tissue invasion may occur, ulcers develop, and dysentery or other symptoms follow. The extent to which the amebas may be equipped with enzymes specifically facilitating tissue penetration may also be a factor, but inconsistent results have been obtained in efforts to demonstrate such enzymes in *E. histolytica.* Braden reported that *E. histolytica* produced a hyaluronidase (spreading factor); Nakamura et al. and Harinasuta et al. in 1959 could not confirm this although proteolytic enzymes were demonstrated.

Tissue-invading forms, perhaps because of their ingestion of red blood corpuscles and their complete immersion in nutritious cellular food, grow larger in size and cease to produce cysts.

Although some workers have claimed that *E. histolytica* is invariably a tissue invader, James, way back in 1928, called attention to the fact that the numbers of amebas often found in the stools makes it incredible that they all came from invaded tissues. Since amebas in tissues do not form cysts, the tissue invasion may not be the usual state of affairs.

Infection in other animals. As noted on p. 59, *E. histolytica* can be transferred to many other animals, and natural infections occur commonly in monkeys, occasionally in rats, and probably frequently in pigs. Rabbits, guinea pigs, hamsters, rats, kittens, and dogs are all useful laboratory animals for the study of experimental amebic infections, although they usually have to be infected by special methods such as intracecal injection, by duodenal tube, or through the anus. Hegner in 1932 showed that New World monkeys often show severe amebic lesions when inoculated. Recently it was claimed that Old World monkeys were more like man in being relatively refractory, but these monkeys, in contrast to Hegner's, were previously infected ones, and the influence of immunity was overlooked. In dogs and kittens, the amebas do not produce cysts unless the hosts are fed liver.

Pathogenicity. 1. NONDYSENTERIC INFECTIONS. Although amebiasis is usually thought of as the cause of dysentery with blood and mucus-containing stools, or of liver abscesses, these conditions are actually the exception rather than the rule, and some workers have reported that as many as 90% of the dysentery cases in temperate climates are apparently symptomless. Even in the tropics dysentery is exceptional; in an examination by the senior writer of 500 stools of Egyptian villagers, nearly 100% of which had contained amebas with *E. histolytica* morphology, only four of the stools were suggestive of dysentery. Faust (1941) found *E. histolytica* in 13 (6.5%) of 202 autopsies of persons who had suffered sudden accidental death in New Orleans, but in only 5 of these were amebic lesions demonstrated, and these were superficial, confined to the mucosa. Craig, on the other hand, stated that in his experience 65% of so-called carriers have symptoms referable to their infection, which disappear after eradication of the parasite. Sapero (1939), in a study of 216 non-dysenteric cases, found symptoms in 100 of them, in many cases trivial, but often severe enough to require hospitalization.

On the other hand, Miller and Gilani (1951) found that a group of patients in Calcutta who harbored *E. histolytica* and had chronic gastrointestinal symptoms but not dysentery showed no better response

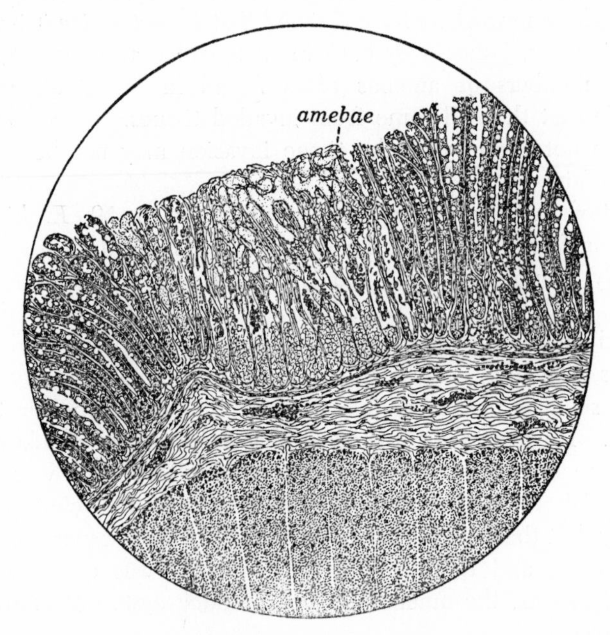

Fig. 6. Section of colon of cat showing an amebic ulcer limited to the mucous membrane. Note broken-down and necrotic epithelium of invaded glands, extravasated blood, and masses of amebae at bottom of glands. (Drawn from slide prepared by H. E. Meleney.)

to treatment with amebicidal drugs than did a group with similar complaints but no amebas. Obviously, further study is necessary to show just how much nondysenteric symptoms are really due to amebic infection; otherwise the discovery of amebas in the stool might be misleading and postpone a correct diagnosis. The symptoms commonly associated with chronic amebiasis are abdominal pain, nausea, flatulence, and bowl irregularity, with headaches, fatigability, and nervousness in a minority of cases. The symptoms resemble those of many other gastrointestinal disorders, particularly appendicitis and peptic ulcer. Appendicitis, or pains simulating it, are frequent enough so that Craig (1944) believed that in all cases of suspected appendicitis an examination for *E. histolytica* should be made, and, if found, amebic treatment should be given before resorting to operation.

2. AMEBIC DYSENTERY. Although the small-race amebas appear to be harmless commensals, at least in the majority of cases, the large race may, under favorable conditions, as noted on p. 66, eat into the tissues, first destroying the mucosa (Fig. 6) and then pushing deeper into the submucosa (Fig. 7), where they spread out and produce flask-shaped ulcers. The abscesses extrude their contents into the intestine

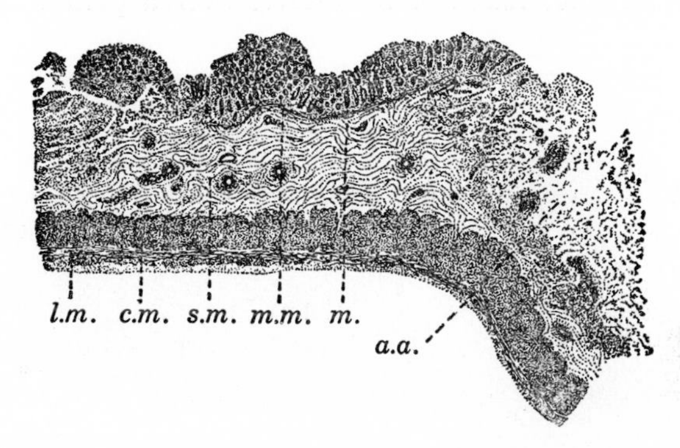

Fig. 7. Section of human colon showing deep amebic ulcer broken through into the submucosa. Note abnormal thickening (edema) of submucosa and pus-like contents of ulcer: *l.m.*, longitudinal muscle layer; *c.m.*, circular muscle layer; *s.m.*, submucosa; *m.m.*, muscularis mucosae; *m.*, mucosa; *a.a.*, amebic abscess. (Drawn from slide prepared by H. E. Meleney.)

as necrosis becomes complete, causing the edges of the injured mucous membrane to cave in, giving a craterlike effect. Sometimes several undermining abscesses coalesce under the surface. Fortunately, the muscular coats of the intestinal wall usually act as a barrier, but sometimes this layer is penetrated by way of the connective tissue sheaths, and the amebas reach the serous membrane, causing extensive adhesions or dangerous perforations. The ulcers vary greatly in number and size; in severe cases almost the entire colon is undermined. When not invaded by bacteria, the ulcers show no signs of inflammation, but invasion by bacteria often occurs. Sometimes amebic granulomata form in the colon and may be confused with cancerous growths.

The ulceration of the bowel, as noted previously, may produce severe dysentery, though it does so in a minority of cases. The reason for the greater frequency of dysentery in the tropics has been the subject of much speculation, and although, as already noted, some authors have attributed it to the existence in temperate climates of fixed races of very low virulence, others, notably Brug, believe that a tropical climate in itself favors the production of the dysenteric symptoms.

In amebic dysentery the stools, usually acid, consist of almost pure blood and mucus, in which swarms of amebas, laden with blood corpuscles, are usually present. The patient is literally "pot-bound," owing to the rectal straining and intense griping pains, with the passage of blood and mucus stools every few minutes. In uncomplicated cases there is little or no fever, a point which is sometimes useful in differentiating amebic from bacillary dysentery. Recurring symptoms sometimes manifest themselves over a period of 30 to 40

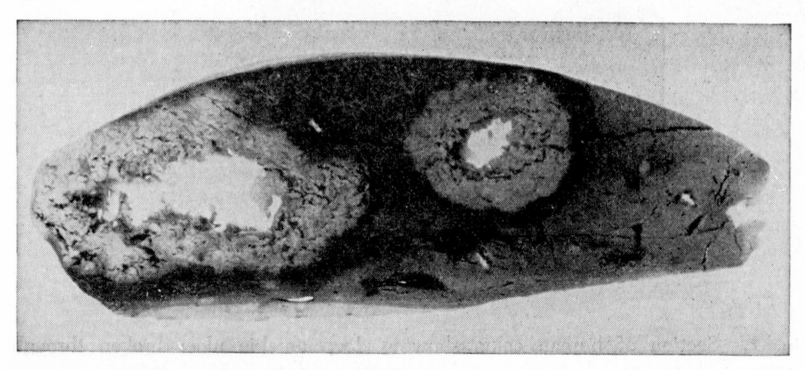

Fig. 8. Multiple amebic abscesses of the liver. (About ½ diameter.) (From photograph by Sir Philip Manson-Bahr.)

years or even longer, and there may be latent periods lasting for at least 6 or 8 years. D'Antoni has suggested that many infections in young adults may have been acquired during early childhood from infected mothers.

3. ABSCESSES IN LIVER, LUNG, ETC. It is clear that invading amebas actively dissolving the tissues may frequently be drawn into the portal circulation. Such amebas are carried to the liver and sometimes settle there, attacking the liver tissue (Fig. 8). In view, however, of the frequency with which amebas are undoubtedly carried to the liver from intestinal ulcers it is evident that this organ must have a high natural resistance to infection. Nevertheless, small or temporary amebic infections of the liver are probably much commoner than we usually think; sometimes liver abscesses develop without any preceding attack of dysentery. The abscesses are usually sterile so far as bacteria are concerned, although they are sometimes secondarily infected; they may become very large and filled with a slimy, bloody, chocolate-colored material resembling pus, but made up of dead amebas, blood, and fibrous tissue left by the amebas, with active amebas in the

enlarging walls. The patient has pain in the liver region, fever, and a high leucocyte count, and his face presents a sorrowful aspect of weariness and apathy, with sallow skin, sunken cheeks, and dark-circled eyes.

Amebas which have escaped into the blood stream are not necessarily halted in the liver but may be carried to any part of the body. Lung abscesses are fairly frequent; these are usually caused by direct extension from a liver abscess through the diaphragm. Such an abscess may rupture into the pleural or pericardial cavity, but it usually works directly into the lung tissue where the lung adheres to the diseased diaphragm. The lung abscess in turn usually ruptures into a bronchial tube and discharges a brown mucoid material which is coughed out with the sputum. Next in frequency are abscesses of the brain. Abscesses elsewhere are rare. Skin infections, however, may develop about the incisions made for surgical treatment of amebic abscesses.

Factors determining pathogenicity. There is no doubt that chronic infections with indefinite symptoms are the rule, with the amebas localized in the large intestine or liver. Whether an amebic infection produces acute dysentery, a chronic state of vague discomfort, or no obvious symptoms at all probably depends on several factors, some of which have already been discussed, e.g., races or strains of amebas, and their possible production of hyaluronidase. The importance of irritation of the intestine by chemical means was demonstrated by Nauss and Rappaport, and the importance of bacterial irritation was well shown by Westphal (1948) in a study of intestinal disturbances in German troops in North Africa. He found evidence that amebic infections tended to remain chronic, nondysenteric, lumen infections until resistance to tissue penetration was lowered by intercurrent bacterial infections. In his experience tissue invasion occurs in about one-third of *E. histolytica* carriers who acquire bacillary intestinal infections. It has been suggested by a number of workers that intestinal bacteria, which ordinarily produce no disease, may cooperate with *Entamoeba histolytica* in the invasion of the intestinal mucosa. Bock and Mudrow-Reichenow (1955) showed in extensive animal experiments that the number of amebas inoculated bore no relation to the degree of ulceration produced and concluded that the course of infection is determined by other factors such as the accompanying bacterial flora. Phillips and his associates (1955) have furnished evidence for this view by studies on the behavior of *E. histolytica* in germ-free guinea pigs and in guinea pigs having a single species of bacterium in the digestive tract. In the germ-free animal, no invasion of intestinal epithelium occurred, but ulcerative amebiasis was produced in animals harboring a single bacterial symbiont. These experiments do not prove that the bacteria are responsible for tissue destruction (also see Neal, 1956). Since it is known that the amebas do not thrive in the presence of oxygen, even in very small amounts, it might be guessed that the bacteria may aid the amebas by reducing the oxygen tension in the region of the mucosa. This line of thought is supported by the observation that if reducing substances, such as thioglycollate or cysteine, are administered with amebas to germ-free hosts, lesions in the intestinal wall are produced (Phillips and Wolfe unpublished, cited by Phillips, 1957). In 1956 Krupp reported that in laboratory animals the migrations of ascarid nematode larvae facilitated the transport of amebas to the liver.

Diet has a great influence. High protein diets are unfavorable for the parasites owing to unfavorable effects on the environment (see p. 22), and adequate protein is necessary to permit the development of immunity (see p. 30). Milk is beneficial in monkey infections, but apparently not in rodent infections. Lack of vitamin C in guinea pigs (Sadun, Braden, and Faust, 1951), and of niacin in dogs (Larsh, 1952), lowers resistance to amebiasis. Raw liver or liver extract (by mouth but not by injection) benefits dogs with amebiasis just as it does dogs with black-tongue, a disease due to niacin deficiency. Human beings suffering from amebic infections sometimes lose their symptoms promptly when put on a diet rich in proteins and vitamins.

Immunity undoubtedly plays an important part, as Swartzwelder and Avant (1952) showed in the case of dogs and of rats. As in other parasitic infections, antibodies undoubtedly help keep the disease in check, at least so far as tissue invasion is concerned. However, it must be said that acquired resistance and the basis for it when it occurs is not well understood.

Mode of infection and epidemiology. Since only the cysts can survive outside the body, these alone are concerned in transmission. The trophozoites are rarely capable of passing through the human stomach and intestine to reach their promised land in the colon. Since dysenteric cases rarely pass cysts, they are not usually concerned in transmission; persons who are cyst-passers with few or no symptoms are principally concerned. Even the cysts probably find the stomach a dangerous hazard and, like typhoid organisms, may cause infection more readily when ingested with water than with food because of the greater rapidity with which they pass the stomach. It also throws light on the relative frequency of amebic infections in individuals with abnormally low stomach acidity.

Since amebic cysts survive for considerable periods outside the body if not desiccated (up to 8 days in soil), it is obvious that if they get into drinking water or moist foods, such as raw vegetables, they are in an advantageous position both from the standpoint of length of life and of opportunities to "thumb a ride" into a human alimentary canal. Polluted water is undoubtedly one of the most important means of transmission, and wherever unprotected or untreated ground water is used for drinking in areas where there is widespread soil pollution, amebic infections will be common. Such conditions prevail over vast portions of the tropics and in the rural areas of our own southern states.

Even when a purified water system prevails, accidents may lead to widespread outbreaks of water-borne infections. One hazard is in defective plumbing. The basements of hotels and public buildings frequently contain a veritable maze of pipes, gradually built up, repaired, and replaced throughout a generation or more, and it is not as surprising as it seems at first that errors in plumbing should be made. This was strikingly demonstrated by an outbreak during the Chicago World's Fair in 1933 in which defective plumbing caused almost 1000 known cases of amebiasis and 58 deaths scattered over 206 cities. The plumbing hazards included back siphonage from sanitary fixtures into water lines, leakage of sewer pipes into basements, and even cross connections between sewer pipes and water pipes made by careless and muddled plumbers. Such faults in plumbing seem to be surprisingly common, but only exceptionally do they cause explosive epidemics. A recent water-borne outbreak occurred among the employees of a factory at South Bend, Indiana, in 1953. More than half of about 1500 persons became infected and several died. In this instance very excellent chlorinated drinking water was contaminated with sewage at a leaky joint in a pipe under suction (Le Maistre, et al., 1956). Chlorination of water in most cities prevents sewage-tainted supplies from causing typhoid or other bacterial infections, but it has no effect on protozoan cysts. Sand filtration, properly carried out, seems to remove the cysts very well. It seems probable that amebic infections might be acquired from dirty swimming pools if very much of the water is swallowed.

Although it has been shown that cysts survive for only a few minutes on hands, except under long, closely fitting fingernails, and that the chances are against transmission by soiled hands of food handlers on any single occasion, nevertheless there is no doubt that oft-repeated exposure of one's food to handling by careless, infected food handlers is dangerous. Although cyst-passing cooks, dairy workers, icemen, waitresses, etc., must all occasionally transmit amebic infection, probably housewives and mothers are most important, since the frequency of exposure is greatest. However, the fact that food handlers do not transmit infection as readily as was once supposed is comforting news for those whose gastronomical needs are ministered to by native servants, public food handlers, or rural southern hospitality.

Animal carriers probably do not contribute to human infection, but flies and roaches may do so. Pipkin (1949) showed that filth flies are capable of passing viable cysts in their vomitus up to an hour after ingesting them, and in their feces for over four hours. Flies probably play some part in transmission in unsanitary communities, and may occasionally be an important factor in local outbreaks; Craig described one such outbreak in some troops in El Paso, Texas, in 1916.

Diagnosis. This should usually be based on the finding of *Entamoeba histolytica* in the stools by the methods described on pp. 61–62. Complement fixation and cultural methods were discussed on pp. 62, 59–60. In examining dysenteric stools for trophozoites, special attention should be paid to flakes of blood or mucus. When clinical manifestations of infection are present, a high percentage of cases can be diagnosed by a single examination of a stool or of the rectum by a proctoscope. The presence of whetstone-shaped "Charcot-Leyden" crystals in feces is usually indicative of *E. histolytica* infection.

Experience in identifying Protozoa is required. As great a danger lies in making a false positive diagnosis as a false negative one, for to inexperienced workers an ameba is an ameba, and often even epithelial cells and other objects are amebas. Many a patient unfortunate enough to have an undiscriminating technician mistake leucocytes in a bacillary dysentery stool or find an innocent *E. coli* or *Endolimax nana* has had to submit to a course of treatment which was useless if not injurious to himself and quite innocuous to the amebas. Dobell in 1917 wrote: "The errors committed by an examiner with little or no previous experience are such as I could not have believed possible if I had not actually encountered them; and in cases where the health of the patient is at stake, it is, I believe, almost better that no examination at all should be made, than that it should be made by an incompetent and inexperienced person."

Treatment. Like many other protozoan diseases, amebiasis if left untreated tends to become chronic and to persist indefinitely. Since the amebas are found both on the surface of the mucosal cells in the lumen and buried in the tissues, permanent cure requires a drug or drugs that will reach them in either situation. Different drugs are needed (1) to stop acute dysentery promptly, (2) to eliminate acute or chronic infections from the intestine, and (3) to cure hepatic, pulmonary, or other extra-intestinal infections.

For prompt relief of acute or subacute dysentery, emetin was long the drug of choice in spite of its toxicity and the fact that it has to be given by injection, but certain antibiotics—Fumagillin, Terramycin, Erythromycin, and Aureomycin—are more effective, and can be given by mouth.

For eradication of intestinal infections after the dysentery is controlled, or in chronic cases, certain arsenic compounds (Carbarsone, Thiocarbarsone, Milibis) and a number of iodine compounds (Chiniofon (Yatren), Diodoquin, and Vioform) are effective when given over a period of 7 to 10 days (for doses of Carbarsone and iodine compounds see Craig, 1944 or Faust, 1954). Milibis is a very insoluble compound; 0.5 grams can be given to adults three times daily after meals for 7 days. Half this amount can be given to children. Except where there is acute dysentery the arsenic and iodine compounds are still the drugs of choice even though the antibiotics may also be effective, because the latter sometimes cause serious disturbances, with elimination of the normal bacterial flora of the intestine. Like many other upsets of the balance of nature, this may have very undesirable effects.

For amebiasis of the liver or lungs, or other extra-intestinal amebic infections, emetin, long the only effective drug known, has been almost entirely superseded by the anti-malarial drug, Chloroquine. The results are dramatic; apparent cures occur in a week, though treatment should be continued with reduced dosage for 2 or 3 weeks. Conan (1948) got good results with 0.3 gram of the base twice daily for 2 days, followed by sustaining doses of 0.3 gram daily for 12 to 19 days more. One recommended regimen is one 0.25 gram tablet of Chloroquine diphosphate every 6 hours for 4 days, then one twice daily for 7 days, and then one daily to the twenty-first day, given after food.

Because intestinal infections may be cured without affecting incipient liver infections, Berberian et al. (1952) tried treatment and subsequent prophylaxis with a combination of Milibis and Chloroquine, and found it very successful. Martin et al. (1953) experimented with a number of drugs singly and in combination in a group of 538 acute infections in Korea. They found that Terramycin alone or in combination with emetin, Carbarsone, Chiniofon, or Chloroquine; Milibis and Chloroquine; and Aureomycin and Chloroquine all gave excellent initial responses and low relapse rates. They did not use Fumagillin,

Erythromycin, or Puromycin which have been used by later workers with satisfactory results.

Diodoquin, a nontoxic drug, was recommended by Craig as a prophylactic for travelers in places where there is danger of infection. A nonmetallic organic compound, Camoform, is useful for both intestinal and extra-intestinal amebiasis.

Prevention. The essentials in the prevention of amebic infection are sanitation and protection of water and vegetables from pollution. Soil pollution, especially by use of night soil, is dangerous in places where unfiltered water is used for drinking, even if it is chlorinated. Clark showed that there was a great falling off of amebic dysentery in Panama after a good water system was installed in 1914–1915. In view of the plumbing hazards discovered in Chicago in 1933 and in South Bend in 1953 it seems evident that public health officials and city governments should spend sufficient money for inspection of hotels and public buildings, but for the most part they have not done so.

Vegetables such as lettuce, radishes, and strawberries, grown in ground fertilized by night soil or even in ground subject to ordinary pollution, are dangerous. Beaver and Deschamps (1949) recommend immersion in 5% acetic acid (or vinegar) for 15 minutes at 30°C. or in 2.5% for 5 minutes at 45°C. It is customary for Europeans in the Orient to soak uncooked vegetables in a potassium permanganate solution for an hour, but usually several cooks have to be discharged before one is found who will actually carry out what he considers a silly notion rather than risk being caught not doing it. Even then he feels that if he sets a head of lettuce in an inch of "red water" he has sufficiently carried out instructions. Immersion for 30 seconds in water at about 150°F. has also been recommended.

Although transmission by the hands may not be so easy as was once supposed, Craig believes that food handlers constitute the most important means of transmission in sanitated cities. The tendency for the infection to spread in families indicates transmission from person to person, probably as a rule from the servant or housewife who prepares the food. Continually repeated exposure may be dangerous when occasional exposure is not. A careful washing of the hands with soap and water after using a toilet would probably eliminate most of the danger. James recommended that Europeans in the tropics should insist on all servants cleaning their hands thoroughly with scrubbing brush, antiseptic soap, and water several times a day, especially before preparing or serving food. This is excellent advice, but in some tropical countries one would have to stand over each servant with both eyes wide open during the entire process of each washing and would very likely have to render assistance!

Other Intestinal Amebas

The other amebas which inhabit the human intestine, with the exception of *Dientamoeba fragilis,* would be of very little consequence if it were not for the danger of confusion between them and *Entamoeba histolytica.* They are never tissue parasites, and there is no good evidence that a human being is any worse off for harboring these guests in his intestine. They ordinarily live free in the lumen of the intestine, where they feed on bacteria, small cysts, starch grains, and all sorts of debris found in the semifluid medium in which they live. In other respects, such as life cycle, cultural characteristics, mode of transmission, transferability to other kinds of animals, etc., they appear to be similar to *E. histolytica.*

Entamoeba coli. This is the commonest species of ameba in the human intestine and has been stated to occur probably in 50% of human beings; its distribution is world-wide. According to Dobell "no race, nor any country, has yet been discovered in which infections with this species are not common." In the United States it has roughly about three times the incidence of *E. histolytica.* The motile forms are found especially in the upper part of the large intestine, and the precystic and cyst forms lower down.

The outstanding characteristics of *E. coli* (Fig. 4) have been mentioned in connection with its differentiation from *E. histolytica,* but they may advantageously be summarized again. The living forms are usually 20 to 30 μ in diameter and are never as small as the smallest races of *histolytica.* The body usually has very little ectoplasm, and even the ponderous pseudopodia are usually composed mainly of endoplasm, although clear ones are occasionally produced. Unlike *histolytica* this ameba tends to move about sluggishly in one place without making much headway in any one direction. The body is usually crammed with food vacuoles, for it is a voracious and undiscriminating feeder. The nucleus is visible in living specimens as a refractile ring.

In stained specimens the contained food and the nucleus distinguish it from *histolytica.* The nucleus has a coarser peripheral layer of chromatin, a larger and eccentrically placed endosome, and usually dots of chromatin strung on the linin network.

The precyst stages are the most difficult to distinguish from those of *histolytica:* the distinction can be made only by observation of the nuclear structure in good specimens; the 2-nucleated stage (Fig. 4*B*) usually has a very large glycogen vacuole, which nearly fills the cyst, lying between the nuclei, but this begins to become diffuse even by the time the cyst becomes 4-nucleated. The mature cysts, which are most commonly found in fresh stools, are 15 to 22 μ in diameter, have 8 nuclei of the typical *coli* type, more granular cytoplasm than in *histolytica,* and either no chromatoid bodies or else a few flakes like splintered glass, but never the heavy bars found in young *histolytica* cysts. According to Hegner, the cysts hatch as entire 8-nucleated amebas.

Endolimax nana. This little ameba is almost as frequent an inhabitant of the human intestine as is *Entamoeba coli,* and is frequently found in 15 to 30% of cases in routine examinations in this country. Its principal characteristics are those given under the genus *Endolimax* on p. 58 and shown in Fig. 5*A* and *A′*. It is a very small ameba, varying from 6 to 12 or 15 μ in diameter, but usually averaging only about 7 to 9 μ. It creeps sluggishly like *E. coli* and often contains numerous food vacuoles filled with bacteria. The 4-nucleated cysts (Fig. 5*A′*) are distinguishable by their small size (usually 6 to 10 μ by 5 to 8 μ), their usually oval shape, and the peculiar structure of the nuclei, described on p. 58. Like *E. coli,* it cannot be eliminated by drugs although it temporarily disappears during emetin treatment. *Endolimax nana* also occurs in monkeys, and probably identical forms occur in rats and pigs; a form from a guinea pig differing only in its smaller size has also been described. Other probably different species occur in frogs, lizards, and fowls.

Iodamoeba bütschlii. This small ameba is usually larger than *Endolimax nana* and smaller than the entamebas. Usually the amebas average about 9 to 11 μ in diameter, but specimens varying from 4 to 19 μ have been found, and Wenrich (1937) believes that there are large and small races. The characteristic features of the nucleus are mentioned on p. 58, and shown in Fig. 5*B* and *B′*. The living trophozoites are sluggish but move about by the extrusion of clear ectoplasmic pseudopodia; the nucleus is not usually visible, but there are usually ingested food particles. In stained specimens the body does not usually show any clear ectoplasm and often has a vacuolated appearance.

The cysts of this ameba are peculiar in several respects. They are about the same size as the trophozoites and are of irregular shape, as if formed under pressure. The endosome moves to an eccentric position almost in contact with the nuclear membrane, and the granules between it and the periphery usually cluster into a cresent-shaped mass on the inner side of it. Usually the nucleus remains single, but Wenrich states that occasionally cysts with two or even three nuclei are formed. The most striking feature of the cysts, however, is a large, sharply defined vacuole filled with glycogen and therefore staining brown in iodine. In fixed and stained specimens the glycogen dissolves out and leaves a large cavity.

Iodamoeba bütschlii infests a very high percentage of monkeys and pigs. Cauchemez estimated that 50% or more pigs in France are infected, and Feibel found 20% of pigs slaughtered in Hamburg harboring it. The pig may, in fact, be considered the normal host in temperate climates. This is another example of the close parasitological relations between pigs and man. This ameba is not so common in man as those hitherto described; in most surveys in this country it occurs in 2 to 6%, especially of adults. Much higher incidences have been reported in Mexico and China, and in some Egyptian villages the incidence is over 50%, in spite of the fact that pigs are absolutely taboo.

There is one unique and amazing instance of fatal generalized amebiasis in a Japanese soldier, captured in New Guinea. There were amebic ulcers in his digestive tract from stomach to colon, and also in his lymph nodes, lungs, and brain (but not in his liver), in which the amebas were definitely not *Entamoeba histolytica,* but either *Iodamoeba bütschlii* or one closely resembling it (Derrick, 1948).

Dientamoeba fragilis. This is another small ameba, usually ranging in size from 3.5 to 12 μ; the average is usually around 9 μ, but in diarrheic stools it may be 11 μ. This ameba (Fig. 5*C* and *C′*) is peculiar in that, in most populations of it, about 80% have two nuclei, and occasionally supernucleate forms appear. This is because, unlike other amebas, the nucleus divides shortly after cell division instead of just before it. The nucleus has in its center a cluster of four to eight deep-staining granules, one of which is an endosome. The cytoplasm has a granular or frothy appearance, with or without food vacuoles, and often the ectoplasm is sharply demarcated from the endoplasm by a deep-staining zone. This and other characters of the nucleus and cytoplasm have led both Dobell and Wenrich to suggest relationships to the flagellate *Histomonas* (Fig. 15); Dobell says that "*Dientamoeba* is, indeed, a typical flagellate except for the important circumstance that it possesses no flagella." No cysts are formed by *Dientamoeba.* In living or iodine-stained specimens the nuclei are not visible, and they stain poorly in MIF. In saline suspensions the trophozoites remain in a sort of dazed, immobile state for 5 or 10 minutes, appearing rounded and granular; they then extrude broad, flat, leaflike pseudopodia of clear ectoplasm. In tap water they swell up and explode, leaving a hollow shell of ectoplasm.

Dientamoeba has a world-wide distribution, and its incidence is probably much higher than commonly reported, partly because methods suitable for its detection are not used and partly because many technicians fail to recognize it when they see it. Wenrich and his co-workers found it in 4.3% of 1060 students at the University of Pennsylvania and in 3.9% of 190 Philadelphia food handlers, when only a single stool was examined; with more examinations an incidence of 7.4% was obtained in the latter group. In some institutions incidences of 36, 42, and 50% have been obtained. In some Egyptian villages the senior writer found it in 11% of the population by a single 10-minute stool examination in MIF. Yet Wenrich (1944) reported that, in 19 surveys in the United States between 1934 and 1944, 65,253 persons were examined without this elusive organism being found in any of them! The organism is best recognized in smears fixed with picro-formol-acetic in the proportions of 75:15:10, and stained with iron hematoxylin. In the group of 1060 students mentioned above, over 60% of the *Dientamoeba* infections were recognized only on stained slides.

There is strong evidence that *Dientamoeba* may sometimes be pathogenic, so, unlike the other amebas except *Entamoeba histolytica,* it is of interest to the medical practitioner as well as to the parasitologist in an ivory tower. Wenrich et al. in 1936 reported a higher incidence of gastrointestinal disturbances among students playing host to this ameba than among those harboring *Entamoeba histolytica,* and Sapero (1939) recorded that 27% of *Dientamoeba* cases had complaints as compared with 43% of those with *E. histolytica,* and 7% of those devoid of intestinal protozoa or harboring only *Entamoeba coli, Endolimax,* or *Iodamoeba.* Wenrich calls attention to frequent eosinophilia in *Dientamoeba* cases. The commonest symptoms associated with it are diarrhea, colicky pains, fatigue, and weight loss. Burrows et al. (1954) found this ameba in 4 of 581 appendices; in every case they were ingesting red corpuscles (though not invading tissues), and they appeared to be the cause of a fibrosis of the walls.

How *Dientamoeba* is transmitted is still an unsolved problem, since the organism usually dies in from a few hours to a day or two in feces, and explodes in water. Dobell failed to infect either himself or two monkeys by means of swallowed cultures, although the infection does occur in monkeys. Dobell's belief that *Dientamoeba* is related to *Histomonas* suggests the possibility that it may be transmitted by the eggs of parasitic worms, since *Histomonas* is transmitted in this manner (see pp. 104–105). Burrows and Swerdlow (1956) suggested that *Dientamoeba* enters the host in the eggs of the pinworm, *Enterobius*, but this has not been demonstrated by experiment.

Dientamoeba infections usually respond to Carbarsone and other drugs effective against *E. histolytica*. This in itself is suggestive of its being a pathogenic parasite, for the species of amebas that are content to live a saprophytic life in the lumen of the intestine are unaffected by these drugs.

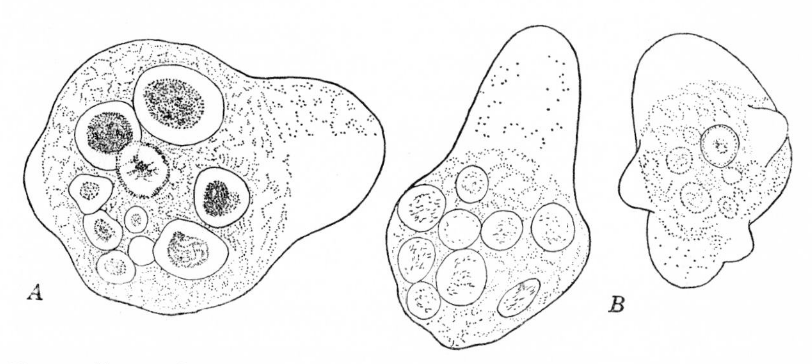

Fig. 9. *Entamoeba gingivalis: A*, stained specimen; *B*, living specimens; nucleus near center of stained specimen, and visible in unstained specimen at right. (Adapted from Kofoid and Swezy, *Univ. Calif. Publ. Zool.*)

Mouth Amebas (*Entamoeba gingivalis*)

In contrast to all other amebas living in man or animals, there is one species of *Entamoeba, E. gingivalis,* which inhabits the mouth instead of the large intestine. The same or a similar species has been found in pyorrheal pus from the mouths of dogs and cats, around the teeth of horses, and in a high percentage of captive monkeys. Dogs with inflamed gums or pus pockets can be infected with the human mouth ameba. In man it can be found in a high percentage of individuals, increasing with advancing age until, according to Kofoid, 75% or more of people over 40 harbor it.

E. gingivalis markedly resembles *E. histolytica*. It is about 12 to 20 μ in diameter and has crystal clear ectoplasm (Fig. 9). The vacuolated endoplasm is usually crowded with food particles which seem to float in the center of large, fluid cavities. The pseudopodia are normally broad and rounded, like large blisters, and the ameba normally progresses rapidly in various directions. The nucleus has the peripheral chromatin in rather uneven granules, and the endosome consists of several closely associated granules. Whereas the *E. histolytica* nucleus has a clear halo around the endosome and a finely granular outer zone between the halo and the nuclear membrane, *E. gingivalis* has a granular, cloudy halo, especially dense around the endosome, and a clear outer zone, through which a few spokelike strands of linin run.

Unlike its close relatives this species fails to form cysts; apparently the ease and rapidity with which infections can spread from one human mouth to another do away with the biological necessity for cysts. As would be expected of an organism inhabiting the mouth, it is rather more adaptive than the intestinal amebas to changing environmental conditions, e.g., heat and cold, hydrogen-ion concentration, and chemical constituents of a culture medium.

The food vacuoles of *Entamoeba gingivalis* sometimes contain bacteria, but they most often contain the nuclei of leucocytes in various stages of digestion (Fig. 9). Although *E. gingivalis* seems to occur more commonly in diseased than in healthy mouths, there is no convincing evidence that the organism causes oral disease. It has been found on a few occasions to multiply abundantly in bronchial mucus and appear in sputum, and this could easily lead to a false diagnosis of pulmonary *E. histolytica* infection.

REFERENCES

Anderson, H. H., et al. 1952. Fumagillin in amebiasis. *Am. J. Trop. Med. Hyg.*, 1: 552–558.

Andrews, J. 1942. The transmission of *Endamoeba histolytica* and amebic disease. *Southern Med. J.*, 35: 693–699.

Beaver, P. C., and Deschamps, G. 1949. The effect of acetic acid on the viability of *Endamoeba histolytica* cysts. *Am. J. Trop. Med.*, 29: 189–197.

Beaver, P. C., et al. 1956. Experimental *Entamoeba histolytica* infections in man. *Am. J. Trop. Med. Hyg.*, 5: 1000–1009.

Benham, R. S., and Havens, I. 1958. Some cultural properties of *Endamoeba histolytica*. *J. Infectious Diseases*, 102: 121–142.

Berberian, D. A., et al. 1952. Drug prophylaxis of amebiasis. *J. Am. Med. Assoc.*, 148: 700–704.

Bock, M., and Mudrow-Reichenow, L. 1955. Experimentelle Untersuchungen über *Entamoeba histolytica*. *Z. Tropenmed. u. Parasitol.*, 6: 344–347.

Brooke, M. M., et al. 1955. Studies of a water-borne outbreak of amebiasis, South Bend, Indiana. III. Investigation of family contacts. *Am. J. Hyg.*, 62: 214–226.

Brooke, M. M., and Goldman, M. 1948. Polyvinyl alcohol-fixative as a preservative and adhesive solution for staining Protozoa in dysenteric stools and other liquid materials. *J. Parasitol.*, 34: Suppl. 12.

Brown, J. A. H., and Whitby, J. L. 1955. An immobilization test for amoebiasis. *J. Clin. Pathol.*, 8: 245–246.

Burrows, R. B., and Klink, G. E. 1955. *Endamoeba polecki* infections in man. *Am. J. Hyg.*, 62: 156–167.

Burrows, R. B., and Swerdlow, M. A. 1956. *Enterobius vermicularis* as a probable vector of *Dientamoeba fragilis*. *Am. J. Trop. Med. Hyg.*, 5: 258–265.

Burrows, R. B., et al. 1954. Pathology of *Dientamoeba fragilis* infections of the appendix. *Am. J. Trop. Med. Hyg.*, 3: 1033–1039.

Chinn, B. D., Jacobs, L., Reardon, L. V., and Rees, C. W. 1942. The influence of the bacterial flora on the cultivation of *Endamoeba histolytica*. *Am. J. Trop. Med.*, 22: 137–146.

Conan, N. J. 1948. Chloroquine in amebiasis. *Am. J. Trop. Med.*, 28: 107–110.

Craig, C. F. 1944. *Etiology, Diagnosis and Treatment of Amebiasis.* Williams and Wilkins, Baltimore.

Derrick, E. H. 1948. A fatal case of generalized amoebiasis due to a protozoan closely resembling, if not identical with, *Iodamoeba bütschlii*. *Trans. Roy. Soc. Trop. Med. Hyg.*, 42: 191–198.

Dobell, C. 1919. *The Amoebae Living in Man.* Bale, Sons and Danielsson, London.

1928–1938. Researches on the intestinal Protozoa of monkeys and man. *Parasitology,* 20: 357–412; 21: 446–468; 23: 1–72; 25: 436–467; 28: 541–593; 30: 195–238.

Dobell, C., and O'Connor, F. W. 1921. *The Intestinal Protozoa of Man.* Bale, Sons and Danielsson, London.

Faust, E. C. 1941. The prevalence of amebiasis in the western hemisphere. *Am. J. Trop. Med.*, 22: 93–105.

1954. *Amebiasis.* Thomas, Springfield, Ill.

Faust, E. C., Sawitz, W., Tobie, J., Odom, V., Perez, C., and Lincicome, D. R. 1939. Comparative efficiency of various technics for the diagnosis of protozoa and helminths in feces. *J. Parasitol.*, 25: 241–262.

Freedman, L., and Elsdon-Dew, R. 1958. Size variation in *Entamoeba histolytica*. *Nature,* 181: 433–434.

Hakansson, E. G. 1936. *Dientamoeba fragilis,* a cause of illness. *Am. J. Trop. Med.*, 16: 175–185.

Hoare, C. A. 1958. The enigma of host-parasite relations in amebiasis. *Rice Inst. Pamphlet,* XLV (1): 23–34.

Hoare, C. A. 1952. The commensal phase of *Entamoeba histolytica*. *Exp. Parasitol.*, 1: 411–427.

Hussey, K. L., and Brown, H. W. 1950. The complement fixation test for hepatic amebiasis. *Am. J. Trop. Med.*, 30: 147–154; discussion, 154–157.

James, W. M. 1928. Human amebiasis due to infection with *Endamoeba histolytica*. *Ann. Trop. Med. Parasitol.*, 22: 201–258.

Kenney, M. 1952. Micro-Kolmer complement fixation test for amebiasis. *Am. J. Trop. Med. Hyg.*, 1: 717–726.

Kofoid, C. A. 1929. The protozoa of the human mouth. *J. Parasitol.*, 15: 151–174.

Larsh, J. E., Jr. 1952. The effect of a blacktongue-producing diet on experimental amebiasis in dogs. *Am. J. Trop. Med. Hyg.*, 1: 970–979.

Le Maistre, C. A., et al. 1956. Studies of a water-borne outbreak of amebiasis,

South Bend, Indiana. I. Epidemiological aspects. *Am. J. Hyg.*, 64: 30–45.

McConnachie, E. W. Modification and elimination of the bacterial flora in cultures of *Entamoeba invadens* Rodhain, 1934. *Parasitology,* 46: 117–129.

McDearman, S. C., and Dunham, W. B. 1952. Complement fixation tests as an aid in the differential diagnosis of extra-intestinal amebiasis. *Am. J. Trop. Med. Hyg.*, 1: 182–188.

Martin, G. A., Garfinkel, B. T., Brooke, M. M., Weinstein, P. P., and Frye, W. W. 1953. Comparative efficacies of amebacides and antibiotics in acute amebic dysentery, used alone and in combination in 538 cases. *J. Am. Med. Assoc.*, 151: 1055–1059.

Meleney, H. E., and Frye, W. W. 1936. The pathogenicity of *Endamoeba histolytica. Trans. Roy. Soc. Trop. Med. Hyg.*, 29: 369–379.

Michael, K. M. M., and Cooray, G. H. 1952. A culture method as an aid for routine diagnosis of amebic infection in Ceylon. *Am. J. Trop. Med. Hyg.*, 1: 543–547.

Miller, M. J., and Gilani, A. 1951. The clinical significance of non-dysenteric intestinal amoebiasis. *Trans. Roy. Soc. Trop. Med. Hyg.*, 45: 131–136.

Neal, R. A. 1956. Strain variation in *Entamoeba histolytica.* III. The influence of the bacterial flora on virulence to rats. *Parasitology,* 46: 183–191.

Neal, R. A. 1953. Studies on the morphology and biology of *Entamoeba moshkovskii* Tshalaia. *Parasitology,* 43: 253–268.

Phillips, B. P., et al. 1955. Studies on the ameba-bacteria relationship in amebiasis. Comparative results of the intracecal inoculation of germfree, monocontaminated, and conventional guinea pigs with *Entamoeba histolytica. Am. J. Trop. Med. Hyg.*, 4: 675–692.

Phillips, B. P. 1957. The pathogenic mechanisms in amebiasis. *Am. J. Proctology,* 8: 445–450.

Phillips, B. P., and Wolfe, P. A. 1959. The use of germfree guinea pigs in studies on the microbial interrelationships in amebiasis. *Ann. N. Y. Acad. Sci.*, 78(1): 308–314.

Rees, C. W. 1955. *Problems in Amebiasis.* Thomas, Springfield, Ill.

Reeves, R. E., Meleney, H. E., and Frye, W. W. 1957. Bacteria-free cultures of *Entamoeba histolytica* with chick embryo tissue juice. *Z. Tropenmed. Parasitol.*, 8: 213–218.

Rendtorff, R. C. 1954. The experimental transmission of human intestinal protozoan parasites. I. (*Entamoeba coli*). *Am. J. Hyg.*, 59: 196–208.

Sadun, E. H., Braden, J. L., Jr., and Faust, E. C. 1951. The effect of ascorbic acid deficiency on the resistance of guinea pigs to infection with *Endamoeba histolytica* of human origin. *Am. J. Trop. Med.*, 31: 426–437.

Sapero, J. J. 1939. Clinical studies in non-dysenteric intestinal amebiasis. *Am. J. Trop. Med.*, 19: 497–514.

Sapero, J. J., and Johnson, C. M. 1939. An evaluation of the role of the food handler in the transmission of amebiasis. *Am. J. Trop. Med.*, 19: 255–264.

Sapero, J. J., and Lawless, D. K. 1953. The MIF stain-preservation technique for the identification of intestinal Protozoa. *Am. J. Trop. Med. Hyg.*, 2: 613–619.

Sawitz, W. G., and Faust, E. C. 1942. The probability of detecting intestinal protozoa by successive stool examinations. *Am. J. Trop. Med.*, 22: 131–136.

Shaffer, G. G., Ryden, F. W., and Frye, W. W. 1949. Studies on the growth requirements of *Endamoeba histolytica IV. Am. J. Hyg.*, 49: 127–133.

Swartzwelder, J. C., and Avant, W. H. 1952. Immunity to amebic infection in dogs. *Am. J. Trop. Med. Hyg.*, 1: 567–575.

Tobie, J. E., et al. 1951. Laboratory results in the efficacy of terramycin, aureomycin and bacitracin in the treatment of asymptomatic amebiasis. *Am. J. Trop. Med.*, 31: 414–419.

Wenrich, D. H. 1944. Studies on *Dientamoeba fragilis* (Protozoa)—IV. Further observations with an outline of present-day knowledge of this species. *J. Parasitol.*, 30: 322–328.

1937. Studies on *Iodamoeba bütschlii* (Protozoa) with special reference to nuclear structure. *Proc. Am. Phil. Soc.*, 77: 183–205.

Wenrich, D. H., Stabler, R. M., and Arnett, J. H. 1935. *Endamoeba histolytica* and other intestinal Protozoa in 1060 college freshmen. *Am. J. Trop. Med.*, 15: 331–345.

Westphal, A. 1948. Zur Epidemiologie und Pathogenese der Amöbenruhr in Nord-Afrika. *Z. Hyg. Infektionskrankh.*, 128: 73–86.

World Health Organization. 1955. *Epidem. and Vital Stat. Rept.*, 8: 127–131.

Chapter 5

INTESTINAL FLAGELLATES AND CILIATES

Flagellates in General

The flagellates (subphylum Mastigophora) surpass all other Protozoa in numbers of individuals and in variety of environments successfully occupied. Free-living forms range from the "red snows" of Alpine summits to the ooze of the ocean's depths, and they abound in natural waters, soil, decaying organic matter, and, either as commensals or parasites, in the bodies of the majority of species of animals and many plants, where few tissues or organs are immune to their invasions. In their varied forms of nutrition and metabolism they afford valuable material for the study of fundamental biological problems.

Classification. As noted on p. 48, the Mastigophora are divided into two classes, the Phytomastigophorea and the Zoomastigophorea, containing the green algalike forms (and their allies) and the animal-like forms, respectively. All the flagellates that are parasitic in higher animals belong to the Zoomastigophorea, and there are representatives in four of the five orders. The fifth one, Hypermastigida, contains cellulose-digesting symbiotes of termites and roaches, enabling these insects to utilize such things as telegraph poles and old books as food.

Although there is no general agreement on divisions into families, the following arrangement, following Jahn and Jahn, and Hall, is as logical as any:

Class **Zoomastigophorea.**

Order 1. **Rhizomastigida.** Body ameboid, with pseudopodia, and 1 flagellum.

Order 2. **Protomastigida.** One or two flagella; body plastic but not ameboid; no axostyle. Several families, including the following:

Family **Trypanosomidae.** One flagellum ending in a blepharoplast, with a kinetoplast (see p. 41) near it (rarely absent). Includes the hemoflagellates of vertebrates.

Family **Cryptobiidae.** Two flagella, one trailing and adherent to body; large kinetoplast. Parasites of mollusks and digestive tract of fishes, *Cryptobia* and, in blood of fishes, *Trypanoplasma.*

Family **Bodonidae.** Two flagella, one usually trailing in swimming. Some in digestive tract of amphibians and reptiles; mostly saprozoic. *Bodo* and *Cercomonas* often found in stale feces or urine, and sometimes in urinary bladder.

Order 3. **Polymastigida.** A heterogeneous group of flagellates with 3 to 8 flagella (2 in *Retortamonas*) and 1 to several nuclei. No costa (see p. 42), parabasal body, or axostyles, except in Hexamitidae; some free-living, some parasitic; includes 8 families, including the following:

Family **Tetramitidae.** Flagella 3 or 4, with 1 or 2 trailing; 1 nucleus; no cytostome. Includes *Enteromonas* and *Tricercomonas* (if a distinct genus [see p. 103]).

Family **Retortamonadidae.** Flagella 2 or 4, 1 trailing, 1 nucleus; cytostome present with supporting fibrils. Includes *Retortamonas* and *Chilomastix.*

Family **Hexamitidate.** Flagella 6 or 8; 2 nuclei; parabasal bodies and axostyles in some. Includes *Giardia* and *Hexamita.*

Order 4. **Trichomonadida.** Axostyle and costa present. Flagella in one or more groups (mastigonts) of 3 to 6 each, 1 trailing; each mastigont associated with 1 nucleus. In insects, especially termites, and vertebrates.

Family **Trichomonadidae.** Have axostyle, costa, and undulating membrane connecting trailing flagellum to body; 3 to 5 anterior flagella; 1 nucleus. Includes *Trichomonas.*

Order 5. **Hypermastigida.** Numerous flagella and multiple axostyles and parabasal bodies, but 1 nucleus. Parasites of termites and roaches.

For convenience we can divide all the flagellates found in man and domestic animals into two groups, the hemoflagellates and the intestinal flagellates. The hemoflagellates live in the blood, lymph, and tissues of their vertebrate hosts and usually pass one phase of their life cycle in the gut of insects. All these belong to the family Trypanosomidae and will be considered in the next chapter. The intestinal flagellates will be considered in the present chapter, along with the allied forms found in the mouth and vagina. At the end of the chapter we shall also present a brief discussion of intestinal ciliates, only one species of which is a true parasite of man, though many are commensals in the alimentary canals of large herbivorous animals. The coccidians, some of which are important parasites of the digestive tracts of many birds and mammals, will be reserved for consideration in Chapter 10, along with other Sporozoa.

INTESTINAL FLAGELLATES

The human "intestinal" flagellates which are commonly recognized belong to five genera, of which *Trichomonas* lives in the mouth, large intestine, and vagina; *Chilomastix,* and probably the rarer *Retortamonas* and *Enteromonas,* lives in the large intestine; and *Giardia* lives in the small intestine. In addition to these genera we shall briefly consider *Histomonas,* a parasite of the intestine, ceca, and liver of turkeys, and *Hexamita,* in the intestine of various birds. Most of these intestinal flagellates form cysts, but neither *Histomonas* nor any of the trichomonads do so. None of these flagellates require an intermediate host except *Histomonas,* which makes use of the eggs of an intestinal worm (*Heterakis*) for this purpose.

All the intestinal flagellates except *Giardia* are easily cultivated in artificial media and are less fastidious about their culture media than are the amebas, although any medium satisfactory for amebas will also grow the intestinal flagellates. Unlike the amebas, however, the flagellates do not require accompanying bacteria. A simple and successful culture consists of a long slant of 1.5% nutrient agar, without a butt, half covered with a sterile Ringer solution with one-twentieth part of horse serum added. For *Trichomonas vaginalis* the slant should be three-quarters covered, the *p*H lowered to 5.5 to 6, and 0.2% dextrose added. It has been customary to subculture every few days, but Wenrich found that, if the nutrients and evaporated water are replaced as needed, cultures will live possibly for years.

Some of the intestinal flagellates appear to be harmless commensals; such are *Chilomastix, Retortamonas,* and *Enteromonas. Trichomonas* and *Giardia,* on the other hand, unquestionably have pathogenic propensities, although some authors tend either to minimize or exaggerate them.

There is still much doubt as to the extent to which intestinal protozoans are confined to particular hosts. Some workers believe that each animal has its own species peculiar to it, and that these species do not normally infect other hosts. However, evidence is accumulating to show that many intestinal protozoans of man are able to live in such animals as monkeys, rats, and hogs.

Naturally these parasites are seldom discovered except when there is some intestinal ailment, since in normal health feces are seldom submitted for examination. Where routine examinations have been made regardless of physical condition, it has been found that a large percentage of people in unsanitary places are infected. According to

Russian workers, there is some evidence that one penalty of collectivization in the Soviet Union and her allied countries is enhanced transmission of intestinal protozoa. In examinations of children under a year of age in Egypt (unpublished data by Lawless), flagellates, particularly *Trichomonas* and *Giardia,* are common, whereas amebas usually do not appear until near the end of the children's first year.

It is important to remember that free-living, coprozoic flagellates not infrequently appear in stale specimens of feces or urine, and may be a cause of confusion to unsuspecting technicians. Especially common are species of *Bodo* and *Cercomonas,* 5 to 10 μ long, both of which have two flagella, one anterior and one trailing. *Bodo* has an indistinct cytostome and a parabasal body, which *Cercomonas* lacks.

Trichomonas

General morphology. The trichomonads (Fig. 10) are all spindle- or pear-shaped organisms easily recognizable by their free anterior flagella, which are three to five in number, and their undulating membrane. The latter has a flagellum and an accessory fibril along its outer margin, giving it a double appearance, and a deep-staining basal rod or costa along its attachment to the body. The body is supported by a stiff axostyle often protruding posteriorly like a tail spine. The anterior nucleus is round or oval with varying amounts of chromatin. In most species there is a sausage-shaped parabasal body anteriorly, close to the nucleus, with a posteriorly directed parabasal fiber. In *T. hominis* the parabasal seemed to be lacking, but Kirby (1945) noted a small rounded one in a slightly different position. The anterior flagella, when three or four in number, arise together from an anterior blepharoplast; when a fifth is present it arises separately and is posteriorly directed. Trichomonads do not have a cytostome, although some authors refer to such a structure in these animals. As a matter of fact, most food particles are ingested at the posterior end of the body by phagocytosis. All these structures are shown in Figs. 10 and 11.

Species. Many vertebrates, including fish, frogs, reptiles, birds, and mammals, harbor species of *Trichomonas.* Some of these species habitually have three, some four, and some five anterior flagella. These have been placed by some authors in separate genera, *Tritrichomonas, Trichomonas,* and *Pentatrichomonas,* respectively. There has long been a belief that the human intestinal species, *T. hominis,* unlike other forms, many have either three, four, or five flagella, but there is still some uncertainty about it, since two or more of the clustered

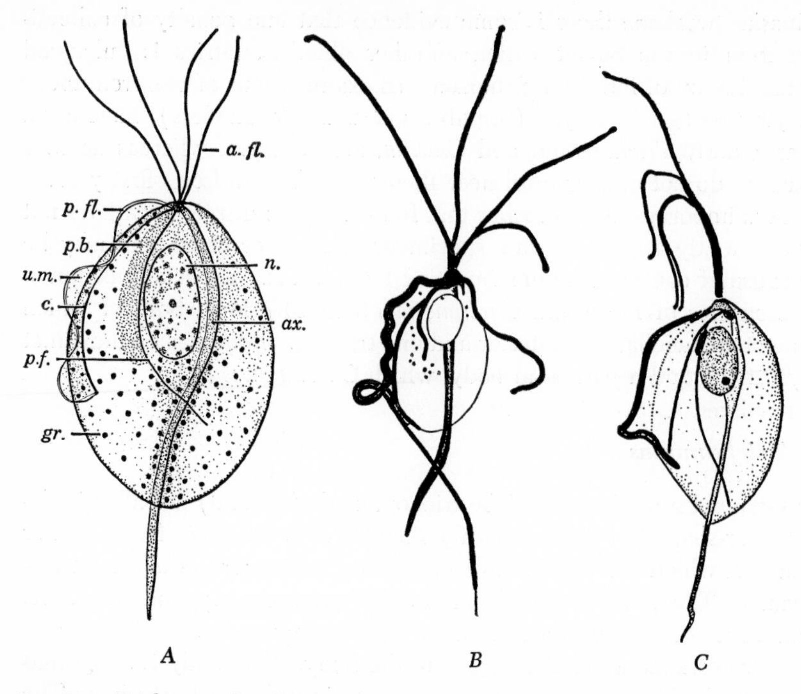

Fig. 10. Trichomonads of man: *A, T. vaginalis; B, T. tenax; C, T. hominis.* Abbreviations: *a.fl*, anterior flagellum; *ax.*, axostyle; *c*, costa; *gr.*, metachromatic granules; *n.*, nucleus; *p.b.*, parabasal body; *p.f.*, parabasal fibril; *p.fl.*, posterior flagellum; *u.m.*, undulating membrane. (*A* after Wenrich, *Am. J. Trop. Med.*, 1944; *B* after Honigberg, *Am. J. Hyg.*, 1959; *C* after Kirby, *J. Parasitol.*, 1945.)

anterior flagella tend to adhere to each other. In 1954, Flick showed that, in cultures of *T. hominis*, 77% of the cells have five flagella. The common five-flagellated form differs from other species not only in having the additional independent flagellum, but also in having a full-length undulating membrane and costa, free posterior flagellum, and a different type of parabasal body. These characteristics seem sufficient to warrant separating these five-flagellated intestinal forms into a separate genus, *Pentatrichomonas.* They occur not only in man but also in monkeys, cats, dogs, and rats. Wenrich, however, prefers to be conservative and retain the name *Trichomonas hominis* for the intestinal group until the status of four- and five-flagellated forms is determined.

Many forms of *Trichomonas* show distinctive morphological and physiological characters which warrant their recognition as distinct species. There is, however, no justification for recognizing new species simply because they are found in new hosts, since, to the annoyance of those who adhere to a belief in fairly close host specificity for intestinal Protozoa, many trichomonads are remarkably promiscuous about their hosts. *T. gallinae,* for instance, a common pathogen of pigeons, can establish itself in chickens, turkeys, hawks, parakeets, and sparrows, and *T. hominis* can be established in monkeys, cats, and rats.

Trichomonads are more finicky about their habitats in the body than they are about their hosts. *T. hominis* and *T. gallinarum* inhabit the lower alimentary canal; *T. gallinae* the throat, esophagus, and crop; *T. tenax, canistomae,* and *equibuccalis* the gums about the roots of the teeth; *T. vaginalis* the vagina and prostate; and *T. foetus* the vagina and uterus of cows and the preputial cavity of bulls. One three-flagellated species, *T. faecalis,* recovered repeatedly from the feces of a single human being, grew in fecal and hay infusions and was successfully established in frogs and tadpoles. Wenrich suspects that this species may be identical with *T. batrachorum* of Amphibia.

Trichomonads vary in pathogenicity from the harmless coprozoic form *T. faecalis* to highly pathogenic species like *T. foetus* and *T. gallinae* (see pp. 95–96). Fortunately the pathogenicity of the species found in man is relatively low.

There has been much dispute as to whether the three species in man —*vaginalis, hominis,* and *tenax*—are distinct species, but it is now definitely established that they are, since they differ in both morphology and physiology, and are not transferable from one habitat to another. *T. vaginalis* and *T. tenax* resemble each other more than they resemble *T. hominis,* but *T. tenax* will not grow in culture media in which *T. vaginalis* flourishes.

Miss Bonestell in 1936 succeeded, as have others, in establishing *T. hominis,* but not *vaginalis* or *tenax,* in the large intestines of kittens, and she could establish *tenax,* but not the others, in the mouths of kittens. *T. vaginalis* has not been established elsewhere than in the human vagina, probably because here alone it finds suitably high acidity (*p*H 4 to 5).

Biology. Trichomonads swim with a characteristic wobbly or rolling motion; sometimes they use their flagella to whirl their bodies about while anchored to a bit of debris by the axostyle. In worming their way through devious passages they can squeeze their bodies, especially the fore part, into distorted shapes. The intestinal forms feed extensively on bacteria and debris, but the vaginal and buccal forms taken from their natural environment seldom contain any solid food except leucocytes or their remains, although in cultures they contain bacteria. All species feed in part by absorption of dissolved substances, since they can be grown in liquid media containing serum without accompanying bacteria. Johnson, Trussel, and Jahn (1945) were the first to succeed in obtaining bacteria-free cultures of *T. vaginalis,* which they did with the help of penicillin.

Multiplication is by simple fission, but when it is rapid the division of the cytoplasm may fail to keep pace with growth and nuclear division, so that large multinucleate bodies are occasionally formed. No sexual phenomena have been observed.

No evidence exists that any of the species encyst. The trophozoites are apparently hardy enough to live outside the body long enough to be transferred to new hosts. *T. hominis* lives in undiminished numbers for several hours, and in some individuals for days, in the feces, and will survive a day or two in water or milk. Satti and Honigberg (1959) reported that *T. gallinae* can withstand freezing in tap water. *T. tenax* will live for several days in tap water at room temperature; in mixed material Stabler et al. found *T. tenax,* but not *T. vaginalis,* to survive when held at 16° to 18°C. for 48 hours before incubating at 37°C. In contrast, Satti and Honigberg (1959), working in the junior author's laboratory, found that *T. vaginalis* will survive at least 24 hours on damp cloth. *T. vaginalis* survives less readily than the others, and its means of transfer from host to host is somewhat of a mystery. It is known, however, that it is often transmitted venereally, but there is reason to believe that transmission may occur by contaminated toilet seats, on towels, etc. Quantitative estimates of the most frequent mode of transmission are not available and there seems to be little justification for assuming that there is a *usual* method. The occurrence of the infection in children from families where adult females are infected suggests that more than one means of transmission exists.

Trichomonas vaginalis (Fig. 19A). This is a very common human parasite. Various authors in many parts of the world have reported it in 20 to 40% or more of women where unselected series of examinations have been made, whereas in series of cases with leucorrheic conditions the organism is commonly found in 50 to as many as 70% of the patients examined. The incidence is nearly twice as high in Negro women as in white. It also occurs in 4 to 15% of men.

This is the largest of the trichomonads found in man; it varies in length from about 10 to 30 μ, but most individuals are usually between 15 and 20 μ long. There are four anterior flagella and a short undulating membrane which seldom reaches beyond the middle of the body. The axostyle projects as a slender spike at the posterior end, and the organism is frequently seen to anchor itself to debris by this structure.

The nucleus is oval and contains rather scanty chromatin scattered in granules. Deep-staining granules are also abundant in the cytoplasm, many of them in rows beside the axostyle or along the costa. The cytostome is very inconspicuous, and the body contains few food vacuoles. The parabasal apparatus is a sausage-shaped, rather faintly staining body lying beside the nucleus, and a more slender but deeper-staining fibril reaching to near the middle of the body.

T. vaginalis inhabits the vagina primarily, but also invades Skene's glands in the urethra; it is only occasionally found in other parts of the female urinogenital system. In males it occurs in the urethra and prostate. Repeated reinfection from the sexual partner has frequently been found to account for infections in women that seemed refractory to treatment. *T. vaginalis* often grows in abundance in the upper part of the vagina around the cervix but seems to show no tendency to invade the uterus as does *T. foetus* in cattle (see p. 95). It occasionally occurs in the urinary bladder, but care must be taken not to confuse it with coprozoic flagellates (see p. 89), which are frequently found in carelessly collected or stale urine.

The presence of *T. vaginalis* in the vagina is associated with a characteristic acid, creamy-white, frothy discharge which may be very abundant, and which to the experienced eye is usually sufficient for a diagnosis of the infection. The discharge often persists for months or years. The vulva becomes red and chafed, and the mucosa of the vagina and cervix is congested, with a deep red mottling. Some patients complain of severe itching or irritation in the genital region, but many seem to have no symptoms other than the discharge. That *Trichomonas* is actually the cause of these symptoms has been proved by inoculation with bacteria-free cultures of 29 women; 9 became infected and 7 showed symptoms after incubation periods of 5 to 20 days. According to Moore and Simpson (1956) emotional disturbance may be an important complicating factor in *T. vaginalis* infections.

Bland, Wenrich, and Goldstein in a series of 250 cases found a significantly higher morbidity rate in childbirth in infected than in uninfected women, and they think that pregnant women with obvious infections should be treated and, if possible, freed of the parasites in the prenatal period.

Karnaky believes that *T. vaginalis* infections are associated with a lowered acidity of the vagina, along with a thinner epithelium and less glycogen in the cells. The normal high acidity of the mature human vagina is due to the presence of a flourishing culture of Döderlèin bacilli, which are probably identical with *Bacillus acidophilus.* The vagina, however, is not highly acid in children. Since *T. vaginalis* does not thrive in a normally acid vagina, treatment involves efforts to restore the acidity as well as to kill the organisms. For this purpose douches of dilute vinegar, powders containing boric acid, or acid creams or jellies have been found useful, and also application of lactose to stimulate the growth of the acid-producing bacteria, with or without cultures of *Lactobacillus.* Frequently these methods alone are sufficient.

Drugs found useful for killing the trichomonads include the arsenic and iodine compounds used against amebas, Argyrol (an organic silver compound), or, more recently, the antibiotics Aureomycin and Terramycin. The Terramycin is given in vaginal suppositories with addition of parasepts to prevent complicating fungus infections. The other drugs, mixed with kaolin or cornstarch, are given as insufflations after drying out the vagina, or in gelatin capsules inserted high up. Male infections are best treated by Carbarsone by mouth—0.25 gram three times a day for 7 to 10 days—together with urethral instillations of 1:3000 acriflavine twice a week. Systemic treatment of vaginal infections has not proved very successful.

Trichomonas tenax. This form of *Trichomonas* (Fig. 10*B*) resembles *T. vaginalis* very closely in most respects, but is smaller, usually only 6 to 10 μ in length. The nucleus has much more chromatin and often stains almost solid black, and the granules in the cytoplasm are scattered and less conspicuous. Formerly this was regarded as a rather uncommon parasite, but Hinshaw, using cultural methods, found it in 40% of the people whom he examined who were above 30 years of age, but most of these had pyorrheic conditions. Beatman (1933) found it in more than 22% of 350 examinations of adults in Philadelphia: 26.5% in diseased mouths and 11.4% in apparently normal mouths. It is probably this species which is occasionally found in bronchial and pulmonary infections. The same or similar forms are found in the mouths of monkeys and also dogs.

Although this parasite has been found suspiciously associated with advanced inflammatory pyorrhea, its pathogenicity has not yet been proved. Along with *Entamoeba gingivalis* (see p. 81) it may well play some role, even if a minor one, in this disease. There are no special means of treatment or prevention of this parasite; only oral cleanliness is of any value.

Intestinal trichomonads. The question of whether there is more than one species of *Trichomonas* inhabiting the human intestine has not yet been settled to the satisfaction of all, but most parasitologists are now coming to the view that there is only one, *Trichomonas* (or

Pentatrichomonas) *hominis,* which usually, if not always, has five anterior flagella.

T. hominis (Fig. 10C) is easily separable from *T. vaginalis* and *T. tenax* by the fact that the undulating membrane extends the full length of the body, the flagellum along its margin continuing free at the posterior end. The parabasal apparatus and chromatic granules are not usually in evidence. In size this species is intermediate, being commonly 8 to 12 μ in length.

T. hominis occurs in a rather low percentage of people in temperate climates, but may affect 10% or more of children in the tropics. Although the pathogenicity of *T. hominis* has not been proved to the satisfaction of all, the infection is often associated with persistent diarrhea, for which some investigators believe it responsible. Kessel in 1928 reported pathogenic effects in naturally and artificially infected kittens in China, but Hegner and Eskridge failed to confirm this in experiments in the United States. Hegner (1924) found that a diet rich in carbohydrates favored an abundance of intestinal trichomonads in rats, whereas a protein diet inhibited them. Ratcliffe (1928) concluded that the number of *Trichomonas* was inversely proportional to the abundance of proteolytic anaerobic bacteria, which are favored by a protein diet. Flagellate infections do not exist in strictly carnivorous animals, whereas they are abundant in herbivorous ones, and also occur in omnivorous ones.

Trichomonads in domestic animals. Three important *Trichomonas* infections occur among domestic animals. *Trichomonas foetus* (Fig. 11) is a world-wide common and injurious parasite in the genital tract of cattle; it can be experimentally established in sheep and hamsters also.

It is a venereal disease, transmitted from infected bulls to heifers, in which it attacks the mucous membrane of the vagina and invades the uterus, causing abortions, stillbirths, delayed conceptions, and other damage. After a number of months the animals overcome the disease and are immune to further infection. A *T. foetus*-like organism has recently been described from the nose of pigs, but its relationship to bovine trichomoniasis is not yet clearly established. Deer are susceptible to *T. foetus* and many be rendered sterile by an infection. Bulls are usually infected in the preputial cavity and remain infected for life. Morgan in 1947 reported promise in the treatment of bulls with sodium iodide, but it has unpleasant effects, and after several injections both the bulls and the owners are uncooperative, so other chemotherapeutic agents are being tested. In Europe considerable success has attended the spraying and injection of the prepuce with

3% H_2O_2 with a wetting agent. Three treatments at weekly intervals are said to cure most *Trichomonas foetus* infections.

Trichomonas gallinae of young pigeons and other birds (Fig. 11*B*) attacks the mucous membranes of the throat region and esophagus,

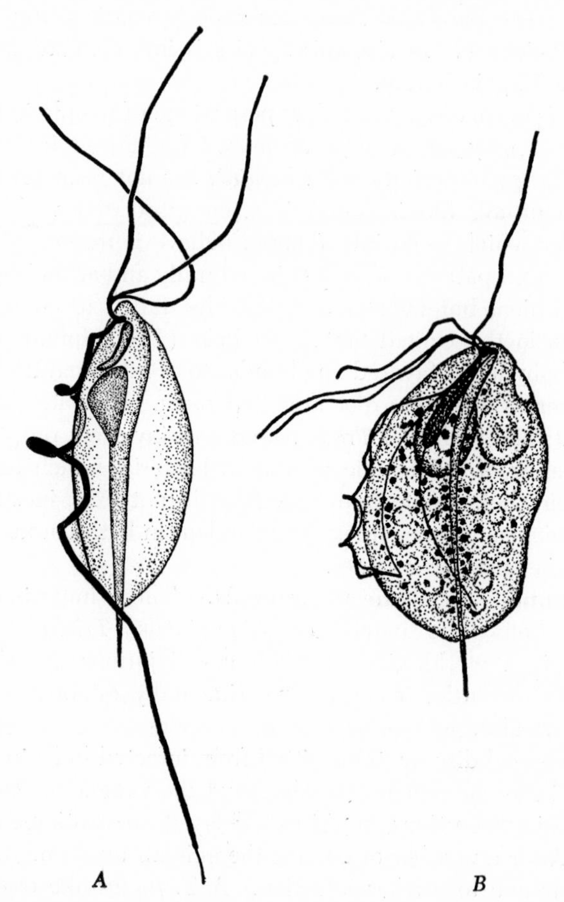

Fig. 11. *A*, *Trichomonas* (*Tritrichomonas*) *foetus* of cattle; *B*, *T. gallinae* of pigeons and other birds. (*A* after Kirby, *J. Parasitol.*, 1951. *B* after Stabler, *J. Parasitol.*, 1947.)

and occasionally of ducts in the liver and pancreas, and causes a considerable mortality. Pigeons are ideal hosts for this parasite, since they feed their squabs by regurgitation of "pigeon milk" and transfer the parasites at the same time. Many other birds are susceptible, but chickens and pheasants are usually refractory. In an active state the infection causes caseation and necrosis of tissues in the mouth and throat and is called "canker." Birds which do not die continue to harbor the organisms for a long time. Stabler (1947) found a wide variation in the virulence of different strains in pigeons and obtained a high degree of immunity to severe strains by inoculation with relatively harmless ones. Honigberg and Read (1960) recently demonstrated transformation of an avirulent strain of *T. gallinae* to virulence by treating the cells with a cell-free preparation from a virulent strain. The transformation was blocked by treatment with DNA-ase, suggesting a similarity to bacterial transformations. Enheptin, a nitrothiazole, was shown by Stabler and Mellentin (see Stabler, 1954) to be very effective in treatment.

T. gallinarum affects the lower digestive tract of galliform birds but is especially injurious to turkeys. The parasite affects particularly the liver and ceca. It causes droopiness and liquid yellow droppings, and is often fatal to young turkeys. A 1 : 2000 solution of copper sulfate substituted for drinking water for 2 or 3 days is said to be helpful in treatment.

Spindler, Shorb, and Hill (1953) reported that a disease of the nose of pigs, called atrophic rhinitis, is due to infection by a species of *Trichomonas*, but Levine et al. (1954) as well as other workers have not been able to produce the disease by introducing bacteria-free trichomonads isolated from sick pigs. Another species of *Trichomonas* lives in the ileocecal region of the intestine of pigs.

Chilomastix mesnili

This organism, often confused with *Trichomonas* by careless observers, inhabits the large intestine of about 3 to 10% or more of human beings; in Egyptian villages the senior author found 13% infected. Closely similar forms are found in all groups of vertebrates. They are common in both rats and frogs. *Chilomastix* (Fig. 12) is an unsymmetrical, pear-shaped animal which has its posterior end drawn out into a sharply pointed tail. It varies in length from 6 to 20 μ, but the usual length is 10 to 15 μ. The body is less plastic than in *Trichomonas*, so there is less variability in shape. It has three slender anterior flagella which, like those of *Trichomonas*, function as two groups, two of them lashing back against the left side of the body and one against the right. The relatively enormous cytostome is an oval groove half or more the length of the body, the lips of which are supported by a complicated system of fibers. Lying in this groove is a fourth flagellum, attached to the left lip by an undulating membrane. By its flickering movements this "tongue" wafts food particles into the depths of the groove, where they pass into the body to be enclosed in food vacuoles, with which the body is often literally crammed. The nucleus lies in the fore part of the body just behind the free flagella.

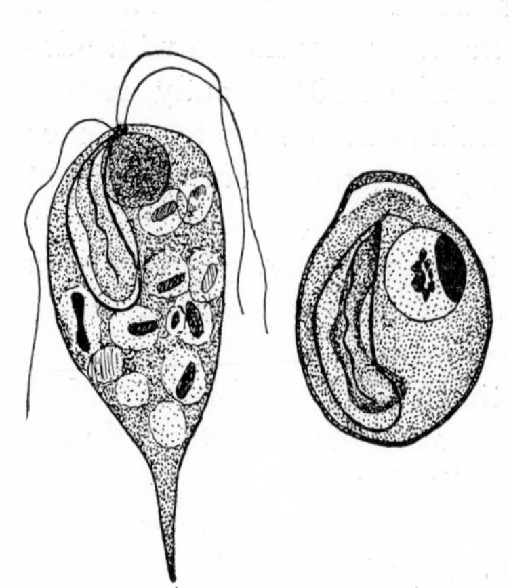

Fig. 12. *Chilomastix mesnili. Left,* trophozoite; *right,* cyst. ×3000. (After Boeck, *J. Expl. Med.,* 1921.)

The animals do not move as rapidly as *Trichomonas,* but proceed by a sort of jerky spiral movement unlike the continuous wobbly progression of *Trichomonas.*

The ordinary multiplication is by simple fission, but sometimes large multinucleate forms are produced. Unlike *Trichomonas, Chilomastix* forms lemon-shaped cysts, narrower at the anterior end. The cysts are usually about 7 to 9 μ long; they have thin walls except where thickened at the anterior end, and the fibers of the cytostome, practically unaltered in form, lie alongside of or overlapping the nucleus. Occasionally the nuclei and cytostomal fibers are duplicated in the cysts, which then presumably give rise to two individuals when they hatch. The cysts are very resistant and live for months in water at room temperature, and for several days in the intestine of flies. Boeck found that a temperature of 72°C. was necessary to kill them.

There is little evidence that *Chilomastix* is pathogenic. Westphal (1939), in experiments on himself, found this parasite and also *Enteromonas hominis* to fluctuate with the condition of the intestine, and considers their presence a result rather than the cause of intestinal ailments with which they may be associated. Cerva and Vetrovska in Czechoslovakia and Mueller (1959) in this country have recently reconsidered the possibility that *Chilomastix* may be mildly pathogenic. It may be that this flagellate is an opportunist, flourishing in the presence of intestinal disease but incapable of initiating it.

Giardia

Giardia (Fig. 13), once known as *Lamblia,* contains flagellates which are remarkable in a number of ways. They are odd-looking creatures, which have their nuclei and other organelles reduplicated like closely bound Siamese twins. They inhabit the upper part of the small intestine instead of the large intestine favored by all the other intestinal Protozoa; they attach themselves to the surface of the mucosal cells where they presumably absorb nourishment directly from the host. They have such close specificity that one recent writer on them (Ansari, 1952) thinks that when one species is found in more than one kind of animal, it is only because it was a parasite of a common ancestor from which the present-day hosts evolved, the parasites having undergone evolution more slowly than the hosts. They are the only intestinal Protozoa which cannot be cultured in artificial media. They inhabit the intestines of all kinds of vertebrates from fish to man; Ansari (1952) listed 38 named and described species, as well as 8 or 10 others noncommittally referred to as "*Giardia* sp."

The species found in man, usually called *G. lamblia* by American writers and *G. intestinalis* in Europe, is also reported from both Old and New World monkeys. It is hard to go along with Ansari's idea that this single species has failed to undergo any evolutionary change since the millions of years ago when these groups of Primates began their evolution, so we must assume either that the monkey strains are not actually identical with the human ones, or that the parasites have enough adaptability to pass from one Primate species to another. Haiba (1956) reported the transmission of *Giardia* from humans to laboratory rats.

Morphology. In appearance *Giardia* is a fantastic little animal. It is bilaterally symmetrical, with two nuclei analogous to a pair of *Chilomastix*-like flagellates fused together in the middle line. A dead giardia trophozoite gives the impression of a wizened monkey face looking up at you. The outline of the body is strikingly like that of a tennis racket without the handle. In side view it is shaped like a pear split lengthwise in two parts, with the flat surface in the broadest part gouged out as a large concave sucking disc, with slightly raised margins. The finely tapering posterior end is usually turned up over the convex back. There are eight flagella, arranged as shown in Fig. 13. They may be thought of as corresponding more or less to eyebrows, moustaches, and beard. The body is 8 to 16 μ in length by 5 to 12 μ in width, the mean being about 12 by 8.5 μ. There appear to be two races which differ in size and in shape of the parabasal body.

The two nuclei have large central endosomes. Between them are two slender rods, the axostyles, to which the nuclei are anchored by slender fibrils. There is a complicated system of basal granules and fibrils connecting with the flagella and the rods supporting the sucking disc, as shown in Fig. 13*A*. The cysts (Fig. 13*C*) are thick-walled and oval and about 8 to 14 μ in length, commonly about 10 μ. Incomplete division takes place in the cyst; when mature the cyst usually contains 4 nuclei, either clustered at one end or lying in pairs at opposite poles, and it also shows deep-staining axostyles and fibrils lying diagonally in the cyst, and 2 or 4 curved bodies.

Biology. As already noted, *Giardia* makes its home in the small intestine, especially in the duodenum, occasionally invading bile ducts, etc. It frequently develops in the large intestine in dogs infected

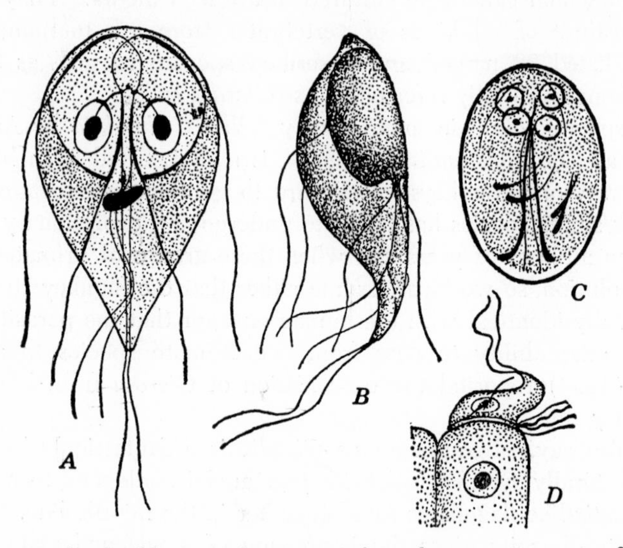

Fig. 13. *Giardia lamblia.* *A*, face view of trophozoite; *B*, semiprofile view; *C*, cyst; *D*, position of trophozoite resting on epithelial cell. *A*, *B*, and *C*, ×3000; *D*, ×1000. (*A*, after Simon, *Am. J. Hyg.*, 1921; *B* and *D*, after Grassi and Schewiakoff, *Ztschr. Wissensch. Zool.*, 1888; *C*, original.)

via the rectum. Hegner, however, found giardias of both rat and human origin to localize only in the upper part of the small intestine of rats, and showed that they were attracted by bile salts. Although *Giardia* infections are found in people of all ages, they are commoner in children. In Egyptian villages, where there is constant exposure to infection, Chandler (1954) found this parasite in 16% of people below the age of puberty and only 3% of those above. On the other hand, Rendtorff (1954) was able to infect 100% of adult men in the United States when 100 or more cysts were fed, although most of the infections were very light, and all of them disappeared spontaneously, usually within 1 to 6 weeks. A possible explanation for these apparently discordant results is that the parasites stimulate some degree of resistance; this allows new infections to become established only infrequently when there is constant exposure, but otherwise it wears off and allows new infections to develop which last until the resistance is restimulated.

In life these grotesque little creatures fasten themselves by their hollow faces to the convex surfaces of epithelial cells in the small intestine, their flagella streaming like the barbels of a catfish (Fig. 13*D*). Sometimes large areas of epithelium are practically covered with them, each one perched on a separate cell. Their vast numbers can be judged from the fact that in one instance Miss Porter estimated the number of cysts in a single stool to exceed 14,000,000,000. The number of cysts in an average stool in a case of moderate infection she estimated at over 300,000,000. The motile forms are not normally found in the stools, but in cases of diarrhea dead ones may be present in considerable numbers. They do not ingest solid food, nor do they appear to dissolve tissue cells; possibly they feed on the abundant secretion of mucus which their presence stimulates, and on a variety of amino acids, vitamins, and other substances which are constantly passing in and out of the intestinal mucosal cells (see Read, 1950).

Multiplication occurs by division into two in a plane parallel with the broad surfaces, and occasionally multiple fission occurs as in other intestinal flagellates. The cysts are formed intermittently; enormous numbers may be found on one day and then none for several days, when a shower of them again appears. Occasionally fecal examination fails to reveal them even when they are present in the duodenum in large numbers. The cysts remain alive in feces for 10 days or more and survive many days in the gut of roaches. The parasite is a very persistent one; infections sometimes last for many years, possibly in some cases for life.

Pathogenicity. There is no longer any doubt of the pathogenicity of *Giardia.* Véghelyi in 1939 found evidence of mechanical interference with absorption, particularly of fats, from the intestine by the layer of parasites adhering to its wall. It is obvious that this might lead to vitamin deficiencies, particularly of the fat-soluble ones. The presence of large amounts of unabsorbed fats in the stools causes a persistent or recurring diarrhea, often with large amounts of yellow mucus. The symptoms may resemble those of celiac disease, sprue, or chronic gall bladder disease. Epigastric pains, vague abdominal discomfort, loss of appetite, apathy, headache, and diarrhea alternating with constipation are common. An allergic dermatitis sometimes occurs. In many cases, on the other hand, there are no evident symptoms. Occasionally the parasites are found in the bile ducts and even in the gall bladder. It is possible that they may cause some irritation in the bile ducts and predispose them to chronic infection, but the evidence for this is inconclusive.

Giardia infections are very susceptible to the anti-malarial drugs, atebrin, Chloroquin, and Camoquin, given at the rate of 0.1 to 0.2 gram three times a day for about 5 days. Cysts cease to be passed after the second or third day. One *Giardia*-infected young man whom the senior writer treated with atebrin was extremely ill, had had no appetite for weeks, and was very emaciated. After treatment he recovered his appetite within 24 hours, gained several pounds a week, and was restored to normal health in a month. Patients tend to benefit from administration of fat soluble vitamins, particularly vitamin A.

Other Intestinal Flagellates

A few other flagellates may be residents of the human intestine, but they are relatively rare and of little importance.

Retortamonas intestinalis. This little slipper-shaped animal, formerly called *Embadomonas intestinalis,* though rare, has been found in many parts of the world. The fact that members of the same genus

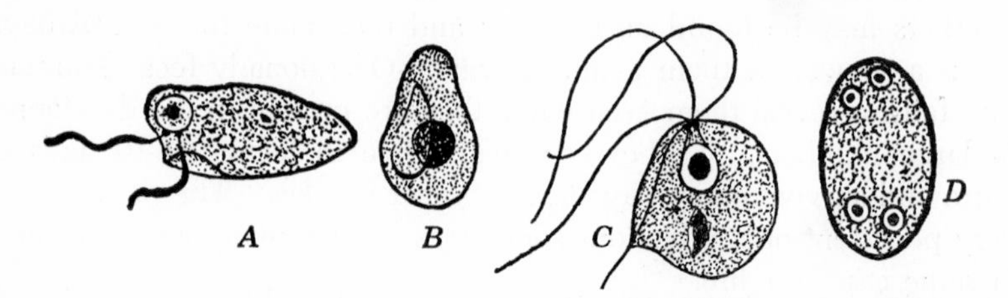

Fig. 14. *A* and *B*, *Retortamonas* (= *Embadomonas*) *intestinalis,* trophozoite and cyst; *C* and *D*, *Enteromonas* (= *Tricercomonas*) *intestinalis.* ×3000. (After Wenyon and O'Connor, 1917.)

occur in various insects, especially aquatic ones, and in frogs and turtles, suggests that the infections of man and other mammals in which they have been found may perhaps be derived from the swallowing of cysts of some insect or aquatic species with water. Its rarity makes it doubtful that it is normally a human parasite. It is very small, only 4 to 9 μ long by 3 to 4 μ in breadth; it has two flagella, a long, slender anterior one and a shorter, thicker one which lies partly in the large elongated cytostome, the borders of which have supporting fibers (Fig. 14*A* and *B*). The nucleus is anterior in position. The cysts are whitish, opalescent, pear-shaped bodies, 4.5 to 6 or 7 μ long when living. When stained they show what appears to be the endosome of the nucleus, sometimes dumbbell-shaped, and fibers which Wenyon interprets as the marginal fibers of the cytostome.

Faust described another species, *R. sinensis*, from China; it is larger and is said to have the two flagella alike, but Wenyon believes it to be identical with *R. intestinalis*. It was found in nine cases with diarrheic stools, and was again reported from two cases in China by Watt in 1933. It has been successfully cultivated, and seems to be a valid species.

Enteromonas hominis. This flagellate (Fig. 14*C* and *D*) which is believed by Dobell to be identical with another flagellate reported from man, *Tricercomonas intestinalis*, is an extremely small oval or pear-shaped organism, 4 to 10 μ long by 3 to 6 μ broad, slightly flattened on one side, where a flagellum is attached until it becomes free at the posterior end. There is also a cluster of three anterior flagella. Small oval cysts 6 to 8 μ long are formed which have well-developed cyst walls, giving them a double outline, and one to four nuclei, visible only when stained. In cysts with two or four nuclei these are arranged at opposite ends. The parasite has been reported from many parts of the world, but is usually considered rare. In an Egyptian village Lawless (personal communication) found it in 74% of 100 people, 80 of whom were examined six times over a period of 30 months, and Kessel et al. found 20% incidence in Tahiti. Its small size makes it very easily overlooked even if searched for with a high-dry lens. There is no evidence that it is pathogenic, and infections sometimes persist only for a short time.

Intestinal Flagellates of Domestic Animals

Although all species of vertebrate animals are probably parasitized by a number of species of flagellates, pathogenic effects are produced in only a few cases. Certain species of *Trichomonas* (see pp. 95–96) are exceptions. *Giardia* has been reported as sometimes causing severe damage to rabbits and dogs, and the writers have seen dogs with intermittent attacks of diarrhea of the type associated with *Giardia* infections. Other important flagellates are *Histomonas meleagridis* and *Hexamita* spp.

Histomonas meleagridis. This important parasite of galliform birds (Fig. 15*A*, *B*) causes infectious enterohepatitis or "blackhead" in turkeys. It is found in both the ceca and the liver, and occasionally

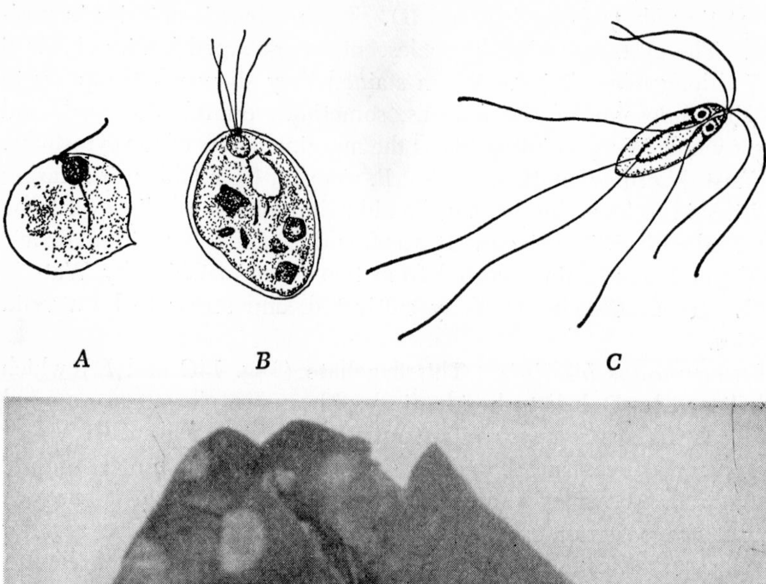

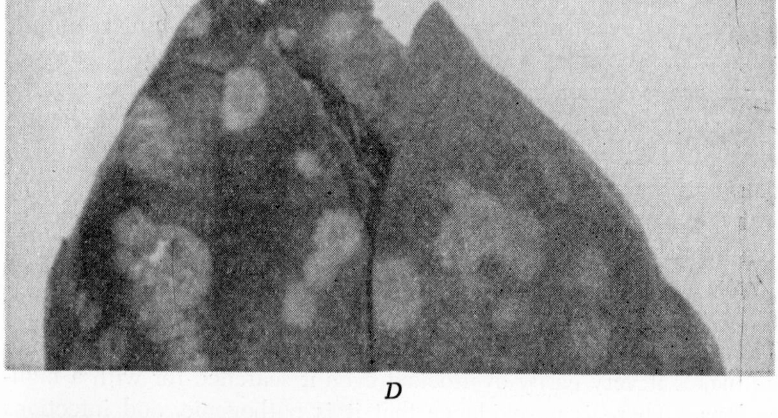

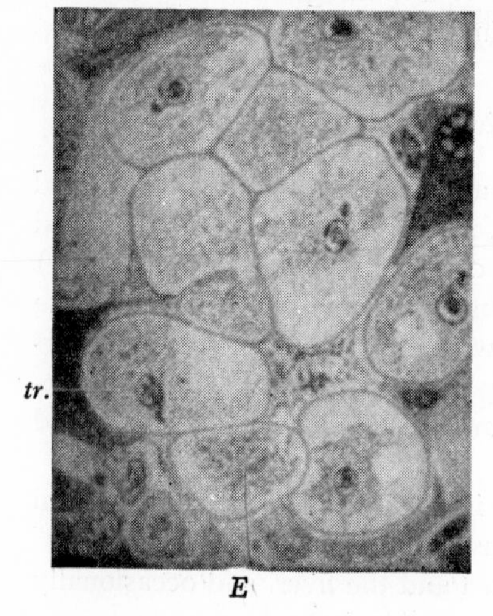

Fig. 15. *A–D*, *Histomonas meleagridis*: *A*, tissue form with one flagellum and cytoplasmic fibril; *B*, lumen form with four flagella; *C*, *Hexamita meleagridis*; *D*, liver of a turkey with the characteristic lesions of histomoniasis; *E*, section of a liver lesion showing the tissue form (*tr.*) of *Histomonas*. (*A* and *B*, after Tyzzer, *Proc. Am. Acad. Arts Sci.*, 69, 1934; *C* and *D*, after Lapage, *Vet. Parasitol.*, 1956; *E*, after McNeil, Hinshaw, and Kofoid, *Am. J. Hyg.*, 34, 1941.)

in the kidney and spleen. It is an ameboid organism 8 to 10 μ in diameter, with a small eccentric nucleus with a blepharoplast on or near the nuclear membrane. From this arise one to four flagella in the intestinal forms, often, however, not extending beyond the cell wall. Organisms in the tissues have no flagella. The organism fails to produce cysts and lives for a very short time when passed in the droppings, which do not cause infection when swallowed. The organism has, however, a very clever means of transfer to a new host by becoming enclosed inside the egg shells of cecal worms, *Heterakis* (see p. 459). When embryonated eggs of *Heterakis* are fed to turkey poults a high mortality from blackhead results. In nature, turkeys are infected by worm eggs passed from healthy chicken carriers; this is the principal reason why it is usually disastrous to try to raise chickens and turkeys togeher. The disease can be induced experimentally by rectal injection of infected material; the ceca are first attacked, and then the parasite migrates to the liver by way of the blood stream (Farmer et al., 1951). A nitrothiazole, called Enheptin T, when mixed with food at 0.05%, is a highly effective prophylactic, and is curative at 0.1% (see Horton-Smith and Long, 1951).

Hexamita spp. A number of species of *Hexamita* occur in various vertebrates. They are more or less elongated flagellates with two anterior nuclei, four anterior flagella in pairs, and two which arise anteriorly but pass posteriorly through the body to emerge near the posterior end (Fig. 15*C*). These parasites cause a severe diarrhea in young turkeys and pigeons. Quail, partridges, and chicks suffer less. According to McNeil, Hinshaw, and Kofoid (1941), *H. meleagridis* of turkeys and *H. columbae* of other birds are two distinct species. *H. columbae* was not transferable to turkeys, and *H. meleagridis* caused only temporary infections in chickens and ducks.

INTESTINAL CILIATES

All the intestinal ciliates of warm-blooded animals belong to the subclass Euciliatia (see p. 49). Amphibia, on the other hand, have the rectum inhabited by many species of Opalinidae, which belong in the subclass Protociliatia.

The Euciliatia are classified as follows:

Order 1. **Holotrichida.** No adoral zone of flattened cilia or membranelles. Includes *Paramecium* and many coprozoic ciliates, but no parasites of higher vertebrates.

Order 2. **Spirotrichida.** Adoral zone of membranelles winding clockwise to cytostome, the peristome (mouth region) not protruded. Two suborders contain parasites of vertebrates: **Heterotrichina,** with body covered with cilia, includes *Balantidium,** **Oligotrichina,** with body non-ciliated, but with adoral and other zones of membranelles, includes numerous species in stomach of ruminants and colon of horses.

Order 3. **Chonotrichida.** Like order 2, but peristome protruding like a funnel. No vertebrate parasites.

Order 4. **Peritrichida.** Body not entirely ciliated; anterior region disclike with counterclockwise adoral zone of membranelles. Example, *Vorticella.* None parasitic in vertebrates.

The stomachs of ruminants and the large intestine of horses harbor numerous species of commensal ciliates belonging to the suborder Oligotrichina. They may play some part in the digestion of cellulose in these animals. One of the species from ruminants is shown in Fig. 1.

Balantidium coli. This is a parasite of the large intestine of man, monkeys, and pigs. A parasite in rats identical with *B. coli* was reported from Moscow; and rats and guinea pigs can be experimentally infected. McDonald, in 1922, believed that the pig harbors another species, a *B. suis,* which is not infective for man, but Hegner, in 1934, doubted this.

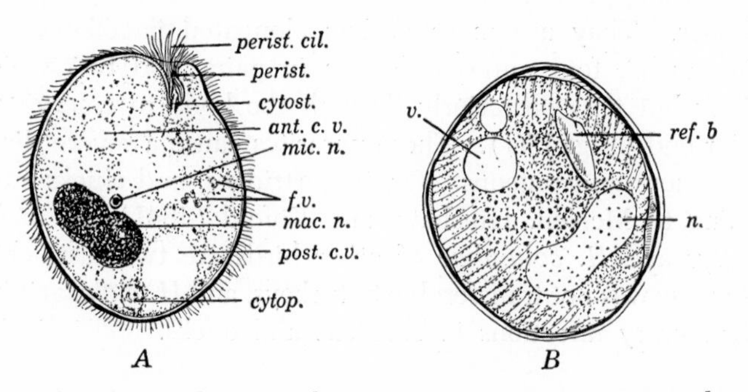

Fig. 16. *Balantidium coli:* *A*, trophozoite; *ant.c.v.*, anterior contractile vacuole; *cytost.*, cytostome; *cytop.*, cytopyge; *f.v.*, food vacuoles; *mac.n.*, macronucleus; *mic.n.*, micronucleus; *perist.*, peristome; *perist.cil.*, peristomal cilia; *post.c.v.*, posterior contractile vacuole. *B*, cyst: *n.*, macronucleus; *ref.b.*, refractile body; *v.*, vacuole. About ×500. (*A*, original; *B*, adapted from Dobell and O'Connor, *Intestinal Protozoa of Man,* 1921.)

B. coli (Fig. 16), as found in man, is much larger than any of the other protozoan inhabitants of the human intestine and usually measures 50 to 80 μ in length, with a breadth between two-thirds and three-fourths as great. In pigs it sometimes reaches a length of 200 μ.

* Corliss recently classified *Balantidium* as a Holotrich.

It is shaped like an egg or pear, and has at the anterior end an obliquely arranged depression, the peristome, which may appear wide open or slitlike, and in the bottom of which is the cytostome. The whole body is covered with fine cilia arranged in rows, with a special row of longer "adoral" cilia surrounding the peristome. The macronucleus is only very slightly curved, usually with a slight concavity on either side. It usually lies obliquely near the middle of the body and is about two-fifths the length of the body. The micronucleus is very small and inconspicuous. There are two contractile vacuoles, and food vacuoles circulate in the endoplasm. Like other ciliates, *Balantidium* divides by transverse fission, a new cytostome being formed by the posterior daughter.

A process of conjugation occurs, similar in its general features to that of *Paramecium*. Thick-walled cysts are formed in which single individuals are usually enclosed. Slow-moving cilia are at first visible on encysted ciliates, but later all structures except the nuclei and sometimes one or more refractile bodies disappear. No multiplication takes place in the cysts.

Pigs are usually regarded as important sources of human infection. Such infections are rather infrequently reported, but they may be locally common. Young (1939) reported 7 cases, all with marked diarrhea, among 142 insane hospital patients examined in South Carolina. Among 3000 Puerto Ricans in New York, 20 cases were found (Shookhoff, 1951) whereas in Puerto Rico there is about twice that incidence. Of the 20 cases in New York, all in children, 18 had a history of contact with pigs. Apparently large or repeated doses of cysts or special susceptibility is necessary for human infection, for Young, using moderate numbers of parasites, was unable to produce infection experimentally in two volunteers. Large inocula are also necessary to get cultures going.

In man *B. coli* is known to be a pathogenic parasite, though in pigs it appears to be harmless. In man it may cause ulceration of the large intestine and invade the tissues of the walls; the colon is sometimes ulcerated from end to end. Nevertheless, the majority of cases suffer only from diarrhea and may show no symptoms at all; only a small number develop severe or fatal dysentery. In the 20 Puerto Rican cases in New York, 10 admitted having symptoms (diarrhea or dysentery) but probably more did, for many people in the tropics consider diarrhea more or less a normal part of life, as indeed it is if universality constitutes normality.

Ciliates of the genus *Nyctotherus* have on rare occasions been recorded from human feces, but it is probably coprophagous (see

Wichterman, 1938). Like amebas and flagellates, coprozoic ciliates are common and have misled more than one parasitologist.

Balantidium infections have been treated with varying success with some anti-amebic drugs, particularly Carbarsone, but there is now evidence that Aureomycin and Terramycin are very effective. These antibiotics are given for 10 to 15 days, the total dosage being from 7 to 28 grams. The ciliates disappear within 2 to 4 days, and no relapses occur for at least 2 weeks or so after treatment.

REFERENCES

Ansari, M. A. R. 1952. Contribution à l'étude du genre *Giardia* Kunstler, 1882 (Mastigophora, Octomitidae) I. *Ann. de Parasitol.*, 27: 421–484.

Arean, V. M., and Koppisch, E. 1956. Balantidiasis. A review and report of cases. *Am. J. Pathol.*, 32: 1089–1115.

Ball, G. H. 1932. Observations on the life history of *Chilomastix*. *Am. J. Hyg.*, 16: 85–96.

Beatman, L. H. 1933. Studies on *Trichomonas buccalis*. *J. Dental Research*, 13: 339–347.

Bland, P. B., Goldstein, L., Wenrich, D. H., and Weiner, E. 1932. Studies on the biology of *Trichomonas vaginalis*. *Am. J. Hyg.*, 16: 492–512.

Boeck, W. C. 1924. Studies on *Tricercomonas intestinalis*. *Am. J. Trop. Med.*, 4: 519–535.

Breuer, A. 1938. Die Symptomatologie und die Behändlung der Lamblien-Infektion des Menschen. *Arch. Schiffs- u. Tropen-Hyg.*, 42: 201–222.

Burrows, R. B., and Jahnes, W. G. 1952. The effect of aureomycin on balantidiasis. *Am. J. Trop. Med. Hyg.*, 1: 626–630.

Dobell, C., 1934–1935. Researches on the intestinal protozoa of monkeys and man, VI, VII. *Parasitology*, 26: 531–577; 27: 564–592.

Dobell, C., and O'Connor, R. W. 1921. *The Intestinal Protozoa of Man*. Bale, Sons and Danielsson, London.

Farmer, R. K., Hughes, D. L., and Whiting, G. 1951. Infectious entero-hepatitis (blackhead) in turkeys and a study of the pathology of the artificially induced disease. *J. Comp. Pathol.*, 61: 251–252.

Geiman, Q. M. 1935. Cytological studies of the *Chilomastix* of man and other animals. *J. Morphol.*, 57: 429–459.

Haiba, M. H. 1956. Further study on the susceptibility of murines to human giardiasis. *Parasitenk.*, 17: 339–345.

Hegner, R. W. 1924. A comparative study of the giardias living in man, rabbit, and dog. *Am. J. Hyg.*, 2: 442–454.

1924. The relations between a carnivorous diet and mammalian infections with intestinal protozoa. *Ibid.*, 4: 393–400.

Hegner, R. W., and Chu, H. J. 1930. A comparative study of the intestinal protozoa of wild monkeys and man. *Am. J. Hyg.*, 12: 62–108.

Hess, E. 1949. Behändlung Trichomonadeninfizierte Zuchtstiere. *Schweiz. Arch. Tierheilk.*, 91: 481–491.

Honigberg, B. M., and Lee, J. J. 1959. Structure and division of *Trichomonas tenax* (O. F. Müller). *Am. J. Hyg.*, 69: 177–201.

Honigberg, B. M., and Read, C. P. 1960. Virulence transformation of a trichomonad protozoan. *Science,* 131: 352–353.

Horton-Smith, C., and Long, P. L. 1951. Enheptin T in the treatment of histomoniasis (blackhead) in turkeys. *Vet. Record,* 63: 507.

Johnson, G., Trussell, M., and Jahn, F. 1945. Isolation of *Trichomonas vaginalis* with penicillin. *Science,* 102: 126–128.

Kirby, H. 1945. The structure of the common intestinal trichomonad of man. *J. Parasitol.,* 31: 163–175.

Kofoid, C. A. 1929. The Protozoa of the human mouth. *J. Parasitol.,* 15: 151–174.

Levine, N. D., Boley, L. E., and Hester, H. R. 1941. Experimental transmission of *Trichomonas gallinae* from the chicken to other birds. *Am. J. Hyg.,* 31(C): 23–32.

Levine, N. D., Marquardt, W. C., and Beamer, P. D. 1954. Failure of bacteria-free *Trichomonas* to cause atrophic rhinitis in young pigs. *J. Am. Vet. Assoc.,* 125: 61–63.

McNeil, E., Hinshaw, W. R., and Kofoid, C. A. 1941. *Hexamita* sp. nov. from the turkey. *Am. J. Hyg.,* 34: 71–82.

Moore, S. F., and Simpson, J. W. 1956. Trichomonal vaginitis: an emotionally conditioned symptom. *Southern Med. J.,* 49: 1495–1501.

Morgan, B. B. 1944. *Bovine Trichomoniasis.* Burgess, Minneapolis.

1947. A summary of research on *Trichomonas foetus. J. Parasitol.,* 33: 201–206.

Mueller, J. F. 1959. Is Chilomastix a pathogen? *J. Parasitol.,* 45: 170.

Nelson, E. C. 1935. Cultivation and cross-infection experiments with balantidia from pig, chimpanzee, guinea pig, and *Macacus rhesus. Am. J. Hyg.,* 22: 26–43.

Read, C. P. 1950. The vertebrate small intestine as an environment for parasitic helminths. *Rice Inst. Pamphlet,* 37: No. 2, iv + 94 pp.

Read, C. P. 1957. Comparative studies on the physiology of trichomonad protozoa. *J. Parasitol.,* 43: 385–394.

Rendtorff, R. C. 1954. The experimental transmission of human intestinal protozoan parasites. II (*Giardia lamblia*). *Am. J. Hyg.,* 59: 209–220.

Satti, M. H., and Honigberg, B. M. 1959. Observations on thermal resistance of *Trichomonas gallinae* and *T. vaginalis. J. Parasitol.,* 45 (Suppl.): 51.

Shookhoff, H. B. 1951. *Balantidium coli* infection with special reference to treatment. *Am. J. Trop. Med.,* 31: 442–457.

Stabler, R. M. 1947. *Trichomonas gallinae,* pathogenic trichomonad of birds. *J. Parasitol.,* 33: 207–213.

1954. *Trichomonas gallinae:* A review. *Exp. Parasitol.,* 3: 368–402.

Stabler, R. M., Feo, L. Q., and Rakoff, A. E. 1941–1942. Implantation of intestinal trichomonads (*T. hominis*) into the human vagina. *Am. J. Hyg.,* 34(C): 114–118. Survival time of intravaginally implanted *Trichomonas hominis. Am. J. Trop. Med.,* 22: 633–637. Inoculation of the oral trichomonad (*T. tenax*) into the human vagina. *Ibid.,* 22: 639–642.

Trussell, R. E. 1947. *Trichomonas vaginalis and Trichomoniasis.* Thomas, Springfield, Ill.

Tyzzer, E. E. 1934. Studies on histomoniasis, or "blackhead" infection in the chicken and turkey. *Proc. Am. Acad. Arts Sci.,* 69: 189–264.

Webster, B. H. 1958. Human infection with *Giardia lamblia*. *Am. J. Digest. Diseases,* 3: 64–70.

Weinstein, P. P., et al. 1952. Treatment of a case of balantidial dysentery with terramycin. *Am. J. Trop. Med. Hyg.,* 1: 980–981.

Wenrich, D. H. 1943. Observations in the morphology of *Histomonas* (Protozoa, Mastigophora) from pheasants and chickens. *J. Morphol.,* 72: 279–303.

Westphal, A. 1939. Beziehungen zwischen Infektionsstärke and "Krankheitsbild" bei Infektionen mit *Chilomastix mesnili* und anderen Dickdarmflagellaten. *Z. Hyg. Infektionskrankh.,* 122: 146–158.

Westphal, A. 1957. Experimentelle Infektionen des Meerschweinchens mit *Balantidium coli*. *Z. Tropenmed. Parasitol.,* 8: 288–294.

Wichterman, R. 1938. The present status of knowledge concerning the existence of species of *Nyctotherus* living in man. *Am. J. Trop. Med.,* 18: 67.

Wilhelm, R. E. 1957. Urticaria associated with giardiasis lamblia. *J. Allergy,* 28: 351–353.

Chapter 6

HEMOFLAGELLATES
I. LEISHMANIA and LEISHMANIASIS

The Trypanosomidae

The term "hemoflagellates" is used for those flagellates which habitually live in the blood or tissues of man or other animals. There are only two kinds of these which occur in man, namely, the leishman bodies, belonging to the genus *Leishmania,* and the trypanosomes, belonging to the genus *Trypanosoma.* These two types of organisms, however, are only two of a number of genera which all belong to one family, Trypanosomidae, in the order Protomastigida (see p. 86). Other members of the family occur as gut parasites of insects or lizards, and still others as parasites of plants. Since both the hemoflagellates and the plant parasites undergo cycles of development in the gut of insects, it is safe to presume that this entire group of flagellates was originally and primitively parasitic in the gut of insects.

Four distinct morphological types of these parasites are found in the bodies of insects:

1. The leptomonas type (Fig. 17*A*). This is the most primitive type, in which the body is more or less elongate or pear shaped: it contains a nucleus near the center, a kinetoplast near the anterior end, and a single long slender flagellum which arises from a basal granule closely associated with the kinetoplast. All the other types of Trypanosomidae may be considered as having arisen from this.
2. The crithidia type (Fig. 17*B*). This differs in that the flagellum arises from a kinetoplast which has shifted back to a position just in front of the nucleus and is connected with the body, up to the anterior end, by an undulating membrane.
3. The trypanosoma type (Fig. 17*C*). In this the kinetoplast has moved far behind the nucleus to a point near the posterior end of the

body, and the flagellum is attached to the body for most of its length, with or without an undulating membrane.

4. The leishmania type (Fig. 17*D*). This is a rounded-up form which contains a nucleus and a kinetoplast, but is entirely devoid of a flagellum. Any of the other three types may assume this form and, conversely, may be developed out of it.

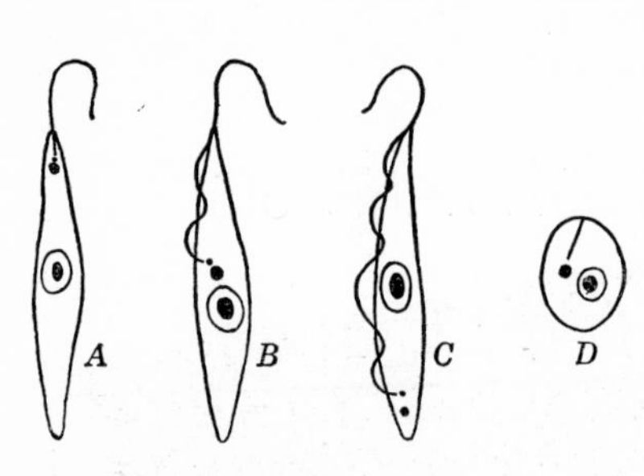

Fig. 17. Diagram of forms assumed by Trypanosomidae either as adults or as developmental forms. *A*, Leptomonas; *B*, Crithidia; *C*, Trypanosoma; *D*, Leishmania. (After Wenyon, *Protozoology*, 1926.)

Any or all of these forms may occur in the digestive tracts of insects, but only the leishmania and trypanosome forms occur in the blood of vertebrates.

The fact that some flagellates never develop farther than the leptomonas form, and others never, so far as known at present, farther than the crithidia form, whereas the trypanosomes go through all the stages, makes a study of this group of flagellates very confusing. When a leptomonas or crithidia type is found in an insect gut, it is impossible to say, without further investigation, whether it is an adult animal which never undergoes any further development or is only a developmental phase of a trypanosome of a vertebrate animal. A number of crithidias which were supposed to be purely insect parasites with no trypanosome stage have been found to develop into trypanosomes in the blood of certain vertebrates, so it may be that most of the crithidias are really developmental stages of these parasites.

The Trypanosomidae are divided into a number of genera on the basis of the morphological forms they assume and on whether they are transmissible to vertebrate animals or to plants. The following genera are usually recognized:

1. Genus ***Leptomonas.*** Species having only leptomonas and leishmania stages, and confined to invertebrate hosts. They are common in various kinds of bugs, larvae and adults of fleas, various Diptera, and other insects. They live in the hindgut, where they attach themselves to the epithelial cells by their flagellar ends, the free flagella being very short or lacking (Fig. 18). Often they occur in rosettes of dozens of individuals. They produce resistant cyst-like forms resembling ordinary leishmania forms but apparently protected by cyst walls.

2. Genus ***Leishmania.*** Species having only leptomonas and leishmania stages, but transmissible to vertebrates. Unlike *Leptomonas*, they develop mainly in the stomach and foregut, and form no resistant cystlike bodies.

In vertebrates they develop intracellularly and entirely in the leishmania phase, except one species, *Leishmania chameleonis,* which retains the leptomonas form in the intestine of lizards. In artificial cultures or in insects they assume the leptomonas form and are extracellular.

3. Genus ***Phytomonas.*** Similar to *Leptomonas,* but transmitted to plants, particularly *Euphorbia* and milkweeds, where they multiply in the latex. In some, at least, the organisms are said to be inoculated by the bites of insects, and cystlike forms are not found in the feces.

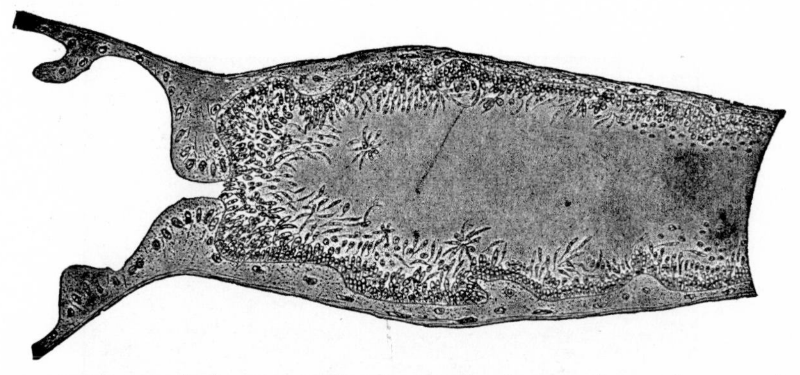

Fig. 18. Longitudinal section of the intestine of a dog flea, showing leptomonads lining the hindgut. ×170. (After Wenyon, *Protozoology,* 1926.)

4. Genus ***Crithidia.*** Strictly insect parasites in which leptomonas, leishmania, and crithidia stages occur, and in which cystlike forms are voided in the feces of the host. As noted, many of these have proved to be developmental stages of trypanosomes.

5. Genus ***Herpetomonas.*** Strictly insect parasites having leptomonas, leishmania, and crithidia stages, and also a stage in which the kinetoplast is at the posterior end of the body as in trypanosomes, but with the flagellum passing along the body like a rhizoplast, instead of being attached to an undulating membrane as in true trypanosomes. Cystlike forms are produced in the feces of the host.

6. Genus ***Endotrypanum.*** A single poorly known species, *E. schaudinni,* occurring as an intracellular parasite in the red blood cells of sloths. Leptomonas and trypanosome stages are known, but the developmental stages are not. Cell division has not been observed in the vertebrate host.

7. Genus ***Schizotrypanum.*** Species which go through all the stages of development and have both vertebrate and invertebrate hosts. In the vertebrate nondividing trypanosome and dividing intracellular leishmania forms occur and in the invertebrate they multiply in the hindgut like typical insect flagellates. Transmission from invertebrate to vertebrate is thought to occur by contamination rather than by biting.

8. Genus ***Trypanosoma.*** Species which have both vertebrate and invertebrate hosts and may go through all stages of development in the invertebrate but only occur in the trypanosome form in vertebrates. Devel-

opment may take place in the anterior or posterior gut of the invertebrate and the trypanosome stage in the vertebrate reproduces by simple fission.

In the tissues of mammals parasites of the genus *Leishmania* invade cells and multiply in them, particularly cells of the reticulo-endothelial system in the skin, mucosa, lymph nodes, spleen, liver, bone marrow, etc. They are taken up by leucocytes, especially large mononuclears, and thus enter the blood stream. They are very small, round or oval bodies, usually about 1.5 to 4 μ in diameter, possessing a round nucleus

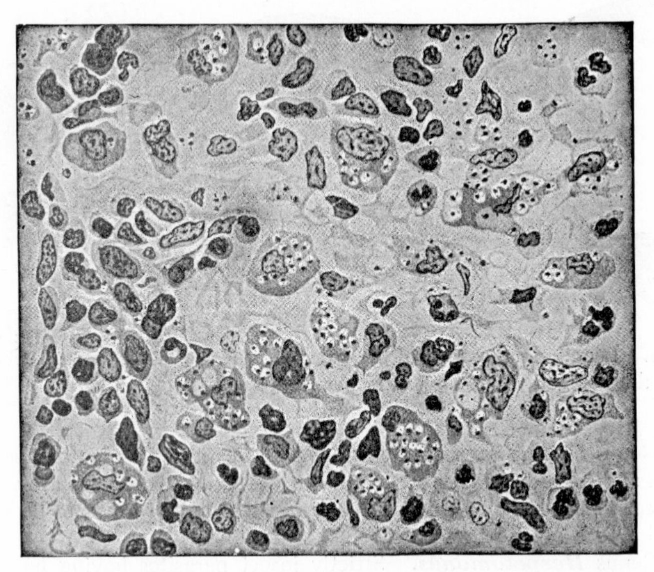

Fig. 19. Section of human spleen showing numerous leishman bodies in the cells. $\times 750$. (After Nattan-Larrier, from Wenyon, *Protozoology,* 1926.)

and a dot- or rod-shaped kinetoplast which stain well with Giemsa or related stains (Fig. 19). Torpedo-shaped forms are sometimes seen. Inside the cells the organisms multiply by simple fission, producing large clusters of 50 to 200 parasites, which distend and finally rupture the cells, setting the minute parasites free to invade, or be taken up by, other cells. They may also pass from cell to cell along protoplasmic processes. This parasitization of cells of the reticulo-endothelial system is an interesting reversal of the usual course of events. The parasites are probably picked up by these cells with phagocytic intent, but instead of being digested they grow and multiply, and ultimately destroy the would-be destroyer.

Cultural forms. The parasites develop readily in cultures containing blood when kept at room temperatures. In such cultures they

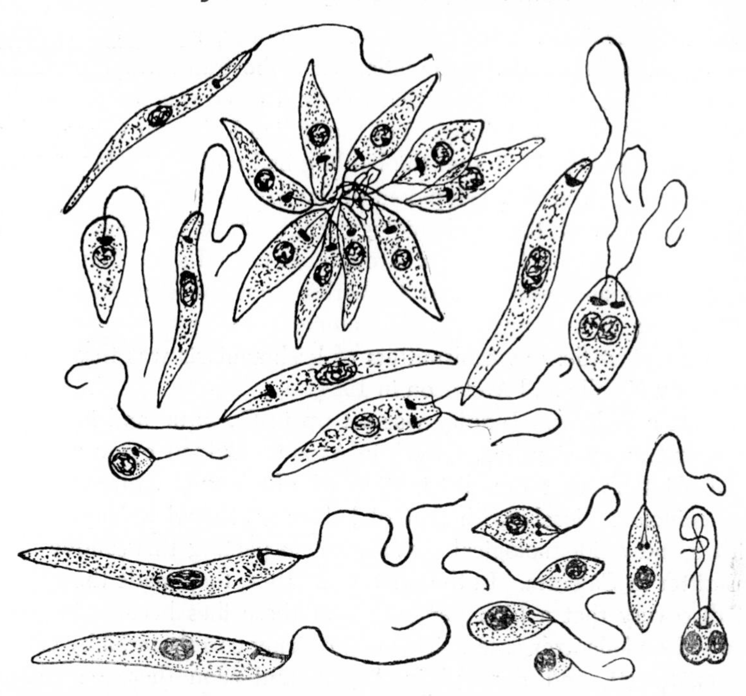

Fig. 20. *Leishmania donovani. Upper figures,* forms seen in cultures (original). *Lower figures,* forms found in midgut of *Phlebotomus argentipes; at left,* forms found in lumen; *at right,* forms found attached to walls. ×1600. (Sketched from figures by Shortt, Barraud, and Craighead, *Ind. J. Med. Research,* 1926.)

transform into active, flagellated leptomonas forms (Fig. 20). Typically these are spindle shaped, 14 to 20 μ long and 1.5 to 3.5 μ broad. The flagellum is as long as, or longer than, the body. The round or oval nucleus is near the center of the body, and the oval kinetoplast lies transversely near the anterior end. Division is by longitudinal fission; sometimes rosette clusters develop. In young cultures many stumpy, pear-shaped, or oval forms are found, but longer and more slender forms predominate later.

Development in insects and transmission. When ingested by certain insects the parasites undergo development into leptomonas forms just as they do in cultures, and may produce leishmania forms also. This development was first observed by Patton in India (1907) to occur in bedbugs fed on kala-azar patients. This was an unfortunate discovery, since it started investigators on a false trail in search for the transmitting agent and led to more than a dozen years of futile work. As a matter of fact, few problems in parasitology have caused more fruitless effort, more blasted hopes, more false conclusions, or more unfounded speculation than the transmission of leishmaniasis; it was not until 1942 that the final piece was fitted into the puzzle. Shortt and his colleagues of the Indian Kala-azar Commission finally concluded, in 1925, that the bedbug had nothing to do with the transmission in spite of development of the flagellates in its gut, as any careful observer of the epidemiology might have guessed. In the Mediterranean region fleas were strongly suspected since they fed on infected dogs and were found to harbor leptomonads; these were finally shown to be species of *Leptomonas* peculiar to the fleas, and the case against fleas as transmitters of leishmaniasis was thrown out of court by Nicolle and Anderson in 1924.

In 1921 a Kala-azar Inquiry was set up in Calcutta, and in 1924 a Kala-azar Commission began work in Assam. Guided by an observation of Sinton's that the distribution of *Phlebotomus argentipes* in India coincides closely with that of kala-azar, Knowles, Napier, and Smith found epidemiological reasons for suspecting this sandfly as a transmitter in Calcutta. In the same year, 1924, they made the important discovery that a high percentage of these flies became infected when fed on kala-azar cases. This was quickly corroborated by Christophers, Shortt, and Barraud in Assam, and soon thereafter many important details were added concerning the development of the flagellates in the sandfly, including demonstration of occasional massive infections of the pharynx and proboscis.

Then followed years of patient but largely fruitless effort to prove actual transmission by sandflies. In the course of hundreds of trials, only four successful infections were obtained, all in hamsters; transmission to human volunteers always failed. In 1939 Smith, Halder, and Ahmed made the interesting discovery that if sandflies, after an infective meal, were fed on raisins instead of additional blood meals, the flagellates frequently grew so numerous that they blocked the pharynx as do plague germs in fleas (Fig. 21). These authors then subjected five hamsters to bites of flies fed on raisins after their infective blood meal, and every one developed kala-azar; of five others fed on by flies given repeated blood meals, at least four failed to become infected (one escaped). In confirmation of this remarkable result, Swaminath, Shortt, and Anderson (1942) then succeeded in infecting each one of five human volunteers in Assam. Thus to a successful end came 20 years of patience, perseverance, labor and ingenuity.

This work, taken in conjunction with the epidemiology and the success of experimental infection of sandflies in various parts of the world, leaves no further doubt that sandflies are an important factor in the transmission of all forms of leishmaniasis, although other methods are also possible, e.g., by excretions of infected individuals in kala-azar, and by means of flies in the cutaneous forms of the disease.

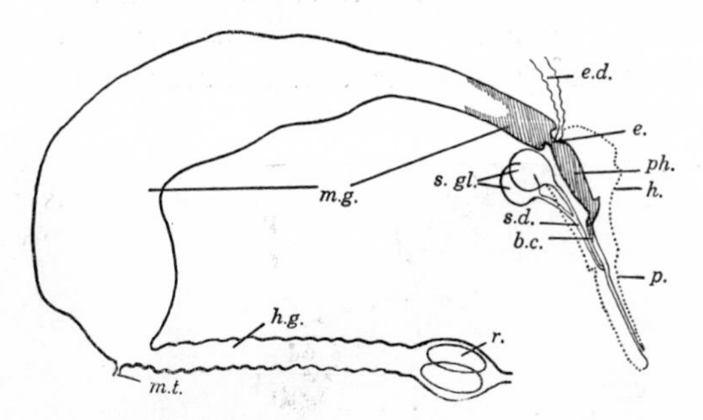

Fig. 21. Gut of sandfly, showing "blocking" of pharynx and forepart of midgut (shaded area) by *Leishmania donovani; b.c.*, buccal cavity; *e.*, esophagus; *e.d.*, esophageal diverticulum; *h.*, head; *h.g.*, hindgut; *m.g.*, midgut; *m.t.*, Malpighian tubule; *p.*, proboscis; *ph.*, pharynx; *r.*, rectum; *s.d.*, salivary duct; *s.gl.*, salivary glands. (After Shortt, Barraud, and Craighead, *Ind. J. Med. Research*, 1926.)

Types of leishmaniasis. *Leishmania* infections are usually classed in two general types, visceral and cutaneous, but there are several types of each, and intermediate conditions exist. Visceral leishmaniasis, or kala-azar, is a generalized and often fatal disease, accompanied by fever and enlargement of spleen and liver. Cutaneous leishmaniasis is limited to development of one or more local sores, usually without fever or generalized symptoms. These sores may be confined to the skin, as in Oriental sore of the Old World, or may spread to mucous membranes of the nose and mouth, as in espundia of tropical America. This mucocutaneous form of the disease is evidently caused by parasites that are intermediate in invasive power.

It is probably best to recognize only two distinct species of *Leishmania: L. donovani,* causing the various visceral forms of the disease; and *L. tropica,* causing the various cutaneous and mucocutaneous forms, of which several subspecies are recognized, based on clinical manifestations (see the following). Possibly the vectors are different also. The parasite of the severe mucocutaneous form of the disease in Brazil (espundia) is sometimes recognized as a distinct species, *brasiliensis*, but since there are intermediate forms between this and typical *L. tropica,* all are best considered subspecies, as suggested by Biagi (1953), who gives an excellent review of the subject. The various subspecies show no constant morphological, cultural, or serological differences, nor do they show clear-cut differences in their pathogenicity for laboratory animals.

Of visceral leishmaniasis or kala-azar, Biagi (1953) recognizes five types: (1) Chinese, with some preference for children; affects dogs, and is transmitted by *Phlebotomus chinensis* and *P. sergenti;* (2) Indian, occurring chiefly in adults, does not attack dogs, and is transmitted by *P. argentipes;* (3) Mediterranean, almost confined to children, with dogs an important reservoir; transmitted by *P. perniciosus, P. major,* and *P. longicuspis;* (4) Sudanese, characterized by frequent oral lesions and by unusual refractoriness to antimony treatment; and (5) South American, attacking all ages, with reservoir in wild and domestic dogs and cats; vectors *P. intermedius* and *P. longipalpis.*

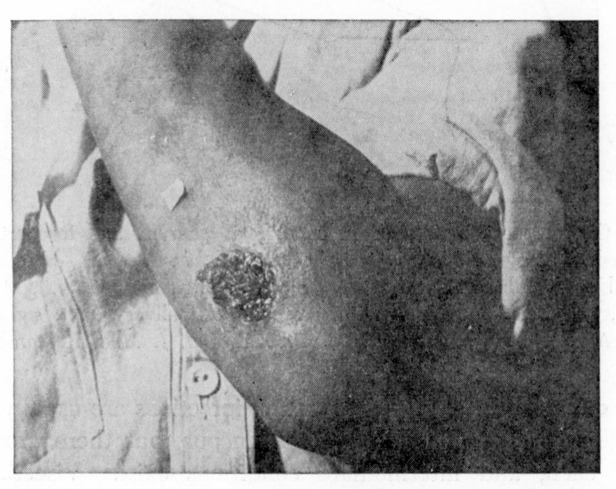

Fig. 22. Oriental sore on arm. (From Army Institute of Pathology, photograph 79107.)

In the Old World there are two forms of cutaneous leishmaniasis, the parasites of which the Russians have designated as *L. tropica minor* and *L. t. major*, respectively: (1) Classical Oriental sore, caused by *L. t. minor*, common from the Mediterranean to central and northern India; produces circumscribed "dry" sores (Fig. 22), with abundant parasites in them; incubation several months; subject to spontaneous cure; no metastatic lesions, and no mucous membrane involvement; involvement of lymph glands in 10%, dogs susceptible; urban; transmitted by *P. papatasii, P. sergenti, P. perfiliewi* (in Italy) and *P. longicuspis* (in Algeria). (2) "Wet" sores, caused by *L. t. major,* with a short incubation of 1 to 6 weeks; sores quick to ulcerate; lymph glands commonly involved; no cross-immunity with Oriental sore; rural, with reservoir in wild rodents (gerbils and ground squirrels), and transmitted by *P. caucasicus*, which lives in the burrows of these rodents.

In tropical America, Biagi recognizes four types of cutaneous leishmaniasis as follows: (1) Mucocutaneous (Fig. 23A) in rain forests of Brazil; spreading, chronic, cutaneous lesions, tending to invade mucous

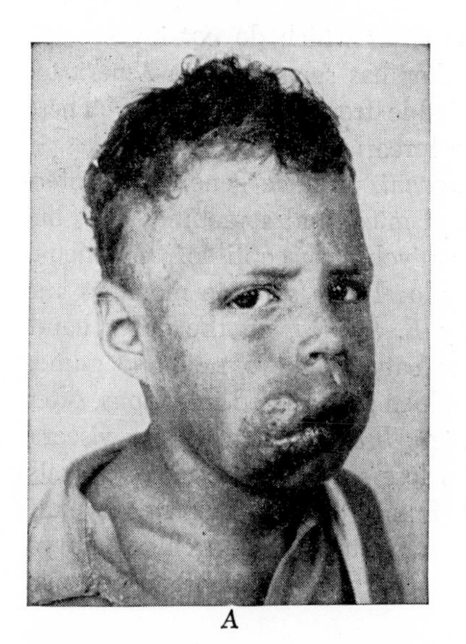

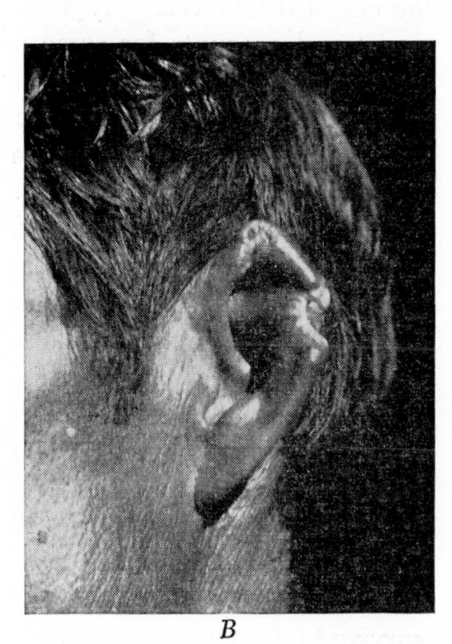

A *B*

Fig. 23. American leishmaniasis. *A*, a case of the mucocutaneous form from Brazil. (Photograph supplied by Dr. Nery-Guimaraes.) *B*, chiclero ulcer of the ear; a case of moderate duration as commonly seen in southern Mexico. (Photograph supplied by Dr. F. Biagi.)

membranes either by extension or (more often) by metastasis; much destruction of tissue; parasites relatively scarce in lesions, and deep; spontaneous recovery rare; lymph gland involvement infrequent. (2) Uta, in mountains of Peru; small, numerous skin lesions, causing little destruction of tissue; benign. (3) Leishmaniasis or "buba" or "pian bois" of Panama, Costa Rica, and the Guianas, showing moderate ulceration tending to spontaneous cure after a few years except when in the nose; mucous membrane lesions in 5%, always by extension and not metastases; lymph gland involvement in 10%. Floch (1954) proposed the subspecies name "*guianensis*" for this variety, and considered the parasite of uta in Peru to be the same. (4) Chiclero ulcer of Guatemala, Belize, and southeastern Mexico, characterized by small, nondestructive skin lesions which get well in a few weeks or months except when on the ear (Fig. 23*B*), for which the parasites have a special predilection and where they cause chronic, disfiguring, subcutaneous, nodular ulcers lasting many years; no metastasis to mucous membranes, rarely cutaneous metastases, and lymph gland involvement rare (2%); parasites scanty in lesions. For this variety Biagi proposed the name *L. tropica,* subspecies *mexicana.* In addition to these recognizably different clinical forms it is quite possible that the classical Old World type of Oriental sore has been established in Brazil and accounts for some of the cases which do *not* involve the mucous membranes. The junior author has seen cases in Americans in Panama which were indistinguishable from Oriental sore. These were most frequently lesions on the forearm or wrist.

Visceral strains, caused by *L. donovani,* produce generalized infections in monkeys, dogs, hamsters, and mice, and sometimes rats, but cats, rabbits, and guinea pigs are relatively insusceptible. Cutaneous inoculations sometimes produce only local skin sores in monkeys and dogs. Cutaneous strains, caused by *L. tropica,* on the other hand, produce only local infections in dogs, cats, monkeys, rats, and guinea pigs, whereas in mice they often produce generalized infections, often with skin lesions as well. The variety *brasiliensis* produces visceral infections in hamsters, but produces less severe effects in mice, rats, and cotton rats. Geiman (1940) found that *L. tropica* develops readily in the chorio-allantoic fluid of a 5- to 9-day-old chick embryo, whereas *L. brasiliensis* does not, though the original organisms may survive to a second passage.

The difference between visceral and cutaneous infections seems clearly to be one of virulence of the parasites. The body defenses, except in mice, are capable of localizing the cutaneous strains of *Leishmania,* thus confining them to the skin or testicles where inoculated, whereas they are unable to exert a similar restraining action on the visceral strains.

Leishmania donovani and Kala-azar

Kala-azar is a disease that is insidious in origin, slow in development, and fearful in effects. In 1890–1900 an epidemic swept Assam which depopulated whole villages and reduced populations over large areas. In 1917 another epidemic started in Assam and Bengal, reached its height about 1925, and then mysteriously subsided until, by 1931, it was almost gone. In 1937 a new outbreak began in Bihar. In other parts of the world it is less subject to such vacillations. A few decades ago kala-azar brought terror and persecution in its path. Today, knowledge of its epidemiology, diagnosis, and treatment has shorn it of much of its power for evil.

In the Old World typical kala-azar occurs in India, particularly in Assam, Bengal, and Bihar; in North China; in Turkestan; and in Sudan and many other places in tropical Africa; around the Mediterranean and in western and middle Asia; and from Venezuela to northern Argentina in South America.

In kala-azar the parasites are widely distributed in the body, but the special habitat seems to be the large endothelial cells of blood vessels and lymphatics. They are especially abundant in the spleen, liver, and bone marrow, but they are by no means confined to these organs. They are found both inside and outside of the tissue cells, and are present in limited numbers in the circulating blood, usually inside of monocytes, but occasionally free.

Kala-azar is usually a house and site infection, and for this reason it was once thought that the infection spread by contaminated soil. Although now established that sandflies (*Phlebotomus*) (see pp. 668–672) are the principal vectors, the parasites may sometimes be transmitted by nasal secretions, urine, and feces; infection by mouth is possible in experimental animals. Archibald and Mansour (1937) infected monkeys by swabbing or spraying the nose with infected nasal secretion and also by confining them in an insect-proof room with infected comrades. As noted on p. 118, dogs are important reservoirs of the disease in the Mediterranean area and in South America, are susceptible in China, but seem not to be involved very much in India or Sudan. In the Mediterranean area it is probable that transmission from dog to man is more frequent than from man to man. A similar situation exists between cutaneous leishmaniasis of dogs and man. Cats are sometimes infected, and occasionally horses, sheep, and bullocks. Experimentally monkeys, mice, hamsters, and ground squirrels are susceptible. In Sudan and Kenya, cases are sporadic, often occurring in persons traveling in uninhabited regions. This suggests that wild animals may be a reservoir. Stauber (1958) has made an exhaustive study of host susceptibility and course of infection of *L. donovani* from Sudan in a variety of small mammals. The differences in susceptibility, acquisition of resistance, and outcome of infection are striking. Stauber's studies indicate the futility of predicting with any confidence the probability of a particular animal serving as a host without careful laboratory investigation.

All the known or suspected transmitters of kala-azar in the Old World (see pp. 668–674) belong to the *P. major* group of *Phlebotomus* flies. These are listed on p. 673. Adler and Theodor suggested that the frequent occurrence of kala-azar in dogs and infants in the Mediterranean region, and not in India or China, may be due to the fact that the Mediterranean vectors may infect their victims very frequently by direct inoculation into the skin at the time of biting, whereas the Indian and Chinese vectors less frequently inoculate the parasites by their bites but cause infection by being crushed. Since dogs and babies are not so adept at slapping the flies as are adults, they escape infection.

Whether any species of *Phlebotomus* in the United States can serve as transmitters is not yet known, but in 1955 Thorson found *L. donovani* in the United States in a dog imported from an endemic area.

The disease. In kala-azar the reaction to parasitization of the reticulo-endothelial cells in internal organs leads to a great increase in their number, especially in the spleen and liver, which may become grotesquely enlarged.

The disease often comes on with symptoms suggestive of typhoid, malaria, or dysentery, and may actually be precipitated by these diseases, for there is now evidence that there is a high natural resistance to kala-azar and that probably the parasites are held under control in many latent infections, and no symptoms appear until resistance is lowered. In a case that Adler experimentally infected by inoculation of a massive dose of cultured *Leishmania,* no symptoms appeared over a period of 9 months, although numerous parasites were found post-mortem. The incubation period is usually at least several months.

After onset there is an irregular fever with enlargement of spleen and liver, rheumatic aches, anemia, and a progressive emaciation. The leucocytes are reduced in number, and the skin is often edematous. Untreated cases usually die in a few weeks to several years, usually from some intercurrent disease which the patient cannot fight with his macrophage system converted into a *Leishmania* breeding ground. Often, in patients who have been treated and have recovered from the systemic disease, whitish spots develop in the skin and eventually grow into nodules the size of split peas; they occur mainly on the face and neck. This condition is called post-kala-azar dermal leishmanoid. Apparently the parasites are able to survive in the skin after the viscera have become too "hot" for them. A number of cases of extensive lesions in the mouth have been seen in the Sudan, in which the parasites were found in abundance in the oral lesions, although they could not be found in the enlarged liver and spleen. Such cases probably represent intermediate conditions of parasite virulence and host resistance between typical kala-azar and cutaneous leishmaniasis.

The Mediterranean type of the disease in infants and dogs runs a similar course but may be of shorter duration.

Diagnosis. Although the clinical symptoms are highly suggestive in endemic localities, diagnosis should be confirmed either by finding the parasites or by serological tests. Puncture of liver, spleen, or lymph glands is useful in finding the parasites. Some workers recommend sternal puncture, but Shortt considers this less effective and more unpleasant for the patient. Shortt et al. have been able to find parasites in over 75% of cases by examination of a thick edge left after making a blood smear. Another method is to make a smear from the dermis exposed with as little bleeding as possible. Inoculation of NNN culture medium with spleen juice, blood, or bits of excised dermis is a reliable procedure. Shortt particularly recommends seeding 3 or 4 NNN culture tubes with the top of the sediment obtained by centrifuging 2 to 5 cc. of blood added to four times its volume of citrated saline. The tubes are incubated at 22 to 24°C., and flagellates appear in 7 days or later in 90% of untreated cases.

A number of simple serological tests have been recommended. One of the first was Napier's aldehyde test, in which a drop of strong formalin is added to 1 cc. of serum; in positive cases the serum gels and turns milky white; a mere gel is not diagnostic. Precipitates are also formed with organic antimony compounds, resorcinol, alcohol, peptonate of iron, lactic acid, and even distilled water, under conditions in which they are not formed by normal serum. The multiplication of apparently unrelated serum tests was becoming very confusing until Chorine (1937) showed that most of them are due to increase in euglobulin and decrease of albumin in kala-azar serum.

Treatment. Before the discovery of the striking effectiveness of antimony compounds for all forms of leishmaniasis, the death rate in kala-azar cases was about 95%; now it is less than 5%. Two groups of compounds are used: trivalent ones such as sodium and potassium antimonyl tartrates and sodium antimonyl gluconate, and pentavalent ones, the most extensively used being Neostibosan, Neostam, Solustibosan, and urea stibamine. Some trivalent compounds, such as Anthiomaline and Fuadin, which are very useful in schistosomiasis and filariasis, are less effective against leishmaniasis. The pentavalent compounds have the advantage of being less toxic, more quickly effective, and most of them injectable intramuscularly as well as intravenously, but they are more expensive.

Mediterranean and Sudanese forms of kala-azar do not respond to antimony treatment as well as the Indian form, but they respond well to one of the aromatic diamidines (see p. 145), Stilbamidine. Unfortunately this drug is quite toxic; it causes more or less alarming symptoms after injection, and sometimes dangerous ones weeks or even months later. Also, unlike the antimony compounds, the diamidines and related drugs do not immediately relieve the symptoms of kala-azar, but do so several weeks later, when both patient and physician are getting discouraged.

The great trouble with treatment is the long time that has been required for complete cures. Of the tartrates at least 25 or 30 doses daily or on alternate days, totaling at least 2500 mg., are needed, and of the pentavalent compounds, about 10 or 12 doses, totaling 2700 to 4000 mg., are required. Kirk and Sati, however, reported excellent results in Sudan kala-azar cases using sodium antimonyl gluconate in large daily doses for only 4 days, and then 2 to 6 more doses, usually after an interval of 2 weeks. They got immediate clinical response; usually gland and spleen punctures were negative after the first four injections. They claimed toxic effects to be negligible.

Prevention. Protection against kala-azar involves avoidance and control of sandflies, which is discussed on page 674. Infected houses and people should be avoided after dusk, when sandflies are biting, unless repellents are used. Habitations where cases have occurred should be sprayed with DDT or Dieldrin. Some control can be obtained locally by the establishment of free clinics and treatment of all cases. In endemic regions where the canine disease occurs, Sergent et al. recommend destruction of all dogs showing evidence of infection by symptoms or blood tests, and of all stray dogs; control of movement of dogs into and out of infected areas; and prevention of contacts between children and dogs. Destruction of the majority of dogs in Canea on the island of Crete in 1933 led to a markedly lower incidence of human kala-azar in the following year.

Oriental Sore (Old World Cutaneous Leishmaniasis)

One of the commonest sights in many tropical cities, particularly those of the eastern Mediterranean region and southwestern Asia, is the great number of children, usually under three years of age, who have on the exposed parts of their bodies unsightly ulcerating sores, upon which swarms of flies are constantly feeding. In some cities infection is so common and so inevitable that normal children are expected to have the disease soon after they begin playing outdoors, and visitors seldom escape a sore as a souvenir. Since one attack usually gives immunity, Oriental sores appearing on an adult person in Baghdad brands him as probably a new arrival, and the same is true in many other tropical cities. Dogs frequently suffer from cutaneous leishmaniasis also, especially on the nose and ears, and undoubtedly constitute an important reservoir. Many other animals develop local and sometimes visceral lesions when inoculated (see p. 120).

The disease is more or less prevalent from the shores of the Mediterranean to central Asia and the drier parts of central and western India, and also in parts of China and in many parts of Africa. It is possible that true Oriental sore has been introduced into South America also, but here it is obviously difficult to distinguish it from the native South American infection.

The parasites are found in the dermal tissues of the sores, where greatly increased numbers of large monocytes and other reticuloendothelial cells are literally packed with them. Torpedo-shaped parasites are more commonly found than in kala-azar.

The disease. Classical Oriental sore (Fig. 22) begins as a small red papule, like an insect bite, which gradually enlarges to a diameter of an inch or more. The covering epithelium eventually breaks down and granulation tissue is exposed, but no pus is evident unless the sore is secondarily infected by bacteria. In uncomplicated cases the ulcer remains shallow and sharply defined by raised edges. It persists from a few months to a year or more. The incubation period varies from a few days to several months. In an outbreak among fresh troops in Quetta the incubation period was over 3 months in more than half the cases.

There may be one sore or several, sometimes many, probably due to multiple infective bites. Neighboring lymph glands may be invaded and become large and painful, but general invasion of the body does not occur; generalized symptoms and changes in the blood are lacking unless there are secondary infections. The "wet" type of cutaneous leishmaniasis, which is rural with ground squirrels and gerbils as reservoirs, occurs along with typical Oriental sore in Turkestan. Its special characters are listed on p. 118.

Transmission. Although the parasites may occasionally be inoculated into broken skin by contact or by flies that have just fed on other sores, the disease is usually transmitted by *Phlebotomus* flies. The pupiparous fly, *Hippobosca canina* (see p. 706), may act as a mechanical transmitter among dogs.

Either *Phlebotomus papatasii* or *P. sergenti* or both occur in most places where Oriental sore occurs. Both are readily infected after feeding on infective material, are frequently found naturally infected, and produce infection when crushed and rubbed into scarified skin. According to Adler (1957), some strains of *L. tropica* seem better adapted to *P. sergenti* and others to *P. papatasii.* In Italy *P. perfiliewi,* although primarily zoophilic, is believed to be a vector, and in Algeria *P. longicuspis;* as noted on p. 118, the rural "wet"-sore disease of central Asia is transmitted by *P. caucasicus.* For further details concerning *Phlebotomus* flies see pp. 668–674.

Treatment and prevention. If only one or a few sores are present, local treatment is best. Since secondary infections are common, the scabs must be removed and the sores cleaned and antiseptically treated with powders or ointments. Injections of atebrin or berberine sulfate around the sores is said to have good effects. Multiple or chronic sores are best treated by injections of antimony compounds as for kala-azar, although in such cases intramuscular injections of the milder compound, Fuadin, are satisfactory. Usually, if the sores are protected, they heal in 15 to 30 days. Other local treatments with carbon dioxide snow, x-rays, and various antiseptic ointments have favorable influence but are not as effective as the methods mentioned above.

Control probably lies largely in keeping the sores on either man or dog protected so that sandflies or other biting insects cannot get at them. It is not likely that insects can become infected from sucking blood elsewhere, since blood cultures are never positive. Inoculation with cultures into unexposed parts of the body is recommended in endemic areas.

American Cutaneous and Mucocutaneous Leishmaniasis

As noted on p. 119, there are several clinically different types of cutaneous leishmaniasis in tropical America, ranging from Yucatan and Campeche in Mexico to northern Argentina. The disease is called chiclero ulcer in Mexico and Guatemala, Bay sore in British Honduras, Bosch yaws, forest yaws, and pian bois in the Guianas, espundia in Brazil and eastern Peru, uta in other parts of Peru, buba in Paraguay, and papalomoyo in Costa Rica. Shattuck (1936) pointed out that heat and moisture characterize the climate of all the endemic foci, with the possible exception of the mountain valleys in Peru where uta occurs.

It is almost always contracted, as is jungle yellow fever, in virgin forests, usually among men gathering chicle, rubber, or maté, or constructing railways through the forests. It is common among chicle gatherers in low-lying rain forest areas in Yucatan, especially from August to January when the collecting season is at its height. Biagi found 62% of these people positive to the Montenegro skin test for leishmaniasis. Although usually acquired in rain forests, outbreaks of leishmaniasis have occurred in residential parts of Rio de Janeiro, where there were gardens and shrubbery, and in Peruvian villages at elevations of 4500 to 7500 ft., where the mild form of the disease known as uta occurs.

Dogs are sometimes found naturally infected, but not as commonly as with Oriental sore in the Old World. Monkeys and dogs can be experimentally infected, and Fuller and Geiman in 1942 found that squirrels, especially Texas ground squirrels, are susceptible to cutaneous but not to intraperitoneal inoculation, and develop ulcerating sores. Hamsters are less easily infected but may develop nodular skin lesions that do not ulcerate, and also visceral infections. Milder infections can be induced in mice.

Another species, *L. enriettii,* described by Muniz and Medina in 1948, produces mucocutaneous leishmaniasis in the guinea pig. Interestingly, this form first appeared in a laboratory animal in South America. The cutaneous lesions found in a wild-caught Brazilian paca by Forratini and Santos in 1955 may well be *L. enriettii.*

Mucocutaneous leishmaniasis (espundia). This severe form of the disease occurs from Brazil to Paraguay and northern Argentina. In typical cases the infection begins precisely as in Oriental sore and frequently follows a similar course, but there is a greater tendency for the sores to spread over extensive areas and for more numerous sores to appear. In one instance 248 sores were reported. The ears, face, forearms, and lower legs are the favorite sites for the original lesions, but laborers naked to the waist may get ulcers on the trunk, and occasionally on the genitals or elsewhere. Sometimes the sores show a mass of raw granular tissue raised above the surface; at other times they become extensively eroded, with sharply defined, raised, purplish edges and a surrounding red inflamed area. The foul-smelling fluid which exudes sometimes crusts over, but may be inoculated into abrasions elsewhere and cause secondary ulcers. Secondary infections with bacteria, spirochetes, fungi, or maggots are frequent. The rarity of leishmanias in late stages suggests that secondary infections may play an important role, though the prompt healing which follows antimony treatment shows that leishmanias still play a leading part. There is nothing about the sores to distinguish them with certainty from others caused by blastomycosis, syphilis, tropical ulcers, or even in some cases yaws, so it is little wonder that there has been much confusion about their distribution and etiology.

The most striking feature of the disease is the secondary development, sometimes by extension but more frequently by metastasis, of ulcerations in the nasal cavities, mouth, and pharynx, which may occur in 20% or more of the cases, though much more commonly in some geographic regions than in others. In rare cases ulcers occur in the vagina. According to Villela et al., small incipient lesions can be found in the nose in many cases where no obvious lesions are present, and scrapings of the mucosa frequently reveal *Leishmania* even when it is perfectly normal in appearance. The mucous membrane ulcerations may appear before the skin lesions have healed, but usually they develop from several months to several years later. Ordinarily they start as tiny itching spots or swellings of the mucous membrane, usually in the nose, the infected membrane becoming inflamed and marked either with small granular sores or with blister-like swellings. The lymph glands in the infected regions become swollen and turgid. A granular ulceration begins in a short time, invading all the mucous membranes of the nose and spreading by means of infective fluid which flows down over the upper lip into the mouth cavity, attacking the hard and soft palate.

Advance of the lesion is obstinate and slow, and gives rise to serious complications. The nostrils become too clogged to admit the passage of sufficient air and the patient has to keep his mouth constantly open to breathe. His repulsive appearance and fetid breath help to make his life miserable. Affections of the organs of smell and hearing, and even sight, may supervene, and the voice is weakened or even temporarily lost. The digestive tract becomes upset from the constant swallowing of the exudations mixed with saliva or food. A spreading of the nose due to the eating away of the septum is a characteristic feature. Although in late stages of the disease the entire surface of the palate and nasal cavities is attacked and the septum between the nostrils destroyed, the bones are left intact, a feature which readily distinguishes a leishmanian ulcer from a syphilitic one. Usually the victim of espundia, if untreated, dies of some intercurrent infection, but he may suffer for years and eventually succumb to the disease itself.

Diagnosis is usually made by finding the leishmanias in the lesions, but a skin test described by Montenegro in 1926 is sometimes helpful. Dead cultured flagellates are injected into the skin; in positive cases an allergic inflammation develops within 48 hours.

Transmission. Little is definitely known about the transmission of the disease, though by analogy with other forms of leishmaniasis it is highly probable that bites of *Phlebotomus* flies are usually responsible. Support for this view is provided by instances where typical sores developed at the site of bites of *P. lutzi* in Brazil. Aragão in 1922 found leptomonads in some wild *P. intermedius* captured in a locality in Rio de Janeiro where a local outbreak occurred, and in five of the flies that had fed on espundia sores 3 days before, he found similar flagellates. When emulsions of these flies were inoculated into the nose of a dog, an ulcerating sore containing leishmanias developed 3 days later. Natural or experimental infections of this species, a semidomestic one, have been reported in many South American countries, from Argentina to Venezuela. Natural and experimental infections have been reported in at least six other species in various parts of South America (Bustamente, 1948). For further details see pp. 673–674. In Peru Townsend obtained some experimental evidence that two species of midges of the genus *Forcipomyia* (see p. 676) are the transmitters of uta, but this work needs confirmation.

Treatment. Most cases respond well to injections of antimony compounds, but some respond better to Neosalvarsan, and in some cases resistant to antimony and arsenic compounds, Lomidine, one of the diamidines (see p. 145), has proved effective. Treatment should be accompanied by removal of scabs from ulcers, even on nose, lips, or mouth, and cleansing to get rid of bacterial infections. In lesions confined to the skin local treatments are helpful, using antimony tartrate applied as a powder or in 1 or 2% solutions, or the methods described for Oriental sore can be employed. For chiclero ulcers the local treatments are satisfactory except for the chronic ear lesions, which requires systemic treatment with tartar emetic, Fuadin, or Lomidine.

REFERENCES

Adler, S., and Ber, M. 1941. The transmission of *Leishmania tropica* by the bite of *Phlebotomus papatasii*. *Ind. J. Med. Research*, 29: 803–809.

Adler, S., and Theodor, O. 1931–1938. Investigations on Mediterranean kala-azar. *Proc. Roy. Soc. London, B108:* 447–453; 110: 402–412; 116: 494–504; 125: 491–516.

Adler, S., and Theodor, O. 1957. Transmission of disease agents by phlebotomine sand flies. *Ann. Rev. Entomol.*, 2: 203–226.

Archibald, R. G., and Mansour, H. 1937. Some observations on the epidemiology of kala-azar in the Sudan. *Trans. Roy. Soc. Trop. Med. Hyg.*, 30: 395–406.

Beltran, E. 1944. Cutaneous leishmaniasis in Mexico. *Sci. Monthly*, 59: 108–119.

Beltran, E., and Bustamente, M. E. 1942. Datos epidemiológicos acerca de la "Ulcera de los chicleros" (leishmaniasis americana) en México. *Rev. Inst. Sal. Enf. Trop.*, 3: 1–28.

Biagi, F. F. 1953. Algunos comentarios sobre las leishmaniasis y sus agentes etiológicos. *Leishmania tropica mexicana,* nueva subespecie. *Medicina Rev. Mex.,* 33: 401–406.

Bustamente, M. E. 1948. Epidemiológia de la Leishmaniasis en América. *Bol. Ofic. Sanit. Panamer.,* 27: 611–618.

Chorine, V. 1937. Les réactions sérologiques dues aux euglobulines. *Ann. inst. Pasteur,* 58: 78–124.

Dotrovsky, A., and Sagher, F. 1946. The intracutaneous test in cutaneous leishmaniasis. *Ann. Trop. Med. Parasitol.,* 40: 265–269.

Geiman, G. M. 1940. A study of four Peruvian strains of *Leishmania brasiliensis.* *J. Parasitol.,* 26: Suppl., 22–23.

Hoare, C. A. 1944. Cutaneous leishmaniasis (critical view of recent Russian work). *Trop. Diseases Bull.,* 41: 331–345.

1948. The relationship of the haemoflagellates. *Proc. 4th Intern. Congr. Trop. Med. Malaria,* 2: Sect. VII: 1110–1118.

Hoeppli, R. 1940. The epidemiology of kala-azar in China. *Chinese Med. J.,* 57: 364–372.

Kirk, R., et al. 1940–1945. Studies in leishmaniasis in the Anglo-Egyptian Sudan. *Trans. Roy. Soc. Trop. Med. Hyg.,* 33: 501–506, 623–634; 34: 213–216; 35: 257–270; 38: 61–70, 489–492.

League of Nations. 1935. On the diagnosis, treatment and epidemiology of visceral leishmaniasis in the Mediterranean basin. *Quart. Bull. Health Organisation,* 4: 789.

Montero-Gei, F. 1956. Contribución al estudio de *Endotrypanum schaudinni* (Trypanosomidae). *Rev. biol. trop.,* 4: 41–68.

Napier, L. E., et al. 1942. The treatment of kala-azar by diamidine stilbene. *Indian Med. Gaz.,* 77: 321–338.

Pessôa, S. B., and Pereira Barretto, M. 1948. *Leishmaniose tegumentar americana.* Impr. Nacional, Rio de Janeiro.

Reports of the Indian Kala-azar Commission. (1924–1932.) No. 1, *Indian Med. Res. Mem.* 4; No. 2, *ibid.,* 25. (Contains numerous papers by Calcutta and Assam workers.)

Shattuck, G. C. 1936. The distribution of American leishmaniasis in relation to that of Phlebotomus. *Am. J. Trop. Med.,* 16: 187–205.

Smith, R. O. A., Halder, K. C., and Ahmed, J. (Transmission of Kala-azar.) 1940–1941. *Indian Med. Gaz.,* 75: 67–69; Editorial, *ibid.,* 97–98; *Ind. J. Med. Research,* 28: 575–591; 29: 783–802.

Stauber, L. A. 1958. Host resistance to the Khartoum strain of *Leishmania donovani.* *Rice Inst. Pamphlet,* XLV (1): 80–96.

Swaminath, C. S., Shortt, H. E., and Anderson, L. A. P. 1942. Transmission of Indian kala-azar to man by the bites of *Phlebotomus argentipes.* *Ind. J. Med. Research,* 30: 473–477.

Chapter 7

HEMOFLAGELLATES
II. TRYPANOSOMES

One of the blackest clouds overhanging the civilization of tropical Africa is the scourge of trypanosome diseases which affect both man and domestic animals. The ravages of sleeping sickness, which is the final phase of trypanosome infection in man, were well known to the old slave traders, and the presence of Africans lying prostrate on wharves and docks with saliva drooling from their mouths, insensible to emotions or pain, was a familiar sight. It did not take these astute merchants long to find that death was a frequent outcome of the disease, and they soon recognized swollen glands in the neck as an early symptom and refused to accept as slaves Negroes with swollen glands. Nevertheless sleeping sickness must often have been introduced with its parasites into various parts of North and South America, as it frequently is even at the present time, and only the absence of a suitable means of transmission has saved the western hemisphere from being swept by it.

In Africa the most obvious effect of sleeping sickness is depopulation by death, induced sterility, high infant mortality, and displacement of population, with the result that the land is relinquished to wild animals and tsetse flies. Important as human infection has been in Africa, the effect of trypanosomiasis on domestic animals has, over vast areas, had a tremendous influence on the economy and development of that continent. As Hornby (1950) remarked, "Trypanosomiasis is unique among diseases in that it is the only one which by itself has denied vast areas of land to all domestic animals other than poultry. The areas of complete denial are all in Africa and add up to perhaps one-quarter of the total land surface of this continent." Animal trypanosomiasis, by making it impossible to keep cattle or other animals, causes loss of soil fertility by the absence of manure, diminishes vitality of the people because of deficiencies in the diet, and deprives them of beasts of burden.

Fossil remains of tsetse flies, belonging to the Oligocene period, have been found in Colorado, and it has been suggested that the extinction of prehistoric camels and horses in North America, which cradled them in the early days of their evolution, may have been brought about by tsetse-borne trypanosome diseases.

History. Although trypanosomes were first discovered in 1841, which is very ancient history in parasitology, the first connection with disease was the discovery in 1880 that they were the cause of surra in horses and other animals in India. In 1895 Bruce showed that nagana of domestic animals in Africa was caused by trypanosomes. In 1902 Forde and Dutton discovered the presence of trypanosomes in human blood in a case of "Gambia fever," the preliminary stage of sleeping sickness. In 1903 Castellani found trypanosomes in the cerebrospinal fluid of cases of sleeping sickness in Uganda. Kleine, in 1909, showed that the tsetse fly is no mere mechanical transmitter but a true intermediate host. In that same year there was discovered a new type of human sleeping sickness in Rhodesia, and Chagas described an entirely different human trypanosome infection in South America.

Trypanosomes have played an interesting role in the development of certain concepts in biology and medicine. Observations on trypanosomes led Ehrlich to the idea that specific drugs acting on disease-producing agents could be found, ultimately resulting in the discovery of drugs for treating syphilis. The first observations of drug-resistance in a microorganism were made on trypanosomes. These little animals have continued to fascinate research workers who are interested in a variety of biological problems.

The parasites. The general relationships of trypanosomes have been discussed on pp. 111–113. They may be regarded as having developed in the course of evolution from the crithidias of invertebrates, adapted to living in the blood of vertebrates on which the invertebrates habitually feed. They thus bear the same relation to *Crithidia* that *Leishmania* bears to *Leptomonas.*

Trypanosomes exist as parasites in all sorts of vertebrates—fish, amphibians, reptiles, birds, and mammals—living in the blood, lymph, or tissues of their hosts. A great number of different species have been named; usually any trypanosome found in a new host is named after the host as a tentative label until more is found out about it. Though this procedure is not in accordance with rules of naming animals, it is better than the alternatives of having numerous nameless trypanosomes to deal with, or of identifying them with species from which they may subsequently be found to differ.

In form most trypanosomes are active, wriggling little creatures somewhat suggesting diminutive dolphins or eels, according to their slenderness (Fig. 24). They swim in the direction of the pointed end of the body, being propelled by the wave motions of the undulating membrane. The body is shaped like a curved, flattened blade, tapering to a fine point anteriorly, from which a free flagellum often continues forward. This flagellum continues nearly to the posterior end of the body, and is connected with the body by an undulating membrane, like a long fin or crest; in some species it is thrown into numerous graceful ripples; in certain others (e.g., *Schizotrypanum*

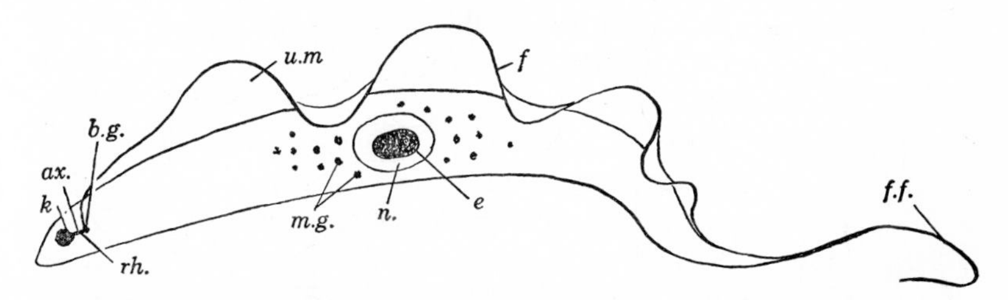

Fig. 24. *Trypanosoma gambiense*, slender form: *ax.*, axoneme; *f.*, flagellum; *f.f.*, free flagellum; *k.*, kinetoplast; *m.g.*, metachromatic granules; *n.*, nucleus; *rh.*, rhizoplast; *u.m.*, undulating membrane.

cruzi) it is only slightly rippled. The body contains a nucleus which varies in its position in different species and under different circumstances. Near the posterior end, or sometimes at the tip, there is a kinetoplast and a basal granule from which the flagellum arises. The kinetoplast is lost in *T. equinum* and some strains of *T. evansi*, species which have become independent of intermediate hosts; apparently, as Hoare has pointed out, the kinetoplast is necessary for development in tsetse flies, but not for multiplication in vertebrate hosts. Many species also contain scattered, deep-staining granules in the cytoplasm. Trypanosomes are commonly spoken of as polymorphic or monomorphic. Polymorphic forms are those in which some individuals have a free flagellum and others do not (e.g., *T. gambiense, brucei,* and *rhodesiense*), whereas monomorphic forms may always have a free flagellum (e.g., *lewisi, vivax, evansi, equinum, equiperdum,* and S. *cruzi*), or may always lack one (e.g., *T. congolense, simiae*) (see Fig. 25).

Life cycles. In the vertebrate hosts most trypanosomes usually multiply by simple fission. The blepharoplast is the first structure to divide; next a new flagellum begins to grow out along the margin of the undulating membrane; then the nucleus divides; and finally the body splits from the anterior end backwards.

The African polymorphic trypanosomes of man and animals are mainly parasites of the lymphatic and intercellular fluids, but multiplication also occurs in the blood stream; *T. equiperdum* thrives in edematous fluid of sex organs and skin; and some trypanosomes of birds apparently live mainly in the bone marrow. The members of

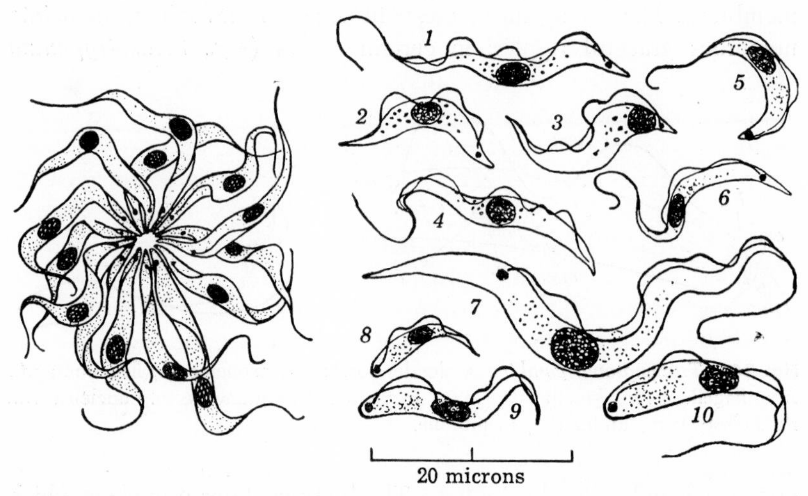

Fig. 25. *Left,* agglomeration of trypanosomes, *T. lewisi,* in blood of immunized rat. (After Laveran and Mesnil.) *Right,* mammalian trypanosomes: *1, brucei* or *gambiense* with free flagellum; *2,* same, without free flagellum; *3, brucei* or *rhodesiense,* form with posterior nucleus; *4, equinum; 5, cruzi; 6, lewisi; 7, theileri; 8, congolense; 9, simiae; 10, vivax.*

the genus *Schizotrypanum,* however, multiply intracellularly in a leishmania form (Fig. 30), changing to the trypanosome form before being liberated from the cells; the free trypanosome forms get into the lymph and blood but do not multiply there (see p. 113). Because of these peculiarities many South and Central American workers have long recognized *Schizotrypanum* as a genus separate from *Trypanosoma* and *Leishmania.* Most European and North American workers have not accepted *Schizotrypanum,* but it seems apparent to the authors that it is a logical separation. In many cases the favorite habitat of trypanosomes is unknown, only the blood forms having been recognized. In some species at least, e.g., *T. lewisi* of rats and *duttoni* of mice, an immune response which inhibits reproduction soon develops, making further multiplication impossible.

At least one trypanosome, *T. equiperdum,* has become completely independent of its ancestral invertebrate hosts and is transmitted directly from horse to horse during copulation, and other trypanosomes can live and multiply indefinitely in vertebrate hosts if artificially injected by the soiled proboscis of biting flies. The majority of them, however, when they reach a suitable invertebrate host, hark back to the traditions of their remote forebears and go through a cycle of development more or less like that of typical crithidias. Some, such as S. *cruzi* and *T. lewisi,* finding themselves in the ancestral home, revert almost completely. After being sucked into the stomach of an insect they assume the crithidial form, attach themselves to the epithelial cells or enter them, and multiply. Gradually they move backwards towards the rectum, and the infective forms are voided with the feces. Infection occurs either by contamination of the bite with the feces, which is probably the usual way in S. *cruzi;* by ingestion of the feces of the insect when licking the bites, as in the case of *T. lewisi;* or by ingestion of the whole insect.

Those trypanosomes which develop thus in the hindgut of invertebrates are said to develop in the posterior station; they are the conservatives. The trypanosomes of this group use a variety of invertebrates as intermediate hosts; for example, *T. lewisi* of rats develops in fleas, *T. melophagium* of sheep in sheep ticks (keds), *T. theileri* of cattle in tabanids, S. *cruzi* of man and other animals in triatomid bugs, and African reptilian trypanosomes in tsetse flies. There are other trypanosomes, however—the radicals—which after ingestion by their insect vectors develop in the anterior part of the alimentary canal and infect by way of the proboscis. They form an evolutionary series: first the *T. vivax* group, which develop only in the proboscis; then the *congolense* group, which develop first in the stomach and then move forward to the pharynx; and finally the *brucei* group, which, after some development in the stomach, move forward and invade the salivary glands, where the cycle of development is completed. So far as is known at present, this specialized procedure occurs only in tsetse flies, which serve as transmitters of African mammalian trypanosomes, and in leeches, which transmit the trypanosomes of aquatic animals. Such trypanosomes are said to develop in the anterior station. The infective trypanosomes that appear at the end of the cycle in insects, whether in the anterior or posterior station, are called metacyclic forms; they resemble the blood forms but are smaller.

Identification of mammalian trypanosomes. The following key indicates the principal differences between some of the commoner mammalian trypanosomes:

I. Polymorphic forms; undulating membrane convoluted; kinetoplast usually not terminal; body with metachromatic granules. Invade salivary glands of tsetse flies.
   1. Nucleus nearly always central or slightly posterior; low virulence for domestic and laboratory animals (Fig. 25, *1* and *2*) *T. gambiense*
   2. Nucleus sometimes posterior, especially in small laboratory animals (Fig. 25, *3*); highly virulent in laboratory and domestic animals.
      a. Man not susceptible *T. brucei*
      b. Man infectible *T. rhodesiense*

II. Monomorphic forms with free flagellum.
   1. Undulating membrane only slightly convoluted.
      a. Kinetoplast small, not terminal; nucleus anterior; length about 25 μ; parasite of rats; develops in fleas (Fig. 25, *6*) *T. lewisi*
      b. Kinetoplast large, egg-shaped, usually terminal; nucleus central; body stumpy; length about 20 μ; parasite of various small mammals and man; develops in triatomids; does not undergo reproduction in the trypanosome stage (Fig. 25, *5*) *Schizotrypanum cruzi*
   2. Undulating membrane moderately or strongly convoluted.
      a. Size large, usually 50–70 μ long; kinetoplast distant from posterior end.
         (1) In cattle; develops in tabanids (Fig. 25, *7*) *T. theileri*
         (2) In sheep; develops in sheep-tick *T. melophagium*
      b. Size moderate (18–36 μ).
         (1) Posterior end swollen and rounded; undulating membrane moderately convoluted; kinetoplast terminal or nearly so; highly pathogenic for domestic animals, but laboratory animals insusceptible; develops in tsetse fly in proboscis only (Fig. 25, *10*) *T. vivax*
         (2) Posterior end pointed.
            (*a*) Nucleus consistently anterior (three-eighths length of body from anterior end); kinetoplast small, subterminal; extremely long and slender forms develop in culture; develops in *Rhodnius* (Fig. 33) *T. rangeli*
            (*b*) Nucleus central or slightly posterior; kinetoplast small, subterminal; undulating membrane well convoluted, closely resemble flagellated phase of *brucei* (Figs. 24, 25, *1*).
               (1) Highly pathogenic for horses, dogs and camels, causing surra, milder in cattle; transmitted by biting flies mechanically *T. evansi*
               (2) Milder in dogs, causes dourine in horses; venereally transmitted, with no insect intermediary *T. equiperdum*
            (*c*) Similar to (*b*) but kinteoplast absent; causes mal-de-caderas in horses in South America (Fig. 25, *4*) *T. equinum*

III. Monomorphic forms with no free flagellum.
   1. Small, 9–18 μ long; kinetoplast terminal or nearly so; produces

chronic wasting disease in domestic animals, especially destructive to cattle; not highly virulent for laboratory animals; develops in stomach and pharynx of tsetses, not in salivary glands (Fig. 25, 8) *T. congolense*

2. Larger, 14–24 μ long; highly virulent for pigs; monkeys, sheep, and goats also susceptible, but usually not other domestic or laboratory animals (Fig. 25, 9) *T. simiae*

Pathogenicity and immunity. The very name trypanosome suggests deadly disease, yet at least the majority of trypanosomes are harmless to their hosts. Wenyon in 1926 went so far as to say: "As a general statement, it is safe to regard all trypanosomes as nonpathogenic to their natural hosts." The so-called pathogenic trypanosomes of man and domestic animals he regards as owing their injuriousness to their being in unnatural hosts; in the wild game animals of Africa, which he regards as the natural hosts, they are harbored without ill effects. *T. gambiense* has undoubtedly arisen from a *brucei*-like ancestor; it has not yet reached a stage of equilibrium with its new host where it can exist without creating a disturbance. It is significant that where human infections have existed longest the disease tends to assume a mild chronic form. *T. rhodesiense,* a more recent offshoot from *T. brucei,* is much *more* pathogenic for man.

The pathogenicity of trypanosomes depends largely on ability of the hosts to develop trypanocidal antibodies and in some cases reproduction-inhibiting antibodies (Taliaferro, 1926). Vitamin deficiencies may increase their harmfulness; Becker et al. in 1947 showed that the usually benign *T. lewisi* may become pathogenic in pantothenate-deficient rats. The serum of a recovered animal contains protective antibodies against the particular trypanosome involved, and shows the usual immune reactions, such as complement fixation and lysis. It also causes the trypanosomes to clump together in rosettes, attached by their posterior ends (Fig. 25, left), and to adhere to leucocytes and platelets in the blood. Serum of naturally immune animals protects against infection when injected but is not destructive *in vitro.*

African Trypanosomiasis and Sleeping Sickness

Two distinct types of trypanosomes cause human disease, one type in Africa, the other in South and Central America. The African trypanosomes belong to a group of closely related polymorphic forms. One of these, *Trypanosoma brucei,* is found in many African wild animals, is highly virulent for domestic animals, especially horses and camels, is infective for almost every kind of mammal except baboons and man, and is transmitted by *Glossina morsitans.* This is without doubt the parent form from which two species or strains capable of infecting man have arisen, namely, *T. gambiense* and *T. rhodesiense.* Some authorities consider both of them distinct species, some think that *gambiense* but not *rhodesiense* is distinct from *brucei,* and some think that all of them are mere strains of a single species.

It seems evident that here again, as in the case of spirochetes, amebas, and leishmanias, we have run up against the difficulty in classification that comes from the fact that pathogenic organisms are not immutable things that can be described like simple chemical compounds, but are constantly undergoing adaptation and change. Their evolutionary possibilities are great because of their rapid multiplication, and are further enhanced by the isolation and variety of environmental conditions afforded them by life in a variety of intermediate and definitive hosts.

Aside from minor and inconstant differences in behavior in tsetse flies and in effects on laboratory animals, the only difference between *T. gambiense* and *T. rhodesiense* is the fact that the latter, like *T. brucei,* when developing in small laboratory animals produces a small percentage of forms in which the nucleus is displaced to the posterior end of the body. Even this is not a constant difference, and it is possible that it results from unusually rapid multiplication, and may be only an indication of virulence of the parasite or susceptibility of the host.

Morphology and habits. The African polymorphic trypanosomes vary in length from about 15 to 30 μ, with exceptionally longer or shorter forms. They show the characteristic slender forms with free flagellum, stumpy forms without a free flagellum, and intermediate forms, in any single blood or gland smear (Fig. 26). They are nearly always sparse in human blood, and are usually more abundant in the juice of enlarged lymph glands. They also occur in the spleen, which is often enlarged. Later, usually after 3 months, they appear in the cerebrospinal fluid and even in the tissues of the brain and spinal cord. Throughout the infection they live between the cells and are only found inside the cells when they have been ingested by phagocytes. These trypanosomes in the past have not been easily cultured in artificial media, but Weinman has succeeded with a medium containing blood, peptone, and beef-heart infusion. The cultured forms resemble those that develop in tsetse flies; they are not infectious but they *are* antigenic. The trypanosomes can also be cultured in chick embryos.

Trypanosoma gambiense is readily inoculated into certain kinds of monkeys and less readily into small laboratory animals, unless first passed through a monkey. Various antelopes and other herbivorous animals, and also dogs, are susceptible. The Situtunga antelope is an important natural reservoir in some places, and domestic animals, particularly pigs, may also serve in this capacity.

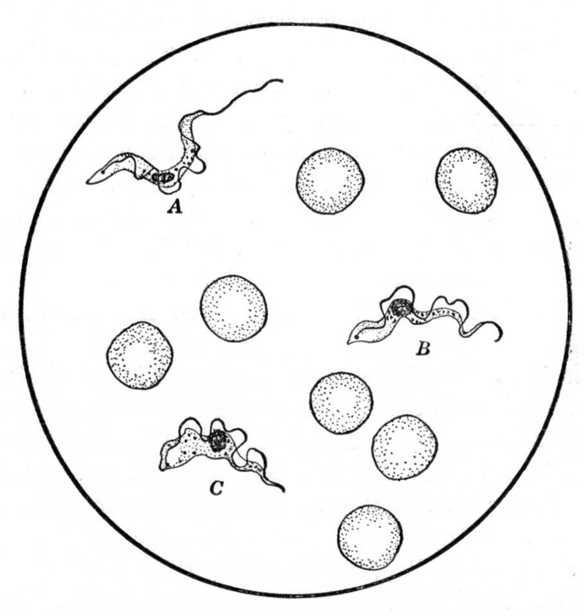

Fig. 26. *Trypanosoma gambiense* showing *A*, long form with free flagellum; *B*, intermediate form; and *C*, short form without free flagellum. About $\times 1200$.

Until quite recently *T. rhodesiense* had never been actually isolated from wild game animals although there was strong experimental and epidemiological evidence favoring the idea that wild animals are sources of human infection. In 1958, the transmission of *rhodesiense* to man from a bush buck was demonstrated. Normal human serum is toxic to all the African trypanosomes except *T. gambiense* and *T. rhodesiense,* but whereas *T. gambiense* apparently never loses this immunity, *T. rhodesiense* may do so after being kept in culture or in laboratory animals for a long time. This suggests that *T. rhodesiense* is a strain that has not become as thoroughly acclimated to human blood as *T. gambiense,* which is also indicated by its greater virulence.

Distribution. Human infections with *Trypanosoma gambiense* occur in a wide area in tropical western Africa from Senegal to Loanda, extending inland along the rivers, particularly the Congo and

Niger. The affected areas have been extended to Lake Tanganyika and southern Sudan by white settlement and consequent movement of infected natives. Stanley's expedition to reach Emir Pasha in 1888 may have introduced the disease to virgin territory in Uganda and the Great Lakes region, where it gave rise to a terrible epidemic that in one district reduced the population from 300,000 in 1901 to 100,000 in 1908. Some whole villages and islands were depopulated. Buxton (1948), however, suggested that this outbreak may have been due in part to *T. rhodesiense,* which was unknown at that time.

In more recent years the severity of the disease has been reduced in many parts of Africa by preventive measures and treatment, together with a natural decrease in virulence in many areas. In Nigeria alone, from 1931–1943, over 4,000,000 examinations were made, with detection and treatment of over 300,000 cases. The infection rate in Nigeria is now only one-tenth of what it used to be. However, in many endemic areas in Africa, 5 to 30% or more of the natives are still infected.

T. rhodesiense infection was first reported in Rhodesia in 1909. Since then many more cases have occurred over a fairly wide area in eastern Africa from Kenya to southern Rhodesia and northern Mozambique, and inland across Tanganyika to Uganda and eastern Congo. A few outbreaks that might be termed epidemics have occurred, including one in Uganda in 1940–1943, which has continued to smolder, and one in Belgian Congo in 1954, but in general the infection in *T. rhodesiense* areas is markedly sporadic. This species has a tendency to recur in certain spots, sometimes after intervals of many years. In spite of its greater pathogenicity and high case mortality rate, it affects such a small percentage of the population as to be a relatively minor public health problem. Animal trypanosomiasis has far more serious effects on the welfare of East Africa.

Transmission and epidemiology. The entire group of polymorphic trypanosomes is transmitted by tsetse flies; they differ from other tsetse fly-transmitted trypanosomes by invading the salivary glands of the flies. Although animal trypanosomiasis is often transmitted mechanically by biting flies, such as tabanids and *Stomoxys,* this method of transmission is negligible as far as man is concerned, although mechanical transmission by tsetse flies may not be infrequent.

The species of tsetse flies involved vary with the strains. The principal vector of *brucei* and *rhodesiense* is *Glossina morsitans,* although in one part of Tanganyika *G. swynnertoni* is also an important carrier. In the Uganda epidemic of *T. rhodesiense* in 1940–1943, *G. pallidipes* seemed to be the principal transmitter. The principal vector of *gam-*

biense is *G. palpalis,* but in northern Nigeria and Cameroons *G. tachinoides* seems to be the important transmitter. These species of *Glossina* are discussed on pp. 698–706. Experimentally *T. brucei* and *rhodesiense* can also be transmitted by *G. palpalis* and other tsetses, and *T. gambiense* by *G. morsitans.*

When cyclical development occurs (Figs. 27 and 28), the ingested parasites multiply first in the middle intestine. After the tenth to fifteenth day long slender forms are developed which move forward to the proventriculus. After several more days the trypanosomes make

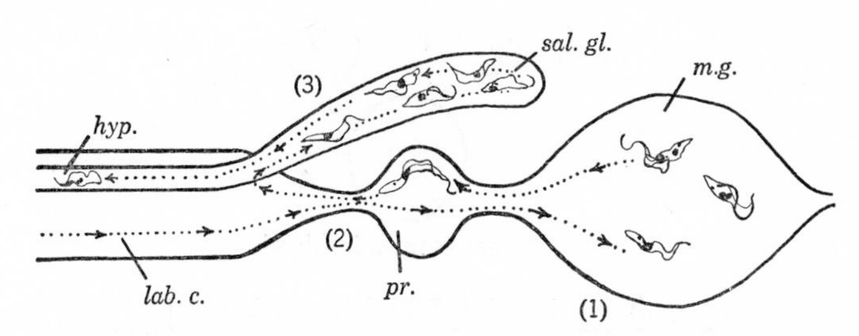

Fig. 27. Course of development of *T. gambiense* in tsetse fly. (1) multiplication in midgut in trypanosome form; (2) long, slender trypanosomes in proventriculus; (3) passage to salivary glands and development there of crithidial and then infective trypanosome forms; hypopharynx and labial cavity used for passage only; *hyp.*, hypopharynx; *lab.c.*, labial cavity; *m.g.*, midgut; *pr.*, proventriculus; *sal.gl.*, salivary gland.

their way to the fly's salivary glands, to the walls of which they attach themselves by their flagella and, rapidly multiplying, undergo a crithidial stage. As multiplication continues, free-swimming "metacyclic" trypanosome forms are produced which very closely resemble the parasites in vertebrate blood and are now capable of infecting a vertebrate host. The whole cycle in the fly usually occupies 20 to 30 days. A temperature between 75°F. and 85°F. is necessary for full development of the parasite in the fly, ending in invasion of the salivary glands.

In nature not more than 1 or 2 per 1000 wild tsetses are found infected with trypanosomes of the *brucei* group, although 20 to 30% of the game on which they feed are infected. Many flies appear to be completely refractory to infection, as Huff has shown to be true of mosquitoes and malaria. On the other hand, individuals gifted with susceptibility are often found to be infected with more than one species of trypanosome. Strains of trypanosomes vary in their ability to infect the salivary glands of tsetses, and some lose their power of cyclical transmission entirely after prolonged cultivation or passage through animals. In 1956 Willet showed that, in the case of *T. rhodesiense,* about 80,000 must be introduced into a man to establish infection.

As just noted, game animals may serve as reservoirs for the human trypanosomes, particularly *T. rhodesiense.* The recurrence of *rhodesiense* infections in the same spots raises the question whether man

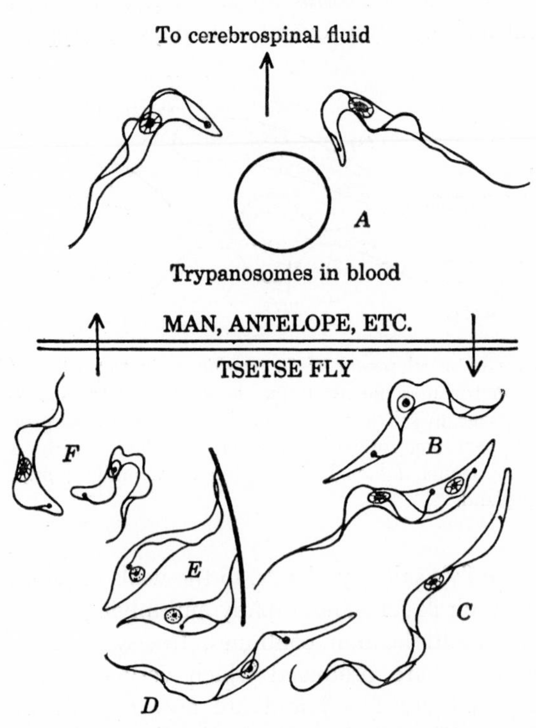

Fig. 28. Life cycle of *T. gambiense.* *A*, long and short forms in human blood; *B*, trypanosomes in midgut of tsetse 48 hours after blood meal; *C*, slender proventricular forms, 10th to 15th day; *D*, trypanosome newly arrived in salivary glands, 12th to 20th day; *E*, crithidias in salivary glands, 15th to 25th day; *F*, infective trypanosomes in salivary glands, 20th to 30th day. (Constructed from figures by Robertson, *Proc. Roy. Soc. London,* 1912.)

or some animal is the more important reservoir (Buxton, 1948). Healthy human carriers exist and might be responsible for keeping the disease alive, but it is also known that this trypanosome remains capable of infecting man even after a long succession of cyclically transmitted passages to sheep or antelopes, although virulence for man then diminishes. The *rhodesiense* infection usually occurs on the fringes of thinly populated back country abounding in big game, whereas *gambiense* infections prevail in well-populated areas with sparse mammalian fauna. It is reasonable to believe that man-to-man transmission plays an important part in *rhodesiense* infections. Outbreaks of *gambiense* infections tend to be widespread and prolonged, whereas *rhodesiense* outbreaks are limited and brief. In part, at least, this is because natives infected with *rhodesiense*, by the time the trypanosomes appear in their blood, are far too ill to venture into the "bush" where *Glossina morsitans* could suck up their parasites, whereas the mild, prolonged course of the *gambiense* disease enables infected people to carry on with their usual occupations for months or years, and to serve as sources of infection for tsetses. Studies on persons who have migrated from African trypanosomiasis areas to Europe have shown that the protozoans may reside quietly in the body for several years, symptoms appearing when the individual suffers some other infection, nutritional deficiency, or even physical fatigue.

The habits and characteristics of the vectors of the two infections also contribute. The *gambiense* vectors and man both require water; therefore in dry seasons in poorly watered areas the concentration of flies and man in limited areas provides ideal conditions for the spread of a parasite that does not easily infect tsetses and is not easily transmitted by them. The people spread the infection along trade routes, by local travel, or by movement away from a decimated locality. This, according to Morris (1951), accounts for the great epidemics in the interior savanna woodlands of West Africa and their absence in the wetter forested coastal areas, where the flies, though abundant, are free to wander, and may only occasionally bite another person after picking up an infection. In East Africa the vectors of *rhodesiense* infections are not so dependent on water, and concentrations of man and flies in the same place is of less frequent occurrence, hence the lesser frequency and more limited nature of the outbreaks.

The disease. The course of the disease caused by trypanosome infection is insidious and irregular. The Gambian and Rhodesian diseases are essentially alike in their symptoms and in the course they run, except that the latter is usually more rapid in development and more virulent in effect, as a rule causing death within 3 or 4 months after infection, often without the enlarged glands so characteristic of *gambiense* infections. The variety of the Gambian disease found in Nigeria is comparatively mild and of long duration.

The bite of an infected tsetse fly is usually followed by itching and irritation near the wound, and frequently a local, dark-red, buttonlike lesion develops, occasionally increasing to considerable size. After a few days, fever and headache develop, recurring at irregular intervals for weeks or even months, accompanied by increasing weakness, enlarged glands, and usually some edema, and a markedly lowered resistance to other diseases. Often a peculiar tenderness of the muscles is complained of also. Usually an irritating rash breaks out during the early stages of the disease. Sometimes for long intervals trypanosomes are so sparse in the blood that they can be detected only by animal inoculation. Loss of ambition and vitality usually figures prominently, and childbirth is seriously interfered with.

It is possible that after weeks or months or years of irregular fever and debility the disease may spontaneously disappear and never become more than trypanosome fever. Usually, however, the parasites ultimately succeed in penetrating to the cerebrospinal fluid of the brain and spinal cord, and "sleeping sickness" results.* In the Rhodesian disease the central nervous system is usually invaded early, but in the Gambian disease this is a late manifestation and may appear from a few months to at least seven years after the onset of symptoms. The invasion is accompanied by a striking accumulation of round cells in and around the walls of vessels in the brain and by characteristic increases in the cells of the cerebrospinal fluid.

Sleeping sickness is ushered in by an increase in the general physical and mental depression. The victim wants to sleep constantly and lies in a stupor; his mind works very slowly, and even the slightest physical exertion is obnoxious. Eventually the sleepiness gets such a hold on him that he is likely to lose consciousness at any time and even neglects to swallow his food. After weeks of this increasing drowsiness his body becomes emaciated, a trembling of the hands and other parts of the body develops, with occasional muscular convulsions and sometimes maniacal attacks. He finally passes into a state of total coma ending in death, or death may end the unhappy condition earlier during an unusually intense convulsion or fever, or through the agency of some complicating disease. If untreated, death is probably the inevitable outcome. A large percentage of infections occurs among people of middle age; old people are significantly few in number in sleeping sickness districts.

Diagnosis. Diagnosis should be confirmed by finding the parasites in blood or gland juice in early cases, or in cerebrospinal fluid when symptoms or increased cells suggest involvement of the central nervous system. Often the parasites are very scanty, and inoculation of laboratory animals may have to be resorted to. Blood can be examined

* It should be noted that the so-called sleeping sickness of the United States is a totally different disease, caused by a filtrable virus.

by centrifuging it twice, just enough to throw down the red cells each time, removing the supernatant and leucocyte cream, and recentrifuging this at high speed for a long time, then examining the sediment. Sometimes the parasites are detectable by their movement in fresh, fairly thick preparations, or they may be found in Giemsa-stained thick smears.

Examination of gland juice obtained by puncturing an enlarged gland with a dry needle is usually more reliable in early cases, and in late cases centrifuged cerebrospinal fluid is more frequently positive than is the blood. Marrow obtained by sternal puncture is sometimes positive, but less often than in leishmaniasis.

Treatment. Arsenic and antimony compounds were until recently the standard drugs for treatment of trypanosomiasis, but now they are rarely used except for late stages when the parasites have invaded the central nervous system. Two drugs, Bayer 205 (also called Antrypol, Germanin, or Suramin), and pentamidine or a closely related drug, Lomidine, which are aromatic diamidines, are now most widely used for both treatment and prophylaxis of human infections; they are low in toxicity, effective in treatment, and prevent reinfection for several months.

For treatment of blood and lymph infections, five intramuscular injections of pentamidine or Lomidine at 2-day intervals, at the rate of 4 mg. of the base per kilogram of weight, have been found to produce no damage to the nervous system, liver or kidneys, and to be followed by very few relapses in over 12,000 cases in French West Africa, although the immediate transient reactions characteristic of the diamidines were observed. Antrypol has a buffer effect on the diamidines, preventing their toxic effects, so a combination of the two drugs may permit larger doses and effective treatment in 2 or 3 days, possibly in 1 day. Lourie (1953) thinks that a combination of these two drugs might be particularly advantageous in mass prophylaxis (see below). This case of two drugs, apparently acting together therapeutically, but one annulling the other's toxic effects, seems to be unique in therapeutics.

The prophylactic effect of Antrypol is due to the fact that this combines with globulin in the blood and remains in the circulation. Presumably the long-lasting effects of the diamidines and also Antrycide (see p. 158) may be similarly explained. Single injections of pentamidine or Lomidine, or two injections of Antrypol (0.5 gram followed by 1.5 grams after 2 days), prevent reinfection for about 6 months or more. In French West Africa and Belgian Congo, where prophylactic doses have been given to hundreds of thousands of people, infection rates have been reduced in the treated people by 90 to 99% after 3 to 5 treatments at 6-month intervals, and even in untreated people there has been 50 to 65% reduction.

After invasion of the central nervous system, an arsenic compound, Tryparsamide, is an effective drug but very toxic, and strains of trypanosomes resistant to it have cropped up all over Africa. Tryparsamide treatment requires a series of once-a-week injections of 0.04 gram per kilogram of body weight for 10 or 12 weeks, usually repeated once or even twice after 3-month intervals; thus a completed treatment takes as long as a year. The drug sometimes injures the optic nerve and may cause blindness.

As an alternative for cases that do not respond to Tryparsamide, melarsen oxide, a trivalent arsenic compound, was suggested by Weinman in 1946, and a somewhat detoxified derivative of it, Mel B, by Friedheim in 1949. These compounds have had extensive trial and have been shown to be highly effective. However, both drugs are very toxic and obviously unsuitable for mass treatment. They can be used effectively if the patient is hospitalized and the drug administered under careful supervision. They are most useful in cases that have failed to respond to Tryparsamide and are sure to die if left untreated. Nitrofurazone compounds and the antibiotic, Puromycin, administered with other drugs, have shown promise in treating advanced cases of *gambiense* infection. Thus far, no method of dealing with drug-fast strains of trypanosomes has been found except the use of a variety of different drugs, for the parasites obstinately retain their resistance even after passage through tsetse flies.

Prevention. Therapeutic and prophylactic treatment, including the establishment of sleeping sickness dispensaries, can certainly reduce human infection to a low level, and possibly eradicate it in the case of *gambiense* infections. However, application of such methods has its problems. In Nigeria it has been found that when the incidence of human trypanosomiasis drops there is increased difficulty in obtaining the cooperation of the native population in mass surveys and treatment. They no longer fear the disease, and native authorities slacken their efforts at persuasion. (This is reminiscent of the laxity of Americans who failed to obtain polio immunization after development of a vaccine.) Control through treatment of cases is more difficult in *rhodesiense* areas because the disease is more rapidly fatal and wild animals play a much larger role in its epidemiology. Aside from mass treatment and prophylaxis, control of human trypanosomiasis resolves itself into getting rid of tsetse flies, or reducing contacts between them and the human population, by one or more of the methods discussed on pp. 704–705. Wholesale evacuation of human inhabitants of an area, as was done in the great Uganda epidemic at the beginning of the century, is seldom warranted now, for the day has come when the fly and/or the disease, instead of man, can be ousted from a stricken area if the population is not so sparse and the land so poor that it is not worth fighting for.

One of the best demonstrations of what can be accomplished was the eradication of tsetses in the Anchau corridor, a strip 70 miles long and 10 miles wide, inhabited by 60,000 people, and its subsequent rehabilitation (see Nash, 1948).

American Trypanosomiases

In the western hemisphere, where there are no tsetse flies, certain species of trypanosomes parasitic in domestic animals, which have become independent of their ancestral hosts, have become established. These include *T. equiperdum, T. equinum,* and some closely related forms which may be identical, and *T. vivax* (see pp. 156–157). In addition, however, there are at least two trypanosomes of different type which are parasitic in man and many other animals. These are *Trypanosoma rangeli* (see p. 155), found in northern South and Central America and transmitted by triatomid bugs of the genus *Rhodnius;* and *Schizotrypanum cruzi,* found from southern United States to Argentina and transmitted by many kinds of triatomids.

SCHIZOTRYPANUM CRUZI AND CHAGAS' DISEASE

Chagas, in 1909, found that in villages in the state of Minas Gerais, Brazil, the thatched roof houses were infested with large bloodsucking bugs, *Panstrongylus megistus,* called barbeiros, and that these bugs were infected with flagellates which, when inoculated into monkeys and guinea pigs, caused acute infections. On further investigation he found that in the infested houses there were frequent cases, especially among infants and young children, of an acute disease characterized by fever, enlarged glands, anemia, and disturbances of the nervous system. In one of these cases trypanosomes were found in the blood, and in others they were demonstrated by injection of animals.

These trypanosomes, named *Schizotrypanum cruzi* (see p. 134), have since been found to parasitize many small mammals and to be common in bugs of the subfamily Triatominae all the way from the pampas of Argentina to the deserts and canyons of Southern California and Arizona, and the scrub woods and farms of Southern Texas. It has recently been reported from raccoons in Maryland, and a similar parasite has been found in Asiatic monkeys. Fortunately, human infection is less widely distributed.

A closely related species, S. *vespertilionis,* occurs in bats all over the world (Zeledon and Vieto, 1958).

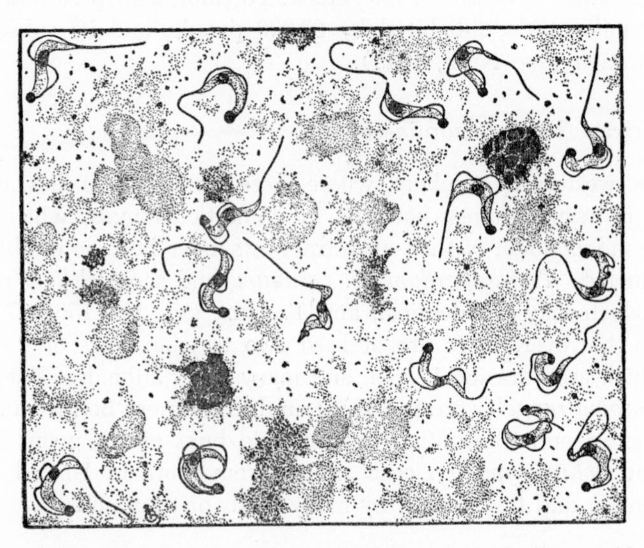

Fig. 29. *Schizotrypanum cruzi* in dehemoglobinized thick drop of blood from experimentally infected mouse. (After Brumpt, *Presse méd.,* 1939.)

The parasite. *Schizotrypanum cruzi* is a curved, stumpy trypanosome about 20 μ long, with a pointed posterior end, an elongated nucleus in the center of the body, a large egg-shaped kinetoplast close to or at the posterior end, a narrow and only slightly rippled undulating membrane, and a moderately long free flagellum (Fig. 29).

This species as found in the blood never exhibits stages in division. Greatly swollen cells, enclosing a mass of rapidly dividing parasites, varying in their number from just a few to many hundreds, are found in tissues of infected man and animals. The parasites multiply thus, especially in cells of the heart and voluntary muscles, central nervous system, and various glands. During the stage of rapid intracellular multiplication the parasites are round in form and resemble leishman bodies (Fig. 30*A*), but as they grow older a flagellum grows out, and crithidial and eventually trypanosome forms develop (Fig. 30*B*). Then the loaded cell ruptures, liberating the parasites, which invade neighboring cells or are distributed to other parts of the body by the lymph or blood system, unless destroyed by immunological reactions before they reach the safety of another cell. It is only in the early, acute stage of the disease that the parasites can live in the blood for long. In the chronic stage, when antibody reaction has occurred, the

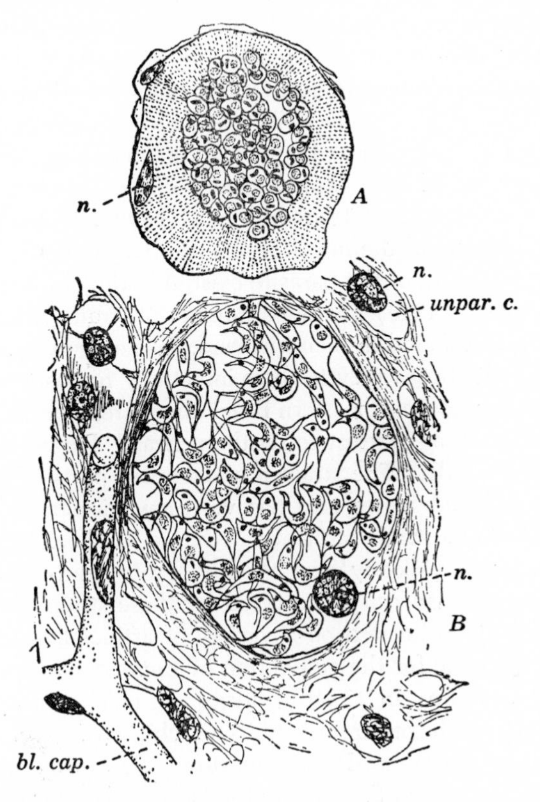

Fig. 30. *Schizotrypanum cruzi.* *A*, cyst containing *Leishmania* forms in muscle fiber of guinea pig, cross section; *n.*, nucleus of muscle fiber. *B*, older cyst, containing trypanosome forms, in neuroglia cell in gray matter of cerebrum; *n.*, nucleus of parasitized cell; *bl.cap.*, blood capillary; *unpar.c.*, unparasitized cell. ×1000. (After Vianna, *Mem. Inst. Oswaldo Cruz,* 1911.)

blood forms are rarely seen and can be demonstrated only by animal inoculation or culture, although parasites may still be abundant in tissue cells.

S. *cruzi* is a very versatile parasite. Its natural hosts appear to be armadillos, opossums, bats, racoons, and rodents; it can thrive in monkeys, marmosets, guinea pigs, rats, rabbits, cats, dogs, and other mammals, but not birds. Borsos and Warren (1959) showed that chicken serum contains an antibody that lyses S. *cruzi*. Large herbivorous animals such as horses, ruminants, and pigs are little affected, in contrast to their susceptibility to the African trypanosomes, but Brumpt et al. (1939) found that sheep, goats, and pigs could be experimentally infected, and he thinks that pigs, as well as dogs and cats, may constitute a reservoir of infection, a view later supported by the finding of a naturally infected pig in southern Brazil. Man seems to be somewhat more resistant to infection than many other mammals. Dias et al. (1945) think that wild animals are not important reservoirs of infection as far as man is concerned, although cats and dogs may be of some importance. Where human infection is common, the transmitting triatomid bugs have adapted themselves to living in houses and the infection is essentially a residential one.

A number of strains of the parasite exist, differing in their infectivity for man. Mazzoti (1940) isolated a number of strains from triatomids in Mexico which differed in their virulence for guinea pigs. In French Guiana, different strains vary in their infectivity for particular species of triatomids. In arid northeastern Brazil there are many chronic cases, but no acute ones, in contrast to southern Brazil, Paraguay, Uruguay, Argentina, and Chile. The strain found in arid parts of Mexico and southwestern United States also has very low virulence for man, so much so that only a single authenticated case has been reported from the United States (Woody and Woody, 1955) and only a few in Mexico, although man *can* be infected by these strains (see p. 152). S. *cruzi* is readily cultivated on a variety of artificial media, and is a satisfactory associate for the cultivation of *Entamoeba histolytica* (see p. 59). It develops in tissue cultures of chick embryos and in cultures of macrophages from rabbits or chickens.

Intermediate hosts, and transmission. The usual intermediate hosts and transmitters of S. *cruzi* are the large and often highly colored bugs of the subfamily Triatominae, of the family Reduviidae (see p. 608), but this parasite seems not to be very fastidious about the arthropod hosts in which it will develop. Experimentally it develops not only in all species of Triatominae in which it has been tried, but also in bedbugs, *Melophagus*, ticks, and even the body cavity of a caterpillar. In nature, however, the triatomids are probably the only transmitters of any importance. At least thirty-six species of these bugs have been found naturally infected. In southwestern United States and Mexico, about 20 to 25% of bugs are infected, 40 to 60% in Central and South America.

Even of the triatomids, there are relatively few species that are important as transmitters to man; these are the ones that habitually invade houses. These transmitters and their distribution are discussed on p. 611.

The development of *S. cruzi* in the intermediate host (Fig. 31) is similar to that of *T. lewisi* of rats, and takes place in the "posterior station." Within 24 hours the trypanosomes may pass into the intestinal portion of the midgut, where they transform into crithidia and multiply abundantly. Eventually crithidial forms pass to the rectum where small ones are found attached to the epithelium. In the rectum

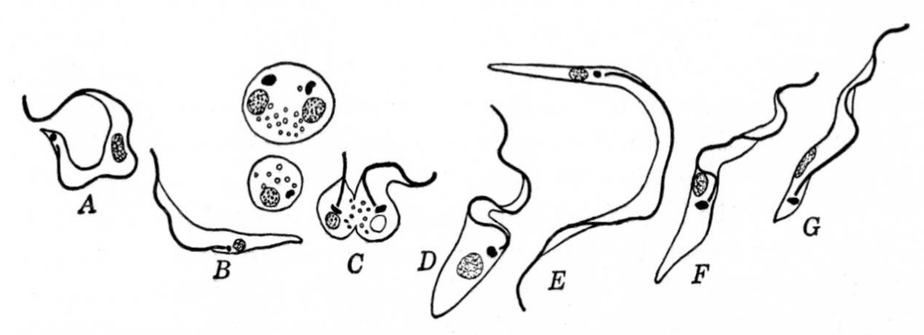

Fig. 31. Development of *Schizotrypanum cruzi* in digestive tract of *Rhodnius prolixus.* *A,* freshly ingested trypanosome; *B,* crithidia, 6 to 10 hours after ingestion; *C,* leishmania in midgut, 10 to 20 hours after ingestion; *D,* redevelopment of flagellum and undulating membrane, 21 hours after ingestion; *E* and *F,* crithidia in hindgut, 25 hours after ingestion; *G,* metacyclic trypanosome in rectum, 8 days or more after ingestion. (*A, B,* and *G* adapted from Wenyon, *Protozoology,* 1926; *C–F,* after Brumpt, *Nouveau traité de méd.,* 1925.)

they give rise to "metacyclic" trypanosomes which resemble those in vertebrate blood and are the infective forms. These appear about the sixth day in larval bugs, but not until the tenth to fifteenth in adults. As many as 3500 of these per cubic millimeter may be voided with the insect's excreta.

Normally the salivary glands do not become infective, but S. F. Wood in 1942 found trypanosomes in great numbers in the body cavity of dead bugs. Probably accidents of some sort caused the body-cavity invasions, which resulted in death of the bugs, but such bugs *might* transmit the infection by their bites before dying.

Transmission usually comes from contamination of mucous membranes or skin with infected excreta. Human infection seems most frequently to come from rubbing the eyes after a bite on the lids, presumably contaminating the mucous membranes or conjunctiva with feces deposited by the bug while feeding, or squeezed out by slapping. Animals can become infected by eating the bugs or licking their bites.

Cats can be infected by eating infected rodents. The infection can also spread from bug to bug by cannibalism or ingestion of liquid feces (see pp. 613–614). Once infected, a bug remains so probably for the rest of its life.

Human infection. Human infection with Chagas' disease occurs all the way from Mexico to northern Argentina but is by no means evenly distributed. It is an odd fact that only one human infection has been observed in the United States, in spite of the fact that S. *cruzi* has been found in many species of triatomids in the Southwest and frequently in a high percentage of individual bugs even in houses (see p. 612). That the Texas strain is infective for man is indicated by a successful experimental transmission by Packchanian (1943). In South America the disease is most prevalent in areas where species of triatomids live like bedbugs in houses and habitually suck human blood. Since the bites of triatomids are not infective unless contaminated by the bugs' feces, only constant exposure to bites, especially during sleep, when they are unconsciously rubbed, is likely to lead to infection, and then only if the bugs regularly defecate during or immediately after a meal, as the "domestic" South American species do. It seems probable that the species of triatomids in North America are not as quick to defecate after a meal and are therefore less prone to transmit infection. This is known to be true of *T. protracta* and Shields and Walsh in 1956 reported that *T. sanguisuga* does not defecate for 20 to 30 minutes after feeding.

Only a few cases occur in Mexico and Central America, and little is heard of the disease in Colombia, Ecuador, or Peru. It is common, however, in many parts of Venezuela, French Guiana, Brazil, Bolivia, northern Chile, Paraguay, Uruguay, and northern Argentina. In some areas in Brazil, Bolivia, Chile, and Argentina 10 to 20 per cent of the inhabitants show evidence of infection. In Venezuela Torrealba estimated that there were a million human infections, constituting the principal cause of myocarditis in rural parts of that country, as it apparently does in Brazil. In the northern part of Chile 12% give positive Machado reactions.

There seems to be a rather high natural resistance to acute infections; these occur most frequently in infants and children in whom resistance is weakened by a goiterous condition, malnutrition, or chronic malaria or other chronic diseases. However, acute cases do occur in individuals who are otherwise apparently healthy. The occasional finding of parasites in the blood or by xenodiagnosis in unsuspected cases, high incidence of positive Machado reactions (see p. 154), and frequency of characteristic electrocardiographic changes indicate that the infection is much more frequent than was formerly thought.

When Chagas first discovered human trypanosome infections in the regions of endemic goiter in Minas Geraes, he believed that the goiter, with all its sinister consequences—myxedema, infantilism, cretinism, etc.—was caused by the trypanosome infection, through a supposed toxic effect on the thyroid gland. But it is now evident that Chagas got the cart before the horse—acute trypanosome infections were the result of the goiter, not vice versa.

Transmission of S. *cruzi* in blood transfusions can occur. Addition of gentian violet to blood is reported to render it safe for transfusion. Prenatal infection has been reported in humans and dogs and a few cases of the disease have apparently been acquired with the mother's milk.

The disease. Acute cases of Chagas' disease are especially common in infants or young children. Frequently the disease begins with an edematous swelling of the eyelids and conjunctiva (Fig. 32) and sometimes other parts of the face, usually only on one side. This is accompanied by inflammation of the tear gland and swelling of lymph glands of the neck. These symptoms suggest that the eye may be the usual site of inoculation, the insect biting the lids, and the victim then rubbing infected excreta from the bug into his eye. The swelling, called a primary "chagoma," is caused by an inflammatory exudation in the locality where the parasites are colonizing in tissue cells, particularly in subcutaneous fat cells. Later other chagomata may appear in distant parts of the body; they may be conspicuous or only detectable by palpation. During early days of the disease there may be severe headache and marked prostration, with more or less continuous fever. After this acute stage subsides the disease goes into a chronic stage, which many authors believe persists for life, but there may first be a long latent period with few or no symptoms, during which the disease is insidiously progressing. Characteristic symptoms in the chronic stage are extensive hard edema, inflamed lymph glands, and enlarged liver and spleen. In protracted cases there is a progres-

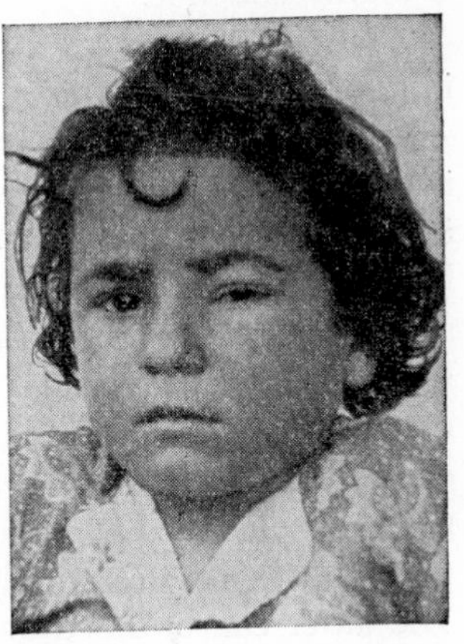

Fig. 32. Acute case of Chagas' disease in 5-year-old Brazilian girl, with trypanosomes easily seen in blood. (Photograph supplied by Emmanuel Dias.)

sive anemia and sometimes nervous disturbances. In severe cases death may occur in two or three weeks.

Disturbance of the heart is very common and according to Chagas is the commonest chronic manifestation in man. Nearly all fatal cases show injury to the heart muscle, which is one of the favorite tissues attacked by the parasite. The injury causes separation of the cells, inflammatory infiltration by phagocytic cells, and increase of fibrous tissue, which weakens the heart in chronic cases. Electrocardiograms show characteristic types of heart blocks and other abnormalities. In Brazil Dias (1953) found heart abnormalities in 16% of electrocardiograms in a place where positive serological tests ran over 50%, but in only 6% in a nonendemic area. According to Koberle, dilatation of the gut (megacolon or mega-esophagus) is a common complication in Chagas' disease, described as being due to destruction of nervous tissue associated with maintenance of smooth muscle tonus. Old-age security is a minor problem in areas where Chagas' disease is highly endemic.

Diagnosis. In acute cases a diagnosis can usually be made by finding the parasites in direct blood smears, but in some acute and most subacute cases such slow or tiresome methods as blood cultures, large inoculations of blood into susceptible animals, or xenodiagnosis (feeding of uninfected bugs on the patient) have been resorted to. Muniz, however, showed in the late 1940's that a precipitin test is helpful in such cases, using a polysaccharide extract of cultured trypanosomes as antigen; this method is simple, rapid, and reliable, since no cross-reactions are obtained in leishmaniasis cases. In chronic cases a complement fixation test, called the Machado reaction, is useful; negative reactions rule out all but very early stages of infection, and strongly positive results occur only in S. *cruzi* and *Leishmania* infections. The latter can be ruled out by clinical examination or by the formol-gel test (see p. 123), which is not positive in Chagas' disease. The large size of the kinetoplast of S. *cruzi* makes it distinguishable from other trypanosomes in blood smears, and from *Leishmania* in fixed tissues.

Treatment and prevention. No successful method of treatment is known. Mazza et al. say that Bayer 7602 relieves acute cases, and Pizzi et al. (1953) found Primaquine and an antibiotic, Puromycin, to be helpful by suppressing the parasites temporarily, thus enabling the host to develop resistance. For prevention, good results have been obtained by dusting or spraying with Lindane or dieldrin. The latter has good residual action. Rebuilding of better houses is being urged everywhere in Brazil and with good results. The town of Belo-

Horizonte, for example, is said to have been nearly freed from Chagas' disease by remodeling of the houses. People accidentally bitten by triatomids, if conscious of it, should avoid possible contamination of the bite by the feces of the bug.

Trypanosoma rangeli

A trypanosome which is infective for man has been found to be harbored by a considerable percentage of *Rhodnius prolixus* in Venezuela, Colombia, and Guatemala. It was named *T. rangeli* when discovered in *R. prolixus* in Venezuela by Tejera in 1920. Later a parasite which is now considered identical was found in four babies in Guatemala and named *R. guatemalensis*. Later, infections were found in a considerable number of human cases, and also in dogs, in Venezuela, along with *S. cruzi* (Pifano, 1948). In 1951 Groot et al. described what is apparently the same parasite in a large number of people (67 of 183 examined) in the Ariari Valley in Colombia, most of them without *S. cruzi*. They described the trypanosome under the name *T. ariarii* which Groot later concluded was identical with *T. rangeli*. Subsequently, the trypanosome was found in humans in Salvador and Costa Rica. Zeledon (1954) considered *T. cebus*, described by Floch and Abonnenc from monkeys in French Guiana, to be identical with *T. rangeli*.

These trypanosomes, as they occur in the blood, are characterized by a small kinetoplast, rippled undulating membrane, anteriorly placed nucleus, and large size (see key, p. 136). Herbig-Sandreuter (1957) made an exhaustive study of the tissues of animals experimentally infected with *T. rangeli* and was unable to find developmental stages in the tissues nor was there any evidence of pathology. Contrary to statements by some authors, there is no reliable evidence that *T. rangeli* has a leishmania stage in the vertebrate host nor has any other pathology in the vertebrate host been attributed to it.

These parasites develop in the gut of *Rhodnius prolixus* and in at least one other species of *Rhodnius*, later invading the hemocoele and salivary glands of the insect. They produce pyriform, crithidial, and metacyclic forms, distinguishable from comparable stages of *S. cruzi* by the small kinetoplast and the extraordinary length attained by some of the crithidias—as much as 100 μ (Fig. 33). These trypanosomes are uniquely different from *S. cruzi* in invading the hemocoele and salivary glands of the host bug and in being commonly transmitted to the vertebrate by the bite of the insect. Since they are found in the posterior gut, as well as in the salivary glands, transmission may also occur by fecal contamination of the bite as in *S. cruzi*. Astonish-

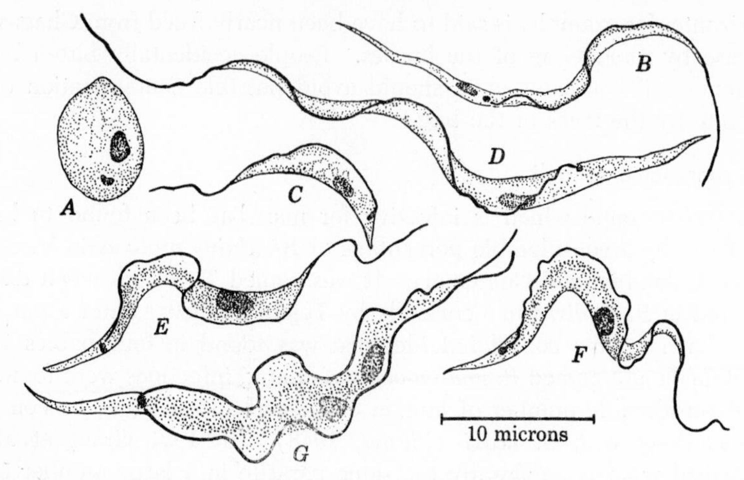

Fig. 33. *Trypanosoma rangeli* (= *ariarii*). *A* and *B*, from cultures; *C*, from anterior part of gut of *Rhodnius; D*, from posterior part of gut of *Rhodnius; E* and *F*, from human blood; *G*, dividing form from blood of experimentally infected mouse. (After Groot, Renjifo, and Uribe, *Am. J. Trop. Med.*, 1951.)

ingly, Grewal (1957) showed that *T. rangeli* is highly pathogenic in the intermediate host. This seems to be the only known instance in which a trypanosome is apparently nonpathogenic in the vertebrate but is pathogenic in the invertebrate host. *T. rangeli* is infective for dogs, opossums, anteaters, and baby mice, as well as monkeys and man.

Trypanosomiasis of Animals

As noted at the beginning of this chapter, tsetse-borne trypanosome infections, by making it impossible to keep domestic animals, particularly cattle, where even small numbers of tsetse flies are present, have been a tremendous impediment to the development of Africa. Some trypanosomes of animals have become partly or completely independent of tsetse flies, and have become a scourge in parts of the world far distant from the home of the tsetses.

Species

There are four groups of trypanosomes that affect domestic animals. (For morphological characters, see key, p. 136, and Fig. 25.)

1. The *brucei* group includes *brucei, gambiense,* and *rhodesiense,* which invade the salivary glands of the tsetse fly (Fig. 27) (see p. 141); and also three species that have become independent of their

ancestral tsetse hosts: *evansi* and *equinum,* which are transmitted mechanically by *Stomoxys* and tabanids, or even vampire bats, and are found in all tropical countries; and *equiperdum,* which has gone all the way in becoming independent of insects and depends on venereal transmission.

2. The *congolense* group includes *congolense, simiae,* and *dimorphon,* which in the tsetse fly pass forward from the stomach to the labial cavity but do not invade the salivary glands.

3. The *vivax* group includes *vivax* and *uniforme,* which develop only in the proboscis of tsetse, and do not survive in the stomach.

4. *T. theileri* is a large, nonpathogenic species in cattle which develops in the posterior station in tabanids, and *T. melophagium* is a similar parasite of sheep and sheep "ticks," *Melophagus ovinus.*

These species differ in their morphology (see Fig. 25), in their effects on various animal species, and in their response to drugs, as well as in their development in tsetses. The disease caused by the tsetse-borne African species is called nagana; by the mechanically transmitted *evansi,* surra; by *equinum,* mal-de-caderas; and by the venereally transmitted *equiperdum,* dourine.

T. congolense is the most destructive to cattle. It causes a slowly progressing anemia which is nearly always fatal to adult cattle, though calves frequently recover and have "cryptic" infections. Horses, sheep, goats, camels, and dogs are also susceptible, but pigs are very little affected. Strains of the parasite and adequacy of nutrition of the host are important factors in determining the outcome of this as of the other trypanosome infections. This species lives its whole life in the capillaries, producing its effects almost entirely by blocking capillaries and causing anemia.

T. simiae causes acute, rapidly fatal infections of pigs and camels, is chronic in monkeys, but is of little importance to other animals. This species produces disease in the same way as *congolense.*

T. brucei causes a very serious and usually fatal disease of horses, sheep, goats, camels, and dogs, but a very mild disease in cattle. This parasite injures the walls of capillaries and passes through into the intercellular tissue spaces; no organ is immune from its damage.

T. vivax, unlike those mentioned previously, although cyclically transmitted in tsetses, can survive outside tsetse fly areas, where it depends on mechanical transmission by *Stomoxys* and tabanids; it has become established not only in tsetse-free areas in Africa, but also in tropical America. It causes disease in all large domestic animals except dogs, but different strains vary greatly in their pathogenicity, from rapidly fatal infections to chronic or cryptic ones. Like *brucei,* it leaves the capillaries and invades the tissues. The related *T. uniforme* is very much like *vivax,* but seems to be relatively harmless to sheep and goats.

T. evansi, the cause of surra, is essentially a disease of camels, but it also causes acute and fatal disease in horses and dogs, chronic in donkeys, and transient in cattle. Its effects are quite similar to those of *brucei,* from which it and *equiperdum* have doubtless both been derived. *T. equinum* and some related or identical forms in South America are probably mere strains of *evansi* which have lost the kinetoplast (see p. 133).

T. equiperdum is the cause of dourine, a venereal disease of horses. In addition to the anemia, edema, and wasting caused by all forms of trypanosomiasis in animals, this one causes acute inflammation and edema of the penis and vagina, and plaquelike skin eruptions.

Treatment and prophylaxis. Formerly animals were treated, with only fair success, with antimony compounds and antrypol, but in recent years two new drugs, dimidium bromide (phenanthridium) and Antrycide in the form of methyl sulfate or chloride salts, have greatly brightened the picture. Antrycide is not equally effective for all the species of trypanosomes, being much more effective against *congolense* than against *vivax,* somewhat less against *brucei, evansi,* and *equiperdum,* and rather ineffective against *simiae.* However, in tolerated doses of 5 mg. per kilo of the more soluble methyl salt it is curative for cattle and horses, and in the less soluble chloride form a single dose gives protection against infection for 2 to 8 months. A mixture of the two salts, called a "pro-salt," given subcutaneously combines therapeutic and prophylactic virtues; it has been claimed, however, that the effect of the drug is suppressive rather than prophylactic. Dimidium bromide is more expensive but is effective in smaller doses (1 to 1.5 mg. per kilogram), but in some animals it causes photosensitization and is ineffective against *brucei, evansi,* and *simiae.*

These drugs are especially valuable when animals are temporarily exposed to fly infections, e.g., in transit through fly belts, dry-season grazing in fly areas, or in connection with clearance schemes where tsetses are seasonally present.

One of the darkest clouds on the horizon now is the fact that Antrycide-resistant strains of trypanosomes, particularly of *congolense,* can and do develop, which threatens the dream of Africa as a vast new home for contented cows.

REFERENCES

Brand, T. von. 1951. Physiology of blood flagellates. Chapter 7 in *Parasitic Infections in Man,* H. Most (Editor). Columbia University Press, New York.

Brumpt, E. 1939. La maladie de C. Chagas. *Presse méd.,* 47: 1013–1015, 1081–1085.

Buxton, P. A. 1948. *Trypanosomiasis in Eastern Africa, 1947.* Colonial Office, London.

Chagas, C. 1909. Ueber eine neue Trypanosomiasis des Menschen. *Mem. inst. Oswaldo Cruz,* 1: 158–218.

Chandler, A. C. 1958. Some considerations relative to the nature of immunity in *Trypanosoma lewisi* infections. *J. Parasitol.,* 44: 129–135.

Davey, D. G. 1950. Experiments with "Antrycide" in the Sudan and East Africa. *Trans. Roy. Soc. Trop. Med. Hyg.,* 43: 583–616.

Davey, T. H. 1948. *Trypanosomiasis in British West Africa.* Colonial Office, London.

Dias, E. 1951–1953. Doença de Chagas nas Americas. *Rev. Brasil. Malariol. e Doenças Trop.,* 3: 448–472 (U. S.), 555–570 (Mexico); 4: 75–87 (Central America), 255–280 (Colombia, Venezuela and Guianas), 319–325 (Ecuador and Peru); 5: 11–16 (Bolivia and Paraguay), 131–136 (Chile).

Dias, E., and Chandler, A. C. 1951. Human diseases transmitted by parasitic bugs (Portuguese and English). *Mem. inst. Oswaldo Cruz,* 47: 403–441.

Dias, E., and Laranja, F. S. 1948. Chagas' disease and its control. *4th Intern. Congr. Trop. Med. Malaria,* 2: 1159–1167.

Grewal, M. S. 1957. Pathogenicity of *Trypanosoma rangeli* Tejera, 1920, in the invertebrate host. *Exp. Parasitol.,* 6: 123–130.

Groot, H. 1952. Further observations on *Trypanosoma ariarii* of Colombia, South America. *Am. J. Trop. Med. Hyg.,* 1: 585–592.

Herbig-Sandreuter, A. 1957. Further studies on *Trypanosoma rangeli* Tejera, 1920. *Acta Trop.,* 14: 193–207.

Hornby, H. E. 1950. *Animal Trypanosomiasis in Eastern Africa,* 1949. Colonial Office, London.

Laranja, F. S., et al. 1956. Chagas' disease: A clinical, epidemiologic, and pathologic study. *Circulation,* 14: 1035–1060.

Lourie, E. M. 1951. Combined treatment by Suramin (Moranyl) and Pentamidine, providing a situation unique in therapeutics and with potential advantages in the prophylaxis of sleeping sickness. *Bur. Perm. Interafr. de la Tsétsé et de la Tryp.,* No. 160/0.

Mazza, S., et al. 1939–1942. Investigaciones sobre enfermedad de Chagas. *Univ. Buenos Aires: Missión de estudios de patologia regional argentina* (Jujuy), *Pub.,* 42–63.

Mazzoti, L. 1940. Variations in virulence for mice and guinea pigs in strains of *Trypanosoma cruzi* Chagas from different species of bugs (Triatomidae) from different localities in Mexico. *Am. J. Hyg.,* 31:(C) 67–85.

Morris, K. R. S. 1949. Planning the control of sleeping sickness. *Trans. Roy. Soc. Trop. Med. Hyg.,* 43: 165–198.

1951. The ecology of epidemic sleeping sickness. *Bull. Entomol. Research,* 42: 427–443.

Nash, T. A. M. 1948. *The Anchau Rural Development and Settlement Scheme.* Colonial Office, London.

Noble, E. R. 1955. The morphology and life cycles of trypanosomes. *Quart. Rev. Biol.*, 30: 1–28.

Packchanian, A. 1939–1940. Natural infection of *Triatoma gerstakeri* with *Trypanosoma cruzi* in Texas. *Public Health Repts.*, 54: 1547–1554; Natural infection of *Triatoma heidemanni* with *Trypanosoma cruzi* in Texas. *Public Health Repts.*, 55: 1300–1306.

1943. Infectivity of the Texas strain of *Trypanosoma cruzi* to man. *Am. J. Trop. Med.*, 23: 309–314.

Pifano, C. 1948. Nouvelle trypanosomiase humaine de la région neotropicale produite par le *Trypanosoma rangeli* Tejera, 1920. *Bull. Soc. Pathol. Exotique*, 41: 671–681.

Pizzi, T., Prager, R., and Knierim, F. 1953. Ensayos quinoterapia de la enfermedad de Chagas experimental. XII. Acción de la puromycina sola y associada a la primiquina. *Bol. inform. Parasitol. chilenas,* 8: 77–79.

Scott, D. 1957. The epidemiology of human trypanosomiasis in Ashanti, Ghana (Gold Coast). *J. Trop. Med. Hyg.*, 60: 205–215; 238–249; 257–274; 302–315.

Taliaferro, W. H. 1926. Host resistance and types of infections in trypanosomiasis and malaria. *Quart. Rev. Biol.*, 1: 246–269.

Veatch, E. P., Bequaert, J. C., and Weinman, M. D. 1946. Human trypanosomiasis and tsetse-flies in Liberia. *Am. J. Trop. Med.*, Suppl., 26(5): 1–105.

Weinman, D. 1950. The Trypanosomiases of Man. Reprinted from Oxford Medicine, Vol. 5, Chap. 35.

Wood, S. F. 1951. Importance of feeding and defecation times of insect vectors in transmission of Chagas' Disease. *J. Econ. Entomol.*, 44: 52–54.

Wood, F. D., and Wood, S. F. 1941. Present knowledge of the distribution of *Trypanosoma cruzi* in reservoir animals and vectors. *Am. J. Trop. Med.*, 21: 335–345.

Woody, N. C., and Woody, H. B. 1955. (Chagas' disease) American trypanosomiasis. First indigenous case in the United States. *J. Am. Med. Assoc.*, 159: 676–677.

Zeledon, R. 1954. Tripanosomiasis rangeli. *Rev. Biol. Trop.*, 2: 231–268.

Zeledon, R., and Vieto, P. L. 1958. Comparative study of *Schizotrypanum cruzi* Chagas, 1909, and *S. vespertilionis* (Battaglia, 1904) from Costa Rica. *J. Parasitol.*, 44: 499–502.

Chapter 8

THE SPOROZOA

I. MALARIA

The Sporozoa include a large and varied assemblage of Protozoa which have little in common except a parasitic mode of life, the lack of any organs of locomotion during most stages of their development, and the evolution of a complicated life cycle usually involving an alternation of generations and the production of resistant stages which in some cases are called spores. There is little foundation to support the inclusion in a single group of such diverse parasitic organisms as the Coccidia, haemosporidians, and gregarines, constituting the class Telosporidea; the Myxoporida and Microsporida, parasites of fishes and arthropods, constituting the class Cnidosporidea; and the Haplosporidia, inhabiting many hosts from rotifers to fishes. However, since most protozoologists prefer a single conglomerate mess to several groups of very uncertain relationships, we shall follow along and include all these varied organisms in the subphylum Sporozoa (see p. 49).

Classification

All the parasites which concern us here fall in the class Telosporidea, which, following Jahn and Jahn, and Hall, we shall classify in this way:

Class **Telosporidea** (for characteristics, see p. 49).

Subclass 1. **Gregarinidia.** Typically parasites of digestive tract and body cavities of invertebrates, where trophozoites usually live free, generally as elongated, spindle-shaped organisms. These eventually form gametocytes either by direct transformation or after one or more schizogonic multiplications. Gametocytes associate in pairs, undergo syngamy, and then usually encyst, forming an oocyst in which sporozoites are produced. The oocyst or "spore" is the usual means of transfer.

Subclass 2. **Coccidiida.** Typically parasites of epithelial cells of invertebrates and vertebrates, intracellular throughout most of the life

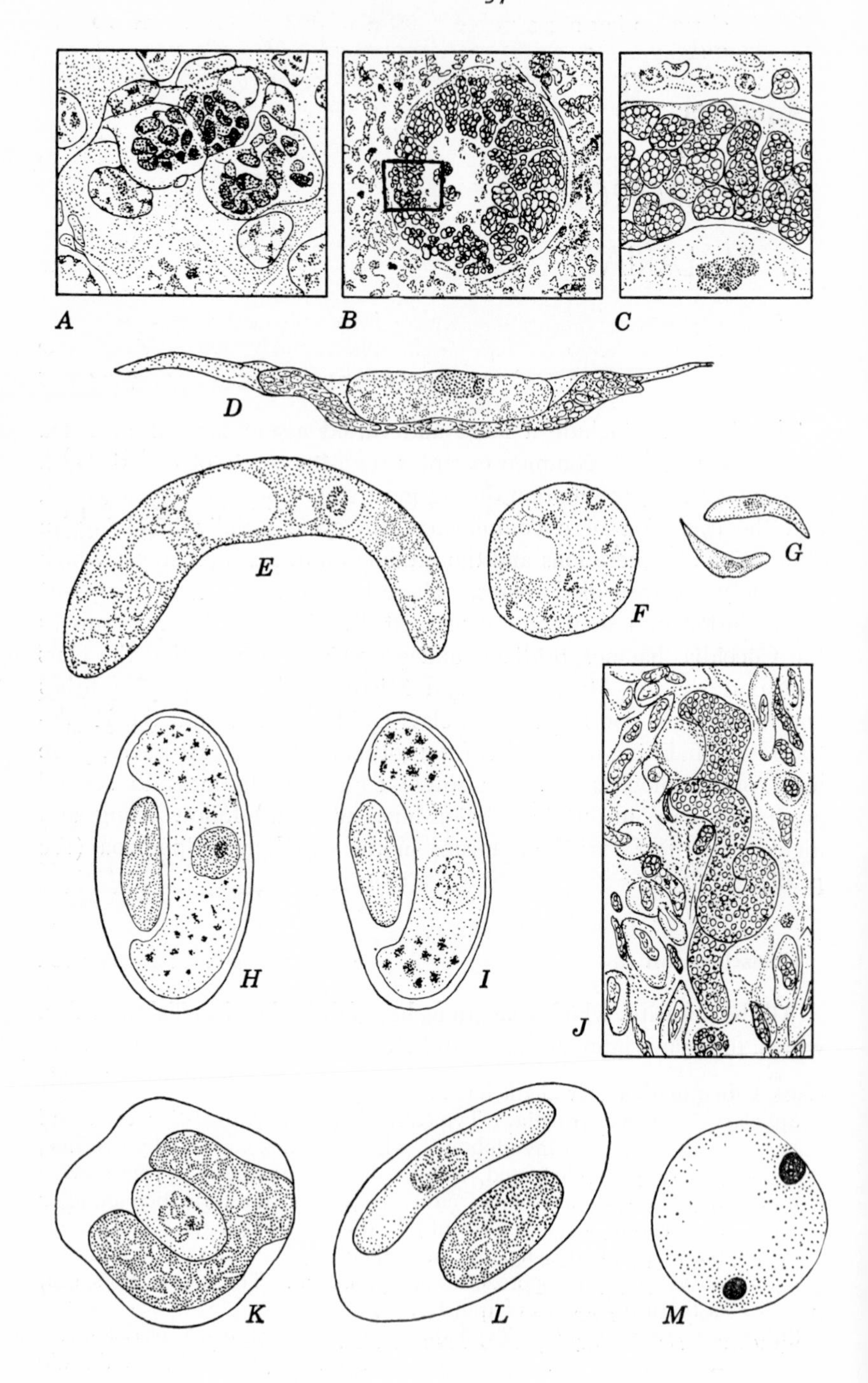
A
B
C
D
E
F
G
H
I
J
K
L
M

cycle. Sporozoites, naked or in sporocysts or oocysts, are ingested and enter tissue cells, where they multiply by repeated schizogony. Last generation of merozoites develop into gametocytes. These form dissimilar gametes which undergo syngamy, forming zygotes enclosed in cysts (oocysts) which do not grow in size. In these sporozoites are produced, usually in intermediate products of multiplication, the sporocysts. Intermediate host may or may not be involved.

Order 1. **Adeleida.** Gametocytes dissimilar, associated during development; microgametes few. Two suborders: (1) *Adeleina*, with inactive zygote and nongrowing oocysts; includes parasites of invertebrates and one, *Klossiella*, of mice and guinea pigs. (2) *Haemogregarinina*, with motile zygotes and growing oocysts in an intermediate host. *Haemogregarinina* undergoes schizogony in fixed cells and gametocytes (Fig. 34*L*) enter erythrocytes of turtles or snakes, sexual phase in leeches and arthropods. In *Hepatozoon* the gametocytes enter leucocytes of birds or mammals (Fig. 34*K*), and the sexual phase is in arthropods.

Order 2. **Eimeriida.** Gametocytes similar, not associated during development; microgametes numerous. Fertilization in oocysts; latter do not grow, but produce sporozoites, usually in sporocysts. Most have no intermediate hosts. Includes the Coccidia of vertebrates (family Eimeriidae) (see pp. 205–212).

Subclass 3. **Haemosporidia.*** Schizogony in fixed tissue cells, with or without subsequent schizogony in erythrocytes. Gamonts or gametocytes in red or white blood cells. Fertilization in arthropod host. Zygote motile; oocyst grows and produces many sporozoites. Asexual phase in lizards, birds, or mammals, sexual phase in arthropods.

Order 1. **Plasmodiida.** Gametocytes in blood cells of vertebrates; formation and syngamy of gametes in arthropod hosts; pigment granules deposited when red cells invaded. Includes true malaria parasites (*Plasmodium*), and allied genera such as *Haemoproteus* and *Leucocytozoon.*

* Some recent classifications place the Haemosporidia in the subclass Coccidiida.

Fig. 34. Miscellaneous blood-inhabiting organisms. (Adapted from various authors.)

Leucocytozoön simondi, A–G: A, early developing schizont in liver of duck, ×1640; *B*, later stage of megaloschizont in liver of duck, ×410; *C*, enlarged portion of *B*, ×1180; *D*, gametocyte in duck blood, ×1640; *E*, oökinete in stomach of *Simulium* (magnification not given); *F*, oöcyst in stomach, ×2500; *G*, sporozoites, ×2500, also in stomach.

Haemoproteus columbae, H–J: H and *I*, ♀ and ♂ gametocytes, respectively, in pigeon erythrocytes; *J*, schizont in lung capillaries of pigeon.

Hepatozoön muris, K: gametocyte in mononuclear leucocyte of rat.

Haemogregarina sp., *L:* gametocyte in erythrocyte of turtle (differs from *Haemoproteus* in having no pigment granules).

Anaplasma marginale, M: in erythrocyte of cattle.

Order 2. **Babesiida.** Unpigmented forms in red corpuscle, dividing into groups of 2 or 4, rarely more, these probably being gamonts, which develop into gametocytes and produce gametes in ticks; with or without schizogony in various vertebrate tissues. Life cycles imperfectly known. Contains the family Babesiidae with the genus *Babesia*, the family Theileridae with the genus *Theileria*, and the family Gonderidae including the genera *Gonderia* and *Cytauxzoon*, all parasitic in mammals and birds (see p. 196).

Of uncertain status: *Toxoplasma, Besnoitia, Sarcocystis.*

Formerly another group of minute parasites of blood corpuscles—*Bartonella, Anaplasma, Eperythrozoön,* and a few others—was included in the Protozoa, attached as "riders" to Haemosporidia, but these are no longer regarded as Protozoa (see p. 218).

Classification of Malarialike Parasites

In this chapter we shall consider only the malaria parasites and their allies, i.e., the order Plasmodiida as previously given. This group includes numerous parasites which occur in lizards, birds, and certain mammals. Primates, rodents, shrews, bats, and hippopotami are the mammals thus far known to harbor them. Only members of the genus *Plasmodium,* which includes the human malaria parasites, undergo schizogony in red blood cells; the others undergo this multiplication only in fixed or (in bat malaria) in fixed *and* wandering macrophages. Before the exo-erythrocytic (e.e.) stages of *Plasmodium* became recognized as essential parts of the life cycle, this genus was segregated into a separate family of its own (Plasmodiidae) and the others grouped in a family, Haemoproteidae, but there no longer seems justification for this. Attempts to classify these organisms on the basis of type of exo-erythrocytic development or of their arthropod hosts have not been successful either.

At the present, we may recognize the following genera:

1. ***Plasmodium:*** parasites of lizards, birds, and mammals; gametocytes in red cells; erythrocytic schizogony; e.e. forms in various tissues, but always solid or at most vacuolated bodies (Fig. 38); arthropod hosts mosquitoes (*Anopheles* for mammalian species; culicines, or rarely *Anopheles*, for bird and lizard species).

2. ***Leucocytozoon:*** parasites of birds; no erythrocytic schizogony; gametocytes in round or elongated lymphocytes or red cells (Fig. 34*D*); e.e. forms (Fig. 34*A–C*) in parenchyma of liver, heart, or kidney, forming large bodies divided into cytomeres; arthropod hosts, as far as known, blackflies (*Simulium*), with development in stomach (Fig. 34*E–G*).

3. ***Haemoproteus:*** parasites of birds and reptiles; no erythrocytic schizogony; gametocytes halter-shaped in red cells (Fig. 34*H, I*); e.e. forms in

endothelium of blood vessels, especially in lungs (Fig. 34*J*); arthropod hosts, as far as known, Pupipara.

4. ***Hepatocystis:*** parasites of lower monkeys, squirrels, bats, and hippopotami; no erythrocytic schizogony; gametocytes in red cells; e.e. forms cyst-like, very large, in liver parenchyma; arthropod hosts unknown.

5. Species assigned to ***Plasmodium*** until more is known about them: parasites of bats and elephant shrews; no erythrocytic schizogony; gametocytes in red cells; e.e. forms in one bat species in fixed and wandering macrophages and reticular cells of bone marrow; arthropod hosts unknown.

Malaria and Malarialike Diseases in Animals Other than Man

Haemoproteus infections. The elongated, halter-shaped gametocytes (halteridia) (Fig. 34*H, I*) of species of *Haemoproteus*, curving around the nuclei of the red blood cells, have been found in the blood of numerous species of birds (see Herman, 1944). One species, *H. columbae,* is very common in pigeons, although it does not do them very much harm. The life cycle of three species have been worked out. In two cases, pupiparous flies of the family Hippoboscidae and in the other flies of the genus *Culicoides* serve as intermediate hosts.

Leucocytozoon infections. The gametocytes of *Leucocytozoon* (Fig. 34*D*) also are found in the blood of many species of birds. Two species, *L. simondi* (=*anatis*) of ducklings and *L. smithi* of turkeys, cause serious losses. The schizogony stages are of two kinds—small schizonts in parenchyma cells of the liver (Fig. 34*A*) and "megaloschizonts" (Fig. 34*B,C*) up to over 150 μ long, and containing over a million merozoites, in the heart, liver, and spleen in macrophages and heart muscle. The intermediate hosts are species of blackflies (Simuliidae). In the case of *L. simondi,* at least, the whole sexual cycle, including development of the sporozoites (Fig. 34, *E–G*), usually occurs *in* the stomach instead of on its outer wall, and the salivary glands are not commonly invaded.

***Plasmodium* infections of lizards and birds.** The bird parasites of the genus *Plasmodium* are of particular interest because of their extensive use in the study of the biology and treatment of malarial parasites in general. Until the discovery of *P. berghei* in rodents in 1948, bird malaria had to be depended upon to a large extent for screening anti-malarial drugs, since monkeys were too few and too expensive to be used on such a large scale. It was in *P. gallinaceum* of chickens that exo-erythrocytic stages were found to be an essential part of the life cycle of malaria parasites.

The majority of the species of *Plasmodium* of birds are, primarily at least, parasites of passerine birds; such are *relictum, cathemerium, elongatum, circumflexum,* and many others, but *gallinaceum* is primarily a parasite of galliform birds and *lophurae* of some galliform birds and of ducks. Huff et al. (1947) showed that the behavior of these various parasites differs considerably in different hosts, so that it may not be easy to decide which are the natural ones. In unnatural hosts there may be no development at all; more or less suppression of either the blood or exo-erythrocytic stages or both; failure to produce gametocytes; or lack of sexual potency of the gametocytes. Interesting examples of these conditions have been given by Huff and Coulston (1946) and by Huff, Coulston, Laird, and Porter (1947). The bird species are transmitted by various species of *Aëdes* and *Culex,* and rarely by *Anopheles.*

There are two types of development of e.e. forms: the commoner *gallinaceum* type in which they develop in the fixed cells (macrophages and endothelial cells) of the brain, spleen, lung, and many other organs; and the *elongatum* type in which they develop in cells of the blood-forming organs and circulate in both red and white blood cells. In lizards, which harbor a number of species, both types of development occur, at least in some species.

Mammalian species of *Plasmodium*. Typical species of *Plasmodium,* i.e., those which undergo schizogony in the red blood cells and are transmitted by mosquitoes, are found among mammals only in man and monkeys with the exception of two related species, *P. berghei* and *P. vinckei,* found in African rats and transmissible to various rodents. Both the primate and rodent parasites are transmitted by species of *Anopheles* only.

Apes and monkeys harbor a number of species, some of them so closely related to the species found in man that there is question about their distinctness. It is very likely that *P. rodhaini of* chimpanzees is identical with *P. malariae* of man.* *P. cynomolgi,* which is very similar to *P. vivax,* has proved very useful in studies on the e.e. cycle, and also on drugs. *P. knowlesi,* which causes rapidly fatal infections in rhesus monkeys, causes relatively mild and temporary infections in man, and was once used for malaria therapy of syphilis. *P. berghei* is now being used experimentally in the study of mammalian malaria, but it does not show as close an affinity to the human species as the monkey parasites do.

An interesting light was thrown on host-parasite relations of malaria parasites by the observation that whereas rats are easily infected by either blood or sporozoite inoculation with *P. berghei,* mice can be

* Eyles et al. (*Science* 131: 1812) have recently reported the transmission of a *P. vivax*-like parasite from the macaque monkey to man.

infected only by blood inoculation, showing that the conditions for the erythrocytic and exo-erythrocytic cycles may not be parallel in different hosts.

Human Malaria

History and importance. Malaria, meaning bad air, was so named because of association of the disease with the odorous air of swamps, particularly at night, and fear of damp night air still exists, even in the United States. Although historians and economists have largely failed to recognize it, malaria must have played a large part in the history of the world and the progress of nations.

The malaria parasite was discovered in the blood by Laveran in 1880. In 1898 Ross experimentally proved the mosquito transmission of the disease, and worked out the details in the case of bird malaria. Immediately afterward Grassi and his pupils, working independently, described the cycle of human malaria in *Anopheles*.

Although malaria still ranks, as of this writing, as a very important human disease, present prospects are that future editions of this book will treat it as a relatively uncommon infection. The biggest public health program in human history is in progress with a battle cry: Eradicate malaria. The program has moved with jet-age speed.

A program for the eradication of malaria in the United States came into operation in 1947 and was successful by 1950. An integrated international malaria eradication program was suggested by Soper in 1950 and by 1955 a plan was accepted by the World Health Organization. By the end of 1956 considerable progress had been made in several countries. By 1958 seventy-six countries were planning, carrying out, or had completed the eradication of malaria, a truly remarkable course of events. The results of this ambitious program have been impressive. Deaths attributed to malaria numbered about 3 million in 1946, declined to 2 million in 1955, and were perhaps 1 million in 1958. A most important aspect is the economic advantage of malaria control. In India, for example, the total loss of income due to malaria amounts to 500 million dollars per year. The World Health Organization estimates that malaria eradication will cost about 114 million dollars. It is obvious that eradication is a worthwhile business venture.

Countries or territories now at or near malaria eradication are Argentina, British Guiana, Chile, France (Corsica), Cyprus, French Guiana, the Gaza strip, Italy (including Sardinia and Sicily), Martinique, Mauritius, Netherlands, Puerto Rico, Reunion, Romania, U. S. A., and Venezuela.

Even in such notoriously malarious areas as parts of Africa, India, and Ceylon, vast strides have been made in the control of this disease. In Ceylon, reduction has had the dramatic effect of arousing pessimistic warnings about liberation from disease without preparation for a balanced economy, which, they say, may as well lead to overpopulation, unemployment, communism, and war, as to peace, prosperity, and contentment. But as Russell ably pointed out, withholding public health cannot possibly restore a balance that never existed in the past and cannot in the future *without* public health.

During World War II malaria was the number 1 problem of nonimmune troops and caused tremendous havoc in the Mediterranean, India-Burma, and South Pacific theaters of operation. At the fall of Bataan, when quinine ran out, 85% of every regiment had acute malaria. In the South Pacific campaign, malaria caused more than five times as many casualties as did combat. In April 1942, the case rate on Efati was 2678 per 1000 per year, and on Guadalcanal in November 1942, two months after occupation, the rate was 1800 per 1000 per year. That was before the importance of malaria was realized, and before strict malaria discipline was developed. When it became apparent that malaria had to be licked before the Japanese could be, the job was successfully done by the combined efforts of entomologists, sanitarians, engineers, physicians, and occasional courts-martial, and malaria rates dropped to 10 to 50 per 1000 in some of the world's worst malarial areas.

Malaria in the United States. Opinions differ as to whether malaria existed in America in pre-Columbian days, but it developed soon thereafter and was a scourge during colonial and pioneering days. Of 45,713 patients admitted to Charity Hospital in New Orleans from 1814 to 1847, 43% were classed as "fevers" and 20% as "intermittent fevers." In the Civil War one command of 878 men below Savannah had 3313 cases of malaria in 14 months!

During earlier days in this country malaria was prevalent in the northern and midwestern states as well as in the South, but it died out there about the turn of the century, largely as the result of agricultural drainage, treatment, screening, better hygiene, and household spray guns. These factors were reducing the disease in the South, but there the process was greatly hastened by the development of a DDT residual-spraying program. Civilian malaria mortality decreased over 90% between 1935 and 1945, and in the early 1950's malaria ceased to exist as an endemic disease in continental United States. In 1951 the National Malaria Society voluntarily ceased to exist because its goal had been attained—surely a unique occurrence in public health history!

Species of *Plasmodium* in man. Four species of *Plasmodium* are capable of causing malaria in man. One, *P. ovale,* is very rare, though it has been found in such widely separated parts of the world as West Africa, South America, Russia, Palestine, and New Guinea. The other three species, *P. vivax, P. malariae,* and *P. falciparum,* are common and of wide distribution.

P. vivax is the commonest and most widely distributed species, being prevalent in both tropical and temperate zones. It is the cause of "tertian" or "benign tertian" malaria, though a better term is "vivax" malaria. It has a 48-hour cycle of development in man and is particularly likely to cause relapses. *P. malariae* is also widely distributed in both tropical and temperate climates, but it has a spotty distribution and is usually much less common than either *vivax* or *falciparum.* It is the cause of "quartan" malaria. It has a 72-hour cycle and causes infections of many years' duration. *P. falciparum* is very prevalent in the tropics but does not thrive as far north as *vivax* does. It has a 40- to 48-hour cycle of development and is the cause of "malignant tertian," "subtertian" or "aestivo-autumnal" malaria, now preferably called "falciparum" malaria. It causes a much more dangerous disease than the other species but runs a shorter course without relapses, seldom lasting more than 8 to 10 months without reinfection. The principal morphological and physiological differences between the human species are indicated in the table on p. 173 and in Figs. 35 and 36.

In at least two of the human species, *vivax* and *falciparum,* there are races or strains differing in their clinical course, geographic distribution, and ability to produce immunity to each other.

There is no animal reservoir for any of these human parasites except possibly chimpanzees for *P. malariae.* Malaria cannot, therefore, be acquired in uninhabited regions; for malaria to thrive there must be infected human beings and plenty of man-biting *Anopheles,* and easy contact between the two.

Life cycle. Although the general nature of the life cycle of all species of *Plasmodium* is similar, there are some marked differences between them which have very important practical consequences. The general course of the life cycle is shown in Fig. 37. Although partly hypothetical, the circumstantial evidence for its correctness is very strong.

EXO-ERYTHROCYTIC STAGES. When sporozoites are injected by a mosquito they do not enter erythrocytes at once to start their develop-

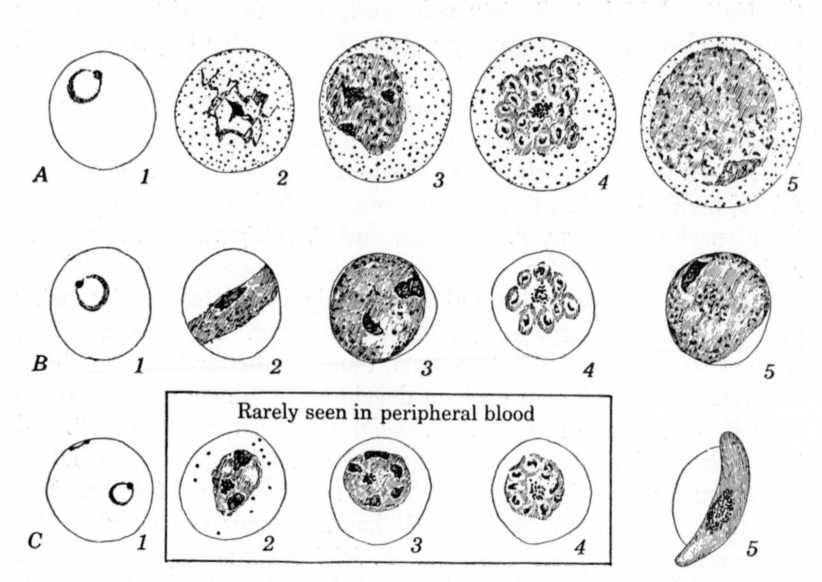

Fig. 35. Comparison of three common species of malaria parasites, illustrating diagnostic characteristics in each stage. *A, Plasmodium vivax; B, P. malariae; C, P. falciparum; 1,* "ring" stages; *2,* growing schizonts; *3,* grown schizonts with dividing nucleus; *4,* segmenting parasites nearly ready to leave corpuscle; *5,* female gametocytes.

ment. Within a few minutes after injection they leave the blood stream and enter tissue cells where they go through at least two schizo-

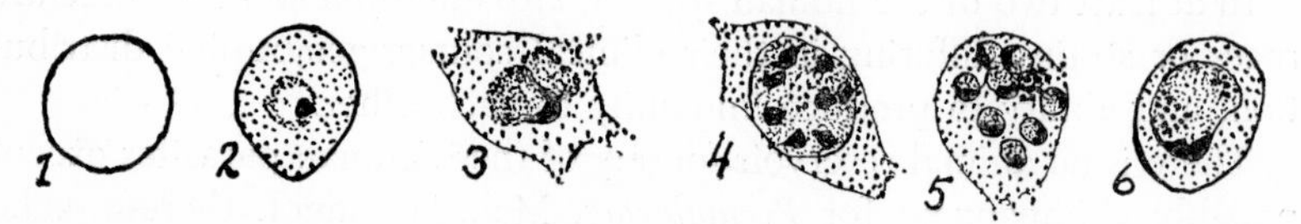

Fig. 36. *Plasmodium ovale: 1,* normal corpuscle. *2,* ring stage; note corpuscle already full of Schüffner's dots. *3,* young schizont; note irregular shape of corpuscle and "fimbriated" appearance, also seen in *4* and *5*. *4,* dividing schizont; note oval, fimbriated corpuscle and round parasite. *5,* segmenter; note only 8 merozoites around a central clump of pigment. *6,* female gametocyte; note similarity to quartan gametocyte except for shape of corpuscle and presence of Schüffner's dots. (Drawn from figures by James, Nicol, and Shute, *Parasitology,* 25, 1922.)

gonic cycles before invading the blood. The first generation of these preerythrocytic tissue parasites are called by Huff "cryptozoites"; the following ones he calls "metacryptozoites." The parasites then invade

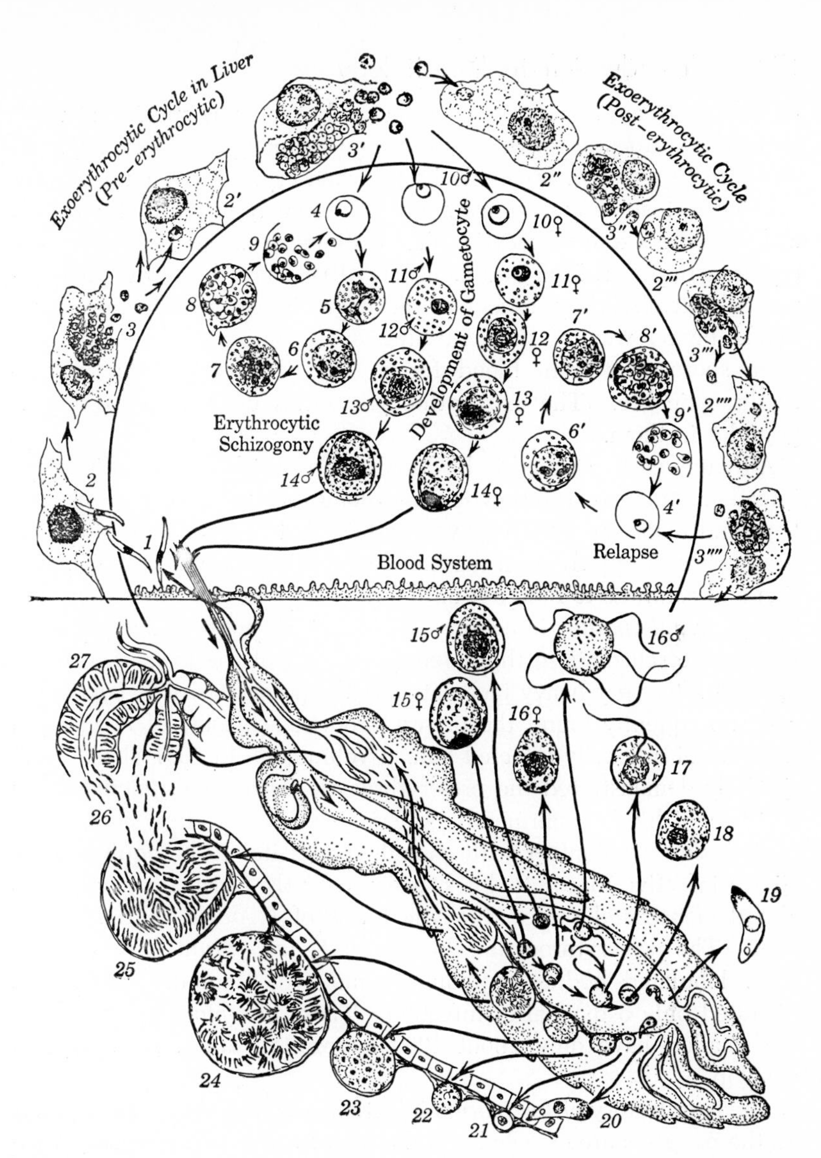

Fig. 37. Life cycle of *Plasmodium vivax,* partly hypothetical. *1,* sporozoites injected; *2,* sporozoites entering liver cell; *3,* exo-erythrocytic schizogony; *4–9,* erythrocytic schizogony, repeated indefinitely; *10–14,* development of ♂ and ♀ gametocytes; *4′–9′,* reinvasion of blood by exo-erythrocytic merozoites, causing relapse; *15,* ♂ and ♀ gametocytes in stomach of *Anopheles; 16*♀, female gamete; *16*♂, formation of male gametes by exflagellation; *17,* fertilization; *18,* zygote; *19,* ookinete; *20,* same, penetrating stomach wall; *21–25,* development of oocyst (*23* shows sporoblasts, *24* and *25* development of sporozoites); *26,* sporozoites liberated into body cavity when oocyst bursts; *27,* sporozoites collecting in salivary gland cells and ducts. Although the ookinete is pictured as penetrating the stomach wall, recent evidence indicates that the zygote may be mechanically pressed into the stomach wall and that the ookinete is not involved (see p. 176).

the blood stream and enter erythrocytes, but in some species, at least, the exo-erythrocytic forms may continue to multiply indefinitely. In some bird species in favorable hosts, when erythrocytic parasites are injected instead of sporozoites, some of these may penetrate the tissues and become exo-erythrocytic. To distinguish these from the pre-erythrocytic forms Huff and Coulston (1946) use the term "phanerozoites," but obviously, except at the beginning of an infection, there would be no way of distinguishing exo-erythrocytic forms derived from sporozoites from blood forms that had reverted to an exo-erythrocytic life. The exo-erythrocytic forms do not have pigment granules in them.

Although easily studied in bird malaria, especially in *P. gallinaceum* infections, the exo-erythrocytic stages long escaped detection in monkeys and man. It was, however, certain that they existed, since the sporozoites disappear from the blood stream 30 minutes after inoculation and continue to hide out until 6 days later in *vivax* infections and 8 days in *falciparum* infections. Shortt et al., in 1948 succeeded in finding the schizonts in the parenchyma cells of the liver of a rhesus monkey after very heavy inoculation with sporozoites of *P. cynomolgi* and subsequently found the pre-erythrocytic schizonts of *P. vivax* in the liver of a heavily infected human volunteer. In 1951 Shortt et al. similarly demonstrated the pre-erythrocytic forms of *P. falciparum,* a finding which was confirmed by American workers (Jeffery et al. 1952) after dosing a volunteer with enough sporozoites to be equivalent to 20,000 infective mosquito bites. More recently Garnham et al. have also demonstrated pre-erythrocytic forms of *P. ovale,* in the human liver. The pre-erythrocytic schizonts of *P. falciparum* (Fig. 38) are large lobulated bodies containing some 40,000 minute merozoites, liberated only 6 days after infection. Those of *cynomolgi* and *vivax* are smaller, have only about 1000 merozoites, and take 8 days to develop.

The later history of *falciparum* infections suggests that in this species all the parasites are expelled into the blood within a few weeks; when the blood infection is destroyed by drugs no relapse occurs. Bray has carried out extensive studies on the development of the human plasmodia in chimpanzees. In *vivax* and *malariae* infections, on the other hand, the exo-erythrocytic forms seem to persist, causing relapses when a falling off in immunity or drug treatment makes it possible for them to reinvade the blood stream. In some strains of *P. vivax,* however (see p. 180), there is a long latent period of many months between the primary attack, which sometimes fails to develop at all, and the first relapse, apparently due to failure of the parasites to try

COMPARISONS BETWEEN ERYTHROCYTIC FORMS OF DIFFERENT SPECIES OF HUMAN MALARIA PARASITES

	P. vivax	*P. malariae*	*P. ovale*	*P. falciparum*
Rings	Coarse, about 1/3 to 1/2 diameter of corpuscle; rarely more than one in corpuscle	Similar to *vivax*	Similar to *vivax*; Schüffner's dots may be present even in this stage	Fine, about 1/6 to 1/5 diameter of corpuscle; doubly infected corpuscles common; often situated at edge of corpuscle; two nuclei frequent
Corpuscles infected with schizonts	Enlarged and pale; contain Schüffner's dots	Size and color normal; no Schüffner's dots	Oval in shape, often fimbriated; not much enlarged; normal in color; contain Schüffner's dots	Not seen in peripheral circulation; size normal; color dark, brassy; have reddish clefts called Maurer's dots, and may have bluish stippling
Growing schizonts	Very irregular, sprawled out over cell; pigment in small brown granules, usually collected in a mass	More compact and rounded, or drawn out band-like across cell; pigment blacker and in coarser granules	Usually round, not ameboid; pigment brownish, coarse, somewhat scattered	Usually compact, rounded; pigment coarse and blackish; not seen in peripheral circulation
Segmenters	Nearly fill enlarged, dotted corpuscle; 15 to 20 merozoites, occasionally up to 32, irregularly arranged	Nearly fill normal-sized corpuscle; 6 to 12 merozoites, commonly 8 or 9, arranged like daisy head	Occupy 3/4 of dotted oval corpuscle; 8 to 10 merozoites, arranged like bunch of grapes	Occupy 2/3 to 3/4 of corpuscle; number of merozoites very variable, from 8 to 32, not seen in peripheral circulation
Gametocytes	Rounded, larger than corpuscles (10–14 μ in diameter); pigment granules fine, brown, evenly peppered throughout cytoplasm	Rounded, smaller, nearly filling a corpuscle of normal size; pigment blacker and coarser than in *vivax* and more or less concentrated at center and periphery	Rounded, filling 3/4 of enlarged dotted corpuscle; pigment coarse, black, evenly peppered	Crescent-shaped or bean-shaped with pigment granules clustered about nucleus at center; remnants of corpuscle often not in evidence
Distinctions between microgametocytes and macrogametocytes	Microgametocytes with pale cytoplasm; nucleus large, pink, often stretched across center of body. Macrogametocytes slightly larger, with deeper blue cytoplasm; nucleus small, red, usually lying at or near one side of body	Same differences as in *vivax*	Same differences as in *vivax*	Microgametocytes short and squatty with pale blue cytoplasm and pink nucleus; pigment granules scattered except at poles. Macrogametocytes longer and more slender, deeper blue, with small red nucleus; pigment more concentrated near center
Interval between sporulations	48 hours	72 hours	48 hours	48 hours

to invade the blood stream rather than to any barrier caused by immunity.

Apparently in *vivax* infections the erythrocytic forms do *not* revert to exo-erythrocytic life as they do in *gallinaceum* infections in birds, for in *vivax* infections induced by blood inoculation no relapses occur once the blood parasites are destroyed.

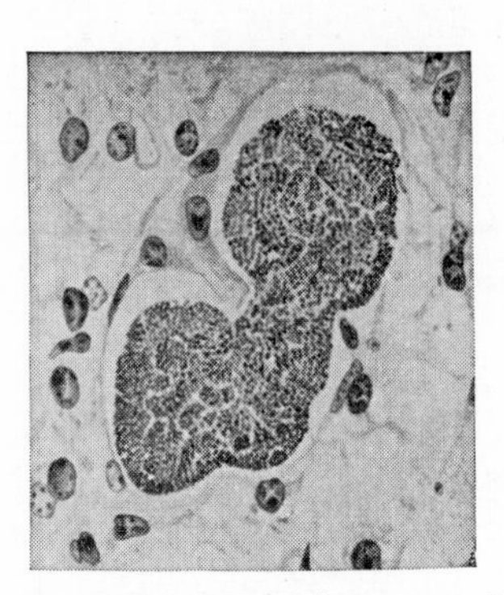

Fig. 38. *Left:* Pre-erythrocytic schizont of *Plasmodium falciparum* in human liver 6 days after infection by mosquito bites. *Right:* Reconstruction of a mature schizont. (From figures by Shortt et al., *Trans. Roy. Soc. Trop. Med. Hyg.*, 44, 1951.)

On the basis of the lack of further exo-erythrocytic development in *falciparum* infections Bray feels that this species should not be retained in the genus *Plasmodium* but in a separate genus, *Laverania,* which many workers have regarded as a synonym of *Plasmodium.* Bray's suggestion seems to have considerable merit, but for the present the authors have retained *falciparum* in *Plasmodium.*

ERYTHROCYTIC STAGES. About a week or ten days after mosquito infection the parasites invade erythrocytes and begin a process of schizogony. The earliest form of the parasites seen in the corpuscles is the "ring stage" (Fig. 35, *1*). It appears like a little signet ring caused by the presence of a transparent or vacuolated area in the center of the parasite, surrounded by a delicate ring of cytoplasm and a tiny nucleus at one side, like the setting in a ring. With the usual blood stains (Giemsa or Wright) the cytoplasmic ring stains blue and the nucleus ruby-red. The rings of *vivax, malariae,* and *ovale* are about one-third the diameter of the blood cells and are indistinguishable, but those of *falciparum* are only about half this size, have hair-like rings, and tend to be perched on the periphery of the corpuscles; corpuscles containing two or more rings are common.

As the parasite grows larger it becomes rounded or irregular in shape. In *Plasmodium falciparum* the infected corpuscles at this stage become viscid and clump together in internal organs and are not seen in the peripheral circulation, but those of the other species continue to circulate in the peripheral blood in all stages. As noted in the table on p. 173, *vivax* infections are distinguishable in all stages beyond the rings by the enlarged, pale corpuscles which they occupy, studded with red-staining granules called "Schüffner's dots" (Fig. 35A, 2–5). Similar dots appear in *ovale* infections, but the infected corpuscles are oval and not enlarged (Fig. 36). No such dots appear in *malariae* infections, and the infected cells fail to enlarge or grow pale. In *falciparum* infections the infected cells (located in internal organs) may have large, irregular reddish clefts called "Maurer's dots"; they have a darker "brassy" color and may also have a bluish stippling.

As the parasites grow, the nucleus divides into two, then four, and eventually more parts. As maturity approaches, the nuclei tend to take up peripheral positions in the schizont, and a small portion of the cytoplasm concentrates around each. These *segmenters* eventually break free from the corpuscles in which they have developed, and the individual *merozoites* thus liberated attack new corpuscles and repeat the process. The number of merozoites in the different species is shown in the table.

The merozoites of *vivax* attack almost exclusively the young immature corpuscles (reticulocytes) and those of *malariae* the older ones, but *falciparum* indiscriminately enters any that are handy. The result is that *vivax* parasites are seldom found in even 1% of the corpuscles and *malariae* in seldom more than 1 in 500, whereas *falciparum* may infect 10% or more, with dire results.

The pigment and other waste products left behind when the parasite breaks up are released into the blood stream and deposited in the spleen or other organs, or under the skin, causing the sallow color so characteristic of malarial patients. It is at the time of the bursting of corpuscles and release of waste products that the characteristic paroxysms of chills and fever are felt.

The species differ in the time they take to mature. *P. vivax, ovale,* and *falciparum* take about 48 hours in which to complete the schizogonic cycle, whereas *P. malariae* takes 72 hours, whence the names tertian and quartan. These are derived from the old Roman method of figuring, which counts the day on which something happens as the first day, the second day following being therefore the third (tertian), and the third following day the fourth (quartan). Although the schizogonic cycles of the human species are all 48 or 72 hours, the paroxysms, particularly in early stages of *vivax* and *falciparum* infections, commonly occur daily (quotidian), presumably because the exoerythrocytic forms enter the blood at various times and consequently different broods mature on different days. The liberation of merozoites does not take place at all hours, however, but is timed by some physiological condition in the host and is largely concentrated within a few hours on each day. In *falciparum* infections the sporulation of all the parasites is less closely synchronized than in the other species resulting in longer drawn out paroxysms of chills and fever.

After a few generations of schizonts have been produced in the blood some of the merozoites have a different destiny. They grow more slowly, produce more pigment, and develop into large single-nucleated organisms. These are the gametocytes, which continue to circulate in the blood for at least a number of weeks, but undergo no further development within the human body. The gametocytes of *falciparum* are crescent shaped, whereas those of the other species are rounded. Distinguishing characters of the gametocytes of the different species and of the males (microgametocytes) and females (macrogametocytes) are shown in the table on p. 173, and Figs. 35 and 36.

MOSQUITO CYCLE. When removed from the warm blood by being sucked up by a mosquito, or even if placed on a microscope slide, the microgametocytes undergo a striking development. The nucleus quickly divides, and within 10 or 15 minutes six to eight long flagella-like structures are extruded; the parasite is then known as a "flagellated body." This process of exflagellation is in reality the formation of microgametes. These slender structures break free and swim actively among the corpuscles ingested by the mosquito, in search of a macrogamete. The macrogametes meanwhile undergo little change except that the *falciparum* crescents become rounded (Fig. 37, *15, 16*).

The result of the union of the filamentous microgamete with the inactive female gamete is a "zygote." If this rounded zygote is on the periphery of the blood meal it may be lodged between the cells of the mosquito stomach wall, these cells becoming extremely flattened when the stomach is distended by the meal of blood. Those unfortunate individuals who do not reach the stomach wall elongate into worm-like bodies which have been called ookinetes. The ookinete, long thought to be an invasive form, is apparently a dying parasite which is voided in the feces of the mosquito. Those zygotes which are pressed into the intercellular areas of the mosquito stomach actually become surrounded by the gut cells as digestion of the blood meal progresses and the stomach wall cells become first cuboidal and finally elongate. These events have recently been demonstrated by Howard (1960).

The long held assumption that the ookinete is a motile invasive zygote is an excellent example of a concept based on observations which were never made. Under the limiting outer membrane of the mosquito stomach, rapid growth of the parasite takes place, and a cyst wall develops. The oocyst thus formed protrudes like a little wart on the outer surface of the stomach and grows until it is 50 to 60 μ in diameter.

Meanwhile its contents undergo important changes. The nucleus divides repeatedly, and a number of faintly outlined cells called sporoblasts are formed, varying in size and number (Fig. 37, *23*). As further nuclear division occurs, dots of refractile chromatin arrange themselves around the periphery of each sporoblast. Granular streaks appear in the protoplasm, and slender spindle-shaped sporozoites develop, each with a chromatin dot as a nucleus. The sporoblasts, meanwhile, enlarge and coalesce, vacuoles form in them, and a sponge-like mesh is produced (Fig. 37, *24*). Eventually the sporozoites, each about 15 μ in length, break loose from their moorings and form a tangled mass in the oocyst, which is crammed with them to the bursting point (Fig. 37, *25*). Such an oocyst may contain more than 10,000 sporozoites, and there may be as many as 50 oocysts on one mosquito's stomach. In about 10 days to 3 weeks, according to temperature, after the mosquito has sucked blood containing gametocytes, the oocyst becomes mature and bursts, releasing the sporozoites into the body cavity of the mosquito. From here they make their way to the three-lobed salivary gland lying in the fore part of the thorax and connecting with the proboscis. They assemble in the cells lining the salivary glands (Fig. 39); there may be up to 200,000 in one mosquito. The sporozoites now invade the lumen of the ducts and are discharged with the saliva when the mosquito bites. Many more may be discharged at one bite than at another, but it takes 20 bites or more to discharge them all.

James in 1926 reported one remarkable mosquito which was caught on August 5 and was finally dissected on November 16 of the same year, with active sporozoites still in its salivary glands. In the meantime it had spent a hectic life in incubators, refrigerators, hospitals, railway trains, etc., and had successfully infected more than forty general paralysis patients as a means of treatment. Humidity does not affect the cycle of development in the mosquito if the mosquito itself can survive.

Usually the number of oocysts which develop on a mosquito's stomach is proportional to the number of gametocytes in the blood sucked, but only a small percentage actually develop. The number of sporozoites in the salivary glands may have little relation to the number of oocysts; sometimes the development never goes beyond the oocyst stage. The oocysts of various species of malaria parasites are distinguishable by the number and character of the pigment granules, according to Shute and Maryon (1952).

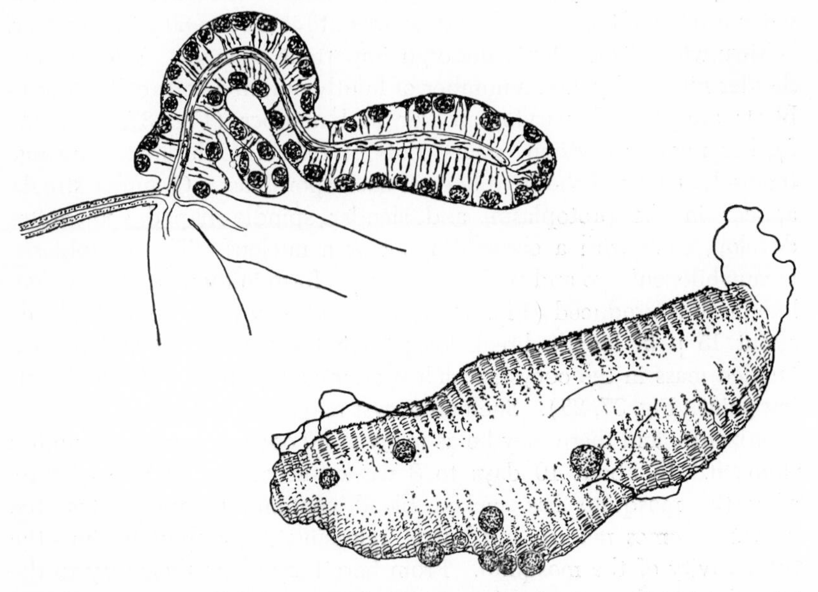

Fig. 39. *Upper:* One lobe of 3-lobed salivary gland of an infected mosquito, showing sporozoites in cells and lumen. *Lower:* Stomach of *Anopheles quadrimaculatus,* showing oocysts of *Plasmodium vivax.*

The differences in the mosquito development of the different species are only minor. Infective gametocytes of *vivax* may appear in the blood of a patient within 4 days after the first appearance of the parasites. As few as 10 gametocytes per cubic millimeter of blood may be enough to infect *Anopheles quadrimaculatus.* Infective gametocytes of *falciparum* are not observed until 10 days after appearance of the parasites, and no infections of *A. quadrimaculatus* are successful with less than about 100 gametocytes per cubic millimeter. However, there is no close correlation between the number of gametocytes in the blood and infectiveness for mosquitoes; the gametocytes in some patients seem to be quite worthless, although the reason is unknown. Sometimes when bird malaria is inoculated into unfavorable hosts the gametocytes lack sexual potency although normal in appearance. The effect of drugs on viability of gametocytes is considered in the section on treatment, p. 187.

Temperature affects the time required for development in mosquitoes. At 85° to 90°F. both *falciparum* and *vivax* may form sporozoites in 7 or 8 days, though the mosquito mortality is high; 98°F. for even 4 hours prevents development entirely after an infective meal, and that temperature for 18 to 21 hours kills growing oocysts. At 77°F. *vivax* oocysts develop in 9 days, *falciparum* in 10 days, *ovale* in 15 days, and *malariae* in 15 to 21 days. *Vivax* stands low temperatures better than *falciparum;* the former has its development stopped below 60°F., the latter below 66°F. Although development is stopped by cold weather, the parasites may remain alive and resume development later, but the minimum temperature at which microgametes are formed is said by Grassi to be 65°F., so no mosquito infections would be expected below that point.

The disease. Although the course of a typical initial case of malaria, with its recurring chills and fever, is very easy for any physician to diagnose, even if he never saw one before, the symptoms may be profoundly modified by treatment, immunity, etc., particularly in old or repeated infections.

In some cases the gastrointestinal tract is affected and symptoms resembling cholera or dysentery develop, either due to the malarial infection alone or to the lighting up of a chronic dysentery infection. In some the symptoms are suggestive of influenza or bronchopneumonia; in others, of dengue; in others, of encephalitis or meningitis. Sometimes the only symptoms are jaundice, anemia, albuminuria, malaise, or digestive disturbances. Any individual who has lived in a malarial locality and shows symptoms of a chronic infection not otherwise diagnosed should be suspected of malaria and his blood should be examined for it. Such blood examinations should be as routine as Wassermann tests or urinalysis.

In typical cases of primary *vivax* infections, after completion of one or two pre-erythrocytic cycles in the tissues, the parasites invade the blood stream about 7 to 14 days after infection, and are demonstrable in the blood a few days later. The time of appearance of blood parasites is usually a little shorter in *falciparum* infections and longer in *malariae* infections. In infections caused by temperate-zone strains of *P. vivax* this early invasion of the blood, resulting in an early primary attack, sometimes either fails to develop, or produces such mild attacks as to be unnoticed. Characteristic symptoms appear when the parasites have reached a concentration of about 200 per cubic millimeters of blood, or about one billion in the entire body.

The characteristic recurrent chills and fever of malaria are correlated with the liberation of successive broods of merozoites from disrupted blood corpuscles. As noted on p. 169, the interval between sporulations of the parasites and consequent paroxysms is at first irregular, tending to be quotidian before it eventually assumes the typical 48- or 72-hour cycle. In relapses and in subsequent infections the appearance of the paroxysms on every second or third day may be apparent from the beginning.

Each attack begins with a shivering chill, sometimes accompanied by convulsions so severe that the teeth chatter, gooseflesh stands out, and the bed rattles. Yet the temperature will be found to be several degrees above normal and still going up. In the wake of the chill comes a burning and weakening fever, with violent headache and nausea and a temperature up to 106° or even higher. The fever stage in turn is followed by a period of sweating so profuse that the clothes or bedding may become wringing wet. The sweating gradually subsides, the temperature drops rapidly, often below normal, and after 6 to 10 hours the patient rests fairly easily until the next attack. The fact that the attacks most commonly occur between midnight and noon, instead of in the evening, is often useful in distinguishing malaria from other intermittent fevers.

In *P. vivax* infections the paroxysms of chills and fever continue every other day for 8 or 10 days to 2 weeks or more. Then they become less pronounced, the parasites become sparse, the patient feels well, and his temperature becomes normal. Sooner or later there are relapses, in which the intermittent chills and fever begin again. In temperate-zone strains of *P. vivax* (e.g., St. Elizabeth), after the primary attack (if any) has subsided, the exo-erythrocytic parasites usually remain quiescent for 6 to 14 months, commonly about 9 months. An interesting example of this occurred in a nonendemic area in California (Brunetti, 1954). A Korean veteran suffered a relapse of *vivax* malaria while camping near an encampment of Campfire Girls and was apparently responsible for 34 cases among the girls; 9 of these had incubation periods of 12 to 38 days, but 25 showed no symptoms until 7 to 10 months later.

When the relapses begin after the long latent period they occur repeatedly at intervals of a month or so. After these have subsided some strains (e.g., in Korea) apparently die out, but others may persist for years. The long latent period between primary and relapse attacks has been observed in strains of *P. vivax* from United States, Europe, Madagascar, and Korea. In at least some *vivax* strains that are of tropical origin, on the other hand, e.g., the "Chesson" strain from New Guinea, there is a series of short-period relapses, gradually spaced further apart and ending after about 18 months. The spacing and probability of relapses are, however, apparently influenced by dosage of sporozoites, acquisition of immunity, and administration of drugs (see Coatney and Cooper, 1948). It is commonly accepted that relapses after long periods may be brought on by such physiological shocks as exhaustion, childbirth, operations, alcoholic binges, etc., but there is a dearth of controlled evidence.

After repeated attacks the patient's vitality is lowered, he becomes anemic, his spleen enlarges, and he finally reaches a chronic run-down condition, at least when there is poor nutrition.

P. ovale infections are not prone to relapse and are more susceptible than any of the other species to drug treatment.

In *P. malariae* infections the paroxysms occur at 72-hour intervals, are milder and of shorter duration, recur more regularly, and the infections persist for a longer time. The milder nature of the disease often results in failure to seek treatment, and this, together with its long duration and tendency to relapse, is believed to explain the frequent kidney disease which is found in quartan cases. Lambers found nephritis in nearly 50% of such cases in a hospital in Dutch Guiana as compared with 4 or 5% in *vivax* and *falciparum* infections; one-sixth of all the nephritis cases were due to quartan malaria.

In *falciparum* malaria we have to deal with quite a different disease. In natives of hyperendemic localities, primary infections are seldom seen except in very young children but are common in visitors. The paroxysms of chills and fever are less well defined, last 12 or 14 to 36 hours, are severe in nature, and often occur daily, a fresh attack sometimes beginning before the previous one has entirely subsided. On days intervening between attacks the patient is sick and does not have a "well" day as in *vivax* infections. As already noted, the parasites frequently become excessively numerous and the spleen becomes very large. The temperature is likely to rise above 105°F., and is often accompanied by vomiting and delirium. The attacks usually last 8 or 10 days, and then the temperature slopes off. In just a few days, however, there is a second series of paroxysms, perhaps even more severe, and these recrudescences then continue in declining severity every 10 or 12 days for about 6 or 8 weeks, after which they become more irregular, although the blood continues to be infective. In the absence of reinfections the disease usually dies out completely in 6 to 8 months. In malarial countries in the tropics, however, no such course is seen, since reinfection is more or less continuous. Under these circumstances the infected persons become "carriers" harboring a few parasites, possibly too few to be found in blood smears, and showing few symptoms or none at all. In subtropical regions, on the other hand, as in southern United States and Italy, the infections die out in cold weather and fresh outbreaks occur every year.

A number of pernicious conditions may develop, usually in *falciparum* malaria. The tendency of corpuscles infected with *P. falciparum* to cling together results in clogging capillaries and preventing the proper flow of blood in vital organs. In the brain this, as well as a direct toxic effect, leads to numerous symptoms, among them total loss of consciousness, or coma, and sometimes sudden death by a "stroke." This "cerebral malaria" causes a large fraction of malarial deaths. In some cases violent gastrointestinal symptoms resembling cholera, typhoid, or dysentery develop, and in others, heart failure or pneumonia. *Falciparum* malaria is always an accompaniment of blackwater fever, but its exact relation to that disease is still uncertain (see p. 191).

Prenatal infection is not infrequent, especially in *falciparum* malaria. Gastrointestinal and pulmonary forms of malaria are especially common in infants.

Immunity. The nature of the acquired resistance to an existing malaria infection and immunity to superinfections has been extensively studied. From the very beginning of an infection there is considerable destruction of the parasites. Knowles pointed out that a single parasite producing twenty merozoites at each successive multiplication, if unchecked, would have increased in 20 days to the point where there would be about four parasites to every blood corpuscle, if the patient could live that long.

Much information, most of it probably applicable to human malaria, has been obtained about the development of immunity in malaria of birds and monkeys. Huff and his colleagues showed that in bird malaria immunity to blood forms does not protect against exo-erythrocytic forms, and they think it possible that two different antibodies may be involved. In human malaria where the e.e. forms are far less abundant than in some bird species, it is doubtful that complete immunity to e.e. forms ever develops.

The mechanism on which the blood immunity depends has been demonstrated by Taliaferro and his colleagues to be mainly phagocytosis by cells of the reticulo-endothelial system, particularly in the spleen, liver, and bone marrow. The phagocytosis begins probably at once, and consists of the engulfing and destruction of the entire parasitized blood corpuscles and not merely the free merozoites. This destruction of invaded corpuscles is believed to be an important factor in malarial anemia. As the disease progresses the activity of these voracious cells is gradually increased, and they begin multiplying in number until a climax is reached when the rate of destruction of parasites greatly exceeds their production. At this time the liver and spleen are enlarged and show great activity of the phagocytic cells in them. There is also a marked increase in lymphoid tissue to build a mesenchymal reserve for the rapid production and mobilization of more macrophages.

This condition gradually declines during the latent period, but rapid mobilization against fresh invaders of the same species may occur for years, especially if the infection has not entirely died out. The immunity is, however, highly specific, and not only fails to protect against attack by other species but sometimes even by other strains of the same species. In well-nourished and healthy individuals, as long as the parasites persist in the body in an exo-erythrocytic reservoir, as in *vivax* and *malariae* infections, or are constantly being reinoculated, as in *falciparum* infections in the tropics, the immunity, once developed, is restimulated as soon as it begins to fail because of scarcity or lack of parasites, and no clinical relapses can occur. Only unhealthy individuals, who are unable to provide for rapid formation of antibodies or phagocytes in response to a renewed stimulus, suffer a relapse before the body reacts.

The presumed absence of an exo-erythrocytic reservoir in *falciparum* infections, and the wiping out of the blood infection in 6 or 8 months, before a high enough degree of immunity is built up to persist very long, results in the tendency of this infection to break into epidemics in areas where conditions are not suitable for continued reinfection, e.g., on the fringes of the tropics (see p. 189). Under such conditions it may be dangerous to control malaria partially in a hyperendemic area or to create a nonmalarial oasis in its midst unless facilities for prompt treatment of cases are available.

Infants in hyperendemic areas do not suffer as much as is sometimes thought from malaria infections since, as in the case of a number of other diseases, they are passively protected by the mother's antibodies until they begin to develop some of their own.

Negroes have a higher degree of tolerance to *P. vivax* than do non-Negroes. Many are refractory to infection even with exotic strains to which they could not have developed specific immunity. Watson and Rice in a study in the Tennessee Valley in 1946 found that although the parasitemia rate in Negroes was five times that in Caucasians, the latter experienced about ten times as many reported cases of malaria and over fifteen times as many sick days.

Epidemiology. As a result of the work of Ross and Grassi, which set such an important milestone in the progress of preventive medicine, malaria is now known to be transmitted naturally, except in some prenatal infections, only by the bites of certain species of mosquitoes, all belonging to the genus *Anopheles*.

The only important exception is in the case of heroin addicts, who frequently pass infections around by means of hypodermic needles.

More than a hundred species of *Anopheles* have been described, but less than two dozen species are of any real importance in the transmission of malaria, except perhaps in local areas. This matter is discussed in the chapter on mosquitoes (p. 715). As shown there, some species are eliminated because they do not readily nurse the malaria parasites through their sporogonic cycle; some are eliminated because of their habits; and others are of no importance on account of their rarity. Local conditions may influence the importance of particular species of mosquitoes in transmitting malaria, and therefore local epidemiological surveys to determine the prevalent transmitters are important. Since the different species vary greatly in their breeding habits, control measures must depend on the habits of the particular species involved.

Malaria does not become endemic wherever suitable *Anopheles* mosquitoes occur. A small deviation above or below the critical point may mean the difference between ultimate extermination and permanent establishment. MacDonald (1952) speaks of a "critical density," meaning the average number of *Anopheles* bites per person per night below which there would be progressive reduction of the disease until it became extinct, as has now happened in the United States. The critical density is determined by a number of factors: number of mosquitoes, fondness for human blood, life expectancy of the mosquitoes, frequency of feeding, extent of development of parasites to the sporozoite stage, etc. For example, where malaria is transmitted by such mosquitoes as *A. gambiae, A. funestus,* and *A. minimus,* the critical density is low (0.029 for *A. gambiae*) and we have stable malaria; where transmitted by such species as *quadrimaculatus* and *albimanus,* which have high critical densities, malaria is unstable, i.e., easily influenced by environmental and climatic changes.

Food preferences of various different species or varieties of *Anopheles* are of prime importance. The presence of abundant *Anopheles* in certain localities in Europe without accompanying malaria when neighboring localities with fewer *Anopheles* might be highly malarious was a mystery until it was found that the European *A. maculipennis* really consists of several distinct races, some of which are "zoophilic" and only exceptionally bite man, whereas others show no discrimination against human blood (see pp. 725–728). Tendency to enter houses, and particularly to rest in them after feeding, is of great importance in connection with the effectiveness of DDT residual spraying; the latter causes great reduction of malaria where such species as *quadrimaculatus* and *darlingi* are the transmitters, but has little effect where *aquasalis* or *bellator* are principally involved.

Climate is a factor, for it may lower the life expectancy of most of the mosquitoes to a point where there may not even be time for the sporogonic cycle of the malaria parasites to be completed. In Punjab, for instance, there is little transmission, even with abundant potential transmitters, when the season is hot and dry.

The species or strains of malaria parasites are also important. Some mosquitoes become infected much more easily with some species or strains of parasites than others. The susceptibility of *A. quadrimaculatus* to *P. falciparum* as compared with *A. crucians* is in the ratio of 64 to 15. *A. maculipennis atroparvus* in England proved refractory to Indian and African strains of *falciparum*, but not to European ones. An infected *A. quadrimaculatus* usually transmits *P. falciparum* by a single bite, whereas it often requires several bites to transmit *vivax*, but this mosquito seems to be susceptible to most if not all strains of both *vivax* and *falciparum*. *A. albimanus* from Cuba or Panama, though highly susceptible to malaria strains from its own region, is refractory to Florida strains. Huff, working with *Culex* vectors of bird malaria, found that refractoriness of mosquitoes to malarial infection is hereditarily transmitted.

As noted on p. 183, epidemics of *vivax* malaria rarely occur since the disease often continues to exist in the host for several years, keeping up immunity by occasional relapses. Only if a new strain of *P. vivax* were introduced from foreign parts could an epidemic occur. In hyperendemic tropical regions there are no *falciparum* epidemics either, the reason being constant reinfection.

In subtropical areas where climatic conditions are such as to cause marked seasonal reduction in anopheline density, true epidemics of *falciparum*, and to a lesser extent of the other species, may occur; their violence is largely dependent upon the interval between seasons of highly favorable conditions for infection. An epidemic may occur of such extraordinary severity as to involve almost the entire population and to cause a mortality of several hundreds per thousand. Such devastating epidemics are nearly always of the *falciparum* type.

Local epidemics may also arise from the bringing in of a new strain of parasite, from the introduction of a new species of *Anopheles*, e.g., *A. gambiae* into Brazil and during World War II into Egypt, or from the development of more favorable conditions for the breeding of dangerous species of *Anopheles.* In parts of Europe inhabited by zoophilic strains of *maculipennis,* malaria disappears as animal husbandry develops. In Java improvement and reconstruction of houses for protection against plague have led to a serious increase in malaria; this is apparently due to a combination of several factors, such as tile roofs, borrow pits for building material, and importation of new parasite strains with laborers.

Diagnosis. In acute cases of malaria the clinical symptoms are usually sufficient for a diagnosis. In more chronic cases a combination of anemia and enlarged spleen, where kala-azar and certain less common conditions can be ruled out, almost unmistakably advertises malaria infection. Nevertheless all diagnosis should be confirmed by blood examination whenever possible. If not possible, failure of a test course of an anti-malarial drug to relieve the symptoms indicates that the fever is not due to malaria.

Accurate diagnosis is made by examinations of blood smears stained by a Romanowsky stain, preferably Giemsa's. In thin smears, made by spreading a film on a slide by drawing a drop across it in the acute angle behind the line of contact of the film slide and the spreading slide, the infected corpuscles are spread in a single layer, and the parasites are stained in their natural positions in the corpuscles. Characters which differentiate the species and stages are easily recognizable in such films. Thick smears, however, are far more valuable for detection of cases where the parasites are sparse, though the identification is more difficult. These smears are made by thoroughly drying thick drops, dehemoglobinizing before or during staining, and then examining for the more concentrated parasites free from the corpuscles. An injection of adrenalin a few minutes before taking the blood for a smear is helpful in finding parasites when they are sparse.

The degree of malariousness of a district can be determined fairly accurately by finding the percentage of children between about 2 and 10 years of age who have enlarged spleens. The "spleen rate" in adults is of little value in highly malarial places because of a reduction in spleen enlargement with continued immunity; it is of use only as an indication of the number of active cases. Any spleen that can be felt below the last rib when a child is lying down may be classed as enlarged and is usually indicative of malaria; in extreme cases the spleen may reach the pubis.

Serological tests for malaria have been recommended but are not generally considered reliable enough to replace examination for parasites.

Treatment. The treatment of malaria has come a long way from the time a Countess Chinchona, returning to Europe from Peru in 1640, introduced a native Indian remedy, "fever bark," containing quinine and allied alkaloids. These alkaloids, especially quinine, remained the standard treatment for malaria for 300 years. Now, however, it is still used only by physicians who have not kept up with the times.

Since quinine had little effect on gametocytes, especially of *P. falciparum,* and did not prevent relapses of *vivax* and quartan malaria, search was made for better drugs. During World War II this search was so intensified that more research was directed toward finding better anti-malarial drugs than toward any other project except the atom bomb. Prior to the war two drugs had been found that were helpful; one, Plasmochin (Pamaquin), was found in 1926 to kill the gametocytes that quinine did not, and to be helpful in preventing relapses, but it was too toxic to use on a large scale; the other, atebrin (Quinacrine, Mepacrine), was a substitute for quinine which was, as a rule, better tolerated and more effective, its chief disadvantage being that it temporarily turns the skin a bilious saffron color. It was, however, an incalculable boon during World War II when the quinine supply was cut off by the Japanese at a time when it was needed more than ever before.

Meanwhile, the discovery of the exo-erythrocytic forms of *P. gallinaceum* and evidence of their occurrence in human malaria, together with the fact that drugs which affect the schizonts in the blood have little or no effect on the e.e. forms, made it evident why these drugs, although suppressing *vivax* and quartan malaria and rapidly controlling the symptoms by killing the blood parasites, failed to bring about complete cures and therefore to prevent relapses. From the fact that *falciparum* malaria is not only suppressed but also cured by the "schizonticides," it has been deduced that in this species the exo-erythrocytic schizogony does not continue long, if at all, after the pre-erythrocytic development, and therefore when the blood parasites are destroyed by drugs, or by immune reaction, the infection dies.

The search for anti-malarial drugs during and after World War II therefore resolved itself into: (1) Finding better schizonticides to destroy the blood parasites and therefore, in *vivax* and quartan malaria, to cure clinical symptoms in developed cases and prevent development of symptoms from new cases as long as the drug is being administered, i.e., act as a suppressant; and in *falciparum* malaria, radically to cure developed cases and prevent new infections from ever producing clinical symptoms even after discontinuance of the drug, i.e., act as a causal prophylactic. (2) Finding drugs which would destroy the e.e. forms of the parasite in the sporozoite, pre-erythrocytic or later e.e. stages, and, therefore, act as a causal prophylactic for all forms of malaria, and radically cure all forms.

Great progress has been made in both quests. Two effective schizonticides, Chloroquine and some related 4-aminoquinolines and Paludrine, were discovered during the war. In the postwar years some 8-aminoquinolines, of which Primaquine is best, were found which are very effective against the tissue parasites, as well as another drug, Daraprim, which has astonishing suppressive activity.

Chloroquin (Resochin, Nivaquine B) and some related drugs (Camoquin, Sontochin, Propoquin, etc.) suppress *vivax* malaria and protect against *falciparum* infections in a single oral dose of 0.3 gram of the base once a week. In South American countries, the incorporation of Chloroquin in table salt has been used for malaria suppression. Even for treatment of clinical cases a single oral dose of 0.8 to 1 gram or less of the base is usually adequate and can be used in mass treatments. The temperature becomes normal within 18 hours, and the parasites disappear within 25 hours. When feasible, a 3-day regimen is preferred—3 doses of 0.3 gram first day, then a single 0.3-gram dose for two more days.

Paludrine (chlorguanide) is also good, but it is slower in action, is less effective against some strains of malaria, and tends to produce drug-resistant strains. Its advantages are cheapness, lack of toxicity, and activity against gametocytes; the latter are slow in disappearing but are rendered noninfective for mosquitoes in a day or two.

Although the tissue forms of malaria parasites are affected to some degree by sulfonamides, some antibiotics, and a few other drugs, only certain 8-aminoquinolines and a pyrimidine (Daraprim) have been found sufficiently active to warrant extensive use as radical cures and causal prophylactics of the relapsing forms of malaria (*vivax* and *malariae*). Unfortunately, the 8-aminoquinolines have rather narrow margins of safety between the effective therapeutic doses and toxic ones; they cause acute hemolytic anemia in some individuals due to an intrinsic susceptibility of their older corpuscles. Plasmochin, the first one used, had a very narrow margin of safety. Then Pentaquine and Isopentaquine were tried, each less toxic than the other, and effective in smaller doses, but still not safe enough. Then about 1950 another drug, Primaquine, was found not only to be superior to the other drugs, but also to be very well tolerated except in a few cases. Its pharmacological properties and effectiveness against malaria were thoroughly studied by Schmidt and his colleagues on *P. cy-*

nomolgi infections in monkeys, and it was then tested on a large number of American troops returning from Korea. A dose of 15 mg. of the base daily for 14 days, plus 1.5 grams Chloroquine in divided doses in 3 days (see above), caused complete cures of *vivax* malaria, with no relapses. Larger doses sometimes produce toxic symptoms. Even in Negroes, who develop anemia more readily than non-Negroes, double this dose (30 mg.) daily for 14 days produced mild anemia in only 17 out of 105, and severe anemia, making it necessary to discontinue treatment, in only 5 Negroes.

Daraprim, a different type of anti-malarial drug, is the most potent one yet discovered, and is unique in being highly effective against *both* blood and tissue forms. Single doses of 25 mg. cause schizonts of *vivax* and *malariae* to disappear, and prevent them from reappearing for a month; and even doses as small as 5 mg. will temporarily clear the parasites. Less than 1 mg. a week suppressed them. In acute attacks, however, Daraprim does not clear parasites or fever as rapidly or effectively as Chloroquine. Early indications were that the drug also destroyed the e.e. forms and was therefore a true causal prophylactic and means of cure; this proved not always to be true, even after 17 weekly doses of 25 mg., but there are long delays after cessation of treatment before relapses occur. In a dose of 25 mg. weekly, or at even longer intervals, Daraprim is a completely reliable suppressant of all forms of malaria, and at that dosage is nontoxic. Although it does not kill gametocytes, it prevents their normal development in mosquitoes, thus preventing transmission. Daraprim-resistant strains of *Plasmodium* have been observed, first in Africa and later elsewhere. At least some of these strains remain sensitive to other anti-malarial drugs. It has been suggested that resistance to Daraprim is likely to develop when low dosages are given. However, Young (1957) described the sudden appearance of resistant strains of *P. malariae* in patients receiving four times greater doses than is usually recommended for suppression. Its slow action in active cases, failure to radically cure *vivax* infections, and tendency of parasites to develop resistance to it are disadvantages; its greatest usefulness is for routine suppression. In some African villages, weekly doses of 25 mg. for 15 weeks virtually eliminated malaria parasites from mosquitoes as well as from the blood of the human population. It was suggested that an annual course of treatment early in the rainy season might help to eradicate malaria over wide areas.

In addition to treatment with specific drugs, there is no doubt, as already pointed out, that nutrition is an important factor in determining the severity of malaria in a community. Of particular interest is the observation that a pure milk diet suppresses *P. berghei* in rats and mice; evidently some factor necessary for the parasites is lacking, since milk *ad lib.* with a normal diet is not suppressive. That milk may have a similar effect on human malaria parasites is suggested by the rarity of malaria in infants under 3 months of age, but some experiments with adults failed to support this view.

Prevention. Although the control of malaria has been relatively simple in principle ever since the discovery by Ross and Grassi of its means of transmission, in practice it has not been easy. It requires community rather than individual effort. Even with our relatively new tools—DDT and suppressive drugs—Ross's statement made many years ago is still true: "It [malaria] is essentially a political disease—one which affects the welfare of whole countries; and the prevention of it should therefore be an important branch of public administration. For the state as for the individual health is the first postulate of prosperity. And prosperity should be the first object of scientific government." Often in the past, however, private industrial organizations have been more active in malaria control than have the governments.

A brilliant example of malaria control was the elimination, in 1939 to 1940, of the disease in northeastern Brazil by the complete eradication of *Anopheles gambiae,* made possible by a cooperative project financed by the Brazilian government and the Rockefeller Foundation. Introduced from Africa to Natal, Brazil, in 1930, *A. gambiae* spread hundreds of miles to northeastern Brazil and created a malaria epidemic that has probably never been equaled in intensity. The need for constant guard against reintroducing of this species is evident, however, from the fact that in 1943 living specimens were found several times in planes arriving from Africa and once in homes near the Natal airport.

With the advent of residual spraying with DDT and related chemicals, and to a lesser extent their use as mosquito larvicides, together with easy and safe prophylactic and suppressive treatment, previously undreamed of progress has been made in the control, or even complete elimination, of malaria, even from large areas in some of the most malaria-ridden countries in the world. The elimination from the United States and the reduction or local eradication in a number of other parts of the world have already been mentioned (p. 167).

The great value of residual spraying with DDT or related compounds (see p. 532) is that a high percentage of the mosquitoes which bite malarial patients in or around houses, privies, etc., by resting on the sprayed walls after a feed, will die before they can transmit the disease.

The prospect for control of malaria is not as good in places where the principal vectors do not enter houses, or enter only to bite and run (see p. 725). Here it will be necessary to destroy the breeding places or kill the larvae of the vectors, or use suppressive drugs (Daraprim or Chloroquine), a procedure which might lead to ultimate eradication.

The control of malaria by anti-mosquito methods is not a problem of general mosquito control, or even *Anopheles* control, but *species* control directed against one or two important local vectors in the particular locality. It is of paramount importance to identify the principal malaria carriers of a region, to study their habits, and then to institute measures directed specifically against these. There are less than two dozen dangerous malaria-carrying species of *Anopheles* in the whole world, but sometimes the habits of individual species vary from place to place, so local studies of them are required.

Except in the case of imperfectly adjusted immigrant species of mosquitoes (e.g. *Anopheles gambiae* in Brazil and Egypt) or under special conditions, such as *A. pseudopunctipennis* in the separated mountain valleys of Chile, complete elimination of malaria-carrying species of mosquitoes is difficult, expensive, and unnecessary. Reducing them to the critical density (see p. 184) is enough. After complete stoppage of malaria transmission over a large area for several years, control operations may be discontinued.

An important fact that is often overlooked is that much malaria is man-made, e.g., by engineers who carelessly leave borrow pits, dam up streams, etc., by irrigation, flooding, impounding, etc., or even, as in Trinidad, by planting forest shade trees that harbor aerial plants (bromeliads) in which *A. bellator* breeds.

During military operations bomb craters, shell holes, foxholes, ditches, vehicle ruts, etc., made many an *Anopheles* happy, for under natural conditions even mosquitoes are often troubled by what corresponds to a housing shortage. Man is an occasionally rational being who in the past has probably done as much to help malaria as he has done to eradicate it.

Although there will be some areas in which malaria will continue to thrive for a while, or perhaps even increase a little, the concept that this disease can be eradicated is not a pipe dream.

In many parts of the world where severe malignant tertian malaria is present, but not in all, a disease occurs which is known as blackwater fever, about the real nature of which there has been much argument but little definite knowledge. It is a scourge in many parts of Africa and in some parts of India, Malaya, and the East Indies, and formerly in the southeastern United States. The disease is characterized by a fever accompanied by an intense jaundice and a tremendous destruction of red corpuscles and excretion of hemoglobin in the urine. In severe cases 60 to 80% of the red blood corpuscles may be destroyed within 24 hours. The disease is usually accompanied by a contraction of the spleen. Severe attacks are usually fatal; cases that recover are prone to subsequent attacks if they remain in an endemic area. Often a blackwater fever attack wipes out the malarial infection. The disease is suggestive of the course of *P. knowlesi* infections in rhesus monkeys.

As yet the reason for the common occurrence of blackwater fever in *falciparum* malaria cases in some areas but not others is unknown; a review of current theories was made by Maegraith (1946). To the senior writer the evidence strongly suggests some antigen-antibody mechanism which reacts against the corpuscles and hemolyzes them, possibly involving something comparable to the Rh factor. Debility, attacks of malaria in adults who have developed no immunity, malnutrition, exhaustion, and exposure to cold seem to be predisposing factors.

Treatment with quinine or the 8-aminoquinolines is contra-indicated, but atebrin or Chloroquine can be given safely. Patients should stay in bed, keep the skin warm and carefully protected from drafts, and drink plenty of warm, salty, alkaline fluids. About half the deaths are due to kidney failure. Complete quiet, diuretics, and a milk diet are recommended, with intravenous saline or glucose injections in some cases, as well as blood transfusions. Enough alkali should be given to make the urine alkaline. Some authors have found injections of liver extract to be of great benefit if given along with atebrin.

REFERENCES

Malaria

Aberle, S. D. 1945. Primate Malaria. *Natl. Research Council Monogr.* 171 pp.

Allison, A. C. 1957. Malaria in carriers of the sickle-cell trait and in newborn children. *Exp. Parasitol.,* 6: 418–448.

American Association for the Advancement of Science. 1941. *A Symposium on Human Malaria.* Edited by F. R. Moulton. Publ. 16 (43 papers by 42 authors). Science Press, Lancaster, Pa.

American Medical Association Council on Pharmacy and Chemistry, Report. Status of Primaquine. 1952. *J. Am. Med. Assoc.,* 149: 1558–1570. (See also series of 5 papers on Korean *vivax* malaria in *Am. J. Trop. Med. Hyg.*, 2: 958–988, 1953.)

Boyd, M. F. 1947. A review of studies on immunity to *vivax* malaria. *J. Natl. Mal. Soc.*, 6: 12–31.

1949. *Malariology.* Saunders, Philadelphia.

Bray, R. S. 1957. Studies on the exo-erythrocytic cycle in the genus *Plasmodium. Mem. London Sch. Hyg. and Trop. Med.* No. 12. 192 pp.

Brunetti, R. 1954. Outbreak of malaria with prolonged incubation period in California; a non-endemic area. *Science,* 119: 74–75.

Coatney, G. R. 1949. Chemotherapy of malaria. *Bull. Pan-Am. Sanit. Bur.*, 28: 27–37.

Coatney, G. R., and Cooper, W. C. 1948. Recrudescence and relapse in *vivax* malaria. *Proc. 4th Internatl. Congr. Trop. Med. Malaria,* 1: 629–639.

Coatney, G. R., et al. 1953. Studies on human malaria. XXII. The protective and therapeutic effects of pyrimethamine (Daraprim) against Chesson strain *vivax* malaria. *Am. J. Trop. Med. Hyg.*, 2: 777–787.

Covell, G., Coatney, G. R., Field, J. W., and Singh, J. 1955. Chemotherapy of malaria. *World Health Organization,* Geneva. 123 pp.

Cowan, A. B. 1955. The development of megaloschizonts of *Leucocytozoon simondi* Mathis and Leger. *J. Protozool.*, 2: 158–167.

Daggy, R. H. 1959. Malaria in oases of eastern Saudi Arabia. *Amer. J. Trop. Med. Hyg.*, 8(2): 223–291.

Fairley, N. H. 1947. Sidelights on malaria in man obtained by subinoculation experiments. *Trans. Roy. Soc. Trop. Med. Hyg.*, 40: 621–676.

1952. The chemoprophylaxis and chemotherapy of malaria in man with special reference to the life cycle. *Australian Ann. Med.*, 1: 7–17.

Fallis, A. M., et al. 1951. Life history of *Leucocytozoon simondi* Mathis and Leger in natural and experimental infections and blood changes produced in the avian host. *Canad. J. Zool.*, 29: 305–328.

Fallis, A. M., Anderson, R. C., and Hennett, F. 1956. Further observations on the transmission and development of *Leucocytozoon simondi. Canad. J. Zool.*, 34: 389–404.

Faust, E. C. 1945. Clinical and public health aspects of malaria in the United States from an historical perspective. *Am. J. Trop. Med.*, 25: 185–201.

1951. The history of malaria in the United States. *Am. Scientist,* 39: 121–130.

Faust, E. C., and Hemphill, F. M. 1948. Malaria mortality and morbidity in the United States for the year 1946. *J. Natl. Malaria Soc.*, 7: 285–292.

Field, J. W., and Shute, P. G. 1956. The microscopic diagnosis of human malaria. II. A morphological study of the erythrocytic parasites. *Studies Inst. Med. Res., Federation of Malaya,* No. 24. Government Press, Kuala Lumpur. 251 pp.

Findlay, G. M. 1947. The toxicity of Mepacrine in man. *Trop. Diseases Bull.*, 44: 763–779.

Fourth International Congress on Tropical Medicine and Malaria. 1948. *Proceedings,* Sect. V, 601–945.

Garnham, C. C. 1948. Exoerythrocytic schizogony in malaria. *Trop. Diseases Bull.*, 45: 831–844.

Hackett, L. W. 1936. Biological factors in malarial control. *Am. J. Trop. Med.*, 16: 341–352.

1937. *Malaria in Europe.* Oxford University Press, London.

1952. The disappearance of malaria in Europe and the United States. *Riv. di Parasitol.,* 13: 43–56.

Howard, L. 1960. Studies on the mechanism of infection of mosquitoes by malaria parasites. D. P. H. Thesis, Johns Hopkins University.

Huff, C. G. 1942. Schizogony and gametocyte development in *Leucocytozoön simondi* and comparisons with *Plasmodium* and *Haemoproteus. J. Infectious Diseases,* 71: 18–32.

Huff, C. G., and Coulston, F. 1944. The development of *Plasmodium gallinaceum* from sporozoite to erythrocyte trophozoite. *J. Infectious Diseases,* 75: 231–249.

1946. The relation and acquired immunity of various avian hosts to the cryptozoites and metacryptozoites of *Plasmodium gallinaceum* and *Plasmodium relictum. J. Infectious Diseases,* 78: 99–117.

Huff, C. G., Coulston, F., Laird, R. L., and Porter, R. J. 1947. Pre-erythrocytic development of *Plasmodium lophurae* in various hosts. *J. Infectious Diseases,* 81: 7–13.

Macdonald, G. 1957. *The Epidemiology and Control of Malaria.* Oxford University Press, N. Y., 201 pp.

Maegraith, B. G., et al. 1952. Suppression of malaria (*P. berghei*) by milk. *Brit. Med. J.,* Dec. 27, 1382–1384, 1405.

Malaria and Other Insect-Borne Diseases in the South Pacific Campaign, 1942–1945. 1947. (Four papers by various authors.) *Am. J. Trop. Med.,* 27(3), Suppl.: 1–128.

National Institutes of Health. 1948. Symposium on exoerythrocytic forms of malaria parasites. *J. Parasitol.,* 34: 261–320.

National Malaria Society. 1951. Symposium: Nation-wide malaria projects in the Americas. *J. Natl. Malaria Soc.,* 10: 97–196.

O'Roke, E. C. 1934. A malaria-like disease of ducks caused by *Leucocytozoon anatis* Wickware. *Univ. Mich. Sch. Forestry and Conserv. Bull.* 4.

Pampana, E. J. 1951. Malaria control with residual sprays. Results in major campaigns. *Bull. World Health Org.,* 3: 557–619.

Pipkin, A. C., and Jensen, D. V. 1956. Avian embryos and tissue culture in the study of parasitic protozoa. I. Malarial parasites. *Exp. Parasitol.,* 7: 491–530.

Russell, P. F. 1951. Malaria and society. *J. Natl. Malaria Soc.,* 10: 1–7.

1952. The present status of malaria in the world. *Am. J. Trop. Med. Hyg.,* 1: 111–123.

Russell, P. F. 1958. Man against malaria—progress and problems. *Rice Inst. Pamphlet,* XLV (1): 9–22.

Rudzinska, M. A., and Trager, W. 1957. Intracellular phagotrophy by malaria parasites: An electron microscope study of *Plasmodium lophurae. J. Protozool.,* 4: 190–199.

Sapero, J. J. 1947. New concept in the treatment of relapsing malaria. *Am. J. Trop. Med.,* 27: 271–283.

Shannon, R. C. 1942. Brief history of *Anopheles gambiae* in Brazil. *Caribbean Med. J.,* 4: 123–129.

Shortt, H. E., Fairley, N. H., Covell, G., Shute, P. G., and Garnham, C. C. 1951. The pre-erythrocytic stage of *Plasmodium falciparum. Trans. Roy. Soc. Trop. Med. Hyg.,* 44: 405–419.

Shortt, H. E., Garnham, C. C., Covell, G., and Shute, P. G. 1948. The pre-erythrocytic stage of human malaria, *Plasmodium vivax*. *Brit. Med. J.*, No. 4550, 547.

Shute, P. G., and Maryon, M. 1952. A study of human malaria oöcysts as an aid to species diagnosis. *Trans. Roy. Soc. Trop. Med. Hyg.*, 46: 275–292.

Swellengrebel, N. H., and de Buck, A. 1938. *Malaria in the Netherlands.* Scheltema and Holkema, Amsterdam.

Taliaferro, W. H., and Cannon, P. R. 1936. The cellular reactions during primary infections and superinfections of *P. brasilianum* in Panamanian monkeys. *J. Infectious Diseases*, 59: 72–125.

Thurston, J. P. 1953. *Plasmodium berghei*. *Exptl. Parasitol.*, 2: 311–332.

Wilcox, A. 1943. Manual for the microscopical diagnosis of malaria in man. *Natl. Insts. Health Bull.*, 180.

Williams, L. L., Jr. 1958. Malaria eradication—growth of the concept and its application. *Am. J. Trop. Med. Hyg.*, 7: 259–267.

Young, M. D. 1957. Resistance of *Plasmodium malariae* to pyrimethamine (Daraprim). *Am. J. Trop. Med. Hyg.*, 6: 621–624.

Young, M. D. 1953. Malaria during the last decade. *Am. J. Trop. Med. Hyg.*, 2: 347–359.

Young, M. D., et al. 1945–1946. Studies on imported malarias, 1–5. *J. Natl. Malaria Soc.*, 4: 127–131, 307–320. *Am. J. Hyg.*, 43: 326–341; *Am. J. Trop. Med.*, 26: 477–482.

Blackwater Fever

Foy, H., and Kondi, A. 1935–1936. Researches on Blackwater fever in Greece, I–IV. *Ann. Trop. Med. Parasitol.*, 29: 383–397, 497–515; 30: 423–433.

Maegraith, B. G. 1946. Blackwater fever—modern theories. A critical review. *Trop. Diseases Bull.*, 43: 801–809.

Ross, G. R. 1932. Researches on blackwater fever in southern Rhodesia. *Mem. London Sch. Trop. Med.*, No. 6.

Chapter 9

SPOROZOA OTHER THAN MALARIA

Haemosporidia

As noted on p. 163, the subclass Haemosporidia contains (1) the order Plasmodiida, considered in the last chapter; (2) the order Babesiida, containing the families Babesiidae, Theileridae, and Gonderidae; and (3) some other parasites of uncertain affinities, e.g. *Toxoplasma, Besnoitia.* These as well as other Sporozoa, of interest as parasites of man or domestic animals, will be considered in the present chapter.

BABESIIDA

The general characteristics of this group of parasites were given on p. 164. The life cycles are rather imperfectly known, even in the most important species, and most of them are entirely unknown. Eight or ten genera have been described. Members of the genera *Babesia* (=*Piroplasma*) and *Theileria* cause important and highly destructive diseases of all the large herbivorous animals and also of dogs, and species of *Aegyptianella* are injurious to poultry. All species of Babesiidae as far as is known are transmitted by ticks. In the genus *Babesia* the parasites are known to multiply only in the red blood corpuscles; they are usually pear-shaped, and grouped in two's or four's (Fig. 40, *A–G*). In the genus *Theileria,* the organisms are smaller, of variable shape, and with no tendency to be grouped in pairs (Fig. 40, *K–M*). In this family (Theileridae) schizogony occurs in lymphocytes. On the other hand, in the family Gonderidae (*Gonderia* and *Cytauxzoon*) schizogony occurs in lymphocytes or histiocytes.

Although about 20 species of *Babesia* have been named, many of these are probably not valid species, and only a few are important

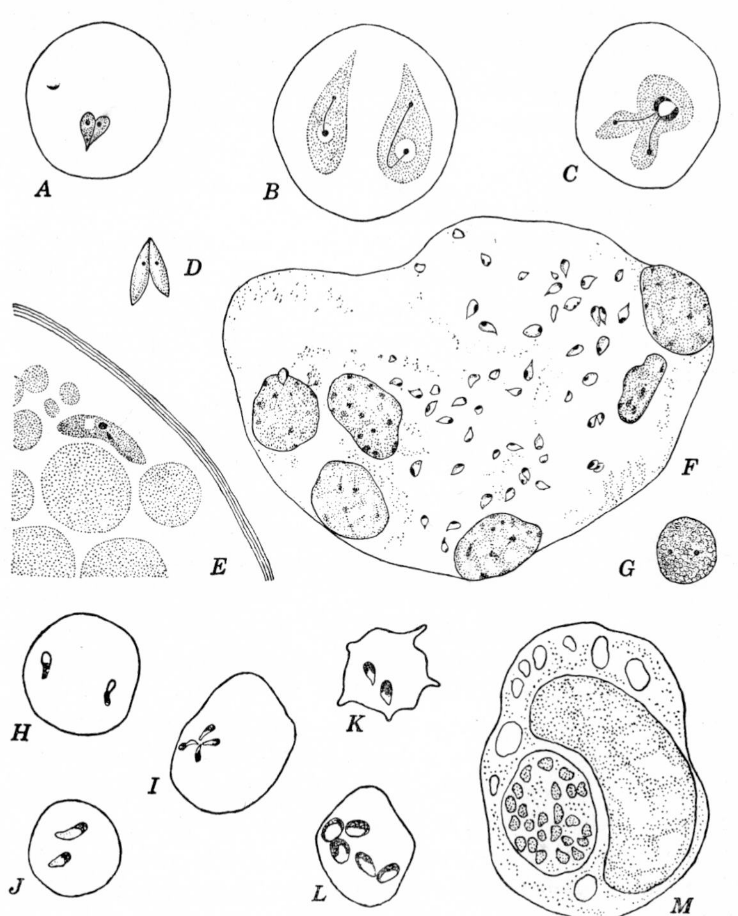

Fig. 40. *A–G, Babesia bigemina; A–C,* in cattle; *D–G,* in tick. *A,* trophozoite in red cell of cattle, showing binary fission; *B,* large pyriform pairs in red cell; *C,* division stage in red cell; *D,* associated isogametes in gut of tick; *E,* ookinete in section of ovum of tick; *F,* sporozoites in a single focal plane of salivary anlage of tick; *G,* late stage of syngamy of isogametes. (Adapted from Dennis, *Univ. Calif. Publ. Zool.,* 33, 1930; 36, 1932.) *H–J, Gonderia mutans* in cattle blood; *K–L, Theileria parva* in cattle blood; *M,* leucocyte containing *T. parva* schizont. (Adapted from Brumpt, *Bull. Soc. Pathol. Exotique,* 13, 1920, and *Ann. Parasitol.,* 1, 1923; and 2, 1924.)

parasites of domestic animals (Neitz, in Cole et al., 1956). These include 5 species in cattle (*bigemina,* cosmopolitan; *argentina* in South America; *bovis* in Europe; and *berbera* and *major* in North Africa); 2 species in sheep and goats (*ovis* and *motasi* in Europe and

Africa); 2 species in pigs (*trautmanni* in Eastern Europe and Africa; *perroncitoi* in Sardinia); 2 in horses (*caballi,* around the Mediterranean and in Central America; *equi* in Europe and Africa); 2 in dogs (*canis* in Europe, Asia, and Africa; *gibsoni* in Asia and Africa); and 1 in cats (*felis* in Africa). It is obvious that most of the Babesiidae are limited to the Old World, but one species, *Babesia bigemina,* has a world-wide distribution in cattle.

The diseases caused by *Babesia,* often called piroplasmosis from the name *Piroplasma,* long used instead of *Babesia,* are characterized by destruction of red blood corpuscles and elimination of hemoglobin with the urine, hence the name "redwater fever." They produce fever, anemia, jaundice, and injury to the liver and kidneys. *B. bigemina* (Fig. 40, *A–G*) is the cause of Texas fever or redwater fever in cattle, of world-wide distribution. The disease is transmitted by ticks of the genus *Boophilus* (see p. 595) in which the parasites, after fertilization in the hindgut, invade the reproductive organs. They become enclosed with the eggs and subsequently undergo extensive multiplication and migration to all the tissues of the developing tick embryo. Some of the parasites enter the salivary glands and can then be transmitted by the seed ticks when they feed (Dennis, 1932). The adult ticks do not transmit the infection. The developmental cycle of *Babesia* in the vertebrate host is not fully known; the only forms known are small oval or pear-shaped bodies which bud to produce clusters of two, or in some species four, parasites (Fig. 40, *A–C*). Some writers, however, have described what they interpret as schizogony in red cells. By quarantine and anti-tick methods this one-time scourge of cattle has been completely eliminated in the United States. *Babesia canis* causes a form of redwater fever in dogs in Brazil.

The genus *Theileria* (Fig. 40, *K–M*) contains a single species, *T. parva,* which causes the deadly East Coast fever of cattle in Africa. After about 15 days of fever, infected animals seem to be convalescing and then often suddenly fall over and die, apparently from edema of the lungs. In adult animals the mortality may be 95%. During recent years the disease has occurred in sudden sporadic outbreaks. This may be related to the fact that the reservoirs are wild bovine animals and that the ticks involved apparently do not transmit the disease agent to their offspring. Aureomycin has been reported to kill the schizonts of *T. parva* in lymphocytes but has no effect on erythrocytic stages. The latter are sensitive to the anti-malarial drug, plasmoquin.

The species belonging to the Gonderidae produce very mild infections. *Gonderia mutans,* for example, causes a practically symptomless infection of cattle over most of the Old World. The cycle of development of these parasites in the tick hosts has been worked out for only one species, *G. annulata* (=*Theileria dispar*), by Sergent et al. in 1936. Neitz (in Cole et al., 1956) reviewed the classification, transmission, and biology of the species which infect domestic animals.

Toxoplasma

This organism is still one of the least understood of the parasites of man and domestic animals, although it is extremely common and under some circumstances very dangerous. Its taxonomic relations are so poorly known that it has even been suspected of being a fungus instead of a protozoan, and one man has argued for inclusion in the Trypanosomidae.

The parasites (Fig. 41) are crescent shaped or oval, 6 to 12 μ long, with a discrete central nucleus, and occur in pairs or groups of pairs. It has been generally assumed that *Toxoplasma* multiplies by binary fission; however, Goldman, Carver, and Sulzer (1958) reported a peculiar reproduction, involving internal budding with two new individuals produced inside the parent cell. Using the electron microscope, Bringmann and Holz (1954), Gustafson et al. (1954), and several later workers demonstrated an organelle at the anterior pole which may be a cell mouth or a hold-fast and fibrils extending over the outside of the anterior two-thirds of the body; the latter may account for the gliding movement of the organism. Similar external fibrils occur in malaria sporozoites.

The parasites are found free in the blood stream or in tissues, or inside cells of many different types, particularly those of the reticuloendothelial system, white blood cells, and epithelial cells. After experimental inoculations in laboratory animals the organisms appear free in the blood plasma and in various organs within 4 hours; they subsequently disappear from the blood, but in rats they are irregularly demonstrable in the liver, lungs, spleen, etc., for several weeks and are uniformly present in the brain for at least 2 years. Here, and sometimes in other parts of the body, in chronic cases they occur in large intracellular "pseudocysts" containing up to 50 or more organisms (Fig. 41), apparently in an inactive resting state, since they cause no inflammatory reaction as do the free forms. It is believed that it is due to rupture of these pseudocysts, from unknown causes, that the infection sometimes flares up again in chronic cases.

As far as known there is only one species of *Toxoplasma, T. gondii,*

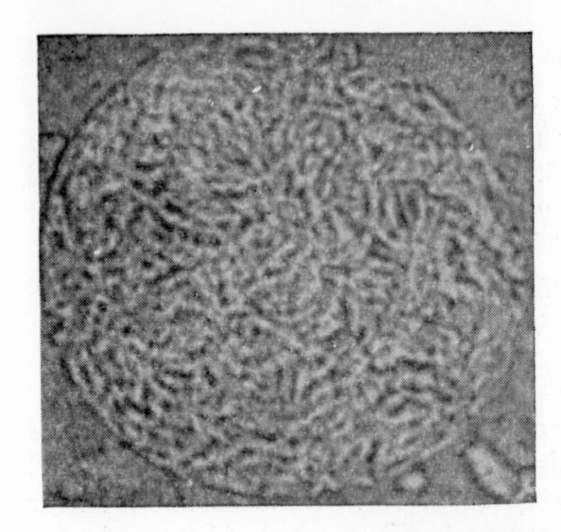

A

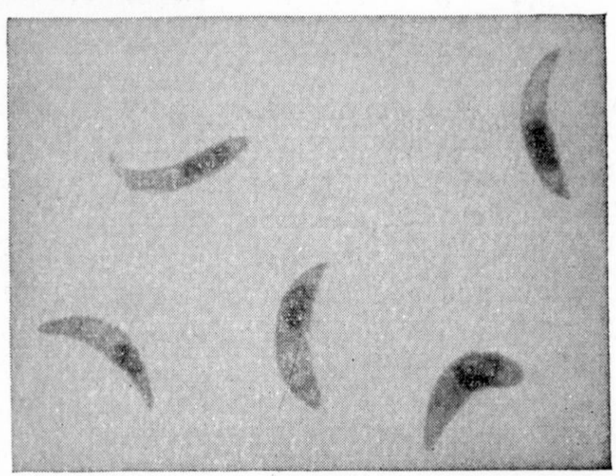

B

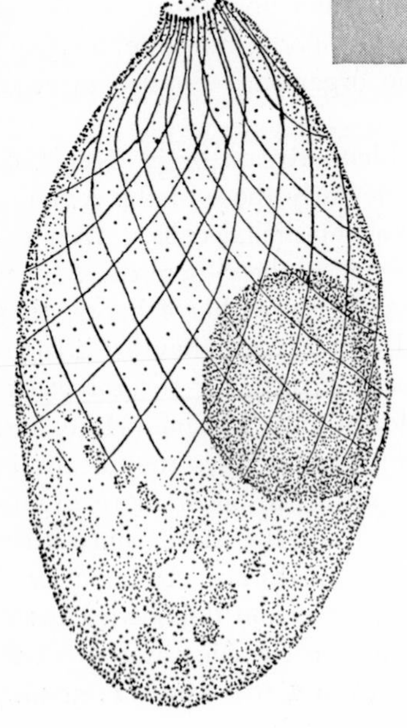

D

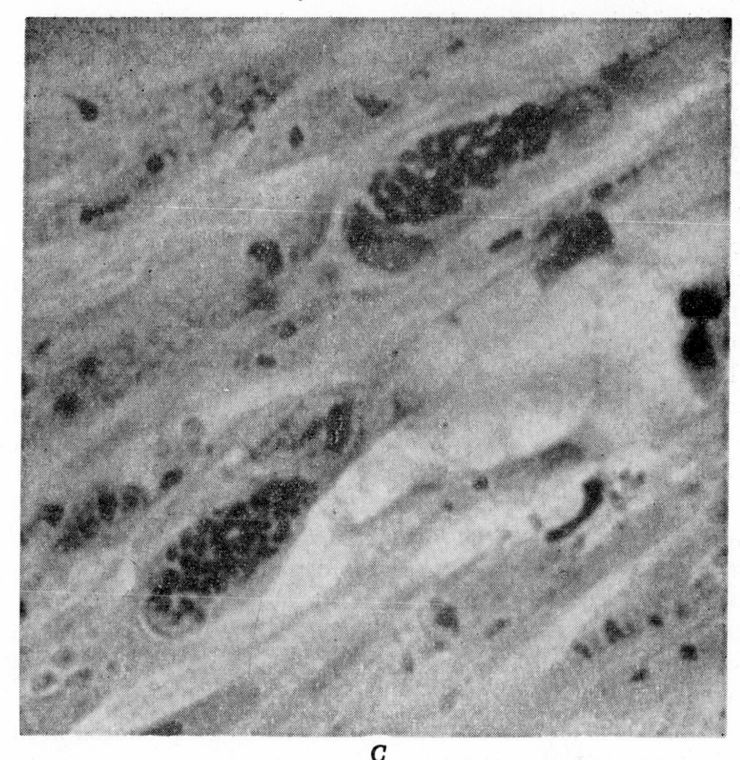

C

Fig. 41. *Toxoplasma gondii.* *A*, pseudocyst in the brain of a mouse; *B*, from peritoneal exudate of a mouse; *C*, toxoplasmas in tissue culture of chick embryo liver. Note pairs of parasites and agglomerations within cells; *D*, as seen by electron microscopy. (*A*, after Frenkel, *Ann. N. Y. Acad. Sci.*, 64, Art. 2, 1956; *B* and *C*, after Jacobs, *Ann. N. Y. Acad. Sci*, 64, Art. 2, 1956; *D*, after Bringmann and Holz, *Z. Tropenmed. Parasitol.*, 5, 1954.)

which shows practically no choosiness about its hosts, apparently being able to infect all kinds of birds and mammals and even surviving in lizards and turtles for at least a month. As indicated by antibody tests, and in many cases by transmission to laboratory animals as well, this parasite has been reported to occur in 59% of dogs, 34% of cats, 48% of goats, 30% of pigs, 3 to over 20% of rats, and 10 to 12% of pigeons. In man the incidence of significantly positive antibody tests ranges from 0% in Eskimos to 68% in Tahiti; in American cities the incidence ranges from 17 to 35%. The prevalence of human infection is highest in warm, moist areas, less in warm, dry areas, and least in the very cold areas of the world. It will be observed that man's closest animal friends, dogs and cats, must constitute important reservoirs of the disease, and that rats, in addition to their long-known relation to plague and typhus, are important reservoirs of this disease as well; infective *Toxoplasma* was demonstrated in the brains of 14 of 160 wild rats in Savannah, Georgia, and in over 3% in Memphis, Tennessee.

There is no direct evidence as to the usual method of transmission, but prenatal infections occur in man and animals, and animals can be infected by eating infected flesh or feces. Several workers have reported that animals become infected when fed pseudocysts but not when fed free toxoplasmas. This suggests that the pseudocyst may be a resistant stage, perhaps comparable in epidemiological significance to the cystic stages of intestinal amebas. There is no evidence of seasonal infection and person-to-person transmission seems rare. There is a high incidence of antibodies in rabbit handlers. Blood-sucking arthropods may occasionally play a part; Laarman in 1956 reported the mechanical transmission of *Toxoplasma* by the stable fly, *Stomoxys*, and fleas, after feeding on infected animals, are infective when eaten. Ticks have been reported to harbor *Toxoplasma* for at least 60 days after feeding on infected animals.

In animals, toxoplasmosis may produce anything from rapidly fatal infections to unapparent ones. Mice, rabbits, and birds often suffer severely, but rats usually have mild and transient infections. According to Eyles and Coleman a fatal infection can be produced in mice by the inoculation of one parasite.

In man, the symptoms as a rule become less and less pronounced with advancing age. After the maturation of defense mechanisms (see p. 26) of human hosts, the parasites are usually got under control before they do serious damage. Few fatal cases have been reported in human adults. Adults and children alike only rarely show symptoms. The symptoms may involve a typhuslike rash, fever, enlargement of lymph glands, and eye disturbances (chorioretinitis, uveitis). Fatal encephalitis attributed to *Toxoplasma* has been reported. In infants infected before birth, fatal results are usual, the acute infection being severe, commonly with effects in the brain and eye, which are often almost completely destroyed. In congenital cases there is almost always severe eye injury (chorioretinitis), hydrocephaly or microcephaly in two-thirds, and severe mental and nervous disturbances and convulsions in nearly all of them.

Undoubtedly, as in animals, these prenatal human infections are at first generalized, but as the mother develops antibodies and shares them with the fetus, the parasites are inhibited in the visceral organs but continue their devastating activity in the brain. As Frenkel and

Friedlander (1951) point out, this "blood-brain barrier," as it is called, effective in all infectious diseases that penetrate into the central nervous system, is probably related to the fact that there is only about 0.3 to 0.5% as much protein in the cerebrospinal fluid as in the blood, and a correspondingly small amount of antibody content. Hence parasites are able to persist in the central nervous system long after antibody reaction has made it too "hot" for them in other parts of the body.

Evidence indicates that infections in newborn babies result from prenatal infection from mothers that acquire infections, usually symptomless, during pregnancy, *not* to old chronic infections. Even if a baby survives it is often so severely damaged that it can lead only a vegetative existence. Children sometimes have transitory attacks of chorioretinitis due either to new infections or to relapses from the rupture of pseudocysts in the brain.

Aside from the finding of parasites by smears or animal inoculation, diagnosis of present or past infections may be made by several immunological tests. One is intradermal inoculation of an extract of the parasites called toxoplasmin; it is not very reliable. Another is complement fixation, but this is slow in development, and fades quickly. The most valuable has been a "dye test" devised by Sabin and Feldman (1948). When living parasites in mouse peritoneal fluid are placed in serum containing antibodies against *Toxoplasma,* along with a methylene blue dye and some normal human serum (preserved by freezing) to serve as "activator," the parasites lose their affinity for the dye, whereas in normal serum containing no antibodies they round up and have both the nucleus and cytoplasm stained. In serum dilutions of 1 : 16 or greater this dye test appears to be a reliable index to present or past infection; it remains positive for as long as 25 years, although the titer gradually drops. This test becomes positive early in an infection, usually during the second week, quickly rises to a titer of 1 : 1000 or more, and then remains high for a year or more. A rising titer by either the dye test or complement fixation indicates an existing or very recent infection, but a single test may only mean past infection. The dye test has the disadvantage that it requires the maintenance of *Toxoplasma* in the laboratory. Jacobs and Lunde (1957) described a hemagglutination test which they regard as highly promising for routine laboratory use. Inoculation of suspected material into animals has been a useful diagnostic method for research purposes. Goldman (1957) has described a specific staining procedure for *Toxoplasma* using a fluorescein-labeled antibody as the reagent. This may be useful as a means of locating the organism in tissues.

A combination of the anti-malarial drug, Daraprim (see p. 189) and sulfonamides has given the best results in treatment of acute cases, but, in rats at least, the parasites are less easily eliminated after 3 or 4 months, when pseudocysts presumably have formed. Antibiotics have proven to be useless.

Organisms Resembling *Toxoplasma*

There are several genera of *Toxoplasma*-like organisms occurring in a wide variety of animals. One of these, *Besnoitia besnoiti,* produces a generalized infection with a dermatitis in horses and cattle. In the acute stage of the infection the organisms are almost indistinguishable from *Toxoplasma* but in its chronic phase produces large cysts (up to 2 mm. in diameter) very unlike the pseudocysts of *T. gondii.* A relative, *B. jellisoni,* is found in deer mice in the United States. Experimentally, animals can be infected by feeding infected material (Jellison et al., in Cole, 1956).

Sarcocystis may also be related to *Toxoplasma.* A number of species have been named from herbivorous birds and mammals, including man. Only cyst stages from cardiac and skeletal muscle are well known. Virtually nothing is known of the life history of *Sarcocystis.* Some workers have reported transmission by feeding of infected muscle, but Frenkel (in Cole, 1956) considers this as poorly substantiated. The fact that carnivores are rarely infected would argue against meat-eating as the important mode of transmission.

Encephalitozoon occurring in cysts in the brains of rodents may also belong in this group. *Besnoitia, Sarcocystis,* and *Encephalitozoon* are immunologically different from *Toxoplasma.*

Hepatozoon and Hemogregarines

These parasites, although classified in the subclass Coccidiia, resemble the Haemosporidia in requiring an intermediate host. The forms found in cold-blooded animals may belong to either of two genera: (1) *Haemogregarina,* in which the schizogonic cycle and the gametocytes (Fig. 34*L*) may both occur in blood corpuscles, and oocysts containing free sporozoites which develop in leeches or arthropods; or (2) *Karyolysus,* in which schizogony occurs in endothelial cells, producing merozoites which enter red corpuscles and become gametocytes, the sporogonic cycle occurring in mites. The hemogregarines of birds and mammals are placed in the genus *Hepatozoon;* schizogony occurs in the reticulo-endothelial cells of liver, spleen, or bone marrow, and the gametocytes (Fig. 34*K*) develop in circulating mononuclear leucocytes. The sporogonic cycle of *Hepatozoon muris* of rats and *H. griseisciuri* of squirrels occurs in dermanyssid mites (see p. 555), whereas that of *H. canis* of dogs occurs in ixodid ticks. In these intermediate hosts there develop large coccidium-like oocysts containing numerous sporocysts, each with about sixteen sporozoites. The resemblance to coccidians is further indicated by the fact that the sporocysts cause infection only via the alimentary canal when swallowed with the mites or ticks in which they develop. *H. canis* causes a serious and sometimes fatal illness in dogs in India and Africa.

The gametocytes of hemogregarines in red cells are distinguishable from those of *Haemoproteus* by the lack of pigment.

Coccidia

General account. Although negligible as human parasites, Coccidia cause a greater economic loss among domestic and game animals in temperate climates than any other group of Protozoa. They are of major importance to poultry raisers and produce serious disease in rabbits and cattle. Horses, sheep, goats, pigs, dogs, cats, guinea pigs, ducks, geese, pigeons, and even canaries frequently suffer from their attacks. For relationships of the Coccidia to other Sporozoa, see pp. 161–163.

The Coccidia are most commonly parasites of the epithelial cells of some part of the intestine, although some species attack the liver and other organs. The species which cause important infections in domestic animals belong to two genera, *Eimeria* and *Isospora,* which differ from each other mainly in details of development within the oocysts. The life cycle is graphically shown in Fig. 42. After one, two, or more schizogonic cycles the merozoites develop into gametocytes, usually in the same type of cells in which the schizogony occurred. The microgametocytes produce a swarm of minute two-flagellated microgametes which fertilize the macrogametes, usually after the latter have escaped from the cells which mothered them. The macrogametes are provided with cyst walls, but have a small opening called a micropyle at one end through which the microgametes are able to enter. The resulting zygote is a young oocyst, ready for escape from the host in which it was developed, and prepared to withstand conditions in the outside world until opportunity to enter another host is afforded.

In most species the oocysts (Fig. 43A) are undeveloped when they leave the host with the feces and require 30 hours to 2 weeks to

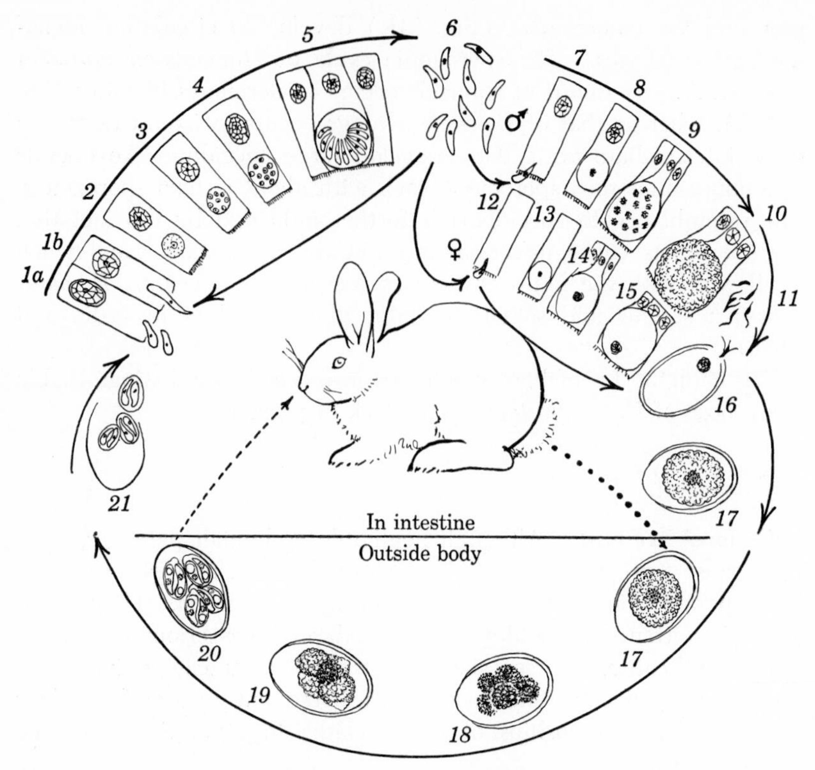

Fig. 42. Life history of *Eimeria perforans* of rabbit. *1a*, sporozoite entering intestinal cell; *1b*, merozoite entering intestinal cell; *2*, developing schizont; *3–5*, multiplication of nuclei and formation of merozoites; *6*, liberated merozoites, re-infecting other intestinal cells; *7–10*, development of ♂ gametocyte; *11*, micro-gametes; *12–15*, development of ♀ gametocyte; *16*, fertilization of macrogamete; *17*, undeveloped oocyst ready to leave body of rabbit, and after escape with feces; *18–19*, formation of sporoblasts; *20*, development of sporocysts and sporozoites (ripe oocyst); *21*, escape of sporozoites after ingestion of ripe oocyst. (Adapted from figures by Becker and by Wetzel.)

develop, depending upon the species and the temperature. Development takes place in two steps, (1) a division of the nucleus and cytoplasm into a number of parts called sporoblasts, often leaving a residual mass of cytoplasm which may subsequently disappear, and (2) the further development of these sporoblasts into sporocysts with resistant cyst walls, and the division of their contents into a number of sporozoites; sometimes each sporocyst has a residual mass of cytoplasm of its own. The sporocysts are cysts within cysts, and in some species may be liberated from the parent oocysts before re-entering a host. In the genus *Eimeria* each oocyst produces 4 sporocysts each with 2 sporozoites (Fig. 44*A*), whereas in the genus *Isospora* each oocyst produces 2 sporocysts each with 4 sporozoites (Fig. 44*B*).

The oocysts are easily destroyed by a temperature of about 50°C., by desiccation, and by extreme cold, but are highly resistant to chemicals. They can be cultured in the laboratory in a 2 to 5% dichromate or a 1% chromic acid solution. It is startling to find them undergoing development in fixed and stained slides!

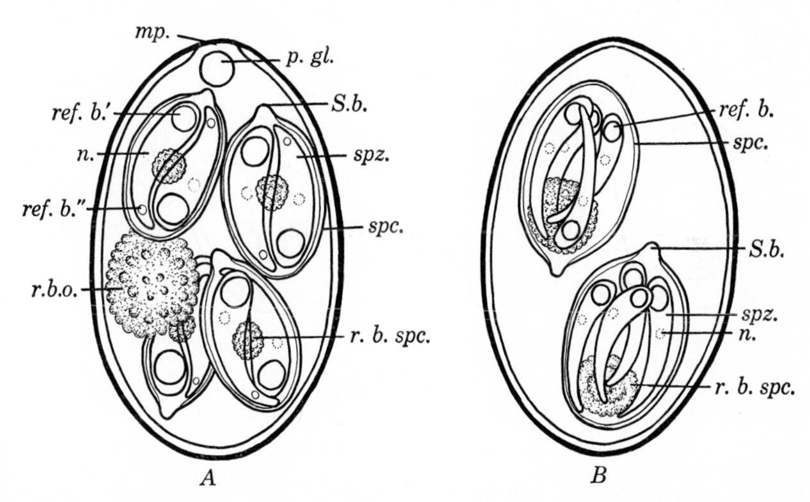

Fig. 43. Diagrammatic representation of oocysts of *Eimeria* (*A*) and *Isospora* (*B*); *mp.*, micropyle; *n.*, nucleus of sporozoite; *p.gl.*, polar globule; *r.b.c.*, residual body of oocyst; *ref.b.*, refractile bodies of sporozoites; *r.b.spc.*, residual body of sporocyst; *S.b.*, Stieda body or "plug" of sporocyst; *spc.*, sporocyst; *spz.*, sporozoite. (Adapted from Boughton and Volk, *Bird Banding*, 9, 1938.)

The sporozoites liberated from ingested oocysts penetrate cells in their chosen sites of development, grow into schizonts, and then divide into a cluster of spindle-shaped merozoites, usually about 16 to 30, but in the deadly *E. tenella* of chicks about 900, and in *E. bovis* of cattle over 100,000. In most coccidians two or more generations of schizonts are produced before the sexual forms are developed, but in most if not all species the number of schizogonic generations is limited, and therefore continuation of the disease depends on repeated reinfections.

Eimeria bovis of cattle differs from typical species of *Eimeria* in producing a single generation of huge schizonts, 250 to 400 μ in diameter, easily visible to the naked eye, and containing over 100,000 merozoites.

The sporozoites begin their development in endothelial instead of epithelial cells; the schizonts mature in 14 to 18 days and then occupy the outer portion of the interior of badly bulged villi. Obviously, with such prolific schizonts, one generation of them is enough. Although the schizonts of this species occur in the small intestine, the gametocytes and oocysts develop in the cecum and colon; the pathological effects and symptoms are associated only with the latter.

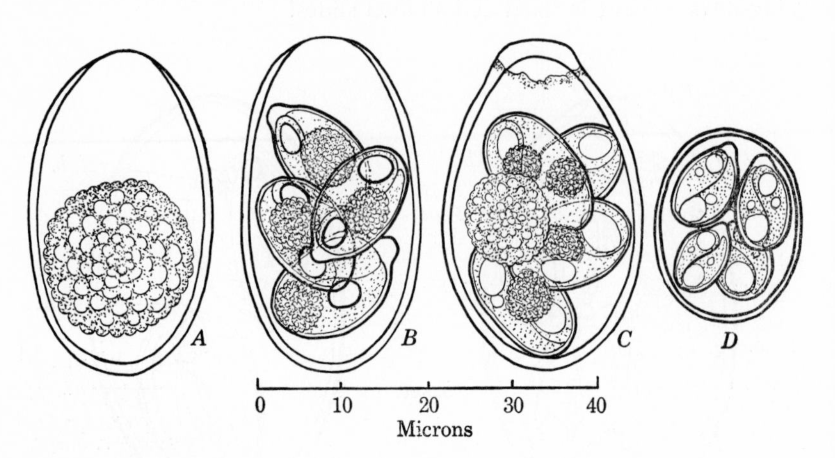

Fig. 44. Various types of oocysts of *Eimeria*. *A*, unsporulated oocyst of *E. stiedae* of liver of rabbit; *B*, ripe oocyst of *E. stiedae* with residual bodies in sporocysts but not in oocyst; *C*, ripe oocyst of *E. magna* of intestine of rabbit, with residual bodies in both oocyst and sporocysts; *D*, *E. tenella* of cecum of chickens, with no residual bodies. (*A* to *C* after Kessel and Jankiewicz, *Am. J. Hyg.*, 14, 1931.)

The true nature of these huge schizonts has only recently been determined (Hammond et al., 1946). Similar organisms occur in horses, sheep, and camels and were formerly recognized under the names *Globidium* or *Gastrocystis*. When the life cycles of more of these are determined, it may be desirable to separate them from the genus *Eimeria*.

Species. The Coccidia comprise numerous species, most of which show marked host specificity; *Isospora* shows more laxity in its choice of hosts than *Eimeria*. Not only do most species inhabit only a single kind of animal or a few closely related ones, but also a single animal may harbor several different species of Coccidia. Members of the genus *Eimeria* have made themselves at home in almost every kind of vertebrate, especially herbivorous ones, and in some invertebrates. In cold-blooded animals the oocysts mature before leaving the host; in warm-blooded ones they usually mature afterwards. Intestinal forms are very common in rodents, pigs, ruminants, and poultry, and one species is common in the liver of rabbits. *Isospora* is common in small birds (Fig. 45*C*); Boughton reported references to its occurrences in 173 species, mostly passerines, but whether these represent one or many species is unknown. English sparrows show a very high incidence of infection, especially in the southern states. Canaries suffer from the infection but chickens do not; the latter are afflicted only by *Eimeria.* Among mammals species of *Isospora* are especially frequent in carnivores but are also reported from man, pig, and hedgehog.

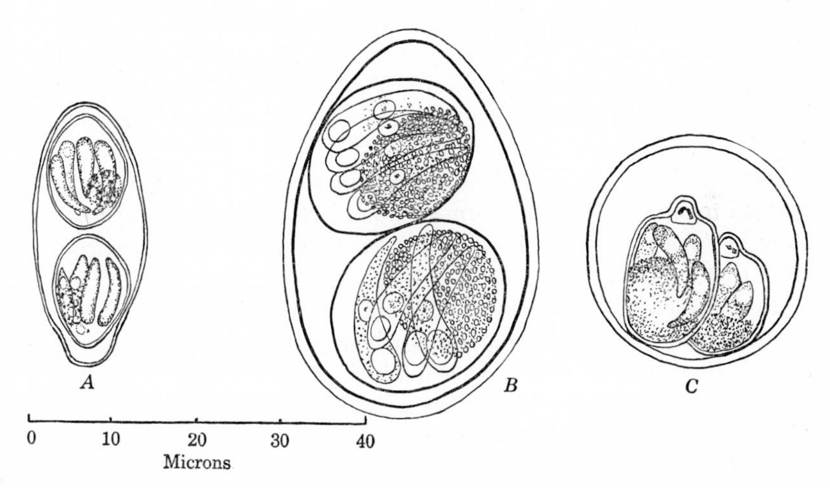

Fig. 45. Oocysts of three species of *Isospora.* *A, I. belli* of man (after Haughwout, *Philip. J. Sci.,* 18, 1921); *B, I. felis* of cats (after Wenyon, *Protozoology,* 1926); *C, I. lacazii* of sparrows (after Becker, *Coccidia and Coccidiosis,* 1934).

Chickens harbor at least 8 species of *Eimeria,* cattle 6, rabbits 6, and pigeons 1; dogs harbor 3 species of *Isospora,* all of which are shared by cats and foxes. Only *Isospora* occurs in man, 3 species having been reported.

The various species of Coccidia vary in the site and developmental details of their schizogonic cycle but are nearly always identifiable by the oocysts alone, which is fortunate since these are the only forms commonly seen. The principal characters used in distinguishing the oocysts of different species are size and shape of entire oocysts, size and shape of sporocysts, the distinctness of the micropyle, the presence or absence of residual bodies in the oocyst and in the sporocysts, and the thickness of the oocyst wall.

Ingestion of oocysts of "foreign" coccidians, passing through the alimentary canal intact, may lead to errors of interpretation. Species which are parasitic in the liver of herrings and in the testes of sardines, for example, have been mistakenly described as human parasites, and it is possible that some of the reported *Eimeria* infections in man may have been pseudo-infections from eating rabbit livers infected with *E. stiedae* (Fig. 44A, *B*). Passage of the oocysts through insusceptible animals may serve as a means of distribution.

Human infections. Although only a few hundred cases of *Isospora* infections in man have been reported, the infection is undoubtedly much commoner in many unsanitary countries than that would suggest. Most of the cases have been reported from the Middle East, Africa, Southwest Pacific, Japan, and South America. In São Paulo, Brazil, *Isospora* oocysts were found in 0.1% of the stools examined, and a somewhat similar incidence occurs in Egyptian villages near Cairo.

Isospora hominis and *I. belli,* both reported from man, were long merged together under the former name, but recent work by Elsdon-Dew in Africa and by Meira and Correa in São Paulo supports the view that these are quite distinct species. Both occur in the United States. *I. hominis* has oocysts that are ovoid and measure about 16 by 10 μ, with developed sporocysts when passed; it closely resembles *I. bigemina* of dogs and may actually be that species. *I. belli* has larger oocysts, undeveloped when passed, that measure 25 to 33 by 13 to 16 μ, and are definitely narrower at one end (Fig. 45*A*). This seems to be much the commonest species in man. A third species, *I. natalensis,* with oocysts about 30 by 25 by 21 μ, and somewhat resembling *I. rivolta* of dogs, was reported once from an African in Natal by Elsdon-Dew. After leaving the body the oocysts of *I. belli* develop 2 sporocysts each with 4 sporozoites, packed like sardines in a tin, within 24 to 72 hours. There is evidence that both these species develop in the duodenum or perhaps even in the bile duct. In 28 cases in São Paulo, 13 had large, immature oocysts of *I. belli,* whereas 15 had small, mature oocysts of *I. hominis.* In Chile all of 11 cases studied were *I. belli.*

Mild diarrhea with light-colored, fatty stools, abdominal distress, and eosinophilia are commonly described in connection with human infections, but some infections seem to be without symptoms. In experimental infections diarrhea and fever appear in about a week. Jefferey (1958) recently reported an explosive outbreak of diarrheic disease attributed to *I. belli* in South Carolina. In general, the infections appear to be self-limited and last only a few weeks.

Infections in animals. In animals the pathogenic effect varies considerably with different species and with the severity of the infection. In light cases there are often no symptoms, but in severe attacks by pathogenic species there is extensive destruction of the epithelium in the chosen sites, with sloughing of the walls and severe hemorrhage. The symptoms are loss of appetite, emaciation, weakness, pallor, diarrhea, bloody feces, and sometimes fever. Animals develop immunity from repeated sublethal infections, but much more quickly and permanently against some species than against others. The immunity is local in nature, with no relation to antibodies in the blood.

Chickens harbor eight species of *Eimeria;* fortunately most of them are mildly pathogenic, but *E. tenella* (Fig. 44*D*) and *E. necatrix* are very harmful, the former causing severe injury to the ceca in young chicks, with often fatal bloody diarrhea, the latter a more chronic intestinal disease in older birds, characterized by leg weakness, pallor, and general unthriftiness. In cattle several species are pathogenic, including *E. zurnii, E. bovis,* and *E. alabamensis. E. bovis* has a cycle of development in the host of 20 days and produces huge schizonts visible to the naked eye, with merozoite families numbering over 100,000 (see p. 207). *E. alabamensis* has a cycle of 8 to 12 days and produces modest merozoite families of 16 or less. Almost all calves in the southern states are infected; if heavily infected they suffer red or watery diarrhea, weakness, and emaciation, which is often fatal. Other species of Coccidia harmful to domestic animals are *E. meleagrimitis* in turkeys, *E. truncata* in geese, *E. debliecki* in pigs, *E. solipedum* and *E. ungulata* in horses, and *Isospora bigemina* in dogs.

Treatment and prevention. A number of drugs are capable of preventing infection if given with food or water prior to establishment of the infection, e.g., sulfur, certain organic sulfur compounds, and many sulfa drugs (sulfonamides), especially Sulfaquinoxaline. The last has therapeutic effect against *E. bovis* in cattle but not against *E. alabamensis,* presumably because the abundant merozoites of the former are more exposed while making their way to the large intestine where the sexual stages develop, whereas the few merozoites of *E. alabamensis* immediately enter neighboring cells (Boughton, 1943).

Sulfa drugs with a pyrimidine nucleus (Sulfadiazine, Sulfamethazine, and Sulfamerazine especially) have been found to have curative as well as preventive action against cecal coccidiosis in chickens (*E. tenella*) if given as soon as symptoms appear, or before. However, since symptoms do not usually appear until the fifth day after infection, and 90% of the mortality occurs by the end of the seventh day, there is not time for much procrastination. These drugs, added to the food at the rate of 0.5 to 2% for 3 days, beginning within 4 days after infection, greatly reduce mortality in chicks (Swales, 1946). Addition of 2 grams of sodium Sulfamerazine per quart of drinking water for 3 days is also effective and perhaps better, since infected chicks lose their appetites but not their desire for water. Sulfaquinoxaline at the rate of 0.05% of the food, given intermittently, or one-fourth that dose continuously, effectively controls *Eimeria tenella* and *E. necatrix* infections in chicks, reducing deaths from over 17% to 1 or 2%. It prevents development of later stages if given up to the fourth day after infection.

About all a harassed poultryman or animal raiser can do to hold in check the ravages of coccidiosis is to try by sanitary means to limit the ingestion of oocysts to a number that will lead to immunity rather than death, and to utilize the sulfa drugs in food or water as soon as evidence of infection appears. Methyl bromide applied at the rate of 0.15 to 0.3 cc. per square foot of litter or soil inactivates oocysts. It can also be used as a space fumigant in brooder houses; 2 lb. per 1000 cubic foot prevents infection.

REFERENCES

Sporozoa (Including *Toxoplasma*)

Anonymous. 1949. Coccidiosis (papers by various authors). *Ann. N. Y. Acad. Sci.*, 52: 429–624.

Barksdale, W. L., and Routh, C. F. 1948. *Isospora hominis* infections among American personnel in the southwest Pacific. *Am. J. Trop. Med.*, 28: 639–644.

Becker, E. R. 1934. *Coccidia and Coccidiosis.* Iowa State College, Ames, Iowa.

Clark, G. M. 1958. *Hepatozoön griseisciuri* n. sp.; a new species of *Hepatozoön* from the grey squirrel (*sciurus carolinensis* Gmelin, 1788), with studies on the life cycle. *J. Parasitol.*, 44: 52–63.

Cole, C. R., et al. 1956. Some protozoan diseases of man and animals: Anaplasmosis, babesiosis, and toxoplasmosis. *Ann. N. Y. Acad. Sci.*, 64 (Art. 2): 25–277.

Cook, M. K., and Jacobs, L. 1958. Cultivation of *Toxoplasma gondii* in tissue cultures of various derivations. *J. Parasitol.*, 44: 172–182.

Dennis, E. W. 1932. The life cycle of *Babesia bigemina* (Smith and Kilbourne) of Texas cattle-fever in the tick, *Margaropus annulatus* (Say). *Univ. Calif. Publ. Zool.*, 36: 263–298.

Dodds, S. E., and Elsdon-Dew, R. 1955. Further observations on human coccidiosis in Natal. *S. African J. Lab. Clin. Med.*, 1: 104–109.

Eyles, D. E. 1953. The present status of the chemotherapy of toxoplasmosis. *Am. J. Trop. Med. Hyg.*, 2: 429–444.

Eyles, D. E., and Frenkel, J. K. 1952. A bibliography of toxoplasmosis and *Toxoplasma gondii. Public Health Serv. Publs.,* 247. Washington. First supplement, Memphis, 1954. (Contains over 1300 references.)

Garnham, P. C. C., et al. 1957. Symposium on Toxoplasmosis. *Trans. Roy. Soc. Trop. Med. Hyg.,* 51: 93–122.

Goldman, M. 1957. Staining *Toxoplasma gondii* with fluorescein-labeled antibody. I. and II. *J. Exptl. Med.,* 105: 549–573.

Goldman, M., Carver, R. K., and Sulzer, A. L. 1958. Reproduction of *Toxoplasma gondii* by internal budding. *J. Parasitol.,* 44: 161–171.

Gustafson, P. V., Agar, H. D., and Cramer, D. I. 1954. An electron microscope study of *Toxoplasma. Am. J. Trop. Med. Hyg.,* 3: 1008–1014.

Hammond, D. M., Bowman, G. W., Davis, L. R., and Simms, B. T. 1946. The endogenous phase of the life cycle of *Eimeria bovis. J. Parasitol.,* 32: 409–427.

Herman, C. M. 1944. The blood Protozoa of North American birds. *Bird-Banding,* 15: 90–112.

Hoare, C. A. 1956. Classification of coccidia (Eimeriidae) in a "periodic system" of homologous genera. *Rev. brasil. Malariol.,* 8: 197–202.

Hornby, H. E. 1935. Piroplasms of domestic animals. *12th Intern. Vet. Congr.,* 3: 314–324.

Jacobs, L., and Jones, F. E. 1950. The parasitemia in experimental toxoplasmosis. *J. Infectious Diseases,* 87: 78–89.

Jefferey, G. M. 1958. Epidemiologic considerations of isosporiasis in a school for mental defectives. *Am. J. Hyg.,* 67: 251–255.

Liebow, A. A., Milliken, N. T., and Hannum, C. A. 1948. *Isospora* infections in man. *Am. J. Trop. Med.,* 28: 261–273.

Ludvik, J. 1956. Vergleichende elektronenoptische Untersuchungen an *Toxoplasma gondii* und *Sarcocystis tenella. Zent. f. Bakteriol. I. Abt. Orig.,* 166: 60–65.

Magath, T. B. 1935. The Coccidia of man. *Am. J. Trop. Med.,* 15: 91–129.

Neveu-Lemaire, M. 1943. Piroplasmidea, in *Traité de Protozoologie Médicale et Vétérinaire,* pp. 458–515.

Perrin, T. L., Brigham, G. D., and Pickens, E. G. 1943. Toxoplasmosis in wild rats. *J. Infectious Diseases,* 72: 91–96.

Ray, H. N. 1950. Hereditary transmission of *Theileria annulata* infection in the tick, *Hyalomma aegyptium. Trans. Roy. Soc. Trop. Med. Hyg.,* 44: 93–104.

Sabin, A. B., and Feldman, H. A. 1948. Dyes as microchemical indicators of a new immunity phenomenon affecting a protozoan parasite (Toxoplasma). *Science,* 108: 660–663.

Sergent, E., Donatien, A., Parrot, L., and Lestoquard, F. 1936. Cycle évolutif de *Theileria dispar* du boeuf chez la tique, *Hyalomma mauritanicum. Arch. inst. Pasteur Algérie,* 14: 259–294.

Swales, W. E. 1944–1946. On the chemotherapy of cecal coccidiosis (*Eimeria tenalla*) in chickens, I. *Canad. J. Research,* D. 22: 131–140; II. *Canad. J. Comp. Med. Vet. Sci.,* January, 1946, 3–13.

Tyzzer, E. C., Theiler, H., and Jones, E. E. 1932. Coccidiosis in gallinaceous birds, II. *Am. J. Hyg.,* 15: 319–393.

Chapter 10

ARTHROPOD–BORNE ORGANISMS OTHER THAN PROTOZOA

We shall briefly discuss the organisms or disease agents other than Protozoa that are commonly transmitted biologically by arthropods in order to give a more comprehensive view of them and a better understanding of their relations to each other and to their arthropod hosts than could be gained from the discussions of them in the chapters dealing with various vectors. Four groups will be considered: the rickettsias and related forms; the *Bartonella-Anaplasma* group of organisms; the bacteria among which certain spirochetes and species of *Pasteurella* are biologically transmitted; and the filtrable viruses. As far as is known no fungus infections are habitually transmitted by arthropods.

Rickettsia and Related Organisms

General characters. Rickettsias, which are minute organisms resembling nothing more than exceptionally tiny bacteria, are strictly intracellular parasites and, like filtrable viruses, have not yet been cultivated except in the presence of living cells in tissue cultures or chick embryos. This is evidently because, again like filtrable viruses, they are degenerate parasites which have become dependent on host cells for essential enzymes. They vary in form, being coccuslike, rod-shaped, or filamentous, and they do not stain readily with aniline dyes as do bacteria, but respond well to Giemsa's stain as do Protozoa. Their primary habitat is in the tissues of arthropods, where some have become necessary commensals; in the hosts to which they are well adapted they cause no apparent disturbance and may be transovarially transmitted generation after generation, e.g., in ticks, but in hosts to which they are poorly adapted they may cause fatal infections, e.g., rickettsia of epidemic typhus in lice.

Vertebrate infections are undoubtedly a secondary development, resulting when rickettsias in a blood-sucking arthropod are inoculated into a vertebrate and are able to multiply there. Sometimes the infections may then be passed to other individuals without the arthropods being further concerned, as in Q fever infections. In vertebrate hosts the organisms multiply principally in endothelial cells of blood and lymph vessels and in cells lining serous cavities.

The rickettsias are typically placed in a family, Rickettsiaceae, the classification of which has been reviewed by Philip (1956). They are arranged in five principal genera: *Rickettsia, Coxiella, Cowdria, Neorickettsia,* and *Ehrlichia.*

The genus *Rickettsia* contains nonfiltrable coccuslike or rod-shaped organisms, less than half a micron in diameter, which cause typhus-like diseases in man. Those of one group, recognized by Philip as a subgenus *Dermacentroxenus,* multiply in the nuclei as well as in the cytoplasm of cells in both the mammal and arthropod hosts; they are the cause of a group of diseases known as spotted fever, tick typhus, etc., which are transmitted by ticks, and also one species, *R. akari,* that causes a disease known as rickettsialpox, transmitted by dermanyssid mites. *R. pavlovskyi* is reported to be transmitted by ticks, mites, and fleas! Species of another group, subgenus *Rickettsia,* multiply only in the cytoplasm of cells, and are transmitted by fleas and lice, in which there is no transovarial transmission. To this group belong *R. typhi,* which causes the flea-borne, endemic or murine typhus, and *R. prowazekii,* the cause of louse-borne epidemic typhus, and perhaps also *R. quintana,* a species which causes trench fever, but which differs in regularly multiplying extracellularly in the intestine of lice (see p. 629). A third group, subgenus *Zinserra,* contains one species, *R. tsutsugamushi,* which causes scrub typhus and is transmitted by the larvae of trombiculid mites (redbugs), being transovarially transmitted in these mites. In addition to its different mode of transmission it differs from the typical ricekttsias in other minor characters.

The genus *Coxiella* contains the important species, *C. burnetti,* which causes Q fever of man and animals. It multiplies in the cytoplasm of mammalian cells, but *outside* the cells in arthropods, and is filtrable. It multiplies in all kinds of ticks, but also in fleas, lice, bugs, and dermanyssid bird mites. It is not transovarially transmitted by arthropods and is commonly transmitted by such agents as dust, milk, meats, etc. It has been reported to occur in chickens and to be transmitted in eggs.

The genus *Cowdria* also contains only one species, *C. ruminantium,* which causes "heartwater" of ruminants, and is transmitted by ticks, not, however, transovarially.

In addition to these rickettsias there are other organisms which have been included with them, but they (and their affinities) are less well known. These include a few species of *Ehrlichia* pathogenic in mammals and transmitted by ticks (*E. bovis* in cattle; *E. canis* in dogs); *Neorickettsia helmintheca* which causes salmon-poisoning in dogs and is transmitted by a trematode (see p. 323); and a number of species from arthropods, some of which are probably commensals.

Typical rickettsial diseases. The typical rickettsias cause typhus or typhuslike diseases which are transmitted by ticks, mites, fleas, or lice. The diseases of this group all have similar general pathology and clinical character and show various degrees of cross-immunity. In all of them the serum of an immunized animal causes agglutination of certain strains of bacilli of the genus *Proteus,* although these bacteria are in no way related to the rickettsias; this is called a Weil-Felix reaction. In the tick-borne, flea-borne, and louse-borne types of typhus, as well as rickettsialpox, strains called OX19 and OX2 are agglutinated in various titers; in the mite-borne scrub typhus, strain OXK is agglutinated but the others are not.

Characteristic features of the typhus group of diseases are fever, an eruptive rash or purplish spotting of the skin a few days after onset, nervous and often gastrointestinal symptoms, microscopic nodules around arterioles, and a marked reduction in leucocytes (leucopenia). In some, e.g., scrub typhus and boutonneuse fever, an ulcer develops at the site of the infective bite.

The various forms of tick typhus (see p. 589) caused by the *Dermacentroxenus* group are certainly closely related, although several groups are sufficiently distinctive so that their rickettsias are dignified by separate species names, e.g., *R. rickettsii* of spotted fever in North and South America; *R. conorii* of boutonneuse fever in the Mediterranean region, and similar if not identical tick-borne, typhuslike diseases of Central and South Africa and India; and *R. australis* of North Queensland tick typhus. The parasite of rickettsialpox also belongs to *Dermacentroxenus,* although its vectors are mites instead of ticks (see p. 553).

There is likewise a close relationship between flea-borne endemic or murine typhus (see p. 653) and louse-borne epidemic typhus (see p. 628); they differ principally in their effects on laboratory animals. There is some evidence that either one of these types may revert to the other. The flea-borne type is undoubtedly the primitive one, from which the louse-borne type developed when lice originally acquired the rickettsias from human beings who were accidentally infected by rat fleas, for lice invariably die from typhus infections, whereas fleas do not appear to be inconvenienced. However, transovarial transmission does not occur even in fleas, so one may speculate that possibly a tick- or mite-borne type was the parent of the flea-borne type and the grand-parent of the louse-borne type.

Scrub typhus (see p. 553), transmitted by larval trombiculid mites (redbugs) and called by various names in different parts of the Orient, is immunologically a distinct type, though clinically it resembles other typhus diseases. In 1946 a rickettsia resembling that of scrub typhus, and thought to be transmitted by *Trombicula microti*, was found infecting numerous meadow mice (*Microtus*) on an island in the St. Lawrence near Quebec. No human infections have been observed.

Unlike most virus diseases, the rickettsial diseases are susceptible to treatment with certain antibiotics, particularly Terramycin, but also chloramphenicol (Chloromycetin) and Aureomycin to a somewhat less degree. Patients who are treated with these drugs 7 days or more after onset of the disease never relapse, whereas those treated earlier sometimes do so temporarily. This suggests that the antibiotics do not directly kill the organisms but suppress them for a number of days until the patient has time to mobilize his own defensive mechanism. A high degree of protection against the typhus-like diseases is obtained by vaccination with killed rickettsias grown in chick-embryo yolk sacs, in ticks or lice, or in lungs of intranasally inoculated animals.

Other rickettsial diseases. Q fever (see p. 215), as noted above, is caused by a filtrable rickettsia, *Coxiella burnetii*. This disease was probably originally tick-borne, and thus transmitted among wild animals, but it is secondarily transmitted, particularly to man, by contact with cattle or their feces or dried urine, by unpasteurized milk, or by inhaled dust or droplets. In this disease pneumonic symptoms are prominent and there is no rash. Q fever is susceptible to treatment with Terramycin but not with the other antibiotics.

Salmon poisoning, long thought to be a fluke-transmitted virus (see p. 323), was shown by Cordy and Gorham (1950) to be caused by a rickettsialike organism, *Neorickettsia helmintheca*.

Heartwater fever, caused by *Cowdria ruminantinum*, that differs somewhat from other rickettsias morphologically, is a disease of cattle, sheep, and goats in South Africa, transmitted by ticks. The most characteristic symptom is accumulation of fluid in the pericardium.

In addition to these forms, rickettsias or rickettsialike organisms have been described from Bullis fever (see p. 593) and trench fever (see p. 629) in man. Elementary bodies of a number of filtrable viruses (see p. 227) have also been interpreted by some workers as rickettsias. Actually, the borderline between rickettsias and filtrable viruses with visible elementary bodies is becoming increasingly obscure, as Philip (1943) pointed out. All the arthropod-borne rickettsial diseases are discussed further under their respective vectors.

Bartonella-Anaplasma Group (Bartonellaceae)

These minute organisms, like the rickettsias, stain poorly with aniline dyes, but well with Giemsa's stain. They parasitize the red blood cells of man and other mammals, and some, at least, have arthropod vectors. Some were once classed with Protozoa, but they show no distinct cytoplasm or other typical protozoan characters. The included organisms are as follows:

1. ***Anaplasma,*** which causes severe and often fatal disease of cattle, characterized by fever and intense anemia. Appears as minute round, deep-staining dots, about 1 μ in diameter in red corpuscles (Fig. 34*M*). Is transmitted by ticks (see p. 597) but also mechanically by instruments in dehorning, interrupted feeding of biting flies, etc.
2. ***Eperythrozoon,*** species of which cause somewhat less severe disease in a variety of animals, including ictero-anemia of pigs. Somewhat resembles *Anaplasma,* but stains blue or violet instead of red. One species, in rodents, is known to be transmitted by a louse.
3. ***Grahamella,*** bacterialike bodies that stain light blue with darker areas at the poles, in corpuscles of moles; the same or similar forms reported in many other animals. Pathogenic to moles and rodents.
4. ***Haemobartonella,*** minute coccus or bacillary forms scattered on surface of red cells of many animals; cause disease only after splenectomy, are eliminated by arsenical drugs, and are cultivated on artificial media with difficulty. One species, *H. muris,* is practically universal in wild and most strains of laboratory rats, but demonstrable only after splenectomy. Transmission by fleas, lice, bugs, etc.
5. ***Bartonella,*** which multiplies in endothelial cells as well as in red blood cells. Cultivable on cell-free media. Includes one species, *B. bacilliformis,* the cause of Oroya fever and verruga peruviana of man in South America (see p. 674). Transmitted by sandflies (*Phlebotomus*).

Of these organisms, *Anaplasma* was the last to be evicted from the Protozoa, where it was long thrown in with the Haemosporidia for want of a better place to put it. De Robertis and Epstein in 1951 showed by electron microscopy that it reproduces by budding off numerous particles of submicroscopic size; these scatter in the corpuscles and presumably are liberated in the blood to infect other corpuscles.

Spirochetes

The spirochetes, once included in the Protozoa with which they have some affinities, particularly in their staining reactions, are now considered an Order of the bacteria. They lack a distinct nucleus, they divide by transverse fission, they are not oriented into anterior or posterior ends, and they have extraordinarily flexible and motile bodies, although devoid of true flagella. All are slender, elongated

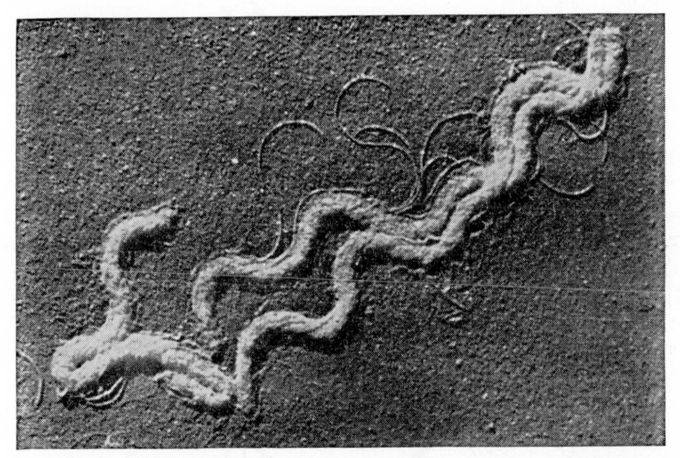

Fig. 46. Electron micrograph of *Borrelia vincenti*, showing "flagella." (From R. W. G. Wyckoff, *Electron Microscopy*, copyright 1949, Interscience Publishers, New York, London.)

organisms, spirally twisted, swimming by means of spiral waves passing through the body, and suggestive of minute, very animated corkscrews with an urge to go, no matter where (Fig. 46).

The spirochetes may be divided into two families: Spirochaetaceae, containing three genera of large, coarse-spiraled organisms (Fig. 47, *A–C*), 45 to 500 μ long, which live in stagnant water or (genus *Cristispira*) in the crystalline style of bivalve mollusks; and Treponemataceae containing small, extremely slender organisms, 6 to 18 μ in length and only one-fourth of a micron or less in diameter. Only these are of concern to us here. There are three genera of the Treponemataceae: *Borrelia, Treponema,* and *Leptospira.* Some are free-living, aquatic saprophytes, some are normally harmless commensal inhabitants of mouth, intestines, sores, etc., and some rank among the most highly pathogenic parasites of man and animals.

Borrelia. These spirochetes have smooth, graceful coils when alive, although these become distorted after death (Fig. 48). Unlike those of the other genera, they stain readily with aniline dyes as well

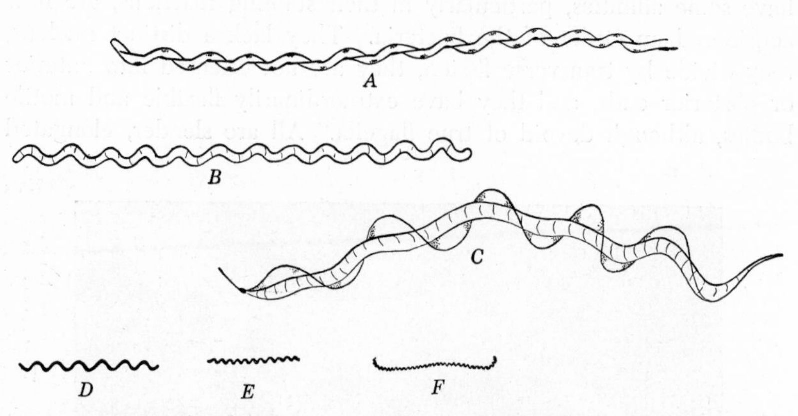

Fig. 47. Types of spirochetes, drawn to scale. *A, Spirochaeta plicatilis* (500 × 0.75 μ); *B, Saprospira grandis* (50 × 0.8 μ); *C, Cristispira anodontae* (60 × 1.8 μ); *D, Borrelia recurrentis* (15 × 0.25 μ); *E, Treponema pallidum* (10 × 0.25 μ); *F, Leptospira icterohaemorrhagiae* (15 × 0.2 μ). (Adapted from various authors.)

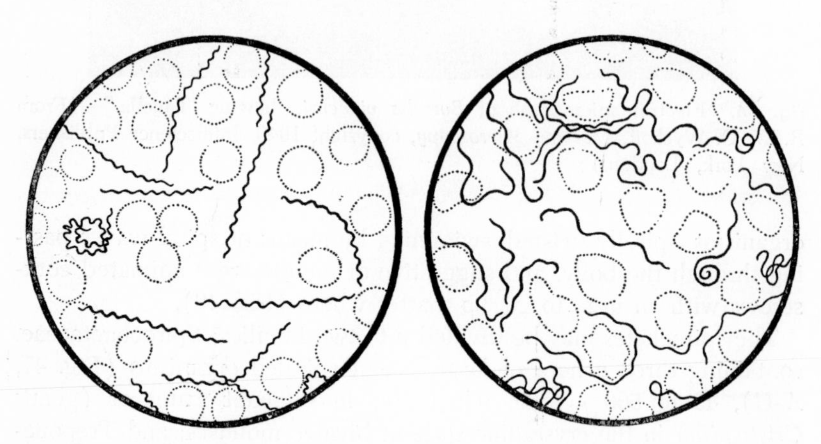

Fig. 48. *Borrelia duttoni,* showing appearance of spirochetes when living (*left*) and when dried and stained. ×1000. (After Wenyon, *Protozoology,* London, 1926.)

as with Giemsa stain. Some species are found on mucous membranes or in sores, where they ordinarily live as harmless commensals, although either they or organisms that closely resemble them become

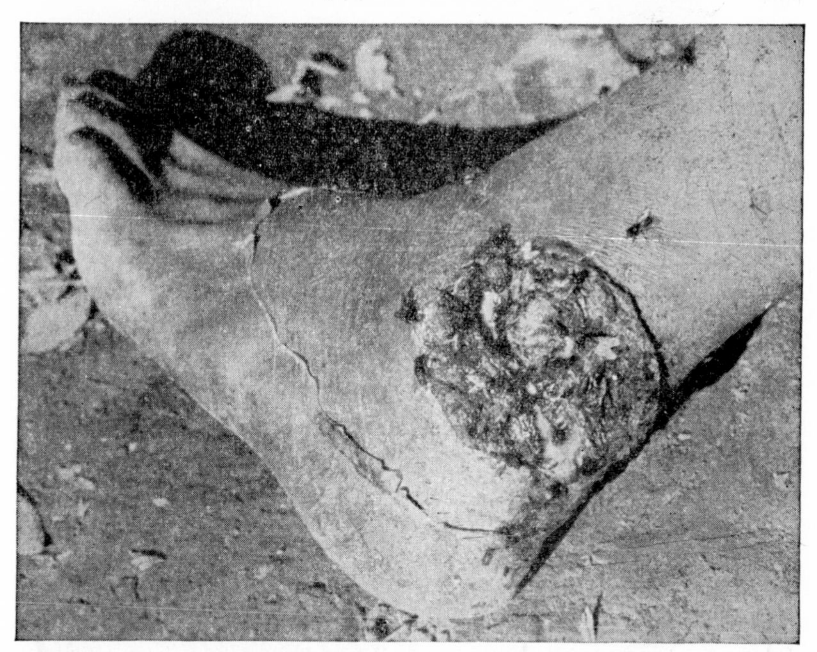

Fig. 49. A tropical ulcer. (U. S. Army Institute of Pathology photograph C43-1.)

pathogenic under favorable circumstances, and swarm in local lesions in the mouth or throat (Vincent's angina, "noma," etc.), in the lungs (bronchospirochetosis), and in nasty, sloughing skin sores [tropical ulcer (Fig. 49), phagedena, Naga sore, "jungle rot," etc.]. In these situations the spirochetes are never in pure culture, so the extent to which they are the causative agents is always open to question. They are commonly accompanied by a large cigar-shaped bacillus, suggesting some symbiotic relationship.

The most important species of *Borrelia* are those that live and multiply in the blood of birds and mammals causing relapsing fever. They should probably be regarded as primarily parasites of argasid ticks (see pp. 579–580), numerous species of which harbor their own specific strains that they transmit transovarially with various degrees of efficiency, generation after generation. One strain (*B. recurrentis*) has adapted itself to lice, but is not transovarially transmitted by them (see p. 630). In their arthropod hosts they live in the hemocoele and invade all the tissues. They are transmitted by ticks either (or both) by bite and coxal fluid (see p. 587), but in lice their only escape is by their host being crushed or broken. In their vertebrate hosts these spirochetes are capable of undergoing antigenic changes, so that as antibodies appear against one form another develops and causes a relapse. This process may be repeated five or six times in the tick-borne strains, but the louse-borne strain usually produces only one relapse.

Although, like other spirochetes, *Borrelia* is susceptible to both arsenic compounds and certain antibiotics, particularly penicillin and terramycin, relapses tend to occur because of the failure of spirochetes in the brain to be destroyed.

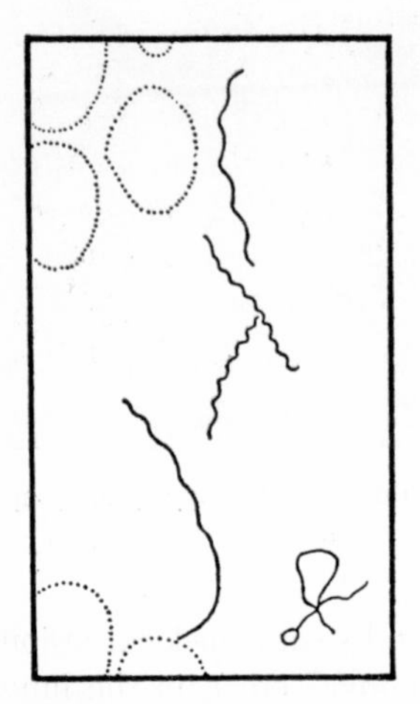

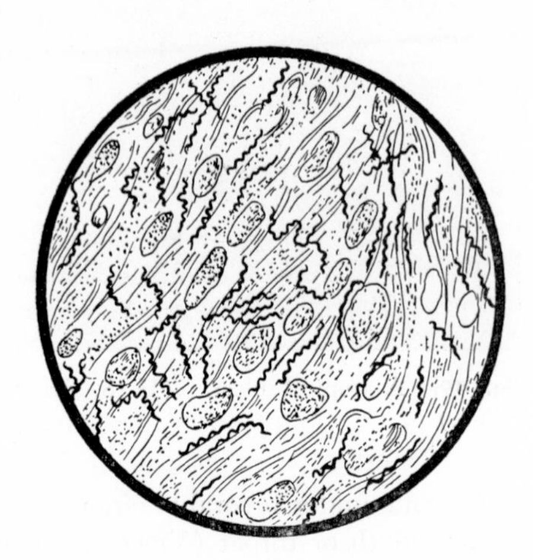

Fig. 50. *Left,* spirochetes from a syphilitic lesion. The two in the center are *Treponema pallidum;* the others a saprophyte, *T. refringens. Right, T. pallidum* in liver tissue of a syphilitic fetus.

Treponema. These are slender, kinky-coiled spirochetes that do not stain readily with aniline dyes but can be stained with Giemsa or by silver impregnation methods. Eight species have been recognized, of which four are pathogenic and will not grow in artificial culture. Three of the latter—*T. pallidum* of syphilis, *T. pertenue* of yaws, and *T. carateum* of pinta are, in nature, strictly human parasites; the fourth, *T. cuniculi,* causes a syphilislike disease of rabbits. The other species of *Treponema* are saprophytic forms that thrive in the mouth, genitalia or sores, or (one species) in pond water. All the treponemas are morphologically similar; (Fig. 50) the body consists of 6 to 14 short, sharp coils each occupying about 1 μ.

Probably all the pathogenic forms have been derived from a common ancestor (see Hudson, 1946), which under diverse climatic and sociological conditions has differentiated into forms that differ in their clinical manifestations, geographical distributions, usual modes of transmission, and predilection for tissues (other than the skin). All, however, respond alike to the serological tests (Wassermann, Kahn, etc.), in which nonspecific lipoidal extracts of normal tissues are used as antigens, and to treatment with arsenicals and with penicillin and other antibiotics.

Syphilis, caused by *T. pallidum,* is a pitifully common and widely distributed disease in temperate climates. In most places it is primarily a venereal disease, since the parasites are very fragile and are successfully transmitted only by very intimate and direct contact. However, in certain countries where living standards are low, there are nonvenereal forms of the disease, attacking children primarily. The spirochetes have shown no tendency, as have so many organisms, to develop resistance to penicillin; in earlier stages of the disease a single injection of a slowly absorbed form of this is usually adequate for cure, and even in neurosyphilis most cases respond to a series of six injections. As a consequence this disease is far less formidable a menace than it was before and during World War II.

Yaws, caused by *T. pertenue,* is a disabling and disfiguring disease especially prevalent in tropical America, Africa, and the Far East. (See Fig. 51.) It attacks almost exclusively the dark-skinned races. In many areas practically 100% of the children acquire the disease, and many adults continue to suffer from some of its horrible later manifestations, e.g., crab yaws, ulcerated lesions, gangosa, goundou, and bone deformities. *T. pertenue* attacks primarily the epidermis; later it may attack the bones or other tissues, but rarely the nervous or vascular tissues which are so prone to attack by late syphilis. Yaws is transmitted through abrasions in the skin by direct contagion or through the agency of flies, particularly eye flies (see p. 709), that feed avidly on the spirochete-loaded skin lesions. Yaws is usually easily diagnosed by its lesions, and it responds to treatment with penicillin even more readily than syphilis. Prior to penicillin there were estimated to be over 50 million cases of yaws, half of them in Africa, but nation-wide campaigns sponsored by WHO are making tremendous inroads on this. In Haiti, for example, where over one-third of the rural population was afflicted a few years ago, yaws is now practically wiped out.

Pinta or mal-del-pinto, caused by *T. carateum,* occurs in tropical America from Mexico to Ecuador and Bolivia, being particularly common in southern Mexico and in Colombia; cases have also been reported in the Old World. Like yaws it affects dark-skinned people almost exclusively. Its principal and usually only serious effect, other than psychological, is the eventual development of unsightly white patches on the skin, and often a thickening of the palms and soles late in the disease. It is transmissible by contagion, and possibly by ectoparasites, through minute abrasions of the skin, but not by blood or pre-

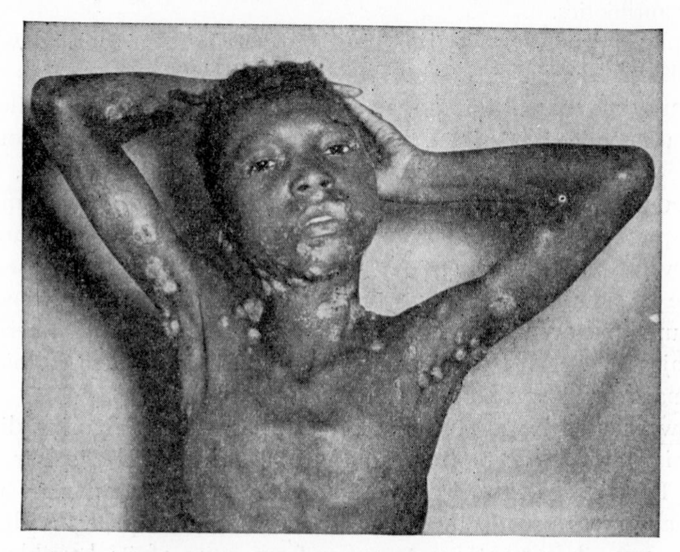

Fig. 51. A case of yaws. (U. S. Army Institute of Pathology photograph.)

natal infection. The Wassermann and other serological tests are slower in developing than in syphilis or yaws. There is no cross-immunity with syphilis or yaws. For an excellent review of this disease, see Marquez et al. (1955).

Leptospira. These are extremely slender organisms with spirals so fine that they are usually visible only by dark-field or phase contrast microscopy. The cells are hooked at one or both ends (Fig. 47*C*), and usually show a few gross undulations of the body. They react to stains as do the treponemas.

There are numerous (20 or 30) strains or "serotypes" of *Leptospira*, differentiated primarily by their serological reactions. Many may occur in a single geographical area, e.g., at least 10 occur in Panama. Many of these serotypes differ also in their pathogenicity for various animals, the clinical symptoms produced, and the principal reservoir hosts (see Davis, 1948; Babudieri, 1958). One strain (*L. biflexa*) seems to be a saprophytic water dweller, but even the parasitic forms may survive in neutral or alkaline water for two or three weeks. Rodents are the primary and reservoir hosts of most if not all of the strains (Stoenner, 1957), although dogs, cats, and pigs are important for some. Birds appear to be of minor importance. The leptospiras enter through mucous membranes or minute skin abrasions, live and multiply for a while in the blood and tissues, especially the liver, kidneys, and meninges, and eventually may persist for life in the kidney tubules, where they are protected from antibodies, and whence they are continually excreted in the urine. Most if not all of the

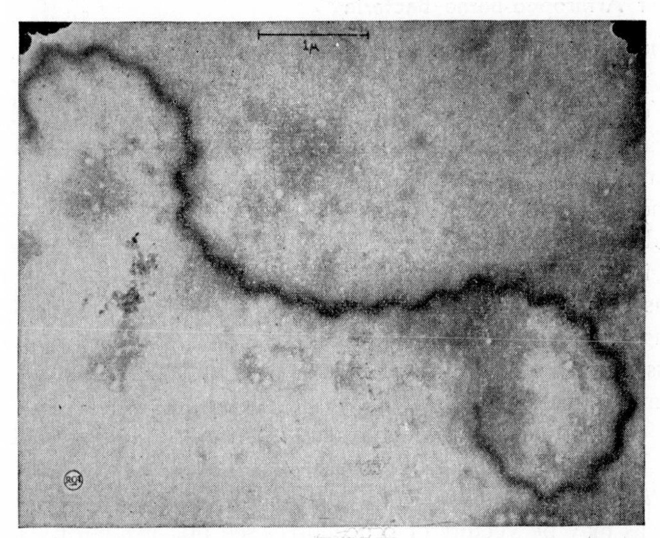

Fig. 52. Electron micrograph of *Leptospira icterohaemorrhagiae* by Harry E. Morton. (From *J. Bacteriol.*, 45: 144, 1943.)

strains are capable of infecting man. Contamination of water, food, or other objects is the principal means of transmission. Bathers or waders in polluted water or mud are particularly vulnerable, and the sniffing habit of dogs undoubtedly contributes to their exposure to infection. It has recently been demonstrated that ticks may serve as conservators or transmitters of leptospiras. Diagnosis can be made by dark-field examinations of blood (early), by inoculation of cultures with blood, or by inoculation of animals with blood, urine, etc. When a particular serotype of leptospira is mixed with homologous serum, the organisms are immobilized, agglutinated, and lysed. No good specific treatment is known, although some antibiotics have some effect in dogs and pigs.

Some of the commoner and more widespread serotypes of *Leptospira* (Fig. 52) are (1) *L. icterohaemorrhagiae,* causing Weil's disease or infectious jaundice, chief reservoir and international distributor, house rats; (2) *L. canicola,* chief reservoir, dogs; and (3) *L. pomona,* common in pigs, cattle, and horses, and especially harmful for cattle. *L. grippotyphosa* of swamp fever in Europe, *L. hebdomadis* of 7-day fever in Japan, *L. australis* A and B of Mossman fever in Australia, and *L. bataviae* of infectious jaundice in the Far East (none *limited* to those localities), as well as many other strains, have their chief reservoirs in wild or commensal rats, mice, or voles.

Other Arthropod-Borne Bacteria

Although a few bacteria are able to cause disease in arthropods, the latter, particularly some of the blood suckers, are remarkably resistant to the establishment and multiplication of most bacteria other than the atypical spirochetes and rickettsias. In many species, e.g., bedbugs, lice, and some ticks, the intestine contains potent bactericidal properties and may be completely sterile, whereas the intestines of others are packed with particular kinds of bacteria (see Steinhaus, 1946). In many insects, e.g., flies, roaches, and ticks, the intestine may harbor various bacteria (such as those of anthrax and the intestinal group causing typhoid, dysentery, and food poisoning) long enough for them to pass through and be voided in the feces, sometimes perhaps with a little multiplication en route; however, such infections do not persist. In the case of ticks such infections are often transmitted while the tick is biting. *Ornithodoros* and fleas have been found to harbor *Salmonella enteritidis* and to transmit it to experimental animals; ticks, bedbugs, and fleas may be vectors of brucellosis (see pp. 595, 607). The tick *Dermacentor albopictus* has been shown to harbor a bacterium, *Klebsiella paralytica,* that causes a disease of moose.

The only typical bacteria which establish themselves, multiply, and persist in arthropods, so that the arthropods act as true biological transmitters, are two species of the genus *Pasteurella: P. pestis,* the cause of plague, and *P. tularensis,* the cause of tularemia. These are small, nonmotile, nonspore-forming bacilli which stain deeply at the ends (bipolar staining). *P. pestis,* when ingested by fleas with the blood of an infected animal, multiplies so prodigiously in the esophagus, proventriculus, and stomach that it frequently blocks the alimentary canal, but it does not invade other parts of the body and does not multiply in any other insects. *P. tularensis,* on the other hand, can be transmitted by a variety of arthropods, including ticks, deerflies, lice, bedbugs, and fleas. However, there is little doubt that ticks are the primary arthropod hosts. Unlike *P. pestis* in fleas, *P. tularensis* invades the hemocoele in ticks and is transovarially transmitted. Plague is discussed further on pp. 650–653, and tularemia on pp. 594–595.

Filtrable Viruses

A considerable number of important human diseases have causative agents so small that they are beyond the range of visibility and will pass through filters which will hold back any microscopically visible organisms. These filtrable viruses are probably not primitive forms of life but highly degenerate forms that have become more and more dependent on the cells in which they live to provide them with enzymes and materials that free-living organisms provide for themselves. The intracellular rickettsias and some of the largest viruses, e.g., vaccinia, are only a little less complex than bacteria, but the process of loss and of concomitant decrease in size continues until in the smallest and simplest viruses, like those of tobacco mosaic and poliomyelitis, little or nothing is left but naked nucleoprotein molecules which have inherent in them the power of reproduction. The host cells provide all the necessary enzymes and materials. In many ways the simplest viruses are comparable with genes both in size and properties, differing mainly in the ability to move from cell to cell. This difference loses a good deal of its meaning in the light of recent work on transformation of bacteria by deoxyribonucleic acid preparations. Like genes the viruses undergo mutation.

Although not a natural group, all viruses have certain features in common. They are more resistant to antiseptics than bacteria. They all tend to stimulate the cells in which they grow to increase multiplication, followed by death of the cells, but the necrosis may occur too rapidly for the growth stimulus to be apparent, and this is the only observed effect in nerve cells, which cannot multiply. Many produce "inclusion bodies" which are in reality masses of the elementary filtrable bodies. As noted on p. 218, some of the larger of these elementary bodies are not unlike rickettsias, and those of trachoma, psittacosis, and lymphogranuloma have been so interpreted. Sometimes the masses of elementary bodies are surrounded by a mantle of amorphous material produced by cellular reaction, suggesting Protozoa, for which a special class "Chlamydozoa" (mantled animals) was once proposed. All viruses stimulate formation of neutralizing antibodies which inactivate them and thereby provide the most commonly used method for identifying them.

An attempt has been made by Holmes to classify viruses and give them genus and species names as for other organisms, but this has not been widely accepted and virologists continue to refer to them by somewhat informal names.

The viruses of vertebrates that are primarily arthropod-borne fall reasonably well into two principal groups. One contains forms that are primarily "visceral"; these produce fevers, sometimes a rash, and a marked decrease in leucocytes (leucopenia). Yellow fever, Rift Valley fever, dengue, sandfly fever, and Colorado tick fever belong to this group. The second group contains forms that are neurotropic, at least to the extent that their most serious pathogenic effects are on the central nervous system, although in a high percentage of cases, as in poliomyelitis, these neural symptoms may not appear at all; to this group belong the encephalomyelitis viruses (see pp. 752–754). The distinction between these two groups is not, however, very sharp, for neurotropic strains of yellow fever and dengue have been produced in the laboratory, and there are many viruses in tropical Africa and South America which show no sharp tendency to be either viscerotropic or neurotropic, or whose affinities are very imperfectly known, if at all. Good examples are the West Nile and Ntaya viruses of Africa (see p. 755).

In addition to the groups of viruses discussed in the last paragraph there are a number of others that may be transmitted by arthropods but are not primarily dependent on them, such as fowlpox of birds and swamp fever of horses.

It is still uncertain what role arthropods play in transmission of poliomyelitis, but the epidemiology strongly suggests their implication. The disease is prevalent in summer, is more frequent in small towns and edges of cities than in the centers of them, and does not spread in crowds. Flies harbor the virus, which they acquire from human feces, and monkeys eating food contaminated by infected flies show evidence of infection by developing neutralizing antibodies, as do a large percentage of human beings when an outbreak occurs, but there is no paralysis. Furthermore, fly-controlled areas in south Texas had as high an incidence of polio cases as did uncontrolled areas. Might it not be possible that infection via the digestive system by small doses of virus deposited on food by flies may lead to immunity without paralysis? The absence of the virus from the blood argues against arthropods acquiring it by sucking blood, although they might possibly inoculate it if infected in some other way. Mosquitoes like *Culex* might acquire the virus from sewage-polluted water as larvae, and subsequently inoculate it after becoming adults.

Of the first group of primarily arthropod-borne virus diseases just mentioned, yellow fever is the most severe and is the only one causing mortality. It is transmitted by mosquitoes, although an instance of experimental transmission to a monkey by a mite has been reported.

The disease is discussed in more detail on p. 742. The virus is present in the blood of patients for only 3 days and requires about 10 days for a mosquito to become infective, after which it persists for life, although it is not transovarially transmitted. Dengue fever (see p. 750), of which there are at least three distinct antigenic types, is also transmitted only by mosquitoes, but the evidence is conflicting as to how long it takes for mosquitoes to become infective and whether transovarial transmission can occur (see p. 751). Sandfly fever is very similar to dengue, but is transmitted only by *Phlebotomus papatasii*, in which it is transovarially transmitted; at least two antigenic types exist. Colorado tick fever (see p. 596) is also very similar to dengue, but is transmitted by ticks; transovarial infection occurs. Rift Valley Fever of east central Africa is a severe disease of sheep but mild in man; it also affects cattle, monkeys, and rodents. It is believed to be transmitted by mosquitoes.

Of the group of arthropod-borne viruses causing encephalomyelitis or encephalitis, sometimes erroneously called sleeping sickness, there are many mosquito-transmitted strains (see p. 752). Three strains are common in the United States and Canada—eastern and western equine and St. Louis encephalomyelitis; several in South America, including Venezuelan and Argentinian; two in the Far East, Japanese B, and Russian spring-summer encephalitis; one in Australia, Murray Valley; and about a dozen, distinct from all the others, in Africa, of which West Nile and Ntaya are commonest. All appear to be primarily mosquito borne and are discussed further on pp. 752–757, except the Russian spring-summer disease, which is transmitted by ticks and is discussed on p. 597. As Warren pointed out in 1946, there is a very close relationship between the virus of this disease and that of louping ill in sheep; the former commonly attacks man in Siberia, and the latter is a disease of sheep in Russia and northern Britain.

The North American encephalitides can also infect mites, which for a while were thought to be important as reservoirs to carry them through winters and between epidemics. Eklund (1954) called attention to the fact that temperate-zone encephalitis viruses are primarily transmitted by *Culex* and involve birds extensively, whereas tropical viruses, including yellow fever and dengue as well as the neurotropic ones, are primarily transmitted by *Aedes* or other mosquitoes and not primarily by *Culex*, and that the principal vertebrate hosts are always mammals rather than birds. It has been suggested that the temperate-zone viruses may depend for survival on migration of their bird reservoirs.

Little is known about transmission of viruses by helminths, but this is a subject worthy of further investigation. The virus of swine influenza is transmitted by a nematode lungworm (see p. 446), and that of lymphocytic choriomeningitis has been shown to be transmissible by *Trichinella* larvae. It is possible that such diseases as poliomyelitis, ornithosis (psittacosis), and other viruses which are abundant in the intestine, might be transmitted by eggs of intestinal worms.

REFERENCES

Alcock, A. 1929. *Bartonella muris-ratti* and the infectious anemia of rats. *Trop. Diseases Bull.*, 26: 519–524.

American Association for the Advancement of Science. 1942. *A Symposium on Relapsing Fever in the Americas.* Publication 18, edited by F. R. Moulton. Science Press, Lancaster, Pa.

American Association for the Advancement of Science. 1948. *A Symposium on the Rickettsial Diseases of Man.* Edited by F. R. Moulton. Washington.

Babudieri, B. 1958. Animal reservoirs of leptospires. In *Animal Diseases and Human Health. Ann. N. Y. Acad. Sci.*, 70, Art. 3; 393–413 (and other papers, pp. 391–444).

Baltazard, M. 1954. Sur le classement des spirochètes récurrents. *Ann. Parasitol. humaine et comparée*, 29: 12–32.

Beadle, L. D. 1952. Eastern equine encephalomyelitis in the United States. *Mosq. News*, 12: 102–107.

Bell, E. J., and Philip, C. B. 1952. The human rickettsioses. *Ann. Rev. Microbiol.*, 6: 91–118.

Berge, T. O., and Lennette, E. H. 1953. World distribution of Q fever: human, animal and arthropod infections. *Am. J. Hyg.*, 57: 125–143.

Blank, H. 1947. Tropical phagedenic ulcer (Vincent's ulcer). *Am. J. Trop. Med.*, 27: 383–398.

Borman, F. von. 1952. Die in New York und Boston unter der Namen "Brillsche Krankheit" bekannten Fleckfieberkrankungen und der Frage der Spätrucksfalle beim klassischen Fleckfieber. *Z. Hyg. u. Infektionskr.*, 135: 448–471.

Boyton, W. H., and Woods, G. M. 1944. Some information on anaplasmosis for the veterinarian. *Mich. State Coll. Vet. Sci.*, 25.

Brumpt, E., and Brumpt, L.-Ch. 1942. Étude épidémiologique concernant l'apparition de la verruga du Pérou en Colombie. *Ann. parasit. humaine et comparée*, 19: 1–50.

Brumpt, E., Mazzotti, L., and Brumpt, L. C. 1939. Étude épidémiologique de la fièvre récurrent endémique des hauts plateaux mexicains. *Ann. parasitol. humaine et comparée*, 17: 275–286.

Burnet, F. M., Derrick, E. H., Smith, D. J. W., et al. 1940–1942. Studies in the epidemiology of Q fever. *Austral. J. Exptl. Biol. Med. Sci.*, 18: 1–8, 99, 108–118, 119–123, 193–200, 409; 20: 105–110, 213–217, 295–296.

Chandler, A. C., and Rice, L. 1923. Observations on the etiology of dengue. *Am. J. Trop. Med.*, 3: 233–262.

Chung, H. L., and Wei, Y. L. 1938. Studies on transmission of relapsing fever in North China. II. Observations on mechanism of transmission of relapsing fever in man. *Am. J. Trop. Med.,* 18: 661–674.

Coffin, D. L., and Stubbs, E. L. 1942. Observations on canine leptospirosis in the Philadelphia area. *Univ. Penn. Vet. Ext. Quart.,* 87: 3–14.

Cox, H. R. 1941. Cultivation of the Rocky Mountain spotted fever, typhus, and Q fever groups in the embryonic tissues of developing chick. *Science,* 94: 399–403.

Davis, G. E. 1948. The spirochetes. *Ann. Rev. Microbiol.* for 1948, 281–334.

Day, M. F. 1957. The relation of arthropod-borne viruses to their invertebrate hosts. *Trans. N. Y. Acad. Sci.,* Ser. II, 19: 244–251.

Dickmans, G. 1948. Anaplasmosis. *4th Intern. Congr. on Trop. Med. and Malaria,* 2: 1404–1411.

Eklund, C. M. 1953. The ecology of mosquito-borne viruses. *Ann. Rev. Microbiol.,* 8: 339–360.

1954. Mosquito-transmitted encephalitis viruses. A review of their insect and vertebrate hosts and the mechanisms for survival and dispersion. *Exp. Parasitol.,* 3: 285–305.

Florio, L., and Miller, M. S. 1948. Epidemiology of Colorado tick fever. *Am. J. Public Health,* 38: 211–213.

Francis, E. 1937. Sources of infection and seasonal incidence of tularemia in man. *Public Health Repts.* 52: 103–113.

Gochenour, W. S., et al. 1952. Leptospiral etiology of Fort Bragg fever. *Fed. Soc. Proc.,* 11, Part 1, 469.

Gsell, O. 1952. *Leptospirosen.* Huber, Bern.

Guthe, T., and Willcox, R. R. 1954. Treponematoses. A world problem. *Chronicle World Health Organization,* 8: 37–114.

Hackett, C. J. 1953. The natural history of yaws. *Trans. Roy. Soc. Trop. Med. Hyg.,* 47: 318–320.

Hammon, W. McD. 1948. The arthropod-borne virus encephalitides. *Am. J. Trop. Med.,* 28: 515–525.

Hertig, M. 1942. *Phlebotomus* and Carrion's Disease. *Am. J. Trop. Med.,* 22: No. 5, Suppl.

Hudson, W. H. 1946. Treponematosis. *Oxford Loose Leaf Medicine,* Chapter 27-C, 9–122. (Reprint.)

Hurlbut, H. S. 1956. West Nile virus in arthropods. *Am. J. Trop. Med. Hyg.,* 5: 76–85.

Jefferson Medical College. 1955. Yellow fever. A symposium in commemoration of Carlos Juan Finlay. (A comprehensive survey of yellow fever.) Philadelphia.

Kemp, H. A., Moursund, W. H., and Wright, H. E. 1933–1935. Relapsing fever in Texas. *Am. J. Trop. Med.,* 13: 425–435; 14: 159–162, 163–179, 479–487; 15: 495–506.

Kirk, R. 1943. Some observations on the study and control of yellow fever in Africa, with particular reference to the Anglo-Egyptian Sudan. *Trans. Roy. Soc. Trop. Med. Hyg.,* 37: 125–150.

Lewthwaite, R. 1952. The typhus group of fevers. *Brit. Med. J.,* 826–828, 875–876.

Livesay, H. R., and Pollard, M. 1943. Laboratory report on a clinical syndrome referred to as "Bullis fever." *Am. J. Trop. Med.,* 23: 475–479.

Marmion, B. P. 1954. Q fever. II. Natural history and epidemiology of Q fever in man. *Trans. Roy. Soc. Trop. Med. Hyg.*, 48: 197–207.

Marquez, F., Rein, C. R., and Arias, O. 1955. Mal del Pinto in Mexico. *Bull. World Health Organization*, 13: 299–332.

Meyer, K. F. 1942. The known and the unknown in plague. *Am. J. Trop. Med.*, 22: 9–37.

Parker, R. R. 1938. Rocky Mountain spotted fever. *J. Am. Med. Assoc.*, 110: 1185–1188, 1273–1278.

Philip, C. B. 1953. Nomenclature of the Rickettsiaceae pathogenic to vertebrates. *Ann. N. Y. Acad. Sci.*, 56: 484–494.

Philip, C. B. 1956. Comments on the classification of the order Rickettsiales. *Canad. J. Microbiol.*, 2: 261–270.

Pinkerton, H., Weinman, D., and Hertig, M. 1937. Carrion's disease (Oroya fever and verruga peruviana) I–V. *Proc. Soc. Exptl. Biol. Med.*, 37: 587–600.

Pratt, H. D. 1958. The changing picture of murine typhus in the United States. *Ann. N. Y. Acad. Sci.*, 70, Art. 3: 516–527.

Reeves, W. C. 1951. The encephalitis problem in the United States. *Am. J. Public Health*, 41: 678–686.

Reinhard, K. R. 1953. Newer knowledge of leptospirosis in the United States. *Exp. Parasitol.*, 2: 87–115.

Rickettsialpox—1946–1947. A newly recognized rickettsial disease, I–VI (various authors). I. *Public Health Repts.*, 61: 1605–1614. II. *J. Am. Med. Assoc.*, 133: 901–906. III. *Am. J. Public Health*, 37: 860–868. IV. *Public Health Repts.*, 61: 1677–1682. V. *Ibid.*, 62: 777–780.

Rivers, T. M. (Editor). 1952. *Viral and Rickettsial Infections of Man.* Lippincott, Philadelphia.

de Roberts, E., and Epstein, B. 1951. Electron microscope study of anaplasmosis in bovine red blood cells. *Proc. Soc. Exptl. Biol. Med.*, 77: 254–258.

Sabin, A. B. 1952. Research on dengue during World War II. *Am. J. Trop. Med. Hyg.*, 1: 30–50.

Schuhardt, V. T., and Wilkerson, M. 1951. Relapse phenomena in rats infected with single spirochetes (Borrelia recurrentis var. turicatae). *J. Bacteriol.*, 62: 215–219.

Simmons, J. S., St. John, J. H., and Reynolds, F. H. K. 1931. Experimental studies of dengue. *Bur. Sci. Monogr.* 29. Manila.

Smith, D. T. 1932. *Oral Spirochetes and Related Organisms in Fusospirochetal Disease.* Wood, Baltimore.

Smith, M. G., et al. 1948. Experiments on the role of the chicken mite, *Dermanyssus gallinae*, and the mosquito in the epidemiology of St. Louis encephalomyelitis. *J. Exptl. Med.*, 87: 119–138.

Smithburn, K. C. 1952. Neutralizing antibodies against certain recently isolated viruses in the sera of human beings residing in East Africa. *J. Immunol.*, 69: 223–234.

Soper, F. L. 1937. Present day methods for the study and control of yellow fever. *Am. J. Trop. Med.*, 17: 655–676.

1937. The newer epidemiology of yellow fever. *Am. J. Public Health*, 27: 1–14.

Soper, F. L., Davis, W. A., Markham, F. S., and Riehl, L. A. 1947. Typhus fever in Italy, 1943–1945, and its control with louse powder. *Am. J. Hyg.*, 45: 305–334.

Splitter, E. J. 1950. *Eperythrozoön suis*, etiological agent of icteroanemia, an *Anaplasma*-like disease of swine. *Am. J. Vet. Research*, 11: 324–330.

Steinhaus, E. A. 1946. *Insect Microbiology.* Comstock, Ithaca, N. Y.

Stoenner, H. G. 1957*a*. The laboratory diagnosis of leptospirosis. *Vet. Med.*, 52: 540–542.

1957*b*. The sylvatic and ecological aspects of leptospirosis. *Vet. Med.*, 52: 553–555.

Stoker, M. G., and Marmion, B. P. 1955. The spread of Q fever from animals to man. The natural history of a rickettsial disease. *Bull. World Health Organization*, 13: 781–806.

Strode, J. K. (Editor). 1951. *Yellow Fever.* McGraw-Hill, New York.

Strong, R. P. 1918. *Trench Fever Report.* Medical Research Committee, American Red Cross. Oxford University Press, New York.

Symposium on Animal Disease and Human Health. 1958. I. Lieberman (Editor). Part I. The viral encephalitides. (5 papers by various authors.) *Ann. N. Y. Acad. Sci.*, 70, Art. 3: 292–341.

van Thiel, P. H. 1948. Diagnosis and treatment of leptospirosis. *Proc. 4th Intern. Congr. Trop. Med. Malaria*, 1: Sect. III, 321–327.

1948. *The Leptospiroses.* University Press, Leiden.

Tigertt, W. D., and Hammon, W. McD. 1950. Japanese B encephalitis: a complete review of experiences in Okinawa 1945–1949. *Am. J. Trop. Med.*, 30: 689–722.

Topping, N. Y., Cullyford, J. S., and Davis, G. E. 1940. Colorado tick fever. *Public Health Repts.*, 55: 2224–2237.

Turner, T. B., et al. 1935. Yaws in Jamaica. *Am. J. Hyg.*, 21: 483–521, 522–539.

Turner, T. B. 1957. Biology of the treponematoses. *World Health Organization Monogr. Ser.* No. 35, Geneva.

Van Rooyen, C. E., and Rhodes, A. J. 1948. *Virus Diseases of Man.* 2nd ed. Nelson, New York.

Weinman, D. 1941. Bartonellosis: a public health problem in South America. *J. Trop. Med. Hyg.*, 44: 62–64.

Welsh, H. H., et al. 1958. Air-borne transmission of Q fever: The role of parturition in the generation of infective aerosols. *Ann. N. Y. Acad. Sci.*, 70, Art. 3: 528–540.

World Health Organization. 1953. First international symposium on yaws control. *Monogr. Ser.* No. 15.

World Health Organization. 1956. Treponematoses. (6 papers by different authors.) *Bull. World Health Organization*, 15: 863–1096.

Yager, R. H., and Gochenour, W. S. 1952. Leptospirosis in North America. *Am J. Trop. Med. Hyg.*, 1: 457–467.

Part II

HELMINTHS

Chapter 11

INTRODUCTION TO THE "WORMS"

Classification. The name "worm" is an indefinite though suggestive term popularly applied to any elongated creeping thing that is not obviously something else. There is hardly a branch or phylum of the animal kingdom that does not contain members to which the term worm has been applied, not excepting even the Chordata. In fact, some animals, such as many insects, are "worms" during one phase of their life history and something quite different during another.

In a more restricted sense the name worm, or preferably helminth, is applied to a few phyla of animals, all of which superficially resemble one another in being unquestionably "wormlike," though in life and structure they are widely different. In fact, man has more in common with a salamander than a nematode has with a tapeworm. The helminths which are of interest to us as parasites of vertebrates belong to four different phyla of the animal kingdom, the Platyhelminthes or flatworms, the Acanthocephala or spiny-headed worms, the Nemathelminthes or roundworms, and the Annelida (or Annulata), the segmented worms including the leeches.

The relative importance of the major groups of helminths may be roughly judged by Stoll's 1947 estimate that there exist in the world today, among some 2200 million people, 72 million cestode, 148 million trematode, and over 2000 million nematode infections. There is little reason to believe that these numbers have decreased in the ensuing 12 years; as a matter of fact they are probably too low now. Of course, like money and brains, these are not evenly distributed among the individuals!

Platyhelminthes. This phylum contains the helminths of lowest organization. The great majority of them are flattened from the dorsal to the ventral side, hence the common name flatworm. Unlike nearly all other many-celled animals they have no body cavity, the organs being embedded in a sort of spongy "parenchyma" or packing tissue. The digestive tract in its simplest form, and as it occurs in the asexual generation of flukes known as rediae, consists of a blind sac with only a single opening, serving both as a mouth and as a vent, but in most adult forms this sac is variously branched and in a few flukes even has an anus. On the other hand, tapeworms get along very well without any digestive tract, food being absorbed through the outer surface, some from nutrients in the host's meals and some from the secretions of the host intestine. The nervous system is very simple, and the primi-

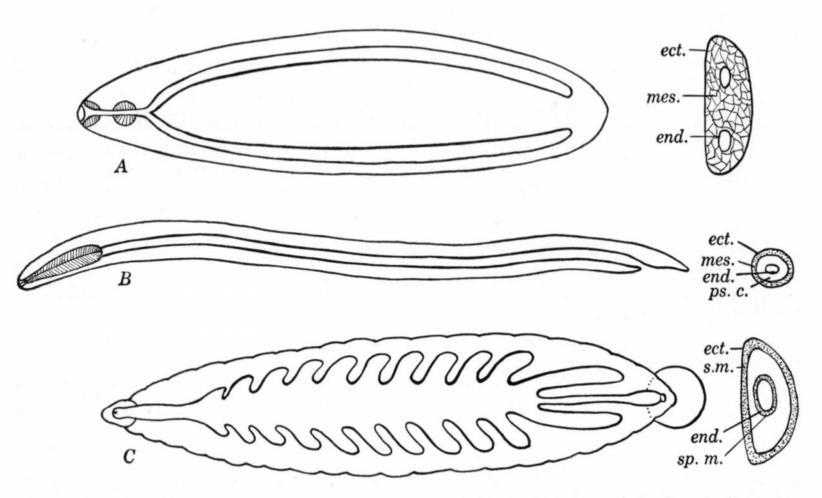

Fig. 53. Diagrams of digestive tracts and cross sections of bodies of representatives of Platyhelminthes (*A*); Nemathelminthes (*B*); and Annelida (*C*). *A*, fluke; gut branched into two ceca, no anus and no body cavity. *B*, nematode; gut a simple tube with only pharynx differentiated, anus present, body cavity a pseudocoelom not lined by splanchnic mesoderm internally. *C*, leech; gut with ceca for surplus food, anus present, body cavity a true coelom lined by splanchnic mesoderm internally and somatic mesoderm externally. Abbreviations: *ect.*, ectoderm; *end.*, endoderm; *mes.*, mesoderm; *ps.c.*, pseudocoelom; *s.m.*, somatic mesoderm; *sp.m.*, splanchnic mesoderm.

tive ganglia which serve as a brain are located in the anterior portion of the worm. Performing the function of kidneys is a system of tubes, the terminal branches of which are closed by "flame cells," so-called from the flamelike flickering of a brush of cilia which keeps up a flow of fluid toward the larger branches of the system and ultimately to the excretory pore, thus conducting the waste products out of the body.

The most highly developed systems of organs, occupying a large portion of the body, are those concerned with reproduction. All but a few of the Platyhelminthes are hermaphroditic, containing complete male and female systems in each individual; in tapeworms both systems are usually complete in each segment and there may even be double sets in each segment. In addition to the ordinary sexual reproduction of the adults, many flukes and tapeworms have special asexual methods of multiplication in the course of their life cycles.

The flatworms are usually divided into three classes, the Turbellaria, the Trematoda, and the Cestoidea, but some zoologists include also the Rhynchocoela, a group of band-shaped marine worms, none of which is of interest in connection with human parasitology.

The Turbellaria are ciliated and for the most part free-living animals; they include the "planarians," which can be found creeping on the underside of stones in ponds. The Trematoda include the flukes, all of which are parasitic, some externally on aquatic animals, others internally on aquatic or land animals. They are soft-bodied, usually flattened animals, commonly oval or leaf-shaped, and furnished with suckers for adhering to their hosts. The flukes that live as external parasites of aquatic animals have a comparatively simple life history, whereas those that are internal parasites have a complex life history, including two or three asexual generations, and involving two, three, or even four different hosts.

The third class, the Cestoidea or tapeworms, with the exception of one primitive family living in the body cavity of ganoid fishes, and the members of one genus which are able to complete their development precociously in annelid worms, are (in the adult stage) invariably parasites of the digestive tracts of vertebrate animals and are profoundly modified for this kind of an existence. Except for a few evolutionarily precocious forms in the genus *Hymenolepis* (see p. 367), all tapeworms begin their development in an alternative host, which may be either a vertebrate or an invertebrate. Sometimes two or even three intermediate hosts are involved in the life cycle. Although in some forms a number of adults may develop from one larva as the result of a budding process, there is never an alternation of generations such as occurs in most trematodes. In one subclass, the Cestodaria, the adult worms are single individuals, suggestive of gutless flukes; in the other subclass, Cestoda, all but the members of a single family (Caryophyllaeidae) consist of chains of segments.

Nemathelminthes. This phylum name has been used in the past to include various assemblages of organisms, of which only the class Nematoda has always been present; the Nemathelminthes once included not only the Acanthocephala but even the gregarines (see p. 161). Because of this, Hyman (1951), who includes six classes in this phylum, prefers the less shopworn name "Aschelminthes." She defines the phylum as mostly wormlike, bilaterally symmetrical, unsegmented animals provided with a body cavity (not a true coelom), with a digestive tract lacking a definite muscular wall and with both a mouth an anus, and no respiratory or circulatory systems. In addition to the Nematoda, five other classes are included, all of them unknown to the average layman, and most of them even to zoologists who have not by some odd chance been attracted to them; they are (1) the rotifers, (2) the gastrotrichs, (3) the kinorhynchs and echinoderans, (4) the priapulids, and (5) the horsehair worms (Gordiacea or Nematomorpha). The last class, Gordiacea, are the only ones likely to attract attention as "worms"; the name "horsehair worm" comes from a popular idea, not yet dead, that they develop out of horsehairs that fall into water. They are very long and slender hairlike worms that live as parasites in insects until almost mature, when they emerge from the insects and reproduce in water or soil. Occasionally they are accidentally swallowed with drinking water and are usually promptly vomited, much to the surprise and horror of the temporarily infested person.

The true nematodes are cylindrical or spindle-shaped worms covered by a very resistant cuticle; they have a simple digestive tract with mouth and anus, a fluid-filled body cavity which is not lined by epithelium as in other animals; and usually separate sexes, with the sex glands continuous with their ducts in the form of slender tubules. There is an excretory system consisting of a glandular apparatus opening through an anteriorly situated excretory pore, in some forms connected with longitudinal lateral canals. The development is always direct and simple but sometimes requires two hosts for its completion.

Acanthocephala. The spiny-headed worms were long included in the Nemanthelminthes, but they have little in common with the nematodes in either structure or development; they constitute a very aberrant and sharply defined group of parasites of vertebrate animals. They are characterized by having a large body cavity, complete lack of a digestive tract, a spiny proboscis retractile into a sac, and separate sexes with reproductive systems of unique character (see Chapter 16). The development, which involves an arthropod intermediate host, more nearly resembles that of flatworms than that of nematodes. For the present it seems best to recognize Acanthocephala as a distinct phylum with affinities closer to the cestodes than to nematodes, as Van Cleave did in 1948.

Annelids. The most highly organized group of helminths is the phylum Annelida, including the earthworms, leeches, etc. In three important respects these worms are the first animals in the scale of evolution to develop the type of structure characteristic of the vertebrate animals, namely, a division of the body into segments, the presence of a blood system, and the presence of "nephridia"—primitive excretory organs of the same fundamental type as the kidneys of higher animals. In addition the digestive system is highly developed and there is a well-developed nervous system with a primitive brain in the head. In some annelids the sexes are separate, though in others both reproductive systems occur in the same individual.

Three classes of Annelida are usually recognized, of which one, the Hirudinea, or leeches, are of interest as bloodsuckers. These differ from other annelids in lacking setae, in the possession of suckers for adhering, and in the fact that the external annulation of the body does not correspond exactly to the true internal segmentation. These animals superficially resemble flukes, so much so that liver flukes are often referred to as liver leeches, but they can be distinguished externally by the segmentation of the body and internally by their totally

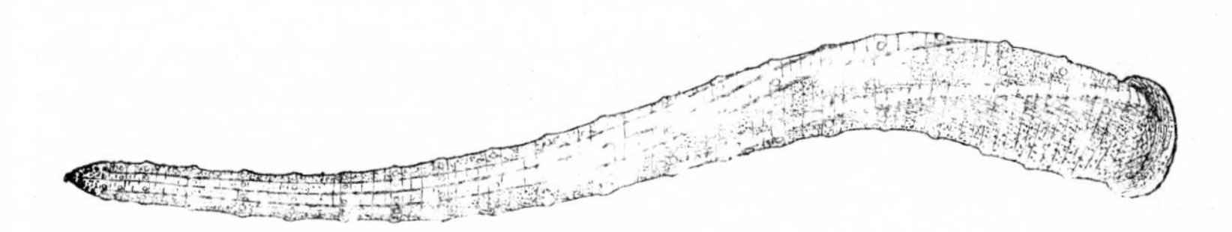

Fig. 54. Japanese land leech, *Haemadipsa,* extended. ×2. (Adapted from Whitman, *Quart. J. Micr. Sci.,* 26, 1886.)

different anatomy. Both sexes are represented in the same individual.

Every boy who has experienced the delights of hanging his clothes on a hickory limb and immersing his naked body in a muddy-bottomed river or pond is familiar with leeches. These are related to the medicinal leeches that were an important stock in trade of medieval physicians, for whom blood-letting was as important as transfusions are to modern physicians. Still more familiar with them is any tourist who has journeyed on foot through the jungles of Ceylon, Sumatra, or Borneo, or through the warm moist valleys of the Himalayas or Andes, for hordes of bloodthirsty land leeches infest these places. In southwest Asia *Haemadipsa zeylandica* causes much loss of blood as well as ulceration and inflammation of the bites, which continue to bleed even after the leeches are filled to repletion and drop off. Clothing impregnated with the U. S. Army repellent (see p. 537) is protective against them.

Thirsty horses, and occasionally men, gulping water from pools or streams in Palestine, North Africa, and China, may suffer severe or even fatal loss of blood, and sometimes strangulation, from the settling of "horse leeches" of the genera *Limnatis* or *Haemopis* in the pharynx or nasal passages or sometimes in urinary passages, where they may hang on for days or months (Masterson, 1908). Another aquatic leech, *Dinobdella ferox,* is a serious pest in southeastern Asia (see Chin, 1949). When animals or human beings drink from water infested by it, the young worms quickly enter the nostrils or mouth and loop their way to the back of the pharynx or larynx, where they attach, grow rapidly, and often do much damage. Leeches lodged in the nasopharynx let go when 5% cocaine is sprayed into the nostrils.

Although land leeches are nasty pests in some places, they are not known to be vectors of any human infection. Aquatic leeches serve as intermediate hosts for trypanosomes and Sporozoa of fish, amphibians, and turtles. They affect water birds by entering the nostrils and trachea, causing asphyxiation; by serving as intermediate hosts of flukes; by transmitting fowlpox and possibly other diseases; and by sucking blood from around the vent. As one writer put it, whether a duck is tickled or worried is hard to say, but not having hemorrhoids they presumably do not benefit from the leeches as did mankind thus afflicted in the Middle Ages. Some large leeches in northern United States feed voraciously on large snails and thereby seem to have rendered some lakes free of swimmer's itch (see p. 294). These species do not suck blood from vertebrates.

Parasitic habitats. Hardly any organ or tissue is exempt from attack by worms of one kind or another. There are flukes parasitic in man which habitually infest the intestine, liver, lungs, and blood vessels, and one species occasionally wanders to the muscles, spleen, brains, and many other organs. In other animals there are species with even more specialized habitats; some inhabit the Eustachian tubes of frogs, the frontal sinuses of polecats, the eye sockets of birds, cysts in the skin of birds, etc. All the adult tapeworms of man are resident in the small intestine, but there are species in sheep and goats and one in rats which habitually live in the bile duct; larval tapeworms are found in a great variety of locations—in the liver, spleen, muscles, subcutaneous tissues, eye, brain, etc.

The majority of the parasitic nematodes of man are resident in the intestine, but the filariae and their relatives inhabit various tissues and internal organs, such as the lymph sinuses and subcutaneous connective tissue. Nematode parasites of other animals, many of which are occasional or accidental in the human body, may occur in all parts of the alimentary canal and in its walls, and in liver, lungs, kidneys, bladder, heart, blood vessels, trachea, peritoneum, skin, eye, and sinuses. None live as adults in the central nervous system, although invasion of it by larvae that have lost their way is a common and sometimes dangerous occurrence. The surface of the body and cavities of the nose and throat of man are not the habitat of any helminth parasites except leeches and the tongueworms; the latter are really arthropods (see p. 562), although they have much more in common with the helminths.

Physiology. Parasitic worms vary in their diets. Tapeworms, having no alimentary canal, absorb carbohydrates from material ingested by the host but certain nitrogenous and probably other substances from the intestinal secretions of the host. There is recent evidence that competition for carbohydrate may be the critical factor concerned in the "crowding effect" in tapeworm infections (Read, 1959; Holmes, 1959). Flukes feed in part on blood and lymph, in part on cells and tissue debris. Some nematodes, e.g., hookworms, feed mainly on blood, but others subsist principally on tissues, either ingesting them and digesting them in the intestine or first liquefying them, presumably through the action of lytic secretions of esophageal glands. Ability to dissolve tissues is also shared by some flukes. As discussed on pp. 28–29, the tissues seem to be capable of developing resistance to digestion by the worms. In hosts immunized by repeated or long-standing infections, intestinal nematodes, unable to feed in the midst of plenty, fail to grow and are frequently eliminated.

Although some parasitic helminths live in blood and tissues where there is an abundance of oxygen, and others in the intestine or bile ducts where oxygen is a scarce commodity, many seem to be facultative anaerobes, showing considerable tolerance to absence of oxygen. Some apparently can get along indefinitely at very low oxygen tensions and are injured when exposed to very much of it, whereas others thrive in its absence for only limited periods. All helminths, however, consume oxygen to some extent when it is available, and produce carbon dioxide. It has generally been assumed that the lumen of the intestine of a vertebrate animal is a practically anaerobic environment, but Read (1950) pointed out that the space next to the mucosa may be quite similar physiologically to intercellular spaces inside the host, and may contain considerable amounts of oxygen and other substances that would not be found in appreciable amounts free in the lumen.

The need for oxygen may be the reason why hookworms have such an insatiable appetite for blood, which is constantly sucked from the host, passed through the body as through a vein, and spilled into the intestine (see p. 431). Rogers has shown that some of the small nematodes of the digestive tract have a pattern of metabolism which is characteristic of aerobic organisms. Most parasite eggs also require oxygen to complete their development.

The intermediary metabolism of helminths apparently involves a variety of enzymes catalyzing chemical reactions identical with those in the host, but Bueding has shown that these enzymes are immunologically different and, in some cases, have different physical properties.

This is a field of parasitology which is still largely unexplored, but in which there is growing interest. Knowledge of the physiology and biochemistry of helminths may make it possible to find effective anthelmintics by a logical process of reasoning instead of the hit or miss method of trial and error used in the past, for a large proportion of modern chemotherapy, as noted on p. 9, is based on interference with enzyme systems of parasites. Some advances along this line have already been made. The recent *in vitro* cultivation of cestodes by Smyth, by Mueller, and by Schiller, Read, and Rothman and of parasitic nematodes by Stoll and by Weinstein and Jones furnish new tools for studying the physiology of animal parasites.

Although helminths are very poorly equipped with sensory organs, they show amazing ability to react when the necessity arises. With no evident specialized sense organs of any kind whatever, a single male *Trichinella* finds a single female in the relatively vast expanses of a rat's intestine, and some miracidia, the larvae of flukes, are attracted by their proper snail hosts as are filings by a magnet.

Life history and modes of infection. The life history and modes of infection of worms vary with the habitat in the body. Every parasitic worm must have some method of gaining access to the body of its host and must have some means for the escape of its offspring, either eggs or larvae, from the host's body in order to continue the existence of its race. Many species utilize intermediate hosts as a means of transfer from one host to another; others have a direct life history, i.e., either they develop inside the escaped egg and depend on such agencies as food and water to be transferred to a new host, or they develop into free-living larvae which are swallowed by or burrow into a new host.

Most of the intestinal helminths enter their host by way of the mouth, and the eggs escape with the feces. Many species enter as larvae in the tissues of an intermediate host which is eaten by the final host, e.g., most of the tapeworms, many flukes, and some nematodes (spiruroids). Some nematodes of the intestine, such as the pinworm and whipworm, make their entry as fully developed embryos in the eggs. Others, like the schistosomes, hookworms, and *Strongyloides*, usually reach their destination in an indirect way by burrowing through the skin. All the intestinal worms except *Trichinella* produce eggs or larvae that escape from the body with the feces. In *Trichinella* the larvae encyst in the muscles, and their salvation depends on their host's being eaten by another animal.

Many of the helminths of other organs of the body also enter by way of the mouth and digestive tract, though they have various means of exit for the eggs or larvae. The liver flukes enter and escape from the body as do intestinal parasites; the schistosomes enter by burrowing through the skin, and the eggs escape with either feces or urine; filariae enter and leave the body by the aid of bloodsucking insects; the guinea worm enters by the mouth, and the larvae leave through the skin. The larval tapeworms which infest man enter either by the mouth or by accidental invasion of the stomach from an adult in the intestine, or, in the case of *Sparganum* infections in China, by the bizarre method of burrowing into the eyes from infected frogs used as poultices. Like those of *Trichinella,* larvae of tapeworms are usually permanently sidetracked in man, and escape only if their hosts have the greater misfortune of being eaten by cannibals, leopards, or rats.

Adjustments in life cycles. It is obvious that parasitic worms have a tremendous problem to solve in insuring the safe arrival of their offspring in the bodies of other hosts, on which the survival of the race depends, for sooner or later the body which is affording food and shelter will die, and however immortal the soul may be, the parasites can derive little comfort from it. The problem is difficult enough for worms like *Ascaris, Trichuris,* and hookworms, whose offspring merely have to spend a relatively short time in the great outdoors before being ready to return, either as stowaways in food or water or by their own burrowing, to another host of the same species. But flukes and tapeworms are so hampered by heritage and tradition that they have to undergo a preliminary development in some entirely different but often very particular kind of animal, and sometimes must even spend an apprenticeship in a third kind, before they are ready for their ultimate life of ease and comfort in the definitive host. When one considers the experiences through which a lung fluke, for example, must go in order to live and reproduce its kind, first as a minute free-swimming protozoanlike organism, then as an asexually reproducing parasite of certain species of snails, then as a tissue-invading parasite of crabs, and finally as a human invader that must find its way from the stomach to the lungs, he would be incredulous if he were not confronted with the fact that the lung fluke not only succeeds in accomplishing this, but succeeds so well that in some places it constitutes a serious menace to the health of whole communities.

Since the vicissitudes of life for the offspring of parasitic worms are so great, it is obvious that there must be a tremendous waste of offspring which do not succeed in the struggle, and therefore a sufficiently large number of eggs or young must be produced so that the chances of survival are a little greater than the chances of destruction. The numbers necessary to accomplish this are amazing. The hookworm, *Ancylostoma duodenale,* lays in the neighborhood of 20,000 eggs a day, and it may do this for at least 5 years; the total offspring of such a worm would number over 36,000,000. If the number of hookworms in a community remains about constant, as it usually does, and the percentage of males and females is equal, the chances against a male and female hookworm gaining access to a host, and living for the full period of 5 years, is then 18,000,000 to 1. The hookworm, however, has a comparatively simple time of it. Flukes and tapeworms have an even more difficult problem to face. According to estimates of Penfold et al. (1937), a beef tapeworm produces over 2500 million eggs in 10 years, yet this worm is rare enough so that most practitioners keep specimens in bottles on their shelves!

Flukes and tapeworms owe such success as they have to two special devices in their life cycles. In the first place, they have to a large extent substituted self-fertilization for cross-fertilization; they combine male and female organs of reproduction in a single individual and do not take chances on other individuals of the opposite sex being present to render the eggs viable. In the second place, efficient egg-making machines as they are, they have found the production of sufficient eggs by one body inadequate. A tapeworm overcomes the difficulty by constantly reproducing, sometimes for years, more egg-producing segments, in essence new individuals, by a process of budding. Some, such as *Multiceps* and *Echinococcus,* go even further and produce several or even many thousands of buds while in the larval stage, each of which is capable of developing into a new individual when, if ever, it reaches its final host.

Flukes attain the same end in a different way. Instead of producing a sufficient number of eggs to overcome the chances of destruction through the whole cycle of development, they distribute the risk. They produce enough eggs to overcome the chances against their reaching the mollusk which serves as the first intermediate host; then, in order to overcome the odds against them in the subsequent part of the life cycle, the successful individuals reproduce asexually. This is accomplished by a process of multiplication and separation of the germ cells before they cooperate to form a new individual—what zoologists call polyembryony. A single schistosome embryo, after successfully reaching the liver of a snail, may give birth, by asexual reproduction, to over 100,000 progeny. Without this advantage a schistosome would have to produce thousands of times as many eggs as it does.

Significance of intermediate hosts. One might reasonably ask why some worms adhere to the life cycles which they have, when so much simpler ways of reaching their hosts would seem to be available. A fluke which lives as a parasite in the intestine of a bat, for example, would seem to be very ill-advised to select a snail, on which bats do not feed, as an intermediate host, when an insect would serve so much better. Nature is in this respect strangely inconsistent—she is a peculiar mixture of progressiveness and conservatism. In many instances, as we have seen, she has evolved the most intricate specializations both in life cycle and in structure; there are innumerable instances in the animal kingdom of short cuts and detours in life cycles, devised to meet newly developing conditions. On the other hand, there are some short cuts that nature is too conservative to take. It is one of the fundamental precepts of embryology that ontogeny, i.e., the development of the individual, recapitulates phylogeny, which is evolutionary development of the race. Many unnecessary phases are, however, slurred over or greatly altered, and sometimes entirely new phases are interposed to meet the exigencies of the situation, as, for example, the pupa of insects (see p. 526).

Now intermediate hosts, in which partial development occurs, are unquestionably in many instances ancestral hosts. Mollusks are probably to be regarded as the hosts of the redialike or cercarialike ancestors of digenetic flukes. In the course of evolution these developed further until they reached the condition of modern flukes. Nature, however, has been too conservative to produce flukes in which the mollusk phase of the phylogeny is omitted in the ontogeny; this is apparently too radical a short cut. The result is that all flukes, with the exception of two species with forked-tailed cercariae, *Cercaria loossi* and *C. hartmanae,* which develops in annelids, must first be molluscan parasites regardless of their final destiny, just as a chicken must have gill slits like a fish before it can have lungs like its parents. Therefore, we have the irrational condition of flukes becoming first parasites of snails, then of insects, and only after sojourns in these animals, parasites of bats or birds. Undoubtedly the earliest method of transfer of flukes to their final hosts was by the eating of the infected mollusks, a method still adhered to by many flukes of mollusk-eating animals. In more highly specialized flukes, however, the cercariae leave the mollusk to encyst on vegetation if the host is a vegetarian, in fishes or other animals if it is carnivorous, or, in the case of the schistosomes, to take an active instead of a passive attitude and burrow directly into their final hosts.

A somewhat similar course has been followed by tapeworms, which were probably originally parasites of Crustacea, then became parasites of Crustacea-eaters, and eventually parasites of eaters of Crustacea-eaters. The tapeworms, however, seem a little less hidebound by tradition, for some have substituted insects or even mammals for the ancestral Crustacea, and a very few (e.g., *Hymenolepis nana*) have done away with the need for intermediate hosts altogether.

Not all intermediate hosts can be regarded as heritages from the remote past; some are quite clearly secondarily acquired conveniences in the life cycle to facilitate access to a new host. This is true of the mollusk and earthworm hosts of lungworms (see p. 446) and almost certainly of the insect hosts of spiruroid and filarial worms. But as Rothschild and Clay remark in their delightful book *Fleas, Flukes and Cuckoos,* by whatever paths they have evolved, the life cycles of parasitic worms are today sufficiently complicated and extraordinary to satisfy the imagination of Salvador Dali himself. As they say: "When the flatworms gave up their freedom they certainly began an odyssey compared with which the voyages of Ulysses seem singularly uneventful and commonplace."

Host relations. The effects produced by parasitic worms depend in part on the organs or tissues occupied, in part on the habits of the worms, and in part on the poisonous qualities of their secretions or excretions, in part on the numbers of worms harbored, and to a very large degree on acquisition of immunity.

Worm infections differ radically from bacterial or protozoan infections in that, in most cases, the worms do not multiply in the body of the host, and thus the infections are quantitative in nature. The bite of a single lightly infected mosquito may produce as severe a case of malaria as numerous bites by heavily infected mosquitoes, but the acquisition of a few hookworms, liver flukes, or filariae produces in a given individual a very different effect from the oft-repeated acquisition of large numbers of these worms. The term "infestation" instead of "infection" is frequently used to distinguish nonmultiplying invaders from multiplying ones.

In some instances even single worms may cause a serious disturbance. Thus a single *Dibothriocephalus latus* may cause severe anemia; a single gnathostome may cause a fatal perforation of the stomach wall; a single *Ascaris* may block the bile or pancreatic duct; and a single guinea worm creeping under the skin may lead to an infection causing loss of a limb. In the majority of cases, however, the pathogenicity of worms is proportional to the number present.

Some investigators tend to minimize the damage done by helminths, especially intestinal ones, whereas others undoubtedly overestimate it. Improved facilities for discovering infection have demonstrated the presence of intestinal parasites in so many unsuspected cases that we are likely to incriminate them in nearly every morbid condition for which we cannot, with equal readiness, discover another cause. Differences in the effects of worm infestations are due in part to the variable susceptibility of different races and individuals; in part to presence or absence, and degree, of malnutrition or other debilitating influences; in part to number of worms harbored and rate at which the worm burden is increased; and in part to acquired immunity. If the rate of acquisition of worms is slow enough in a well-nourished individual, immunity can develop before serious damage has been done (see p. 26).

Effects of parasitism. The principal ways in which helminths harm their hosts are by mechanical damage, devouring of tissues or loss of blood, and by toxic effects. Some large worms, such as the larger tapeworms, may rob the host of enough food, especially proteins or vitamins, to cause malnutrition or mild vitamin deficiencies in hosts that are on a suboptimum intake, especially in the case of young, growing individuals. This is especially striking in infestations with *Dibothriocephalus;* this worm has a special affinity for vitamin B_{12}, and as a result precipitates pernicious anemia in persons who are on the borderline of it (see p. 351).

The mechanical injuries are almost as numerous as the kinds of worms. Some, such as the hookworms, bite the intestinal wall and cause hemorrhages, which are intensified by a secretion which prevents the blood from coagulating; some, such as the lung flukes and guinea worms, cause tissue damage and inflammation by burrowing; some, such as schistosomes and numerous Spirurata (*Gnathostoma, Gongylonema, Onchocerca, Spirocerca,* etc.), cause the formation of tumors, and in some cases—either by irritation or toxic action—of true cancerous growth; some, such as *Ascaris,* may block ducts or even cause intestinal obstruction; some, such as gnathostomes and occasionally *Ascaris,* may cause perforation of the walls of the digestive tract and consequent peritonitis; some, such as the liver flukes, may choke up the bile passages of the liver; some, such as Bancroft's filaria, may interfere with the normal flow of lymph and divert it into abnormal channels; some, such as hydatid cysts, may interfere with the proper functioning of neighboring organs by pressure; some, such as the schistosomes, may produce profound irritation of the tissues by extruding their eggs into them; and some, such as hookworms and spiny-headed worms, open up portals of entry for bacteria. We have awakened to the importance of a "whole skin" and the danger which accompanies the piercing of it by the unclean proboscides of biting insects. We have not yet fully awakened to the importance of an uninjured mucous membrane. As has been pointed out by Shipley, the intestinal worms play a part within our bodies similar to that played by bloodsucking arthropods on our skins, except that they are *more* dangerous since, after all, only a relatively small number of biting insects have their proboscides soiled by organisms pathogenic to man, whereas the intestinal worms are constantly accompanied by bacteria that are capable of becoming pathogenic if they gain access to the deeper tissues. Weinberg found that, whereas he was unable to infect unparasitized apes with typhoid bacilli, apes infested with tapeworms or whipworms readily contracted typhoid fever, the bacteria presumably gaining entrance through wounds in the mucous membrane made by the worms.

Some effects of worms which were once ascribed to toxic products liberated into the tissues are now known to be caused in other ways, e.g., the primary anemia in hookworm infections, and the pernicious anemia in *Dibothriocephalus* infections. Probably all worms that live in the blood or tissues, however, sensitize the body to their secretions or excretions, or to their body substance, and this causes allergic reactions. These are largely responsible for the symptoms in infestations with such adult worms as the filarias, schistosomes, and guinea worms, with the migrating larvae of such worms as *Trichinella, Strongyloides,* hookworms, and *Ascaris,* and even with such adult worms in the intestine as *Ascaris* and *Enterobius.*

A characteristic feature of an allergic reaction is an increase in the number of eosinophiles. These are white blood corpuscles containing granules that stain red with eosin. Such an increase in eosinophiles (called eosinophilia) is one of the most characteristic features of helminthic infection, particularly during the period when the helminths are developing or migrating in the body. From a normal of 1 or 2% of the leucocytes, the eosinophiles may rise to 5% or more, sometimes to 75%, in infections with *Trichinella,* schistosomes, *Echinococcus* cysts, etc. There is evidence that ACTH or cortisone may alleviate severe allergic reactions sufficiently to be a life saver in severe *Trichinella* infections.

Diagnosis. The diagnosis of infection with various species of worms depends principally on the identification of their eggs or larvae as found in the feces or other excretions by microscopic examination. Nearly every species of parasite has recognizably distinct characteristics of the eggs, the chief variations being in size, shape, color, thickness of shell, stage in development, appearance of the embryo if present, and presence or absence of an operculum or lid.

In many instances whole groups of worms have egg characteristics in common; for example, the eggs of flukes, except schistosomes, have an operculum at one end; those of schistomes have spines; those of most tapeworms of warm-blooded animals have no operculum and contain a fully developed six-hooked embryo. Those of *Dibothriocephalus*, however, and most of the tapeworms of cold-blooded animals, have undeveloped operculated eggs like those of flukes; eggs of the tapeworm family *Taeniidae* are characterized by thick inner shells (embryophores) which have a striated appearance in optical section; eggs of ascarids are thick shelled, bile stained, and with surface markings; those of whipworms and their allies are brown with an opercular plug at each end; those of oxyurids are colorless and flattened on one side; and those of the hookworms and all their allies of the suborder Strongylata have thin-shelled, unstained eggs without either opercula or surface markings. Some eggs of the commoner worms are shown in a comparative way in Fig. 55.

Most worm infections of the digestive system can be diagnosed by finding eggs or larvae in the feces, but this cannot be relied on in some kinds of infections. Tapeworms, other than *Dibothriocephalus* and its allies, have no natural exit for the eggs from the segments, but in most species the eggs are easily released from ruptures of the uterus at the ends of the detached ripe segments. *Hymenolepis* eggs are probably always released inside the body and become mixed with the feces. *Taenia saginata* eggs are more frequently found on the skin near the anus, since the detached segments often squirm out of the rectum. However, infections with the larger tapeworms are best discovered by examining the surface of freshly passed stools for the segments, which squirm actively like flukes (for which they are sometimes mistaken). Pinworms (*Enterobius*) do not ordinarily deposit the eggs in the feces at all; the females crawl out of the anus and deposit the eggs on the perianal skin, where they can be picked up by special devices described on p. 462. *Trichinella*, since its embryos do not normally leave the body at all, cannot often be diagnosed by fecal examination, though sometimes some of the adult worms can be expelled by violent purges or anthelmintics. For this and some other helminthic diseases—filariasis, infestations with larval tapeworms, and schistosomiasis—skin tests or other immunological tests must often be resorted

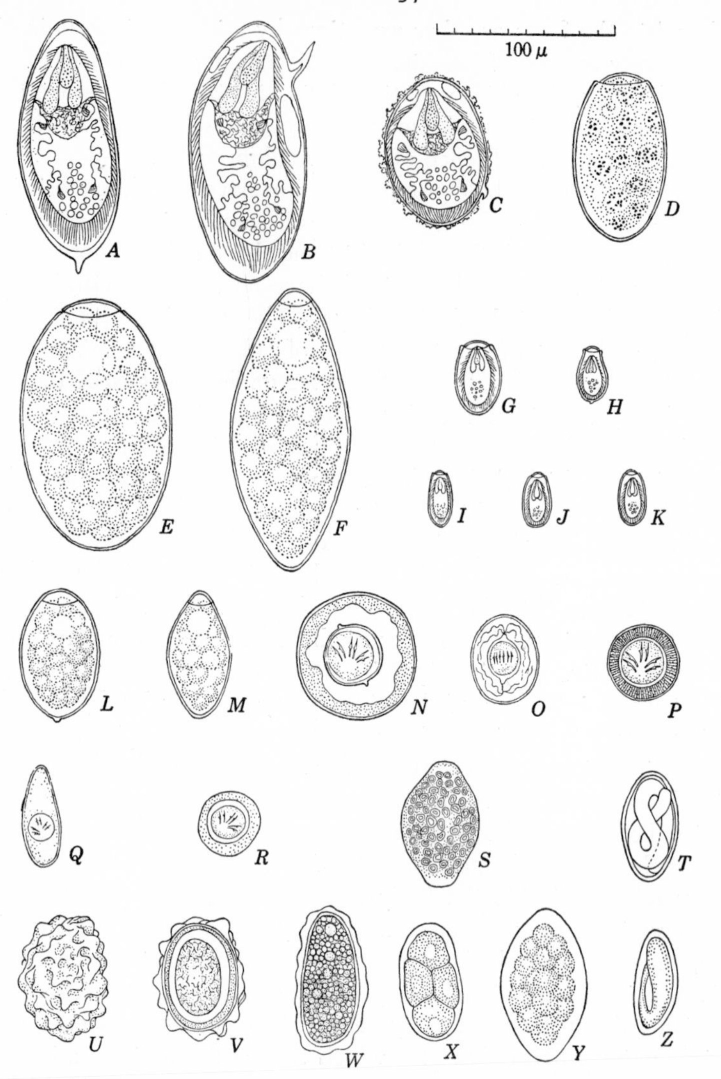

Fig. 55. Eggs of parasitic worms, drawn to scale (×250). Flukes: *A*, *Schistosoma haematobium;* *B*, *Schistosoma mansoni;* *C*, *Schistosoma japonicum;* *D*, *Paragonimus westermanni;* *E*, *Fasciolopsis buski;* *F*, *Gastrodiscoides hominis;* *G*, *Dicrocoelium dendriticum;* *H*, *Clonorchis sinensis;* *I*, *Opisthorchis felineus;* *J*, *Heterophyes heterophyes;* *K*, *Metagonimus yokogawai.* Tapeworms: *L*, *Dibothriocephalus latus;* *M*, *Spirometra mansonoides;* *N*, *Hymenolepis diminuta;* *O*, *Hymenolepis nana;* *P*, *Taenia* sp. or *Echinococcus;* *Q*, *Raillietina madagascariensis;* *R*, *Dipylidium caninum.* Nematodes: *S*, *Dioctophyma renale;* *T*, *Gongylonema* sp.; *U*, *Ascaris lumbricoides* (surface view); *V*, Same, optical section; *W*, Same, unfertilized; *X*, *Necator americanus;* *Y*, *Trichostrongylus* sp.; *Z*, *Enterobius vermicularis.*

to; these will be discussed in connection with the diseases concerned.

Intestinal worm infestations can usually be diagnosed by finding eggs or larvae, though the number present may vary considerably from day to day. In heavy infestations microscopic examination of a simple smear in water, thin enough to read newspaper print through, is sufficient, but many light infestations escape detection by this method and concentration methods are necessary.

Flotation methods in heavy salt or sugar solutions are valuable for eggs of most kinds of nematodes and some tapeworms but fail to float the eggs of schistosomes, the operculated eggs of flukes or *Dibothriocephalus,* the porous eggs of tapeworms of the family Taeniidae, the eggs of Acanthocephala, or the unfertilized eggs of *Ascaris.*

One of the best techniques for demonstration of these eggs, particularly those of intestinal schistosomes, is a modification of the old Telemann acid-ether technique, known as the AMS III method (Hunter et al., 1948), performed as follows:

1. Comminute 2 grams of stool with 5 cc. of an equal mixture of HCl (40 cc. concentrated HCl in 60 cc. of water) and sodium sulfate (sp. gr. 1.08).
2. Strain through gauze moistened with this mixture, pour into a 15 cc. centrifuge tube, and wash two or three times by centrifuging for about 1 minute at 2000 r.p.m., decanting the supernatant fluid each time and mixing the sediment with fresh mixture.
3. Add 5 cc. of fresh mixture plus 3 drops of Triton NE (a detergent) plus 5 cc. refrigerated ether, shake for 30 seconds, and centrifuge for 1 minute.
4. Remove tube, loosen ring of debris at interface of ether and HCl-Na_2SO_4 mixture with applicator stick, and decant.
5. Swab tube down to the sediment with cotton swab.
6. Add physiologic saline to the 0.4 cc. mark, mix sediment, pipette onto 1 or 2 slides, apply cover glass, and examine.

Another modification of the Telemann technique, which is good for protozoan cysts as well as eggs, consists of comminution of the stool in saline followed by straining and washing as above, addition of 10 per cent formalin to the sediment, then addition of ether after 5 minutes, followed by steps 4 to 6 above (Ritchie, 1948).

For flotation saturated NaCl (sp. gr. 1.200), or $ZnSO_4$ (sp. gr. 1.180), which also brings up protozoan cysts, is most frequently used, though a sugar solution is preferred by some. The simplest flotation method is that of Willis (1921), in which a 1-oz. or 2-oz. tin container for collecting fecal samples is left one-sixth to one-tenth full of feces, and is then stirred gradually with salt solution until brimful. A 2 by 3 in. glass slide is then placed over it in contact with the fluid; in 10 minutes the slide is carefully lifted by a straight upward pull, inverted, and examined.

Lane (1928) devised a method of direct centrifugal flotation (DCF). About 1 cc. of stool is thoroughly mixed with water in a centrifuge tube with a ground top, centrifuged, and the supernatant poured off. The residue is then mixed with saturated NaCl or $ZnSO_4$ (sp. gr. 1.180), the tube filled to the top, covered with a No. 2 cover glass, and placed in centrifuge buckets provided with four projecting horns to prevent the cover glass from sliding off during the centrifuging. After centrifuging for 1 minute at 1000 rpm, the cover is removed and the adhering fluid examined as a hanging drop or by dropping the cover on a slide. The special apparatus is unnecessary if the last few drops of solution to fill the tube are added after centrifuging, and the surface film is removed by touching a cover to it or by means of a 4-mm. bacteriological loop. The DCF method demonstrates a high percentage of the eggs present, and in a small area.

Egg counts. About 1920 Darling called attention to the importance of quantitative diagnosis of worm infections. Stoll in 1923 devised a satisfactory method of estimating eggs per gram of feces by diluting a measured quantity of feces in a measured volume of 0.1 *N* NaOH, counting the eggs in a measured fraction, and multiplying by the proper factor. Stoll and Hausheer in 1926 recommended the use of a special narrow-necked flask filled to a 56-cc. mark with 0.1 *N* NaOH, and then to a 60-cc. mark with feces, thus diluting 4 cc. fifteen times. After thorough shaking, the eggs in a 0.075-cc. drop of this are counted under a 25-mm. square cover glass. The eggs counted, multiplied by 200, represent the eggs per gram. The method will not do for very light infections and is unreliable in individual counts, but when averaged, even for small groups, is useful in estimating the worm burden of a community, and in determining the relative number of light, medium, and heavy infections. Adjustments should usually be made to bring the results of the examinations to a uniform basis of formed stools. Egg counts in "mushy" stools are multiplied by two, and in liquid stools by four. Beaver (1949) devised a method by which adjustments for density of stools are made with the help of a light meter (photoelectric cell).

Treatment. Treatment of the various worm infections is considered under the head of the different kinds of worms, but a few general principles should be noted here.

Drugs which are used for expelling worms are known as anthelmintics. An ideal anthelmintic is one which effectively kills or expels the particular worms for which it is used, is not injurious to the host in the dose required, is easily administered, and is cheap.

There is little probability that any anthelmintic will be found that is effective against all kinds of worms, even all intestinal ones. Although nematodes and trematodes, which have digestive tracts, respond to many of the same drugs, tapeworms are affected very little by most of these drugs, but respond to entirely different ones which have little effect on the other worms. Even among intestinal nematodes susceptibility to drugs varies greatly. Hookworms are easily killed by tetrachlorethylene, but *Necator* more readily than *Ancylostoma,* and *Ascaris* is merely annoyed by it; the reverse holds true to a considerable degree for oil of chenopodium. Phenothiazine is highly effective against *Haemonchus* and moderately so against species of *Trichostrongylus,* but is without significant effect against *Nematodirus;* yet all of these belong to the same family, Trichostrongylidae. Dithiazanine comes closest to being a broad spectrum human anthelmintic, acting against hookworms, *Ascaris, Trichuris, Enterobius,* and *Strongyloides.*

Little is known about the mechanisms by which most anthelmintics exert their effects. Some, e.g., ficin and papain, which are proteolytic enzymes in fresh fig latex and papaya, respectively, digest the tissues of nematodes. Many anthelmintics probably act, as do other chemotherapeutic drugs, by interfering with enzyme systems, e.g., the antimony drugs used against schistosomes, but for the most part the mode of action is unknown. Some drugs exert their effects slowly, and may have sterilizing effects by gradually destroying the reproductive organs, particularly the ovary, e.g., antimony compounds used against schistosomes, and arsenic compounds against filariae. Some antifilarial drugs quickly kill the microfilariae (embryos) and only slowly affect the adult worms; others have the opposite effects. Dithiazanine may act by inhibiting oxidative energy metabolism. It is a cyanine compound, a class of drugs which Bueding showed to inhibit the respiration of certain helminths. It is quite possible that some of the drugs used against intestinal worms might have similar effects in preventing reproduction by slowly destroying the sex organs if used in small doses over a period of time, but anthelmintics against intestinal worms are usually discarded and considered worthless if they are not immediately lethal.

REFERENCES

Baylis, H. A. 1929. *A Manual of Helminthology, Medical and Veterinary.* Baillière, Tyndall and Cox, London.

Beaver, P. C. 1949. Quantitative hookworm diagnosis by direct smear. *J. Parasitol.,* 35: 125–135.

Brand, T. von. 1952. *Chemical Physiology of Endoparasitic Animals.* Academic Press, New York.

Bueding, E. 1954. Some aspects of the comparative biochemistry of *Ascaris* and of schistosomes. In *Cellular Metabolism and Infections* (Edited by E. Racker). Academic Press, New York.

Bueding, E., and Most, H. 1953. Helminths: metabolism, nutrition and chemotherapy. *Ann. Rev. Microbiol.,* 7: 295–326.

Chandler, A. C. 1935–1938. Studies on the nature of immunity to intestinal helminths, I-VI. *Am. J. Hyg.,* 22: 157–168, 243–256; 23: 46–54; 26: 292–307, 309–321; 28: 51–62.

1939. The nature and mechanism of immunity in various intestinal infections. *Am. J. Trop. Med.,* 19: 309–317.

1953a. The relation of nutrition to parasitism. *J. Egyptian Med. Assoc.,* 36: 533–552.

1953b. Immunity in parasitic diseases. *J. Egyptian Med. Assoc.,* 36: 811–834.

Chitwood, B. G., and Chitwood, M. B. 1937–1942, 1950. An introduction to Nematology. Sect. I, Pts. I-III, and Sect. II, Pts. I and II, 1937–1942; Sect. 2, revised 1950. Nematology and Co., Marquette, Mich.

Fairbairn, D. 1957. The biochemistry of *Ascaris. Exp. Parasitol.,* 6: 491–554.

Faust, E. C. *Human Helminthology.* 3rd ed. Lea and Febiger, Philadelphia.

Harwood, P. D. 1953. The anthelmintic properties of phenothiazine *Exptl. Parasitol,* 2: 428–455.

Holmes, J. C. 1959. Competition between the rat tapeworm *Hymenolepis diminuta* and the acanthocephalan *Moniliformis dubius* in the rat. Ph.D. Thesis, Rice Institute.

Hunter, G. W., et al. 1948. Studies on schistosomiasis. II. Summary of further studies on methods of recovering eggs of *S. japonicum* from stools. *Bull. U. S. Army Med. Dept.,* 7: 128–131.

Hyman, L. 1951. *The Invertebrates,* Vol. 2, *Platyhelminthes and Rhynchocoela;* Vol. 3, *Acanthocephala, Aschelminthes and Entoprocta.* McGraw-Hill, New York.

Kükenthal, W. (Editor). 1928–1933. *Handbuch der Zoologie,* Bd. II, 1 u. 2, Vermes, Berlin.

Lapage, G. 1937. *Nematodes Parasitic in Animals.* Cambridge University Press, London.

1951. *Parasitic Animals.* Cambridge University Press, London.

LaRue, G. R. 1951. Host-parasite relations among the digenetic trematodes. *J. Parasitol.,* 37: 333–342.

Mönnig, H. O. 1949. *Veterinary Helminthology and Entomology,* 3rd ed. Williams and Wilkins, Baltimore.

Morgan, B. B., and Hawkins, P. A. 1949. *Veterinary Helminthology,* Burgess, Minneapolis.

Neveu-Lemaire, M. 1936. *Traité d'helminthologie médicale et vétérinaire.* Vigot Frères, Paris.

Penfold, W. J., Penfold, H. B., and Phillips, M. 1937. *Taenia saginata;* its growth and propagation. *J. Helminthol.,* 15: 41–48.

Rausch, R. 1951. Biotic interrelationships of helminth parasitism. *Public Health Repts.,* 66: 18–24.

Read, C. P. 1950. The vertebrate small intestine as an environment for parasitic helminths. *Rice Inst. Pamphlet,* 37: No. 2, 94 pp.

1951. The crowding effect in tapeworm infections. *J. Parasitol.,* 37: 174–178.

Read, C. P. 1959. The role of carbohydrate in the biology of cestodes. VIII. Some conclusions and hypotheses. *Exp. Parasitol.,* 8: 365–382.

Ritchie, L. S. 1948. An ether sedimentation technique for routine stool examinations. *Bull. U. S. Army Med. Dept.,* 8: 326.

Rothschild, M., and Clay, T. 1952. *Fleas, Flukes and Cuckoos.* Philosophical Library, New York.

Stoll, N. R. 1947. This wormy world. *J. Parasitol.,* 33: 1–18.

Stunkard, H. W. 1937. The physiology, life cycles, and phylogeny of the parasitic flatworms. *Am. Mus. Novitates,* No. 908.

1953. Life histories and systematics of parasitic worms. *Systematic Zool.,* 2: 7–18.

Van Cleave, H. J. 1948. Expanding horizons in the recognition of a phylum. *J. Parasitol.,* 34: 1–20.

Wardle, R. A., and McLeod, J. A. 1952. *The Zoology of Tapeworms.* University Minnesota Press, Minneapolis.

Whitlock, J. H. 1947. *Illustrated Laboratory Outline of Veterinary Entomology and Helminthology.* Burgess, Minneapolis.

Yorke, W., and Maplestone, P. A. 1926. *The Nematode Parasites of Vertebrates.* J. and A. Churchill, London.

LEECHES

Chin, T. H. 1949. Further note on leech infestation of man. *J. Parasitol.,* 35: 215.

Masterman, E. W. G. 1908. Hirudinea as human parasites in Palestine. *Parasitology,* 1: 182–185.

Neveu-Lemaire, M. 1938. Hirudinea, in *Traité d'entomologie médicale et vétérinaire.* Vigot Frères, Paris.

Whitman, C. O. 1886. Leeches of Japan. *Quart. J. Microscop. Sci.,* 26: 317–416.

Chapter 12

THE TREMATODES OR FLUKES

I. GENERAL ACCOUNT

The trematodes are animals of a very low order of development in some respects and of very high specialization in others; all of them are parasitic. There are two principal groups: the monogenetic flukes with no asexual generations, which are primarily external or semi-external parasites of aquatic animals; and the digenetic flukes with two or more asexual generations and an alternation of hosts, which are internal parasites of all kinds of vertebrates. Since only the digenetic flukes are of interest as parasites of man and domestic animals, the following account refers to this group except when otherwise specified.

In shape the flukes are usually flat and often leaflike, with the mouth at the bottom of a muscular sucker, usually at the anterior end; in most groups there is a second sucker, for adhesion, on the ventral surface. The monogenetic flukes have highly specialized compound suckers at the posterior end, usually supplemented with hooks. The thin cuticle of trematodes, often spiny, is apparently secreted by mesodermal cells, since the true ectoderm is lost during development of the cercariae. Under the cuticle are layers of circular, longitudinal and diagonal muscles, and inside of this the loose mesh of the parenchyma.

The development of the nervous system is of low grade; a small ganglion at the forward end of the body gives off a few longitudinal nerves. Sense organs are almost lacking. There is no blood or blood system, the result being that the digestive tract and excretory system are branched, often to a surprising extent, in order to carry food to all parts of the body and to carry waste products out from all parts. The digestive system (Fig. 56) usually has a muscular *pharynx* near the mouth, and then branches into two blind pouches, the *intestinal ceca.* In some of the larger flukes, e.g., *Fasciola hepatica,* these ceca have numerous branches and subbranches, whereas in the schistosomes the ceca reunite posteriorly to form a single stem. Only in a few aberrant species do the ceca open posteriorly.

The excretory system consists of a complicated arrangement of branched tubules. At the ends of ultimate fine branches are flame cells which keep up a flow of fluid towards the excretory pore. The

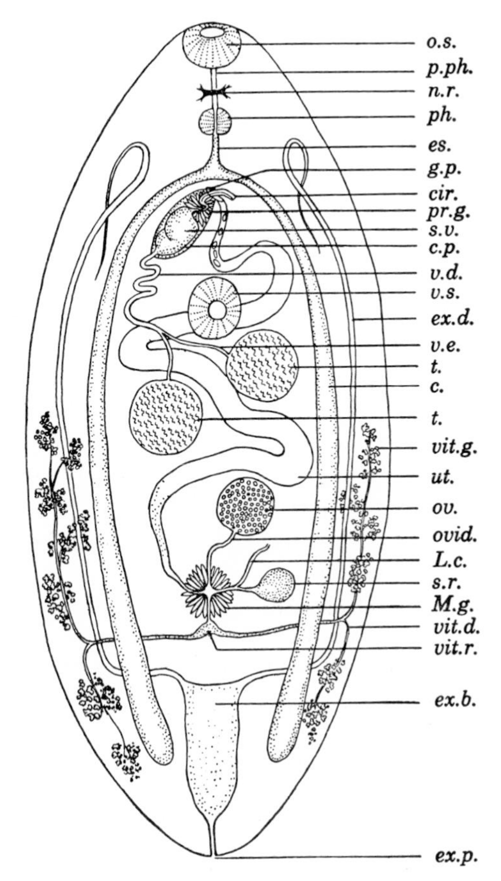

Fig. 56. Diagrammatic fluke to illustrate principal morphological characteristics. Abbreviations: *c.*, cecum; *cir.*, cirrus; *c.p.*, cirrus pouch; *es.*, esophagus; *ex.b.*, excretory bladder; *ex.d.*, excretory duct; *ex.p.*, excretory pore; *g.p.*, genital pore; *L.c.*, Laurer's canal; *M.g.*, Mehlis' gland; *n.r.*, nerve ring; *o.s.*, oral sucker; *ov.*, ovary; *ovid.*, oviduct; *ph.*, pharynx; *p.ph.*, prepharynx; *pr.g.*, prostate glands; *s.r.*, seminal receptacle; *s.v.*, seminal vesicle; *t.*, testis; *ut.*, uterus; *v.d.*, vas deferens; *v.e.*, vas efferens; *vit.d.*, vitelline duct; *vit.g.*, vitelline glands; *vit.r.*, vitelline reservoir; *v.s.*, ventral sucker.

finer branches unite in a definite manner, varying in different groups. In the monogenetic flukes the excretory system may open by two anterior pores (Fig. 58), but in the digenetic flukes the main collecting

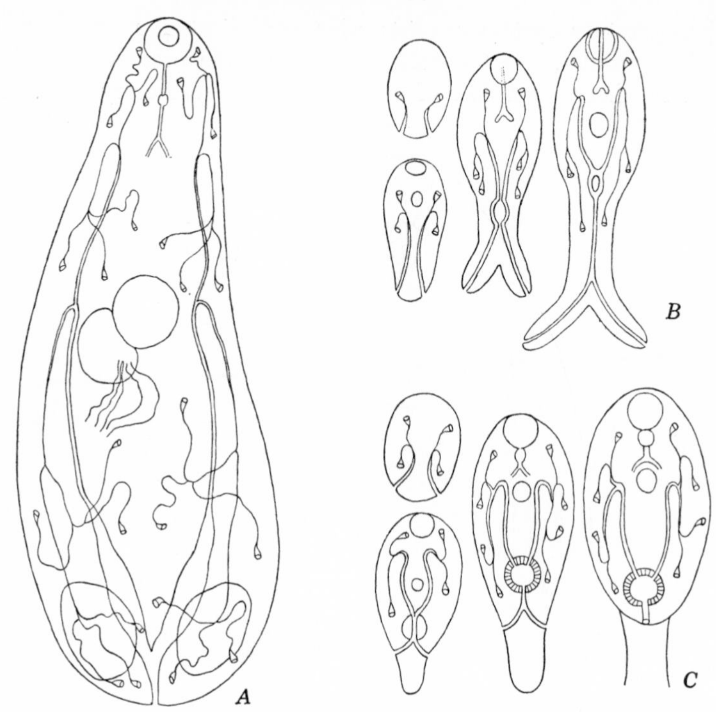

Fig. 57. Excretory system of digenetic flukes. *A*, excretory system of *Heterophyes heterophyes.* Note Y-shaped bladder, and division of main collecting vessels in mid-region of body ("Mesostomate," in contrast to stenostomate condition in which main vessels extend far forward and then turn backward), and two groups of three flame cells each on both anterior and posterior branches, giving a flame-cell formula of 2 [(3 + 3) + (3 + 3)]. *B*, development of a schistosome cercaria, showing formation of primitive, thin-walled excretory bladder by fusion of two principal ducts (characteristic of Anepitheliocystida; see p. 273), and primary excretory pores on sides or ends of tail, which is formed partly by molding of body and partly by growth of tissues so molded. *C*, development of stylet cercaria, showing thick-walled epithelial bladder replacing primitive bladder (characteristic of Epitheliocystida; see p. 274), and absence of excretory system in tail, which is formed entirely by backward growth of tissue lying between primary excretory pores. (*A*, after Looss from Stunkard, *J. Parasitol.,* 1929; *B* and *C* adapted from unpublished sketches by LaRue.)

tubules on each side open into a posteriorly situated bladder which in turn opens to the exterior by a single pore (Fig. 57). In digenetic flukes the type of branching of the excretory system is of value in classification but is difficult to determine in the adults; group differences are more readily determined in the living cercariae, and are even present in the ciliated embryos or miracidia. The flame cell arrangement is commonly expressed in a formula; the arrangement in Fig. 57*A* is $2[(3 + 3) + (3 + 3)]$, indicating that on each side there are two groups, of three flame cells each, on the anterior main branch of the excretory tubule, and two groups, of three each, on the posterior main branch. The excretory bladder in some flukes is formed by fusion of the two lateral tubules in a developing cercaria and expansion into a bulb (Fig. 57*B*), and in others from a mass of cells (Fig. 57*C*).

Some flukes, especially the amphistomes (see p. 315), have a lymphatic system of much-branched, delicate tubules in the parenchyma; this seems to function as a primitive circulatory system.

Few animals have more intricate and highly specialized reproductive systems, and their life histories are so marvelously complex as to tax our credulity. Many flukes, especially those living as internal parasites in land animals, pass through four and sometimes even five distinct phases of existence, during some of which they are free-living, and during others may parasitize successively two, three, or even four different hosts.

In all flukes except those of the family Schistosomatidae and one other family, both male and female reproductive systems occur in the same individual and occupy a large portion of the body of the animal.

In the female system there are separate glands for the production of the ova proper, the yolk and shell material, and the fluid in which the eggs are carried. In a typical digenetic fluke the organs are arranged as diagrammatically shown in Fig. 56, but there are many variations. The *ovary* has a short duct, the *oviduct,* which is joined by a *yolk* or *vitelline duct,* a short duct from a *seminal receptacle* if one is present, and by a duct which leads to a pore on the dorsal surface of the fluke, called *Laurer's canal.* Close to where these various ducts meet there is a slight bulblike enlargement called an *oötype,* surrounded by a cluster of unicellular glands called *Mehlis' gland* and formerly thought to be a shell gland; it is now known that the shell material comes from granules in the yolk cells.

The oötype is an assembly plant for the production of finished eggs; in some flukes the daily output is estimated at 25,000! The yolk and shell material are provided by clusters of little *vitelline glands* usually situated in the lateral parts of the fluke but occasionally posteriorly or anteriorly. These clusters of glands are connected by ducts to one main transverse duct from either side; these right and left ducts come together to form a common duct shortly before entering the oötype, often with a small *vitelline reservoir* at their junction. Sometimes there is no separate seminal receptacle; instead the sperms are stored in the region of the oötype or lower part of the uterus. Laurer's canal, sometimes connected with the seminal receptacle or its duct instead of directly to the oviduct, is believed to be a vestigial vagina, but in many species it fails to reach the dorsal surface; like the human appendix, it is a useless heirloom. In most flukes it is probable that the sperms make their way down through the uterus before this becomes jammed with eggs.

When the eggs are fertilized, supplied with yolk cells to provide nourishment, and surounded by shell material which gradually hardens and darkens, they enter the *uterus.* This, usually much coiled and convoluted, leads to the *genital pore,* where it opens in common with the male reproductive system. The terminal part of the uterus is often provided with special muscular walls and is called the *metraterm.*

The male system consists of two or more *testes* for the production of the sperms; two sperm ducts which meet to form a *vas deferens,* usually with an enlargement, the *seminal vesicle,* for the storage of sperms; a cluster of *prostate glands;* and a retractile muscular organ or *cirrus* which serves as a copulatory organ. The seminal vesicle, prostate glands, and cirrus are usually enclosed in a *cirrus sac.* All these complex sex organs in a single animal which may be much smaller than the head of a pin!

Important variations that are of taxonomic value occur in (1) the absolute and relative positions of the ovary and testes; (2) the position of the uterus; (3) the position and arrangement of yolk glands; (4) the presence or absence of a seminal receptacle; (5) the position of the genital pore; (6) the presence or absence of a cirrus sac and the nature of the cirrus; and (7) the presence of a seminal vesicle inside or outside the cirrus sac. The Schistosomatidae, as already noted, are peculiar in having the male and female systems in separate individuals.

The Monogenea differ in their internal structure mainly in (1) having the excretory system of the right and left side separate, opening by two anteriorly situated pores; (2) the testes more variable in number; (3) the presence of a canal connecting the oviduct with the right intestinal cecum in some; and (4) single or paired vaginas or copulation canals in the majority (Fig. 58).

Life cycle. The more primitive monogenetic flukes, belonging to the orders Monogenea and Aspidobothrea, have a direct development involving a simple metamorphosis but no interpolated nonsexual generations. The Monogenea are parasitic externally or in the excretory bladder or on the gills of aquatic vertebrates, whereas the Aspidobothrea are parasitic on or in the soft parts of mollusks or in the intestines of aquatic vertebrates. Some Aspidobothrea develop to maturity in a single molluscan host, whereas others have achieved an alternation of hosts without an alternation of generations. Flukes of the order Digenea, on the other hand, have very complicated life cycles involving several nonsexual generations which, except in two species,

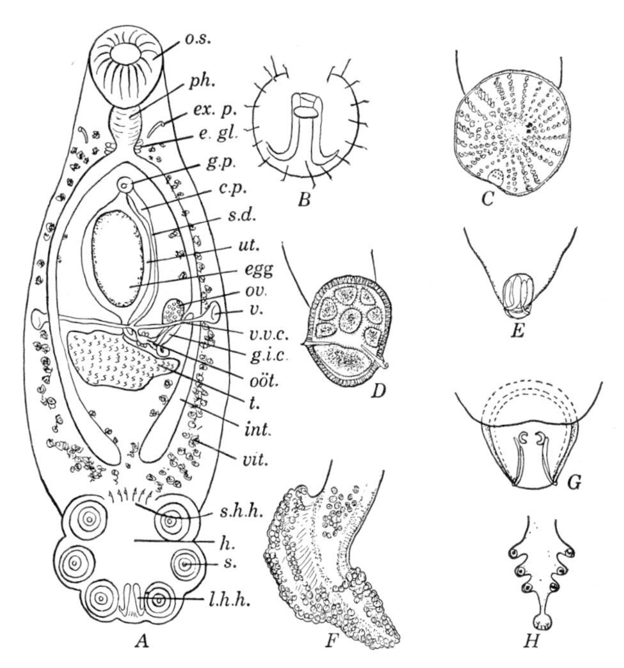

Fig. 58. Monogenea. *A*, *Polystomoidella oblongum* (Polystomatidae) (after Cable, *Illustrated Lab. Manual of Parasitology,* 1947); *B–H*, various types of haptors of Monogenea (adapted from various authors); *C*, Acanthocotylidae; *D*, Monocotylidae; *E*, Microbothriidae; *F*, Microcotylidae; *G*, Capsalidae; *H*, Discocotylidae. Abbreviations: *c.p.*, cirrus pouch; *e.gl.*, esophageal glands; *ex.p.*, excretory pore; *g.p.*, genital pore; *g.i.c.*, genito-intestinal canal; *h.*, haptor; *int.*, intestine; *l.h.h.*, large haptorial hooks; *oöt.*, oötype; *o.s.*, oral sucker; *ov.*, ovary; *ph.*, pharynx; *s.*, sucker; *s.d.*, sperm duct; *s.h.h.*, small haptorial hooks; *t.*, testis; *ut.*, uterus; *v.*, vagina; *vit.*, vitellaria; *v.v.c.*, vitellovaginal canal.

as far as known at present (see p. 247), always develop in snails or bivalve mollusks.

MIRACIDIA. Digenetic flukes produce eggs, often by tens of thousands, which escape from the host's body with the feces, urine, or sputum, according to the habitat of the adults. Either before or after the eggs have escaped from the host ciliated embryos develop within them; these hatch either in water or in the intestines of mollusks which serve as intermediate hosts. The embryos, called miracidia, are free-

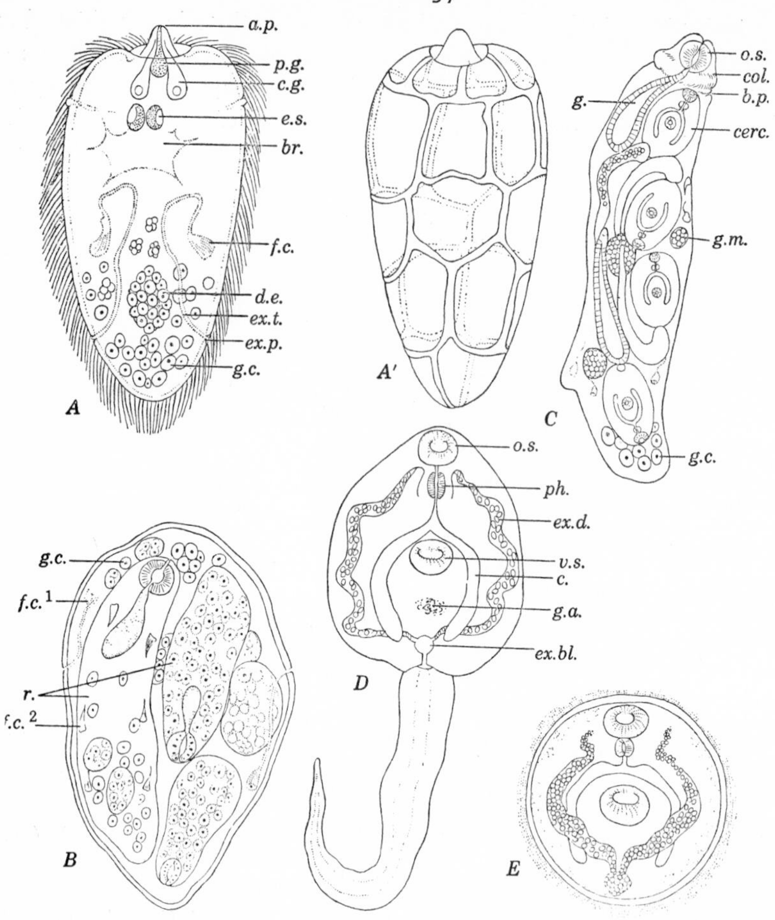

Fig. 59. Stages in life cycle of a fluke. *A*, miracidium showing internal organs, *A′* showing ectodermal cells; *a.p.*, apical papilla; *br.*, brain; *c.g.*, cephalic gland; *d.e.*, developing embryo; *e.s.*, eye spot; *ex.p.*, excretory pore; *ex.t.*, excretory tubule; *f.c.*, flame cell; *g.c.*, germ cells; *p.g.*, primitive gut. *B*, sporocyst; *f.c.*[1], flame cell of sporocyst; *f.c.*[2], flame cell of redia; *g.c.*, germ cells; *r.*, developing rediae. *C*, redia; *b.p.*, birth pore; *cerc.*, developing cercaria; *col.*, collar; *g.*, gut; *g.c.*, germ cells; *o.s.*, oral sucker; *g.m.*, germinal mass. *D*, cercaria; *c.*, cecum; *ex.bl.*, excretory bladder; *ex.d.*, excretory duct; *g.a.*, genital anlage; *o.s.*, oral sucker; *ph.*, pharynx; *v.s.*, ventral sucker. *E*, encysted metacercaria.

swimming animals suggestive of ciliated protozoans. They are covered by a ciliated epithelium of relatively few large flat cells (Fig. 59*A′*) and have a short saclike gut, one or more pairs of penetration glands, one or more pairs of flame cells and excretory tubules, and a cluster of germ cells which are destined to give rise to a new generation of organisms (Fig. 59*A*). Many miracidia have eye spots, but some are blind. The miracidia do not feed, and they die in 24 hours or less if unsuccessful in finding a proper molluscan host.

Free miracidia swim in a characteristic spirally rotating manner, in quest of a mollusk of the particular species which is to serve as an intermediate host. When they come very close to such a mollusk they become greatly excited and make a headlong dash for it, although they ignore other kinds of mollusks. They attach themselves to the soft part of the mollusk by the secretion of their glands and proceed to bore or digest their way into the tissues. Some miracidia, e.g., those of *Clonorchis* and *Dicrocoelium*, hatch only after the eggs have been eaten by the proper snails. It is obvious that only a very small percentage of the embryos are likely to survive the double risk of not reaching water and, if safely in water, of not reaching a suitable mollusk in which to develop.

The miracidia make their way to various tissues of the molluscan host, according to the species, but usually do *not* settle in the digestive gland or liver although that is the commonest site of development for their offspring. The miracidium sheds its coat of flat, ciliated cells and changes in form to become an irregular-shaped, saclike, filamentous or branched body called a mother sporocyst (Fig. 59*B*). This sporocyst is essentially a germinal sac inside of which are the germ cells, which have descended in a direct line from the original ovum from which the miracidium developed. The germ cells may continue to multiply as single cells or may cluster together to form germinal masses. The latter may then differentiate directly into a new generation of germinal sacs with enclosed germ cells, or may bud these off from their surfaces over a considerable period of time.

DAUGHTER SPOROCYSTS OR REDIAE. This new generation may be simple sacs like the mother and are then called daughter sporocysts, or they may have an oral sucker, saclike gut, and rudimentary appendages, and are then called rediae (Fig. 59*C*). These emerge from a birth pore in the mother sporocyst and usually make their way to the digestive gland of the mollusk. Here the enclosed germ cells, with or without multiplication as individual cells, again form germ masses which directly develop into a new generation of organisms, or shed partially formed embryos from their surfaces as the cells in the masses continue to multiply.

In some flukes the germ cells in the daughter sporocysts or rediae develop into a structurally different generation, the cercariae, but in some there is a second or even a third generation of rediae before cercariae are produced. In schistosomes and strigeids the mother sporocysts may still be producing daughter sporocysts long after the latter have started producing cercariae; the only limitation to the polyembryonic capacity of the mother sporocysts seems to be the food and space available in the digestive gland of the snail host, and the daughter sporocysts appear to be able to produce cercariae as long as the snail lives. Thus these reproductive machines may produce up to a million cercariae from a single original egg. There is one record of a snail that gave off cercariae for 7 years at the rate of over 1,000,000 a year. In some flukes, on the other hand, e.g., *Paragonimus*, a miracidium may end up producing only a few hundred cercariae.

CERCARIAE. The cercariae (Fig. 59*D*) are not germ sacs in which the germ cells multiply and produce new embryos, as are sporocysts and redia. These are true larval forms which undergo no further reproduction in the mollusk, but must by some route, sometimes amazingly devious, reach the final vertebrate host, where they will grow to maturity and reproduce by the more orthodox sexual method.

The cercariae are odd mixtures of features characteristic of the adult flukes into which they will grow, and of special adaptive characters which enable at least a few of them to succeed in the often hazardous transfer from the mollusk where they were born to the vertebrate where they will mature. Usually features connected with the digestive tract, excretory system, and suckers, and a few special features such as the circlet of head spines in echinostomes, are prophetic of what type of fluke the cercaria will eventually develop into, but other features are purely adaptive, e.g., the stylet, the tail, the fins, the penetration and/or cystogenous glands, and other features which are sometimes very bizarre. Often certain of these features, together with the few adult characters that may be present, make it possible to predict what the adult of a newly discovered cercaria will be. However, sometimes adult characters that would be expected are absent, e.g., a ventral sucker, or a ventral sucker may be present when absent in the adult.

The cercarial features often give better clues to relationships and correct classification than do the adult flukes. Although nobody would have dreamed of including such widely different adult flukes as schistosomes, strigeids, *Clinostomum*, and gasterostomes (and some others) in one suborder, the strikingly similar features of their miracidia and cercariae (all forked-tailed) make such a grouping unavoidable. Another feature which seems to have been evolved by nature only once is the stylet, a little spine near the oral sucker to help in the penetration of arthropod hosts, so all stylet cercaria belong in one group, even though one would never guess that the adults belong together. Here, however, we run into possible difficulty, for some cercariae without stylets may have secondarily lost them but

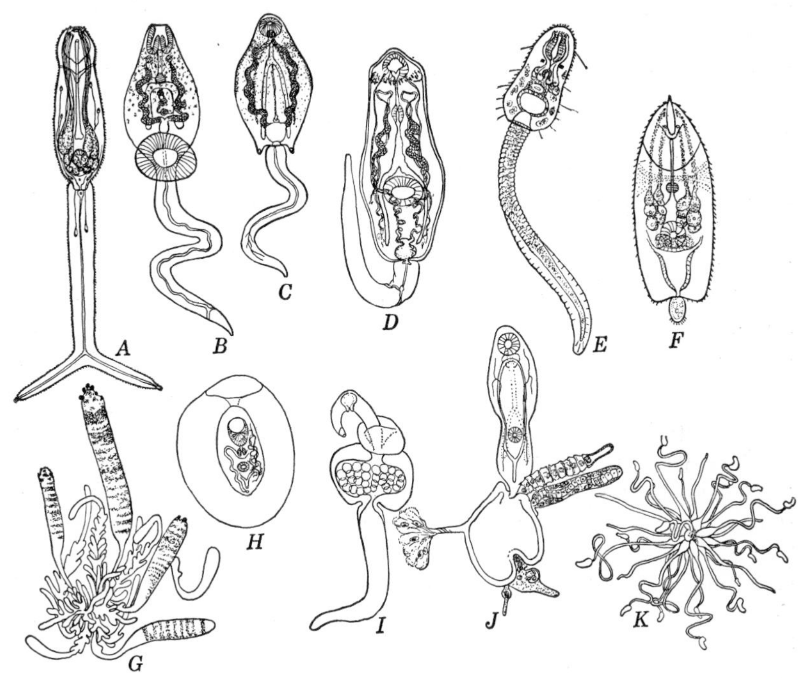

Fig. 60. Some types of cercariae. *A*, furcocercous (*Schistosoma japonicum*); *B*, amphistome (*C. inhabilis*); *C*, monostome (*C. urbanensis*); *D*, echinostome (*Echinostoma revolutum*); *E*, pleurolophocercous (*Opisthorchis felineus*); *F*, stylet, microcercous (*Paragonimus westermanni*); *G*, colored sporocysts of *Leucochloridium paradoxum*, which grow out from tentacles of land snail; *H*, metacercaria from a sporocyst of *Leucochloridium migranum*, enclosed in jelly-like cyst capsule; *I*, cystocercous (gorgoderine) (*C. macrocerca*); *J*, cystophorous (hemiurid), showing cercarial body and various appendages evaginated from tail cyst; *K*, cluster of "rattenkönig" cercariae (*C. gorgonocephala*). (Adapted from various authors.)

should nevertheless not be ostracized. Figure 60 gives some idea of nature's uninhibited ideas as to what a cercaria may look like.

All but a few kinds of cercariae have tails; they may be long or short, single or forked (with or without a stem), and with or without fins, bulbs, or other peculiar features; in some the cercaria's body can be withdrawn into the anterior end of the tail, or the tail may form a cyst from which peculiar appendages protrude. A few cercariae that develop in land snails and could not use tails if they had them, have secondarily lost them—such a cercaria is called a cercariaeum.

Cercariae usually have one or both of two kinds of glands. Penetration glands are conspicuous features in most, and very useful in identification; they are histolytic and are used for escaping from the snail's tissues, and subsequently for penetrating into intermediate hosts or, in the case of the schistosomes, the final hosts. Many species also have cystogenous glands, the product of which is used in forming cyst walls. These glands are particularly conspicuous in such flukes as *Fasciola, Fasciolopsis,* and amphistomes, which encyst on inanimate objects in the water and form thick protective cysts. Most cercariae do not lead a free existence in water for more than a few days at most.

METACERCARIAE. All flukes except the schistosomes and their near relatives undergo some further development before finally growing into adults. On penetrating a host or preparing to encyst, a cercaria nonchalantly flips off its tail, exudes the contents of its glands, and proceeds to become a metacercaria (Fig. 59*E*). In most flukes this metacercarial stage takes place during encystment on vegetation or in an intermediate host, but in some flukes, e.g., some that develop in the humors or lens of the eyes of fishes, they become metacercariae without encysting. In intermediate hosts the cercarial cyst is usually very delicate and transparent, but the host lays down a fibrous or glassy outer cyst, often conspicuous with pigment granules, which becomes thicker with age.

In some flukes the metacercariae are infective for the final host within a few hours, in others they require days or weeks. In one genus of strigeids (*Alaria*) an additional stage, the mesocercaria, is interposed between the cercaria and the metacercaria, so these flukes require four successive hosts—snails, tadpoles, mice, and mink.

TRANSFER TO FINAL HOST. The manner in which the cercariae accomplish the transfer from snail to final host varies greatly. The self-reliant cercariae of schistosomes actively seek the final host and bore directly into it, but most cercariae depend on the host to pick them up. The simplest method of transfer is encystment directly in the primary mollusk host or even in the mother redia, as happens in some echinostomes. *Leucochloridium* has the cercariae encysted in large colored sporocysts that grow out of the tentacles of snails and resemble tempting worms for birds to peck at (Fig. 60*G*). In flukes like the Fasciolidae and amphistomes, which reach maturity in herbivorous animals, the cercariae encyst on vegetation in the water and patiently wait to be eaten by the final host. In flukes which mature in carnivorous or insectivorous hosts, the cercariae penetrate into the tissues of frogs, fish, insects, crustacea, etc., where they encyst and await salvation by the second intermediate host being eaten by the final one. It is for this reason that many human fluke infections are prevalent only in the Orient, where fish or crabs are eaten without thorough cooking. Occasionally metacercariae are progenetic, i.e., they become sexually mature before reaching the final host.

The skin-penetrating cercariae of schistosomes are carried to their final destination in the mesenteric blood vessels by the blood stream, but encysted metacercariae always enter by way of the mouth. Their cyst walls are digested away in the intestine, and the young liberated flukes migrate by various routes to the parts of the body in which they are to mature.

Examination of mollusks for asexual generations and cercariae. When mollusks are collected and brought to the laboratory for examination for immature stages of flukes, those producing cercariae can readily be determined by placing them, first in groups and later individually, in half-pint bottles and leaving them for 12 to 24 hours. The emerged cercariae will then be seen swimming in the water or, in a few instances, crawling on the bottom. In some species all the cercariae emerge almost simultaneously at a certain time of day. Some are attracted to light, some to the upper layers of the water; and some swim almost continuously, others hang motionless most of the time. The cercariae of some flukes appear only for a short season, whereas others continue to emerge for months or even years.

Sporocysts and rediae are found by crushing the mollusks or picking away the shell until the body can be dragged out intact. In most cases they will be found in the brown digestive gland of the mollusk; their presence can often be detected with the naked eye as yellowish mottlings. These must then be dissected out carefully. Cercariae obtained after crushing a mollusk are frequently immature and unlike those escaping naturally. The cercariae, sporocysts, and rediae should be studied in the living state as much as possible, with the help of such intravitam strains as neutral red (1:1000) or nile blue sulfate. Subsequent studies can be made on material fixed and stained by various standard tissue methods.

Molluscan hosts. The mollusks involved as intermediate hosts of flukes include many families of snails and bivalves, but only snails are involved as hosts for the sporocyst and redia generations of flukes parasitic in man and domestic animals. These include several families of fresh-water snails which have gills and can close their shells by means of an operculum (order Prosobranchia), and both fresh-water and land snails (order Pulmonata), which have a respiratory sac instead of gills and no operculum.

In the Prosobranchia are included: (1) Hydrobiidae, small aquatic or more often amphibious conical snails containing *Oncomelania* (Fig. 67), hosts of *Schistosoma japonicum,* and *Bulimus* (Fig. 75), *Parafossarulus* (Fig. 73), and *Amnicola,* hosts of the Opisthorchiidae. (2) Thiaridae (formerly Melaniidae), large, high-spired, rough-shelled aquatic snails, containing *Thiara* and *Semisulcospira* (Fig. 69), hosts of *Paragonimus* and *Metagonimus; Hua,* of *Clonorchis;* and *Goniobasis,* of *Nanophyetus salmincola.* (3) Potamididae, containing *Pirenella,* host of *Heterophyes heterophyes;* and several marine snails that are hosts for bird schistosomes causing swimmer's itch.

In the Pulmonata are included in the fresh-water group (Basommatophora): (1) Planorbidae (Figs. 67, 84), flatly-coiled snails serving as hosts of *Schistosoma mansoni, Fasciolopsis,* and some amphistomes and echinostomes. (2) Lymnaeidae (Fig. 61), dextrally coiled snails serving as hosts of *Fasciola,* and the principal cercariae causing swimmer's itch. (3) Bulinidae (see footnote, p. 287) (Fig. 67), sinistrally coiled snails serving as hosts of *Schistosoma haematobium.* In the terrestrial group, Stylommatophora, which have *two* pairs of head tentacles, are included the hosts of Dicrocoeliidae (*Helicella, Cionella* [Fig. 72], etc.).

Classification. The classification of flukes has undergone a gradual evolution. In 1858 van Beneden divided the class Trematoda into those having direct development, the subclass Monogenea, and those having indirect development, involving two or more asexual generations interposed in the life cycle, the subclass Digenea. This primary division still stands, except that most authors separate the small family Aspidogastridae, parasitic in mollusks, fish, and turtles, into a separate subclass Aspidobothrea. These resemble Monogenea in lacking asexual generations although some have an alternation of hosts, but they resemble the Digenea in most of their anatomical characters and in being endoparasites. In their suctorial apparatus they resemble neither.

The Monogenea fall readily into two orders on the basis of the character of the posterior sucking organs (see below). The Digenea present a greater problem. The earliest classification was, of necessity, based on characters of the adults, at first mainly the number, position, and characters of the suckers, from which arose such familiar names as polystome, monostome, distome, amphistome, and gasterostome. Later more attention was paid to internal organs, at first mainly the reproductive and digestive organs, later the excretory

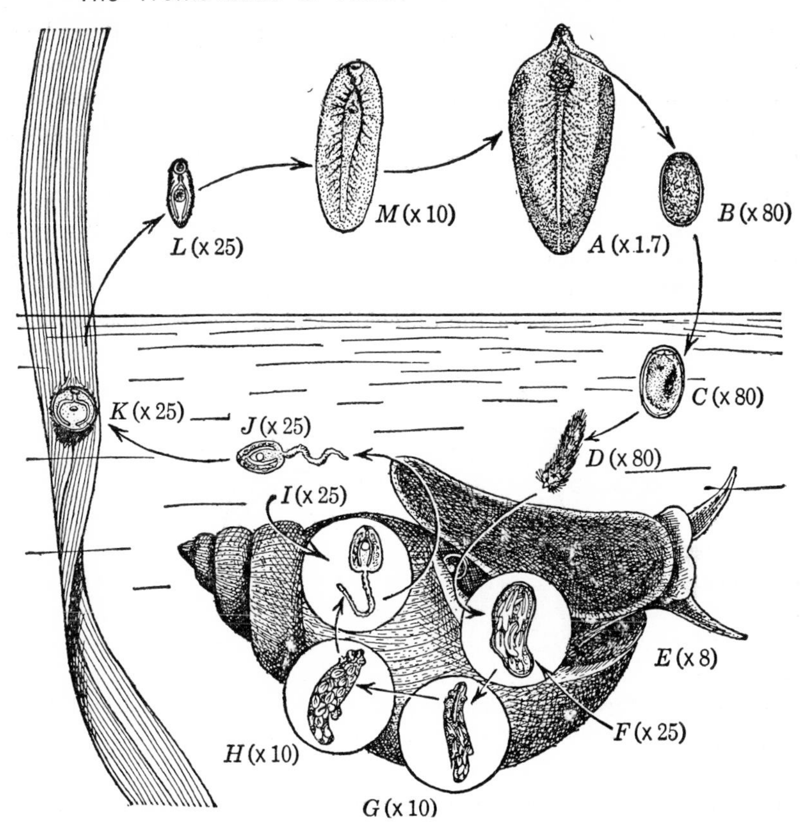

Fig. 61. Life history of liver fluke, *Fasciola hepatica.* A, adult in liver of sheep; *B*, freshly passed egg; *C*, egg with developed embryo, ready to hatch in water; *D*, ciliated embryo in water, about to enter pulmonary chamber of snail (*E*); *F*, sporocyst containing rediae; *G*, redia containing daughter rediae; *H*, redia of second generation containing cercariae; *I*, cercaria; *J*, same, having emerged from snail into water; *K*, cercaria encysted on blade of grass; *L*, cercaria liberated from cyst after ingestion by sheep; *M*, young fluke developing in liver of sheep.

system. With the development of knowledge of the life cycles it became apparent, as La Rue pointed out in 1938, that a taxonomic system which really indicates relationship must be based on comparative anatomy of all the stages of the life cycle, and especially of the miracidia and cercariae. As will be seen, life cycle studies have brought some astonishing revelations of previously unsuspected relationships; in fact, so many unexpected skeletons in family closets have appeared that a complete revision of the classifications accepted 10 or 15 years ago has become necessary. The "Monostomata,"

"Amphistomata," and "Gasterostomata" have entirely disappeared as major groups. Stunkard (1946), when far fewer life cycles were known than at present, although many unexpected relationships had already appeared, concluded that in the state of knowledge then existing the only groups above families in which one could pin any faith were the subclasses Monogenea and Digenea. Since then many new life cycles have been determined, particularly by Cable and his colleagues, and a classification based on true relationships is now emerging, although the proper allocation of some groups is still highly speculative. Characters of the miracidia and of the cercariae appear

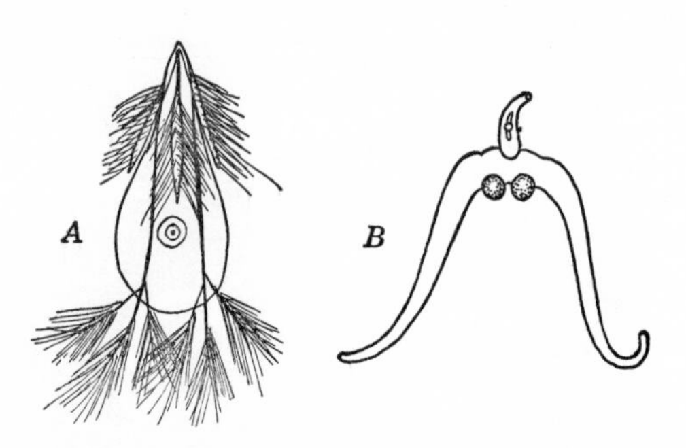

Fig. 62. *A*, Gasterostome miracidium, and *B*, cercaria. (*A* adapted from Woodhead; *B* from Lühe, *Susswasserfauna Deutschlands, Trematoda.*)

in general to be much more significant than do adult characters. In the miracidia the presence of one or two pairs of flame cells, and the presence or absence and distribution of cilia are important (Fig. 62). In the cercariae it is necessary to distinguish between truly phylogenetic and adaptive ones. For example, a forked tail, the presence of a stylet, and the method of formation of the excretory bladder and of the tail appear to be reliable clues to common ancestries, but the absence of a tail and occasionally loss of its forks, and possible loss of the stylet, are characters that must be reckoned with. Adult characters, pertaining particularly to such things as position of reproductive glands and uterus, nature and distribution of vitellaria, presence or absence of a cirrus pouch, details of the excretory system, division into fore- and hind-bodies, etc., are useful mainly as family or generic characters.

The latest classification, although probably not the last, is one proposed by La Rue (1957). As shown below, he divides the Anepitheliocystida into three orders and the Epitheliocystida into two, all of which appear to be natural groups containing phylogenetically related forms, in spite of the vast differences in appearance of the adults, e.g., the schistosomes, strigeids, and gasterostomes (Figs. 64 and 83) in the Strigeatoidea; the Fasciolidae, echinostomes, and amphistomes (Figs. 71, 77, 82) in the Echinostomida; the great variety of forms with stylet cercariae in the Plagiorchiida; and the opisthorchiids and heterophyids (Figs. 73, 74, 79) in the Opisthorchiida. Although the majority of the families of trematodes can now be placed with a fair degree of confidence in one or another of these orders, there are many flukes whose position on the family tree is still very uncertain.

La Rue's Classification of Trematodes

Subclass **Monogenea.** Direct development with no asexual generations; large postero-ventral adhesive apparatus, usually armed with hooks or spines and/or muscular suckers; excretory pores 2, anteriorly situated. External or semiexternal parasites of aquatic animals, usually fish or Amphibia.

Order **Monopisthocotylea.** Opisthaptor a single disc, usually with 1 to 3 pairs of large hooks and 12 to 16 marginal hooklets (Fig. 58*B–E, G*); without genito-intestinal canal connecting oviduct and cecum.

Order **Polyopisthocotylea.** Opisthaptor comprising a number of muscular suckers or clamps, on a disc or free (Fig. 58*A, F, H*); genito-intestinal canal present (Fig. 58*A*).

Subclass **Aspidobothrea** (often incorrectly called **Aspidogastrea**). No asexual generations, but may have alternation of hosts; one or several rows of suckers on ventral surface, usually on a large ventral disc; internal anatomy resembles that of Digenea. One family, internal parasites of mollusks, fish or turtles.

Subclass **Digenea.** Asexual generations interposed in life cycle, nearly always in mollusks. One or two cuplike suckers for adhesion; excretory pore single, posterior. Internal parasites of vertebrates.

Superorder **Anepitheliocystida.** Excretory bladder in cercaria thin-walled, formed by fusion of 2 excretory ducts (Fig. 57*B*). Cercarial tail formed in part by molding of body (see legend for Fig. 57*B*).

Order **Strigeatoidea.** Cercariae forked-tailed (Fig. 60*A*, 64*B*) or derived from that condition (Fig. 60, *G, H*), usually developing in sporocysts. Miracidia in more primitive forms with 2 pairs of flame cells. Includes strigeids (pp. 326 and 266); schistosomes (Chapter 13); *Clinostomum* (pp. 307 and 266); gasterostomes (Bucephalidae), parasites of fishes with the mouth on the mid-ventral surface and a saclike gut, the larval stages of which are injurious to oysters and other bivalves; and many other fish parasites.

Order **Echinostomida.** Cercariae with large bodies and strong unforked tails (Fig. 60*B*, *C*, *D*; 82*A*), and with abundant cystogenous glands; metacercariae encyst on vegetation, in snails, on frog skins, and only occasionally inside aquatic vertebrates; miracidia with 1 pair of flame cells. Includes the Fasciolidae (pp. 317–319, 304–307); echinostomes (pp. 324–326); amphistomes (pp. 315–317); and certain "monostomes" of birds and reptiles.

Order **Renicolida.** Cercariae develop in sporocysts in marine snails; have excretory bladder with numerous diverticula, and large unforked tail with fins. Kidney parasites of birds.

Superorder **Epitheliocystida.** Excretory bladder in developed cercariae with thick epithelial walls derived from a mass of mesodermal cells (Fig. 57*C*).

Order **Opisthorchiida.** Cercarial tail formed as in Anepitheliocystida, so with excretory ducts entering it, at least during development; no stylet. One suborder, Opisthorchiata, contains Opisthorchiidae (pp. 309–314), Heterophyidae (pp. 319–323) and many fish parasites, all with "pleurolophocercous" tail, i.e., long, with lateral fins (Figs. 60*E*, 75*A*), which encyst in lower vertebrates. The other suborder, Hemiurata, contains numerous parasites of fishes, all with nonciliated miracidia, and peculiar "cystophorous" cercariae (Fig. 62*J*), which use copepods as second intermediate hosts.

Order **Plagiorchiida.** Cercarial tail formed entirely by backward growth of tissue from area between primary excretory pores, so no excretory vessels in tail; stylet usually present (Fig. 60*F*, 70*F*); encyst in invertebrates, usually arthropods, as second intermediate hosts. Includes Troglotrematidae (e.g., *Paragonimus,* pp. 299–304 and *Nanophyetus,* pp. 323–324); Dicrocoeliidae (pp. 307–309); Plagiorchiidae (pp. 327–328); Lecithodendriidae, containing many minute parasites of bats; and many families of fish parasites.

The flukes which infect man may be divided for convenience into four groups: (1) the blood flukes or schistosomes, (2) the lung flukes, (3) the liver flukes, and (4) the intestinal flukes. Over forty different species have been recorded as human parasites, but only ten of these are common enough to be more than medical curiosities.

Control. In many cases the most feasible method of control of fluke diseases is destruction of the snails which serve as intermediate hosts. The methods employed depend upon the species of snails involved and on local conditions.

Use of chemical substances is often feasible and offers valuable possibilities. Liver flukes of cattle and sheep do not occur in salty pastures, and a liberal use of salt can under very special conditions be of advantage. The writer (Chandler, 1920) found that all species of snails are destroyed by very high dilutions of copper salts. Since then copper sulfate and copper carbonate have been more extensively employed for destruction of both aquatic and amphibious snails than any other chemicals. Although lime has been advocated for some snails, McMullen and Graham found it of no value against the snail host of *Schistosoma japonicum* in the Philippines, but they found calcium cyanamide a good substitute for copper salts against this snail. Active research is going on to find molluscicides that will be as effective as copper sulfate, but will have longer lasting effects. This matter is discussed further on pp. 293–294.

Other possible methods of controlling fluke diseases would be by mass treatment and, for man, prevention of pollution of water, and protection against cercariae in the case of schistosomes.

Monogenea

The majority of the monogenetic flukes are parasites of the gills, skin, and cloaca of fishes, but some have established themselves in the urinary bladder of amphibians and in the urinary bladder or mouth cavity of turtles, and one genus is found in the eyes of hippopotamuses. Some of the species attacking the gills and fins of fishes cause serious losses in fish hatcheries, where they are of much more importance than their endoparasitic relatives. In nature they seldom cause much trouble since there is less opportunity for heavy infections early in life, and a protective immunity develops. Most Monogenea produce large eggs that hatch into larvae which at once attack their definitive hosts, but those of one family, Gyrodactylidae, give birth to larvae of large size, one at a time. *Polystoma* develops either on the gills of tadpoles or in the urinary bladder of adult toads or frogs. Most Monogenea are fairly specific with respect to hosts.

REFERENCES

Flukes in General

Bhalerao, G. D. 1947. Applied helminthology, its past and future in India. *Proc. 34th India Sci. Congr.*

Chandler, A. C. 1920. Control of fluke diseases by destruction of the intermediate host. *J. Agr. Research*, 20: 193–208.

Cort, W. W. 1944. The germ cell cycle in the digenetic trematodes. *Quart. Rev. Biol.*, 19: 275–284.

Cort, W. W., Ameel, D. J., and Van der Woude, A. 1954. Germinal development in the sporocysts and rediae of digenetic trematodes. *Exp. Parasitol.*, 3: 185–225.

Dawes, B. 1946. *The Trematoda.* University Press, Cambridge.

Faust, E. C. 1949. *Human Helminthology,* 3rd ed. Lea and Febiger, Philadelphia.

Fuhrmann, O. 1928. Trematoda, in *Handbuch der Zoologie* (Edited by W. Kükenthal). Bd. II, Hälfte 1. Vermes Amera.

Hyman, L. H. 1951. *The Invertebrates,* II. *Platyhelminthes and Rhynchocoela.* McGraw-Hill, New York.

Hassall, A. 1908. Index catalogue of medical and veterinary zoology: trematoda and trematode diseases. *Hyg. Lab. Bull.,* 37.

La Rue, G. R. 1957. The classification of digenetic trematodes: A review and a new system. *Exp. Parasitol.,* 6: 306–344.

Reinhard, E. G. 1957. Landmarks of parasitology. I. The discovery of the life cycle of the liver fluke. *Exp. Parasitol.,* 6: 208–232.

Stoll, N. R. 1947. This wormy world. *J. Parasitol.,* 33: 1–18.

Stunkard, H. W. 1946. Interrelationships and taxonomy of the digenetic trematodes. *Biol. Reviews, Cambridge Phil. Soc.,* 21: 148–158.

Yamaguti, S. 1958. *Systema helminthum.* Vol. I. *The Digenetic Trematodes of Vertebrates.* Interscience, New York. 1575 pp.

Monogenea

Hargis, W. J., Jr. 1957. The host specificity of monogenetic trematodes. *Exp. Parasitol.,* 6: 610–625.

Mizelle, J. D. 1938. Comparative studies on trematodes (Gyrodactyloidea) from the gills of North American freshwater fishes. *Univ. Illinois Bull.,* 36: No. 8.

Mueller, J. F. 1937. The Gyrodactyloidea of North American freshwater fishes. *Fish Culture* (N. Y. Conserv. Dept.).

Paul, A. A. 1938. Life history studies on North American freshwater polystomes. *J. Parasitol.,* 24: 489–510.

Sproston, N. G. 1946. A synopsis of the monogenetic trematodes. *Trans. Zool. Soc. London,* 25: 185–600.

Chapter 13

THE TREMATODES OR FLUKES II. SCHISTOSOMES

Human Schistosomiasis

Schistosomiasis is today the most important human disease caused by animal parasites. Although hookworm disease has gradually decreased as a public health problem in many parts of the world, and chloroquine and insecticides have suddenly sharply tipped the balance against malaria, schistosomiasis seems to be on the upgrade. In spite of intensive research there is still no easy cure for it and no easy means of control, and the extension of irrigation projects and concentration of human populations are increasing its distribution and intensity. Stoll (1947) estimated that there are 114,000,000 people infected with schistosomes in the world, 46,000,000 of them in the Orient. In 1950, Wright thought that the number in the Orient, on a conservative estimate, might be no more than two-thirds of Stoll's figure. Maegraith (1958) regarded schistosomiasis as the most serious parasitic disease in Red China and estimated that there were more than 11,000,000 infected persons in the country. Indeed, in a news release in July of 1958 the New China News Agency announced that research groups had been organized to deal with 1089 disease subjects and schistosomiasis led the list. It seems probable that Stoll's estimate for Africa is too low. In lower Egypt, for instance, an incidence of 60% infection was estimated, but more thoroughgoing diagnosis demonstrated 95% in some localities. Instead of 6,000,000 schistosome infections in Egypt, there are probably more like 10,000,000. Schistosomiasis is undoubtedly one of the principal scourges of that country.

The seriousness of schistosomiasis can be judged by Wright's estimate that infection of 1700 American troops during the recapture of Leyte in 1944 caused the loss of over 300,000 man-days and medical care costs of $3,000,000, not to mention subsequent veterans' benefits. Hunter et al. (1952) estimated an annual loss of $3,000,000 in wages and treatment costs in just one of the five endemic areas in Japan (about 90 sq. miles). The geopolitical effects of schistosomiasis have rarely been evaluated. Kierman (1959) has recently described the decisive role of schistosomiasis in preventing a military assault by Red China on the island of Formosa in early 1950. An estimated 30

Fig. 63. *Australorbis glabratus,* the intermediate host of *Schistosoma mansoni* in Puerto Rico. (From *U.S.P.H.S. Community Diseases Center Rept.,* 1957.)

to 50,000 military cases of acute schistosomiasis delayed the communist Chinese attack by at least 6 months. Six months was too long and, at the time of this writing 9 years later, another such opportunity for mounting an attack on Formosa has not presented itself.

The parasites. The human schistosomes and most of the other species in mammals belong to the genus *Schistosoma* (Fig. 64),* in the family Schistosomatidae. This family shows affinity with strigeids

* The name *Bilharzia* has priority over *Schistosoma,* but before this became known *Schistosoma* was approved by the International Commission on Zoological Nomenclature, which should make it inviolable regardless of later findings. Nevertheless many Europeans (and W.H.O.) prefer *Bilharzia,* honoring the discoverer of the parasites.

and certain other flukes (see p. 273) in having miracidia with two pairs of flame cells, daughter sporocysts instead of rediae, and cercariae with forked tails, but is peculiar in having separate males and females, which in the genus *Schistosoma* are morphologically quite different. The male, usually about 8 to 16 mm. long, has a cylindrical appearance but

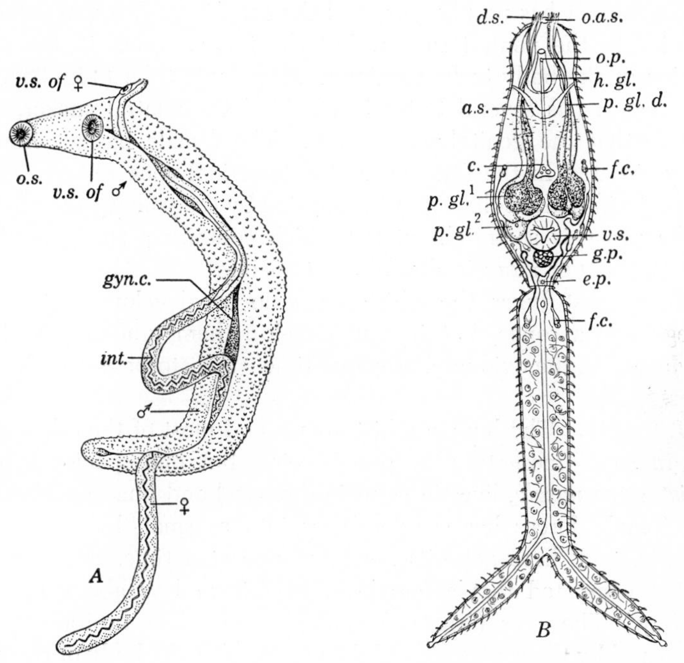

Fig. 64. A, Blood fluke, *Schistosoma mansoni:* male (♂) carrying female (♀) in gynecophoric canal (*gyn.c*); *int.*, intestine of ♀; *o.s.*, oral sucker of ♂; *v.s.*, ventral sucker. ×8. (Adapted from Looss.) *B*, cercaria of *S. japonicum: a.s.*, anterior sucker; *c.*, cecum; *d.s.*, duct spines; *e.p.*, excretory pore; *f.c.*, flame cells; *g.p.*, genital primordium; *h.gl.*, head gland; *o.a.s.*, orifice of anterior sucker; *o.p.*, oral pore; *p.gl.*1, 2 acidophilic penetration glands; *p.gl.*2, 3 basophilic penetration glands; *p.gl.d.*, penetration gland ducts; *v.s.*, ventral sucker. ×275. (Adapted from Faust and Meleney, *Am. J. Hyg.*, Monograph Ser., 1924.)

is actually flat, with the sides of the body posterior to the ventral sucker rolled ventrally to form a groove or "gynecophoric canal," in which the longer and more slender cylindrical female projecting free at each end, but enclosed in the middle, lies safe in the arms of her spouse (Fig. 64A). In most schistosomes they seem to remain permanently wedded and monogamous, the uncoupled females remaining spinsters, but in *Schistosoma mansoni* the union is of more companionate nature.

Oddly enough, the female worms do not become sexually mature until they become associated with males, although the males are able to develop quite independently of the females (see Moore et al., 1954).

Both male and female worms are provided with oral and ventral suckers; in the male the ventral sucker is large and powerful. The digestive tract has no pharynx, and the esophagus forks, as usual, just anterior to the ventral sucker, but the forks reunite in the middle portion of the body to be continued as a single tube (Fig. 65). The male worm has several testes just behind the ventral sucker, and it is here that the genital pore opens. The female has an elongated ovary situated in the fork where the intestinal ceca rejoin. Most of the posterior half of the worm is occupied by the yolk glands. Anterior to the ovary is a straight uterus which contains a small number of eggs, 1 to 50 or more in the different species.

Unlike most flukes, the schistosomes do not develop great numbers of eggs all at once, but instead develop them gradually and have only a few in the oviduct at any one time. Schistosomes live for many years.

Life cycle. The human schistosomes and most of the other species live in small mesenteric or pelvic veins, but one species in cattle, *Schistosoma nasale,* lives in veins in the nasal and pharyngeal mucosa. The female forces her slender body into as small blood vessels as possible, and there deposits her eggs, one at a time. Pesigan et al. (1958) estimated that a female S. *japonicum* lays about 1200 eggs per day. The eggs (Fig. 65) usually retain their position by their spines and by the contraction of the vessels after the body of the parent worm has been withdrawn; presumably aided by histolytic secretions of the embryo inside, they gradually work their way out of the vessels and into the tissues of the walls of the intestine or bladder, and finally into the lumen of these organs, whence they escape with the feces or urine. Some of the eggs, however, are accidentally carried to the liver or lungs where, as in other organs, they set up inflammations. Eventually the irritated tissues become so thickened that most of the eggs are trapped. The eggs of S. *japonicum* take 9 or 10 days to develop mature miracidia while passing through the tissues, and will live for 10 or 12 days longer if not expelled, but many die in the tissues, becoming blackened and calcified. S. *mansoni* eggs remain alive in tissues of experimentally infected mice for 3 to 4 weeks after the parent worms have been killed by drugs.

Cross-fertilization between different species is possible. The sex of the future adult worms is already determined in the miracidium;

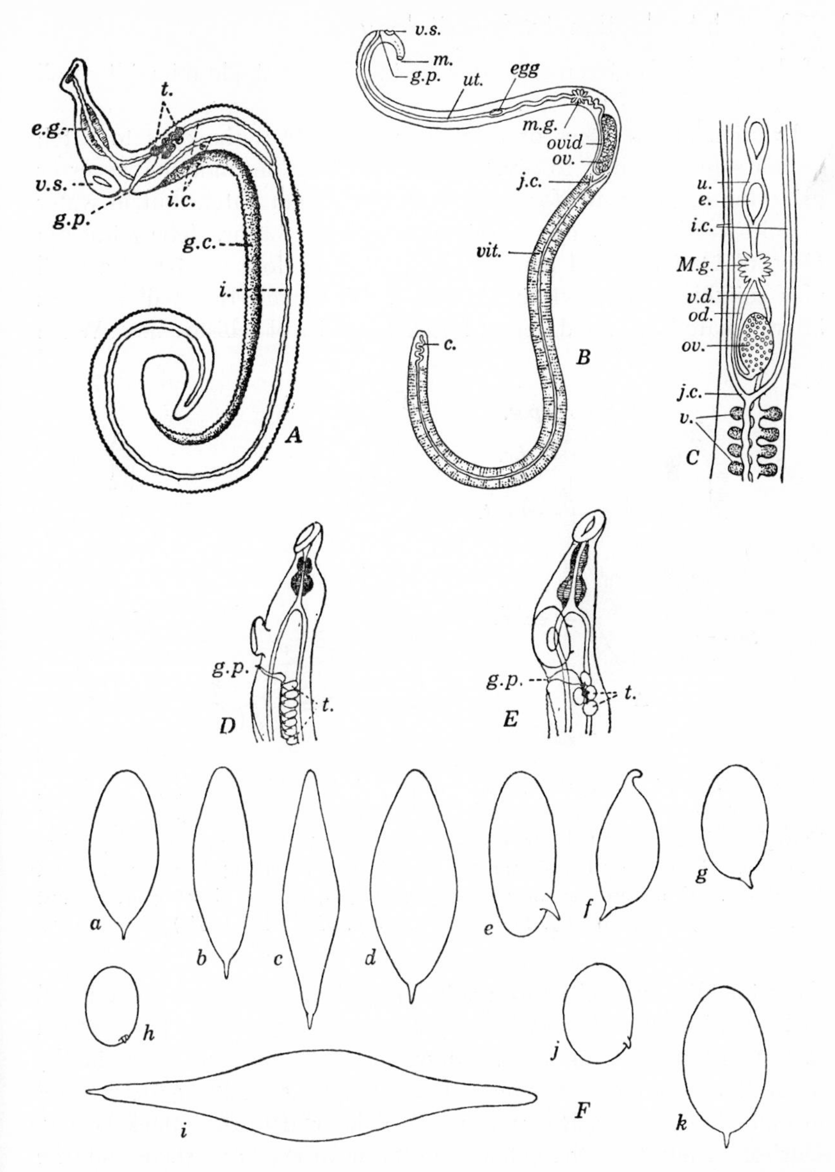

Fig. 65. Schistosome adults and eggs. *A*, ♂ *S. mansoni; B*, ♀ *S. mansoni; C*, ovarial region of *S. haematobium; D*, anterior end of ♂ *S. japonicum; E*, anterior end of ♂ *S. haematobium; e.*, egg; *e.g.*, esophageal glands; *g.p.*, genital pore; *g.c.*, gynecophoric canal; *i.*, intestine; *i.c.*, intestinal ceca; *j.c.*, junction of ceca; *M.g.*, Mehlis' gland; *od.*, oviduct; *ov.*, ovary; *t.*, testes; *u.*, uterus; *v.*, vitellaria; *v.d.*, vitelline duct; *v.s.*, ventral sucker. *F*, *a–k*, eggs of various schistosomes: *a, haematobium; b, intercalatum; c, bovis; d, matheei; e, mansoni; f, rodhaini; g, incognitum; h, margrebowiei; i, spindale; j, japonicum; k, indicum.* (Adapted from various authors; *F* mostly from Schwetz, Baumann, and Fort, *Ann. soc. belge méd. trop.*, 33, 1953.)

of the thousands of cercariae developing from a single miracidium all produce worms of one sex.

Dilution of the feces or urine in water causes the eggs to hatch within a few minutes to several hours or more; in undiluted feces or urine the eggs survive for some time but do not hatch; but in water in cold weather they may survive for several months. The miracidia (Fig. 66*A*) live for 24 hours or less, and therefore must find a snail of a suitable species within this time. The snails that will serve as intermediate hosts are different for each species of schistosome. When

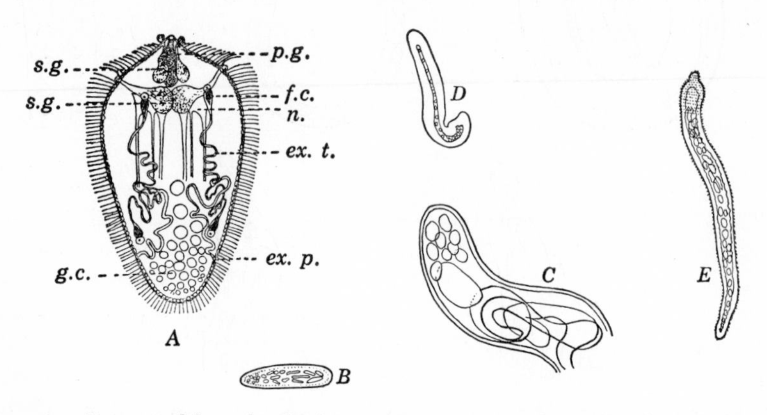

Fig. 66. Stages in life-cycle of schistosomes. *A*, miracidium of S. *haematobium*. Abbreviations as in Fig. 59A; ×300. (From *Human Helminthology*, by Ernest Carroll Faust, Lea and Febiger, Philadelphia.) *B*, primary sporocyst 8 days after infection. *C*, mature primary sporocyst (19 days). *D*, young daughter sporocyst (19 days). *E*, mature daughter sporocyst. (*B*, *C*, *D*, and *E* ×35; after Faust and Hoffman, *Puerto Rico J. Publ. Health Trop. Med.*, 10, 1934.)

the miracidia come close to a suitable snail they become excited and make a dash for it, burrowing into the tentacles or other parts, much to the irritation of the snail. Many miracidia become mired in the tough tissues of the foot or head; those attacking the soft parts succeed in embedding themselves within a half hour after the attack begins. During penetration the ciliated outer coverings are shed and the miracidia elongate and become tubular sporocysts. These make their way through the viscera to the digestive gland at the innermost extremity of the snail.

In S. *mansoni* the sporocysts reach a length of 1 mm. in about two weeks, and begin to produce daughter sporocysts which burst free from the mother sporocyst. These, in turn, reaching a length of 1.5 mm. by 0.09 mm., produce forked-tailed cercariae from germ masses at their posterior ends. The mature cercariae (Fig. 64) begin emerging from a birth pore near the anterior end of the sporocyst, which continues to produce them for several months. S. *haematobium* and S. *mansoni* are said to begin shedding cercariae about 4 to 6 weeks after infection under optimum conditions, but in Leyte S. *japonicum* was found to require 11 weeks. A snail infected by a single miracidium of S. *mansoni* was observed by Faust and Hoffman (1934) to discharge an average of 3500 cercariae a day for a long time; in one instance the total progeny of a single miracidium exceeded 200,000. S. *japonicum* evidently develops less abundantly since Pesigan et al. in 1958 reported that infected snails emit about 5 cercariae in 2 days.

The cercariae of *haematobium* and *mansoni* have a body about 200 μ long with a tail stem slightly longer and forks about 75 μ long; those of S. *japonicum* are slightly smaller. They escape from the snail into the water in "puffs," a number at a time. For two or three days the cercariae alternately swim and rest in the water; if they fail to reach a final host in this time they die. If successful they burrow through the skin, using the histolytic and hyaluronidase-bearing secretions of their penetration glands just as the miracidia do. The natives of some parts of Africa where S. *haematobium* occurs realize that infection may result from bathing, but from the nature of the disease they believe that infection takes place by way of the urinary passages, and therefore vainly employ mechanical devices to prevent infection in this manner. Ruminants, because of the neutral or alkaline nature of parts of the stomach, can become infected with schistosomes by drinking cercaria-infected water, but other animals cannot.

Skin penetration requires several minutes and may or may not be accompanied by a prickling sensation and subsequent dermatitis (see p. 294. If ingested with water the cercariae attach themselves to the mucous membranes of the mouth or throat and similarly bore in. They leave their tails behind them, and can be found in the skin for about 18 hours, but eventually they find their way into the blood system and are carried via the heart to the lungs. Young S. *mansoni* accumulate in the lungs on the second and third days; by the sixth day they appear in numbers in the liver, where they are well established by the fifteenth day. Apparently these larvae feed only on the portal blood, but once in the liver they grow rapidly. Migration of this species to the mesenteric veins begins about the twenty-third day, and mating and egg production about the fortieth day.

Species in man. There are three species of *Schistosoma* which are habitual human parasites of wide distribution—S. *haematobium*, S.

mansoni, and *S. japonicum,* the first affecting primarily the urinary system, the other two the intestine. *S. haematobium* has large, terminal-spined eggs measuring 115 to 170 μ by 45 to 65 μ (Figs. 55*A*, 65*F*, *a*); 4 to 6 testes in the males; the female with the ovary posterior to the middle of the body; 20 to 30 eggs in the uterus; and a tuberculated cuticle. *S. mansoni* has lateral-spined eggs of similar size (Figs. 55*B*, 65*F*, *e*); 8 or 9 testes; the ovary anterior to the middle of the body; only 1 egg in the uterus; and a tuberculated body. *S. japonicum* has smaller and more rounded eggs (70 to 100 μ by 50 to 65 μ) (Figs. 55*C*, 65*F*, *j*), with a rudimentary lateral spine that is often difficult to see; 7 testes; ovary posterior to the middle of the body; 50 or more eggs in the long uterus; and a smooth cuticle. The cercariae of these three species are distinguishable by the number and type of their penetration glands.

S. mansoni is occasionally found in nature in animals other than man, including monkeys and various kinds of rodents. Monkeys have been found naturally infected with *S. haematobium* and a variety of rodents can be infected in the laboratory. In the Orient *S. japonicum,* on the other hand, is a common parasite of cattle, goats, pigs, dogs, and cats among domestic animals and also weasels, meadow mice, moles, and shrews in the wild animals. *S. haematobium* is widely distributed in Africa, the Middle East, western India, and part of Portugal; *S. mansoni* over most of Africa and also in South America from Brazil to Venezuela, and in some of the West Indies; and *S. japonicum* in the Far East in Japan, China, some of the Phillipine Islands, and Celebes.

The intermediate hosts of these three species belong to three distinct families of snails (see p. 270); those of *S. mansoni* to the family Planorbidae, those of *S. japonicum* to the family Hydrobiidae, and those of *S. haematobium* to the family Physidae (or Bulinidae), but only certain species of certain genera of these families are satisfactory hosts. An exception is the recent report that in western India a species of *Ferrissia* (family Ancylidae) may serve as intermediate host for *S. haematobium.* The species of schistomes are not only highly specific in their choice of intermediate hosts, but strains of one species may vary in their ability to serve as hosts, and may be good hosts for one strain of a parasite and not another. For example, Files (1951) found that *Australorbis glabratus* from West Indies and Venezuela is more susceptible to strains of *S. mansoni* from the Western Hemisphere than from Egypt, and one strain of this snail from Brazil was found refactory to *all* strains of *S. mansoni. Biomphalaria boissyi* from Egypt is receptive only to Egyptian *mansoni,* but *B. pfeifferi* in Liberia is a good host for *S. mansoni* from either Egypt or America. Likewise in northern parts of Africa and the Middle East, S. *bovis* of animals develops in *Bulinus truncatus* but not in *B. africanus*, whereas in other parts of Africa S. *bovis* develops in *B. africanus* but not in *B. truncatus*. In nonsusceptible snails there is a tissue reaction around the invading miracidia within 12 to 24 hours which destroys them in a few days, whereas in good hosts there is no apparent reaction (Brooks, 1953).

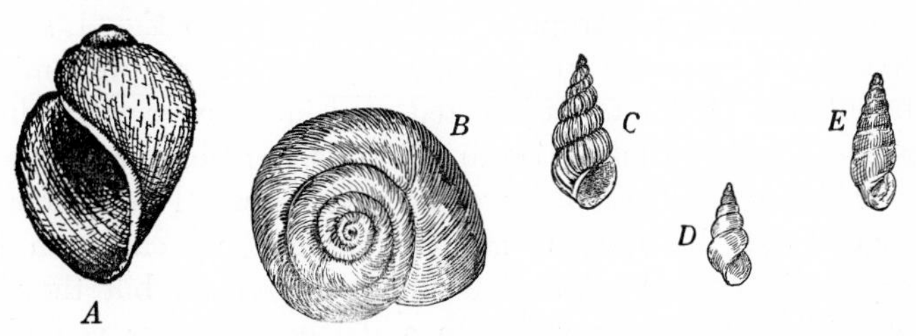

Fig. 67. Intermediate hosts of schistosomes, drawn to scale. *A, Bulinus truncatus*, principal host of S. *haematobium* in Egypt; *B, Australorbis glabratus*, host of S. *mansoni* in tropical America; *C, D,* and *E, Oncomelania hupensis, quadrasi,* and *nosophora,* hosts of S. *japonicum* in China, Philippines, and Japan, respectively. All ×2.

The three widespread human schistosomes just discussed are representative of groups of schistosomes which some writers, e.g., Schwetz et al. (1954), consider species, but which others, e.g., Amberson and Schwarz (1953), consider subspecies. Kuntz (1955) has referred to these as species complexes. The erection of subspecies names among parasitic worms seems to the authors to be a generally questionable procedure. Related to S. *haematobium* are three other African schistosomes, and one in India, which differ in the shape of their eggs (Fig. 65*F, b,c,d,* and *k*) and in their definitive hosts, though all use *Bulinus* as intermediate hosts. These are the widely distributed S. *bovis* of cattle, sheep, and goats; S. *mattheei* of sheep in South Africa; S. *intercalatum* of man in Belgian Congo; and S. *indicum* of horses and other animals in India. All of these are *intestinal* parasites, and even S. *haematobium* is primarily intestinal in monkey hosts. Related to S. *mansoni* is S. *mansoni* var. *rodentorum* of rodents, and S. *rodhaini* of rodents and dogs, which have eggs with a subterminal spine (Fig. 65*F, f*); both of these, like S. *mansoni,* develop in planorbid snails. In 1958, Pitchford reported a *mansoni*-like schistosome to be common in cattle, sheep, and goats and in 5 to 10% of the children in Eastern Transvaal. Hsu and Hsu (1958) and others have shown that S. *japonicum* has at least four recognizably different strains occurring in China, Japan, Formosa, and the Philippines. The Formosan strain does not develop in man but is a parasite of small mammals. A species with eggs suggestive of *japonicum*, *S. margrebowiei* (Fig. 65*F*, *h*), has been found in man in Congo and South Africa. In addition to the above-mentioned species, eggs of a pig schistosome in India, *S. incognitum*, were found by the senior writer in feces believed to be of human origin; these eggs have a subterminal spine (Fig. 65*F*, *g*).

SCHISTOSOMA MANSONI. This is an important human parasite in many parts of Africa and tropical America. In lower Egypt, since the development of perennial irrigation, a majority of the predominantly rural population (fellahin) is infected, and in some irrigated districts in Venezuela Scott estimated up to 90% of the males over 10 years of age to be infected. The adult worms have a special predilection for the branches of the inferior mesenteric veins which drain blood primarily from the large intestine and cecal region, but they occasionally get into the urinary system and have the eggs voided with the urine. A healthy, mated female worm deposits a single egg at a time, about three hundred times a day. Mice, hamsters, gerbils, and cotton rats are the best experimental hosts; rats recover too readily, and in other animals either the eggs fail to be produced, or few worms develop. Egg production begins in about 6 to 7½ weeks in experimental animals.

The intermediate hosts, as noted above, belong to the family Planorbidae. In Africa they are species of the genus *Biomphalaria*. The Egyptian species is usually referred to as *B. boissyi*, and in most other parts of Africa and Madagascar as *B. pfeifferi*, although several other species have been reported as hosts. Amberson and Schwarz consider even *boissyi* and *pfeifferi* to be subspecies of one species, *B. alexandrina*. *B. boissyi* is a snail of ditches and ponds, whereas *P. pfeifferi* is a riverine species. In South America *Australorbis glabratus* (Fig. 67*B*), *A. olivaceus*, and *Tropicorbis centimetralis*, and in the West Indies *A. glabratus* and *A. antiguensis*, serve as hosts. Fortunately the North American planorbids fail to serve as hosts with the exception of *Tropicorbis havanensis* of the southern states, and in only a very few of these do the invading miracidia succeed in producing cercariae.

S. mansoni infections are almost everywhere associated with perennial irrigation, which provides good breeding grounds for the planorbid snails, and brings man into contact with the water. The danger is intensified by the Moslem custom of washing themselves after defecation; as Khali pointed out, a habit directed towards personal cleanliness has become a dangerous contributor to disease. An interesting situation exists in Egypt where both *mansoni* and *haematobium* are prevalent in the delta, but only *haematobium* in Upper Egypt, although both species are again prevalent in Sudan. Helmy showed that this was because *Biomphalaria boissyi* is a quiet-water, surface-feeding snail, and although washed down the Nile from the Sudan it fails to find landing places until the current becomes very reduced, whereas *Bulinus truncatus,* the snail host of S. *haematobium,* settles to the bottom and crawls out. The difference is strikingly demonstrated by putting *Biomphalaria* and *Bulinus* in a glass of water and emptying it—*Biomphalaria* is poured out but *Bulinus* clings to the glass.

SCHISTOSOMA HAEMATOBIUM. This schistosome is very prevalent in many parts of Africa, Madagascar, and southwestern Asia, and a few cases have been reported in India. It is especially prevalent in Egypt and affects 60 to 95% of the fellahin in lower Egypt, where perennial canal irrigation is practiced. It is becoming commoner in Iraq and parts of Syria, Palestine, and Iran as irrigation systems are developed. In Bagdad 25% of the males (60% in some parts of the city) are infected, and in Basrah 33%. In some rural areas 75 to 80% of the males are infected; the females usually show a somewhat lower incidence. Throughout the greater part of Africa, both north and south of the Sahara, S. *haematobium* infections are prevalent wherever local conditions are favorable for the snail vectors and wherever people bathe or work in the water or drink it unfiltered. Kuntz (1952) found open ablution basins in mosques in Yemen to be a breeding ground for snails, and the main source of schistosome infections.

The adult worms live in the pelvic veins of the vesical plexus, and the females normally deposit their eggs in the walls of the urinary bladder, urethra, or ureters, through which the eggs work their way to be voided with the urine; a few eggs, however, often get into the feces also. As with other species, eggs are often swept back to the liver and on to the lungs or other organs. S. *haematobium* can be reared in hamsters experimentally, less readily in mice, and not at all in rats or rabbits.

The intermediate hosts, except in Portugal where a planorbid, *Planorbarius corneus metidjensis,* and in India where an ancylid, *Ferrissia,* have been incriminated, belong to the subfamily *Bulininae*—*Bulinus, Physopsis,* and *Pyrgophysa.** In North Africa, the Mediterranean area, southwestern Asia, the East African highlands, and South Africa the intermediate host is *Bulinus truncatus* (Fig. 67A) or closely related forms. These are snails of moderately warm countries, but not adapted to the heat of low-lying countries in equatorial Africa, where *B. africana* and closely related forms are the hosts. Both these species

* Formerly recognized as a separate family, but now considered a subfamily of Planorbidae.

occur together in many parts of Africa, but only *B. africana* in tropical Central Africa and the Guinea Coast. In the island of Mauritius *B. forskalii* is the host. These same snails also serve as hosts for the other African schistosomes with terminal-spined eggs.

As noted on p. 285, in Belgian Congo another schistosome with terminal-spined eggs, *S. intercalatum,* is found in man. Like *S. haematobium,* it uses *Bulinus* as an intermediate host, but it is an intestinal, not a urinary, species.

SCHISTOSOMA JAPONICUM. This species, as noted previously, occurs naturally in a great variety of reservoir hosts in many parts of the Orient. The adult worms live mainly in branches of the superior mesenteric veins; the eggs, up to 3500 a day, are deposited in the walls of the intestine and make their way into the lumen until, as in other species, the wall becomes too thick and the host reaction too great. Not infrequently eggs become imbedded in the appendix also, and may cause appendicitis. More frequently than in the other species the eggs, which are laid in clusters, are swept back into the liver, and often on to the lungs.

The intermediate hosts of *S. japonicum* are small, operculated, amphibious snails of the genus *Oncomelania* (family *Amnicolidae*) (Fig. 67*C–E*). In Japan the host is *O. nosophora,* a smooth-shelled form; in China it is the rib-shelled *O. hupensis,* and perhaps one or two related species; in the Philippines it is *O. quadrasi,* with a smooth, pointed shell; and in Formosa it is *O. formosana,* but this species will not allow development of *S. japonicum* from Japan. In 1959, Wagner and Chi showed that four Oriental species of *Oncomelania* will readily interbreed in the laboratory. Rey (1959) has recently described a new species, *Oncomelania brasiliensis,* from Brazil. The snails were found in the state of Mato Grosso in an area inhabited by many immigrants from Japan and Okinawa. Rey pointed out that the possibility of establishment of *S. japonicum* in Brazil should be investigated. The nearest relative of *Oncomelania* in the United States is *Pomatiopsis* (see p. 299 and Fig. 69*C*), which is a host for *Paragonimus.* This snail can sometimes produce cercariae of *S. japonicum* but is not a good host.

The snails of this group are only 7 to 10 mm. long with high-spired shells. The young snails live in water but when mature they are amphibious and live in damp places at the edges of water and are commonly found climbing on vegetation, mud, or rocks along irrigation ditches and edges of ponds and streams, especially where the water or soil is enriched with humus or night soil, for they feed on filth. They are frequently submerged with rising or disturbed water and are carried from dirty village ditches to rice fields. While submerged they are attacked by the miracidia, which habitually swim near the surface of the water. In Japan and China irrigation ditches are by far the most important locations for the snails, rice fields less so, but in the Philippines, where agriculture is less highly developed, *O. quadrasi* occurs mainly in small, slow-flowing streams clogged with vegetation.

In northern areas the snails lay their eggs singly in the spring, patting mud or debris on them so that they are hard to detect. The snails get infected in the summer, hibernate in winter, and shed cercariae the following spring. Farther south there are at least two broods a year, and cercaria production is more or less continuous.

Pathology. The diseases produced by various schistosomes are similar in many respects but differ in details. In previously uninfected cases there are three distinct stages of the disease: (1) the period of migration and development to maturity of the young worms; (2) the period of early egg production, when the eggs readily escape to the lumen of the intestine or bladder; and (3) the period of late egg production when the eggs tend to be trapped in the tissues. Of course where there are constant reinfections these distinct stages are not recognizable.

During the pre-egg stage the worms are carried to the lungs, then migrate to the liver, and after 3 weeks or more undertake the last lap of their journey to the mesenteric veins. The first symptoms, if there has been heavy exposure to cercariae, is an irritating bronchial cough lasting a few days. After 2 or 3 weeks there is an itching rash, local dermatitis, fever, aches, and other general toxic symptoms of allergic nature, including eosinophilia.

During the stage of early egg production, which may last one month to a year, the intestinal species produce blood and mucus in the stools, diarrhea, abdominal pain, and slight liver tenderness, whereas *S. haematobium* produces few symptoms except blood in the urine, often only in the last few drops, and sometimes some irritation from urinating, depending on the severity of the infection.

Gradually, as the tissues become sensitized to the eggs passing through the tissues, and tissue reactions cause the walls of the intestine, bladder, and urinary passages to become thickened and inflamed, trapping large numbers of the eggs, the symptoms of the third stage come on. In the intestinal forms there is recurring diarrhea or dysentery, increasing abdominal pain, and enlargement of liver and spleen. As the disease progresses the liver eventually recedes and becomes contracted and cirrhotic, but the spleen continues to enlarge, the abdomen becomes bloated while the rest of the body is pitifully anemic and emaciated, and abdominal pain may become very great. In *haematobium* infections bloody urine continues with increasing pain on urination, often with constrictions of the urinary passages, and eventually inability of the bladder to contract (Fig. 68). Fistulas and even malignant tumors may develop, involving the bladder, prostate, or penis. Usually, however, this infection causes less serious symptoms than does S. *mansoni.* One side effect of the disease is a high incidence of urinary typhoid carriers.

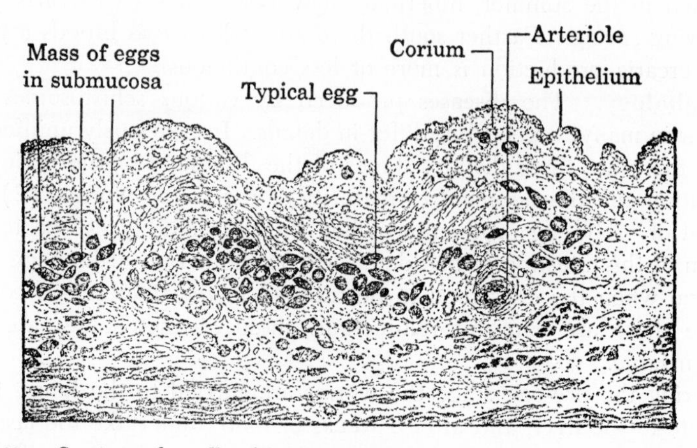

Fig. 68. Section of wall of urinary bladder showing eggs of *Schistosoma haematobium.* (After Brumpt, *Nouveau traité méd.,* 5, 1922.)

In light cases there may be no noticeable symptoms at all except blood in the stools or urine, or there may be only general symptoms such as loss of appetite, weakness, aches, etc. On the other hand, in severe infections there may be a number of serious complications, particularly from eggs being deposited in arterioles in the lungs, causing what is variously called cardio-pulmonary schistosomiasis, "cor pulmonale," or "ayerza disease." Inability of the blood to pass the clogged vessels in the lungs may lead to cyanosis (blueness from inadequate aeration of the blood), congestive heart failure, and aneurisms (blowouts) of the pulmonary artery or other vessels. In South Africa, S. *haematobium* eggs are found in the lungs of people dying of lung diseases twice as frequently as in those dying of other complaints. The more frequent involvement of the lungs in *haematobium* infections is not particularly surprising since the venous drainage of the urinary bladder bypasses the liver and eggs may be carried directly to the heart without being trapped in the liver. There is some evidence that "Egyptian splenomegaly" or Banti's disease, characterized by huge spleen enlargement, edema, and anemia, may also be a manifestation of schistosomiasis.

Lesions in other parts of the body (ectoptic lesions) result from aggregates of tubercles forming around eggs which have escaped and have been carried to distant parts of the body; they vary in size from pinhead lesions in the conjunctiva to lesions the size of an orange in the brain. In *S. japonicum* infections a significant proportion occur in the brain, probably because, according to Faust, the vertebral veins provide a natural channel from the portal and caval veins.

Immunity. Little is known about development of immunity to schistosomiasis by man, but experimental animals develop immunity (see Vogel and Minning, 1953; Meleney and Moore, 1954), and there can be no doubt that man does too. Presumably all the manifestations of immunity discussed on pp. 25–30 come into play, particularly interference with reproduction and resistance to reinfection; if this were not so, people constantly exposed in places like Egypt would soon succumb. There is also little doubt that the nutritional status of a community, and of individuals in the community, has as much to do with the severity of the disease as does exposure to it—in the present writers' opinion probably even more.

Diagnosis. In early or acute cases, after the worms begin ovipositing, diagnosis of urinary schistosomiasis can be made by finding the eggs in sedimented or centrifuged urine; if scanty, addition of water to the sediment will cause the eggs to hatch in 5 to 10 minutes, and the swimming miracidia can be seen with the naked eye or by projection on a screen. Exercising before urinating increases the number of eggs in the urine. In stages of early egg production in *S. mansoni* and *S. japonicum* infections the modified Telemann technique (see p. 253) can be used for stool examination; if scanty, the eggs can be concentrated by repeated sedimentation or centrifuging in 0.5% glycerin water, after which the sediment can be resuspended in water and swimming miracidia observed after 10 or 15 minutes.

In old chronic cases examination of urine or feces is unreliable since so many eggs become encapsulated in the tissues and do not escape regularly. Those that do are often dead or blackened, or surrounded by a fuzzy coat of cells. In these circumstances rectal scraping gives a much larger number of positives than examination of excreta for eggs, even in *S. haematobium* infections, where only a minority of the eggs get into the rectum. In a survey in Egypt Weir et al. (1952) found 60% infection by examination of urine sediment for eggs, but by the production of miracidia method found 31% of the negatives to be positive, and by the rectal scraping method 83% of the negatives were positive; by all three methods combined the incidence was 95%. Immunological tests are also useful in chronic cases, using antigens prepared from cercariae; skin tests, complement fixation, and formation of a precipitate around schistosome cercariae all have their uses. Complement fixation and precipitin reactions are demonstrable within 2 or 3 weeks, skin reactions after 4 to 6 weeks. In a survey in Puerto Rico by Morales et al. (1950) both skin test and biopsy (rectal scraping) proved superior to stool examination, but any one method may miss cases positive by the others. The skin test is easiest to perform and therefore to be preferred in a survey.

Treatment. Tartar emetic (sodium antimony tartrate) and other trivalent antimony compounds, particularly Anthiomaline and Fuadin, have specific effects in schistosomiasis. These drugs cause degenerative changes in the adult worms, especially affecting the reproductive organs. The injured worms lose their hold and are swept into the liver. The drugs do not kill the eggs, but if reproduction of the parent worms is stopped, there will be only dead and blackened eggs in the tissues after about 3 weeks.

Tartar emetic is cheap and effective, but must be given intravenously over a period of about 4 weeks, and it is toxic. Fuadin can be given into the muscles but is more effective intravenously; it is expensive and somewhat less effective but less toxic. Anthiomaline is similar to Fuadin but less expensive and also less efficient. None of these drugs is as effective against S. *japonicum* as against the other species. Pentavalent antimony compounds are comparatively ineffective. To overcome the long time required for cure, attempts have been made to intensify and shorten the treatment, using courses extending 2 to 10 days, with some success (see Alves and Blair, 1946), but also some danger. Friedheim has developed a relatively nontoxic compound, antimony dimercaptosuccinate (TWSb); Salem, Friedheim, and El Sherif (1957) reported a high rate of cure of S. *haematobium* and S. *mansoni* infections with low toxicity. This was confirmed with S. *haematobium* infections by Alves in 1958 and others, and the drug may well replace Fuadin.

There is a group of xanthone derivatives (miracil D, Lucanthone, and others) which are effective against schistosomes when given by mouth. This would be of tremendous advantage if it were not that the drugs produce such unpleasant (though not dangerous) toxic symptoms as nausea, vomiting, mental depression, and pains, so severe that many patients prefer their disease. This drug is most effective against S. *haematobium* infections and no good against S. *japonicum.*

In advanced cases where the rectum or urinary organs have been severely damaged, and in all cases of splenomegaly, surgical treatment is necessary.

Prevention and control. There are three obvious points of attack on schistosomiasis. These are (1) education and sanitation, (2) treatment of infected hosts, and (3) elimination of snails. The first, desirable on general grounds, does not promise to be effective. In Puerto Rico, school children are very familiar with the life history of S. *mansoni,* but the infection persists in this island. In the case of S. *japonicum,* and probably to a certain extent with S. *mansoni,* the reservoir animal hosts are not affected by education and sanitation. In the Orient, two to five people have to live off one acre of land, and the use of night soil (human excreta) is necessary to existence. In countries with low rainfall the same irrigation water has to be used for all purposes without purification—swimming, washing clothes, ablutions, drinking, and irrigation. Treatment of infected persons remains unsatisfactory, particularly on any large scale, and although it is a necessary part of a control program, it can hardly be considered an effective measure.

Ultimate control of schistosomiasis seems to resolve itself into destroying or eliminating the snails that serve as intermediate hosts.

Copper sulfate, found by the senior writer (1920) to be destructive to snails, has been the only really effective and practically feasible chemical for control of snails until recent years; enormous amounts of it have been used in Egypt and other parts of Africa. It has the disadvantage that it is quickly rendered nontoxic to snails by combination with alkalis and organic matter, and has no residual effect. Under some circumstances a mixture of the slowly soluble copper carbonate with copper sulfate increases effectiveness.

Intensive search in the last few years has brought to light several other groups of chemicals that show promise. The pentachlorophenols are one, of which the sodium salt (Santobrite) seems best. This was found by Hunter et al. (1952) to be so effective against the amphibious snail hosts of S. *japonicum* that not only control, but ultimate elimination, may be possible by spring and fall spraying of all irrigation ditches—in the fall to kill snails that become infected during the summer and will shed cercariae in the spring, and in spring to kill infected snails before rice planting begins. Field trials in Egypt by Kuntz and Wells (1951), Wright et al. (1958) and others have shown the efficacy of sodium pentachlorophenate; however, environmental conditions so greatly modify its effectiveness that no standard formula for its application seems practical. Certain mercuric compounds were shown by workers at the National Institutes of Health to be the most effective molluscicides of all in laboratory tests, but adequate field tests have not yet been made. All molluscicides rapidly increase in effectiveness with temperature.

Halawani has called attention to the fact that while complete eradication of snail hosts in Egypt is at present not feasible, very marked reduction in human exposure to infection could be obtained by treating local areas—villages, bridge crossings, washing places, etc.—where the snails are most exposed to miracidia and man most exposed to cercariae. Objection has been raised that snails would quickly invade from untreated parts of ditches, but in Nigeria stretches of streams 1½ to 3 miles long, freed of snails, remained free for 10 to 11 months.

Additional control methods, locally applicable, consist of paving or clearing vegetation from ditches, use of natural enemies of snails, and wholesale treatment. The last might be feasible if a drug could be found which would prevent passage of viable eggs. Attempts have been made to utilize pathogenic bacteria as natural enemies. A snail, *Marisa,* avidly devours eggs of *Australorbis,* and certain leeches are destructive to it. The latter approaches do not impress the authors as ones likely to lead to control.

For personal protection against cercariae, e.g., in military operations, Hunter et al. in 1952, in tests against schistosome dermatitis, found copper oleate ointment to be effective for 6 to 8 hours, and ointments containing dibutyl or dimethyl phthalate or benzyl benzoate for 4 to 6 hours. Hunter et al. (1956) and others have developed a number of other ointments which show great promise in protecting against infection. Clothing impregnated with mixtures of these chemicals, with a detergent added as an emulsifier, is protective even after several soap-and-water washings. Some protection is obtained by carefully wiping the skin after immersion in infected waters, although if given time the cercariae can penetrate while submerged. Protection against cercariae in drinking water and in unfiltered urban water supplies is possible by impounding the water for 48 to 60 hours, or by treatment with enough chlorine to give a residual of 0.5 ppm for 15 minutes or 0.1 ppm for 30 minutes.

Schistosome Dermatitis (Swimmer's Itch)

The penetration of the skin by the cercariae of human schistosomes usually causes a prickling sensation and may or may not cause an itching rash or papules; these skin effects are undoubtedly conditioned by the extent of prior invasion and sensitization or immunity.

As Cort demonstrated in 1928, certain species of "foreign" cercariae, incapable of infecting man, cause a severe dermatitis or "swimmer's itch" when they penetrate the skin of bathers or waders who have become sensitized by repeated exposure. This condition annoys vacationers on sunny bathing beaches in northern United States and southern Canada, from New England and Quebec to Manitoba and Oregon. In the north central states *Cercaria stagnicolae* is the most important since its host, *Stagnicola emarginata,* inhabits the same waters as the human bathers, and the cercaria, like the bathers, swarm near the surface in shallow water on warm, sunny days. McMullen and Beaver (1945) believe that the snails acquire their infections mainly from migrating ducks in the fall when the beaches are otherwise deserted. Swimmer's itch also affects clam diggers and sea bathers on the North Atlantic coast, bathing beauties in Florida and Hawaii, naturalists on rocky shores of southern California and Mexico, fishermen in San Salvador, carp breeders in Germany, rice growers in Japan and Malaya, farmers in Iraq, and lake bathers in Australia and New Zealand.

It may be expected that wherever suitable snail hosts are present in abundance, and birds or other definitive hosts congregate in sufficient numbers and for a long enough time to infect them, and human beings then repeatedly come in contact with the water, swimmer's itch is likely to be a nuisance. Most of the known fresh-water itch-producers, which are cercariae with eye spots belonging to the *Cercaria ocellata* group, develop into species of *Trichobilharzia* in ducks; two others develop into species of *Gigantobilharzia* in passerine birds; one into *Schistosomatium douthitti* in rodents; and one into *Schistosoma spindale* of cattle and goats (see Cort, 1950). One marine form occurring in Rhode Island and Long Island and in Hawaii develops into a *Microbilharzia* of shore birds (see Chu and Cutress, 1954).

Although some cercariae penetrate the skin under water, the annoyance can be much reduced by wiping the skin dry immediately after leaving the water. Children getting alternately wet and dry in shallow water are affected worst. The dermatitis begins with a prickly sensation followed by the development of extremely itchy papules, which sometimes become pustular and may be accompanied by considerable swelling. Some individuals are much more severely affected than others and may lose much sleep and even be prostrated. It usually takes about a week for the condition to subside. In previously unexposed laboratory mammals, and probably in man, bird schistosomes after skin penetration migrate to the lungs and may produce pulmonary hemorrhages, but they fail to go on to the liver (Olivier, 1949). It is only after sensitization that the cercariae are trapped in the skin by tissue reactions. They soon die in the skin and cause allergic irritation.

After penetration soothing and/or antihistaminic applications are said to be helpful. The dermatitis can be effectively controlled in small bodies of water by the use of copper salts to kill the snails. McMullen and Brackett recommend copper sulfate for shallow water, and a 2 to 1 mixture of copper sulfate and copper carbonate for water over 2 ft. in depth, at the rate of 3 lb. of the mixture per 10,000 sq. ft. of bottom. In larger lakes attention to water currents is necessary.

Animal Schistosomes

Cattle, sheep, and goats are severely affected by several species of schistosomes in Africa and Asia, the most important being S. *bovis* in Africa, Southern Europe, and Asia; the nearly related S. *mattheei* of South Africa; S. *spindale* in India, South Africa, and Sumatra; and S. *nasale* (possibly = S. *spindale*) in India. S. *indicum*, confined to India, also affects these animals, but more frequently horses and camels; and S. *incognitum*, the eggs of which the writer first described from feces believed to be human, occurs in pigs and dogs in India. In the Orient many animals are parasitized by S. *japonicum*. These various species are distinguishable by their eggs (Fig. 65*F*). In Africa the intermediate hosts are the same as those of S. *haematobium*, or related species of the family Bulinidae, whereas the commonest host in India is a planorbid, *Indoplanorbis exustus*. In addition to these members of the genus *Schistosoma* there are several species of *Ornithobilharzia* which live in cattle, elephants, and other animals.

S. *nasale* causes a "snoring" disease of cattle; it localizes in the nose and produces cauliflowerlike growths on the nasal septum. All the other species have their eggs voided with the feces, though sometimes with the urine as well. These parasites cause lowered vitality, especially, no doubt, in poorly nourished animals, and cause economic loss from unsalability of the liver.

Ducks sometimes suffer from various species of the subfamily Bilharziellinae, which differ from the typical schistosomes in having both sexes alike in size and form. Cercariae of some of these, as well as those of S. *spindale*, cause swimmer's itch in man.

REFERENCES

Alves, W. 1957. Bilharziasis in Africa, A review. *Central African J. Med.*, 3: 123–127.

Alves, W., and Blair, D. M. 1946. Schistosomiasis; intensive treatment with antimony. *Lancet,* Jan. 5, 1946, 9–12.

Bang, F. B., Hairston, N. G., Graham, O. H., and Ferguson, M. S. 1946. Studies on schistosomiasis japonica, I-V. *Am. J. Hyg.,* 44: 313–378.

Bhalerao, G. D. 1948. Schistosomiasis in animals. *Proc. 4th Intern. Congr. Trop. Med. Malaria,* 2: 1386–1393.

Brumpt, E. 1941. Observations biologiques diverses concernant *Planorbis* (*Australorbis*) *glabratus,* hôte intermédiaire de *Schistosoma mansoni.* *Ann. parasitol. humaine et comparée,* 18: 9–45.

Chu, G. W., and Cutress, C. E. 1954. *Austrobilharzia variglandis* (Miller and Northrup, 1926), Penner, 1953, (Trematoda, Schistosomatidae) in Hawaii with notes on its biology. *J. Parasitol.,* 40: 515–523.

Faust, E. C., et al. 1934. Studies on Schistosomiasis mansoni in Puerto Rico, I-III. *Puerto Rico J. Public Health Trop. Med.,* 9: 154–160, 228–254; 10: 1–47.

Faust, E. C., and Meleney, H. E. 1924. Studies on schistosomiasis japonica. *Am. J. Hyg. Monogr.,* Ser. 3.

Files, V. S. 1951. A study of the vector-parasite relationships in *Schistosoma mansoni.* *Parasitology,* 41: 264–269.

Hernandez Morales, F., et al. 1950. The acid-ether concentration test, the rectal biopsy, and the skin test in the diagnosis of Manson's schistosomiasis. *Puerto Rico J. Public Health Trop. Med.,* 25: 329–334.

Hsu, H. F., and Hsu, S. Y. L. 1958. On the size and shape of the eggs of the geographic strains of *Schistosoma japonicum.* *Am. J. Trop. Med. Hyg.,* 7: 125–134.

Hunter, G. W., et al. 1956. Studies on schistosomiasis, XII. Some ointments protecting mice against the cercariae of *Schistosoma mansoni.* *Am. J. Trop. Med. Hyg.,* 5: 713–736.

Kierman, F. A., Jr. 1959. The blood fluke that saved Formosa. *Harper's Magazine,* 218: 45–47.

Kuntz, R. E. 1952. Schistosomiasis mansoni and S. *haematobium* in the Yemen, Southwest Arabia: with a report on an unusual factor in the epidemiology of *Schistosoma mansoni.* *J. Parasitol.,* 38: 24–28.

Kuntz, R. E. 1955. Biology of the schistosome complexes. *Am. J. Trop. Med. Hyg.,* 4: 383–413.

Kuntz, R. E. 1957. Relationship of temperature to molluscicidal activity. *Am. J. Trop. Med. Hyg.,* 6: 940–945.

Kuntz, R. E., and Wells, W. H. 1951. Laboratory and field evaluation of 2 dinitrophenols as molluscicides for control of schistosome vectors in Egypt with emphasis on the importance of temperature. *Am. J. Trop. Med.,* 31: 784–824.

Maegraith, B. 1958. Schistosomiasis in China. *Lancet,* January 25, 1958: 208–214.

Mandahl-Barth, G. 1957. Intermediate hosts of *Schistosoma.* African *Biomphalaria* and *Bulinus.* I. *Bull. World Health Organization,* 16: 1103–1163; II. *Bull. World Health Organization,* 17: 1–65.

McMullen, D. B., and Beaver, P. C. 1945. Studies on schistosome dermatitis, IX. *Am. J. Hyg.,* 42: 128–154.

McMullen, D. B., and Harry, H. W. 1958. Comments on the epidemiology and control of bilharziasis. *Bull. World Health Organization,* 18: 1037–1047.

McMullen, D. B., et al. 1947. The control of schistosomiasis japonica I-IV. *Am. J. Hyg.*, 45: 259–298.

Meleney, H. E., and Moore, D. V. 1954. Observations on immunity to superinfection with *Schistosoma mansoni* and *S. haematobium* in monkeys. *Exp. Parasitol.*, 3: 128–139.

Meleney, H. E., Moore, D. V., Most, H. and Carney, B. H. 1952. The histopathology of experimental schistosomiasis. I. The hepatic lesions in mice infected with *S. mansoni, S. japonicum* and *S. haematobium. Am. J. Trop. Med. Hyg.*, 1: 263–285; II. Bisexual infections with *S. mansoni, S. japonicum,* and *S. haematobium. Ibid.*, 2: 883–913.

Moore, D. V., Yolles, T. K., and Meleney, H. E. 1954. The relationship of male worms to the sexual development of female *Schistosoma mansoni. J. Parasitol.*, 40: 166–185.

National Institutes of Health (various authors). 1947. Studies on Schistosomiasis. *Natl. Inst. Health Bull.* 189, 212 pp.

Ottolina, C. 1957. El miracidio del *Schistosoma mansoni.* Anatomia-citologia-fisiologia. *Rev. Sanidad y Asistencia Social. Caracas.*, 22: 7–411.

Pesigan, T. P., et al. 1958. Studies on *Schistosoma japonicum* infection in the Philippines. 1. General considerations and epidemiology. *Bull. World Health Organization,* 18: 345–455. 2. The molluscan host. *Bull. World Health Organization,* 18: 481–578.

Rey, L. 1959. Molluscos do gênero *Oncomelania,* no Brazil, e sua possivel importância epidemiólogica. *Rev. Inst. Med. Trop. São Paulo,* 1: 144–149.

Salem, H. H., Friedheim, E. A. H., and El Sherif, A. F. 1957. The treatment of schistosomiasis by antimony dimercaptosuccinate (TWSb). *J. Egyptian Public Health Assoc.*, 32: 313–336.

Schwetz, J., Baumann, H., and Jort, M. 1954. Sur les schistosomes actullement (en 1953) connus en Afrique. *Ann. soc. belge. méd. trop.*, 33: 687–696.

Scott, J. A. 1937. The incidence and distribution of human schistosomes in Egypt. *Am. J. Hyg.*, 25: 566–614.

1942. The epidemiology of schistosomiasis in Venezuela. *Am. J. Hyg.*, 35: 337–366.

Stoll, N. R. 1947. This wormy world. *J. Parasitol.*, 33: 1–18.

Stunkard, H. W. 1946. Possible snail hosts of human schistosomiasis in the United States. *J. Parasitol.*, 32: 539–552.

Vogel, H., and Minning, W. 1953. Über die erworbene Resistenz von *Macacus rhesus* gegenüber *Schistosoma japonicum. Tropenmed. u. Parasitol.*, 4: 418–505.

Weir, J. M., et al. 1952. An evaluation of health and sanitation in Egyptian villages. *J. Egyptian Public Health Assoc.*, 27: 55–114.

Weller, T. H., and Dammin, G. 1945. The incidence and distribution of *Schistosoma mansoni* and other helminths in Puerto Rico. *Puerto Rico J. Public Health Trop. Med.*, 21: 125–165.

Williams, J. E., et al. 1957. Repopulation Control of *Oncomelania nosophora* by molluscicidal applications against juvenile snails through the medium of irrigation water. *Am. J. Trop. Med. Hyg.*, 6: 304–312.

Wright, W. H., Dobrovolny, C. G., and Berry, E. G. 1958. Field trials of molluscicides (chiefly sodium pentachlorophenate) for the control of aquatic intermediate hosts of human bilharziasis. *Bull. World Health Organization,* 18: 963–974.

Chapter 14

THE TREMATODES OR FLUKES
III. OTHER TREMATODES

Lung Flukes (*Paragonimus*)

The lungs of various mammals, including man, carnivores, rats, pigs, and opossums, may be infected with flukes of the genus *Paragonimus*, belonging to the family Troglotrematidae. The members of this family are rather small egg-shaped flukes with a spiny cuticle and with the large testes situated side by side behind the ovary. Besides *Paragonimus* the family includes *Nanophyetus salmincola*, the salmon-poisoning fluke (Fig. 81), and a fluke that lives in cutaneous cysts in birds, *Collyriclum faba.*

Species

Opinion has been divided as to the number of species of lung flukes. In the adults differences occur principally in the body spines and in the size of the eggs, but these are both variable characters. The first form described was *Paragonimus westermanni* from Bengal tigers, whereas the first human specimen, from Formosa, was named *P. ringeri.* A North American form, normally a parasite of mink, has been named *P. kellicotti.* Investigations of the life cycles have demonstrated differences in both the morphology and the behavior of the different developmental stages, not only between Korean and American forms, but also between various Oriental forms. A species parasitizing rats, *P. iloktsuensis,* was found by Chen near Canton which would not develop in carnivores, pigs, or monkeys, and another species, *P. ohirai,* occurs in pigs and sometimes dogs in Japan; the latter species is difficult to distinguish from *P. westermanni* in the adult stage but the larval stages are different. It is still uncertain whether *P. ringeri* and *P. westermanni* are distinct species, but it seems probable that they are identical. The North American form described by Ameel (1934) is a rather common parasite of mink in Michigan, and infected crayfish have been found over a large part of the United States. Since infection is caused by eating raw crabs or crayfish, which serve as second intermediate hosts, human infection is sporadic in most places but is endemic in many parts of the Orient, especially in Korea, Japan, the Philippines, and parts of Indo-China. In some localities 40 to 50% of

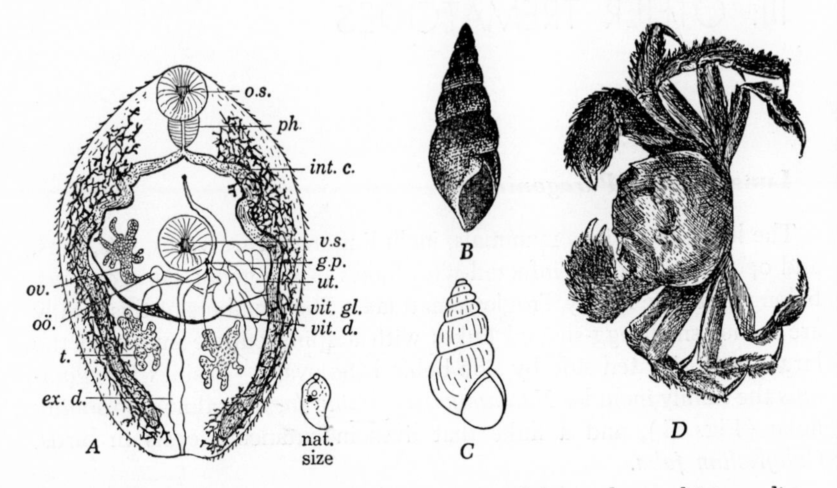

Fig. 69. Lung fluke, *Paragonimus westermanni,* and first and second intermediate hosts. *A,* adult fluke; *g.p.,* genital pore; *oö.,* oötype; other abbreviations as in Fig. 56 on page 259. *B, Semisulcospira libertina,* snail host in Japan and Korea. *C, Pomatiopsis lapidaria,* host of *P. kellicotti* in the United States. *D, Eriocheir japonicus,* a common second intermediate host in Japan. (Adapted from various authors.)

the population are infected. Human infections have also been reported from New Guinea, Indonesia, India, Belgian Congo, Cameroons, Nigeria, Ecuador, and the United States.

The adult flukes are reddish brown, thick, and egg-shaped, about 8 to 12 mm. long and 4 to 6 mm. in diameter. The cuticle is clothed with minute simple or toothed spines. The arrangement of the organs can be seen from Fig. 69A.

Fain and Vandepitte (1957) described a peculiar fluke, *Poikilorchis congolensis,* which produces eggs resembling those of *Paragonimus* and lives in abscesses behind the ears of natives of the Congo. These authors suggested that *Poikilorchis* may be an agent in the lung fluke infections reported from Africa.

Life cycle. The adults live normally in the lungs where, shortly after they have arrived, the host forms cystlike pockets around them, which rupture and liberate the eggs into the bronchial tubes, to be excreted with sputum. These cysts are usually about the size of filberts or larger, and contain commonly two but sometimes as many as six worms, together with infiltrated cells and numerous eggs in a rust-brown semifluid mass. Many of the eggs escape into the tissue, giving it a reddish peppered appearance and causing small tuberclelike abscesses. In some cases the worms apparently get on the wrong track in the body and end up in such places as the spleen, liver, brain, urinary system, intestinal wall, eye, or muscles. In one case more than a hundred mature parasites were found in a muscular abscess.

The eggs (Fig. 55*D*) are yellowish-brown, 80 to 118 μ in length by 48 to 60 μ in diameter; they are commonly found in the feces as the result of being swallowed. Miracidia develop in the eggs slowly after they leave the body, requiring at least 3 weeks, during which time the eggs must be kept moist.

The life cycle of the worm was established in part by several Japanese workers from 1918 to 1921, but the first complete account of the life cycle and of the developmental stages (Fig. 70) was given by Ameel (1934), who studied the American form in Michigan. The miracidia hatch in water and enter suitable snail hosts by burrowing. In the Orient the snail hosts are species of the family Thiaridae (see p. 270), principally *Semisulcospira libertina* (Fig. 69*B*), *S. amurensis*, and *Thiara granifera* (introduced into Florida), all once included in the genus *Melania*. Yogore (1958) found *Brotia asperata* to be the first intermediate host in the Philippines.

Small amphibious snails of the family Hydrobiidae (see p. 270) may also be involved. *Syncera lutea* is a host of a *Paragonimus* found in rats in China, and *Pomatiopsis lapidaria* (Fig. 69*C*) is the host of *P. kellicotti* in Michigan.

In the snails the miracidia change into saclike sporocysts which produce about twelve first-generation rediae; they in turn produce a similar number of second-generation rediae. The last produce 20 or 30 fully developed cercariae. These are 175 to 240 μ long, have a small knoblike tail, spiny cuticle, a stylet, and 14 penetration glands; these cercariae appear 78 days or more after infection of the snail. The cercariae do not swim, but creep in a leechlike manner or float with the current. Those of the American species pierce the cuticle of the crayfish, which is the next host in the series, at vulnerable points and make their way invariably to the heart and pericardium where they become encysted and gradually develop into mature infective metacercariae, a process which takes 6 weeks or more. In China, Japan, and the Philippines various species of fresh-water crabs serve as second

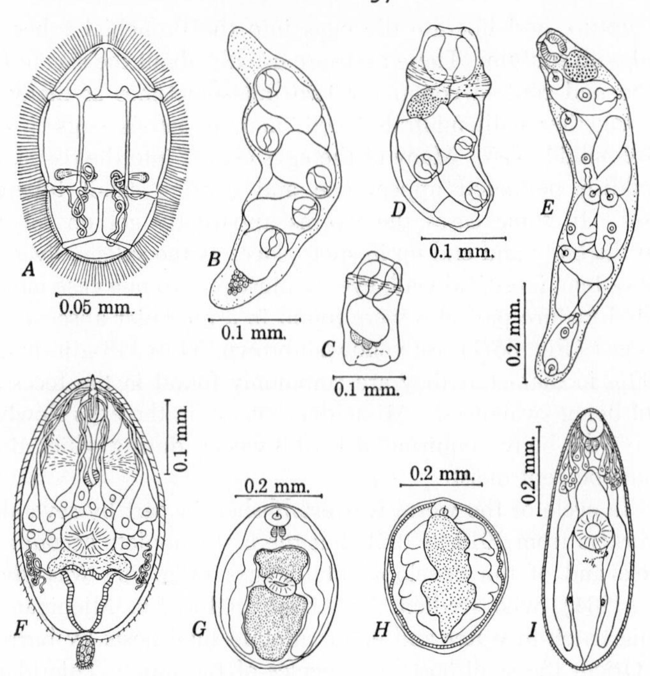

Fig. 70. Stages in life cycle of *Paragonimus kellicotti*. *A*, miracidium, showing ciliated epidermal plates and flame cells; *B*, mature sporocyst containing first-generation rediae; *C*, young first-generation redia; *D*, mature first-generation redia containing second-generation rediae; *E*, mature second-generation redia containing cercariae; *F*, microcercous cercaria; *G*, young encysted metacercaria, 5 weeks old; *H*, mature encysted cercaria; *I*, excysted cercaria, showing excystation glands and beginning of genital organs (refractile granules in large excretory bladder not shown). (After Ameel, *Am. J. Hyg.*, 19, 1934.)

intermediate hosts, and in Korea a crayfish is involved. Most of the Oriental forms do not choose the cardiac region but encyst principally in the gills and the muscles of the body and legs, and sometimes in the liver. In the Philippines Yogore et al. (1958) found metacercariae in the heart or gills of the crab host. The metacercarial cysts (Fig. 70*H*) are nearly round, 0.5 mm. or less in diameter. The enclosed spiny metacercariae lie straight, unlike most encysted forms, and are characterized by the large excretory vesicle filled with refractile granules, with large convoluted intestinal ceca on either side.

Second intermediate hosts. Crabs of the genera *Eriocheir* and *Potamon* are commonly infected in Japan. *Eriocheir japonicus* (Fig. 69*D*) has the highest incidence of infection in Japan—over 90% in some areas in late summer. This crab inhabits rice fields near the sea and small inland streams, and is extensively used as food. The species of *Potamon* are coarse-shelled crabs which abound in shallow water of mountain streams in Japan and Formosa. In the Philippines edible crabs of the genus *Parathelphusa* are intermediate hosts. Another frequently infected crab is *Sesarma*, but this is not an edible form. In the United States probably all the species of *Cambarus* serve as hosts; small sluggish streams, 20 to 30 ft. or less in width, have been found to contain the greatest numbers of infected crayfish, whereas large streams contain few, if any.

Infection usually results from eating the infected crabs or crayfish without cooking. In parts of China and the Philippines, as well as in Japan and Korea, crabs are eaten raw with salt or dunked in wine or vinegar. In some parts of Japan the crabs are not eaten raw but are crushed on a chopping block, which is subsequently used for preparation of other foods that become contaminated by the liberated metacercariae. In some localities the people drink raw juice of crabs or crayfish to reduce fever. Possibly water containing cysts liberated from the gills of dead crabs may also be a source of infection, for such cysts live for some weeks. In 1958 Huang and Chiu reported that an average of 16% of the crabs in markets in Taipei, Formosa, were infected with *Paragonimus*.

Development in final host. When the young flukes are freed from their cysts in the duodenum of their final hosts, they bore through the walls of the intestine, wander about in the abdominal cavity for some time, then go through the diaphragm to the pleural cavity, into the lungs, and finally to the bronchioles, where they remain and grow to maturity in the cysts formed by the host's lung tissue. In a normal host they may reach the pleural cavity in about 4 days and enter the lungs after about 2 weeks, but Ameel found that in white rats they may still be loitering in the abdominal cavity, bereft of ambition or purpose in life, after more than 8 months. Man is probably not the normal host for *Paragonimus;* the frequency with which the worms get lost and find themselves in abnormal localities may be correlated with this fact (see p. 459). The worms in the lungs are long-lived, persisting for at least several years. A German who had become infected in America while enjoying the delicacies provided by a Chinese cook claimed to have had symptoms of lung infection for 10 years before his trouble was diagnosed, and it was not until 13 years later that his symptoms finally disappeared.

The effects produced by *Paragonimus* infection are usually not serious, although they are suggestive of tuberculosis. The most constant symptoms are a cough, which is usually intermittent, blood-stained sputum, mild anemia, slight fever, and weariness. Only rarely are patients incapacitated for work. Positive diagnosis can be made by finding the eggs in the sputum or the feces, where they can be found in about two-thirds of the cases by the AMS III technique (see p. 253). Skin tests using extracts of powdered worms as antigen have been tested but in most cases have not proven highly reliable.

Treatment and prevention. Emetine hydrochloride together with sulfonamides is sometimes effective in treatment. Chloroquine is effective if given relatively early, but after the formation of heavy cysts the drug fails to reach the worms. Prevention of infection consists either in destruction of the snails that serve as intermediate hosts (see pp. 293–294); abstaining from the use of raw crabs or crayfish for food, or of their juices as home remedies; care not to contaminate utensils, etc., during culinary operations on the crabs; and avoidance of use of water for drinking which may possibly contain detached cysts.

Liver Flukes

Fasciolidae

The liver and bile ducts of man and domestic animals are inhabited by flukes of the families Fasciolidae, Dicrocoeliidae, and Opisthorchiidae.

The Fasciolidae include several species of the genera *Fasciola* and *Fascioloides* which are very important liver parasites of cattle, sheep, and goats; *Fasciola* is not infrequently parasitic in man. This family also includes *Fasciolopsis buski* (see p. 317), an important intestinal fluke of man and pigs. The Fasciolidae are large leaflike flukes with branched reproductive organs and usually branched ceca also, with a small coiled uterus lying entirely in front of the sex glands. The eggs are very large; the cercariae (Fig. 61*J*), which have long simple tails, encyst on water vegetation.

Fasciola hepatica is 25 to 30 mm. long, with a small anterior cone, as in other members of this genus, giving it a shouldered appearance. The general arrangement of the organs can be seen from Fig. 71. It is found in cattle, sheep, and goats in nearly all parts of the world and has been found in the livers of many other animals including marsupials, rodents, rabbits, pigs, horses, carnivores, and primates. Olsen in 1948 called attention to the importance of rabbits as reservoirs of infection. In many parts of Africa and the Orient, including Hawaii, *F. hepatica* is replaced by a similar but even larger species, *F. gigantica.* Another related form, *Fascioloides magna,* which lacks the anterior cone, is

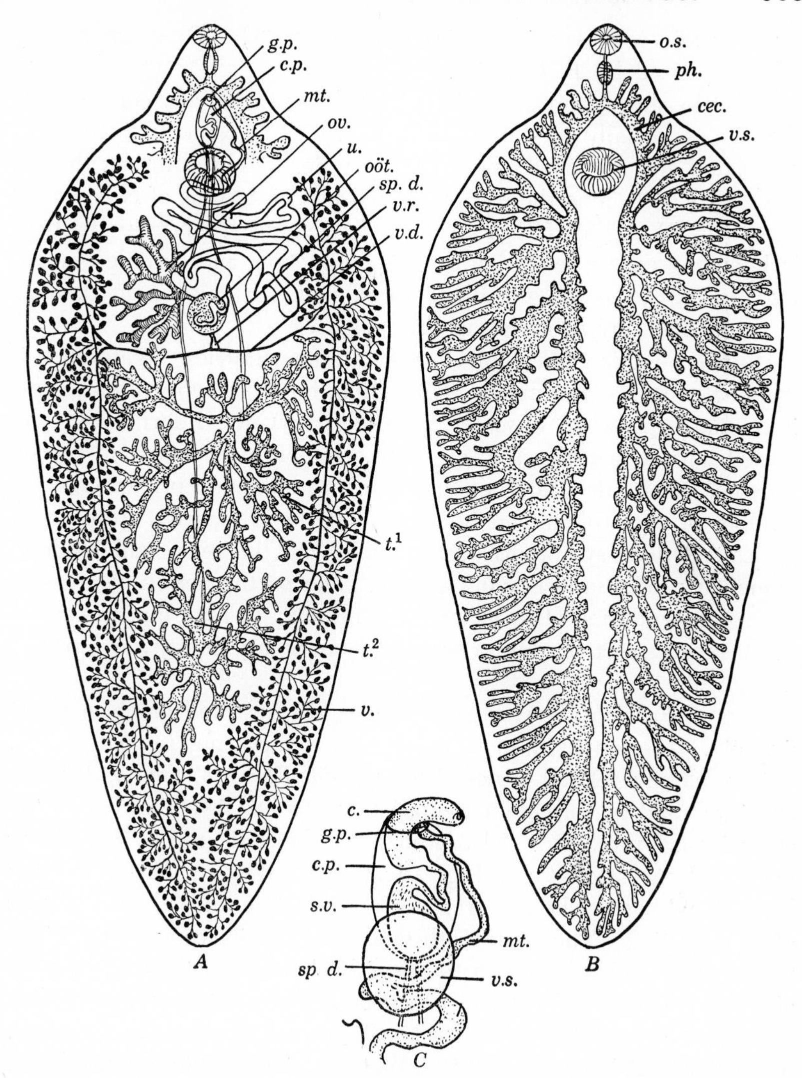

Fig. 71. *Fasciola hepatica.* *A,* showing reproductive systems only; *B,* showing digestive system only; *C,* cirrus pouch region. Abbreviations: *c.,* cirrus; *cec.,* cecum; *mt.,* metraterm; *oöt.,* ootype; *sp.d.,* sperm duct; *t.*^{1}, anterior testes; *t.*^{2}, posterior testes; *v.d.,* vitelline duct; *v.r.,* vitelline reservoir; other abbreviations as in Fig. 56. (Adapted from Leuckart's chart.)

primarily a parasite of deer in North America but also frequently infects cattle; in cattle it commonly becomes encapsulated in the liver tissue, whence its eggs fail to escape from the host. Sheep may be severely affected by this species (Swales, 1935). Deer are usually *not* parasitized by *F. hepatica.*

In cattle, sheep, and goats these liver flukes cause very considerable damage, especially in young animals, which become unthrifty and emaciated and under adverse conditions die. Olsen estimates that on the Gulf Coast alone there is an annual loss of 44 tons of condemned livers (23%) and 58 tons of meat, to say nothing of mortality, particularly among calves, reduction in milk production, and curtailed breeding. In India, according to Bhalerao, *F. gigantica* causes more damage to cattle than any bacterial or virus disease.

Human infection is not infrequent in some countries. Watercress is one of the commonest means of infection, but home-grown watercress is seldom exposed to *Fasciola* cercariae. In Cuba Kourí (1948) reported human fascioliasis to be quite common, particularly in certain provinces, in some years actually reaching epidemic proportions. A number of cases of human infection have been reported from South Africa. Serious symptoms appear, involving the liver, gall bladder, alimentary canal, and nervous system. During the period of invasion there is a syndrome of fever and eosinophilia. A fatal human case of *F. gigantica* infection, due to obstruction of biliary ducts, was reported in Hawaii.

Eggs of *Fasciola* develop after leaving the host and hatch in about 2 weeks. The miracidia develop in snails of the genus *Lymnaea* (Fig. 61) or closely related genera (*Stagnicola, Fossaria, Galba, Pseudosuccinea*) and go through a sporocyst and two redia stages before the cercariae are produced. The latter leave the snail in 5 to 6 weeks or more and encyst on water vegetation, where they remain until eaten by the final host. The cercariae are not infective until about 12 hours after encysting. The cysts withstand short periods of drying. The young flukes normally reach the liver by burrowing through into the abdominal cavity and entering from the surface, but occasionally they get into the circulation and may be distributed to abnormal locations. According to Schumacher (1939), they bore into the liver parenchyma on the second to sixth day after infection but do not enter the bile passages until the seventh or eighth week. The worms live mainly on blood. Egg production begins in about 3 months and lasts for several years. One experimentally infected sheep passed eggs for 11 years. In time the bile passages inhabited by the flukes become very thickened, often with calcified walls, and normal liver function is seriously interfered with.

Olsen in 1943 confirmed the usefulness of hexachloroethane for treatment of cattle; he administered it in a drench with bentonite and water;

at the rate of 10 grams of the drug per 100 lb. of weight he got 90% cures with no ill effects. Kourí recommends emetin as a specific treatment in man.

In the Near East *Fasciola* has been considered the cause in man of a "parasitic laryngo-pharyngitis," or "halzoun," an acute irritation of the throat from temporary attachment of worms eaten with raw food; it is said to come from eating raw livers of sacrificial animals. In 1956 Watson and Kerem found that, in Lebanon at least, halzoun is commonly caused by eating livers containing young *F. hepatica,* although it may also occasionally be caused by leeches. Witenberg believes that *Clinostomum* metacercariae may also cause halzoun. Similar attacks are common in Japan. *Clinostomum* is normally parasitic in water birds. In this country the metacercariae in fish are called "yellow grubs"; their presence ruins vast numbers of fresh-water fish, especially perch, for food. Incidentally, another temporary fluke infection was found by the senior writer to be quite common in the state of Manipur in Assam, caused by eating raw swim bladders of catfish infected with a large flat fluke, *Isoparorchis hypselobagri,* superficially resembling *Fasciolopsis.*

Dicrocoeliidae

The Dicrocoeliidae are small flat flukes, with the testes in front of the ovary and the uterus looped far posteriorly. They have small eggs and stylet cercariae which develop in land snails. The adult worms live, with rare exceptions, in the bile ducts or pancreatic ducts of birds or mammals.

Dicrocoelium dendriticum. This fluke (Fig. 72*A*) is a common liver parasite of sheep and other ruminants in many parts of the world, but particularly in Europe and Asia. Human cases are not infrequent, though often the presence of eggs in human feces is not due to infection but to ingestion of liver of infected animals. The effects are similar to those produced by *Fasciola,* but less severe.

This worm was first discovered in the United States in a cow from upper New York State in 1941; within 10 years it had become an important parasite of sheep and cattle in that area, and had established itself in woodchucks (marmots), deer, and cottontail rabbits as reservoir hosts. The life cycle was worked out by Krull and Mapes (1952, 1953) and found to involve a small land snail, *Cionella lubrica* (Fig. 72*C*) in which the long-tailed cercariae (*C. vitrina*) are continually being produced from germ masses in the daughter sporocysts. These collect in the respiratory chamber of the snail, hundreds of them being rolled together into a "slime ball," formed by secretion

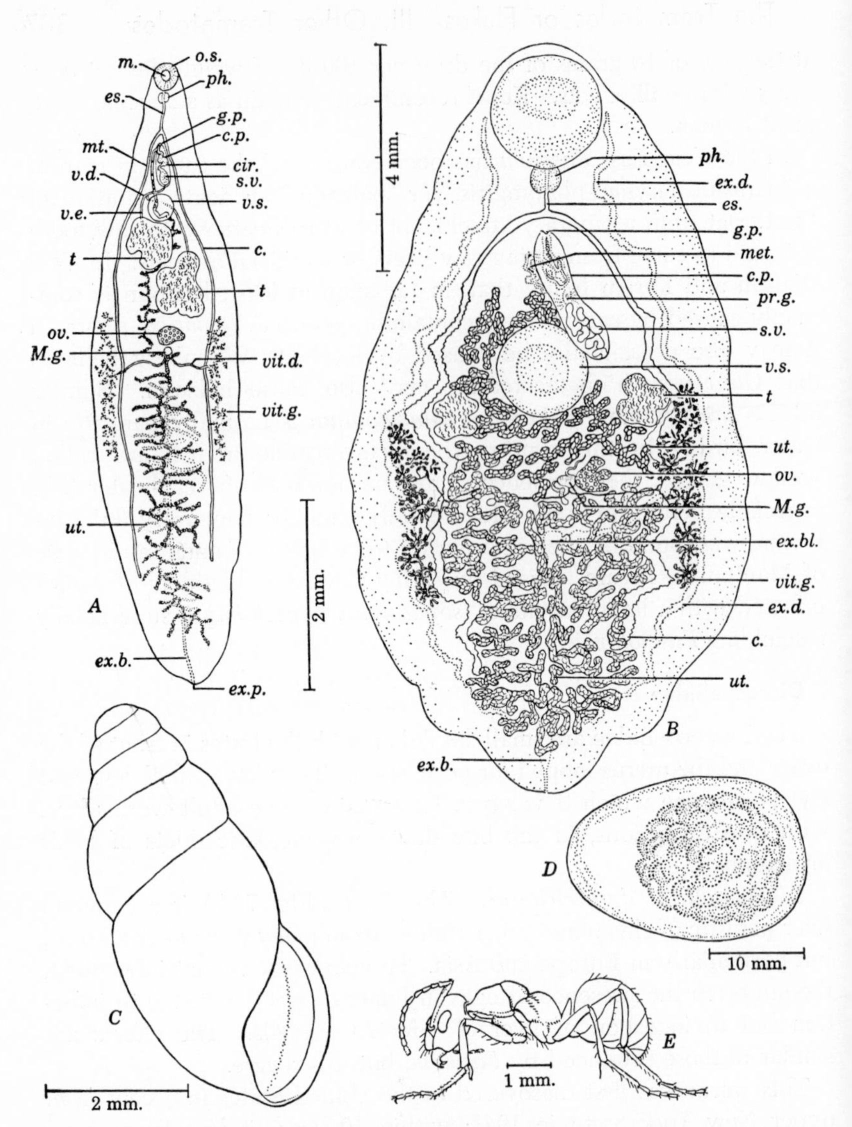

Fig. 72. *A, Dicrocoelium dendriticum. B, Eurytrema pancreaticum* (abbreviations as in Fig. 56). *C, Cionella lubrica,* first intermediate host of *D. dendriticum. D,* slime ball of same. *E, Formica fusca,* second intermediate host of same. Abbrev.: *c.,* cecum; *cir.,* cirrus; *c.p.,* cirrus pouch; *es.,* esophagus; *ex.b.,* excretory bladder; *ex.d.,* excretory duct; *ex.p.,* excretory pore; *g.p.,* genital pore; *m.,* mouth; *M.g.,* Mehlis' glands; *met.,* metraterm; *ph.,* pharynx; *pr.g.,* prostate glands; *o.s.,* oral sucker; *ov.,* ovary; *s.v.,* seminal vesicle; *t.,* testis; *ut.,* uterus; *v.e.,* vas efferens; *v.d.,* vas deferens; *vit.d.,* vitelline duct; *vit.g.,* vitelline glands; *v.s.,* ventral sucker. (*A, C,* adapted from Mapes; *D* from Krull and Mapes, *Cornell Vet.,* 1952; *B* from Looss.)

from the voluminous glands that fill the body of the cercariae, and hardened on the surface to form a sort of community cyst (Fig. 72*D*). These slime balls are dropped by the snail in its wanderings and are regarded by ants (*Formica fusca*) (Fig. 72*E*) as choice food items which they carry to their nests. Metacercariae develop in the ants and these, when eaten with vegetation, infect the definitive hosts, but the latter could not be infected by feeding them either snails or slime balls, contrary to results reported previously in Europe. Work by Vogel and Falcao in 1954 in Germany confirmed the necessary role of ants as second intermediate hosts. The adult flukes, which are 5 to 15 mm. long and only 1.5 to 2.5 mm. broad, live in the bile ducts, which Krull (1958) said they reach directly from the intestine. Krull and Mapes found up to 50,000 flukes in the liver of old sheep.

Eurytrema pancreaticum. This fluke (Fig. 72*B*) lives in the pancreatic ducts of pigs and in the biliary ducts of cattle, water buffaloes, and camels in China. Its thicker body and large oral sucker suffice to distinguish it from *Dicrocoelium.* A few human cases have been recorded from South China. This worm also develops in land snails. Tang (1950) showed that the mother sporocysts continually reproduce daughter sporocysts, but each of these produces only a few almost tailless cercariae, which all mature together and remain, as in a sac, in the thick-walled sporocysts, which are, eventually, shed by the snail. An intermediate host is probably needed. The known or suspected intermediate hosts of the few dicrocoeliids thus far worked out include ants, beetles, isopods, and lizards, but lizards are more likely to be transport hosts.

Opisthorchiidae

The flat, elongate, semitransparent flukes of this family occur in fish-eating animals, particularly in Europe and Asia, but one species, *Metorchis conjunctus,* is very common in Canada, and *Amphimerus pseudofelineus* occurs in cats in the United States. The general arrangement of the organs can be seen from Figs. 73 and 74. The eggs of these flukes are very small and contain miracidia when laid, but the miracidia do not ordinarily hatch until eaten by a suitable snail. The cercariae have long fluted tails and no stylets; they encyst in freshwater fishes, and reach their final hosts when the fish are eaten.

Clonorchis sinensis. This, the most important human parasite in the family, is widely distributed in the Far East from Korea and Japan through China to Indo-China and India. It is common in cats and dogs throughout its range, but human infection is limited to localities where raw fish is esteemed as food. Heavy human infections are common in local areas in Japan, in the vicinity of Canton and Swatow in China, and in the Red River delta in Indo-China. Stoll (1947) estimated about 19,000,000 human cases in all.

The adult flukes vary from 10 to 25 mm. in length and are 3 to 5 mm. wide, with an arrangement of organs as shown in Fig. 73*D*. The deeply branched testes distinguish this genus from the related *Opis-*

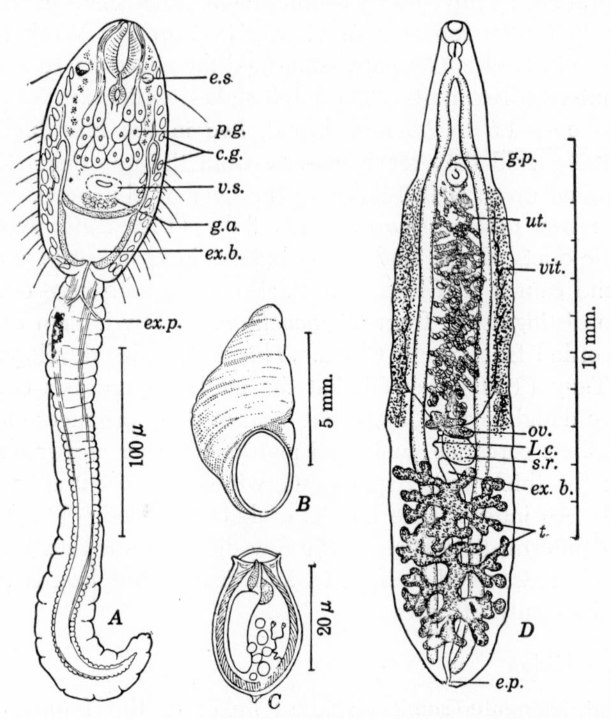

Fig. 73. *Clonorchis sinensis*: *A*, cercaria; *B*, snail host, *Parafossarulus manchouricus; C*, egg containing miracidium (side view); *D*, adult; *c.g.*, cystogenous glands; *e.s.*, eye spot; *ex.b.*, excretory bladder; *ex.p.*, excretory pore; *g.a.*, genital anlage; *p.g.*, penetration glands; other abbreviations as in Fig. 56. (*A*, after Komiya and Tajimi, *J. Shanghai Sci. Inst.*, February, 1940. *B* and *C* adapted from Faust and Khaw, *Am. J. Hyg. Monogr. Ser.*, 8, 1927.)

thorchis, in which the testes are round or lobed. The adults live both in the small biliary ducts of the liver and also in the larger bile ducts leading to the gall bladder, often in hundreds or even thousands.

LIFE CYCLE. The small yellow-brown eggs average 27 by 16 μ in size, the operculum fitting into a thickened rim of the shell like the lid on a sugar bowl (Fig. 73*C*). The miracidia hatch when eaten by small, conical, operculate snails of the subfamily Buliminae (formerly

Bythiniinae) which belong to the family Hydrobiidae (see p. 270). The most important species are *Parafossarulus manchouricus* (=*P. striatulus*) (Fig. 73*B*) and *Bulimus fuchsianus.* One of the Thiaridae (see p. 270), *Hua ningpoensis,* is also an important snail host.

The miracidia develop into rounded sporocysts which produce rediae. The rediae give birth to cercariae (Fig. 73*A*) characterized

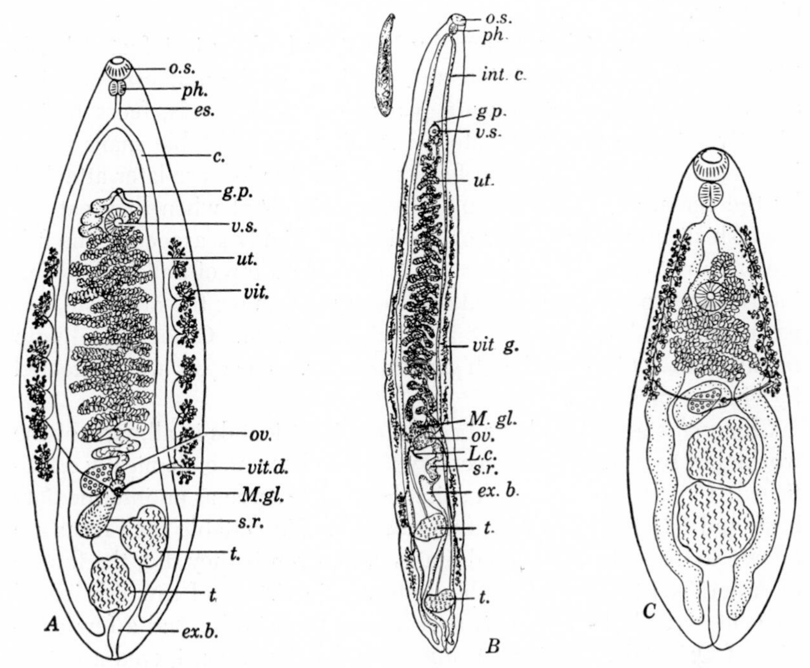

Fig. 74. *A, Opisthorchis felineus,* ×5; *B, Amphimerus pseudofelineus,* ×5; *C, Metorchis conjunctus,* ×20; abbreviations as in Fig. 56 on p. 259. (*B,* adapted from Barker.)

by a long tail with a long dorsal and shorter ventral fins, finely spined cuticle, 7 pairs of penetration glands, 14 cystogenous glands, eye spots, and masses of brownish pigment. The cercariae encyst in the flesh of fresh-water fishes and develop into metacercariae which lose the eye spots and have the saclike excretory bladder filled with coarse, refractile granules. The cysts are oval with thin walls, and average 138 by 115 μ.

Numerous species of fresh-water fish, most of them of the minnow and carp family (Cyprinidae), serve as second intermediate hosts. According to Hsü the metacercariae normally encyst in the flesh and only exceptionally under the scales or in the gills. When infected fish are eaten raw the metacercariae are liberated and enter the bile duct within a few hours after being eaten.

Migration to the liver via the bile duct is probably not the only route, since Wykoff and Lepes (1957) showed that, if the bile ducts of rabbits are tied off, some worms still make their way to the liver. It takes about 3 weeks for the flukes to reach maturity and to begin shedding eggs.

EPIDEMIOLOGY. Observations by Faust and Khaw in infected localities show how *Clonorchis* infections thrive. In the mulberry-growing areas near Canton, latrines are placed over fish ponds, feces falling directly into the water or onto night-soil rafts. Suitable snails occur in the ponds and feed on the fecal material, the fish are later attacked by the cercariae, and the people become infected when they eat the raw fish sliced with radishes or turnips and highly seasoned. The fish are often not eaten entirely raw, but are laid on top of a dish of steaming rice where they are heated sufficiently to remove the raw taste, but not enough to kill cysts in the interior of the flesh. Others merely dip the fish into hot "congee" with similar results. The cysts are unaffected by vinegar or sauces.

Clonorchis infections have been found in Orientals in all parts of the world, but two factors are necessary for it to become endemically established: (1) the presence of a suitable snail to serve as an intermediate host, and (2) the habit of eating raw fish. No suitable snail hosts are known to occur in the United States, and, even if they did, the failure of Americans to appreciate the gastronomic virtues of raw fish would prevent spread of *Clonorchis* as a *human* parasite beyond a few colonies of Orientals. *Bulimus tentaculatus* has been considered a potential intermediate host in northern United States, since a member of this genus, *B. fuchsianus,* serves as a host in the Orient. However, Wykoff was unable to infect *B. tentaculatus* under laboratory conditions.

THE DISEASE AND ITS TREATMENT AND PREVENTION. The flukes injure the epithelium of the biliary ducts, and if numerous they may seriously clog them. The walls of the ducts become thickened, and neighboring portions of the liver tissue may be involved, in severe cases leading to a general cirrhosis. Light infections may show no symptoms at all; more severe infections are accompanied by diarrhea, often with blood, edema, enlarged liver, and abdominal discomfort.

Treatment is uncertain. Some workers have obtained good results with injections of antimony compounds, but complete cures are not usually obtained. Faust and Khaw found that complete cures could be effected in early cases by gentian violet and related dyes given in the form of coated pills, and that even in cases of long standing a proportion of the worms could be reached by a sufficient concentration of the dye to kill them. Recently a fairly high degree of success has been obtained in treatment of clonorchiasis with Chloroquine, giving 0.5 gram daily for several weeks. Success with similar treatment of *Opisthorchis viverrini* infections has been reported.

Prevention would be possible by storing night soil undiluted or adding 10% of ammonium sulfate to kill the eggs before snails got access to them. Susceptibility of the fish to copper sulfate prohibits its use for snail destruction. The best preventive measure is to prohibit the sale of raw fish in public eating places and to educate people to the dangers of eating raw fish. However, it is never easy to suppress well-established tastes in food, and, besides, the cost of fuel for cooking is in some places a real economic factor.

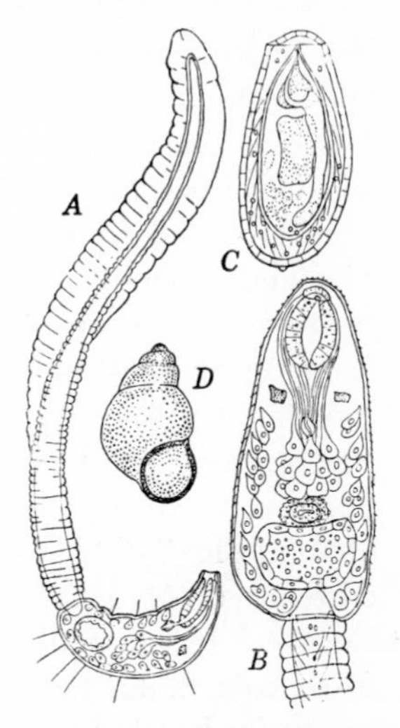

Fig. 75. *Opisthorchis felineus:* A, cercaria hanging in characteristic "pipe stem" manner, ×200; *B*, body of cercaria, ×320; *C*, egg containing miracidium, ×1000; *D*, snail host, *Bulimus tentaculatus* (= *Bythinia leachi*), ×3½. (Adapted from Vogel, *Zoologica*, 33, 1934.)

Opisthorchis spp. The genus *Opisthorchis*, differing from *Clonorchis* in having round or lobed testes (Fig. 74*A*), contains several species of flukes that are parasitic in cats and dogs and related animals, and sometimes in man. One very widespread and common species is *O. felineus*, found from central and eastern Europe to Japan; in some parts of its range it is a common human parasite. It is endemic in human populations in the region of the Dneiper River and its tributaries in the U.S.S.R. It is about 7 to 12 mm. long and 2 to 3 mm. broad, with habits similar to *Clonorchis*. Vogel (1934) found the snail host in East Prussia to be *Bulimus tentaculatus* (=*Bythinia leachi*) (Fig. 75*D*), and the principal fish host the tench. The eggs (Fig. 75*C*) are more slender than those of *Clonorchis*, averaging about 30 by 14 μ. According to Vogel, the miracidia hatch in the gut of the snail and grow into slender sporocysts, which produce numerous rediae over a period of several months; these in turn produce the cercariae. The latter (Fig. 75*A,B*) are born in an undeveloped state and finish their development in the tissues of the snails, eventually leaving the snail after several months.

After penetrating certain fish hosts, for which they show distinct preferences, the cercariae burrow into the tissues and secrete a cyst wall within 24 hours, but it appears to require about 6 weeks of ripening before the metacercariae are infective. During this time they grow to three or four times their original size. Ripe cysts measure about 300 by 200 μ, with a cyst wall about 20 μ thick. As with other members of the family, the liberated metacercariae reach the liver via the bile duct.

Other species, *O. viverrini* in southeastern Asia, *O. noverca* in India, and *O. guayaquilensis* in Ecuador, have similar habits and have also been recorded from man. *O. viverrini* is very common in parts of Thailand, where it is estimated that 1,500,000 people are infected.

Other Opisthorchiidae. The genus *Amphimerus*, distinguished from *Opisthorchis* by having a postovarian division of the yolk glands, contains a species, *A. pseudofelineus* (Fig. 74B), found in cats and coyotes in central United States. The genus *Metorchis* contains flukes that are shorter and broader than *Opisthorchis*, with a rosette-shaped uterus (Fig. 74C). Cameron (1939, 1944) reported the common occurrence of *M. conjunctus* over a wide area in Canada east of the Rockies. It is a small fluke, 1 to 6.6 mm. long, found naturally in dogs, foxes, cats, mink, and raccoons, and occasionally in man, and is injurious to fur-bearing animals. The snail host is *Amnicola limosa porata*, and the metacercariae encyst in the flesh of the common sucker, *Catostomus commersonii*, sometimes in great numbers.

The pathogenic effects, treatment, and epidemiology of these infections do not differ in any way, so far as known, from those of *Clonorchis* (see pp. 312–313).

Intestinal Flukes

The great majority of flukes inhabit the intestine of their hosts, yet there are no flukes that can be considered *primarily* parasites of the human intestine. A few species are very commonly found in man in some localities, though primarily parasitic in other animals, but the majority that have been reported from man are rather rare or accidental infections. On account of the omnivorous and variable food habits of the human being, he is subject to a wide range of such accidental parasites, including species properly belonging to both carnivorous and herbivorous hosts; probably no animal except the pig can compete with man in this respect.

We shall consider the following groups or species of intestinal flukes: (1) amphistomes (families Gastrodiscidae and Paramphistomatidae), normally parasitic in herbivores; (2) *Fasciolopsis* (family Fasciolidae), normally in pigs; (3) Heterophyidae, normally in fish-eating birds and mammals; (4) echinostomes (family Echinostomatidae), commonly parasitic in aquatic birds and mammals; and (5) a few other families which contain important intestinal parasites of lower animals, and sometimes rarely of man—the Strigeidae, Clinostomatidae, Troglotrematidae, and Plagiorchiidae.

Amphistomes

Long considered a distinct suborder, Amphistomata, this group of flukes is characterized by having the ventral sucker near the posterior end. A few are found in cold-blooded vertebrates and birds, but most of them live in the rumen or intestine of herbivorous mammals, literally carpeting considerable areas. Most of those in mammals belong to the families Gastrodiscidae, which have a large ventral disc (Fig. 76*A*), and Paramphistomatidae (Fig. 76*C* and *D*), which are superficially maggot-like in appearance.

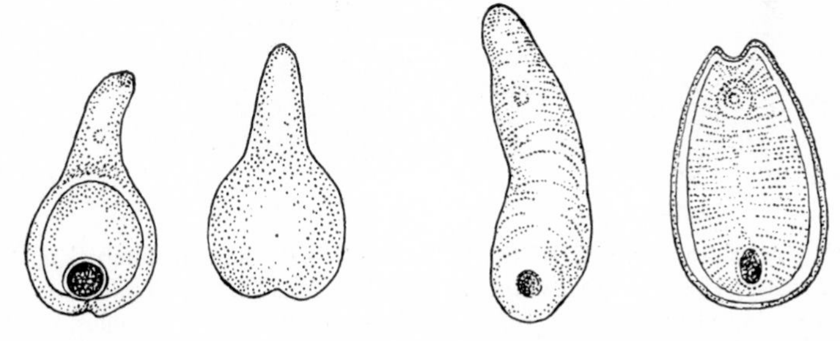

Fig. 76. Amphistome flukes: *A* and *B*, *Gastrodiscoides hominis*, ventral and dorsal views; *C*, *Paramphistomum cervi*, ventral view; *D*, *Watsonius watsoni*, ventral view, about ×4. (*D* after Stiles and Goldberger, *Hyg. Lab. Bull.*, 60, 1910.)

Gastrodiscoides hominis. This member of the Gastrodiscidae is the only amphistome found at all frequently in man; it is a common parasite of pigs in India. Buckley (1939) found it in over 40% of 221 people examined in three villages in Assam, where it is probably widely disseminated. By means of soap-water enemas he obtained nearly 1000 worms from an 8-year-old boy. Although it is present in 50% of pigs in some places in Bengal and Assam, pigs were rare in the locality visited by Buckley and could hardly have served as a reservoir. Human infections have also been reported from Annam.

The worm inhabits the cecum and large intestine of its host, where it causes some inflammation and diarrhea. The adults (Fig. 76*A* and

B), 5 to 7 mm. in length when preserved, have an orange-red appearance when living, caused by a fine network of bright red capillarylike lymphatics in the cuticle, against a flesh-colored background. The body is divided into two parts—a very active, slender, conical or fingerlike anterior portion which has the genital pore on its ventral side, and an almost hemispherical posterior portion, scooped out ventrally in a disc-like manner, with a sucker near its posterior border and a notch at the posterior end. Several closely related species in the genus *Gastrodiscus* occur in the intestines of horses and pigs in Africa.

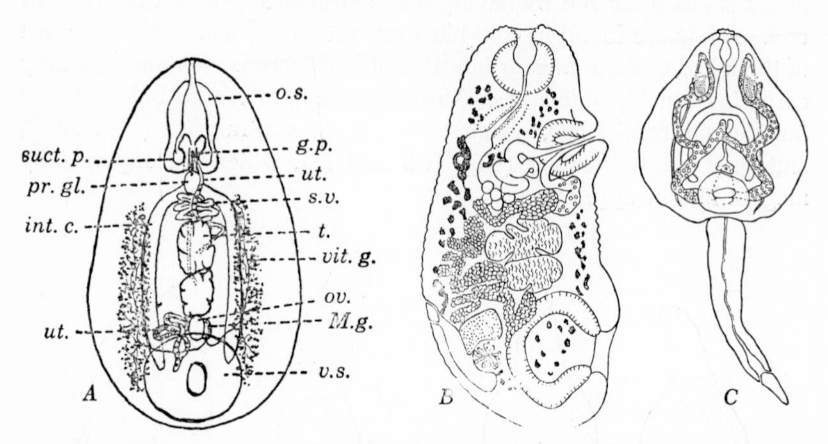

Fig. 77. Amphistomes. *A*, *Watsonius watsoni*, ×6, showing internal anatomy; *suct.p.*, suctorial pouch; other abbreviations as in Fig. 56. *B*, *Cotylophoron cotylophorum*, side view; *C*, same, cercaria (excretory ducts with granules, eye spots lateral to esophagus). (*A*, after Stiles and Goldberger, *Hyg. Lab. Bull.*, 60, 1910. *B*, after Fischoeder from Travassos, 1934. *C*, from Bennett, *Ill. Biol. Mon.*, 14, 1936.)

The eggs (Fig. 55) are very large, as are those of other amphistomes, and rather rhomboidal in shape, tapering rapidly towards each end. The miracidia develop after the eggs have escaped from their host, but nothing is known of the life cycle beyond this point. By analogy with other amphistomes, there is little doubt that the cercariae encyst on water vegetation, and that the life cycle is essentially similar to that of the Fasciolidae. *Gastrodiscoides* is not easily removed by anthelmintics but sometimes responds to soap-water enemas.

Watsonius watsoni. This fluke (Fig. 77A), the only other amphistome thus far found in man, has been recorded but once, from the small intestine of an emaciated Negro who died from severe dysentery in Nigeria; its normal hosts appear to be monkeys, in which the parasite has been found in Africa, Malaya, and Japan. The worm when living is reddish yellow; it is a thick, pear-shaped animal, about 8 to 10 mm. long, slightly concave ventrally, with a translucent gelatinous appearance. It belongs to the family Paramphistomatidae, which contains many species parasitic in the rumen of ruminants. *Paramphistomum cervi* is widespread in the Old World, but *Cotylophoron cotylophorum* (Fig. 77*B*,*C*) is the common species in southern United States. Its life cycle was found by Bennett (1936) to be very similar to that of *Fasciola*. Light amphistome infections are practically harmless, but in heavy infections cattle develop diarrhea, lose weight, and fall off in milk production.

Fasciolopsis buski

Another parasite which man shares with pigs is *Fasciolopsis buski*, a member of the family Fasciolidae (see p. 274). This is a large flat fluke (Fig. 78*K*), creamy pink in color, which reaches a length of 2 to 7.5 cm. When preserved it contracts and thickens, but fresh, relaxed specimens are rather thin and flabby. In general arrangement of organs it resembles *Fasciola*, but it has no thickened cone at the anterior end, and has unbranched intestinal ceca. It is widely distributed in pigs in southeast Asia from central China to Bengal and in many of the East Indian Islands. Stoll (1947) estimated a total of 10 million human infections, most of them in China, though there are a few endemic localities in Assam and Bengal. In some villages near Shaohsing, China, according to Barlow, 100% of the people examined were found to be infected. The eggs (Fig. 78*A*) are large and very variable in size but average about 138 by 83 μ. The miracidia require several weeks to develop after they are passed by the host. The intermediate hosts are members of the family Planorbidae, principally small, flatly coiled, aquatic snails, *Segmentina hemisphaerula* (Fig. 78*J*) and *Hippetis contori*.

In the snails the miracidia change into sporocysts, which are peculiar in possessing a saclike gut like a redia, but no pharynx. Two generations of rediae are produced, the second generation of which produce large heavy-tailed cercariae (Fig. 78*G*), measuring, with the tail, nearly 0.7 mm. in length. These begin leaving the snail after about a month. The free-swimming life is brief, occupying only time enough for the cercaria to get to the plant on which the snail is feeding. In 1 to 3 hours the cercaria has lost its tail and has encysted. The cysts are pearly white and about 200 μ in diameter. The whole development from infection of snails to encystment takes 5 to 7 weeks.

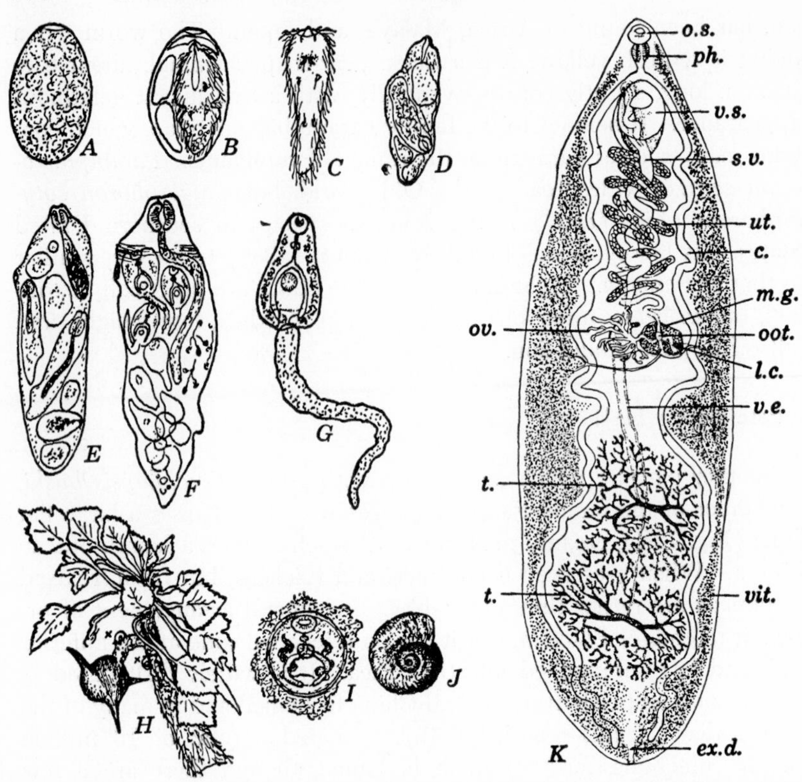

Fig. 78. Stages in life cycle of *Fasciolopsis buski.* *A*, egg as passed in feces, showing yolk balls; *B*, egg containing developed miracidium, with "mucoid plug" at anterior end and oil globules at one side; *C*, miracidium, showing eye spots; *D*, sporocyst containing developing mother rediae; *E*, mother redia containing developing daughter rediae; *F*, daughter redia containing developing cercariae; *G*, cercaria; *H*, Chinese caltrop or water ling (*Trapa natans*) with snails at points marked "X"; *I*, encysted metacercaria; *J*, *Segmentina hemisphaerula*, intermediate host; *K*, adult fluke. *A–D*, ×140; *E*, ×50; *F*, ×40; *G* and *I*, ×70; *H*, ×⅕; *J*, ×1¼; *K*, ×2. (*A–J* sketched from figures by Barlow, *Am. J. Hyg. Monogr. Ser.*, 4, 1925. *K* from Brown, *Synopsis of Medical Parasitology*, 1953.) (Abbreviations as in Fig. 56, p. 259.)

Mode of human infection. In China, human infection has been traced mainly to the eating of the nuts of a water plant known as the red caltrop or red ling (*Trapa natans*) (Fig. 78*H*), on the pods of which the cercariae encyst. These plants are extensively cultivated in ponds in the endemic areas and are fertilized by fresh night soil thrown into the water. The little snails abound in these warm stagnant pools; the plants are fairly alive with snails creeping over their stems and leaves. The nuts are eaten both fresh and dried. When fresh they are kept moist and are peeled with the teeth, during which process the cysts gain access to the mouth and are swallowed. Barlow examined nuts from typical ponds, and found a few to over 200 cysts on each nut. The senior writer (1928) traced some cases of infection in eastern Bengal to the eating of a water nut, *T. bicornis,* closely related to the Chinese nut; there is another focus of infection in India in northern Bihar. Another plant carrying infection is the so-called water chestnut, *Eliocharis tuberosa,* which has tubers like gladiolus bulbs.

Pathology. *Fasciolopsis buski* usually lives in the small intestine, where it causes local inflammation, with bleeding and formation of ulcers. Symptoms develop about 3 months after infection. There is first a period of latency during which there is some asthenia and mild anemia. This is followed by diarrhea, a marked anemia, and usually some abdominal pain. The combination of chronic diarrhea and anemia, together with a distended abdomen, edema of the legs and face, and stunted development, is characteristic of a long-standing infection. In heavy infections the continued diarrhea and edema lead to severe prostration and sometimes death.

According to Barlow, *F. buski* is easily got rid of by a number of different drugs, among which he includes oil of chenopodium, betanaphthol, thymol, and carbon tetrachloride, but some of these drugs would be too toxic for many persons weakened by this infection. Hexylresorcinol crystoids given as for *Ascaris* infections (see p. 456) give excellent results. Probably tetrachloroethylene, given as for hookworm infections, would also be effective.

Prevention consists in educating the people of endemic areas to the danger of eating fresh-water ling, water chestnuts, or other water vegetables unless they are cooked or at least dipped in boiling water. Sterilization of night soil would also be effective, but that presents a vastly more difficult problem.

Heterophyidae

The flukes of this family are extremely small, sometimes only 0.5 mm. in length, and egg-shaped; they are normally parasitic in fish-eating animals. They have the cuticle covered with minute scale-like spines. The genital pore opens into a retractile suckerlike structure which is either incorporated in the ventral sucker or lies to one side of it; Witenberg called this structure a "gonotyl." The arrangement of organs can be seen from Figs. 79 and 80. The life cycle is practically the same as that of the Opisthorchiidae, and a closely related group of snails serve as intermediate hosts.

The eggs (Fig. 79*B*) are very small, being in most species about 20 to 35 μ in length by 10 to 20 μ in diameter. They resemble the eggs of *Clonorchis,* and contain developed miracidia. Hatching occurs when the eggs are eaten by the proper species of snails. As far as known these are species of Thiaridae (see p. 270) in the Far East, but in the Middle East *Pirenella* (family Potamididae) is the host of *Heterophyes heterophyes* and other heterophyids. In the snails two generations of rediae are produced.

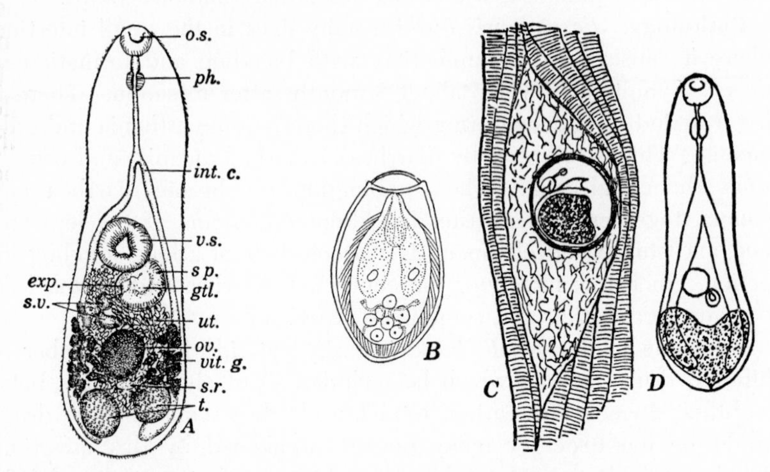

Fig. 79. *Heterophyes heterophyes.* *A,* adult fluke, ×40; *gtl.,* gonotyl; *sp.,* spines of gonotyl; other abbreviations as in Fig. 56. *B,* egg, ×900. *C,* metacercaria encysted in muscles of mullet, ×60. *D,* metacercaria freed from cyst, ×50. (*B,* after Nishigori, *Taiwan Igak. Zasshi,* 1927. *C* and *D,* after Witenberg, *Ann. Trop. Med. Parasitol.,* 23, 1929.)

The cercariae have eye spots and large tails with fluted lateral fins; they are strikingly like those of the Opisthorchiidae but have a special arrangement of spines around the mouth. After leaving the snail host the cercariae usually encyst in fishes, mullets being especially favored, but one species has been found to encyst in frogs as well. Development in the final host is very rapid, maturity being reached in 7 to 10 days.

Host-parasite relations. Numerous species of Heterophyidae have been described. They all seem remarkably versatile with respect to the hosts in which they can mature, but their behavior in abnormal hosts suggests that they feel uncomfortably out of place—in the right pew but in the wrong church, as it were. Faust and Nishigori (1926) found that certain heterophyids of night herons, when experimentally fed to mammals, gradually shifted their position farther and farther back in the gut until finally expelled. Another and more important reaction was observed by Africa, Garcia, and de Leon in 1935. They noted the tendency of various species in the Philippines, when infecting dogs and man, to become buried deep in the mucous membranes. The eggs, instead of escaping normally in the feces, are taken up by the lymphatics or blood vessels and distributed over the body. Often the worms die imprisoned in the tissues. In an American species, *Cryptocotyle lingua,* studied by Stunkard (1941), no actual invasion of the tissues was noted, although there was much tissue damage, especially in abnormal hosts.

Injury to heart and other viscera. Africa et al. (1940) showed that eggs of "foreign" species of Heterophyidae distributed over the body may cause serious injury. The most frequent damage is in the heart, where the eggs are deposited in large numbers. A dropsical condition and acute dilatation of the heart may result, producing symptoms similar to cardiac beri-beri and often fatal. The eggs were also found in the brain and spinal cord, where they are associated with grave nervous symptoms.

The species of Heterophyidae causing these conditions belong to a number of different genera, including *Heterophyes, Haplorchis* (Fig. 80, *left*), and *Diorchitrema. Haplorchis yokogawai,* measuring about 0.7 by 0.28 mm., is the species most frequently causing trouble in man in the Philippines. This species is reported as a common parasite of cats and dogs and occasionally of man. It is possible that the smaller species of Heterophyidae are the most likely to invade the mucosa.

NORMAL INTESTINAL HETEROPHYIDS OF MAN. Two species, *Metagonimus yokogawai* and *Heterophyes heterophyes,* may be regarded as normal parasites of man and other mammals, since they appear to lead an orthodox life in the lumen of the intestine and are very common human parasites in certain localities.

M. yokogawai (Fig. 80, *right*) is a common parasite of dogs and cats in Japan, Korea, China, Palestine, and the Balkans. Human infection is frequent in Japan and in eastern Siberia. Like other members of the family, this tiny fluke is not very particular about its final host, for it infects not only carnivores, pigs, and man, but also pelicans and, experimentally, mice.

The adult worms live in the duodenum, sometimes by thousands. They are only 1 to 2.5 mm. in length and about 0.5 mm. broad. A characteristic feature is the displacement of the ventral sucker to the right side of the body, with the genital opening in a pit at the anterior border of it. The eggs are about 28 to 30 μ by 16 to 17 μ. The snail intermediate hosts are species of Thiaridae (see p. 270). The cercariae attack fresh-water fishes, particularly a species of trout, *Plecoglossus altivelis*, and infection of the final host occurs when the uncooked fish are eaten. The cysts are discoidal and found principally in pockets under the scales.

H. heterophyes (Fig. 79A) is also a very small fluke; relaxed specimens in dogs measure up to 2.7 mm. by 0.9 mm., but in cats they are

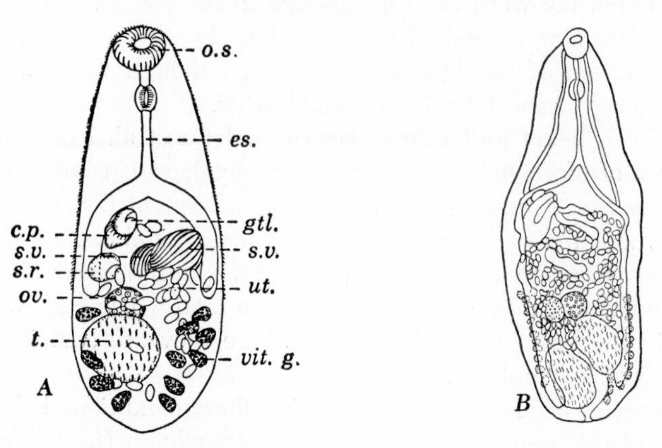

Fig. 80. A, a species of *Haplorchis,* about ×100; *gtl.,* gonotyl; other abbreviations as in Fig. 56. (After Africa and Garcia, *Philip. J. Sci.,* 53, 1935.) *B, Metagonimus yokogawai.* (Adapted from Mönnig, *Veterinary Helminthology and Entomology,* Williams and Wilkins, 1949.)

only about 1.3 by 0.3 mm. They have the ventral sucker on the median line, with a separate spiny genital sucker to the right of it. These flukes are common in cats, dogs, and allied animals in Egypt, Palestine, Yemen, India, and the Far East. In 1956 Wells and Randall found adult worms in kites, domestic cats and dogs, wild cats, foxes, jackals, and house rats in Egypt. Human infections are common both in the Far East and in Egypt and Palestine.

Khalil in 1933 found a common marine and brackish water operculated snail, *Pirenella conica,* to be the snail host in Egypt. The cercariae encyst under the scales and in the flesh of mullets, especially *Mugil cephalus;* in one mullet from the fish market in Jerusalem, Witenberg found over 1000 cysts per gram of flesh. Wells and Randall recently reported metacercariae of *H. heterophyes* in all of 7 species of fishes living in brackish lakes in Egypt. The round cysts (Fig. 79C) lie in spindle-shaped masses of fat globules and measure 0.13 to 0.26 mm. in diameter. The metacercariae, lying folded inside, have the anterior part of the body flattened.

Pathology. In infections with the normal human species the symptoms are usually negligible, though in heavy infections there may be mild digestive disturbances and diarrhea.

Like other intestinal flukes, these species are susceptible to the group of anthelmintics used for nematodes, but their small size and ability to hide away between the villi make treatment unsatisfactory unless the intestine is thoroughly cleaned of contents and mucus beforehand. Prevention consists in eschewing raw infected fish.

Nanophyetus salmincola

These small flukes (Fig. 81), 1 mm. or less in length, belong to the same family as *Paragonimus*, Troglotrematidae. They are common parasites of fish-eating mammals in northwestern United States, the

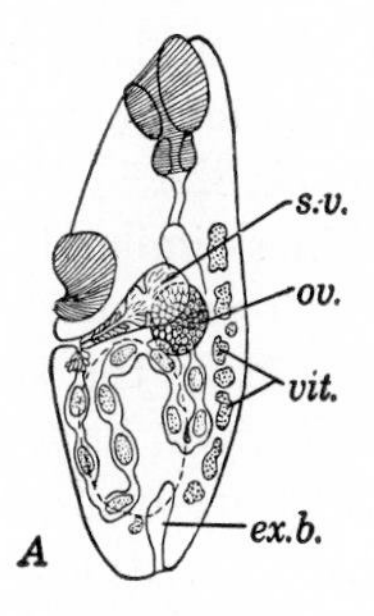

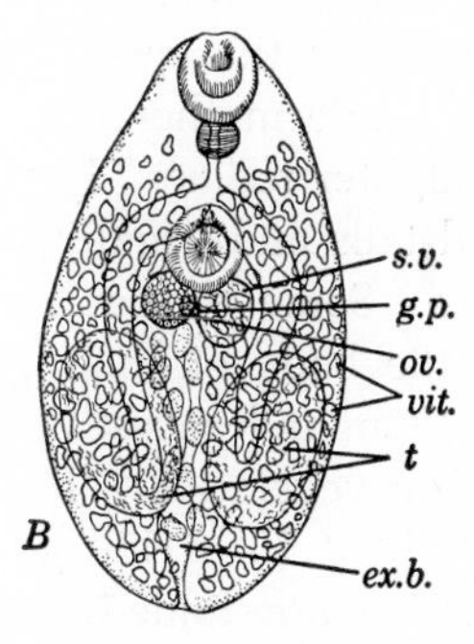

Fig. 81. *Nanophyetus salmincola.* *A*, lateral view; *B*, ventral view; abbreviations as in Fig. 56. (Adapted from Witenberg, *J. Parasitol.*, 18, 1932.)

metacercariae encysting in salmon. The snail host in Oregon, according to Donham, Simms, and Shaw (1932), is *Goniobasis silicula*, a common species in running water. The cercariae resemble those of *Paragonimus.* Dogs and foxes are the best definitive hosts. Human infection has been reported from eastern Siberia.

This parasite is of particular interest because it is associated with a highly fatal disease of dogs called "salmon poisoning." Simms et al. (1932) showed that the disease was caused by an infectious organism transmitted by the metacercariae of *Nanophyetus salmincola* when eaten with the flesh of salmon. It was at first thought to be a virus but was later shown to be a rickettsialike organism, *Neorickettsia helmintheca* (see p. 217). After an incubation period of a week or more there are loss of appetite, fever, and sensory depression, followed by edema, violent vomiting, and dysentery. If diagnosed within 3 hours of onset, 2 to 6 mg. of apomorphine by mouth is protective.

Animals that recover become immune. As yet this disease has not been observed in man.

Echinostomes

The family Echinostomatidae includes numerous species of flukes parasitic in many kinds of vertebrates, particularly aquatic birds. Most species are characterized by a spiny body and a collar of spines near the anterior end. Most of them, like the Heterophyidae, are remarkably promiscuous as to their final hosts, and many are not very particular about their snail hosts either.

The eggs are large, usually over 100 μ long, and contain partly developed embryos when laid; the miracidia have a median eye spot and develop in water. In their snail hosts, usually planorbids, Johnston in 1920 believed the miracidia to develop directly into rediae, omitting the sporocyst stage, but actually, at least in some species, a single mother redia develops in the miracidium and gives rise to daughter rediae. The cercariae (Figs. 60*D*, 82*A* ^{2}) have well-developed tails and usually bear a collar of spines similar to that of the adults. Some species encyst directly in their snail hosts, sometimes in the body of their parent redia; others leave the snail that spawned them and encyst in other snails or in bivalves, insects, frogs, fishes, or on vegetation. The metacercarial cysts are oval or round, and only about 70 to 150 μ in diameter (Fig. 82*A* ^{3}); the contained metacercariae are folded and show two branches of the excretory bladder filled with coarse granules; the collar of spines can be seen on careful examination.

A number of species of echinostomes have been recorded from man, but most are rare and purely accidental parasites.

Echinostoma ilocanum is common in the Ilocanos of the Philippines and was found by Sandground in Java. It is 2.5 to 10 mm. long and 0.5 to 1.5 mm. broad, with 51 collar spines. It is primarily a parasite of field rats, but Chen in 1934 found it common in dogs in Canton. The cercariae of this species after leaving the small planorbid snail, *Gyraulus prashadi,* commonly encyst in a large operculated snail, *Pila luzonica,* which the Ilocanos enjoy eating raw.

E. malayanum, a broader fluke (5 to 10 mm. by 2 to 3.5 mm.) with 43 collar spines, another Far Eastern species, is common in certain tribes who live on the Sino-Tibetan frontier and has also been reported from Malaya and Sumatra.

In central Celebes, Sandground and Bonne (1940) found a high incidence of infection with another echinostome, *E. lindoense* (Fig. 82*A*, *A* ^{1}), which is larger (13 to 16 mm. by 2 to 2.5 mm.) with only

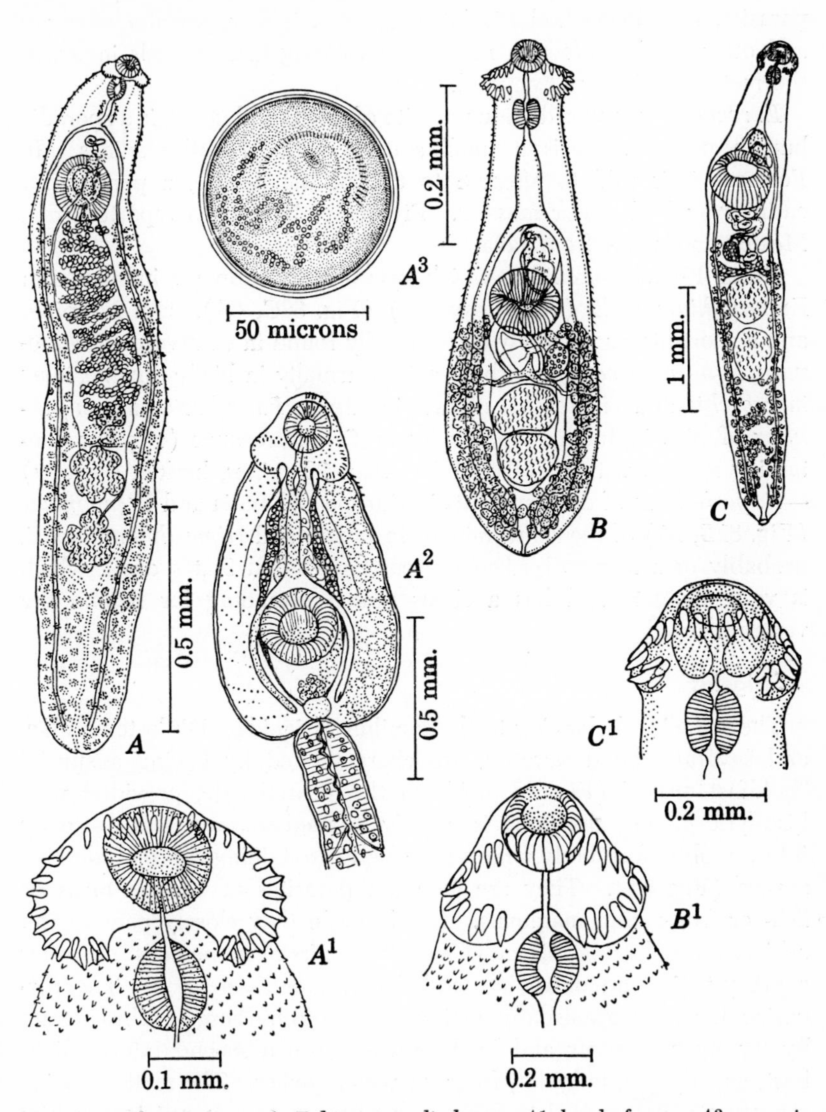

Fig. 82. Echinostomes. *A, Echinostoma lindoense;* A^1, head of same; A^2, cercaria of same. (After Sandground and Bonne, *Am. J. Trop. Med.*, 20, 1940.) A^3, encysted metacercaria of *E. ilocanum.* (After Tubangui and Pasco, *Philip. J. Sci.*, 51, 1933.) *B, Echinochasmus japonicus;* B^1, head of same. (After Yamaguti.) *C, Euparyphium melis;* C^1, head of same. (After Beaver, *J. Parasitol.*, 27, 1941.)

37 collar spines. This was at first thought to be a primary human parasite, but Bonne and Lie later reported it as a parasite of ducks and other fowl. Infection results from eating lake mussels in which the metacercariae are encysted.

E. revolutum, a world-wide parasite of ducks and geese, is a sporadic human parasite. It is a small species with 37 collar spines. In Formosa it is said to affect 3 to 6% of the people, a penalty for eating raw fresh-water mussels. Five cases have been reported from Mexico and a few from Java.

Other Echinostomatidae which occasionally crop up in man are: *Euparyphium melis* (=*E. jassyense*) (Fig. 82*C*, *C*[1]), with 27 spines and a short uterus with few eggs, usually found in carnivores; *Echinoparyphium recurvatum*, with 45 spines, usually in birds; *Echinostoma macrorchis* and *cinetorchis* of rats in Japan; *Paryphostomum sufrartyfex* of pigs in India; two species of *Echinochasmus* (a genus containing species with a collar of less than 20 spines, broken dorsally) —*E. perfoliatus* of dogs and cats in Europe and India and *E. japonicus* (Fig. 82*B*, *B*[1]) of the same animals in Japan; and *Himasthla muhlensi*, probably of a marine bird of the eastern United States coast, which is very elongate and has a cirrus pouch extending far behind the ventral sucker.

Strigeids

The strigeids, belonging to the families Strigeidae, Diplostomatidae, and several related families, are characterized by having a special "holdfast organ" (Fig. 83, *h.f.*) on the ventral side, provided with histolytic glands; the body is usually divided more or less distinctly into a mobile fore body, and a hind body containing the reproductive organs (Fig. 83). They are common parasites of aquatic birds or fish- or frog-eating mammals. In life cycle they closely parallel the schistosomes, having miracidia with two pairs of flame cells, daughter sporocysts instead of rediae, and forked-tailed cercariae. The cercariae, however, are usually distinguishable from those of schistosomes by having a pharynx and by burrowing into a second intermediate host, usually fish, tadpoles, frogs, or water snakes. The metacercariae are often very harmful to fish, since some species encyst in the lens or chambers of the eye, in the spinal cord, or around the heart; some encyst in the skin or muscles, causing "black spot." A typical life cycle has been graphically illustrated by the Hunters (1935).

These parasites when numerous may be very injurious in the intestines of their final hosts. Fortunately, man is rarely parasitized by them, but Nasr (1941) called attention to human infection with an

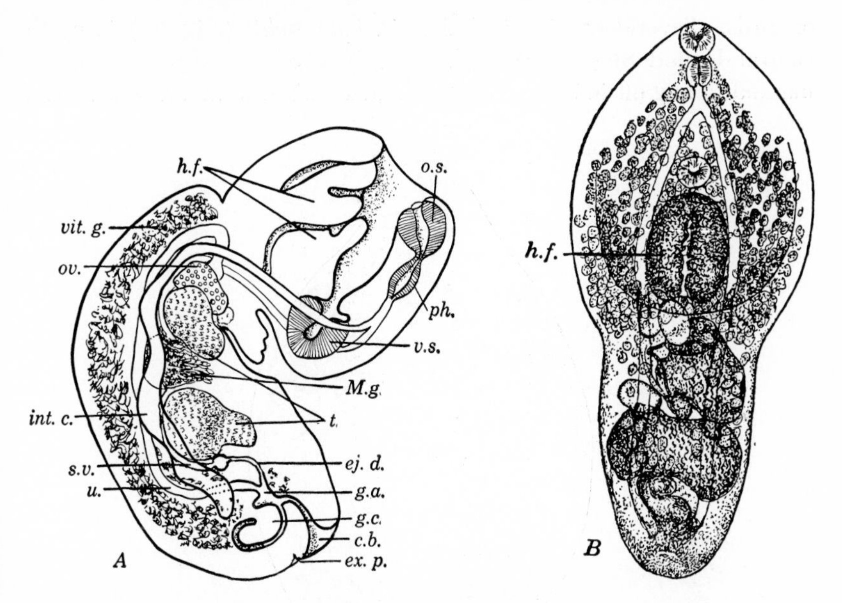

Fig. 83. Two types of strigeids: *A*, *Cotylurus flabelliformis* of ducks, example of Strigeidae; *B*, *Fibricola texensis* of raccoons, example of Diplostomatidae. Note pouchlike character of forebody in *A*, and holdfast organ (*h.f.*) in form of anterior and posterior transverse lips, whereas in *B* the forebody is spatulate and the holdfast organ oval; *c.b.*, copulatory bursa; *ej.d.*, ejaculatory duct; *g.a.*, genital atrium; *g.c.*, genital cone; *h.f.*, holdfast organ. Other abbreviations same as in Fig. 56. (*A*, adapted from Van Haitsma, *Papers Mich. Acad. Sci., Arts, Letters,* 13, 1930. *B,* from Chandler, *Trans. Am. Micr. Soc.*, 61, 1942.)

Egyptian species, *Prohemistomum vivax,* properly a parasite of kites, but also extremely common in dogs and cats which eat, or are fed, raw Nile fishes or tadpoles. One man with 2000 specimens complained of dysenteric symptoms.

Plagiorchiidae

This and a number of closely related families include numerous parasites of various insect-eating vertebrates, especially cold-blooded vertebrates. All have stylet cercariae which encyst in arthropods or vertebrates. Since even a single species, *Plagiorchis muris,* will develop in such a variety of hosts as pigeons, shorebirds, muskrats, mice, and men, it is not surprising that several species of this genus have been reported from man in various parts of the world.

The only important species in domestic animals are members of the genus *Prosthogonimus,* which inhabit the oviduct and bursa fabricii of birds. Several species, including *P. macrorchis* (Fig. 84) in north central United States, are important parasites of poultry, causing a marked falling off in egg production and sometimes fatal disease. The

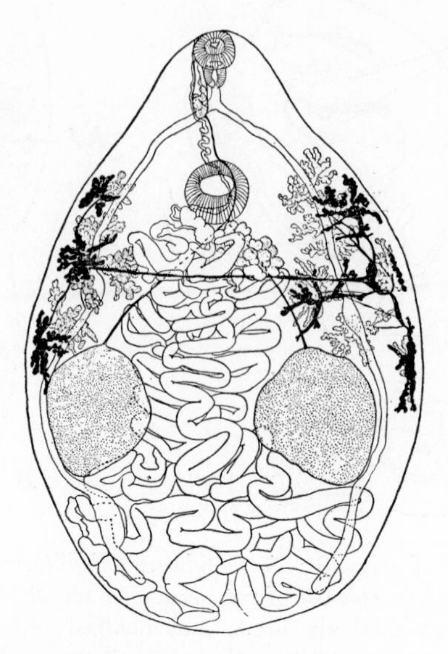

Fig. 84. *Prosthogonimus macrorchis,* oviduct fluke of poultry. ×18. (After Macy, *Univ. Minn. Agric. Exp. Sta. Bull.,* 98, 1934.)

cercariae of these flukes, after developing in snails (*Amnicola*), encyst in dragonfly nymphs or other aquatic insect larvae. Birds become infected by eating either nymphs or adults of dragonflies.

REFERENCES

Paragonimus

Ameel, D. J. 1934. *Paragonimus,* its life history and distribution in North America and its taxonomy. *Am. J. Hyg.,* 19: 279–317.

Bercovitz, Z. 1937. Clinical studies on human lung fluke disease. *Am. J. Trop. Med.,* 17: 101–122.

Fain, A., and Vandepitte, J. 1957. Description du nouveau distome vivant dans des kystes ou abces retroauriculaires chez l'homme au Congo belge. *Ann. Soc. Belge de Med. Trop.* 37: 251–258, 309–315.

Kobayashi, H. 1921. Studies on the lung fluke in Korea. *Mitth. a.d. Med. Hochsch. Z. Keino,* 4: 5–16.

1925. On the development of *Paragonimus westermanni* and its prevention. *Trans. Far Eastern Assoc. Trop. Med.,* 6th Bienn. Congr., 1: 413–417.

Komiya, Y., et al. 1952. Studies on Paragonimiasis in Shizuoka Prefecture, 1 and 2. *Jap. J. Med. Sci. Biol.*, 5: 341–350, 433–445.

La Rue, G. R., and Ameel, D. J. 1937. The distribution of *Paragonimus. J. Parasitol.*, 23: 382–388.

Roque, F. T., Ludwick, R. W., and Bell, J. C. 1953. Pulmonary paragonimiasis: a review with case reports from Korea and the Phillipines. *Ann. Internatl. Med.*, 38: 1206–1221.

Vogel, H., Wu, K., and Watt, J. Y. C. 1935. Preliminary report on the life history of *Paragonimus* in China. *Trans. Far Eastern Assoc. Trop. Med.*, 9th Bienn. Congr., 1: 509–517.

Wu, K. 1938. The epidemiology of paragonimiasis in China. *Far Eastern Assoc. Trop. Med., c.r. dix. congrés.* Hanoi.

Yogore, M. G., Jr. 1958. Studies on paragonimiasis. I. The molluscan and crustacean hosts of Paragonimus in the Philippines. *Philipp. J. Sci.*, 86: 37–45.

Yogore, M. G., Jr., Noble, G. A., and Cabrera, B. D. 1958. Studies on paragonimiasis. II. The morphology of some of the larval stages of Paragonimus in the Philippines. *Philipp. J. Sci.*, 86: 47–69.

Fasciola and Fascioloides

Kourí. P. 1948. Diagnostico, epidemiologia y profilaxis de la fascioliasis hepatica humana en Cuba, I. *Bol. ofic. del. col. med. vet. nat.* Nos. 3 and 4; II and III, *Rev. Kuba de med trop. parasitol.*, 4: Nos. 3, 4, and 5.

Olsen, O. W. 1947. Hexachlorethane-bentonite suspension for controlling the common liver fluke, *Fasciola hepatica,* in cattle in the Gulf Coast region of Texas. *Am. J. Vet. Research,* 8: 353–366.

Sadun, E. H., and Maiphoom, C. 1953. Studies on the epidemiology of the human intestinal fluke, *Fasciolopsis buski* (Lankester) in Central Thailand. *Am. J. Trop. Med. Hyg.*, 2: 1070–1084.

Shaw, J. W., and Simms, B. T. 1930. Studies in fascioliasis in Oregon sheep and goats. *Oregon State Agr. Exp. Sta. Bull.*, 226.

Schumacher, W. 1938. Untersuchungen über den Wanderungsweg and die Entwicklung von *Fasciola hepatica* in Endwirt. *Z. Parasitenk.*, 10: 608–643.

Swales, W. E. 1935. The life cycle of *Fascioloides magna. Can. J. Research,* 12: 177–215.

Dicrocoelium

Krull, W. H., and Mapes, C. R. 1952–53. Studies on the biology of *Dicrocoelium dendriticum* (Rudolphi, 1819) Looss, 1899 (Trematoda: Dicrocoeliidae) including its relation to the intermediate host, *Cionella lubrica* (Muller). I–IX. *Cornell Vet.* 41: 382–432, 433–444; 42: 253–276, 277–285; 339–351, 464–489, 603; 43: 199–202, 389–410.

Neuhaus, W. 1938. Der Invasionsweg der Lanzettegelcercariae bei der Infektion des Endwirtes und ihre Entwicklung zum *Dicrocoelium lanceatum. Z. Parasitenk.*, 10: 476–512.

Tang, C. C. 1950. Studies on the life history of *Eurytrema pancreaticum* Janson, 1889. *J. Parasitol.*, 36: 559–573.

Opisthorchiidae

Cameron, T. W. M. 1944. The morphology, taxonomy, and life history of *Metorchis conjunctus* (Cobbold, 1860). *Can. J. Research,* D, 22: 6–16.

Faust, E. C., and Khaw, O. K. 1927. Studies on *Clonorchis sinensis* (Cobbold). *Am. J. Hyg. Monogr. Ser.* 8.

Hsü, H. F., et al. 1936–40. Studies on certain problems of *Clonorchis sinensis*, I–IV. *Chinese Med. J.*, 50: 1609–1620; 51: 341–356. *Suppl.* II: 385–400; III: 244–254; Festsch. Nocht.: 216–220.

Sadun, E. H. 1955. Studies on *Opisthorchis viverrini* in Thailand. *Am. J. Hyg.*, 62: 81–115.

Vogel, H. 1934. Der Entwicklungszyklus von *Opisthorchis felineus* (Riv.) nebst Bemerkungen über die Systematik und Epidemiologie. *Zoologica*, 33: 86–103.

Wykoff, D. E., and Lepes, T. J. 1957. Studies on *Clonorchis sinensis*. I. Observations on the route of migration in the definitive host. *Am. J. Trop. Med. Hyg.*, 6: 1061–1065.

Amphistomes

Bennett, H. J. 1936. The life history of *Cotylophoron cotylophoron*, a trematode from ruminants. *Illinois Biol. Monogr.* (*n.s.*) 14: No. 4.

Buckley, J. J. C. 1939. Observations on *Gastrodiscoides hominis* and *Fasciolopsis* in Assam. *J. Helminthol.*, 17: 1–12.

Leiper, R. T. 1913. Observations on certain helminths of man. *Trans. Roy. Soc. Trop. Med. Hyg.*, 6: 265–297.

Stiles, C. W., and Goldberger, J. 1910. A study of the anatomy of *Watsonius* (n.g.) *watsoni* of man. *Hyg. Lab. Bull.* 60.

Fasciolopsis

Barlow, C. H. 1925. The life cycle of the human intestinal fluke, *Fasciolopsis buski* (Lankester). *Am. J. Hyg. Monogr. Ser.* 4.

McCoy, O. R., and Chu, T. C. 1937. *Fasciolopsis buski* infection among school children in Shaohsing, and treatment with hexylresorcinol. *Chinese Med. J.*, 51: 937–944.

Vogel, H. 1936. Beobachtungen über *Fasciolopsis* Infektion. *Arch. Schiffs- u. Tropen-Hyg.*, 40: 181–187.

Wu, K. 1937. Deux nouvelles plantes pouvant transmettre le *Fasciolopsis buski*. Revue générale. *Ann. parasitol. humaine et comparée*, 15: 458–464.

Heterophyidae

Africa, C. M., and Garcia, E. Y. 1935. Heterophyid trematodes of man and dog in the Philippines, with descriptions of three new species. *Philippine J. Sci.*, 57: 253–267.

Africa, C. M., de Leon, W., and Garcia, E. Y. 1940. Visceral complications. in intestinal heterophydiasis of man. *Acta Med. Philippina Monogr. Ser.* I.

Chen, H. T. 1936. A study of the Haplorchinae. *Parasitology*, 28: 40–55.

Faust, E. C., and Nishigori, M. 1926. Life cycle of two new species of Heterophyidae, parasitic in mammals and birds. *J. Parasitol.*, 13: 91–128.

Stunkard, H. W. 1941. Pathology and immunity to infection with heterophyid trematodes. *Collecting Net* 16: No. 4.

Nanophyetus salmincola

Cordy, D. R., and Gorham, J. R. 1950. The pathology and etiology of salmon disease in the dog and fox. *Am. J. Pathol.*, 26: 617–637.

Simms, B. T., Donham, C. R., and Shaw, J. N. 1931. Salmon Poisoning. *Am. J. Hyg.*, 13: 363–391.

Simms, B. T., McCapes, A. M., and Muth, O. H. 1932. Salmon poisoning: transmission and immunization experiments. *J. Am. Vet. Med. Assoc.*, 81: 26–36.

Witenberg, G. 1932. On the anatomy and systematic position of the causative agent of so-called salmon poisoning. *J. Parasitol.*, 18: 258–263.

Echinostomes

Beaver, P. C. 1937. Experimental studies on *Echinostoma revolutum* (Froelich), a fluke from birds and mammals. *Illinois Biol. Monogr.* 15.

Sandground, J. H., and Bonne, C. 1940. *Echinostoma lindoensis,* n.sp., a new parasite of man in the Celebes with an account of its lite history and epidemiology. *Am. J. Trop. Med.,* 20: 511–535.

Tubangui, M. A., and Pasco, A. M. 1933. The life history of the human intestinal fluke *Euparyphium ilocanum* (Garrison, 1908), *Philippine J. Sci.,* 51: 581–606.

Strigeids

Dubois, G. 1938. Monographie des Strigeida (Trematoda). *Mém. soc. neuchateloise sci. nat.,* 6: 535 pp.

Hunter, G. W., III, and Hunter, W. S. 1935. Further studies on fish and bird parasites. *Suppl. 24th Ann. Rep. N. Y. State Conserv. Dept.* 1934, No. IX, Rep. of Biol. Surv. Mohawk-Hudson Watershed.

Nasr, M. 1941. The occurrence of *Prohemistomum vivax* infection in man, with a redescription of the parasite. *Lab. and Med. Progress,* 2: 135–149.

Prosthogonimus

Macy, R. W. 1934. Studies on the taxonomy, morphology, and biology of *Prosthogonimus macrorchis* Macy, a common oviduct fluke of domestic fowls in North America. *Univ. Minnesota Agr. Exp. Sta. Tech. Bull.* 98.

Chapter 15

THE CESTOIDEA OR TAPEWORMS

General structure. Except in a few primitive species a mature tapeworm resembles a whole family of animals, consisting sometimes of many hundreds of individuals one behind the other like links of a chain (Fig. 85). The most striking feature is the complete lack of a digestive tract in all stages of development. Larval forms obviously absorb food from the host's tissues through their exposed surfaces, but it has usually been assumed that adult tapeworms in the intestine subsist by absorbing digested but unassimilated foods from the fluid intestinal contents in which they live. It has been shown that this is true for carbohydrates (Chandler, 1943; Read, 1959). However, most other food essentials are obtained from the host, presumably from the secretions. When many worms are present, the crowding forces the worms to compete for food and causes stunting of their growth. Tapeworms are dependent on some constituent of yeast in the host's diet (Addis and Chandler, 1944; Beck, 1951) and also on the level of sex hormones in the host (Addis, 1946; Beck, 1951).

In the subclass Cestodaria no chain of segments is formed, and there is only one set of reproductive organs; this is true also of one order, Caryophyllidea, allied to Pseudophyllidea, but all other tapeworms consist of chains of segments with a "head" or scolex for attachment at one end. Just behind the scolex is a narrow region or "neck" which continually grows and, as it does so, forms partitions, thus constantly budding off new segments. The segments, however, remain connected internally by the musculature and also by nerve trunks and excretory tubes. As the newly formed segments push back the segments previously formed, there is produced a chain of segments called a strobila, each segment being known as a proglottid. The proglottids just behind the neck are the youngest; they are at first indistinct and have no differentiation of internal organs. As they are pushed farther and farther from the scolex, the organs progressively develop, so that it is possible to find in a single tapeworm a complete developmental

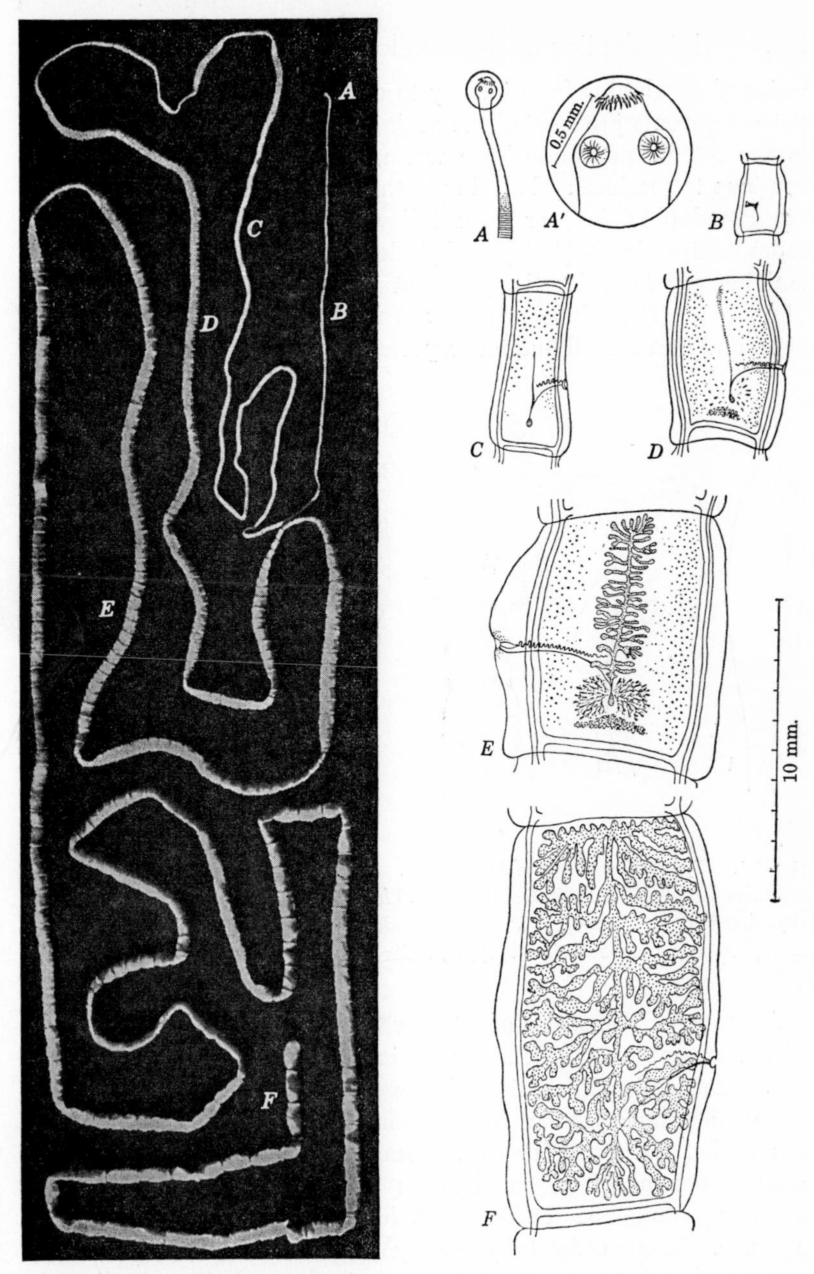

Fig. 85. *Taenia solium,* the pork tapeworm. *Left,* whole worm, about ⅓ natural size. *Right,* enlarged parts of worms from regions indicated on whole worm, showing progressive development of proglottids. In *C* the testes are just beginning to appear; in *D* the male system is fully functional, but the female system is immature; in *E* both systems are fully mature and functional; and in *F* the ripe uterus has usurped the whole segment, only the vagina and sperm duct being still recognizable.

series of proglottids from infancy to old age. The young undifferentiated segments just behind the neck gradually attain sexual maturity in the middle portions of the worm, and then the segments either continue to produce and shed eggs throughout the rest of the strobila (in Pseudophyllidea) or there follows a gradual decadence of the reproductive glands (in Cyclophyllidea) as the segments "go to seed" and become filled by the pregnant uterus with its hordes of eggs (Fig. 85). The whole process can be likened to the development of an undifferentiated bud into a perfect flower and then a seed pod.

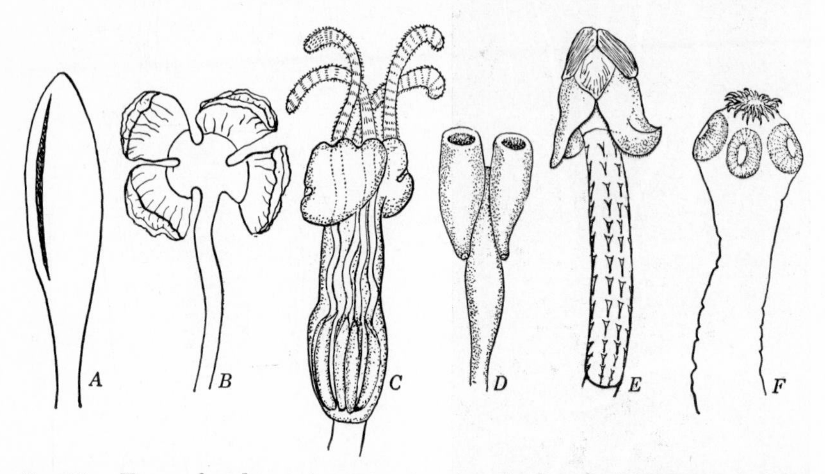

Fig. 86. Types of scoleces of tapeworms. *A,* order Pseudophyllidea (*Dibothriocephalus latus*); *B,* order Tetraphyllidea (*Phyllobothrium* sp.); *C,* order Trypanorhyncha (*Otobothrium* sp.); *D,* order Pseudophyllidea (*Bothridium* sp.); *E,* order Diphyllidea (*Echinobothrium* sp.); *F,* order Cyclophyllidea (*Taenia solium*).

Anatomy. The scolex of a tapeworm serves primarily as an organ of attachment, though it also contains what little brain a tapeworm has. Considering the entire subclass Cestoda, the variety of holdfast organs developed by the scolex is remarkable (Fig. 86), consisting of groovelike, in-cupped, or earlike suckers, and in addition, in some species, crowns of powerful hooks or rows of spines on a fleshy anterior protuberance called a rostellum, in some forms retractile into a pouch. In one order (Trypanorhyncha) there are long, protrusible, spiny proboscides retractile into canals in the neck (Fig. 86*C*). The scoleces of the tapeworms infesting mammals, however, are comparatively monotonous in form.

Tapeworms are covered with a cuticle which recent studies with the electron microscope have shown to be very complex in structure

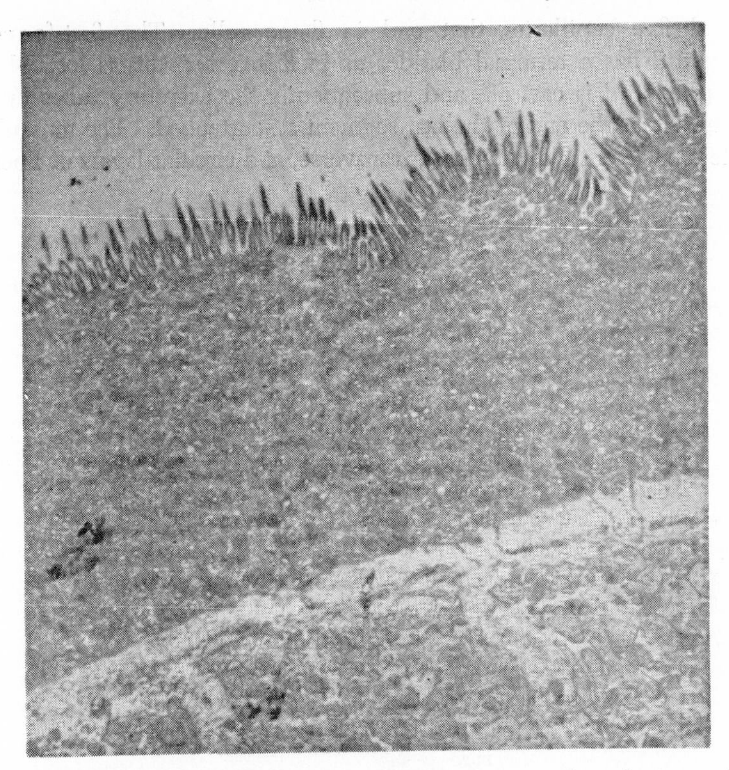

Fig. 87. An electronmicrograph of the "cuticle" of *Hymenolepis diminuta.* The section is about 130 Å in thickness. ×10,940. (Preparation through the courtesy of Dr. F. B. Bang and Mrs. Ilse Vellisto.)

(Fig. 87). The internal organs are imbedded in a spongy mesodermal parenchyma. The nervous system consists of a few ganglia and commissures in the scolex, from which longitudinal nerve cords run through the length of the worm, the largest ones being a pair near the lateral borders. Coordination of movement is very limited, although the whole worm can contract at once, as when dropped into cold water. Individual ripe segments, when detached, show considerable sensitiveness and are often very active. The excretory system or, preferably, osmoregulatory system, according to Wardle and McLeod (1952), is fundamentally of the same type as in flukes and consists typically of two pairs of lateral longitudinal tubes, the ventral pair usually larger than the dorsal and connected by a transverse tube near the posterior end of each proglottid, and sometimes by a network of smaller tubes. The dorsal vessels carry fluid towards the scolex, the ventral ones away from it. From the ventral vessels extend fine capillaries that end in flame cells. The first-formed proglottid has a terminal bladder as in flukes, but this is lost when this proglottid is cast off, and subsequently the excretory tubes open separately at the end of the last segment still attached. The muscular system consists of longitudinal, transverse, and circular layers of fibers,

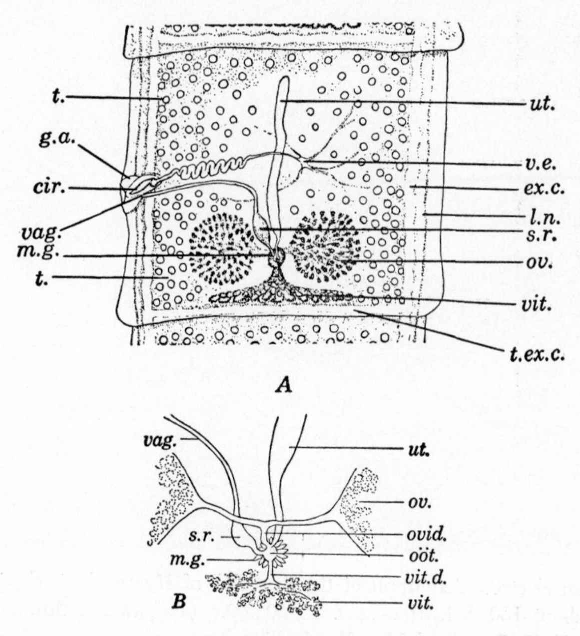

Fig. 88. *Taenia solium: A*, mature proglottid; *B*, region of ootype; *ex.c.*, excretory canal; *l.n.*, lateral nerve; *oöt.*, ootype surrounded by Mehlis' glands; *ov.*, ovary; *s.r.*, seminal receptacle; *t.*, testes; *ut.*, uterus; *vag.*, vagina; *vit.*, vitellaria. *v.e.*, vas efferens. (After Brown, *Synopsis of Medical Parasitology*, 1953.)

some just under the cuticle but mostly in a band which encircles the worm at some distance from the cuticle, dividing the parenchyma into cortical and medullary portions. In thick, fleshy species these muscle layers are well developed, whereas in the semitransparent forms they are not.

As of flukes, the main business of tapeworms is the production of myriads of eggs in order to safeguard the species against extermination in the perilous transfer from host to host. Each proglottid possesses complete reproductive systems of both sexes, fully as complete as in the flukes, if not more so (Fig. 88), and in some species each proglottid has a double set of organs.

The female system consists of an ovary, which may be single or in two more or less distinct lobes; yolk glands, either in a single or bilobed mass, or scattered through the segment; Mehlis' glands around an oötype, where the component parts of the egg are assembled; a vagina for the entrance of the sperms, with an enlarged chamber, the seminal receptacle, for storage of sperms; and a uterus, which may or may not have an exit pore. In the tapeworms which have a pore (order Pseudophyllidea), the development and extrusion of eggs goes on continuously in many segments at once, but in the Cyclophyllidea there is no uterine opening. In these the uterus eventually becomes packed with eggs and may practically fill the segment, which is essentially a seed pod. Such "ripe" segments detach themselves from the end of the chain, subsequently liberating their eggs by disintegration, or by extrusion of the eggs through ruptures during the active contractions and expansions of the segments. The form of the ripe uterus varies greatly in different families, genera, and species, and is often useful in identification (Fig. 89).

The male system consists of a variable number of scattered testes connected by minute tubes with the sperm duct or vas deferens, which is usually convoluted and may have an enlargement, the seminal vesicle, for storage of sperms. The end of the vas deferens is modified into a muscular intromittent organ, the cirrus, which is retractile into a cirrus pouch or sac. In most tapeworms both cirrus and vagina open into a common cup-shaped genital atrium, with a pore on either the lateral border or the mid-ventral surface. Either self-fertilization of a single segment or cross-fertilization between different segments of the same or other worms can occur, but probably fertilization between segments is commonest. As a rule the male reproductive organs mature before the female.

Life cycle. The life cycle is not quite so complicated as in flukes and does not involve asexual generations, although in some species the larval forms multiply by budding. The life cycle of many tapeworms, especially those of fishes, is still unknown; in fact, it was not until the middle of the last century that Küchenmeister proved that the bladderworms in pigs and cattle were in reality the larvae of the common large tapeworms of man; previous to that time they were classified in a separate order, Cystica.

The eggs of tapeworms develop within themselves little spherical embryos characterized by the presence of three pairs of clawlike hooks, whence they are known as oncospheres (Fig. 90*A* and *B*). One or two enclosing membranes inside the egg shell proper form about the developing embryo, the inner one of which is called the embryophore.

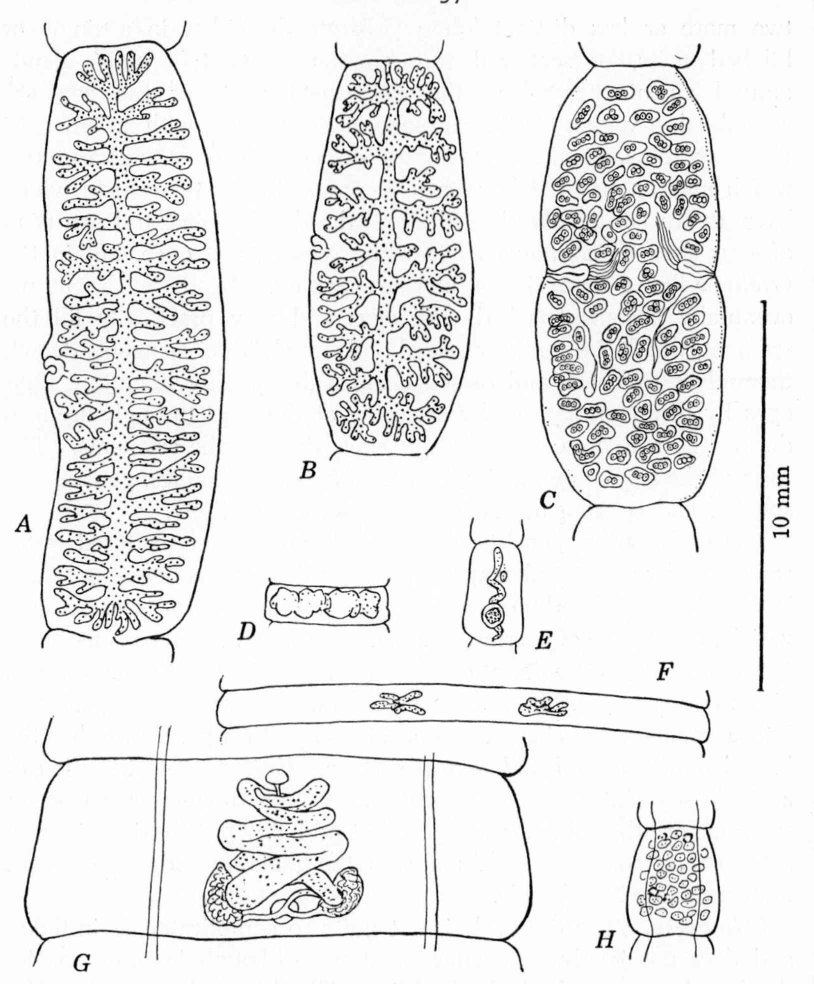

Fig. 89. Ripe proglottids of various tapeworms. *A, Taenia saginata; B, T. solium; C, Dipylidium caninum; D, Hymenolepis diminuta; E, Mesocestoides variabilis; F, Diplogonoporus grandis; G, Dibothriocephalus latus; H, Raillietina* sp.

In the order Pseudophyllidea the embryos, called coracidia (Fig. 94*C*), are covered by a ciliated embryophore. They have a brief free-swimming existence, like miracidia, in which they roll about by means of their cilia long enough to attract the attention of copepods which devour them. In these they shed their ciliated covering and change into elongated oval "procercoids" (Fig. 94*E, F, G*), comparable with sporocysts but solid and incapable of asexual reproduction. The six hooks are still present on a small caudal appendage, the cercomer.

Further development into a "plerocercoid" (Fig. 96) occurs only when the infected copepod is eaten by a fish or other animal. The plerocercoids are solid wormlike larvae with a scolex invaginated at one end. When the animal containing them is eaten by the final host, the scolex turns right side out and attaches itself to the intestinal wall, and the mature tapeworm develops.

In the order Cyclophyllidea, on the other hand, the oncosphere remains passively in the egg, surrounded by the nonciliated embryophore until eaten by the intermediate host. Here it transforms into a bladderlike structure, a part of the wall of which differentiates

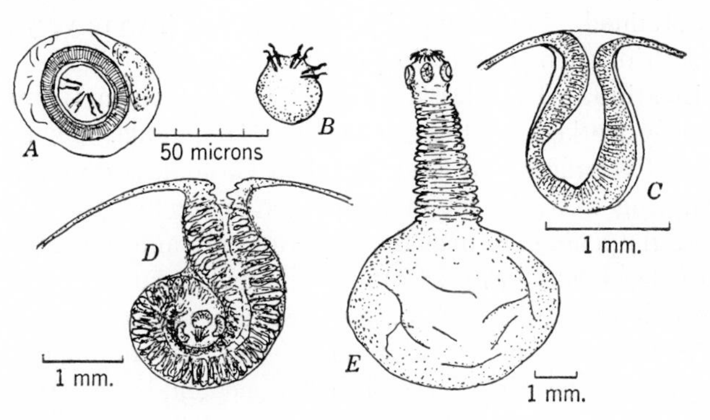

Fig. 90. Stages in life cycle of *Taenia solium.* *A*, egg containing embryophore; *B*, hatched oncosphere; *C*, invagination in wall of developing cysticercus, at bottom of which the scolex will form; *D*, invagination after development of the scolex; *E*, cysticercus with head and neck evaginated. (*A*, *B*, *C*, after Blanchard from Brumpt, *Précis de parasitologie,* 1949.)

into one or more scoleces turned inside out (invaginated) (Fig. 90*C* and *D*). Sometimes the whole embryo becomes hollow and grows into a large bladder, into the spacious cavity of which the relatively small scolex or scoleces are invaginated; such a larva is called a *cysticercus* or bladderworm if there is only one scolex, and a *coenurus* (Fig. 100) if there are a number of them. In one tapeworm, *Echinococcus,* the bladders add a further method of multiplication by budding off daughter and grand-daughter bladders, and the bladder walls, instead of directly producing scoleces, first produce brood-capsules, each of which in turn produces on its wall a number of scoleces, whereby one huge larval cyst, called a *hydatid,* may be the mother of many thousands of tapeworms (Figs. 102, 103). Sometimes the main portion of the body of the embryo remains solid and grows very little, whereas one end of it becomes hollowed out into a small bladder containing the invaginated scolex (Figs. 105, 106). The undeveloped solid portion remains as a caudal appendage. Such a larva, called a *cysticercoid,* is characteristic of those tapeworms which use arthropods as intermediate hosts.

On being eaten by a final host only the scoleces survive; these turn right side out (evaginate), attach themselves to the mucous membrane of the intestine, and grow each into a mature tapeworm. In one progressive genus, *Hymenolepis,* a few species have broken away from the traditional intermediate host idea and may complete their development in one host; the cysticercoids develop inside the intestinal villi and subsequently gain the lumen of the intestine where the mature phase is attained. For a long time parasitologists were very skeptical of the truth of such unorthodoxy on nature's part.

Classification. The classification of the Cestoidea is much more satisfactorily worked out than that of the Trematoda. There are two subclasses, Cestodaria and Cestoda. The Cestodaria do not form segments, having a single set of reproductive organs; the vagina and male genital opening are near the posterior end; and the embryos have ten or twelve hooks (Fig. 91*B*). This subclass contains two orders. The Amphilinidea (Fig. 91*C*) are parasitic in the coelom of fishes or (in one species) a fresh-water turtle, and use amphipods as first intermediate hosts; they are probably neotenic, i.e., precociously mature larvae (plerocercoids) in the second intermediate host, the true definitive hosts having been abandoned. The Gyrocotylidea (Fig. 91*A*), parasitic in the primitive chimaeroid fishes, appear to develop like the monogenetic trematodes, without bothering with an intermediate host. The Cestoda, which alone concern us here, produce reduplicated reproductive organs, usually in distinctly demarcated segments or proglottids, except in one order, Caryophyllidea, which in other anatomical characters resembles the Pseudophyllidea. The Cestoda have a well-developed scolex, genital openings anterior to the ovary, and embryos with only six hooks.

The cestodes are an ancient group which probably evolved with the sharks and rays; four of the eleven orders are still confined to them. It is not surprising that there are a number of aberrant forms which have no near relatives, and which do not fit into any of the large groups. Wardle and McLeod (1952) recognize eleven orders, but only five of these contain numerous forms; the others contain only a few each, one only a single species. The Tetraphyllidea, with numerous and varied forms in elasmobranchs, probably represent a sort of central stock from which in the more or less distant past all the other orders have evolved; the Trypanorhyncha, with their re-

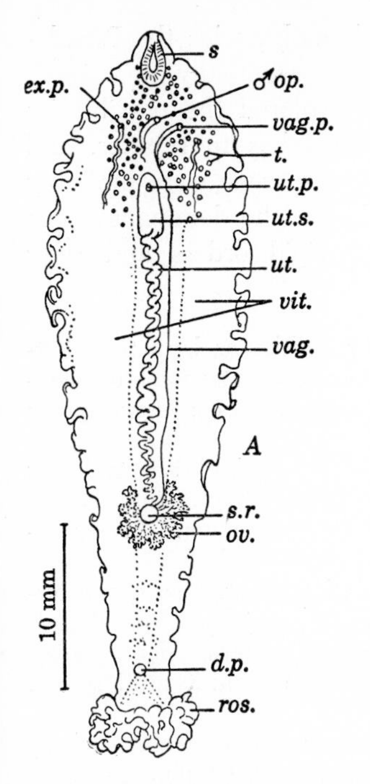

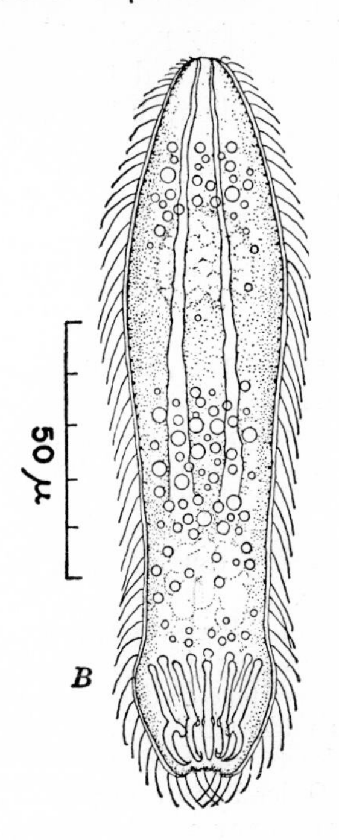

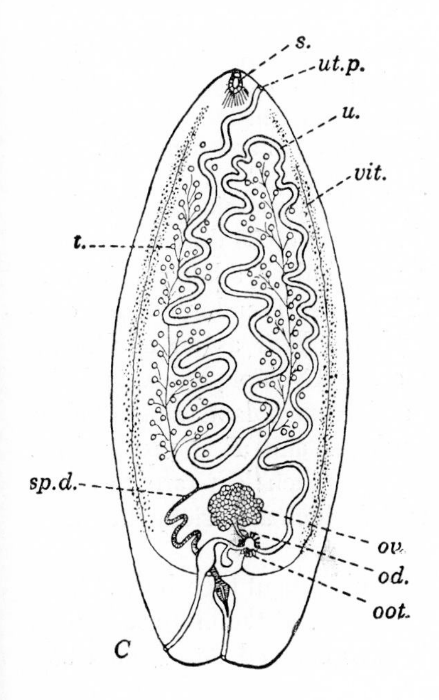

Fig. 91. Cestodaria. *A, Gyrocotyle urna,* adult; *B,* same, newly hatched 10-hooked larva; *C, Amphilina foliacea; d.p.,* dorsal pore; ♂ *op.,* ♂ genital opening; *vag.p.,* vaginal pore; *ut.p.,* uterine pore; *ut.s.,* uterine sac; *s.,* sucker; *ros.,* rosette; *oot.,* ootype; other abbreviations as in Fig. 88. (*A* and *B* adapted from Lynch, *J. Parasitol.,* 1945. *C,* after Wardle and McLeod, *The Zoology of Tapeworms,* University of Minnesota Press, 1953.)

markable spiny proboscides, remained in elasmobranchs; the Pseudophyllidea switched to bony fishes, and some eventually to animals feeding on bony fishes; the Proteocephala explored the possibilities of parasitizing fresh-water vertebrates from fishes to reptiles, and with the advent of birds and mammals onto the scene were probably the group from which came the Cyclophyllidea, the dominant group in these animals. Following are the principal orders and their characters:

1. Order **Tetraphyllidea.** Head with four earlike or lappetlike outgrowths (bothridia) (Fig. 86*B*); proglottids in various stages of development; vitelline glands scattered in two lateral rows or bands; genital pores lateral. In elasmobranchs.
2. Order **Trypanorhyncha.** Head with two or four bothridia and four long evertible proboscides armed with hooks or spines, retractile into sheaths (Fig. 86*C*); yolk glands in continuous sleeve-like distribution. In elasmobranchs.
3. Order **Pseudophyllidea.** Head with two lateral or, rarely, one terminal, sucking grooves (bothria) (Fig. 86*A* and *D*); majority of proglottids in similar stage of development, shedding eggs from a uterine pore; genital pores mid-ventral; vitelline glands scattered in dorsal and ventral sheets. In teleosts and land vertebrates.
4. Order **Proteocephala.** Head with four in-cupped muscular suckers, and sometimes a fifth terminal one; genital pores lateral; vitelline glands in lateral bands; uterus with one or more median ventral openings made by breaks or clefts in the body wall. In fresh-water fishes, amphibians, and reptiles (except one in a shark).
5. Order **Cyclophyllidea.** Head with four in-cupped suckers (Fig. 86*F*); proglottids in all stages of development, ripe ones only near end of chain; no uterine pore; genital pores *usually* lateral; vitelline gland a single or bilobed mass posterior to ovary. Majority in birds and mammals.

Only the Pseudophyllidea and Cyclophyllidea contain species which attack man or domestic animals. Although 25 or 30 different species of tapeworms have been recorded in man, only 4 adult species and 3 larval species are at all common. The order Pseudophyllidea contains one in each group, *Dibothriocephalus latus* as an adult, and *Spirometra mansoni* as a larva; the order Cyclophyllidea includes as adults *Taenia solium, T. saginata,* and *Hymenolepis nana,* and as larvae *T. solium, Echinococcus granulosus* and *E. multilocularis. H. diminuta* and *Dipylidium caninum* are much less rare than the records indicate, but all the others, some of which are briefly described in the following pages, are rare.

Most tapeworms are astonishingly particular about their hosts. They undergo evolution more slowly than the hosts and therefore stick with them, unchanged even after the hosts have evolved into genera and species which may now be separated by oceans and continents. The tapeworms harbored by a bird may be a better indication of its phylogenetic affinities than some of its anatomical characters; in several instances tapeworms have proved more reliable than ornithologists in showing the relationships of birds.

Diagnosis. Tapeworms cannot invariably be diagnosed by examining feces for eggs. The pseudophyllidean tapeworms can be diagnosed in this way, since the operculated eggs are expelled through the uterine pores of many proglottids at a time and are therefore always present in the feces. Like the eggs of flukes, these eggs do not float in saturated salt solution; they can be concentrated by straining and sedimenting or centrifuging in water, or by the AMS III method (see p. 253). *Hymenolepis* infections can also be diagnosed by fecal examination for eggs, even though no birthpore is present, since the segments broken off from the ends of the worms commonly rupture before leaving the body of the host; *Hymenolepis* eggs are easily found by floatation methods. In *H. nana* the partitions between ripe segments break down before the segments are shed, permitting the eggs from several segments to escape from the edges of the segments.

Taenia and many other tapeworm infections cannot be reliably diagnosed in this manner, since the segments often escape from the body uninjured and still alive. Search must be made for the voided segments in the stools; the shape of the segment and form of the gravid uterus serve to identify the species. *Taenia* eggs are present in the feces only when segments rupture; those of *T. saginata* can usually be found on the perianal skin, like those of *Enterobius* (see p. 359). The thick, striated embryophores are porous and therefore cannot be found by flotation.

The eggs of *Dibothriocephalus* (Fig. 55*L*) may be confused with those of flukes, but they are different in size from any common human fluke eggs (60 to 70 μ). All other tapeworm eggs of man are recognizable as such by their six-hooked embryos. *Taenia* and *Hymenolepis* eggs (Fig. 55*N*, *O*, and *P*) cannot be confused when one has once seen them, but many inexperienced physicians, unfamiliar with *Hymenolepis*, take all eggs with six-hooked embryos to be *Taenia*, sometimes with disconcerting results. A physician who found tapeworm eggs in the stool of a Hindu mortally offended him by telling him he had eaten insufficiently cooked beef or pork, when in reality the eggs were those of *Hymenolepis*.

Treatment. For the most part the drugs most useful in expelling tapeworms constitute a group distinct from those effective against nematodes, although carbon tetrachloride and hexylresorcinol have some effectiveness against both groups of worms. Tapeworms are affected by a number of drugs of vegetable origin—filix mas, Cusso, Kamala, pelletierine, arecoline hydrobromide—which have not been found effective against other helminths. Ethereal extract of *Aspidium* (filix mas or male fern) was the standard drug for expelling tapeworms from man for many years, and arecoline hydrobromide is still most extensively used for dogs.

An acridine compound (Acranil), related to atebrin, was first used against tapeworms by Neghme in Chile in 1938, and atebrin by Culbertson shortly after. Atebrin has now largely replaced Aspidium for human tapeworm infections. The usually recommended dose of atebrin for adults is eight to ten 0.1-gram tablets in a single dose—two tablets every 5 minutes. This usually clears 60 to 90% of *Taenia* cases in one treatment. It cures 40 to 75% of *Hymenolepis nana* infections, in all cases removing a high percentage of the worms; in one case reported by Beaver and Sodeman 6000 worms were passed. Neghme and Bertin (1951) found atebrin successful in removing *Dibothriocephalus latus* in all cases tried. The drug is given on an empty stomach in the morning, and followed in 2 to 4 hours by a strong saline purge. Chloroquine seems to be a little less effective than atebrin. Introduction of 0.8 gram of atebrin in 100 cc. of water behind the stomach by duodenal tube is an even more effective treatment; it expelled *all* worms from patients who did not interfere with the treatment by vomiting, which was done more frequently by women. Treatment for species of *Taenia* should not be repeated until segments or eggs again appear in the feces, for sometimes the scoleces are not easily seen when passed. The worms are more easily expelled when there is a considerable length of paralyzed worm to drag on the scolex.

German workers (see Kuhls, 1953) have recently revived the use of tin as an anthelmintic for tapeworms. Tablets containing a mixture of metallic tin, tin oxide, and zinc chloride, given before meals and with no following purge, are reported to be highly effective (90% cures) and nontoxic. Dichlorophen has been reported by Jackson to be effective against tapeworms in South Africans.

Prevention. Prevention varies, of course, with the species of tapeworm and its intermediate host. But since infection with the common human species, with the exception of the species of *Hymenolepis,* results from eating raw or imperfectly cooked beef, pork, or fish in which the larvae have developed, the exclusive use of thoroughly cooked meat and fish is the best preventive measure. Pork and beef bladderworms are killed when heated to 55°C., but it is difficult to heat the center of a large piece of meat even to this point; a ham cooked by boiling for two hours may reach a temperature of only 46°C. in the center. When roasted, pork should always be cut into pieces weighing no more than three or four pounds to insure thorough penetration of heat. Beef which has lost its red or "rare" color is quite safe. At 0° to 2°C. the cysticerci of *T. solium* live for over 50 days, but at —5°C. they die in about a week. Quick-freezing is destructive to cysticerci, as is thorough curing or salting of meat. The meat of sheep, goats, or chickens does not convey any parasites to man, even if uncooked.

Prevention of cysticercus infections in cattle or pigs depends on care to prevent contamination of the animals' water or food by human feces. Eggs of *Taenia saginata* may live in liquid manure for over 10 weeks and on contaminated grass, if not allowed to dry, for over 20 weeks.

The dwarf tapeworm, *H. nana,* and those which develop in arthropods are subject to different means of prevention (see p. 369). No effective method of protection of herbivores against anoplocephalid tapeworms, which utilize free-living mites as intermediate hosts, has yet been devised.

Order Pseudophyllidea

All the members of the Pseudophyllidea which live in man or domestic mammals are members of the family Diphyllobothriidae. These are large worms consisting of long chains of numerous segments and a head provided with a slitlike groove or bothrium on either side. The majority of the segments are mature and functional at one time and deposit eggs through the uterine pores as more are being developed. Eventually, as old age overtakes them, the proglottids cease to produce more eggs; they gradually empty their uteri and then, shrunken and twisted, are sloughed off in long chains. The general type of life cycle, involving a copepod as a first intermediate host and a vertebrate as a second, has already been described on p. 338.

There are about 75 species in the family, living in whales, porpoises, seals, sea lions, fish-eating land carnivores, fish-eating birds, and man. All of these, except a few bird tapeworms that have indistinct or no external demarcation of the segments (*Ligula* and *Schistocephalus*), have been commonly lumped together in the genus *Diphyllobothrium.* Since several well-defined groups occur, with quite well-marked characters, it has become desirable to break up this large group into several genera. Wardle, McLeod, and Stewart (1947) recognized seven genera, five of them parasitic in marine mammals. The species which commonly lives as an adult in man falls into the genus *Dibothriocephalus,* which was created for it before this and all the other species were merged into the genus *Diphyllobothrium.* The latter genus was first erected for a large species with a small oval head and no obvious neck, found in porpoises.

The genus *Dibothriocephalus* (not to be confused with *Bothriocephalus,* which has both its plerocercoid and adult stages in fishes) contains the human species, *D. latus,* and several others reported from fish-eating mammals and birds. They are large, slender, weakly muscular

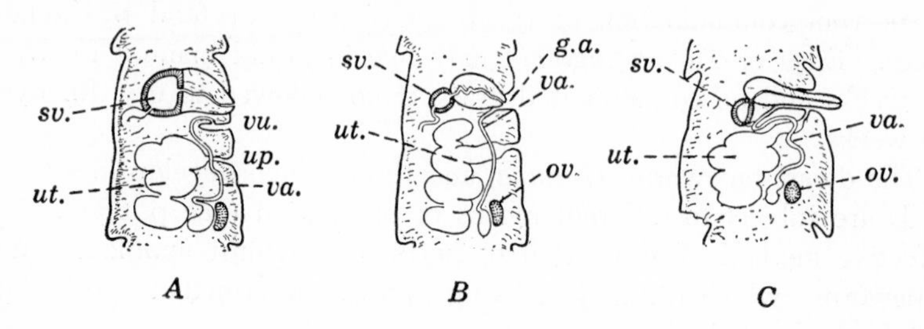

Fig. 92. Arrangement of organs of various genera in the family Diphyllobothriidae as seen in sagittal sections. *A, Spirometra,* with separate openings for cirrus, vagina, and uterus; *B, Dibothriocephalus,* with common opening for cirrus and vagina; *C, Pyramicocephalus,* with common opening for cirrus, vagina, and uterus. Abbreviations: *g.a.,* genital atrium; *ov.,* ovary; *sv.,* seminal vesicle; *up.,* uterine pore; *ut.,* uterus; *va.,* vagina; *vu.,* vulva. (After Mueller, *J. Parasitol.,* 22, 1936.)

forms with an elongated, compressed scolex with slitlike grooves or "bothria" which are narrow and deep; a long slender neck; a rosette-shaped uterus; a common opening for cirrus and vagina (Fig. 92*B*); eggs rounded at the ends; and plerocercoids in fish. A closely related genus, *Spirometra,* contains smaller and weaker worms, found primarily in cats, which have broad and shallower bothria; a uterus with a spiral of close coils; separate openings for cirrus and vagina (Fig. 92*A*); eggs pointed at the ends; and plerocercoids usually in frogs, snakes, birds, or rodents. Some, possibly all, of this group can live in human flesh in the plerocercoid or "sparganum" stage, causing sparganosis. One species, *S. houghtoni,* has been recorded a few times as an adult from a man in China.

There are a few records of human infection with other adult diphyllobothriids, normally found in marine carnivores, and also with worms of the genus *Ligula.* This and the related *Schistocephalus* reach a large size before being transferred from fish to their normal bird hosts. Smyth (1947) was able to get the plerocercoids to mature their reproductive organs in artificial media at 40°C.

Morphology. The arrangement of the organs in mature proglottids of a diphyllobothriid is shown in Figs. 93 and 97. The female system consists of a vagina opening on the mid-ventral surface in the anterior part of the segment and running almost straight posteriorly to an

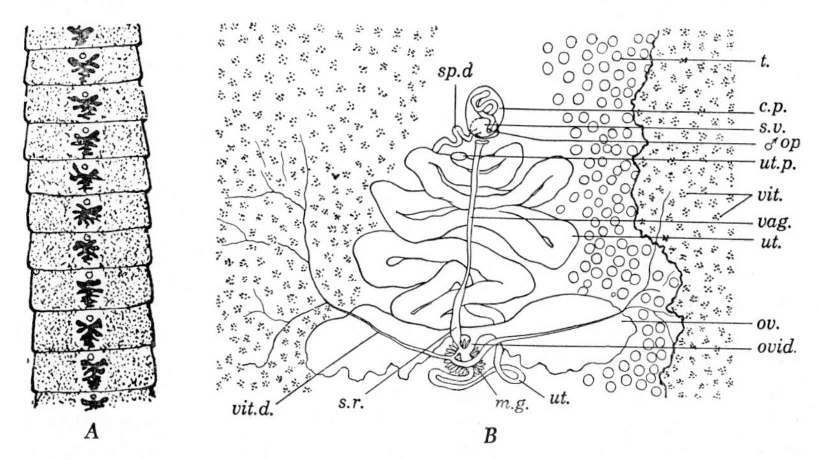

Fig. 93. *Dibothriocephalus latus.* *A*, chain of proglottids, natural size. *B*, middle portion of proglottid; *c.p.*, cirrus pouch; *m.g.*, Mehlis' gland; ♂ *op.*, male genital opening; *ov.*, ovary; *ovid.*, oviduct; *sp.d.*, sperm duct; *s.r.*, seminal receptacle; *s.v.*, seminal vesicle; *ut.*, uterus; *vag.*, vagina; *vit.*, vitellaria; *vit.d.*, vitelline duct. Superficial layer cut away on right to expose testes.

ootype surrounded by a Mehlis gland. The ovaries are paired in the posterior part of the segment. The yolk glands are scattered throughout the lateral fields. The uterus has an inner series of delicate coils as it leaves the ootype, followed by an outer series of large coarse ones. The uterine opening is on the ventral surface but is inconspicuous. The male system consists of a large number of testes scattered in the lateral fields, largely obscured by the vitelline glands which lie dorsal and ventral to them. The cirrus and cirrus pouch are anteriorly situated.

Dibothriocephalus latus. This worm, possibly constituting a group of nearly related species, is called the "broad" or "fish" tapeworm. It is the largest tapeworm found in man. It has long been known in many parts of central Europe; in Finland about 20% of the population harbor this worm, and in some local areas in the Baltics nearly 100% are infested. The worm also occurs in Ireland, Siberia, Palestine,

Japan, Central Africa, Chile, and in Michigan, Minnesota, Wyoming, Manitoba, Alaska, and Florida in North America. It has been thought that the worm was introduced into North America by Scandinavian lumbermen, but it probably also entered by the Bering Straits, or there may have been a native species in our wild carnivores before European or Asian man or dogs arrived. Worms from Manitoba and from Russia *look* different, but there is so much variation in different parts of a worm, in worms of different ages and in different hosts, and in worms differently prepared, that no specifically differential characters have been found.

D. latus reaches maturity in many domestic and wild species of the dog and cat families, in bears, and possibly other fish-eating carnivores. In North America the brown bear may be the normal host; dogs may not be as important since many of the eggs passed by them are not viable.

D. latus is a veritable monster, reaching a length of 10 to more than 30 ft., with a width of 10 to 12 or even up to 20 mm., and with a total of 3000 to 4000 proglottids in large specimens. Tarassov tells of a Russian woman who harbored six worms, aggregating over 290 ft., and of another who supported 143 worms. Fortunately in tapeworm infections the size of the worms usually is in inverse proportion to their number. The proglottids (Fig. 93) for the most part are much broader than long, although the terminal ones become approximately square.

LIFE CYCLE. The broadly oval, operculated eggs, which average about 60 by 42 μ, contain abundant yolk cells (Fig. 55*L*). Ciliated embryos, or coracidia, develop slowly in the eggs, hatching after 8 or 10 days to several weeks, depending on temperature. The coracidia (Fig. 94*C*), 50 to 55 μ in diameter, swim by means of their cilia or creep on the bottom after slipping out of their ciliated coverings, but they must be eaten by certain species of copepods (Fig. 94*D*) in less than 24 hours if they are to continue their development and fulfill their destiny.

The worm is very fastidious about its first intermediate hosts, and in America develops only in certain species of the genus *Diaptomus* (distinguished by having very long first antennae), which live in the open water of lakes. Many species of *Diaptomus* are known to serve as hosts, but other copepods do not, although species of *Cyclops* are the preferred hosts of *Spirometra.*

Soon after the coracidium is ingested by a copepod it loses its ciliated covering, and the naked oncosphere, only 24 μ in diameter, bores through into the body cavity. In 14 to 18 days it develops into a solid, elongate creature. The embryonic hooks persist in a bulblike appendage, the cercomer, which is partially pinched off at the posterior end, and eventually discarded; a cup-shaped depression, into which histolytic glands open, appears at the anterior end; some species also have spines at the anterior end. The worm is now a procercoid, about 500 μ long (Fig. 94*D*, *G*).

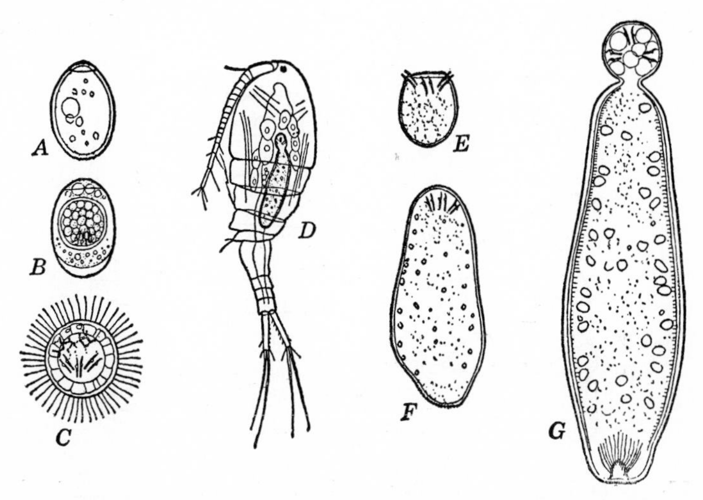

Fig. 94. Developmental stages of *Dibothriocephalus latus:* *A*, undeveloped egg; *B*, egg containing developed embryo; *C*, free embryo or coracidium; *D*, *Cyclops strenuus* containing procercoid; *E*, embryo after shedding ciliated envelope in *Cyclops;* *F*, growing procercoid; *G*, full-grown procercoid. (After Brumpt, *Précis de parasitologie,* Masson, 1949.)

Further development occurs in fish when the infected copepod is eaten. The passage through the intestine and body cavity of the fish is slow, but eventually the larvae reach the flesh of the fish and grow into elongated wormlike plerocercoid larvae, 4 or 5 mm. to several centimeters in length. They are not encysted and are found anywhere in the flesh, in other places only rarely. In this respect *D. latus* differs from some of the related species that develop in birds, since the latter are found encysted on the mesenteries in the abdominal cavity or in the liver. The smaller plerocercoids lie straight, but with growth they become increasingly bent and twisted (Fig. 95). The anterior end has a depression which is the withdrawn and inverted scolex; the remainder of the body is white, somewhat flattened, and marked by irregular wrinkles, but without segmentation. In uncooked fish their opaque white color shows clearly through the translucent flesh, but cysts of flukes or other tapeworms may be confused with them if they are not carefully examined. Cysts of tapeworms of the genus *Proteocephalus* are often present, but these have four or five cup-shaped suckers on the head; the elongate but cramped plerocercoids abundant in tullibee (*Leuciscus*) in some lakes in northern United States are readily distinguished by the tridents on the head; some of the other plerocercoids found in herring, perch, trout, etc., are more difficult to distinguish. Drum, gulf "trout," etc., from the Texas coast

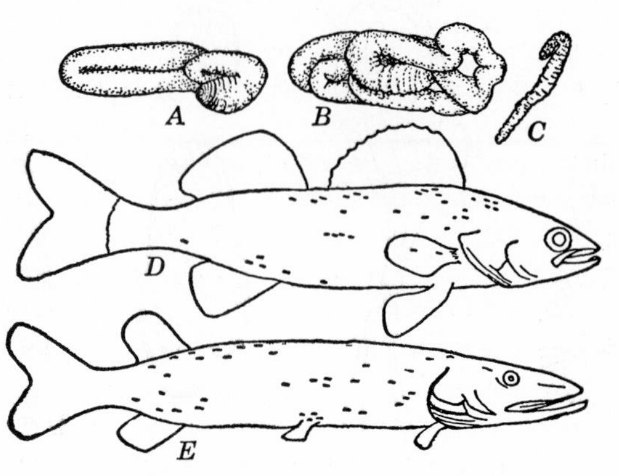

Fig. 95. *A* to *C*, plerocercoid larvae of *D. latus* as they appear in the flesh of fishes, ×3; *D*, wall-eyed pike (*Stizostedion*) showing distribution of 35 plerocercoids in the flesh; *E*, northern pike (*Esox lucius*) with 37 plerocercoids. (After Vergeer, *J. Inf. Dis.*, 44, 1929.)

frequently contain the very elongate "spaghetti worm" plerocercoids of Trypanorhyncha which mature in sharks and rays. None of these become human parasites.

The fish that serve as second intermediate hosts of *Dibothriocephalus latus* are carnivorous species, but they seem to differ in different localities. In northern United States and Canada pike and walleyes (*Esox* and *Stizostedion,* Fig. 95*D* and *E*), are by far the most important hosts; in northern Europe, trout, perch, and burbot; in Lake Baikal, species of *Coregonus* and *Thymallus;* in Chile, Alaska, and the Far East, trout (*Onchorhynchus* and *Salmo*); and in Africa, the barbel (*Barbus*).

In some small lakes in northern United States and Canada 50 to 75% of the pike and walleyes harbor larvae of this worm. These large carnivorous fish do not feed intentionally on copepods and probably ingest them in the stomachs of smaller fish on which they prey. It is a peculiarity of plerocercoid larvae that they are able to reinvade and become re-established in host after host until one is reached in which maturity can be attained in the intestine. Those of *D. latus* can pass from fish to fish, and those of *Spirometra* may be passed about among frogs, reptiles, and mammals.

Infection of the final host comes from eating imperfectly cooked flesh or roe of infected fish or from conveying small plerocercoids to the mouth by the hands, to which they cling while fish is being cleaned. In 3 weeks they may have reached a length of 3 ft. and may begin producing eggs in that time. It has been estimated that one worm produces 36,000 eggs daily. In northern United States many towns pour their sewage directly into lakes, and the inhabitants fish for pike, which harbor the plerocercoids, near the sewage outlets. Summer visitors in camps and hotels often partake of fish hastily prepared, content with a well-done exterior. Dogs and cats are usually given the raw refuse and help to keep the infection alive. Furthermore, millions of pounds of walleyes and pike are annually imported from infected Canadian lakes for the preparation of "gefüllte fish." Many cases develop among Jewish people, presumably as a result of tasting the fish during the preparation, before it is cooked.

PATHOLOGICAL EFFECTS AND TREATMENT. Common effects of *D. latus* infection are abdominal pain, loss of weight, and progressive weakness, similar to the symptoms of *Taenia* infections (see p. 359). This worm is, however, unique among tapeworms in sometimes causing a very severe anemia of the pernicious type. Fortunately, this severe anemia is the exception rather than the rule. In Finland, although the Finlanders are said to be more prone to pernicious anemia than other races, the anemia rate is only 5 to 10 per 10,000 infections. It was once thought that the anemia was produced by toxic substances produced by the worms, but the senior author (Chandler, 1943) suggested absorption of a vitaminlike substance by the worms as a likelier explanation. Subsequently it became known that vitamin B_{12} plays an important role in blood formation, and in 1950 the authors suggested that *Dibothriocephalus* may have an unusual affinity for this vitamin. Work by von Bonnsdorf (reviewed in 1956) and by Nyberg (1958) demonstrated that *D. latus* absorbs ten to fifty times as much vitamin B_{12} as other tapeworms, the fish tapeworm taking up almost half of a single oral dose of radioactive vitamin B_{12} given to the human host. The worm's tissues are so rich in the vitamin that administration of powdered worms along with gastric juice is as effective in treating pernicious anemia as administration of vitamin B_{12}.

CONTROL. Control of *D. latus* infection must depend mainly on more careful cooking of fish. Housewives and cooks preparing "gefüllte fish" should refrain from tasting the raw fish to test their skill in flavoring. Some reduction in the infection of fish could be obtained by education and regulation with respect to pollution of lakes, and the practice of feeding raw fish to dogs and cats should be discouraged.

"*Sparganum*" infections. The plerocercoid larvae of worms of the genus *Spirometra,* for which the name *Sparganum* was given before their adult forms were definitely known, normally develop in frogs, snakes, or amphibious mammals, but when opportunity is afforded can live in man. The first intermediate hosts are *Cyclops,* which abound in shallow water. Galliard and Ngu (1946) showed that *S. mansoni* in Indo-China apparently requires four hosts, the procercoids from *Cyclops* first infecting tadpoles, and later becoming fully developed plerocercoids when these are eaten by frogs, reptiles, or mammals. This recalls the use of minnows as an intermediate step between *Diaptomus* and carnivorous fish by *Dibothriocephalus latus.*

The larval worm as found in man, usually referred to as *Sparganum mansoni,* is a typical plerocercoid, much larger than that of *D. latus,* being 3 to 14 in. in length (Fig. 96). It is a whitish, elastic, wrinkled worm with an invaginated scolex at the broader end. In man it is found in the muscles, subcutaneous connective tissue, or around the eye. The largest number of cases have been recorded from Indo-China, China, and Japan, but scattered cases of this or closely related larvae are known from almost every part of the world. In the Orient, human infection is acquired in a remarkable manner; split fresh frogs are commonly used by the natives as a poultice for sore eyes and wounds, and the spargana then transfer themselves to human flesh. Applied to the eye, they may settle in the lids or go to other parts of the face; they are easier to remove after being encapsulated. As noted on p. 350, spargana are able, when eaten by a host which is not suitable for adult development, to reinvade and become encapsulated over again, ready for another try. The range of hosts in which they can become re-established after development is much larger than that in which they can develop originally.

There is much confusion about the species of *Spirometra,* all of which are primarily cat and dog parasites when mature. Wardle and McLeod (1952) list seventeen species which have been described from all parts of the world. There is no way of distinguishing the spargana, and the adult morphology, as in the genus *Dibothriocephalus,* does not lend itself to easy differentiation of species. Some spargana develop only in frogs, some in mice but not in frogs, and some in both. After becoming spargana, however, some species, at least, will re-establish themselves in fish, frogs, snakes, mice, or men—whatever eats them. It is probable that any species of *Sparganum* of the *Spirometra* group could establish itself in man if swallowed, but it is not known whether the swallowing of infected copepods would

Fig. 96. *Sparganum mansoni,* natural size. (After Ijima and Murata, *Coll. Sci. Imp. Univ. Japan,* 2, 1888.)

result in infection. A few cases of *Sparganum* infection have been recorded in the United States (Read, 1952), but there is no information as to the species to which they belong. *Spirometra mansonoides*

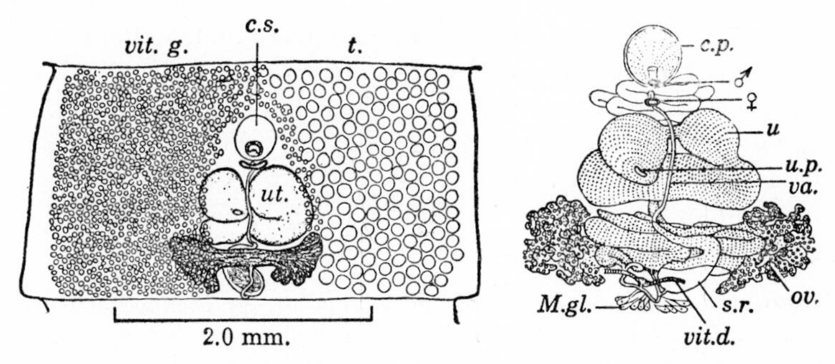

Fig. 97. *Spirometra mansonoides. Left,* young mature proglottid: *c.s.,* cirrus sac or pouch; *t.,* testes; *ut.,* uterus; *vit.g.,* vitelline glands. *Right,* reproductive organs: ♂ and ♀, genital openings; *c.p.,* cirrus pouch; *M.gl.,* Mehlis' gland; *ov.,* ovary; *s.r.,* seminal receptacle; *u.,* uterus; *u.p.,* uterine pore; *vit.d.,* vitelline duct. (After Mueller, *Parasitol.,* 21, 1935.)

(Fig. 97) has a wide distribution in wild and feral cats in eastern United States and Texas, and uses wild species of mice for development of the spargana. Mueller and Coulston (1941) showed that when the young spargana of this species are experimentally implanted in human flesh they grow normally; they thought human cases might be commoner than records indicate and that cysts or fatty tumors removed from under the skin should be suspected.

A few cases have been recorded in which the spargana apparently multiply in the body. Thousands of worms, usually only 3 to 12 mm. in length but sometimes larger, may be present in acnelike nodules in the skin and elsewhere in the body. They apparently proliferate by formation of budlike growths. This so-called *Sparganum proliferum* is now believed to be an abnormal growth in an unfavorable host; it has been found only in man. Mueller (1938), from a careful restudy of specimens of *S. proliferum,* concluded that they are abnormal, degenerate forms without scoleces and without normal orientation of parts.

Order Cyclophyllidea

The vast majority of the tapeworms of mammals and birds belong to the order Cyclophyllidea. These, as noted on p. 342, are distinguished by the presence of four in-cupped muscular suckers on the scolex and often a rostellum armed with hooks, by having the yolk glands concentrated into a single or bilobed mass near the ovary, and by having no uterine pore. The embryos remain passively in the egg or embryophore until eaten by the host in which they are to develop; this may be either a vertebrate or an invertebrate. Reid in 1947 showed that they are provided with a pair of unicellular glands opening between the hooklets and probably helpful in penetration. The larva may be either a cysticercus, a coenurus, a hydatid, or a cysticercoid (see pp. 339–340).

Six families contain species which are habitually or accidentally parasitic in man. These are:

Taeniidae. Medium-sized or large worms, except *Echinococcus,* which is very small. Scolex usually armed with a double row of large hooks but unarmed in *T. saginata;* ripe uterus with a central stem and lateral branches; genital pores lateral on alternating sides; ovaries and yolk gland in posterior part of segment; testes numerous; eggs with thick, striated inner shells. Important genera in man or domestic animals: *Taenia, Multiceps, Echinococcus.*

Hymenolepididae. Medium-sized or small worms, segments usually broader than long; scolex usually with a single row of hooks, but unarmed in *H. diminuta;* ripe uterus sac-like, not breaking up into egg balls; genital pores lateral, usually all on one side; ovary and yolk gland near center of proglottid, and with 1 to 4 testes. Important genus: *Hymenolepis.*

Dilepididae. Medium-sized or small worms; rostellum usually well-developed, retractile into a rostellar sac, and with 1 to 6 or 8 rows of rose-thorn hooks; genitalia single or double; genital pores unilateral or alternating; testes numerous; ripe uterus a transverse sac (subfamily Dilepidinae), or replaced by a paruterine organ (subfamily Paruterininae), or by egg capsules containing one to several eggs (subfamily Dipylidiinae). Important genus: *Dipylidium.*

Davaineidae. Medium-sized or small worms. Scolex with a double row of minute hammer-shaped hooks on rostellum and usually with numerous minute hooklets on the margins of suckers; ovaries and yolk gland near center of segment; uterus breaks up into egg capsules; testes fairly numerous. Important genera: *Davainea, Raillietina.*

Anoplocephalidae. Medium-sized or large worms of herbivorous animals. Scolex unarmed; female genital organs single or double in each segment, situated laterally or near middle; testes numerous; uterus develops a transverse sac or tubular network, in subfamily Thysanosominae later developing paruterine pouches; eggs usually with a pair of hornlike processes (*pyriform apparatus*) on one side of inner shell; cysticeroids, as far as known, develop in oribatid mites. Important genera: *Moniezia, Anoplocephala, Thysanosoma, Bertiella.*

Linstowiidae (formerly included as a subfamily of Anoplocephalidae). Medium-sized or small worms of insectivorous animals; scolex unarmed; sex glands in middle or anterior part of mature segments; testes numerous; uterus breaks down into egg capsules containing one or several eggs; eggs without pyriform apparatus; plerocercoids as far as known develop in beetles. Important genera: *Inermicapsifer, Oochoristica.*

Mesocestoididae. Medium-sized or large worms of carnivorous birds and mammals. Scolex unarmed; genital pore on mid-ventral surface; ripe uterus with or without a paruterine organ; ovaries and yolk glands posterior; testes numerous. Two genera: *Mesocestoides* and *Mesogyna.*

Taeniidae

The family Taeniidae includes for the most part relatively large worms parasitic in mammals. The form of the hooks in the armed species is shown in Fig. 98*B* and the arrangement of organs in the proglottids in Fig. 88. The eggs (Figs. 55*P* and 90*A*) have a very thin outer shell, sometimes provided with a pair of delicate filaments, which is ordinarily lost before the eggs are found in the feces. The inner embryophore has a thick, porous brown shell which on surface view looks honeycombed and in optical section looks striated. The larvae of most species are cysticerci, but in the genus *Multiceps* it is a coenurus and in *Echinococcus* a hydatid.

***Taenia solium* or pork tapeworm.** This worm is common in parts of the world where pork is eaten without thorough cooking, especially in some localities in Europe, but it is rare in the United States. In Jewish and Moslem countries, where the eating of pork is a serious religious misdemeanor, this parasite has little chance of survival and is scandalous evidence of moral turpitude when it does occur, just as is the beef tapeworm in Hindus. It is a remarkable fact that in many parts of the world, e.g., North America, India, the Philippines, human infections with adult worms are so rare that many laboratories are unable to obtain specimens, yet bladderworm infections in pigs are of fairly frequent occurrence. Even human infections with the bladderworm of this species are commoner than infections with the adult. This is one of the unsolved mysteries of parasitology.

MORPHOLOGY. The pork tapeworm (Fig. 85) usually attains a length of 6 to 10 ft.; records of specimens much longer than this are probably due to confusion of parts of more than one worm; there are 800 or 900 proglottids. The scolex (Fig. 98*B*) is smaller than the head of a pin, about 1 mm. in diameter, and has a rostellum armed with 22 to 32 hooks, long ones (180 μ) and short ones (130 μ) alter-

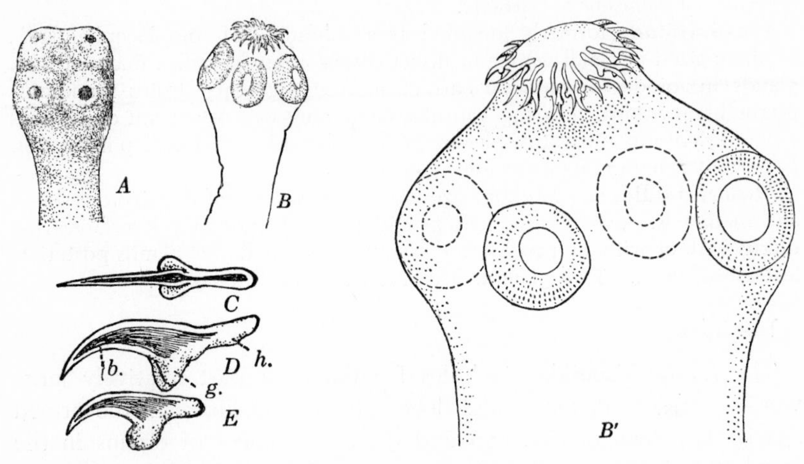

Fig. 98. *A*, unarmed scolex of *Taenia saginata*, $\times 10$; *B*, armed scolex of *T. solium*, $\times 16$; *B'*, same, $\times 60$; *C*, long hook, dorsal view; *D*, same, lateral view; *E*, short hook, lateral view; *b.*, blade; *g.*, guard; *h.*, handle or root; *C–E*, $\times 160$.

nating (Fig. 98*C* and *D*). Behind the head is a thin, unsegmented neck; the younger segments are broader than long, but in the middle part of the worm they become square, and the ripe ones are about twice as long as broad, shaped somewhat like pumpkin seeds and about 12 mm. long. The sexually mature proglottids closely resemble those of *T. saginata* (Fig. 88).

Soon after sexual maturity is reached and sperms for fertilizing the eggs have been received, the uterus begins to develop its lateral branches; in this species there are only 7 to 10 main branches on each side, a fact which is of special value in distinguishing the ripe segments from those of *T. saginata,* which has about twice as many (cf. Fig. 89*A*, *B*). The fully ripe uterus usurps nearly the whole proglottid; most of the other reproductive organs degenerate.

LIFE CYCLE (Fig. 90). A man infested with a pork tapeworm expels ripe segments, singly or in short chains, almost every day.

Several hundred a month are cast off, each loaded with thousands of eggs; the embryophores are nearly spherical and measure 35 to 42 μ in diameter. The shed ripe proglottids, unlike those of *Taenia saginata*, are flabby and inactive and are passed only in the feces, so pigs become infected as a result of coprophagous habits and are likely to have very heavy infections. Free eggs cannot consistently be found in the feces. The eggs probably survive for a considerable time in moist situations, as do those of *T. saginata*. The filthy way in which hogs are usually kept gives ample opportunity for their infection wherever there is human soil pollution or where privies are built in "open-back" style, or so that they leak. Young pigs are especially susceptible. The pig is not, however, the only intermediate host; the bladderworms can also develop in camels, dogs, monkeys, and man.

Upon ingestion by a suitable animal the oncospheres are liberated, bore through the intestinal wall, and make their way, via the blood or lymph channels, usually to the muscles or meat, but they may settle in almost any part of the body. They especially favor the tongue, neck, heart, elbow, and shoulder muscles, and certain muscles of the hams. Having arrived at their destination they grow into bladderworms or cysticerci, named *Cysticercus cellulosae*. The cysticerci are small, oval, whitish bodies with an opalescent transparency, 6 to 18 mm. long (Fig. 99), with a denser white spot on one side where the scolex is invaginated. Pork containing these larvae is called "measly" pork. Sometimes the cysticerci are so numerous as to occupy more than one-half the total volume of a piece of flesh, numbering several thousands to a pound.

When cysticerci in pork are eaten by man, all but the scolex is digested and it, turning right side out and anchoring itself to the wall of the small intestine, grows to maturity in about 2 or 3 months. Man is the only animal known to serve as a final host, though considerable growth takes place in dogs.

PATHOGENICITY. The adult worms in the intestine produce the same effects as *Taenia saginata* (see p. 359). This species, however, is particularly dangerous because the bladderworms as well as the adult can develop in man, causing cysticercosis. Self-infection with the eggs can result either from contaminated hands or by hatching of eggs liberated in the intestine and carried into the stomach by reverse peristalsis. Numerous cases of cysticercosis were diagnosed in British soldiers serving in North Africa or India during World War II.

The effects depend on the location of the cysticerci in the body. A

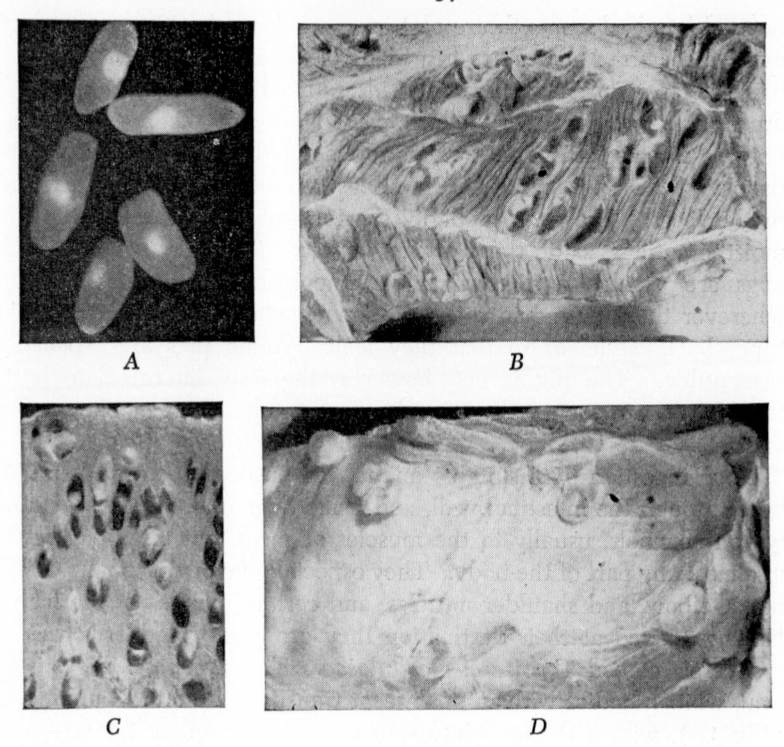

Fig. 99. *Cysticercus cellulosae: A,* freed cysticerci (×1½); *B,* cut pieces of "measly" pork, heavily studded with cysticerci (×½); *C,* cut piece of pig heart, loaded with cysticerci (×½); *D,* same, surface view (slightly reduced).

few in the muscles or subcutaneous tissues are nothing to worry about but, chiefly as the result of mechanical pressure, they may create unpleasant disturbances when they locate in the eye, heart, spinal cord, brain, or other delicate organs.

Eye infections require surgical removal. Brain infections lead to epileptic convulsions, violent headaches, giddiness, local paralysis, vomiting, and optic and psychic disturbances, often hysterialike in nature. Probably many such cases are never correctly diagnosed. Presence of subcutaneous cysticerci should lead to suspicion. Asenjo, in Chile, found 9% of "brain tumors" to be really cysticercosis, and he has devised a method by which even one cysticercus can be identified by ventriculographic x-ray. If the cysticerci are numerous, surgical removal may be impractical, but no other treatment is known.

Prevention and treatment for expulsion of adult worms in the intestine are discussed on pp. 343–345.

***Taenia saginata* or beef tapeworm.** This is the commonest *large* tapeworm of man and is cosmopolitan in distribution. In some localities, e.g., parts of Africa, Tibet, and Syria, where meat is broiled in large chunks over open fires, searing the surface but making the cysticerci in the interior only comfortably warm, it infects 25 to 75% of the people old enough to eat meat. In the Hindu sections of India *T. saginata* is religiously ostracized, since only the lowest outcast will eat the meat of the sacred cow or even of water buffaloes.

MORPHOLOGY. The beef tapeworm ordinarily reaches a length of 15 to 20 ft., but specimens up to 35 to 50 ft. have been recorded; the proglottids of an average worm number 1000 or more. The scolex (Fig. 98*A*) is 1.5 to 2 mm. in diameter and is without hooks. Both mature and ripe segments (Figs. 88 and 89*A*) are larger than those of *T. solium.* The detached terminal segments are about 20 mm. long and 6 mm. wide when relaxed. When freshly passed, usually singly, they are firm and very active, and crawl away like caterpillars; often they creep out of the anus and deposit eggs from the ruptured ends of the uterus on the perianal skin (see p. 343). Several times active specimens from the surface of a fresh stool have been sent to the writers as some new kind of fluke!

LIFE CYCLE. The life cycle is similar to that of *Taenia solium* except that usually the intermediate hosts are cattle or allied animals. However, giraffes, llamas, and pronghorn antelopes are occasionally infected with cysticerci, and lambs and kids have been experimentally infected; two valid human cases have been recorded. In the tropics cattle and buffaloes, habitually coprophagous, often have their flesh thoroughly riddled by the cysticerci. In India cattle, like pigs, frequently follow human beings to the defecation sites in anticipation of a fecal meal. Under favorable conditions the eggs remain viable in pastures for 6 months. Silverman and Griffiths have recently pointed out that *Taenia* eggs in sewage may be distributed by birds, particularly gulls, which frequent sewage disposal works in Britain. The cysticerci (named *Cysticercus bovis*) in measly beef are 7.5 to 10 mm. wide by 4 to 6 mm. long. They are most frequently present in the muscles of mastication and in the heart; these are the portions of the carcass usually examined in meat inspections. They are, however, inconspicuous and can easily be overlooked in raw or rare beef.

PATHOGENICITY. The damage done by adult taenias to their hosts is often either under- or overrated. There are some who believe that the presence of a tapeworm is more or less of a joke, and as such to be got out of the system but not to be taken seriously, whereas others become unnecessarily disturbed over them. They may cause mechanical injury by obstructing the intestinal canal and by injuring the mucous membranes where they adhere, and they may absorb enough nourishment to produce the proverbially ravenous "tapeworm appetite," although much more frequently they cause *loss* of appetite.

Swartzwelder in 1939, in a series of sixty cases in New Orleans, found abdominal pain, excessive appetite, weakness, and loss of weight to be the commonest symptoms. Other symptoms are nausea, difficult breathing, digestive disturbances, dizziness, insomnia, restlessness, false sensations, and occasionally convulsions and epileptic fits. Many of these symptoms might well be due to an induced vitamin deficiency in hosts on a marginal or suboptimal supply, which is deplorably common even in the relatively well-fed United States. Chandler (1943) showed that tapeworms thrive even when there are no vitamins or protein in the diet of the host, and that some, perhaps all, that are needed are acquired directly from the host. Anemia and eosinophilia are rare. The senior author knew of a case in which tuberculosis was suspected; the patient was weak, easily exhausted, and emaciated, with sunken cheeks and staring eyes. A fortnight after two large *Taenia* were expelled he was like a new man. In contrast, a colleague harbored a *Taenia* for years; in spite of a number of unsuccessful efforts to part company with it, "Horace," as he familiarly called his guest, stayed with him, yet there were never any symptoms other than segments in the stools, and the host continued in ruddy and robust health. The latter case is, perhaps, much more common than the former. For diagnosis see p. 343; for treatment and prevention, pp. 343–345.

Other species of *Taenia* and *Multiceps*. The genus *Taenia* and the genus *Multiceps*, distinguishable only by the multiple heads produced in the larvae of *Multiceps*, include many species parasitic as adults in dogs and cats and as larvae in herbivores. Some of the commonest ones in dogs are *T. pisiformis* (= *serrata*), the larvae of which develop in the liver and mesenteries of rabbits; *T. ovis*, developing in the connective tissue in muscles of sheep; *T. hydatigena*, developing in the liver of sheep; *M. multiceps* (Fig. 100), developing as a coenurus in the brain of ruminants and causing gid; and *M. serialis*, developing in subcutaneous connective tissue of rabbits. *T. pisiformis* and *T. hydatigena* occur also in cats, but the commonest form in these animals is *T. taeniaeformis* which develops in the livers of rats and mice. One human case is recorded. The bladderworm of this species, *Cysticercus fasciolaris*, contains a considerable chain of undeveloped segments and is sometimes called a strobilocercus. All these worms resemble *T. saginata* except in minor details; the scoleces differ in the number and size of the hooks.

About a dozen cases of coenurus infection in man have been recorded: several brain infestations with *M. multiceps* (*Coenurus cerebralis*), one of which caused epileptic symptoms; a number of muscular or subcutaneous infections, some identified as *M. serialis* and at least one of the others as *M. glomeratus*, previously described from a gerbil. Crusz in 1948 suggested that these are possibly all

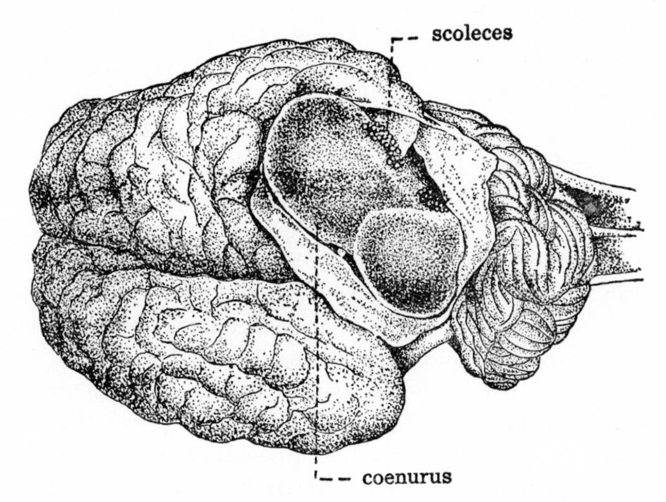

Fig. 100. Brain of giddy sheep with coenurus of *Multiceps multiceps*, showing masses of scoleces. (After Neumann, from Hall.)

one species, but in 1956 Fain showed that subcutaneous human infections in central Africa involve the larvae of *Multiceps brauni* which normally occurs in rodents. The species of *Multiceps* are distinguished mainly by the number, size, and shape of the rostellar hooks.

A few rare species of adult *Taenia* have been found in man. Four cases of "*T. confusa*" have been reported in the United States and three from eastern Africa, but Anderson (1934) believes this form to be only a variant of *T. saginata.* Probably *T. bremneri*, described from a Nigerian, is the same thing. Another species, of which two specimens were obtained from an East African, is *T. africana.* It has segments broader than long, unarmed scolex, and a uterus with unbranched arms.

***Echinococcus* and hydatid cysts.** The genus *Echinococcus* contains several species of very minute tapeworms which live as adults in the intestines of canine and feline animals. One species, *E.*

granulosus, has been recognized as producing hydatid cysts in man and domestic animals. The malignant alveolar form of echinococcosis known from North and Central Europe, and from the subpolar areas of North America, is caused by a distinct species, *E. multilocularis.*

E. granulosis has both a sylvatic and a pastoral epidemiology. The infestation is best known, naturally, in the areas where it is passed back and forth between dogs and sheep or cattle, and more or less frequently to man. It is especially prevalent in such sheep and cattle-raising areas as North and South Africa, the Middle East, Australia, New Zealand, southern South America, and until recently Iceland. Today the infection is becoming rare in Iceland, except in elderly people. In many of the areas mentioned above, one-fourth of the dogs and half of the animals are infected; in the Middle East hydatids occur in about 20% of sheep, 40% of cattle, and 100% of camels. In northern Scandinavia there is a dog-reindeer cycle in which humans become inadvertently involved. Pigs are commonly infected in Virginia, Georgia, and Alabama. Since dogs are not commonly infected in these areas, it seems probable that foxes are involved.

The sylvatic form of *E. granulosus* occurs in circumpolar areas where wolves or foxes and moose or caribou are principally involved, and in Australia, where the parasite oscillates between dingoes and wallabies. Infection is maintained in sled dogs in Alaska by the common native practice of feeding them portions of the viscera of moose or caribou. In Siberia, St. Lawrence Island (Alaska), Central Europe, and continental North America, *E. multilocularis* passes between dogs, foxes or wolves and rodents (*Microtus, Clethrionomys,* and *Citellus*). In most cases, man as a host is a blind alley for the parasite, thus being undesirable for the perpetuation of future generations of worms. Exceptions to this may be found in certain African tribes, e.g., the Masai, or Eskimos who do not bury the dead but abandon their bodies in deserted places to which wild carnivores have access. Schiller (1954) showed that blowflies can carry *Echinococcus* eggs from excreta to food.

MORPHOLOGY. The adult (Fig. 101) is structurally much like a *Taenia,* but is very unlike it in size, and the ripe uterus has a broader central stem with only lobelike out-pocketings, which are often indistinct. The worm is only 2 to 8 mm. long, and consists of a scolex and neck followed by only three or four successively larger segments, one immature, one or two mature, and usually one ripe or nearly ripe. The head has a protrusible rostellum armed with a double row of 28 to 50 hooks, usually 30 to 36. The worms occur by hundreds or even thousands in the intestines of dogs but are usually overlooked on account of their minute size. Each ripe segment contains 500 to 800 eggs. In spite of the small size of the adult worms, they require 4 to 6 weeks to mature in a dog. It is very difficult to distinguish the adults of *E. granulosus* and *E. multilocularis* (Vogel, 1957).

DEVELOPMENT OF HYDATIDS. The eggs, about 30 by 38 μ, are indistinguishable from those of dog taenias. From the feces of dogs, wolves, or foxes they gain access to their intermediate hosts with con-

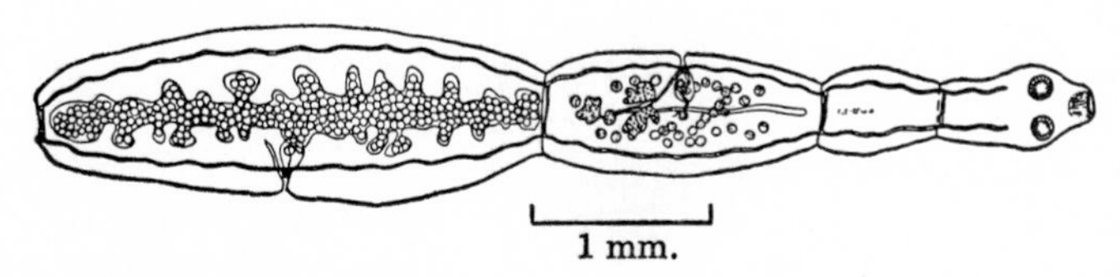

Fig. 101. *Echinococcus granulosus.* (After Mönnig, *Veterinary Helminthology and Entomology,* 1949.)

taminated forage or water. In addition to the animals noted above, pigs, horses, rabbits, and many other herbivores are susceptible. Human infection usually results from too intimate association with dogs; children are especially liable to infection by allowing dogs to "kiss" them or lick their faces with a tongue which, in view of the unclean habits of dogs, is an efficient means of transfer of tapeworm eggs. Transfer of eggs on the hands from an infected dog's fur is also a good means of infection. In moose, deer, and caribou the hydatids develop almost exclusively in the lungs, but in domestic animals the liver and other organs are more frequently sites, and in man about 60 to 75% of the cysts develop in the liver, only about 20% in the lungs. Smaller numbers reach the kidneys, spleen, muscles, bone, heart, brain, and other organs.

Development of the cysts is slow. The young larva changes into a hollow bladder, around which the host adds an enveloping, fibrous cyst wall. At the end of a month these cysts measure only about 1 mm. in diameter; in 5 months they are about 10 mm. in diameter and the inner surface is beginning to produce hollow brood capsules. These ultimately remain attached only by slender stalks and often fall free into the fluid-filled cavity of the mother cyst. As the cyst grows larger more brood capsules form, and the older brood capsules begin to differentiate, on their inner walls, a number of scoleces, usually 3 or 4 to 30 (Figs. 102, 103). Sometimes the mother cyst, as the result of pressure, develops hernialike buds which may detach themselves and continue their development independently as daughter cysts. The fluid of the cysts is nearly colorless; in older cysts there is a granular deposit consisting of liberated brood capsules and free scoleces, called "hydatid sand." A cyst of 2 quarts capacity may produce more than 2 million scoleces.

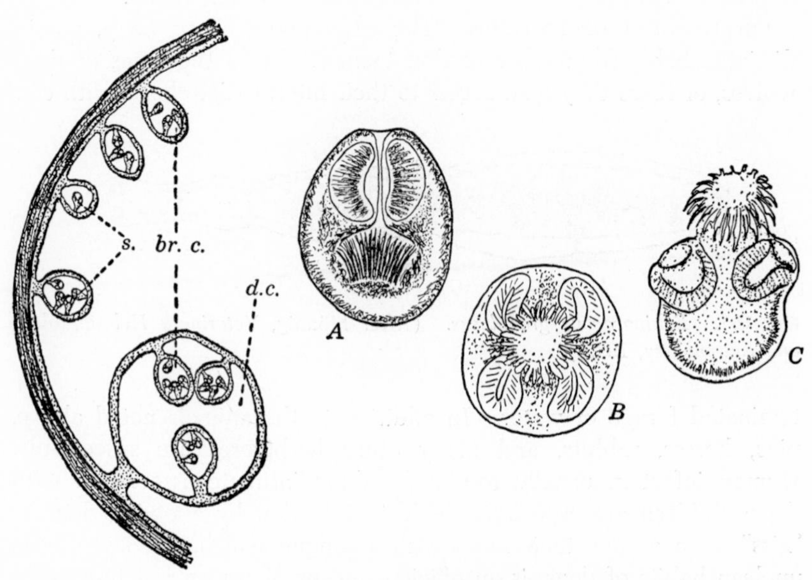

Fig. 102. *Left,* diagram of small hydatid cyst of *Echinococcus,* showing daughter cyst (*d.c.*), brood capsules (*br.c.*), and scoleces (*s.*). Stippled inner wall of cyst is part of parasite; outer fibrous wall is capsule laid down by host. *Right,* scoleces from cyst; *A,* invaginated; *B,* head-on view; *C,* evaginated.

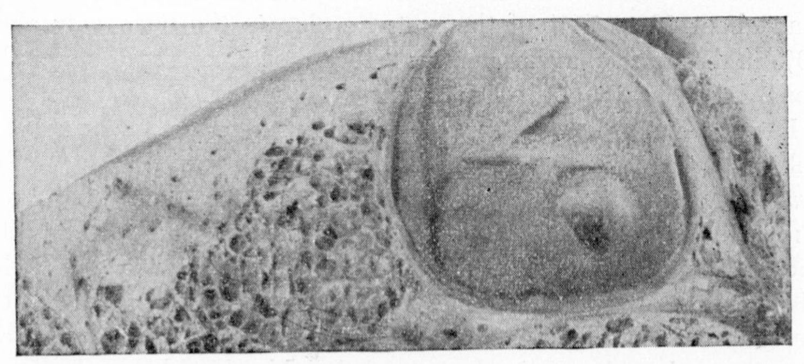

Fig. 103. Cut hydatid cyst showing numerous scoleces like velvet on inner surface.

Eventually the cysts may reach the size of an orange or larger. After 10 to 20 years they may reach enormous size and contain 10 to 15 quarts of liquid, or occasionally even more. When growth is unobstructed the cysts are more or less spherical, but they are often deformed by pressure. When developing in bones they fill the marrow cavities and may cause bone erosion. In 25% of human cases more

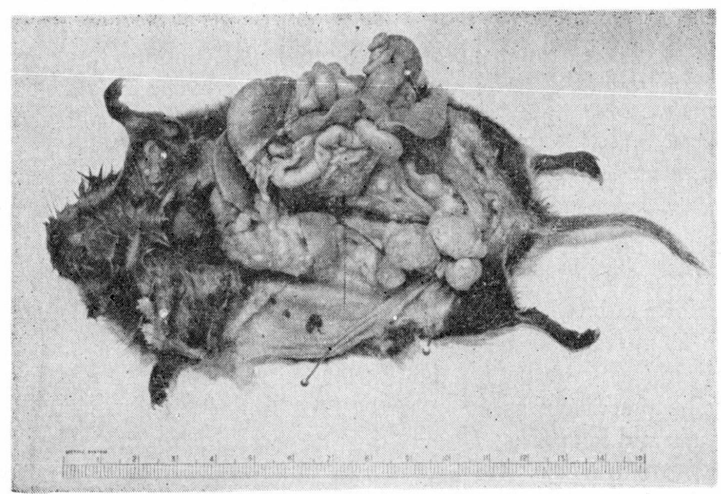

A

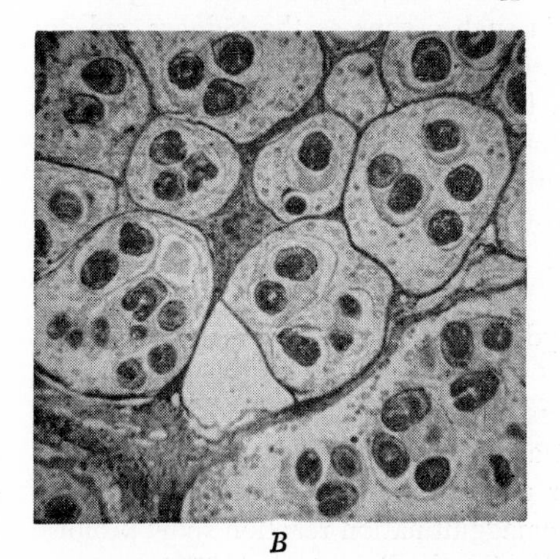

B

Fig. 104. Experimental *Echinococcus multilocularis* infection in a lemming. *A*, the whole animal. *B*, section of the liver from animal shown in *A*. Note the replacement of liver tissue by the growing parasite. (Photographs made in collaboration with Dr. Everett Schiller.)

than one cyst is present, either due to original multiple infection or to development of detached daughter cysts. Not infrequently cysts fail in their primary purpose of producing scoleces and remain "sterile." Possibly this is connected with natural or acquired immunity in the hosts.

MULTILOCULAR CYSTS. Instead of forming a single large vesicle, the larva of *E. multilocularis* forms a spongelike, constantly growing mass of small separate vesicles embedded in a fibrous tissue (Fig. 104). It is not delimited by a capsule formed by the host, and the vesicles contain a gelatinous substance instead of fluid. Roots grow out into neighboring tissues. The central portions degenerate and die while growth continues on the outside, as in a true malignant tumor. Often portions of the growth become separated and continue to grow like the parent; such detached portions may be carried to distant parts of the body. This type of hydatid known as a multilocular or alveolar cyst develops principally in the liver. It occurs in central Europe, Britain, Siberia, northern Japan, and Alaska. Multilocular cysts have been reported from other areas, e.g., Turkey, Africa, South America, but it is not yet clear whether these are *E. multilocularis* or some other species. Vogel has shown that *E. granulosus* may sometimes produce a rather diffuse hydatid in calves, but this can be differentiated from *E. multilocularis* by the lack of septa.

PATHOLOGY. In their natural wild hosts unilocular hydatid cysts may be practically harmless, and this is frequently true in domestic animals and man also. Usually they do serious harm only when they grow to outrageous size in the liver and press on other organs; or liberate their fluid by leaking or rupturing, thus precipitating severe allergic symptoms; or develop in such organs as the kidneys, spleen, brain, or eye. When cysts are ruptured, scattered scoleces and brood capsules develop in other locations; this is especially dangerous if rupture into a blood vessel occurs.

On the other hand, multilocular cysts are probably highly pathogenic to the natural rodent hosts. Growth and actual replacement of liver tissue by parasite tissue are extremely rapid in these animals (Fig. 104).

DIAGNOSIS. Probably most hydatid cysts in either man or animals are detected only at autopsy. In suspected cases precipitation or complement-fixation tests are possible, using hydatid fluid or an extract of adult *Taenia* as antigen, but a skin test with these antigens, called the "Casoni reaction," is better and easier. A bentonite-flocculation test and a recently developed hemagglutination reaction show promise of replacing the Casoni reaction. Garabedian et al. (1957) reported that a hemagglutination test was more sensitive, more specific, and easier to perform than the complement fixation test. Diagnosis by x-ray is often possible, especially for pulmonary cysts, but in the liver the cysts are detectable only when calcified (see Miller, 1953).

TREATMENT AND PREVENTION. Treatment is purely surgical, but this parasite grows fast to the fibrous walls formed by the host and does not "shell out." It is dangerous to withdraw fluid directly, and it is customary to withdraw part of the fluid with a trocar and replace it at once, unless in the lung, with a formalin solution to kill the scoleces, brood capsules, etc. Subsequently the fluid can be drained out. Multilocular cysts can seldom be operated on successfully, and generally lead to death in a few years.

Prevention consists in avoiding too much intimacy with dogs; carefully washing hands after handling them, and also washing dishes from which they have eaten; and avoiding food or water which may have been contaminated by them. Care should also be taken that dogs are not fed, or do not get access to, the entrails or waste parts of slaughtered or dead animals from which they can become infected.

The practical elimination of the disease in Iceland was accomplished by licensing and annual treatment of dogs, and enforcing the burial or burning of infected material. Arecolin hydrobromide, one-sixteenth grain per 10 lb. wt., eliminates 95% of *Echinococcus* from dogs; 3 doses usually eliminate all, and most *Dipylidium* and *Taenia* as well.

Hymenolepididae

This family contains a large number of species of tapeworms parasitic in birds and mammals, particularly in the former. Their characteristics are summarized on p. 354. Three species have been found in man. One, *Hymenolepis nana,* is a very common parasite of man and of rats and mice; another, *H. diminuta,* is abundant in rats and mice but relatively rare in man, though by no means a curiosity; the third, *H. lanceolata,* is a parasite of ducks and geese and has been recorded from man only once.

Hymenolepis nana. The dwarf tapeworm, *H. nana,* is the smallest adult tapeworm found in man, but it makes up for its diminutive size by the large numbers which are often present. It has a world-wide distribution, but it is far commoner in some localities than in others. It is the commonest tapeworm in southern United States, where about 1 to 2% of the population, especially children, are infected. Sunkes and Sellers in 1937 collected data on 927,625 fecal examinations in the southern states and got records of 8085 tapeworm infections; all but 100 of these (98.6%) were *H. nana.* In 1955 Neghme and Silva reported that 8% of 17,219 people examined in Chile were infected. In some parts of India as high as 18 to 28% of the population were found by the senior writer to be infected. In 500 Egyptian villagers examined by the writer, 36 *H. nana* infections were found (7%), but all but 2 of these were in children below the age of puberty.

The adult worm ranges from 7 to over 100 mm. in length. In general the length of the worms is inversely proportional to the number present; in heavy infections it is commonly 20 to 30 mm., with a maximum breadth of only 500 to 600 μ; it is seldom found after treatment, even when diligently sought. The scolex (Fig. 105*A* and *B*) has a well-developed retractile rostellum with a crown of 20 to 30 hooks. All the proglottids are considerably broader than long. The arrangement of the organs in mature proglottids can be seen in Fig. 105*D*. The uterus develops as a sac which practically fills the segment between the excretory vessels. The uterine walls and partitions between segments may break down to allow passage of eggs from segment to segment, and out from the broken posterior end of the worm or between segments.

The eggs have a very characteristic appearance (Fig. 55*O*). The outer shell is oval, thin, and practically colorless; it commonly measures about 40 by 50 μ. The embryophore is lemon shaped, 16 to 20 μ long, with a little knob at either end from which arise a number of long, delicate, wavy filaments which lie in the space between the embryophore and the outer shell. All six embryonic hooks lie approximately parallel in healthy oncospheres.

LIFE CYCLE. *Hymenolepis nana* differs from almost all other tapeworms in being able to complete its entire life cycle in a single host. In this it is radically progressive, having broken away from the age-old tapeworm custom of utilizing intermediate hosts. It can, however, still revert to the habits of its ancestors and develop in fleas or grain beetles. When the eggs are ingested by man, rats, or mice, the oncospheres begin to claw actively inside their shells, and escape in the lumen of the intestine. They burrow into the interior of the villi and there develop into tailless cysticercoids in about 4 days (Fig. 105*G*). On reaching maturity these escape into the lumen of the intestine, the scoleces attach themselves, and the worms grow to maturity in about 15 to 20 days. In grain beetles, however, development of the tailed cysticercoids (Fig. 105*F*) takes 12 to 14 days.

In egg infections, since the worm is parenterally located during development of the cysticercoid, immunity develops, but not after cysticercoid infections.

RELATION OF HUMAN AND RODENT STRAINS. The identity or otherwise of *H. nana* of man and *H. nana fraterna* of mice and rats has been much disputed. The human infection is relatively rare in some localities, especially northern Europe and Canada, where the rodent infections are common, and although eggs of human worms will develop in rodents, and vice versa, this does not occur as readily as when the eggs are ingested by the same hosts as those from which the eggs were derived. The junior author readily infected himself

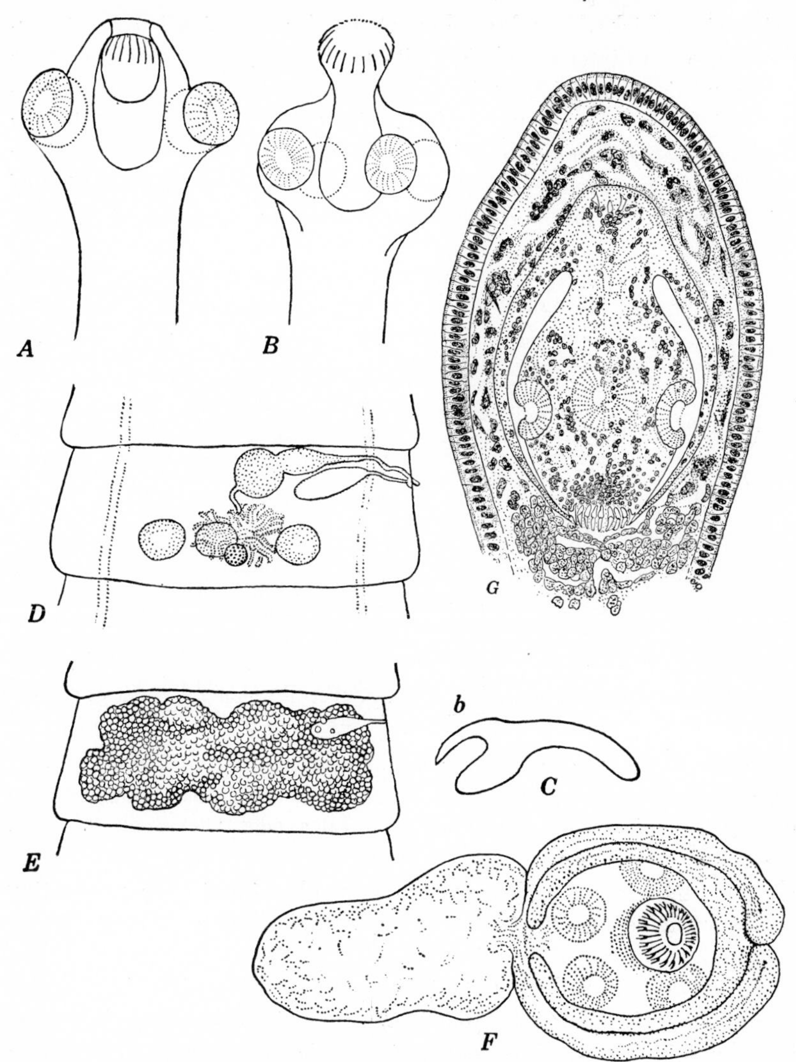

Fig. 105. *Hymenolepis nana: A,* scolex with rostellum retracted; *B,* same, rostellum exserted; *C,* rostellar hook, blade (*b*) up; *D,* mature segment; *E,* ripe segment; *F,* cysticercoid from body cavity of *Tribolium; G,* cysticercoid in villus of mouse intestine. (*G,* adapted from drawing by W. S. Bailey.)

with cysticercoids of worms from mice. Shorb (1933) reported differences between strains from rats and mice.

In India, however, Chandler (1927) found an inverse correlation between the incidence of *H. nana* infections and that of *Ascaris* and

Trichuris, which depend on human fecal contamination for transmission, but a direct correlation with prevalence of household rodents and conditions favoring their access to food, and with such rodent-borne infections as plague and *H. diminuta.* The fact that in our southern states *H. nana* infections are about equally common in cities with sewerage systems and in rural areas is also more suggestive of dissemination by rats and mice than by human contamination. Pre-

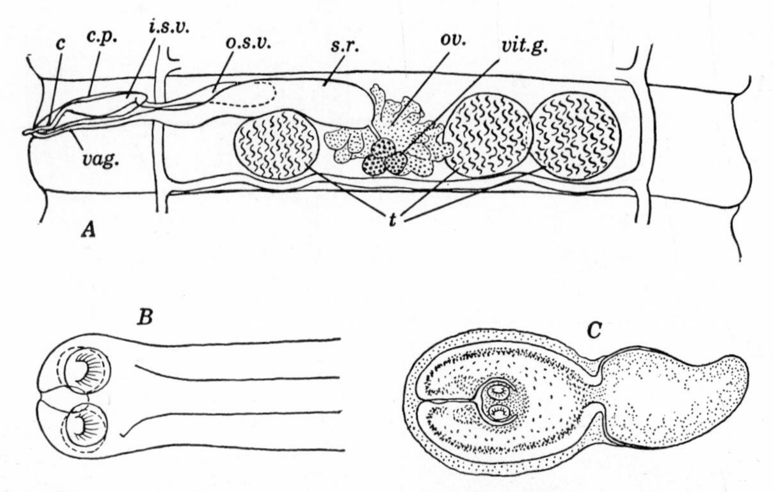

Fig. 106. *Hymenolepis diminuta: A,* mature segment; *B,* scolex; *C,* cysticercoid from *Tenebrio* beetle; *c.,* cirrus; *c.p.,* cirrus pouch; *i.s.v.,* inner seminal vesicle; *o.s.v.,* outer seminal vesicle; *ov.,* ovary; *vag.,* vagina; *vit.g.,* vitelline gland.

vention, therefore, would seem to depend primarily on preventing access of mice or rats to food that is to be eaten without further cooking.

H. nana causes rather severe toxic symptoms, especially in children, including abdominal pain, diarrhea, convulsions, epilepsy, insomnia, and the like. Diagnosis is easily made by finding the eggs in the feces; like nematode eggs, they float in strong salt solutions. Treatment is considered on p. 343.

Other species of *Hymenolepis*. *H. diminuta,* very common in rats and mice in all parts of the world, is much less common in man. It is a much larger worm than *H. nana,* reaching a length of 1 to 3 ft., with a maximum diameter of 3.5 to 4 mm. The head (Fig. 106*B*), unlike that of nearly all other species of *Hymenolepis,* is unarmed, and the segments (Fig. 106*A*) are much broader than long. The structure of mature and ripe segments is much like that of *H. nana.* The eggs

(Fig. 55*N*) are larger (60 to 80 μ in diameter), yellow or yellow-brown, and usually spherical. The oncosphere lacks the knoblike thickenings at the poles, or at best they are rudimentary, and there are no filaments.

Like most kinds of *Hymenolepis* this worm requires an intermediate host for the development of its cysticercoids (Fig. 106*C*). It is satisfied with any one of many grain-infesting insects, including larvae and adults of meal moths (*Pyralis farinalis*), nymphs and adults of earwigs (*Anisolobis annulipes*), adults of various grain beetles such as *Tenebrio* and *Tribolium*, dung beetles, the larvae of fleas, and even myriapods. Human infection results from eating such foods as dried fruits and precooked breakfast cereals in which the grain insects, infected from rat or mouse droppings, are present. Until about 1925 this infection was considered sufficiently rare in man so that every instance was published as an incident worthy of note, but Chandler found 23 cases in about 10,000 fecal examinations in India and found no less than 3 in 50 examinations in one locality where the food habits and rat population were particularly favorable. He also found 9 cases (nearly 2%) in examinations of 500 Egyptian villagers. As is usually true with human tapeworms which belong in another host, this worm is very easily expelled by anthelmintic treatment and is sometimes expelled spontaneously or after a cathartic.

Dipylidiinae

Dipylidium caninum. Although many species of *Dipylidium* have been described, Venard (1938) thinks nearly all of them are really one species, *D. caninum*, an extremely common parasite of flea-infested dogs and cats all over the world. Over 100 human cases, nearly all in children, have been reported, but the actual number of cases is undoubtedly much greater. One doctor from one Texas city has sent the senior writer three specimens from children for identification. *D. caninum* is a delicately built tapeworm, commonly reaching a length of about a foot. The peculiar characteristics of the scolex and proglottids are mentioned on p. 354 and illustrated in Fig. 107. The uterus first develops as a honeycomblike network, but later breaks up into egg balls, each containing 5 to 20 eggs, which remain intact even when the segments disintegrate. The ripe proglottids are the size and shape of elongated pumpkin seeds and are often seen squirming actively in the freshly passed feces of infected animals.

The intermediate hosts are fleas (*Ctenocephalides* and *Pulex*) and dog lice (*Trichodectes canis*). Joyeux (1920) observed that the eggs could not be ingested by adult fleas but are devoured by the larvae.

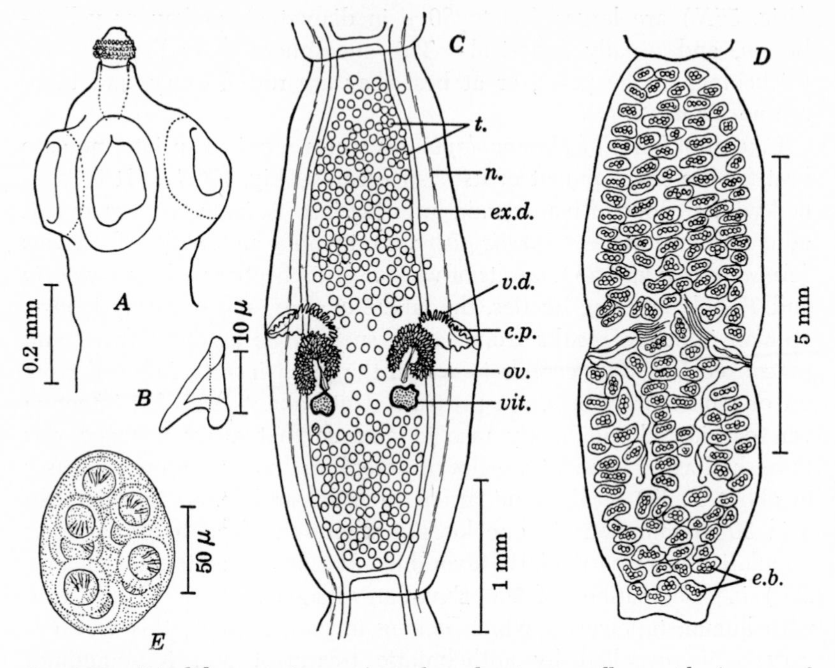

Fig. 107. *Dipylidium caninum: A*, scolex, showing rostellum with 4 rows of hooks; *B*, rose-thorn rostellar hook; *C*, mature proglottid; *D*, ripe proglottid filled with egg balls; *E*, single egg ball. Abbreviations: *c.p.*, cirrus pouch; *e.b.*, egg balls; *ex.d.*, excretory duct; *n.*, lateral nerve; *ov.*, ovary; *t.*, testes; *v.d.*, vas deferens; *vit.*, vitellaria. (*D*, adapted from Hall, *Proc. U.S. Natl. Mus.*, 55, 1919. Others from Witenberg, *Z. Parasitenk.*, 1932.)

The embryos hatch in the intestine and bore through into the body cavity, where they remain very little changed until the flea has transformed into an adult, whereupon it develops into a cysticercoid which infects the final host when the flea is nipped. Children are probably infected by having their faces licked by a dog just after the dog has nipped a flea.

Anoplocephalidae

The Anoplocephalidae, the principal characters of which are mentioned on p. 355, are very common parasites of herbivorous animals, including cattle, sheep, goats, horses, camels, rabbits, rodents, and also apes and pigeons. They are often present in very young animals and the incidence of infection may be very high.

The life cycle of these worms was one of the outstanding mysteries of parasitology until Stunkard (1938) succeeded in developing the cysticercoids of *Moniezia* of cattle and sheep in oribatid mites (see p. 561). The mites (Fig. 108*G*, *H*), living about the roots of grass, are seldom seen but may be very abundant and are undoubtedly often eaten by grazing animals. In a pasture at Beltsville, Md., there were estimated to be 6,000,000 oribatid mites (*Galumna virginiensis*) per acre, nearly 4% of them harboring 1 to 13 *Moniezia* cysticercoids—400,000 potential tapeworms per acre!

Since Stunkard's work a dozen other anoplocephalid tapeworms have been found to develop in oribatid mites, and probably all members of the family do so, now that the former subfamily Linstowiinae has been eliminated and elevated to the rank of a separate family. The development of cysticercoids of typical anoplocephalids in oribatid mites is very slow, taking 5 to 15 weeks.

Many animals harbor anoplocephalid infections. *Moniezia* (Fig. 108*B–D*) are large worms of cattle, sheep, and goats, reaching 10 ft. or more in length, with double sets of reproductive organs in the proglottids. Sheep and goats in western United States commonly harbor the fringed tapeworm, *Thysanosoma actinioides* (Fig. 108*I*) characterized by fringes on the posterior borders of the segments. Horses harbor a number of species, but the most important are two rather short, thick worms of the genus *Anoplocephala: A. magna,* about 10 in. long, in the small intestine, and *A. perfoliata,* only 1 to 2 in. long, in the cecum. Rabbits are commonly afflicted by members of the genus *Cittotaenia.* Young animals are said to suffer digestive disturbances and retarded growth from anoplocephalid infections if their nutrition is poor, as might be expected, but Kates and Goldberg (1951) found no evidence of injury in well-fed lambs.

Human infection with members of this family is limited to *Bertiella studeri* (Fig. 108*A*, *E*. *F*) normally parasitic in apes and monkeys. This worm has been reported from man only 11 times, mostly in children. It is a thick, opaque worm 25 to 30 cm. long and 10 to 15 mm. broad. The arrangement of organs is shown in Fig. 108. The ripe proglottids are very broad, but less than 1 mm. in length; they are shed in blocks of 20 or more. Most of the human cases have occurred around the Indian Ocean, but the infection appears also to have been established in the West Indies—an example of the danger of introducing foreign species of worms with captive animals. Stunkard (1940) developed minute cysticercoids (0.1 to 0.15 mm. in diameter) in oribatid mites but was unsuccessful in infecting man or monkeys with them.

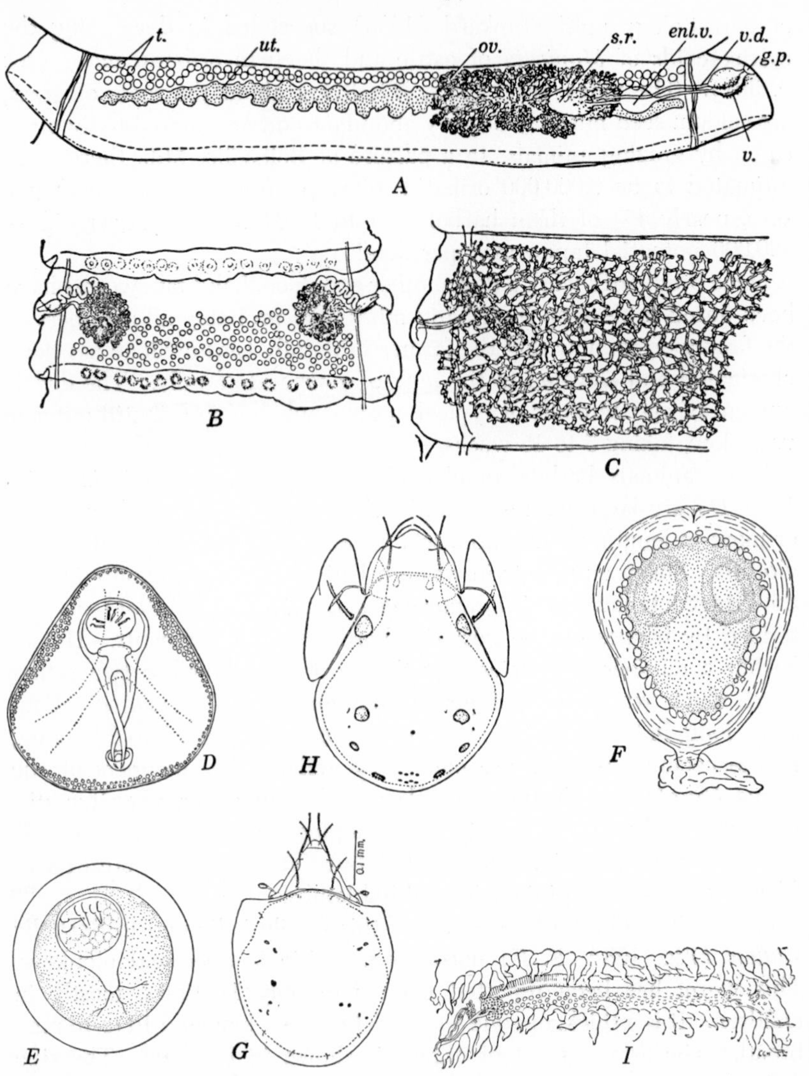

Fig. 108. Anoplocephalids and vectors. *A, Bertiella studeri,* mature segment (from Chandler, *Parasitology,* 17, 1925). *B, Moniezia expansa,* mature segment, and *C,* same, left half of ripe segment (after Fuhrmann from Wardle and McLeod, *The Zoology of Tapeworms,* 1952). *D,* egg of *Moniezia* (adapted from Mönnig, *Veterinary Helminthology and Entomology,* 1949). *E,* egg, and *F,* cysticercoid of *Bertiella studeri* (after Stunkard, *Am. J. Trop. Med.,* 20, 1940). *G, Protoschelobates seghettii,* and *H, Galumna virginiensis,* intermediate hosts of *Moniezia expansa* (after Kates and Runkel, *Proc. Helm. Soc. Wash.,* 15, 1948). *I, Thysanosoma actinioides,* fringed tapeworm (after Fuhrmann in Kükenthal, *Handbuch der Zoologie, Vermes Amera*).

Linstowiidae

This family, which includes the large genus *Oochoristica*, with species in reptiles, birds, and mammals, contains tapeworms which resemble the Davaineidae closely except in having unarmed scoleces. The only life cycles known involve beetles. One member of this family, *Inermicapsifer arvicanthidis* (= *I. cubensis*) (Fig. 109) parasitizes man. This worm was first found in children in Cuba by Kouri, and was named *I. cubensis*. Over 100 human infections have been

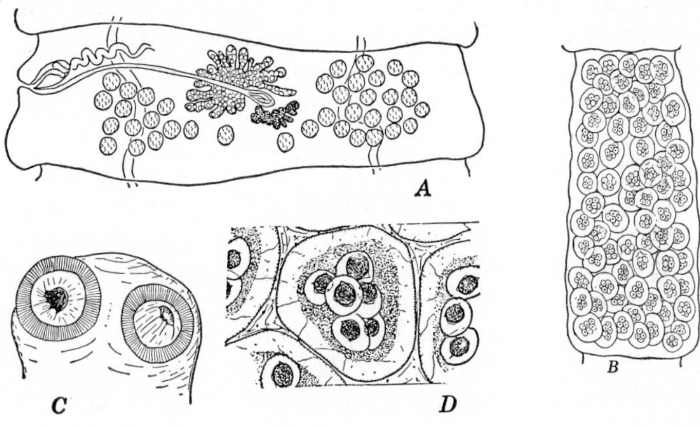

Fig. 109. *Inermicapsifer arvicanthidis* (= *I. cubensis*). *A*, mature proglottid; *B*, ripe proglottid; *C*, scolex; *D*, egg capsules. (From Baer, Kourí and Sotolongo, *Acta Tropica,* 1949.)

reported in Cuba. Later Fain (1950) showed that this worm is the same as *I. arvicanthidis*, common in small rodents in Africa and found a few times in humans in Africa. Baer (1956) reported it from two humans in Mauritius and concluded that the worms had been confused in the past with *Raillietina* (see p. 376) and should be called *I. madagascariensis*. The authors do not feel that Baer has resolved the confusion and prefer to retain the name *I. arvicanthidis*. The worms are 2 to 3 ft. long, the ripe segments being 3 to 4 mm. long and 1 to 2 mm. wide. It seems highly probable that the worm was introduced into Cuba in rodents from Africa, for all the other dozen or so species of *Inermicapsifer* are parasites of hyraxes or rodents in Africa.

Davaineidae

The majority of the tapeworms in this family are parasitic in birds, and several are common and injurious parasites of poultry. The general characteristics are given on p. 354. The cysticercoids of the chicken parasites develop in various intermediate hosts: the minute but injurious *Davainea proglottina* in slugs; *Raillietina tetragona* in maggots of the housefly; *R. echinobothrida,* another particularly pathogenic species, in an ant; and *R. cesticillus* in various beetles. Phenylmercuric compounds in a dose of 50 mg. are effective against *R. cesticillus* but not against the others.

A number of cases of human infection with worms of the genus *Raillietina* have been recorded from various parts of the world—in seaports around the Indian Ocean and South China Sea from Madagascar to Queensland to Japan, in Cuba, and in Guiana and Ecuador in South America. The human cases undoubtedly represent accidental infections with species parasitizing local wild animals.

For many years all the Old World cases were referred to the species *R. madagascariensis,* but Baer and Sandars (1956) concluded that most of these were actually *R. celebensis,* which is primarily a parasite of rats. The South American forms, reported from Guiana, Ecuador, and Cuba were considered by Baer and Sandars to be *R. demerariensis* (Fig. 110), a parasite of howler monkeys and rodents. There is reason to believe that *Raillietina* and *Inermicapsifer* infections in humans have been confused in the past. Chandler and Pradatsundarasar (1957) reviewed the taxonomy of *Raillietina* from man and concluded that most specific identifications have been made from quite inadequate material. These authors described a new species, *Raillietina siriraji,* from children in Bangkok, recognizing the probability that they were dealing with some species previously recovered from humans.

These species of *Raillietina* are slender worms reaching a length of 1 to 3 ft., with a maximum width of 3 to 8 mm. All the genital pores are on one side. The scolex has a double crown of small hooks (Fig. 110, *1*), and the suckers are armed with a number of rows of minute spines. The ripe proglottids are usually squarish or elongate and contain about 100 to 400 egg capsules, each with several elongated eggs.

There is no question but that these worms utilize some arthropod as an intermediate host.

Mesocestoididae

The genus *Mesocestoides* has the peculiar characters listed for Mesocestoididae on p. 354, including a posterior paruterine organ or egg ball (Fig. 110, *4–6*). The number of species has been much disputed, since the worms show considerable variation and there are no good differential characters. The entire life cycle is unknown. *Sparganum*-like larvae called tetrathyridea occur free or encysted in reptiles, birds, and mammals, but these are probably second larval

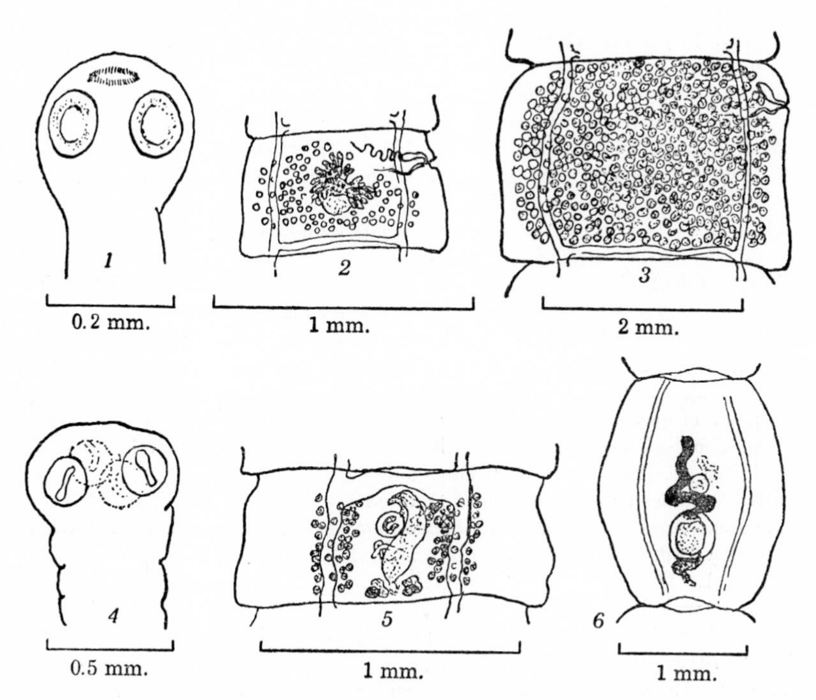

Fig. 110. Upper row, *Raillietina*, from man in Ecuador (*R. demerariensis?*): *1*, scolex, with double crown of about 150 small hooks, and minute spines on suckers; *2*, mature proglottid; *3*, ripe proglottid, with about 200 to 300 egg capsules, each with about 7 to 10 eggs. (After drawings and description by Dollfus, *Ann. Parasitol. hum. et comp.*, 17, 1939–1940.) Lower row, *Mesocestoides variabilis*, from child in Texas; *4*, scolex, showing slit-like openings of suckers; *5*, mature proglottid, showing yolk glands and ovaries posteriorly; cirrus pouch near center; developing uterus; convoluted vagina; and testes on both sides of excretory canals; *6*, ripe proglottid, showing egg ball, remnants of uterus, and cirrus pouch. (After Chandler, *Am. J. Trop. Med.*, 22, 1942.)

stages. The first human infection with a *Mesocestoides* was reported by Chandler (1942) from a child in east Texas. The worms, estimated up to 40 cm. long and about 1.6 mm. wide, are probably *M. variabilis*, previously known from foxes, skunks, raccoons, and dogs in the United States. Subsequently another case was found in a Greenlander in Denmark.

REFERENCES

General

Addis, C. J., Jr., and Chandler, A. C. 1944. Studies on the vitamin requirements of tapeworms. *J. Parasitol.,* 30: 229–236.

Addis, C. J., Jr. 1946. Experiments on the relation between sex hormones and the growth of tapeworms. *J. Parasitol.,* 32: 574–588.

Beaver, P. C., and Sodeman, W. A. 1952. Treatment of *Hymenolepis nana* (dwarf tapeworm) infestation with quinacrine hydrochloride (atebrin). *Am. J. Trop. Med. Hyg.,* 55: 97–99.

Beck, J. W. 1951. Effect of diet upon singly established *Hymenolepis diminuta* in rats. *Exp. Parasitol.,* 1: 46–59.

1952. Effect of gonadectomy and gonadal hormones on singly established *Hymenolepis diminuta* in rats. *Exp. Parasitol.,* 1: 109–117.

Brown, H. W. 1948. Recent developments in the chemotherapy of helminthic diseases. *Proc. 4th Intern. Congr. Trop. Med. Malaria,* 2, Sect. VI, 966–974.

Chandler, A. C. 1939. The effects of number and age of worms on development of primary and secondary infections with *Hymenolepis diminuta* in rats, and an investigation into the true nature of premunition in tapeworm infections. *Am. J. Hyg.,* 29 D: 105–114.

1943. Studies on the nutrition of tapeworms. *Am. J. Hyg.,* 37: 121–130.

Fuhrmann, O. 1931. Cestoidea, in *Handbuch der Zoologie* (ed. by W. Kükenthal). Bd. II, Hälfte 1, Vermes Amera. de Gruyter, Berlin.

Hyman, L. H. 1951. *The Invertebrates.* Vol. 2, Cestoda, pp. 311–422. McGraw-Hill, New York.

Joyeux, C., and Baer, J. G. 1929. Les cestodes rares de l'homme. *Bull. soc. pathol. exotique,* 22: 114–136.

Kuhls, R. 1953. Zinn in der Bandwurmtherapie. *Med. Klin.,* 48: 1511–1514.

Laurie, J. S. The *in vitro* fermentation of carbohydrates by two species of cestodes and one species of Acanthocephala. *Exp. Parasitol.,* 6: 245–260.

Neghme, A., and Bertin, V. 1951. Estado actual de las investigaciones sobre *Diphyllobothrium latum* en Chile. *Rev. Med. Chile,* 79: 637–640.

Read, C. P. 1959. The role of carbohydrates in the biology of cestodes. VIII. Some conclusions and hypotheses. *Exp. Parasitol.,* 8: 365–382.

Read, C. P., Simmons, J. E., and Rothman, A. H. 1960. Permeation and membrane transport in animal parasites. Permeation of amino acids into tapeworms from elasmobranchs. *J. Parasitol.,* 46: 33–41.

Smyth, J. D. 1955. Problems relating to the *in vitro* cultivation of pseudophyllidean cestodes from egg to adult. *Rev. Iber. Parasitol., Tomo Extraord.*: 65–86.

Pseudophyllidea

Bonne, C. 1942. Researches on sparganosis in the Netherlands East Indies. *Am. J. Trop. Med.,* 22: 643–645.

v. Bonsdorff, B. 1956. *Diphyllobothrium latum* as a cause of pernicious anemia. *Exp. Parasitol.,* 5: 207–230.

Galliard, H., and Ngu, D. B. 1946. Particularités du cycle évolutif de *Diphyl-*

lobothrium mansoni au Tonkin. *Ann. parasitol. humaine et comparée,* 21: 246–253.

Mueller, J. F. 1938. The life history of *D. mansonoides* Mueller, 1935 and some considerations with regard to sparganosis in the United States. *Am. J. Trop. Med.,* 18: 41–66.

Mueller, J. F. 1959. The laboratory propagation of *Spirometra mansonoides* (Mueller, 1935) as an experimental tool. III. In vitro cultivation of the plerocercoid larva in a cell-free medium. *J. Parasitol.,* 45: 561–574.

Mueller, J. F., and Coulston, F. 1941. Experimental human infection with the Sparganum larva of *Spirometra mansonoides* (Mueller, 1935). *Am. J. Trop. Med.,* 21: 399–425.

Nyberg, W. 1958. The uptake and distribution of Co^{60}-labeled Vitamin B_{12} by the fish tapeworm, *Diphyllobothrium latum. Exp. Parasitol.,* 7: 178–190.

Read, C. P. 1952. Human sparganosis in south Texas. *J. Parasitol.,* 38: 29–31.

Thomas, L. J. 1938. The life cycle of *Diphyllobothrium oblongatum* Thomas, a tapeworm of gulls. *J. Parasitol.,* 33: 107–117.

Vogel, H. 1929–1930. Studien zur Entwicklung von Diphyllobothrium, I, II. *Z. Parasitenk.,* 2: 213–222, 629–644.

Ward, H. B. 1930. The introduction and spread of the fish tapeworm (*D. latum*) in the United States. DeLamar Lectures, 1929–1930. Baltimore.

Wardle, R. A. 1935. Fish tapeworm. *Bull. Biol. Bd. Canada,* 45: 1–25.

Wardle, R. A., McLeod, J. A., and Stewart, I. E. 1947. Lühe's "Diphyllobothrium" (Cestoda). *J. Parasitol.,* 33: 319–330.

Hymenolepididae

Chandler, A. C. 1927. The distribution of *Hymenolepis* infections in India with a discussion of its epidemiological significance. *Indian J. Med. Research,* 14: 973–994.

Hearin, J. T. 1941. Studies on the acquired immunity to the dwarf tapeworm, *Hymenolepis nana* var. *fraterna* in the mouse host. *Am. J. Hyg.,* 33 D: 71–87.

Hunninen, A. V. 1935. Studies on the life history and host-parasite relations *of Hymenolepis fraterna* (*H. nana* var. *fraterna* Stiles) in white mice. *Am. J. Hyg.,* 22: 414–443.

Read, C. P. 1956. Carbohydrate metabolism of *Hymenolepis diminuta. Exp. Parasitol.,* 5: 325–344.

Schiller, E. L. 1959. Experimental studies on morphological variation in the cestode genus *Hymenolepis.* I, II, III, and IV. *Exp. Parasitol.,* 8: 91–118, 215–235, 427–470, 581–591.

Shorb, D. A. 1933. Host parasite relations of *Hymenolepis fraterna* in the rat and mouse. *Am. J. Hyg.,* 18: 74–113.

(See also references under Chandler, Addis, Beck, and Read under "General.")

Taeniidae

Agosin, M., et al. 1957. Studies on the metabolism of *Echinococcus granulosus.* I. *Exp. Parasitol.,* 6: 37–51.

Anderson, M. 1934. The validity of *Taenia confusa* Ward, 1896. *J. Parasitol.,* 20: 207–218.

Brailsford, J. F. 1941. *Cysticercus cellulosae*—its radiographic detection in the musculature and the central nervous system. *Brit. J. Radiol.,* 14: 79–93.

Dew, H. R. 1928. *Hydatid Disease.* Medical Publishing Company, Sydney.

Dixon, H. B. F., and Smithers, D. W. 1934. Epilepsy in cysticercosis (*Taenia solium*), a study of seventy-one cases. *Quart. J. Med.*, n.s., 3: 603–616.

Dungal, N. 1946. Echinococcosis in Iceland. *Am. J. Med. Sci.*, 212: 12–17.

Garabedian, G. A., Matossian, R. M., and Djanian, A. Y. 1957. An indirect hemagglutination test for hydatid disease. *J. Immunol.*, 78: 269–272.

Hall, M. C. 1919. The adult taenioid cestodes of dogs and cats and of related carnivores in North America. *Proc. U. S. Natl. Museum*, 55: No. 2258.

Henschen, C., and Bircher, R. 1945. Zur Epidemiologie, Pathologie und Chirurgie des *Echinococcus alveolaris. Bull. Schweiz. Akad. Med. Wiss.*, 1: 209–280.

Miller, M. J. 1953. Hydatid infection in Canada. *Can. Med. Assoc. J.*, 68: 423–434.

Penfold, W. J., Penfold, H. B., and Phillips, M. 1937. *Taenia saginata;* its growth and propagation. *J. Helminthol.*, 15: 41–48.

Rausch, R. 1954. Studies on the helminth fauna of Alaska. XXIV. *Echinococcus sibiricensis* n.sp., from St. Lawrence Island. *J. Parasitol.*, 40: 659–662.

Rausch, R. 1956. Studies on the helminth fauna of Alaska. XXX. The occurrence of *Echinococcus multilocularis* Leuckart, 1863, on the mainland of Alaska. *Am. J. Trop. Med. Hyg.*, 5: 1086–1092.

Schiller, E. L. 1954. Studies on the helminth fauna of Alaska. XIX. An experimental study on blowfly (*Phormia regina*) transmission of hydatid disease. *Exp. Parasitol.*, 3: 161–166.

Silverman, P. H., and Griffiths, R. B. 1955. A review of methods of sewage disposal in Great Britain, with special reference to the epizootiology of *Cysticercus bovis. Ann. Trop. Med. Parasitol.*, 49: 436–450.

Swartzwelder, J. C. 1939. Clinical *Taenia* infection: an analysis of sixty cases. *J. Trop. Med. Hyg.*, 42: 226–229.

Viljoen, N. F. 1937. Cysticercosis in swine and bovines, with special reference to South African conditions. *Onderstepoort J. Vet. Sci. Animal Ind.*, 9: 337–570.

Vogel, H. 1957. Über den *Echinococcus multilocularis* Suddeutschlands. I. Das Bandwurmstadium von Stämmen menschlicher und tierischer Herkunft. *Z. u. Tropenmed. Parasitol.*, 8: 404–454.

Wolfgang, R. W., and Poole, J. B. 1956. Distribution of *Echinococcus* disease in northwestern Canada. *Amer. J. Trop. Med. Hyg.*, 5: 869–871.

Anoplocephalidae

Africa, C. M., and Garcia, E. V. 1935. The occurrence of *Bertiella* in man, monkey and dog in the Philippines. *Philippine J. Sci.*, 56: 1–11.

Kates, K. C., and Goldberg, A. 1951. The pathogenicity of the common sheep tapeworm, *Moniezia expansa. Proc. Helminthol. Soc. Wash., D. C.*, 18: 87–101.

Kates, K. C., and Runkel, C. E. 1948. Observations on oribatid mite vectors of *Moniezia expansa* on pastures, with a report of several new vectors from the United States. *Proc. Helminthol. Soc. Wash., D. C.*, 15: 10–33.

Melvin, D. M. 1952. Studies on the life cycle and biology of *Monoecocestus sigmodontis* (Cestoda: Anoplocephalidae) from the cotton rat, *Sigmodon hispidus. J. Parasitol.*, 38: 346–355.

Stunkard, H. W. 1938. The development of *Moniezia expansa* in the intermediate host. *Parasitology,* 30: 491–501.

1940. The morphology and life history of the cestode, *Bertiella studeri. Am. J. Trop. Med.,* 20: 305–333.

Linstowiidae

Baer, J. G., Kourí, P., and Sotolongo, F. 1949. Anatomie, position systématique, et épidémiologie de *Inermicapsifer cubensis* (Kourí 1938) Kourí 1940, cestode parasite de l'homme à Cuba. *Acta Tropica,* 6: 120–130.

Baer, J. G. 1956. The taxonomic position of *Taenia madagascariensis* Davaine, 1870, a tapeworm parasite of man and rodents. *Ann. Trop. Med. Parasitol.,* 50: 152–156.

Fain, A. 1950. *Inermicapsifer cubensis* (Kourí, 1938) présence du cestode *I. cubensis* synonyme de *Inermicapsifer arvicanthidis* (Kofend, 1917) chez un enfant indigène et chez un rat (*Rattus R. rattus* L.) au Ruanda-Urundi (Congo Belge). *Bull. soc. pathol. exotique,* 43: 438–443.

Kourí, P., and Rappaport, I. 1940. A new human helminthic infection in Cuba. *J. Parasitol.,* 26: 179–181.

Dilepididae

Joyeux, C. 1920. Cycle évolutif de quelques cestodes. *Bull. biol. France Belg.,* Suppl. 2.

Venard, C. E. 1938. Morphology, bionomics, and taxonomy of the cestode *Dipylidium caninum. Ann. N. Y. Acad. Sci.,* 37: 273–328.

Zimmerman, H. R. 1937. Life history studies on cestodes of the genus *Dipylidium* from the dog. *Z. Parasitenk.,* 9: 717–729.

Davaineidae

Baer, J. G., and Sandars, D. F. 1956. The first record of *Raillietina* (*Raillietina*) *celebensis* (Janicki, 1902), (Cestoda) in man from Australia, with a critical survey of previous cases. *J. Helminth.,* 30: 173–182.

Chandler, A. C., and Pradatsundarasar, A. 1957. Two cases of *Raillietina* infection in infants in Thailand, with a discussion of the taxonomy of the species of *Raillietina* (Cestoda) in man, rodents, and monkeys. *J. Parasitol.,* 43: 81–89.

Dollfus, R.-Ph. 1939–1940. Cestodes du genre *Raillietina* trouvés chez l'homme en Amérique intertropicale. *Ann. parasitol. humaine et comparée,* 17: 415–442, 542–562.

Joyeux, C. E., and Baer, J. G. 1936. Helminths des rats de Madagascar. Contribution à l'étude de *Davainea madagascariensis. Bull. soc. pathol. exotique,* 29: 611–619.

1949. L'hôte normal de *Raillietina* (R.) *demerariensis* (Daniels, 1895) en Guyane hollandaise. *Acta Tropica,* 6: 141–144.

Mesocestoididae

Chandler, A. C. 1942. First record of a case of human infection with tapeworms of the genus *Mesocestoides. Am. J. Trop. Med.,* 22: 493–497.

Chapter 16

ACANTHOCEPHALA (SPINY-HEADED WORMS)

As noted on p. 240, the Acanthocephala have usually in the past been attached as a rider to the Nemathelminthes for want of a better place to put them, but as Van Cleave (1941) pointed out, they have much more affinity to the Platyhelminthes, particularly the Cestoidea, both in structural characteristics and in life cycle. Van Cleave (1948) raised them to the rank of a phylum. They are all intestinal parasites, found in all classes of vertebrates, though especially common in fishes and birds. They are remarkably uniform in general anatomy, life history, and habits.

Morphology. Like tapeworms, Acanthocephala are devoid of an alimentary canal throughout their lives. The body is divided into a posterior trunk and an anterior presoma, consisting of a spiny proboscis and unspined neck. In some the neck is a short transitional area; in others it may be long and conspicuous The trunk and presoma are demarcated by an infolding of the cuticula and the derivation from the hypodermis of two elongate structures of unknown function called lemnisci (Fig. 111A) which lie in the body cavity. The proboscis, and often the neck also, is in most species retractile into a proboscis sac or receptacle by being turned inside out, and the whole presoma is also retractile, without inversion, into the fore part of the trunk by means of special retractor muscles inserted on the trunk wall (Fig. 111A, *r.m.*). The armature of the proboscis varies from a few to a great many hooks, which are usually in radial or spiral rows; in long proboscides they appear to be in longitudinal rows with quincunxial arrangement (Fig. 113).

The body is covered with a cuticle under which is a syncytial hypodermis or subcuticula. In more primitive forms this has a small number (6 to 20) of large oval or ameboid nuclei; in some forms these break up into numerous nuclear fragments. The hypodermis also has a closed "lacunar" system of longitudinal and transverse

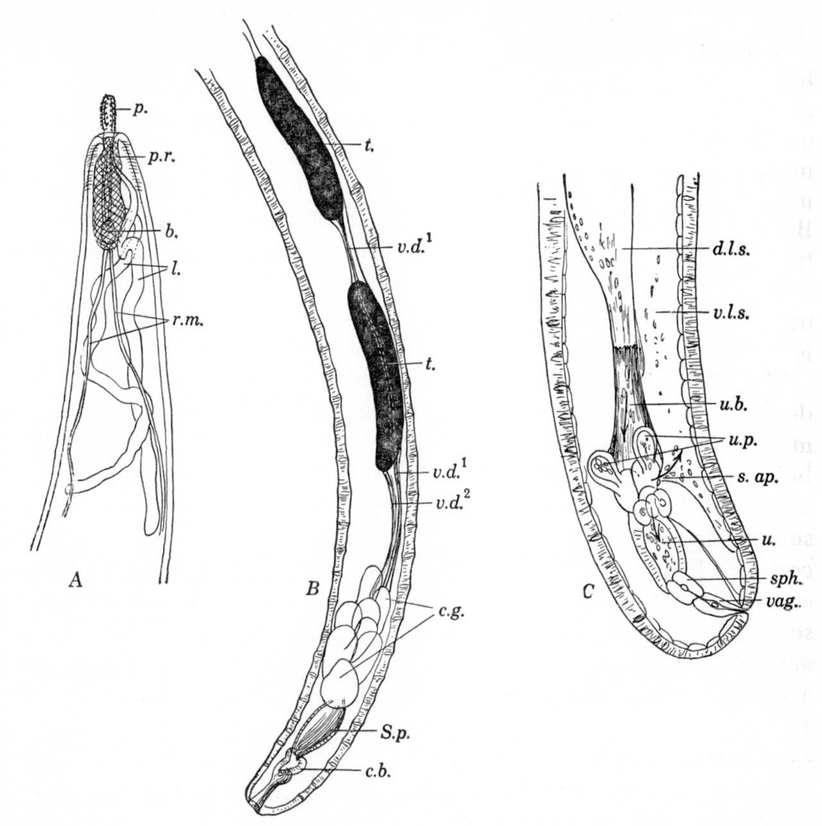

Fig. 111. *Moniliformis dubius: A,* anterior end; *B,* posterior end of ♂; *C,* posterior end of ♀; *b.,* brain; *c.b.,* copulatory bursa; *c.g.,* cement glands; *d.l.s.,* dorsal ligament sac; *l.,* lemnisci; *p.,* proboscis; *p.r.,* proboscis receptacle; *r.m.,* retractor muscles; *s.ap.,* sorting apparatus; *S.p.,* Saefftigen's pouch; *sph.,* sphincter; *t.,* testes; *u.,* uterus, *u.b.,* uterine bell; *u.p.,* uterine pouches; *vag.,* vagina; *v.d.*1 and *v.d.*2, vasa deferentia; *v.l.s.,* ventral ligament sac.

vessels, probably for distribution of nutrients absorbed from the host. These structures are confined to the trunk.

One of the striking things about Acanthocephala is the small number of nuclei; after an acanthocephalan reaches its final host there is no further cell division except in germ cells, even though the body may grow to hundreds of times its original size.

The sexes are separate, and the males are nearly always smaller than the females. The reproductive organs in both sexes are located in the posterior part of the trunk, and are enclosed by connective tissue *ligament sacs.* These are hollow tubes extending most of the length of the body cavity of the trunk, single in males and in females of the order Palaeacanthocephala, but divided into dorsal and ventral ones, communicating anteriorly, in females of the other orders. The males (Fig. 111*B*) have two testes, behind which are cement glands, usually 4 to 8 large unicellular glands but sometimes a syncytial mass. Behind the cement glands in some Acanthocephala is a saclike structure called Saefftigen's pouch, through which, in most Acanthocephala, run the sperm ducts and ducts from the cement glands before they unite at its posterior end. At the posterior end of the worm there is a muscular bursa which can be protruded or retracted into the body.

In females (Fig. 111*C*) an ovary is present only in early stages of development, later breaking up into masses of cells which continue to multiply and produce ova. These float free in the ligament sacs, being retained in the dorsal one in the Eo- and Archiacanthocephala, but liberated into the general body cavity by a disintegration of the single sac in the Palaeacanthocephala. Near the posterior end is a complicated structure called a uterine bell, into the wide-open anterior end of which the eggs are drawn. It acts as a sorting device; the smaller immature eggs are returned to the body cavity or into the ventral ligament sac through lateral openings, while the mature eggs are passed back through an oviduct to the posterior genital opening. The eggs, when ripe, contain a mature embryo called an acanthor surrounded by three envelopes, the outer of which often has shapes or markings useful in identification.

Life cycle. The life cycle involves an intermediate host, which is usually an arthropod: small Crustacea for parasites of aquatic vertebrates; grubs, roaches, etc., for those of land animals. When the embryonated eggs (Fig. 112*A*) are swallowed, the spindle-shaped acanthor (Fig. 112*B*), usually armed with rostellar hooks and small body spines, hatches and bores into the intestinal wall, eventually reaching the body cavity. Meanwhile it grows and undergoes a gradual transformation; as development proceeds, the proboscis, proboscis sac, lemnisci, and rudiments of the sex organs are laid down. For this series of stages (Fig. 112*C*, *D*) leading up to the infective form Van Cleave applied the name "acanthella." A number of workers have applied this name to the fully developed infective larva, and Moore (1946) applied the name "preacanthella" to the earlier pre-infective stages, but Van Cleave insists that his name acanthella should apply only to these pre-infective stages, and that the infective form should be called a "juvenile." Since, however, "juvenile" is also applied to the re-encysted forms in secondary transport hosts, a new name is needed for the fully developed infective form. For this the name "cystacanth" has been proposed. The cystacanth (Fig. 112*E*) is enclosed in a delicate hyaline sheath produced by the larva. The proboscis is fully formed but inverted, and the reproductive organs

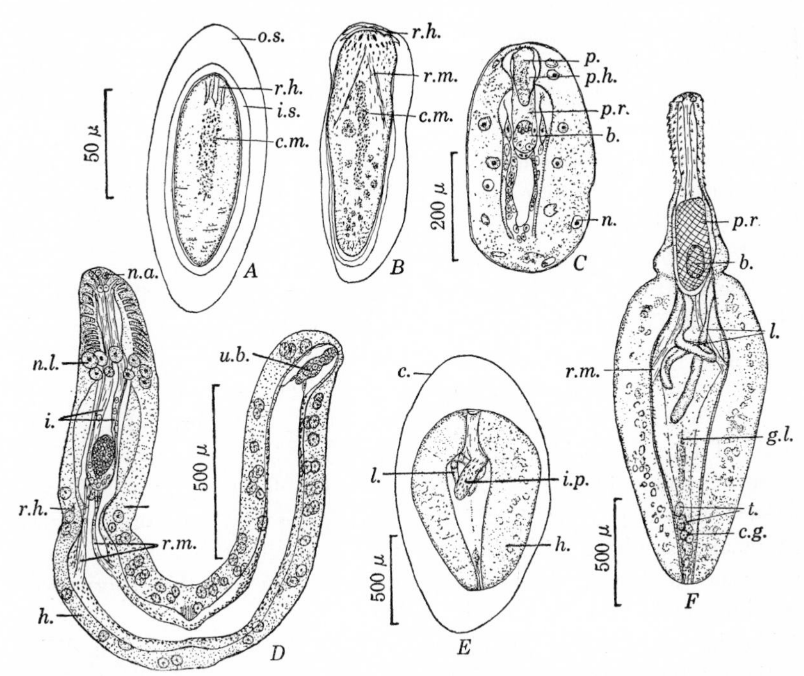

Fig. 112. Life cyle of *Moniliformis dubius: A,* egg; *B,* acanthor in process of escaping from egg shell and membranes; *C,* median sagittal section of larva from body cavity of roach 29 days after infection; *D,* acanthella dissected from enveloping sheath, about 40 days after infection; *E,* cystacanth from body cavity of roach, with proboscis inverted, about 50 days after infection; *F,* cystacanth freed from cyst and proboscis evaginated. Abbreviations: *b.,* brain; *c.g.,* cement glands; *c.m.,* central nuclear mass; *g.l.,* genital ligament; *h.,* hypodermis; *i.,* inverter muscles; *i.p.,* inverted proboscis; *i.s.,* inner shell; *l.,* lemnisci; *n.,* subcuticular nucleus; *n.a.,* nuclei of apical ring; *n.l.,* nuclei of lemniscal ring; *o.s.,* outer shell; *p.,* proboscis; *p.h.,* developing proboscis hooks; *p.r.,* proboscis receptacle; *r.h.,* rostellar hooks; *r.m.,* retractor muscle; *t.,* testes, *u.b.,* uterine bell. (After Moore, *J. Parasitol.,* 1946.)

are sufficiently developed so that the sex is easily recognized. In *Moniliformis* larvae the hypodermis is expanded into broad flanges.

Effects on host. Acanthocephala damage their hosts principally by local injury and inflammation at the point of attachment of the spiny proboscis. When the worm moves and reattaches, the old sore may become infected by bacteria. Occasionally the worms cause perforation of the gut wall and precipitate a fatal peritonitis. In heavy infections, loss of appetite and interference with digestion may lead to unthriftiness. Dogs and coyotes infected with *Oncicola* (see p. 388) are said sometimes to develop rabieslike symptoms, suggesting the possibility of transmission of a virus by the worms. Grassi and Calandruccio in 1888 reported acute pain and violent ringing in the ears experienced by the junior author 19 days after infecting himself with *Moniliformis* larvae.

Burlingame and Chandler (1941) showed that, as with some adult tapeworms, no true immunity to Acanthocephala is developed, resistance to reinfection being primarily a matter of competition for food and for favorable locations in the intestine.

Classification. The Acanthocephala constitute a small group of about a dozen families and about sixty genera which are quite widely divergent from other groups of worms but which are remarkably uniform among themselves, both in morphology and life cycle. Once placed in a single genus, *Echinorhynchus,* they were later (1892) divided into several families, then (1931) into two orders, which were expanded to three in 1936 and finally elevated by Van Cleave (1948) into a phylum containing two classes and four orders: class Metacanthocephala with the orders Palaeacanthocephala and Archiacanthocephala, and the class Eoacanthocephala with the orders Gyracanthocephala and Neoacanthocephala.

To the senior writer the characters used by Van Cleave for differentiating these groups seem trivial. For example, in the table of characters given for distinguishing the orders, the only one in which the Gyracanthocephala and Neoacanthocephala differ is the presence or absence of trunk spines, and even this character is variable in one of the other orders. No good character is presented for differentiating the two classes. For the present, therefore, we prefer to consider the Acanthocephala as constituting a single phylum with three groups, as proposed by Van Cleave in 1936, though we consider even this rather extreme.

1. **Palaeacanthocephala.** Proboscis hooks usually in long rows; spines present on trunk; nuclei in hypodermis usually fragmented; chief lacunar vessels in hypodermis lateral; single ligament sac in ♀ often breaks down; separate cement glands; eggs spindle-shaped, thin-shelled; mostly in fishes and aquatic birds and mammals, cystacanth in Crustacea.

2. **Eoacanthocephala.** Proboscis hooks usually in a few circles; trunk spines present or absent; nuclei in hypodermis few and large; chief lacunar vessels dorsal and ventral; distinct dorsal and ventral ligament sacs in ♀; syncytial cement glands; eggs ellipsoidal, thin-shelled; parasitic in fishes, except one in turtles, cystacanth in Crustacea.

3. **Archiacanthocephala.** Proboscis hooks either in long rows (e.g., *Moniliformis*) or in a few circles (e.g., *Oncicola* and *Macracanthorhynchus*); no spines on trunk; nuclei in hypodermis few and large; chief lacunar vessels dorsal and ventral; dorsal and ventral ligament sacs persist in ♀; separate cement glands; eggs usually oval, thick-shelled; protonephridia present in some; parasitic in terrestrial vertebrates, cystacanth in grubs, roaches, etc.

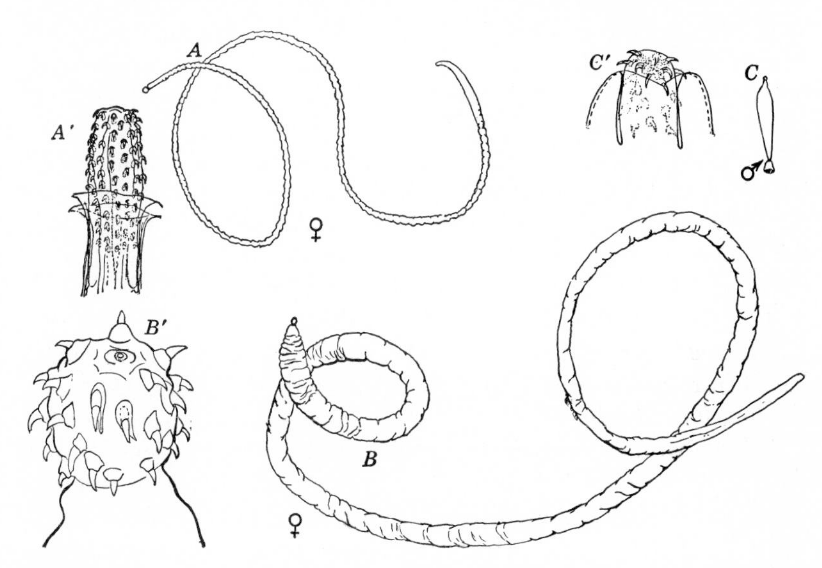

Fig. 113. Various Acanthocephala, adults drawn to same scale: *A*, *Moniliformis dubius* ♀; *A′*, proboscis of same; *B*, *Macracanthorhynchus hirudinaceus; B′*, proboscis of same; *C*, *Oncicola canis; C′*, proboscis of same. (Adapted from various authors.)

The only Acanthocephala found in man, *Moniliformis dubius* and *Macracanthorhynchus hirudinaceus,* belong to the Archiacanthocephala.

Moniliformis. The common spiny-headed worm of house rats, *Moniliformis dubius* (Figs. 111, 113*A*), has been found in man on a few occasions. Its body, 10 to 30 cm. long in females and 6 to 13 cm. in males, has annular rings which give it a tapewormlike appearance.

It has a nearly cylindrical proboscis with 12 to 15 rows of vicious thornlike hooks. It inhabits the small intestine of rats in many parts of the world. The eggs are over 100 μ in length. Cockroaches serve as intermediate hosts; the senior author found over 100 cystacanths in the body cavity of a *Periplaneta americana.* The eggs hatch in the mid-intestine of the roach, the liberated acanthors penetrating into the gut wall. By the tenth day they appear as minute specks on the outside of the intestinal wall, from which they eventually drop into the body cavity (Moore, 1946). The half-grown acanthella lies straight and has very broad ectodermal flanges (Fig. 112*C*), but with further development it bends V-shaped in its cyst (Fig. 112*D*), the body proper elongating and thickening until the flanges become inconspicuous. When fully developed, after 7 to 8 weeks, the cysts are about 1 to 1.2 mm. long and the cystacanths (Fig. 112*F*) 1.5 to 1.8 mm. long. In Europe a beetle (*Blaps*) has been involved as an intermediate host, but the form found in wild rodents in Europe is not identical with that found in rats in the United States and South America.

Sandground (1926) found numerous immature specimens in the intestines of toads and lizards, where they had evidently attached themselves after being eaten with the intermediate hosts. Considering the propensity of Acanthocephala for re-establishing themselves as larvae in abnormal hosts, human infection might be possible without postulating the eating of roaches or beetles.

Macracanthorhynchus. The only other spiny-headed worm which has been recorded from man is the relatively huge species, *Macracanthorhynchus hirudinaceus* (Fig. 113*B*), commonly parasitic in pigs. This large worm, of which the females are 25 to 60 cm. long, though the males are only 5 to 10 cm., is pinkish and has a transversely wrinkled body which tapers from a rather broad, rounded head end to a slender posterior end. The presoma is relatively very small, with a little knoblike proboscis armed by five or six rows of thorns. The eggs are 80 to 100 μ long with sculptured brown shells; they are very resistant to desiccation and cold and remain viable in soil up to 3½ years. White grubs, the larvae of "June bugs," serve as intermediate hosts. The cystacanth is cylindrical and quite different in appearance from that of *Moniliformis.*

Lindemann in 1865 recorded this worm as parasitic in man among the peasants of the Volga Valley in southern Russia.

Other Acanthocephala. Aside from *Macracanthorhynchus* in pigs, the only Acanthocephala of importance to domestic animals are two genera, *Polymorphus* and *Filicollis,* which are injurious to ducks and geese. In both cases the intermediate hosts are Crustacea; *Polymorphus* takes advantage of an amphipod, *Gammarus,* and *Filicollis* of an isopod, *Asellus.* Dogs in Texas are sometimes infected with *Oncicola canis* (Fig. 113*C*), the larvae of which are commonly found in armadillos and sometimes in the walls of the esophagus of turkeys. An arthropod undoubtedly serves as a first intermediate host; it is an open question whether the second intermediate host is obligatory or merely convenient. Monkeys in zoological gardens infested with roaches sometimes suffer from *Prosthenorchis* infections, which are native in South American monkeys but spread to other species.

Treatment. Almost nothing is known about anthelmintics for Acanthocephala. Sodium fluoride given pigs to eliminate *Ascaris* does not affect *Macracanthorhynchus.* Lal performed some experiments on Acanthocephala of fishes *in vitro* and found that CCl_4, $CuSO_4$, and thymol killed at 0.05%, but Santonin did not at 1%. Extract of male fern removed *Moniliformis* in an experimental human infection.

REFERENCES

Burlingame, P. L., and Chandler, A. C. 1941. Host-parasite relations of *Moniliformis dubius* in albino rats, and the environmental nature of resistance to single and superimposed infections with this parasite. *Am. J. Hyg.,* 33: D, 1–21.

Grassi, B., and Calandruccio, S. 1888. Ueber einen *Echinorhynchus,* welcher auch in Menschen parasitiert und dessen Zwischenwirt ein *Blaps* ist. *Zentr. Bakteriol. Parasitenk., Orig.,* 3: 521–525.

Kates, K. C. 1943. Development of the swine thorn-headed worm, *Macracanthorhynchus hirudinaceus,* in its intermediate host. *Am. J. Vet. Research,* 4: 173–181.

Laurie, J. S. 1959. Aerobic metabolism of *Moniliformis dubius* (Acanthocephala). *Exp. Parasitol.,* 8: 188–197.

Meyer, A. 1933. Acanthocephala. In *Bronn's Klassen u. Ordnungen d. Tierreichs,* 4: Abt. 2. Akad. Verlagsgesellsch., Leipzig.

Moore, D. V. 1946. Studies on the life history and development of *Moniliformis dubius* Meyer, 1933. *J. Parasitol.,* 32: 257–271.

Van Cleave, H. J. 1936. The recognition of a new order in the Acanthocephala. *J. Parasitol.,* 22: 202–206.

1941. Relationships of the Acanthocephala. *Am. Naturalist,* 75: 31–47.

1947. A critical review of terminology of immature stages in acanthocephalan life histories. *J. Parasitol.,* 33: 118–125.

1948. Expanding horizons in the recognition of a phylum. *J. Parasitol.,* 34: 1–20.

1952. Some host-parasite relationships of the Acanthocephala, with special reference to the organs of attachment. *Exptl. Parasitol.,* 1: 305–330.

1953. Acanthocephala of North American mammals. *Ill. Biol. Monogr.,* 23, Nos. 1–2.

Witenberg, G. 1938. Studies on Acanthocephala, 3, Genus *Oncicola. Livro jubilar do Prof. Travassos,* 537–560, Rio de Janeiro.

Chapter 17

THE NEMATODES IN GENERAL

The nematodes constitute a large group of worms of comparatively simple organization, nearly all of which are total strangers to everyone but zoologists, although they play extremely important roles in the economy of nature. Popular ignorance of these animals is, as Cobb remarked, easy to understand since they are seldom if ever seen; they do not supply food, raiment, or other valuable material; they are not ornamental; they do not delight our ears with their songs or otherwise amuse us; and they fail even to furnish us with classic examples of industriousness, providence, or other virtues, although they might well be extolled by large-family enthusiasts. Thus avoiding the popular limelight, they do, nevertheless, unobtrusively leave their marks in the world. They have been able to exploit every conceivable aquatic and terrestrial habitat, including ice fields in Spitzbergen and felt mats soaked with beer in German public houses. Most plants are parasitized by nematodes, and probably every species of vertebrate animal on the earth affords harborage for nematode parasites. Stoll (1947) estimated 2000 million human nematode infections in a world harboring 2200 million human inhabitants, a tribute, as he said, to the variety and biological efficiency of nematode life cycles. Only about a dozen species are important human parasites, although over fifty species have been known to make their homes in the human body occasionlly.

Relationships. The Nematoda constitute one of six classes included by Hyman (1951) in the phylum Aschelminthes (see pp. 239–240). The majority of the estimated 500,000 species are free-living in soil or water, including the ocean; others, many closely related to these, have become parasitic in arthropods, mollusks, or plants. Some of the plant-parasitic forms do inestimable damage to crops. Most of the forms that are free-living or that find harborage in invertebrates or plants are barely visible to the naked eye, and are transparent enough so that every structure in the body can be observed as if in a glass model. These forms have very simple life cycles. The species parasitic in vertebrates, on the other hand, are often veritable giants, some up to several feet in length, and may have much more complicated life cycles.

The parasitic forms have without doubt evolved from more than one type of free-living form and do not, therefore, represent a single branch of the class Nematoda, which can properly be classified independently of the free-living forms. This, however, is what has been done in the past, for students of the nematodes parasitic in vertebrate animals had little knowledge or interest in the free-living forms. Only since about 1935 have attempts been made to reconcile these two estranged sections of the nematode clan.

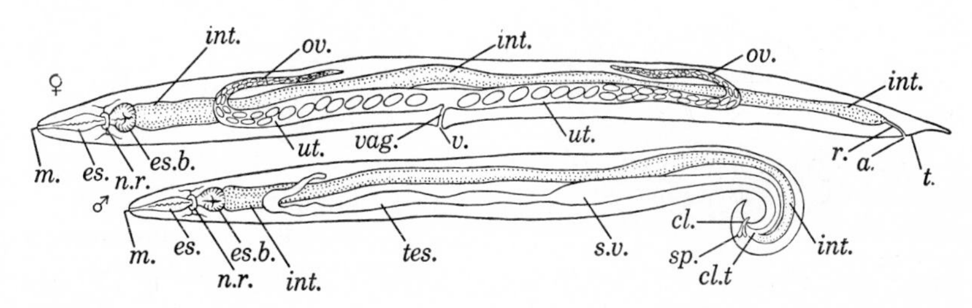

Fig. 114. Diagrams of ♀ and ♂ of free-living nematode of *Rhabditis* type: *a.*, anus; *cl.*, cloaca; *cl.t.*, cloacal tube; *es.*, esophagus; *es.b.*, esophageal bulb; *int.*, intestine; *m.*, mouth; *n.r.*, nerve ring; *ov.*, ovary; *r.*, rectum; *sp.*, spicules; *s.v.*, seminal vesicle and sperm duct; *t.*, tail; *tes.*, testis; *ut.*, uterus; *v.*, vulva; *vag.*, vagina.

General structure. A typical nematode is an elongated, cylindrical worm, tapering more or less at head and tail ends, and encased in a very tough and impermeable transparent or semitransparent cuticle (Fig. 114). This cuticle is not chitin, like the cuticle of arthropods, since it is soluble in potassium hydroxide, but nematodes do have true chitin in the egg shells. Usually the cuticle is marked externally by fine transverse striations; it may have other inconspicuous markings and sometimes has bristles, spines, ridges, or expansions of various kinds. In some parasitic forms there are finlike expansions in the neck region, in others in the tail region of the males, the latter commonly supported by fleshy papillae; they are known respectively as cervical and caudal alae. In the Strongylata there is a bell-shaped expansion at the posterior end of the males supported by fleshy rays conforming in number and arrangement to a definite plan; this is called a bursa. The cuticle is secreted by a protoplasmic syncytial layer called the hypodermis, in which no separate cells can be distinguished. Nuclei are present only in four thickened chords or "lines," one dorsal, one ventral, and two lateral. In these chords run nerve fibers and, in some species, canals connected with the excretory system.

Between the chords there is a single layer of longitudinally spindle-shaped muscle cells of very peculiar structure. In small transparent worms the striated part of the muscle cell is limited to the part of the cell in contact with the hypodermis, and only a few, often only two,

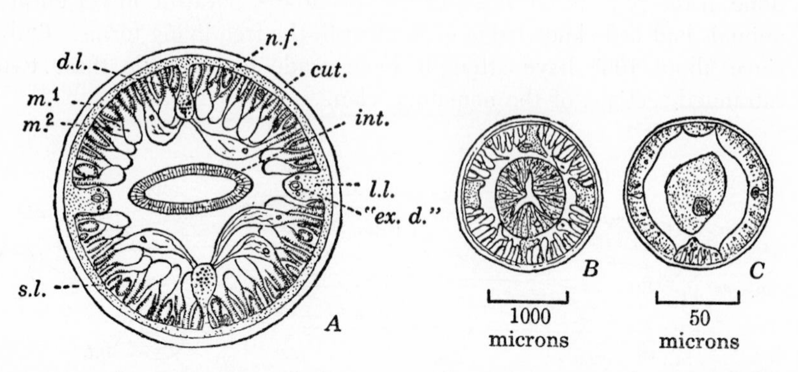

Fig. 115. *A*, cross-section of *Ascaris*, a polymyarian nematode, in prevulvar region; *cut.*, cuticle; *d.l.*, dorsal line; *ex.d.*, excretory duct; *int.*, intestine; *l.l.*, lateral line; *m.*^{1}, striated contractile portion of muscle cell; *m.*^{2}, protoplasmic portion of muscle cell; *n.f.*, nerve fibers. (After Brandes, adapted from Fantham, Stephens, and Theobald.) *B*, cross-section of esophageal region of *Ascaris; C*, of *Trichuris*. In *Ascaris* note the triangular lumen and thick muscular walls of the esophagus, and the numerous muscle cells in body wall. In *Trichuris* note the greatly reduced esophagus imbedded in the protoplasm of a large cell, and broad, flat muscle cells on body wall. (*Ascaris*, original; *Trichuris*, adapted from Chitwood.)

flat muscle cells are in each quadrant of a cross-section. In larger and more opaque worms, however, the muscle cells in each quadrant become very numerous and in cross section have a flask-shaped appearance, with the striations along the "neck" of the flask as well as at the base of the cell (Fig. 115*A*). Worms with these two types of musculature are said to be "meromyarian" and "polymyarian," respectively, but there are all gradations between them. Contraction of these elongated muscles causes a twisting or bending of the body. Special muscles occur in the esophagus, ovejector, etc., and for moving the spicules of the male.

Between the muscles and the gut wall is a relatively spacious body cavity, or pseudocoelom, in which the reproductive organs lie, unattached except at their external openings. This cavity is not lined by an epithelium as is a true coelom. It contains a fluid which serves as a distributing medium for digested food and for collection of waste products. It is provided with a small amount of "mesenterial" tissue and a few large phagocytic cells called coelomocytes.

The nervous system consists of a conspicuous "nerve ring" around the esophagus, from which longitudinal nerve trunks run forward and backward. A few special sensory organs are present; at the anterior end are a pair of supposedly olfactory receptors called amphids, and in some a similar pair, called phasmids, is situated on minute papillae behind the anus. These differ from ordinary tactile papillae in having canals connected with glandlike structures. There are tactile papillae about the mouth, a pair in the neck region of many forms (called deirids), and paired caudal or genital papillae in the males of many forms.

The excretory system is variable. Almost the only constant feature is a pore opening on the mid-ventral surface in the esophageal region; in some forms even this seems to be absent. A well-developed excretory system such as occurs in *Rhabditis* consists of an H-shaped system of tubes, the middle of the crossbar of the H being connected with the pore and the limbs lying in the lateral chords. In addition, two subventral gland cells open into the pore. In *Ascaris* the posterior limbs are well developed; in some the system is reduced to an inverted U or is developed on one side only. In some free-living nematodes the excretory system is reduced to a single glandular cell. Mueller in 1929 expressed the opinion that excretion takes place through the cuticle and that the so-called excretory system is really secretory. There is no circulatory system, and respiration is through the cuticle or possibly through the alimentary canal.

The mouth is variously modified. The primitive type in free-living nematodes is a simple opening surrounded by three lips, one dorsal and two lateroventral. This is retained by many groups of parasitic forms, including *Strongyloides*, oxyurids, and ascarids. In some forms, e.g., the filariae and their allies, the lips have disappeared, but in others two lateral lips, sometimes with a dorsal and ventral one also, have replaced the primitive three. In still others, especially some of the Strongylata, the mouth has been highly modified into a "buccal capsule," which may be supplied with such embellishments as crowns of leaflike processes, cutting ridges, teeth, and lancets.

The mouth, or buccal capsule, leads into the digestive canal. This is a simple tube leading from the anterior mouth to an anus usually a short distance from the posterior end. It consists of two parts, an esophagus and an intestine. The esophagus has a chitinized triradiate lumen usually surrounded by muscle or gland cells (Fig. 115*B*), and ordinarily it has three esophageal glands embedded in its walls. In the suborder Trichurata, however, after a short anterior region the wall is greatly reduced and the lumen of the esophagus appears to pass like a capillary tube through a column of large cells (Figs. 115*C* and 116). Chitwood showed that these cells open by minute ducts into the esophagus, and he interprets them as reduplicated esophageal glands. This column of glandular cells is called a stichosome. Sometimes the posterior end of the esophagus enlarges into a bulb provided with valves (Fig. 114).

The intestine is a flat or cylindrical tube, usually straight, and is lined by a single layer of cells (Fig. 115*A*). In some forms, like the strongyles, it is lined by only 18 to 20 cells in all, whereas in *Ascaris* there are about a million. At the posterior end there is a chitinized rectum. In females the intestine has a separate anal opening, but in males the intestine and reproductive system open into a common cloaca.

Reproductive systems. With rare exceptions parasitic nematodes have separate sexes, which are externally distinguishable; usually the males are smaller, and they differ in the form of the tail. In one instance the male lives as a parasite in the vagina of the female! In both sexes the reproductive system consists primitively of long tubules, part of which serve as ovaries or testes and part as ducts (Fig. 114). In all parasitic nematodes the male system is reduced to a single tubule, but the female system is double with rare exceptions and in a few cases is further reduplicated. The inner ends of the tubules are fine, coiled, threadlike organs closed at the ends, which produce the cells that ultimately become eggs or sperms. These sex glands open directly into a continuous part of the same tube, usually larger in caliber, called the uterus or vas deferens, as the case may be. The walls of the uterus appear to supply the yolk and shell material for the egg.

In the male the single vas deferens usually has an enlargement or seminal vesicle, followed by a muscular ejaculatory duct which opens into the cloaca. Males normally have a pair of sclerotized "spicules," which lie in pouches dorsal to the ejaculatory duct near the cloaca. They are capable of exsertion and are used to guide the sperms into the vagina of the female at the time of copulation. There may be a third smaller sclerotized body or accessory piece called a gubernaculum. The size and shape of the spicules vary greatly in different kinds of nematodes and are often very useful in identification. In a few forms one or both spicules may be missing.

In the females of simple types of nematodes the two uteri come together near the middle of the body and open into a single vulva (Fig. 114). In most parasitic forms, however, the uteri first unite into a common tube, the vagina. Frequently the vagina or the branches of the uteri have enlarged, thin-walled chambers which serve as seminal receptacles, and also muscular ovejectors which by a peristaltic action force the eggs through to the vulva one at a time. The vulva in different species may vary in position from just behind the mouth to a point just in front of the anus.

Development and life cycle. The development of nematodes is a comparatively simple process. The original egg cell, after being enclosed in a membrane or shell, segments into 2, 4, 8, 16, etc., cells, until it forms a solid morula. This then begins to assume a tadpole shape and become hollow inside, and then proceeds to form an elongated embryo provided with a simple digestive tract. After ten consecutive cell divisions, in the later ones of which not all the cells participate each time, the definitive form of the first larval stage is reached. Thereafter development proceeds more slowly and, as in insects, is punctuated by a series of molts, normally four, although in some forms one or two molts may occur in the egg before hatching. Although the successive stages differ in details of structure they are never totally unlike each other.

The state of development at the time the eggs are deposited varies greatly, apparently depending upon different oxygen requirements for development. Some leave the mother's body unsegmented (*Ascaris* and *Trichuris*); some in early stages of segmentation (hookworms and their allies); some in the tadpole stage (*Enterobius*), and some as fully developed embryos (*Trichinella, Strongyloides,* and filariae). Usually no further development occurs until the eggs or embryos have reached a new environment, either outside the body or in an intermediate host, except in the case of *Trichinella,* the embryos of which find *their* new environment in the muscles of the parental host. Having reached this new environment the embryo, either inside the egg or after hatching from it, commonly undergoes two molts, reaching the third stage, before it is infective for another definite host. When it has reached that stage, it ceases to grow or develop until transfer to a new host is accomplished.

The simplest type of life cycle is that in which the embryonated eggs are swallowed by the host. The embryos, usually in the third stage, hatch in the intestine and may develop to maturity there, only burying themselves temporarily in the mucous membranes, e.g., *Enterobius* and *Trichuris,* or they may make a preliminary journey through the host's body via heart, lungs, trachea, and esophagus, and thus back to the intestine, e.g., *Ascaris*. This life cycle may be modified by the first-stage embryos hatching outside the body and growing and developing to the infective stage as free-living larvae, then re-entering the definitive host by burrowing through the skin, e.g., hookworms, or being swallowed with vegetation, e.g., *Haemonchus*.

Strongyloides reproduces parthenogenetically and may intercalate a generation of morphologically different free-living males and females. *Trichinella* produces embryos which penetrate into the host's body and encyst in the muscles to await being eaten by another host, thus substituting the original host for the outside world as a place for preliminary partial development. The filariae and their allies (suborder Spirurata) substitute insects or other invertebrates as a place for partial development, thus requiring a true intermediate host. A few, e.g., *Gnathostoma*, require two intermediate hosts, the larvae developing first in a *Cyclops*, continuing in a fish or other cold-blooded vertebrate, and reaching sexual maturity in a mammal. Some nematodes, after having reached an infective stage, can re-encyst if they get into an unsuitable host.

The methods of escaping from and re-entering a final host vary in accordance with the modifications in the life cycle.

Classification. The classification of nematodes is still in a very unsettled state. This is partly due to the process of promoting nematode groups to higher ranks as more and more species are described. Families or even genera of a few years ago are now superfamilies, suborders, or orders, according to the willingness of helminthologists to recognize the promotions. The classification has been subjected to a veritable earthquake by attempts, which must sooner or later be recognized, to combine the classification of free-living and parasitic groups in a single coordinated whole. Chitwood has done most in reconciling these two estranged groups and has evolved a classification which embraces them both, but Chitwood's conclusions will probably have to undergo some ripening and confirmation before parasitologists in general will accept this nematode classification.

In the first place, he divides the entire class into two subclasses, Phasmidia and Aphasmidia, for the fundamental characters of which the student is referred to Chitwood and Chitwood. The Phasmidia include the majority of soil nematodes, as well as most of the forms parasitic in insects and vertebrates, whereas the Aphasmidia include mainly aquatic forms and a few parasitic ones—the Trichurata, mermithids, and Dioctophymata. The further division of these subclasses into orders involves some unfamiliar names, which may not survive the limelight of publicity, so we omit them and give only the more or less familiar suborders and superfamilies. As an *ad interim* classification the following is suggested:

Subclass **Aphasmidia.** No phasmids (caudal sensory organs); amphids much modified externally except in parasitic forms; excretory system rudimentary or absent; coelomocytes and mesenterial tissue well developed.

1. Suborder **Trichurata.** Esophagus a very long, fine tube embedded for most of its length in a column of glandular cells; females with one ovary; males with one spicule or none. Includes *Trichuris, Trichinella,* and *Capillaria.*
2. Suborder **Dioctophymata.** Large worms; esophagus cylindrical; female with one ovary; male with one spicule and a terminal sucker; no excretory system. Includes kidney worm (*Dioctophyma*).

Subclass **Phasmidia.** Phasmids present; amphids simple pores; excretory system present, not rudimentary; coelomocytes (6 or less) and mesenterial tissue weakly developed.

1. Suborder **Rhabditata.** Small, transparent, meromyarian worms; esophagus usually with one or two bulbs; mouth simple or with 3 or 6 minute lips or papillae; no specialized ovejectors, and vagina transverse; majority free-living, some with an alternating generation of parthogenetic parasitic females. Includes *Rhabditis* and *Strongyloides.*
2. Suborder **Ascaridata.** Esophagus bulbed or cylindrical; vagina elongate; mouth usually with 3 or 6 lips; males usually with 2 spicules; tail of male not spirally coiled but usually curled ventrally; no true bursa, but alae may be present.

   Superfamily 1. **Ascaridoidea.** Cervical papillae present; mostly large, stout polymyarian worms; males with 2 spicules; tail curled ventrally, with or without lateral alae; esophagus muscular, with or without a bulb. Includes *Ascaris* and *Heterakis.*

   Superfamily 2. **Oxyuroidea.** Cervical papillae absent; mostly small or medium-sized transparent meromyarian worms; males with 1 or 2 spicules; esophagus bulbed; tail of female usually slender and pointed. Includes *Enterobius.*
3. Suborder **Strongylata.** Usually meromyarian; males with 2 spicules and with a true bursa supported by 6 paired rays and one dorsal one which may be divided; mouth simple, without lips, or with a buccal capsule; esophagus muscular, club-shaped, or cylindrical; eggs thin-shelled and colorless. Includes hookworms, strongyles, gapeworms, and lungworms.
4. Suborder **Spirurata.** Esophagus cylindrical, often part glandular and part muscular; males usually with 2 spicules and well-developed alae and papillae on spirally coiled tail; mouth either simple with no or rudimentary lips, or with 2 or 4 paired lips; vagina elongated and tubular; posterior part of esophagus with numerous nuclei; require intermediate host.

   Superfamily 1. **Spiruroidea.** Mouth usually with a chitinized vestibule and 2 or 4 paired lips; vulva usually in middle or posterior part of body; males with spirally coiled tail with broad alae supported by papillae; eggs usually escape with feces and are eaten by intermediate host. Includes *Gongylonema, Gnathostoma,* and *Physaloptera.*

Superfamily 2. **Filarioidea.** Slender, delicate worms; mouth usually simple, without lips and rarely a vestibule; females with vulva far anterior; males small with coiled tails with or without alae, but always with papillae; usually give birth to embryos which swarm in blood or skin and develop in bloodsucking insects. Includes filariae (*Wuchereria, Onchocerca,* etc.).

5. Suborder **Camallanata.** Mouth simple or with lateral jaws; posterior part of esophagus with 1 or 3 large nuclei; requires intermediate host.

Superfamily 1. **Dracunculoidea.** Mouth simple, surrounded by circlet of papillae; alimentary canal and vulva atrophied in adult females; males much smaller than females; embryos evacuated through burst uterus and mouth. Includes guinea worm (*Dracunculus*).

REFERENCES

Alicata, J. B. 1935. Early developmental stages of nematodes occurring in swine. *U. S. Dept. Agr. Tech. Bull.* 489.

Baylis, H. A., and Daubney, R. A. 1926. *A Synopsis of the Families and Genera of Nematodes.* British Museum, London.

Bremmer, K. C. 1955. Cytological studies on the specific distinctness of the ovine and bovine "strains" of the nematode *Haemonchus contortus* (Rudolphi). Cobb (Nematoda: Trichostrongylidae). *Australian J. Zool.,* 3: 312–323.

Chandler, A. C. 1939. The nature and mechanism of immunity in various intestinal nematode infections. *Am. J. Trop. Med.,* 19: 309–317.

Chitwood, B. G. 1937. A revised classification of the Nematoda. *Skrjabin Festchr.* Moscow, 69–80.

Chitwood, B. G., and Chitwood, M. B. 1937–1942, 1950. An introduction to Nematology. Sect. I, Pts. I–III, and Sect. II, Pts. I and II so far published. Sect. I, Pts. I–III revised, 1950. Nematology and Co., Marquette, Mich.

Chitwood, M. B. 1957. Intraspecific variation in parasitic nematodes. *Systematic Zool.,* 6: 19–23.

Cram, E. B. 1927. Bird parasites of the nematode suborders Strongylata, Ascaridata, and Spirurata. *U. S. Natl. Mus. Bull.* 140.

Dougherty, E. C. 1951. Evolution of zooparasitic groups in the Phylum Nematoda, with special reference to host-distribution. *J. Parasitol.,* 37: 353–378.

Hoeppli, R. 1927. Ueber Beziehungen zwischen dem biologischen Verhalten parasitischer Nematoden und histologischer Reaktionen des Wirbeltierkorpers. *Arch. Schiffs- u. Tropen-Hyg.,* 31: 207–290.

Hyman, L. B. 1951. *The Invertebrates,* Vol. III, pp. 56–58, 197–455, 480–519. McGraw-Hill, New York.

Lapage, G. 1937. *Nematodes Parasitic in Animals.* Methuen, London.

Rauther, R. 1928–1933. Nematodes in *Handbuch der Zoologie* (Edited by W. Kükenthal). Bd. II, Hälfte 1. de Gruyter, Berlin.

Yorke, W. W., and Maplestone, P. A. 1926. *The Nematode Parasites of Vertebrates.* J. and A. Churchill, London.

Chapter 18

TRICHURIS, TRICHINELLA, AND THEIR ALLIES

SUBORDER TRICHURATA

The worms belonging to the suborder Trichurata differ strikingly from all other nematodes in the appearance of the esophagus, which consists of a fine capillary tube embedded in a long column of single cells which form a structure called a stichosome, and which are believed to function as esophageal glands. The anterior portion of the body, containing only the esophagus, is always very fine and slender and in some forms is sharply demarcated from the relatively coarse posterior part of the body containing the intestine and reproductive organs. The vulva opens either at the end of the esophagus or anterior to this point. The eggs, if produced, are easily recognizable by their barrel shape with an opercular plug at each end. *Trichinella,* however, forms no egg shells, and the embryos hatch before birth.

The families and principal genera of this suborder are differentiated as follows:

Trichuridae. ♀ oviparous; ♂ with protrusible spiny spicular sheath and usually a spicule.

Trichuris. Anterior portion of body much more slender than posterior; whipworms.

Capillaria. Anterior portion slender but not sharply different from posterior; fine, hairlike worms.

Trichinellidae. ♀ viviparous; ♂ with no spicule or spicule sheath; contains *Trichinella* only.

Trichosomoididae. ♀ oviparous; ♂ parasitic in vagina of ♀; in urinary bladder of rodents; 1 genus, *Trichosomoides.*

Trichuris or Whipworms

The whipworm derives its name from its whiplike form, having a thick posterior part of the body containing the reproductive organs and a longer lashlike anterior part occupied only by the slender esophagus. The name *Trichuris* means "thread tail" and was given before it was recognized that the slender part was really a head and not a tail. Someone else more appropriately named the worm *Trichocephalus* (thread head), but since the other name was given first it must be used, in spite of its reflection on the inaccurate observation of its originator.

Whipworms are common inhabitants of the cecum and large intestine of many animals, including dogs, rodents, pigs, and all sorts of ruminants, as well as man and monkeys. Schwartz concluded in 1928 that the whipworms of pig and man are identical and that the whipworm commonly found in apes and monkeys is also the same species. The human species, *Trichuris trichiura,* has a world-wide distribution and is very common in the moist parts of warm countries. It usually inhabits the region of the cecum and appendix but sometimes lives in the sigmoid and rectum also. It buries its slender head in folds of the intestinal wall, occasionally threading it into the mucous membranes.

Morphology. The whipworm has a length of 30 to 50 mm., of which the threadlike esophageal portion occupies about two-thirds. The mouth has no lips but is provided with a minute spear. The males are a little smaller than the females and can be distinguished by the curled tail end of the body (Fig. 116*C*). They have a single long spicule, retractile into a sheath with a spiny, bulbous end. Unlike the condition in most nematodes, the ejaculatory duct (distal part of the sperm duct) joins the intestine a long way from the anus, forming a cloacal tube. This joins the spicular tube, containing the spicule and its sheath, also at some distance from the anus (Fig. 116*B*). The vulva of the female (Fig. 116*A*) is at the junction of the two parts of the body; the single uterus contains many of the barrel-shaped eggs (Fig. 116*D*), which measure about 50 by 22 μ and are unsegmented when they leave the host.

Life cycle and epidemiology. The life cycle is very simple. The eggs develop slowly; even when kept moist and warm they require 3 to 6 weeks for the embryo to reach the hatching point, and under less favorable conditions they may be delayed for months or even years. The eggs are less resistant to desiccation than are those of *Ascaris,* and nearly all die within 12 days when dried on a slide, even in a saturated atmosphere. Epidemiological evidence shows that a high incidence of *Trichuris* infection is always associated with an abundance of moisture in the soil, due either to a heavy and well-distributed rainfall or to dense shade. Infection may result from polluted water or from hand contaminations from polluted moist soil.

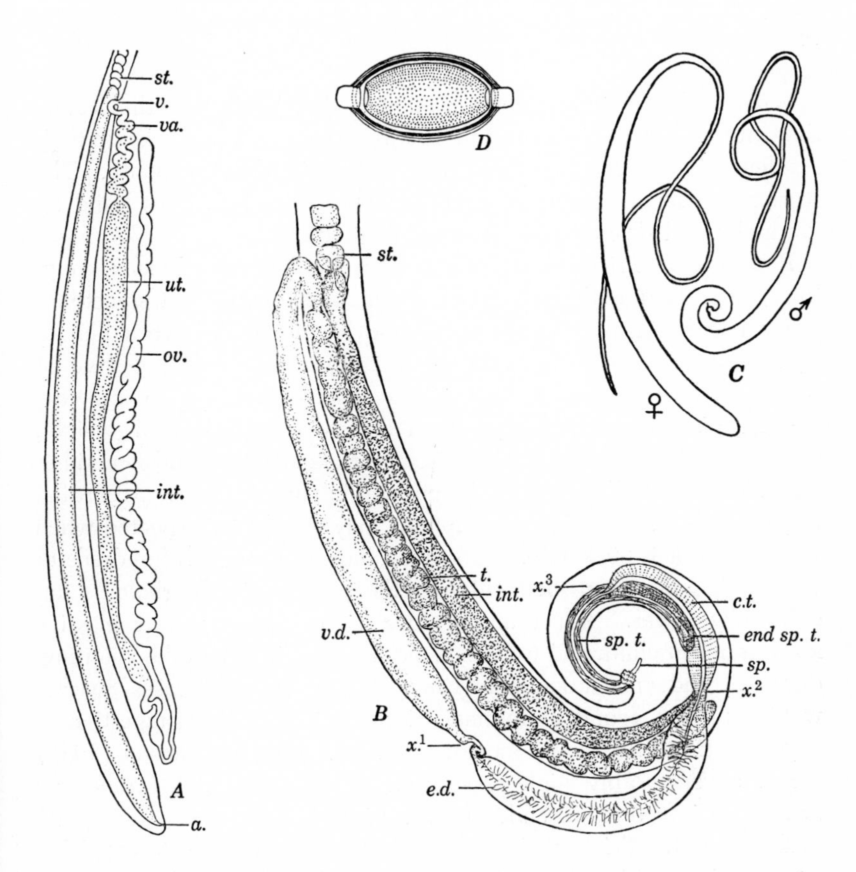

Fig. 116. Human whipworm, *Trichuris trichiura.* *A,* female, and *B,* male, dissected and organs spread, ×15; *C,* body shapes of male and female, ×4; *D,* egg, ×50. Abbreviations: *a,* anus; *c.t.,* cloacal tube; *e.d.,* ejaculatory duct; *int.,* intestine; *ov.,* ovary; *sp.,* spicules; *sp.t.,* spicular tube; *st.,* beginning of stichosome of esophagus; *t.,* testis; *ut.,* uterus; *v.,* vulva; *va.,* vagina; *v.d.,* vas deferens; *x.*[1], junction of vas deferens and ejaculatory duct; *x.*[2], junction of ejaculatory duct and intestine to form cloacal tube; *x.*[3], junction of cloacal tube and spicular tube.

In the United States *Trichuris* infections are more "spotty" in distribution than *Ascaris* and occur abundantly only in places where there is more or less door-yard pollution, dense shade close to the houses, a heavy rainfall, and a dense clay soil to conserve the moisture. These conditions are met in southwestern Louisiana and in the southern Appalachians.

When embryonated eggs are swallowed they hatch near the cecum, the embryos burrow into the villi for a few days, and then take up their residence in the cecum, where they mature in about a month. The worms live for a number of years, and therefore infections build up gradually and do not show seasonal fluctuations.

Pathology. *Trichuris* infections often produce no obvious symptoms, since frequently only a few worms are present, but sometimes the whole lower part of the colon and rectum may have a film of squirming *Trichuris.* Such heavy infections may be suggestive of severe hookworm disease. Symptoms observed are loss of appetite, nausea, diarrhea, blood-streaked stools, weakness, loss of weight, anemia, eosinophilia, abdominal discomfort, emaciation, and sometimes fever. Prolapse of the rectum is common in chronic cases. According to Jung and Beaver (1951) symptoms may always be expected when the egg count is 30,000 per cc. or over, indicating several hundreds of worms; in undernourished children the pathogenic threshold is undoubtedly lower. Kourí and Valdez Dias (1952) report massive infestations in children 1 to 5 years old in Cuba, in whom the vomiting, diarrhea, and emaciation may cause death. In mild cases Swartzwelder (1938) found abdominal discomfort or pains, suggestive of appendicitis, to be the commonest symptom. Jung and Beaver (1951) found a remarkably high association of *Trichuris* and *Entamoeba histolytica* infections in children in Louisiana.

Treatment. *Trichuris* is a particularly difficult worm to expel because of its position in the cecum, remote from either the mouth or the anus. Most anthelmintics given by mouth are relatively ineffective, although good results have been obtained with fresh or refrigerated latex of certain figs (*Ficus*), called by the Spanish name, lêche de higuerón. Jung and Beaver recommended enemas of 0.2% hexylresorcinol in water or glycerin solution; 500 to 700 cc. is given slowly after a cleansing enema, and retained for 30 minutes. Kourí also recommended hexylresorcinol enemas. A simpler but less effective treatment is a mixture of tetrachlorethylene (2.7 cc.) and oil of chenopodium (0.3 cc.) by mouth for adults, less for children, followed by a saline purge in 2 hours. "Enseals" of emetin hydrochloride have also been recommended.

None of these methods of treatment has been satisfactory, particularly in the treatment of populations. A new drug, dithiazanine, shows great promise. It may be given by mouth and also acts against *Ascaris*, hookworm, *Enterobius*, and *Strongyloides* (Swartzwelder et al., 1958).

Other Trichuridae

Species of the genus *Capillaria*, with slender, delicate body and relatively short esophageal portion, are parasitic in a wide variety of vertebrates and exercise a remarkable choice of habitats.

Capillaria hepatica lives in the liver of rats and other rodents where its eggs accumulate in dry, yellow patches. Since the eggs require air to become embryonated, direct eating of the egg-burdened liver does not cause infection; the egg must first be liberated and exposed to air by decomposition of the original host or preliminary passage through the intestine of a predatory animal. Several valid human cases and a number of pseudo-infections in which the eggs were presumably eaten with livers of infected animals have been recorded.

Another species, *C.* (or *Eucoleus*) *aerophila*, occurs in the respiratory system of cats, dogs, etc.; it is an important parasite of foxes, causing more harm than all other infections combined, except distemper. One human case has been reported from Moscow. Other species live in the esophagus and crop or in the intestine of birds, in the stomach of rats, in the urinary bladder of cats and foxes, and in the intestines of many animals.

The life cycles of most species are essentially the same as the life cycle of *Trichuris* except for migration via the blood stream of species living outside the alimentary canal. *C. annulata*, infecting the esophagus and crop, and *C. caudinflata*, infecting the intestine of chicks and turkeys, add an additional chapter, for the eggs fail to become infective until after ingestion by earthworms, which serve as true intermediate hosts (Morehouse, 1944). Consequently the fondness of poultry for earthworms is often penalized by *Capillaria* infection.

Trichinella spiralis and Trichiniasis

The trichina worm, *Trichinella spiralis*, though an intestinal parasite as an adult, is quite different in significance from other intestinal worms. The serious and often fatal results of trichiniasis are due to the offspring of the infecting worms and not to the adult worms in the intestine. It has been said that this worm was responsible for the old Jewish law against the eating of pork. This seems highly unlikely since there is no particular reason for assuming that trichinosis was prevalent in pigs in Moses' day but subsequently has almost disappeared from the Middle East.

Unlike most human helminths, this one is almost entirely absent from the tropics; it is primarily a parasite of Europe, the United States, and arctic regions, with moderate infection in Mexico and southern South America, particularly Chile; it is practically absent from San Francisco to Suez and from Africa and Australia. In the Arctic Rausch found trichinosis prevalent not only in Eskimos, polar bears, and dogs, but also in marine mammals such as seals and white whales, which constitute a large proportion of the natural food of man and dogs. How the fish-eating marine mammals get infected is an unsolved problem. Schiller and Read (1960) recognize a subspecies, *T. spiralis arctica,* which is essentially incapable of infecting rats but readily infects deer mice, and certain carnivores.

Structure and life history. The trichina worm infects many animals. In America hogs are most commonly infected, and infection is common in rats which have access to waste pork. Cats are frequently infected, dogs less often. Man is highly susceptible, and many rodents are easily infected if fed trichinized meat. Birds are very resistant.

The worms gain entrance to the digestive tract as larvae encysted in meat (Fig. 119). They are freed from their cysts in the stomach or intestine and penetrate into the mucosa of the small intestine. Here they undergo a series of molts which bring them to the adult stage. They may reach sexual maturity and copulate as early as 40 hours after being swallowed. The females (Fig. 117) are 3 to 4 mm. long, whitish, slender and tapering from the middle of the body toward the anterior end; the males are only 1.5 mm. long. The long capillary esophagus occupies one-third to one-half the length of the body. In the female the vulva opens near the middle of the esophageal region; the anterior part of the uterus is crowded with embryos, whereas the posterior part contains developing eggs. The males, aside from their minute size, are characterized by the presence of a pair of conical appendages at the posterior end. In both sexes the anus (or cloaca) is terminal. The males have no spicule (Fig. 117*C*).

The adult intestinal worms are essentially short-lived, usually disappearing within 2 or 3 months after infection. Many males pass out of the intestine soon after mating, though some live as long as the females

Trichina embryos develop in the uterus of the mother and are scarcely 0.1 mm. in length when born. The mother worms usually burrow into the mucous membranes far enough so that the young can be deposited in the tissues rather than into the lumen of the

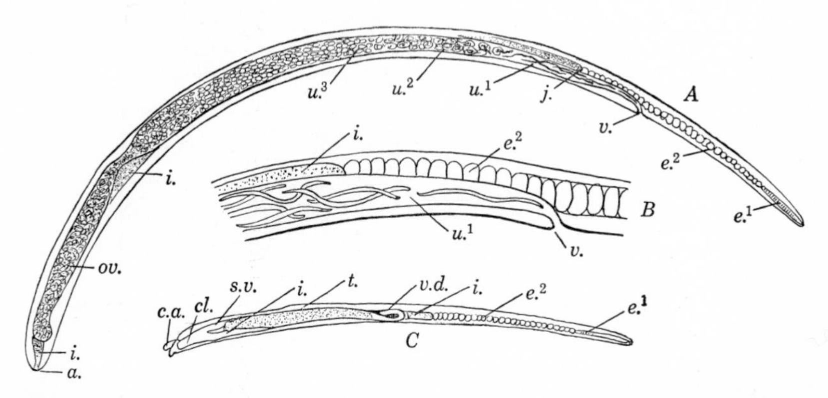

Fig. 117. Adult *Trichinella spiralis.* *A*, female; *B*, vulva region of female; *C*, male; *a.*, anus; *c.a.*, caudal appendages; *cl.*, cloacal tube; *e.*[1], anterior portion of esophagus; *e.*[2], posterior portion of esophagus or stichosome; *i.*, intestine; *j.*, junction of esophagus and intestine; *ov.*, ovary; *s.v.*, seminal vesicle; *t.*, testis; *u.*[1], anterior portion of uterus with free embryos; *u.*[2], middle portion of uterus with embryos coiled in vitelline membrane; *u.*[3], posterior portion of uterus with ova; *v.*, vulva; *v.d.*, vas deferens. Entire worms $\times 40$ in length, width $\times 80$.

intestine. Embryos may be born within a week after the parents have been swallowed by the host and are most numerous in the circulating blood between the eighth and twenty-fifth days after infection.

The embryos enter lymph or blood vessels in the intestinal wall and are distributed over the entire body. They have been found in practically every organ and tissue but undergo further development inside the cells of the voluntary muscles. Active muscles containing a rich blood supply, such as those of the diaphragm, ribs, larynx, tongue, eye, and certain ones in the limbs, are particularly favored, but all the striated muscles in the body except the heart muscle are liable to invasion (Fig. 118). Unlike many tissue-penetrating larvae, however, trichina embryos rarely pass through the placenta and cause prenatal infections.

After entering muscle fibers the worms grow rapidly to a length of 1 mm., ten times their original size, and become sexually differentiated. They finally roll themselves into a spiral and are infective after about 17 or 18 days.

The inflammation caused by the movements and waste products of the worms results in the degeneration of the enclosing muscle fibers and in the formation of cysts around the young worms, beginning about a month after infection. The cysts (Fig. 119), at first very

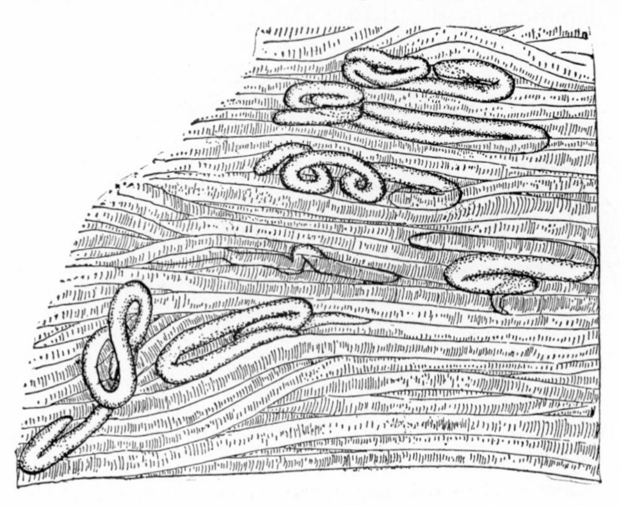

Fig. 118. Larvae of trichina worm burrowing in human flesh before encystment. From preparation from diaphragm of a victim of trichiniasis. ×75.

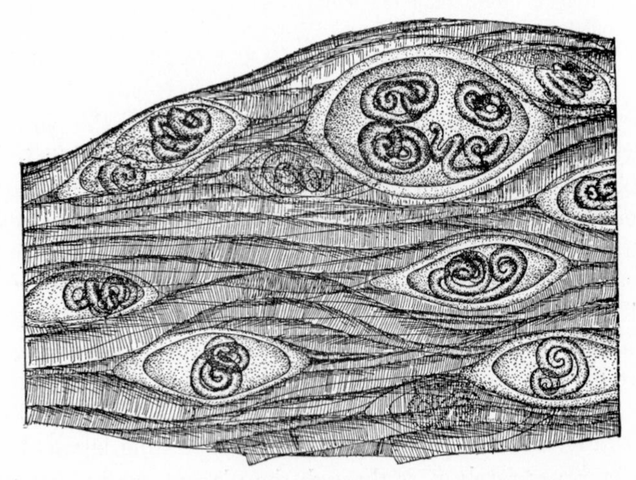

Fig. 119. Larvae of trichina worms, *Trichinella spiralis,* encysted in striated muscle fibers in pork. Camera lucida drawing of cysts in infected sausage. ×75.

delicate but gradually thickening, are lemon-shaped, 0.25 to 0.5 mm. long, lying parallel with the muscle fibers; they are not fully developed until after 7 or 8 weeks. As a rule only one or two worms are enclosed in a cyst but as many as seven have been seen.

After 7 or 8 months or sometimes much later, the cyst walls start to calcify, beginning at the poles. After 18 months or longer the entire cyst becomes calcified and appears as a hard calcareous nodule. Even the enclosed worm, which usually degenerates and dies after some months, becomes calcified after a number of years. At times, however, the trichina worms do not die and disintegrate so soon and the calcification process is much slower. Experimentally the calcification of well-formed cysts can be hastened by administration of calcium and ergosterol or even more by large doses of parathormone.

Estimates of the number of encysted larvae that may be expected per female worm vary greatly, but experimental work with various animals indicates about 1500. An ounce of heavily infected sausage may contain more than 100,000 encysted larvae, over half of which are females, so the eating of it may result in more than a hundred million larvae distributing themselves throughout the body of the unfortunate victim. It has been estimated that for man ingestion of 5 trichina larvae per gram of body weight is fatal, for hogs 10, and for rats 30.

After encysting in the flesh no further development takes place until the flesh is eaten by a susceptible animal, whereupon the worms mature and begin reproducing in a few days. It will be seen that, whereas most worms begin the attempt to find new hosts at the egg or early embryo stage of the second generation, the trichina worm does not make a break from its parental host until it has reached the infective stage for another host.

Mode of infection and prevalence. Obviously man usually becomes infected from eating raw or imperfectly cooked infected meat, in most cases pork. Under modern conditions hogs undoubtedly are most commonly infected by being fed on garbage containing pork scraps, as Hall pointed out in 1937. Nearly 40% of cities of over 4500 population and 50% of cities of over 15,000 dispose of garbage by feeding it to hogs. Stoll suggested calling *Trichinella* the "garbage worm."

Rats appear to play a very minor role in the epidemiology as compared with infected pork scraps, for hogs are not by nature rat-eaters; Hall says that in his experience hogs and rats usually live together on very friendly terms. Rats pass the disease among themselves by cannibalism, but in most cases it is a closed circuit.

Hall showed that the prevalence of the infection in both man and hogs is closely correlated with methods of raising hogs in different parts of the country. It is highest on the North Atlantic seaboard and in California, where hogs are most extensively fed on garbage. In the

Middle West, where a higher percentage are raised on pastures and fed on corn, the incidence is lower and it is still lower in the South where the hogs are generally allowed to roam the fields and woods, competing with the squirrels for acorns and without easy access to kitchen scraps or city garbage.

The incidence of human infection is astonishingly high; where examinations have been made in routine autopsies the infection ranges from about 5% in New Orleans to 18 to 27% in northern and western cities, with a general average of over 16% in the entire United States; even these figures are apparently below the actual incidence. Stoll in 1947 called attention to the fact that the United States has three times as much trichiniasis as all the rest of the world combined. The incidence in hogs in this country is about 1.5%; at this rate, as Gould (1945) pointed out, an average pork-eating American might eat 200 meals of trichinous pork in his lifetime, so the 16% infection is not so surprising. Fortunately, in contrast to this high incidence of infection, outbreaks since 1900 have been mild, with low mortality; less than 600 clinical cases a year are reported, with a mortality of less than 5%.

The most serious outbreaks occur among Germans, Austrians, and Italians who are fond of various forms of uncooked sausage and "wurst." Nearly all serious outbreaks can be traced back to animals slaughtered on farms or in small butchering establishments, since in large slaughterhouses the meat of an infected animal is almost certain to be diluted with the meat of uninfected animals. Moreover, in federally inspected establishments pork destined for raw consumption is refrigerated long enough to destroy the infection (see below). There is a particularly high death rate among rural school teachers and preachers, who are invited by their hospitable neighbors to sample and praise new batches of delicious, freshly made sausage.

Very serious outbreaks have continued to occur in northern Europe. These may affect large numbers of people; e.g., an epidemic in Gdansk in 1952 involved several hundred of them.

The disease. As we have seen, the vast majority of human infections are never diagnosed or suspected unless the diaphragms are examined microscopically or by artificial digestion after death. Hall and Collins call attention to the fact that in not one of 222 infections found post-mortem had a diagnosis of trichiniasis been made, although in some there were almost 1000 worms per gram of muscle, and no person harboring that many worms could by any stretch of the imagination be considered free of symptoms. Clinical symptoms are certainly far commoner than the number of reported cases would indicate. Some cases are mistaken for typhoid, ptomaine poisoning, "intestinal flu," or what not, but in many cases the patients probably just did not feel well. The severity of the symptoms is largely dependent on the number of living worms eaten, although it is undoubtedly influenced also by the general state of health and resistance and by immunity due to prior infections.

The clinical course of trichiniasis is very irregular. Characteristically, the first symptoms are diarrhea, abdominal pains, nausea, and other gastrointestinal symptoms, with or without fever, flushing, etc., caused by irritation of the intestine by the growing and adult worms burrowing into its walls. There is often a sort of general torpor accompanied by weakness, muscular twitching, etc. As the larvae become numerous in the blood and tissues, eosinophilia develops, in extreme cases reaching 50% and even 90%.

The second stage is the period of migration of larvae and penetration of muscles; it is frequently fatal. One of the earliest symptoms in this stage is a marked puffiness under the eyes and in the lids. The characteristic symptoms are intense muscular pains and rheumatic aches. Disturbances in the particular muscles invaded cause interference with movements of the eyes, mastication, respiration, etc. The respiratory troubles become particularly severe in the fourth and fifth weeks of the disease, in fact, sometimes so severe as to cause death from dyspnea or asthma. Profuse sweating and more or less constant fever, though sometimes occurring in the first stage also, are particularly characteristic of the second stage. The fever is commonly absent in children. Eosinophilia and leucocytosis are nearly always present.

The third stage, accompanying the encystment of the parasites, begins about 6 weeks after infection. The symptoms of the second stage become exaggerated, and in addition the face again becomes puffy, and the arms, legs, and abdominal walls are also swollen. The patient becomes anemic, skin eruptions occur, the muscular pains gradually subside, and the swollen portions of the skin often scale off. Pneumonia is a common complication. In fatal cases death usually comes in the fourth to sixth week, rarely before the end of the second or after the seventh.

Numerous variations from this course involve both omissions and additions. In America a more or less persistent diarrhea accompanied by eosinophilia, fever, puffy eyes, and muscular pains should always suggest trichiniasis. Sometimes the characteristic symptoms are overshadowed by others, involving the heart, eye, or nervous system, where the larvae burrow but do not develop. Sometimes even the gastrointestinal symptoms fail to appear, and when there are accompanying bacterial infections there may be no eosinophilia.

Recovery usually does not occur in less than 5 to 6 weeks after infection and often not for several months. Recurrent muscular pains and weakness may continue for a year. Commonly, cases in which a copious diarrhea appears early in the disease are of short duration and mild in type. Young children, owing either to smaller quantities of pork eaten or to greater tendency to diarrhea, are likely to recover quickly.

Diagnosis. To confirm a diagnosis is not easy. Search for adult worms in feces is unreliable, and larvae in blood or cerebrospinal fluid, though present after 8 to 10 days, are difficult to find. The removal of a bit of muscle and examination of it pressed out between two slides are of no use early in the infection but are often diagnostic later.

Bachman (1928) devised a skin test and a precipitin test which have proved helpful. The antigen consists of dried and powdered larvae obtained by artificial digestion of the meat of heavily infected animals. In positive skin tests a blanched wheal appears in 5 minutes and reaches a diameter of 1 to 2 cm. in an hour. This test is seldom positive before about 11 to 14 days, however, and may remain positive for at least 7 years after infection, so might be misleading. It may, however, be put to practical use in the detection of infected hogs (see below). Bozicevich et al. (1951) have developed an easy 15-minute flocculation test in which *Trichinella* antigen is adsorbed on bentonite particles, which then clump on a slide when exposed to serum containing antibodies. More recently, substitution of tanned red blood cells for the bentonite particles has led to the development of a slide hemagglutination test that is rapid and easy to read.

After 2 weeks negative tests are valuable in ruling out trichiniasis, whereas positive ones are valuable as corroborative evidence. Positive reactions sometimes occur in *Trichuris* infections. In view of the difficulty in making a correct diagnosis, it is not surprising that trichiniasis has been mistaken for at least fifty other disease conditions.

Treatment. The search for a good anthelmintic to kill the larvae in the muscles has not yet been very fruitful, and little progress has been made in expelling the adult worms from the intestine. On the basis of success in experimental animals, both Hetrazan and piperazine citrate show promise of effectiveness in removing adult *Trichinella* from man; of piperazine Chou and Brown (in 1954) suggested daily doses of 2 to 3 grams.

Often, however, diagnosis is not made until the critical stage, when millions of embryos are migrating through the body and developing in muscle fibers. The treatment employed then can be only symptomatic.

Cortisone and ACTH have been used to lessen the inflammatory response. However, the inflammatory response in the intestine serves a function in producing a diarrhea and a consequent flushing-out of the adult worms. If cortisone is given during the period when adults are still in the alimentary tract, the drug may allow the adults to persist for a longer time and consequently to produce larger numbers of larvae.

Immunity. Considerable resistance to infection is produced by prior exposure. McCoy showed that this resistance was effective against worms developing in the intestine as well as larvae migrating parenterally. This has been confirmed by Bozicevich and Detre (*NIH Studies in Trichinosis,* VIII, 1940), who found antigen in blood of rabbits 24 hours after they were fed trichina larvae, and by Zaiman (1953, 1954), who found uninfected members of parabiotic twins showing heightened resistance to reinfection when the twins were separated 5 days after infection of one member, before any parenteral larvae were present, and also when the mate was infected with worms rendered incapable of reproduction by x-rays. Roth (1943) obtained some degree of immunity by infecting animals with larvae of one sex only. In immunized mice the worms of subsequent infections show the characteristic evidences of immunity found in other worm infections (see pp. 26–30)—small worms, retarded development, and greatly inhibited reproduction.

A far smaller degree of protection results from injection of vaccines or immune serum. When either adults or larvae are placed in immune serum a precipitate forms at mouth and anus, as in the case of *Nippostrongylus* (see p. 28).

This seems to support the idea that the immune reaction is directed mainly against the products of the worms. This is further supported by Campbell's (1954) and Chipman's (1957) demonstration that resistance of mice to *Trichinella* infection can be enhanced by injection of secretions and excretions of *Trichinella* larvae and adults.

The development of resistance in mice can be altered by behavioral factors. Davis and Read (1959) showed that fighting in mice lowered the capacity of the animals to develop resistance. This was attributed to increased activity of the adrenal glands with enhanced secretion of cortisone or related substances.

From the evidence available it may be presumed that human beings may often be protected from the ill effects of eating heavily trichinized meat by having eaten more lightly infected meat at some earlier date. Data are lacking, however, on how long the immunity is effective.

Prevention. Personal preventive measures against trichiniasis are easy and consist simply in abstinence from all pork that is not thoroughly cooked. Trichinae are quickly destroyed by a temperature of 55°C. (131°F.), but pork must be cooked for a length of time proportionate to its weight in order to insure the permeation of heat to the center. At least 30 to 36 minutes' boiling should be allowed to each kilogram of meat (2¼ lb.). Hurried roasting does not destroy the parasites as long as red or raw portions are left in the center.

Augustine (1933) showed that quick cooling to −34°C., or quick cooling to −18°C., followed by storage at that temperature for 24 hours, or at −15°C. for 48 hours, renders the trichinae noninfective. Cold storage for 20 days at a temperature of −15°C. is required by the U. S. Bureau of Animal Industry for pork products to be used uncooked, unless cured in accordance with certain specified processes, but this appears to be an unnecessarily long time. Salting and smoking are not efficacious unless carried out under certain conditions.

It has been shown that irradiation of trichinous meat with cobalt-60 results in sexual sterilization of the worms, with a failure to produce larvae when such meat is eaten (Gould et al., 1954). This was suggested as a simple method of controlling trichinosis. However, according to Magath and Thompson (1955) such irradiated worms do not stimulate the development of any immunity, and wholesale irradiation of meat might result in a much larger proportion of highly susceptible individuals in the population. A somewhat more practical objection was raised by Otto (1958) who pointed out that under present conditions irradiation of pork in farmyards and small abattoirs could not be enforced; and it is these small establishments which are most often involved in outbreaks of human trichinosis.

Prevention by examination of meat for larvae was thought by Stiles to be too expensive, too incomplete, and inapplicable to the most dangerous sources of infection—hogs butchered on farms or small local establishments. In Chile, however, examination of compressed samples of pork, projected on a screen, has been found inexpensive, rapid, and effective in discovering all but very light infections, which are not dangerous.

The most feasible and practical plan for the control of trichiniasis consists in forbidding the feeding of raw garbage to hogs, as has been done for years in Canada and England. It would be better to stop raising hogs on garbage altogether, but in the United States 1,500,000 hogs are raised annually wholly or in part on commercial garbage.

This practice is responsible for the spread of a number of other hog diseases as well, especially cholera, foot-and-mouth disease, and vesicular exanthema. In 1952 a nationwide outbreak of the last caused heavy losses to hog raisers, and did more than did 25,000,000 human *Trichinella* infections to persuade the swine industry to look upon legislation directed against feeding raw garbage to hogs with a less jaundiced eye. Since that outbreak thirty-seven states either have initiated such legislation or are preparing to do so. For interstate traffic the Public Health Service requires that garbage be steamed or boiled for 30 minutes before being fed to hogs.

Other Aphasmidia

Suborder Dioctophymata. The Giant Kidney Worm

The only aphasmid nematodes other than the Trichurata which are normally parasitic in vertebrates are the Dioctophymata. The females have one ovary, anterior vulva (in *Dioctophyma*), and terminal anus; the males have a terminal bell-shaped bursa without rays, and a single spicule (Fig. 120*A*). The eggs have thick, pitted shells.

The only species of importance is *D. renale,* the giant kidney worm, found in the pelvis of the kidney or in the abdominal cavity of mink, dogs, and occasionally in many other animals, including man (Fig. 120*B*). It is a huge, blood-red worm, the female of which sometimes exceeds 3 ft. in length, with the diameter of a small finger, whereas the male may be 6 to 16 in. long.

It is usually the right kidney that is infected; this swells to several times its size and eventually becomes a mere shell, often with a bony

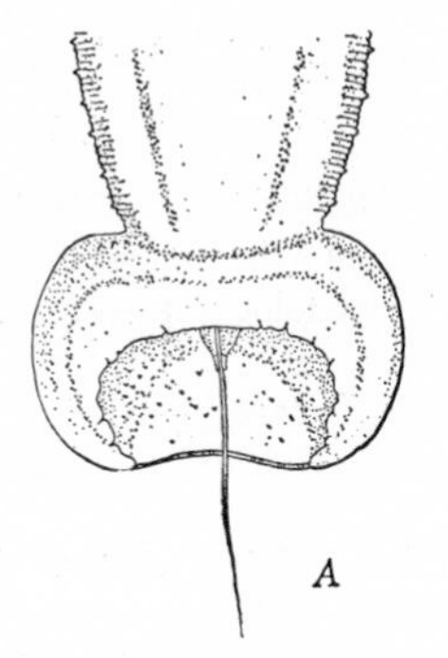

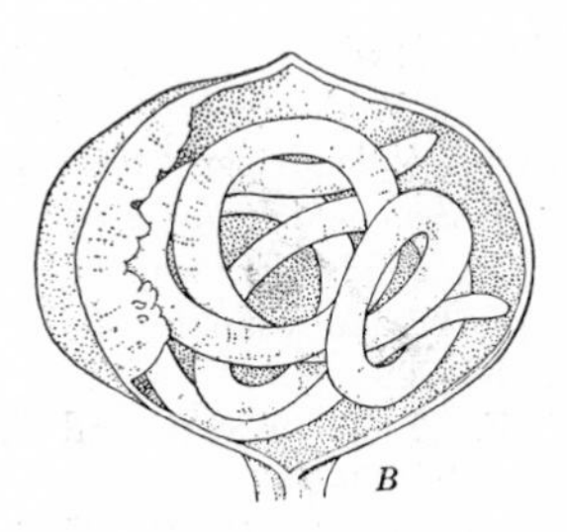

Fig. 120. *Dioctophyma renale,* giant kidney worm. *A,* copulatory bursa of ♂, ventral view, with bristlelike spicule projecting; *B,* adult worm in kidney of a dog. (A, after Stéfansky, *Ann. Parasitol.,* 6, 1928. *B,* adapted from Railliet.)

plate developed on its dorsomedial surface (McNeil, 1948). The left kidney hypertrophies to about twice its normal size. It has generally been assumed that the parasite invades one of the kidneys first and enters the abdominal cavity after the kidney has been more or less destroyed, but Stéfanski and Strankowski (1936) think that it develops in the body cavity and later penetrates the kidney by means of a histolytic secretion from its highly developed esophageal glands.

Woodhead (1945) reported a remarkable life cycle for this worm involving a first stage of development in an annelid symbiote of crayfish and a second one in a fish, *Ameiurus melas* (bullhead), before reaching the adult stage in a mammal. The stages described are strikingly similar to those of the Gordiacea (see p. 240).

REFERENCES

Trichuridae

Allen, R. W., and Wehr, E. E. 1942. Earthworms as possible intermediate hosts of *Capillaria caudinflata* of the chicken and turkey. *Proc. Helminthol. Soc. Wash.*, 9: 72–73.

Brown, H. W. 1948. Recent developments in the chemotherapy of helminthic diseases. *Proc. 4th Intern. Congr. Trop. Med. Malaria,* 2: Sect. VI. 966–974.

Chandler, A. C. 1930. Specific characters in the genus *Trichuris*. *J. Parasitol.*, 16: 198–206.

Christenson, R. O. 1938. Life history and epidemiological studies on the fox lungworm, *Capillaria aerophila*. *Livro jubilar do Prof. Travassos.*, 119–136. Rio de Janeiro.

Fülleborn, F. 1923. Über die Entwicklung von Trichozephalus in Wirte. *Arch. Schiffs- u. Tropen-Hyg.*, 27: 413–420.

Getz, L. 1945. Massive infection with *Trichuris trichiura* in children. Report of four cases, with autopsy. *Am. J. Diseases Children,* 70: 19–24.

Jung, R. C., and Beaver, P. C. 1951. Clinical observations on *Trichocephalus trichiurus* (whipworm) infestation in children. *Pediatrics,* 8: 548–557.

Kourí, P., and Valdez Diaz, R. 1952. Concepto actual sobre el papel patógeno de Tricocéfalo dispar (*Trichuris trichiura*) sintomatologia gastro-intestinale. *Rev. Kuba. med. trop. y parasitol.*, 8: 37–41.

Nolf, L. O. 1932. Experimental studies on certain factors influencing the development and viability of the ova of the human Trichuris. *Am. J. Hyg.*, 16: 288–322.

Otto, G. F. 1932. *Ascaris* and *Trichuris* in southern United States. *J. Parasitol.*, 18: 200–208.

Swartzwelder, J. C. 1939. Clinical *Trichocephalus trichiurus* infection. An analysis of 81 cases. *Am. J. Trop. Med.*, 19: 473–481.

Swartzwelder, J. C., et al. 1958. Therapy of trichuriasis and ascariasis with dithiazanine. *Am. J. Trop. Med. Hyg.*, 7: 329–333.

Thomen, I. J. 1939. The latex of ficus trees and derivatives as anthelmintics: historical account. *Am. J. Trop. Med.*, 19: 409–418.

Trichinella spiralis

Augustine, D. L. 1933. Effects of low temperatures upon encysted *T. spiralis*, *Am. J. Hyg.*, 17: 697–710.

Bachman, G. W. 1928. Precipitin test in experimental trichinosis. *J. Prev. Med.*, 2: 35–48; An intradermal reaction in experimental trichinosis, *ibid.*, 2: 513–523.

Bozicevich, J., and Detre, L. 1940. Studies on trichinosis. VIII. The antigenic phase of trichinosis. *Public Health Repts.*, 55: 683–692.

Bozicevich, J., et al. 1951. A rapid flocculation test for the diagnosis of trichinosis. *Public Health Repts.*, 66: 806–814.

Brandly, P. J., and Rausch, R. 1950. A preliminary note on trichinosis investigations in Alaska. *Arctic*, 3: 105–107.

Campbell, C. H. 1955. The antigenic role of the excretions and secretions of *Trichinella spiralis* in the production of immunity in mice. *J. Parasitol.*, 41: 483–491.

Chipman, P. B. 1957. The antigenic role of the excretions and secretions of adult *Trichinella spiralis* in the production of immunity in mice. *J. Parasitol.*, 43: 593–598.

Coker, C. M. 1955. Effects of cortisone on *Trichinella spiralis* infections in nonimmunized mice. *J. Parasitol.*, 41: 498–504.

Davis, D. E., and Read, C. P. 1958. Effect of behavior on development of resistance in trichinosis. *Proc. Soc. Exp. Biol. Med.*, 99: 269–272.

Culbertson, J. T. 1942. Active immunity in mice against *Trichinella spiralis*. *J. Parasitol.*, 28: 197–202.

Gould, S. E. 1945. *Trichinosis*. Thomas, Springfield, Ill.

Magath, T. B., and Thompson, J. H., Jr. 1955. The effect of irradiation of *Trichinella spiralis* on immunity and its public health importance. *Am. J. Trop. Med. Hyg.*, 4: 941–946.

McCoy, O. R. 1935, 1940. Artificial immunization of rats against *Trichinella spiralis*. *Am. J. Hyg.*, 21: 200–213. Rapid loss of *Trichinella* larvae fed to immune rats and its bearing on the mechanism of immunity. *Ibid.*, 32 D: 105–116.

National Institute of Health (Hall, Wright, Bozicevich, et al.). I–XVI. 1937–1944. Studies on trichinosis. *Public Health Repts.*, 52: 468–490, 512–527, 539–551, 873–886; 53: 652–673, 1086–1105, 1472–1486, 2130–2138; 55: 683–692, 1069–1077; 56: 836–855; 58: 1293–1313; 59: 669–681. *J. Am. Vet. Med. Assoc.*, 94: 601–608; *Am. J. Public Health*, 29: 119–127.

Otto, G. F. 1958. Some reflections on the ecology of parasitism. *J. Parasitol.*, 44: 1–27.

Public Health Reports. 1953. Trichinosis: Highlights from First National Conference on Trichinosis. 1952. *Public Health Repts.*, 68: 417–424.

Reinhard, E. G. 1958. Landmarks of parasitology. II. Demonstration of the life cycle and pathogenicity of the spiral threadworm. *Exp. Parasitol.*, 7: 108–123.

Shookhoff, W. B., Birnkrant, W. B., and Greenberg, M. 1946. An outbreak of trichinosis in New York City, with special reference to intradermal and precipitin tests. *Am. J. Public Health*, 36: 1403–1411.

Stoll, N. R. 1947. This wormy world. *J. Parasitol.*, 33: 1–18.

Zaiman, H., et al. 1953–1954. Studies on the nature of immunity to *Trichinella spiralis* in parabiotic rats. I-V. *Am. J. Hyg.*, 57: 297–315; 59: 39–59.

Dioctophyma

Hallberg, C. W. 1953. *Dioctophyme renale* (Goeze, 1782), a study of the migration routes to the kidneys of mammals and resultant pathology. *Trans. Am. Microscop. Soc.*, 72: 351–363.

McNeil, C. W. 1948. Pathological changes in the kidney of mink due to infection with *Dioctophyma renale* (Goeze, 1782), the giant kidney worm of mammals. *Trans. Am. Microscop. Soc.*, 67: 257–261.

Meyer, M. C., and Witter, J. F. 1950. The giant kidney worm (*Dioctophyma renale*) in mink in Maine, with a summary of recent North American records. *J. Am. Vet. Med. Assoc.*, 116: 367–369.

Stéfanski, W., and Strankowski, M. 1936. Sur un cas de pénétration du strongle géant dans le rein droit du chien. *Ann. parasitol. humaine et comparée*, 14: 55–60.

Underwood, P. C., and Wright, W. H. 1934. A report of the giant nematode, *Dioctophyme renale* from a dog, with a summary of American records. *J. Am. Vet. Med. Assoc.*, 85: 256–258.

Woodhead, A. E. 1945. The life history cycle of *Dioctophyma renale*, the giant kidney worm of man and many other mammals (Abstract). *J. Parasitol.*, 31: 12.

Chapter 19

THE HOOKWORMS AND THEIR ALLIES SUBORDER STRONGYLATA

No group of the nematodes causes more injury to man or greater economic loss through attacks on his domestic animals than the members of the suborder Strongylata, the great majority of which are parasites of mammals. Many are bloodsuckers and cause severe injury to their hosts by loss of blood sucked by them or wasted from hemorrhages; the result is anemia, loss of vitality, and general unthriftiness.

The worms of this suborder have one easily recognizable character which is constant and peculiar to them, namely, a bursa surrounding the cloaca of the male. This is a sort of umbrellalike expansion of the cuticle at the end of the body which is supported by fleshy rays comparable with the ribs of an umbrella. The arrangement of the rays is remarkably constant, and each ray is given a name. Usually the bursa consists of three lobes, two lateral and one dorsal, and it may or may not be split ventrally; it varies in size, and in some of the lung worms (Metastrongylidae) it is vestigial or even absent. The dorsal lobe is supported by a dorsal ray which may be bifurcated only at its tip or may be split almost to the base; from its root there arise a pair of externo-dorsal rays which usually enter the lateral lobes. The latter are supported by three pairs of lateral rays arising from a common root, and two pairs of ventral rays arising from another common root. The names and arrangement of these rays as they occur in hookworms are shown in Figs. 121 and 123*A* and *C*.

Other characteristics of the group are the club-shaped or cylindrical muscular esophagus and the absence of distinct lips; the mouth is either a simple opening at the end of a fine slender head or is provided with a more or less highly specialized buccal capsule. The eggs always have thin transparent shells that do not become bile stained and are therefore colorless. They are in some stage of segmentation or contain embryos when laid. The eggs hatch outside the body into free-living larvae which, after reaching a certain stage of development, enter a new host either by burrowing through the skin or by being ingested with water or vegetation, or in the case of some of the lungworms, with an intermediate or transport host.

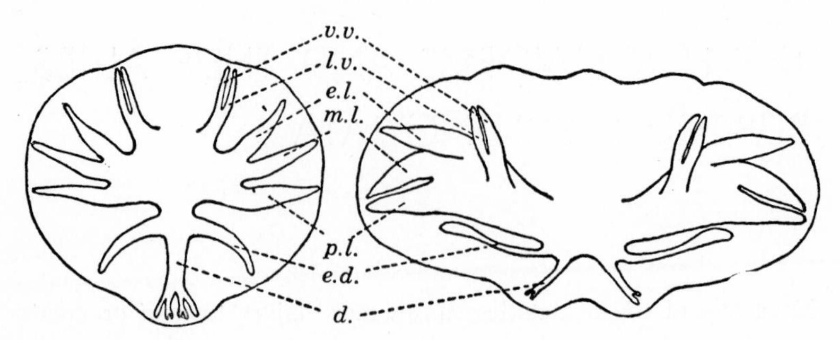

Fig. 121. Diagrams of bursas (spread out flat) of *Ancylostoma duodenale* (*left*) and *Necator americanus* (*right*), showing arrangement of rays: *d.*, dorsal ray; *e.d.*, externo-dorsal; *e.l.*, externo-lateral; *l.v.*, latero-ventral; *m.l.*, medio-lateral; *p.l.*, postero-lateral; *v.v.*, ventro-ventral. (After Chandler, *Hookworm Disease*, 1929.)

The suborder, on the basis of work by Dougherty (1945, 1951), may be classified into families as follows:

I. Eustomatous forms, i.e., with well-developed mouth capsule.
   1. **Ancylostomidae** (hookworms). Capsule with ventral teeth or plates inside opening (Fig. 122).
   2. **Strongylidae.** With crown of leaflike processes (corona radiata) (Fig. 129).
   3. **Syngamidae** (gapeworms). Capsule hooped by large chitinous ring (Fig. 131).

II. Meiostomatous forms, i.e., with reduced or vestigial mouth capsule.
   1. **Trichostrongylidae.** Intestinal forms with ovejectors (Fig. 133*B*); bursa well developed. No intermediate hosts.
   2. **Metastrongylidae.** Lungworms: vulva posterior, vagina long, and uteri parallel (Fig. 133*A*), except in a few forms in carnivores (Skrjabingylinae) and one genus, *Dictyocaulus*, in ruminants; bursa reduced; intermediate hosts (mollusks or annelids) required except by *Dictyocaulus*.

Hookworms of Man

Importance. Until recently hookworm ranked without question as the most important helminthic infection of man, but it has been brought under control in many countries to such an extent in the last few decades that now it is quite likely the schistosomes that deserve first place. Hookworm is never spectacular like some other diseases, but is essentially insidious; year after year, generation after generation, it saps the vitality and undermines the health and efficiency of whole communities. For years the "poor white trash" of some rural parts of our South were considered a shiftless, good-for-nothing, irresponsible people, worthy only of scorn and of the sordid poverty and ignorance which they brought upon themselves as the fruits of their supposedly innate shiftlessness, but the discovery that these unfortunate people were the victims of hookworms which stunted them physically and mentally made them objects of pity rather than scorn. Fortunately, their lot has been enormously improved since the early part of the present century.

Distribution. Hookworm infection exists where local conditions are favorable in most tropical and subtropical parts of the world, bounded approximately by the thirty-sixth parallel in the north and the thirtieth parallel in the south. In North America it occurs in the southeastern United States west to eastern Texas and north to Virginia, Kentucky, southwestern Missouri, and Arkansas; in Mexico, principally on the Gulf Coast from southern Tamaulipas to the base of the Yucatan peninsula; and in most of Central America and some of the West Indies, especially Puerto Rico. In South America it is present over vast areas east of the Andes and south to the River Platte, especially in the Amazon Valley and southeast coast of Brazil. In the Old World it occurs mainly in southern Europe, Egypt, west and central Africa south to Natal, and Madagascar; northern coasts of Asia Minor and irrigated areas in the Middle East; parts of India, Burma, China, Formosa, Japan, Indo-China, Malaya, East Indies, Borneo, New Guinea, some of the Polynesian Islands, and the Queensland coast.

Surveys by Keller, Leathers, et al. (1940) showed a remarkable decrease in both the incidence and intensity of infection in the United States since the first surveys were made about 20 years earlier. In six of eight southern states the average incidence, by techniques that miss very light cases, was 36.6% in 1910–1914 and 11.2% in 1930–1938. In 1940 the incidences varied from 50% in western Florida to 7 to 9% in Tennessee and Kentucky, the areas of important infection being largely localized. The highest incidence is in whites in the 15- to 19-year age group. Today only a small percentage of those infected have enough worms to cause clinical symptoms.

Species. Two species of hookworms are common human parasites, *Ancylostoma duodenale* and *Necator americanus.* They are similar in general appearance and in most details of their life cycle, habits, etc., but *A. duodenale* is much more injurious to its host and is harder to expel by means of anthelmintics. All hookworms, including many species found in dogs, cats, herbivores, and other animals, are rather stocky worms, usually about half an inch in length, with a well-developed bursa, very long, needlelike spicules, and with a conspicuous goblet- or cup-shaped buccal capsule, guarded ventrally by a pair of chitinous plates which either bear teeth, as in the ancylostomes, or have a bladelike edge, as in the necators and their allies (Fig. 122). The human ancylostome, *A. duodenale,* has two well-developed teeth

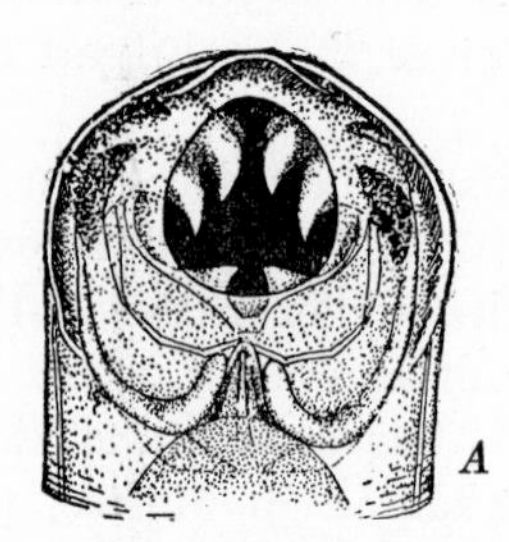

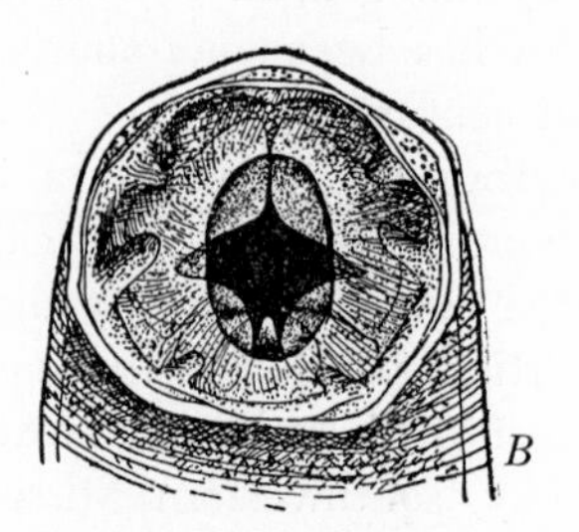

Fig. 122. Mouth and buccal cavity of *Ancylostoma duodenale* (*A*), and *Necator americanus* (*B*), showing teeth in former and cutting ridges in latter. Dorsal view. *A,* ×100; *B,* ×230. (Adapted from Looss.)

on each plate, with a rudimentary third one near the median line (Fig. 122*A*). *A. caninum,* common in dogs and cats, has three pairs of teeth (Fig. 124*C*), and *A. braziliense* and *A. ceylanicum,* also common in cats and dogs, especially in the tropics, have one large tooth and a rudimentary one on each side. These species were confused with each other until Biocca (1951) showed clear differences between them both in mouth capsules and in bursas (Fig. 124).

ANCYLOSTOMA. *Ancylostoma duodenale* (Fig. 123) is primarily a northern species and predominates only in Europe, North Africa, western Asia, northern China, and Japan, but it has accompanied infected mankind to all parts of the world; it is possible that it may have been the original species in at least a part of the American aborigines. It is larger and coarser than *Necator,* the females averaging about 12 mm. and the males about 9 mm. in length. Freshly expelled specimens have a dirty rust color. The vulva of the female is behind the middle of the body, and the tail is tipped by a minute spine. The males are easily recognizable by their broad bursas, which have the rays arranged as shown in Figs. 121 and 123*A*. The single dorsal ray and the nearly equal spread of the three lateral rays are good marks for distinguishing this species from *Necator,* but after a little experience the two genera of either sex can be distinguished with the naked eye by the

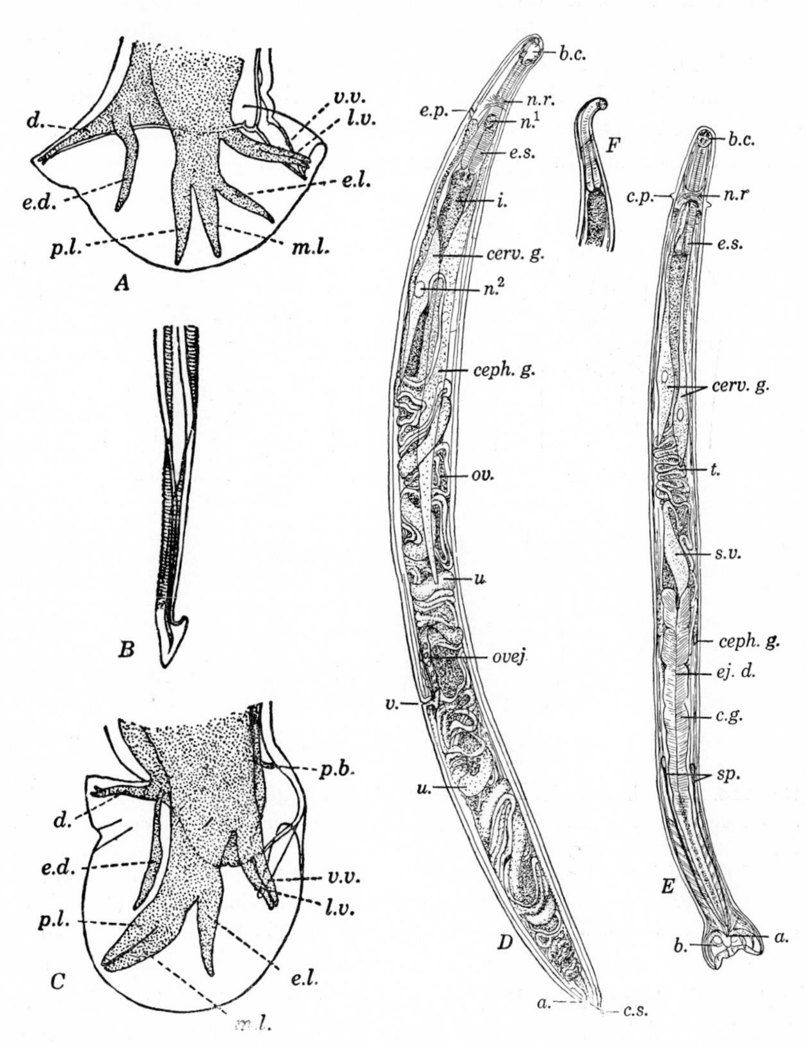

Fig. 123. Hookworms of man. *A*, bursa of *Ancylostoma duodenale; B*, tip of spicules of *Necator; C*, bursa of *Necator americanus* (abbreviations as in Fig. 121); *D*, ♀ *Ancylostoma duodenale; E*, ♂ of same; *F*, head of *Necator americanus*, same scale as *D* and *E*. Abbreviations: *a.*, anus; *b.*, bursa; *b.c.*, buccal capsule; *ceph.g.*, cephalic gland; *cerv.g.*, cervical gland; *c.g.*, cement glands; *c.p.*, cervical papilla; *c.s.*, caudal spine; *e.p.*, excretory pore; *ej.d.*, ejaculatory duct; *es.*, esophagus; *i.*, intestine; *n.*[1], nucleus of cephalic gland; *n.*[2], nucleus of cervical gland; *n.r.*, nerve ring; *ov.*, ovary; *ovej.*, ovejector; *sp.*, spicules; *s.v.*, seminal vesicle; *t.*, testis; *u.*, uterus; *v.*, vulva. (Adapted from Looss.)

form of the head, which in ancylostomes is coarse and only slightly bent dorsally, whereas in necators it is much finer and sharply bent (Fig. 123). The structure of the mouth capsule of this species is shown in Fig. 122. *A. duodenale* is primarily a human parasite but on rare occasions has been found in pigs and experimentally can be reared occasionally in dogs, cats, and monkeys.

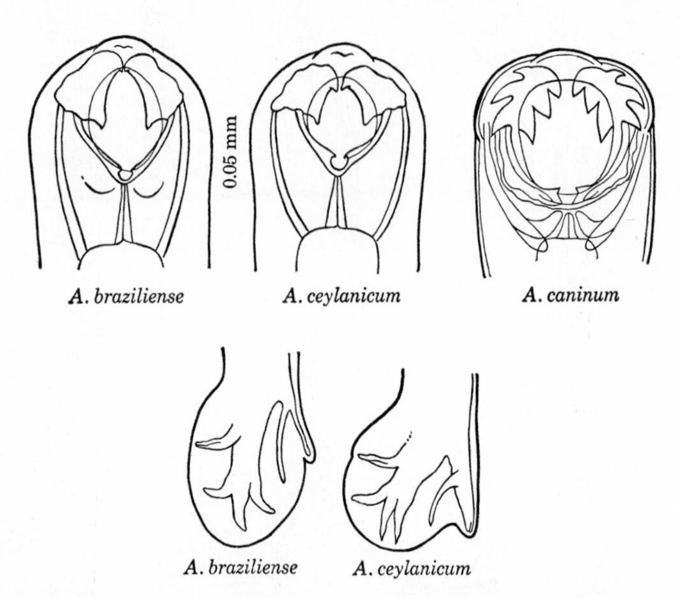

Fig. 124. Mouth capsules and bursas of *Ancylostoma braziliense, A. ceylanicum* and *A. caninum.* Comparing *A. braziliense* and *A. ceylanicum,* note in former smaller accessory tooth in mouth capsule, not directly on contour; longer lateral rays of bursa; longer trunk of lateral rays and these rays all divergent; and more slender externodorsal ray arising nearer base of dorsal ray. (Figures of *braziliense* and *ceylanicum* after Biocca, *J. Helm.,* 1951.)

A. ceylanicum, according to Biocca, is found in Asia and South America, and is an occasional human intestinal parasite in southeast Asia. *A. braziliense* occurs in Africa and South America. It is common in dogs and cats on the Gulf Coast of the United States, and is the cause of creeping eruption in that area (see p. 430).

NECATOR. *Necator americanus* is primarily a tropical worm. It is now the predominant species in all parts of the world except those mentioned in the section on ancylostomes. In our southern states 95% or more of the hookworms are of this species. It is often called the "American" hookworm because it was first discovered here, but it is probably African in origin. Interesting evidence of past and present migrations of mankind can be traced in the hookworm fauna of various countries and islands.

Necator is smaller and more slender than *A. duodenale;* the females average 10 to 11 mm. in length and the males 7 to 8 mm. The vulva of the female is anterior to the middle of the body, and there is no caudal spine. The bursa is longer and narrower than in the ancylostomes (Figs. 121 and 123*C*), and is distinguished by the split dorsal ray and approximation of two of the lateral rays. The structure of the mouth capsule is shown in Fig. 122*B*. *N. americanus* is primarily a human parasite, though capable of development in apes and monkeys, but the same or a very similar form has been found in pigs in tropical America. Other species of *Necator* have been described from chimpanzees.

Life cycle. The adult hookworms of both genera reside in the small intestine, where they draw a bit of the mucous membrane into their buccal capsules and nourish themselves on blood and tissue juices which they suck (Fig. 125). Their main business in life is the production of eggs, and they tend strictly to business! Careful estimates show that each female necator produces 5000 to 10,000 eggs per day, and ancylostomes over twice that many. Yet the bodies of the worms contain on the average only about 5% of this number of eggs at any one time.

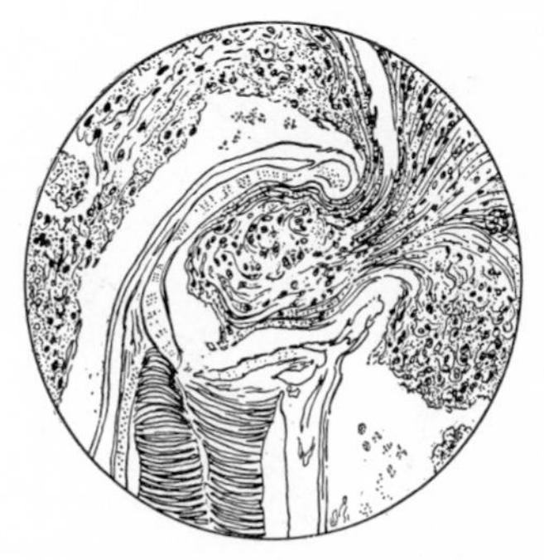

Fig. 125. *Necator americanus;* section showing manner of attachment to intestinal wall. (Adapted from Ashford and Igaravidez, from photograph by W. W. Gray.)

The eggs (Fig. 126, *1* and *2*) average about 70 by 38 μ in necators and 60 by 38 μ in ancylostomes, but the species cannot be identified reliably by eggs in the feces. They are in the four-celled stage when freshly passed and do not develop further until exposed to air. They require moisture and warmth also, and if these conditions are present and there are no injurious substances in the feces, an embryo hatches in less than 24 hours. Usually feces in the tropics are not left undisturbed but are stirred up, aerated, and mixed with soil by dung beetles and other insects, which greatly improves the environment for the eggs and larvae of hookworms.

The hatched larvae (Fig. 126, *8*) are of the "rhabditiform" type, i.e., they have an esophagus with an anterior thick portion connected by a neckline region with a posterior bulb, a character which distinguishes

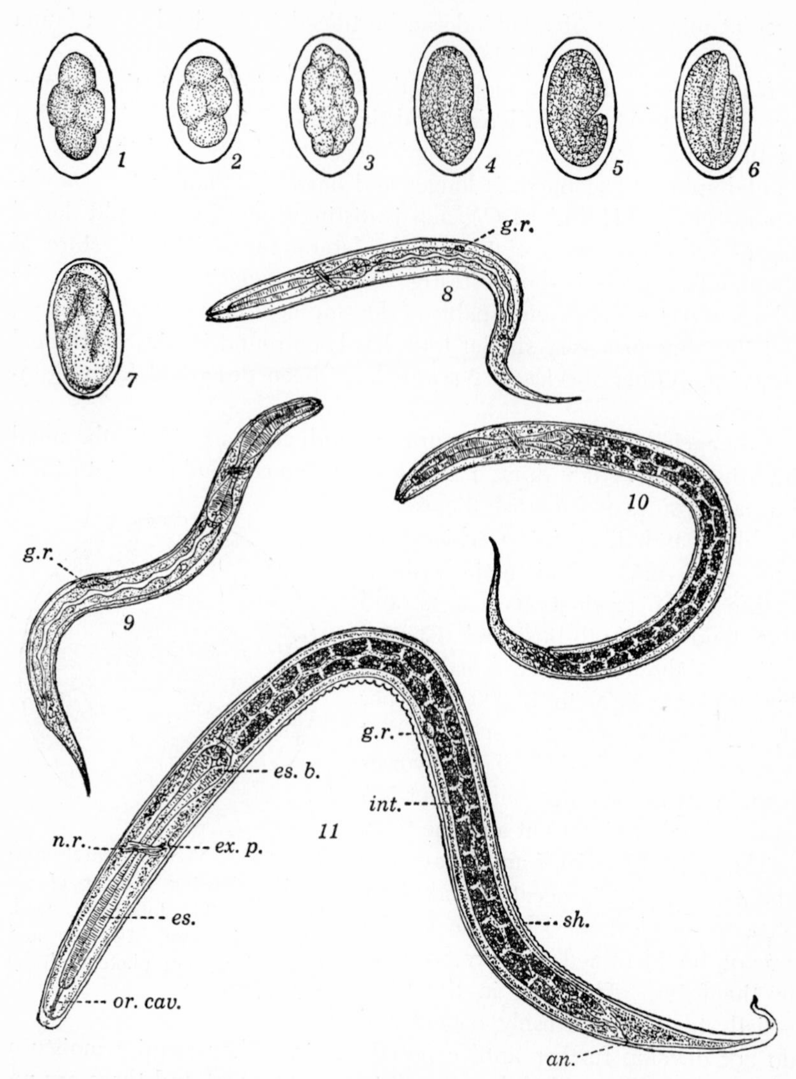

Fig. 126. Stages in life cycle of hookworms from egg to infective larva. *1,* egg of *Necator americanus* at time of leaving body of host; *2,* same of *Ancylostoma duodenale; 3* to *7,* stages in segmentation and development of embryo in the egg; *8,* newly hatched embryo; *9,* same of *Strongyloides* for comparison (note difference in length of oral cavity and size of genital rudiment, *g.r.*); *10,* second stage larva; *11,* third stage (infective) larva; *an.,* anus; *ex.p.,* excretory pore; *g.r.,* genital rudiment; *int.,* intestine; *n.r.,* nerve ring; *es.,* esophagus; *es.b.,* esophageal bulb; *or.cav.,* oral cavity; *sh.,* sheath. ×285. (After Looss from Chandler, *Hookworm Disease,* 1929.)

these larvae from "strongyliform" larvae, which have a long cylindrical esophagus with a terminal bulb that is not sharply demarcated. After the second molt the larvae of hookworms lose their typical rhabditiform esophagus, and become strongyliform. The free-living larvae of hookworms in all stages are distinguishable from the rhabditiform larvae of *Strongyloides* by the long mouth cavity (cf. Figs. 126, *8* and *9*). Larvae of many Strongylata of domestic animals, e.g., esophagostomes and trichostrongyles, are distinguishable by their long, filamentous tails.

The larvae feed on bacteria and perhaps other matter in the feces, and grow rapidly. At the end of about 2 days they molt, grow some more, and at the end of about 5 days they molt again. This time, however, the shed cuticle is retained as a protecting sheath (Fig. 126, *11*), which may remain until the larva penetrates the skin of a host or may be torn or worn away by the movements in the soil. A small oval body, the genital primordium (Fig. 126, *11, g.r.*), is visible near the middle of the body. These larvae are easily distinguishable from the more typically filariform infective larvae of *Strongyloides* (Fig. 137) by the shorter and bulbed esophagus and the pointed tail (notched in *Strongyloides*).

These larvae are now in the infective stage. They eat no more but subsist on food material stored up as granules in the intestinal cells during their 5 days of feasting. The optimum temperature for development is between 70° and 85°F.; lower temperatures retard and finally stop it, and in frosty weather the eggs and young larvae are destroyed; higher temperatures decrease hatching and increase larval mortality. The infective larvae of hookworms are about 500 to 600 μ in length, with characteristic form, color, and movements which make them recognizable after some experience. Minute details of anatomy also make it possible to distinguish necator from ancylostome larvae.

Biology of larvae. Hookworm larvae normally live in the upper half inch of soil, and commonly climb up to the highest points to which a film of moisture extends on soil particles, dead vegetation, etc., and extend their bodies into the air to await an opportunity to apply themselves to a human foot which is unfortunate enough to come in contact with them. When exposed to a hot sun or to superficial drying of the soil they retreat into crevices in the upper layer.

They do not migrate laterally to any great extent, but are dispersed by rain, insects, etc. In loose-textured soil they can migrate vertically, even to the extent of 2 or 3 ft., but in trenches or pits they do not climb the walls, since, from the standpoint of the larvae, the soil walls are very rugged paths. Before they have climbed more than a few inches they reach the top of some projecting particle from which they extend themselves, unaware that they have not reached a vantage point at the top, and are thus trapped. The larvae are strongly attracted by moderate heat and are stimulated to activity by contact with objects; it is these reactions which cause the larvae to burrow into the skin of animals.

Since the larvae have only the stored food granules on which to subsist until a suitable host is reached, the more active they are the sooner their food supply is used up and the sooner they die. Vertical migration through even a few inches of soil is an expensive process for them. Larvae that have had their reserve food greatly depleted, although still alive, may not have enough energy to penetrate the skin of a host.

Mode of infection. Infection normally takes place by penetration of the skin by the larvae; this important discovery was first made by Looss in 1898 when he accidentally spilled some water containing larvae on his hands and acquired an infection. Skin penetration usually results from contact with infested soil, most frequently when the victim is barefooted, but mud containing larvae kicked against the ankles causes numerous light infections in well-shod people. Infection may also come from handling feces-soiled clothing, etc., if they are left damp 4 or 5 days before being laundered. Less frequently, infection results from water or food harboring the infective larvae. Larvae thus swallowed sometimes, though not commonly, develop without migration through the lungs.

The larvae burrow until they enter a lymph or blood vessel and are then carried by the blood stream to the right side of the heart and thence to the lungs, where they are usually caught in the capillaries and again proceed to burrow, this time into the air spaces of the lung. The ciliary movement of the epithelium of the bronchial tubes and trachea carries them to the throat, whence they are either expectorated with sputum or swallowed. If they are swallowed, they go to the intestine and bury themselves between the villi and in the depths of glands for a brief period until the third molt is completed, after which they acquire a provisional mouth capsule and can successfully adhere to the mucosa. The third molt may occur as early as 3 days after infection, but it may be delayed for several days longer. The larvae grow rapidly to a length of 3 to 5 mm. and then molt for the fourth and last time, with the acquisition of the definitive mouth capsule and the development of reproductive organs. In man the eggs first begin to appear in the feces usually about 6 weeks after infection.

Longevity. The length of life of the adult worms in the intestine may be 5 years or more; Palmer recently described an experimental human infection with *Necator* that lasted for 15 years. However, the life span of the majority of a hookworm population is much shorter. There are obviously differences in how long particular humans will harbor the worms. The senior author carried a light experimental infection of *Necator* for over 4 years, whereas the junior author retained a similar infection for less than 6 months.

There is evidence that in natural infections, when repeated reinfections occur, the peak of egg production of newly acquired worms may occur after about 6 months, after which there is a rapid falling off in number of worms. A high percentage of newly acquired worms in repeatedly infected individuals is probably lost within a year. In places where there are prolonged unfavorable seasons little cumulative increase in worms can occur on account of the large annual reduction in worms during the season when reinfection is largely stopped.

Epidemiology. Many environmental factors influence the amount of hookworm infection in a community. Temperature, as already intimated, is a prime controlling factor. Rainfall is also of fundamental importance. Heavy hookworm infections are never common in localities having less than 40 in. of rain a year, and with larger annual rainfall much depends on the seasonal distribution and on the distribution within each month, for hookworm larvae will not withstand complete desiccation.

When soil is drying, e.g., after a shower, or as feces dry up, some of the larvae escape desiccation by retreating under the surface of the soil ahead of the receding zone of free moisture, but many fail to do so and become victims of desiccation. Immediately following a rain the surviving larvae come to the surface and again many fail to escape when the surface dries; they are killed by desiccation long before the soil has lost its moist appearance. Beaver (1953) found that in soil wetted once a day the number of viable larvae fell off about 90% within a week, whereas in soil which was not moistened and became dry and loose at the surface, most of the larvae lived many days longer. Probably extremely few larvae survive in tropical soils for more than 6 or 7 weeks. Excessive rainfall, resulting in saturated soil, may exert an even greater check on hookworm infection. Such local factors as humidity, drainage, and hygroscopic nature of the soil also influence the effect of intershower drying.

The nature of the soil is very influential; hookworm never thrives in regions of heavy clay soil, whereas in adjoining areas with sandy or humus soil it may constitute an important problem. Salt impregnation of soil is also injurious. Vegetation exerts an influence, since dense shade is far more favorable for the development and longevity of larvae than light shade or exposure to sun. Irrigation may make rainless regions favorable for hookworm if moistened soil is selected for defecation.

Animals, such as pigs, dogs, and cattle, which devour feces, especially in the tropics, exert an influence, since in pigs and dogs the eggs in fresh feces pass through the animals uninjured and may be voided with the feces of the animals in places where they are more or less likely to cause infection. In chickens, on the other hand, and probably in cattle also, most of the eggs are destroyed when ingested. Insects play an important role. Dung beetles are allies of hookworms since they mix feces with soil and render the cultural conditions more favorable; cockroaches, on the other hand, destroy most of the eggs in their "gizzards" and were found by the senior author to play an important part in keeping down hookworm infections in Indian mines.

Many human factors also affect the amount of hookworm. Some races are more susceptible than others. The white race is particularly susceptible, and Negroes very slightly so. The other races appear to occupy intermediate positions. Age and sex and the corresponding differences in habits influence the amount of infection and also affect the injury done by a given number of worms, for females are more injured than males and children more than adults.

Occupation is often a determining factor, insofar as it leads to habits which render the acquisition of worms more likely. In most countries agriculture and mining are the main hookworm occupations, but in most places the infection is not strictly agricultural but merely rural. The raising of such crops as coffee, tea, sugar, cacao, and bananas is particularly conducive to hookworm in soil-pollution countries since these crops are grown in moist, warm climates under conditions affording an abundance of shade and suitable soil; cotton and grain raising are much less dangerous since these crops are grown in drier areas, and cotton in unfavorable soil. Raising of rice and jute, mainly on flooded ground, is not associated with heavy hookworm infections. In China and Japan, where night soil is used as fertilizer, hookworm infection is more strictly agricultural and varies greatly with the type of crop produced and the manner in which the night soil is used. Especially heavy infections occur in districts where mulberries and sweet potatoes are raised, since ideal conditions for hookworm propagation are afforded.

Defecation habits are also of great importance. In soil-pollution countries the greater part of the infections are acquired while standing on previously polluted ground during the act of defecation. The concentration of the defecation areas, the extent to which people mingle in common areas around villages, the type of places selected, etc., are all influential factors. Wherever simple soil pollution is modified by the use of standing places or primitive latrines that keep the feet off the polluted ground and bring about an unfavorable concentration of fecal material, hookworm infections are light.

The wearing of shoes also affords a high degree of protection; in southern United States and Queensland heavy hookworm infection is almost entirely limited to children who are less than 14 to 16 years of age, since after the age of 14 shoes are habitually worn. Even simple sandals or wooden soles without uppers, as worn in parts of India, are effective.

Expectoration habits also have an important effect; those individuals who habitually spit out phlegm collecting in the mouth get rid of many of the hookworms which invade the body. From Suez to Singapore the Orient is polka-dotted with the red expectorations of betel-nut chewers; this prevalence of chewing, and consequent expectoration, is probably an important factor in keeping hookworm infections at a relatively low level in Far Eastern countries. A notion was formerly prevalent in our southern states that tobacco chewing is conducive to health. There is no virtue in tobacco juice *per se,* but the constant spitting entailed would have a salubrious effect.

Effect of diet. Diet is of profound importance. As we shall see, most of the injury done by hookworms is the result of blood loss, for the replacement of which large amounts of iron, protein, and vitamins are required in the diet. If these are present in adequate amounts the host soon develops a partial immunity which results both in resistance to reinfection and in inhibition of reproduction and ultimate loss of worms already harbored. In other words, a good diet, as Foster and Cort demonstrated long ago (1932), does not permit serious damage from hookworm infection to occur except when there are overwhelming initial infections with the blood loss so great that *no* diet can compensate for it. A diet deficient in protein, plus loss of protein resulting from the bloodsucking of worms already present, permits no protein reserve for the production of new gamma globulin in the form of antibodies, so no immune response can develop, or if developed will break down. Provision of iron under such circumstances does little good, for protein as well as iron is necessary for manufacture of hemoglobin, and the iron alone does nothing towards development of immunity. As Cruz (1948) pointed out, hookworm disease is essentially associated with malnutrition. The severity of hookworm disease in a community is measured more by the adequacy of the diet than it is by the average number of hookworms harbored or the degree of exposure to infection.

Pathology. PREINTESTINAL PHASE. When human hookworms enter the skin they may cause "ground itch" or "water sore," characterized by itching and inflammation and often development of pustular sores from secondary bacterial invasion. Unless secondarily infected, the skin reaction is of short duration. Similar reactions may occur when *Strongyloides* larvae penetrate the skin (see p. 468) and there is evidence that all these skin reactions are allergic in nature (see Beaver, 1956).

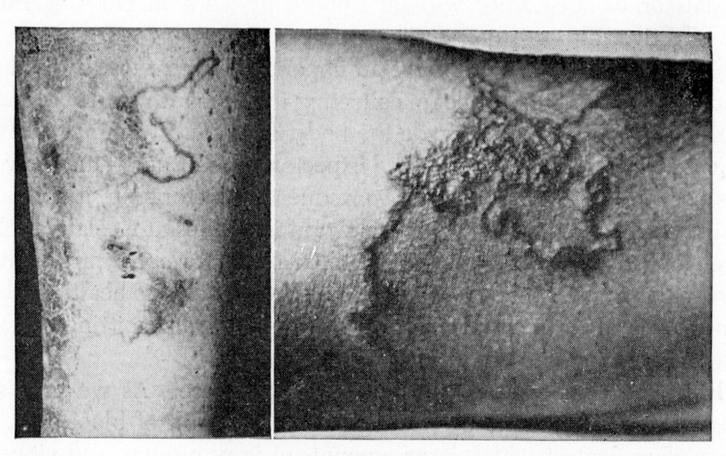

Fig. 127. Creeping eruption; *left,* typical lesions, including results of scratching an infected leg; *right,* lesions experimentally produced by application of a pure culture of *Ancylostoma braziliense* to the forearm. (From photographs by W. E. Dove.)

Some "foreign" species of hookworms, particularly *Ancylostoma braziliense,* commonly fail to find their way below the germinative layer of skin, thus failing to reach blood or lymph vessels. They then wander aimlessly just under the surface, sometimes for 3 months or more, causing tortuous channels—a condition known as cutaneous larva migrans or more popularly as "creeping eruption" (Fig. 127). The severity of the reaction is conditioned by previous sensitization and the allergic responsiveness of the victim. A similar effect is produced by the European dog hookworm, *Uncinaria stenocephala,* but it usually lasts only 2 to 4 weeks; and Mayhew has observed a still less extensive and less durable eruption from the larvae of the cattle hookworm, *Bunostomum phlebotomum.*

In highly allergic individuals *A. caninum* may cause severe skin reactions, though not typical creeping eruption. Creeping eruption is common on the coasts of southern United States and tropical America, where children play in sandpiles or adults on bathing beaches that are the chosen defecation sites of dogs and cats infected with *A. braziliense.* The nonhuman hookworms that produce cutaneous larva migrans may eventually migrate to the interior of the body and persist to produce visceral larva migrans (see p. 459). These infections could be prevented by excluding dogs and cats from such places, or bathers from beaches that are not washed by tides. Millspaugh and Sompayrac (1942) reported that in Florida creeping eruption incapacitated a considerable number of naval personnel. The most effective treatment is local freezing, shortly ahead of the end of the inflamed burrows, by application of ethyl chloride sprays or CO_2 snow. Control of dog and cat hookworms is considered on pp. 436–437.

The next effect of *human* hookworms is in the lungs, where the burrowing larvae may predispose to pulmonary infection or even cause pneumonia symptoms themselves, if numerous. Rodents, in which hookworm larvae migrate to the lungs, die of extensive pulmonary hemorrhages if large doses are administered.

In repeatedly infected cases many larvae are frustrated in their migration through the body by the development of immunity. Immunized serum causes precipitates to form about the mouth, intestine, anus, and excretory pore of migrating larvae, probably directed against digestive enzymes or metabolic products of some kind. The larvae are also immobilized, and many are destroyed by encapsulation and phagocytosis in the skin, lymph glands, or lungs. The eosinophilia and occasional leucocytosis associated with hookworm infection probably result from liberation of proteins from such captured worms in partially immune persons. This larval phase of infection was emphasized by Ashford, Payne, and Payne (1933).

INTESTINAL PHASE. After several weeks, as the worms are maturing in the intestine, there may be some nausea, abdominal discomfort, and sometimes diarrhea, but the principal effects are due to anemia, resulting from the constant sucking of blood. Wells (1931) calculated that 500 *A. caninum* in a dog may suck nearly a pint of blood per day. The permanent loss of most of the iron, as well as large amounts of protein, causes reduction in number and size of corpuscles and in their hemoglobin content, unless *both* the iron and the protein are adequately replaced in the diet. In children the diversion of nutriment to keep the hemoglobin up to standard interferes with normal growth, thus causing a stunting in size.

When the repair cannot keep pace with the damage and immunity fails to develop, symptoms appear. In severe cases the hemoglobin may be reduced to 30% or less, with 2,000,000 or less corpuscles per cubic millimeter. The most noticeable symptoms are a severe pallor; extreme languor and indisposition to play or work, popularly interpreted as laziness; a flabbiness and tenderness of the muscles; breathlessness after slight exertion; enlargement and palpitation of the heart, with weak and irregular pulse; edema, making the face puffy and the abdomen "pot-bellied"; a fishlike stare in the eyes; reduced perspiration; more or less irregular fever; and heartburn, flatulence, and abdom-

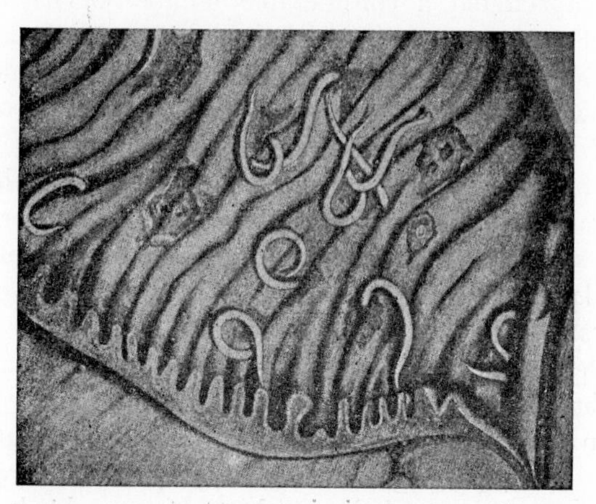

Fig. 128. Hookworms on wall of intestine, showing lesions. (After International Health Board, chart.)

inal discomfort. The appetite is capricious, and frequently there is an abnormal craving for coarse "scratchy" substances such as soil, chalk, and wood. Severe hookworm cases in our southern states a few decades ago were often "dirt-eaters," though they rarely admitted it.*

Children may suffer several years' retardation in physical and mental development, with puberty long delayed. The mental retardation results in stupidity and backwardness in school, and there are sometimes other nervous manifestations, such as dizziness, insomnia, optical illusions, general nervousness, and fidgety movements.

The effects of hookworm infection are particularly severe during pregnancy, when the demand for protein and iron by the developing fetus puts an extra drain on the mother. Hookworm is the cause of a tremendous number of stillbirths and is believed by some to be a

* Dirt-eating or "pica" also occurs in individuals *not* infected with hookworm. The relationship between "pica" and hookworm infection needs further clarification.

more serious complication of pregnancy than even syphilis or eclampsia. The reduction of labor efficiency from hookworm infection may amount to 25 or even 50%, and there may be additional loss from sickness and death.

Grades of infection. In earlier days much emphasis was put on the mere presence or absence of infection. Darling in 1918 was the first to emphasize the importance of the *number* of hookworms harbored, though now we know that even the number may not mean much without considering the adequacy of the diet as well. The first efforts to estimate the number of hookworms harbored were based on worm counts after treatment, but the technical difficulties were too great. In 1923 Stoll devised an easy method of counting the eggs per gram in the feces of infected people and demonstrated that there was a rough correlation between eggs per gram (epg) and number of worms harbored. This method, or modifications of it, has been a valuable yardstick not only for measuring the hookworm burden of communities and the efficacy of control measures, but also for demonstrating the effect of different degrees of infection under varying conditions.

Slight infections with 50 worms or less are practically harmless except in a person who needs additional food much more than he needs hookworm treatment; ordinarily such infections can safely be ignored. Even several hundred worms may produce no measurable symptoms in a person on a good diet, not pregnant, and not suffering from overwork or chronic disease, but Hill and Andrews (1942) in Georgia found a falling off in hemoglobin in the group with 2000 to 4000 eggs per gram (about 60 to 120 worms), which became marked in the 4000 to 8000 group and severe in cases with over 15,000 epg. It is not possible to set any definite limits to these grades of injury in the case of any individual, but in communities the percentage falling into different egg-count groups gives a useful index to the hookworm burden, and correlation with hemoglobin percentages gives valuable information on susceptibility to injury under existing dietary and environmental conditions.

The erroneousness of judging hookworm infection by the percentage of people infected is nowhere better demonstrated than in Bengal, where an average of at least 80% of the 46,000,000 inhabitants are infected, a condition which some years ago was spoken of as "staggering." But egg counts show that in 90% of the area of Bengal the average number of worms harbored per person is less than 20, and not more than 1% of the people are estimated to have over 160 worms and almost none over 400. In other words, instead of being a staggering problem involving the health of over 35,000,000 people, it is negligible from the public-health point of view.

Diagnosis. Hookworm infection can rarely be diagnosed with certainty by symptoms, but a positive diagnosis is easily obtainable by modern flotation methods of finding eggs in the stools. If it is only desired to find infections which need treatment, the simple smear method suffices, but more accurate diagnosis can be made by the methods discussed on p. 253. Community diagnosis, i.e., the hookworm burden of a community, can be ascertained by means of egg counts as described on p. 254.

The collection of fecal samples for diagnosis on a large scale can be made in bottles containing a few cubic centimeters of antiformin as described by Maplestone in 1929, or with 1% NaCl added in the proportion of 30 : 1 as suggested by Maplestone and Mukerji in 1943. In either case the specimens can be sent to a central laboratory for examination, thus eliminating the necessity for a moving field laboratory, for the specimens are useful for both diagnosis and egg counts even after several days.

Treatment. The treatment of hookworm infection has undergone an interesting evolution. Thymol, introduced in 1880, was the classical treatment for many years but was superseded by oil of chenopodium during World War I. The latter drug is more effective against *Ascaris* than against hookworms, and is often combined with tetrachlorethylene when both worms are present. It has the advantage of cheapness and of being administered in liquid, but it is too toxic for general use. In 1921 carbon tetrachloride, previously best known as a fire extinguisher, was introduced and within a few years became widely used all over the world; now, however, it is seldom used because of the damage it does to the liver, and its dangerousness when there is a calcium deficiency.

In 1925 tetrachlorethylene, and in 1930 hexylresorcinol, were found to be highly effective and relatively nontoxic, and these two drugs are the ones most frequently used at present. Tetrachlorethylene is given in soft gelatin capsules in a dose of 0.5 to 0.6 cc. per pound of body weight, up to a maximum of 4 or 5 cc. This removes about 75% of the necators, with complete cures in about two-thirds of the cases, but it is considerably less effective, as are other drugs, against ancylostomes. The only side effects are a brief burning sensation in the stomach, slight nausea, and a drunken sensation. Best results are obtained if the patients are treated before eating in the morning and *not* given a subsequent purge (Carr et al., 1954).

Hexylresorcinol in crystalline form in capsules (crystoids) eliminates about 60 to 75% of the worms harbored and has the added advantage of eliminating 90 to 100% of *Ascaris* and 50% of *Trichuris*. The dose for adults is five 0.2-gram pills (1 gram), three to four for a school child, and one for each 2 years of age up to 6; a sodium sulfate purge is given 2 hours later. This less toxic drug is preferable in pregnancy and illness, and for delicate children. Both tetrachlorethylene and hexylresorcinol have the disadvantage of having to be administered in capsules.

In places where *Ascaris* infections are common a good plan is to give hexylresorcinol first to eliminate the *Ascaris* and most of the hookworms, following it a week or 10 days later with a treatment of tetrachlorethylene. The latter drug irritates but does not kill *Ascaris*, and sometimes causes them to tangle themselves in knots that block the intestine and must be removed by operation.

Copp et al. (1958) reported that the biphenium compounds are effective against certain nematodes and Young et al. (1958) found them to have considerable effectiveness in human hookworm infection. Dithiazanine, which is effective against *Ascaris, Trichuris,* and *Strongyloides,* has some effect against hookworms also. This drug approaches the broad spectrum anthelmintic drug that has been sought for many a year.

Improvement after deworming is very slow, whereas administration of iron in the form of ferrous sulfate or gluconate at the rate of 1 gram per day for adults, with a rich diet in protein and vitamins, causes rapid improvement. In weak, anemic cases, and in pregnancy, the blood should be built up by iron and protein therapy before an anthelmintic is given.

Mass treatment. Mass treatment, first advocated by Darling, greatly speeds up hookworm campaigns. By this is meant the treatment, without preliminary diagnosis, of an entire community at one time, when the great majority of the individuals are found to be infected. The diagnosis itself does not require so much time, but the difficulty in obtaining fecal samples from primitive people is well known to anyone who has tried it; in many cases it is quite impossible. If all the members of a community are treated at once, preferably in a dry or cold season when rapid reinfection from an already badly infested soil cannot occur, the reduction in infection is striking and durable. In Fiji practically the entire population was treated in two years, a feat which could not have been accomplished in *any* length of time by the older methods, for, long before even a fair percentage of the people could have been covered by diagnostic measures, those first treated would again have been infected from their untreated neighbors. The original mass treatment in Fiji was made in 1922 and 1923 and was followed by improvement in the sanitary conditions of the soil. In 1935 Lambert reported that clinical hookworm disease was still rare in Fiji; the people were healthier, happier, and more prosperous, and hookworm had been eliminated as an important economic factor. The Red Chinese have recently reported that they are carrying out a mass antihookworm campaign, but details allowing an evaluation of its extent do not seem to be available.

Prevention. Theoretically, few if any diseases can be as simply, as certainly, and as easily controlled as hookworm. Diagnosis is easy and accurate, treatment reduces existing infections to a negligible point, and reinfection can be prevented by stopping soil pollution, for no other animals, except possibly pigs and apes, and these only in some localities, harbor human hookworms. But in the prevention of soil pollution the sanitarian runs into a snag. The difficulties involved in this seemingly simple procedure are infinitely greater than the average inhabitant of a civilized sanitary country would suspect. It involves an attempt to induce hundreds of millions of people in tropical and subtropical countries to abandon habits which have been ingrained in them for countless generations and in some instances dictated by religion, and to adopt in their place unfamiliar habits that seem to them obnoxious and undesirable and the reasons for which they cannot readily grasp.

Even in our own southern states a survey in the early part of the present century showed that in the hookworm belt about 68% of the rural homes were not provided with privies of any kind. Fortunately there has been much improvement in this respect, but in many rural districts where privies do exist, their use is restricted to the women and children or to the family of the manager. Among the "jibaros" or plantation laborers of Puerto Rico, of 61 hookworm patients who were questioned, 55 never had used privies of any kind, and of the 6 who did occasionally use them, only 2 lived in rural districts.

Five weapons are available for use in the control of hookworm: treatment, dietary supplements, protection of the feet, disinfection of feces or soil, and prevention of soil pollution. Mass treatment gives immediate relief and slows up the rate of reinfection on account of the great reduction in number of eggs reaching the soil; but treatment alone, unless consistently repeated, is inadequate, since it has never yet been and probably never will be found feasible to eliminate all the worms, and reinfection inevitably follows. In Puerto Rico, Hill in 1927 treated 1000 people in an isolated valley and eliminated

97.5% of the worms. In one year the residual infection increased to 500% and was nearly 20% of the infection before treatment.

Cruz and de Mello (1945), recognizing the extent to which hookworm disease is influenced by diet, suggested the wholesale addition of iron to food as preventive measure, just as iodine is added to water to prevent goiter. It is more difficult, since the iron used must be in a cheap, stable form and must not markedly discolor food or give objectionable tastes. In Brazil they suggested ferrous sulfate added to cassava meal, or iron and ammonium citrate to beans. For people with hookworm anemia they recommended 1 gram per day until the hemoglobin is normal, then 0.5 gram for 80 days, and then 0.25 gram for 80 days more. Provided that there is adequate protein in the diet, such a program not only would eliminate symptoms but also would permit development of immunity, reduce the number of eggs reaching the soil, and eventually lead to a much lower hookworm burden even without improvement in sanitary conditions.

Wearing of footgear is a valuable measure when it can be consistently enforced; it is essential for individuals in infected areas who desire to protect themselves. However, the wearing of footgear is often as difficult to enforce in the tropics as is sanitary disposal of feces, and it is far less effective in ultimate control. It is a valuable temporary measure, comparable with the use of mosquito screens for the control of malaria, but it does not get at the root of the trouble.

Disinfection of soil or feces is difficult. Salt can be effectively used under certain conditions, especially in mines, and lime added to feces is an effective method of killing hookworms in night soil. Sodium borate, calcium cyanamide, and urea, especially the first, have been found fairly promising in eliminating hookworm larvae from limited areas, e.g., in dog or cat yards. Methyl bromide applied to the ground under an airtight covering of glue-coated paper (1 lb. to 64 sq. ft.) kills all worm larvae and eggs, and protozoan cysts as well, but it is too expensive except for small areas, for valuable breeding stock, zoos, etc. Further experiments are desirable in adding small amounts of such substances as sodium borate and phenothiazine to the diet of animals. Low-level doses of phenothiazine (2 grams daily to horses) reduces the number of strongylid eggs passed and affects their fertility.

Prevention of soil pollution, then, remains as the only dependable method of control of human hookworms under most conditions. The efforts of sanitarians in this direction are not so prone to be too little and too late as to be too much and too soon; we are likely to try to force on tropical natives our own ideas of sanitary arrangements, just as we try to force on them our ideas of ethics, religion, clothing, and food habits. It is better as a beginning to teach the coolie to defecate into a trench, from a log over a ditch, or from a low branch or root of a tree or even from a projecting stone, than to build an enclosed flyproof latrine, which he promptly befouls and which prejudices him against latrines in general. Meanwhile education will gradually alter prejudices, and eventually really sanitary privies to control such diseases as dysentery and typhoid as well as hookworm will be possible.

In the United States Andrews (1942), in Georgia, pointed out that hookworm work should be directed toward the detection, prevention, and control of clinical infections. Most of these could be found by examination of large, low-income white families living on sandy or sandy-loam soil, without sanitary conveniences, and showing evidence of anemia. Attention should, he thinks, be concentrated on these families, omitting work where there is good sanitation, clay soil, good income, or a Negro population. The bulk of hookworm morbidity would then be revealed at a minimum of time and expense.

The campaign against hookworm disease, which has been sponsored especially by the International Health Board, although a most worthy end in itself, leads to even greater benefits, for the work, while bringing relief to hundreds of thousands of suffering people, is at the same time serving the more useful purpose of creating a popular sentiment in support of permanent agencies for the promotion of public health. In the United States it led to rapid advances in rural hygiene and the establishment of county health organizations all over the country, and similar local organizations have been brought to life in many other countries. Schools of hygiene have been established in various parts of the world to provide trained men to carry on the work. The ultimate results which may come from the simple beginnings centered on the eradication of hookworm disease are impossible to estimate, but in the light of the tremendous accomplishments which we have seen realized since the inception of the work of the International Health Board about 1910 the outlook for the future is bright indeed.

Other Strongylata

Other hookworms. *Ancylostoma caninum, A. ceylanicum,* and *A. braziliense,* important parasites of dogs and cats, were discussed on pp. 422 and 430. In Europe dogs commonly harbor *Uncinaria stenocephala,* a hookworm related to *Necator* but with only one pair of lancets in the depth of the mouth capsule. The hookworms of cattle, sheep, and other ruminants belong to the genus *Bunostomum,* also related to *Necator* but with the dorsal lobe of the bursa asymmetrical.

Family Strongylidae. The members of this family have globular, goblet-shaped or cylindrical deep or shallow mouth capsules, with a

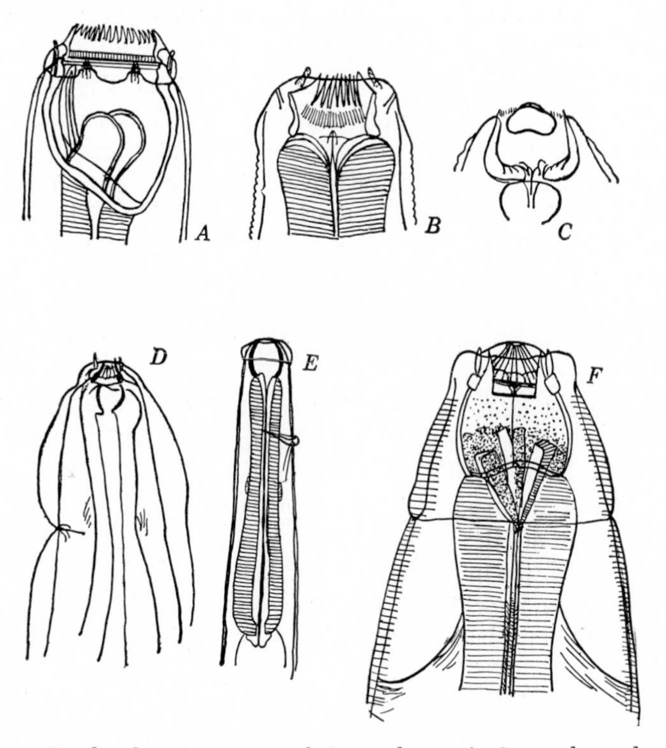

Fig. 129. Heads of various types of Strongylata. *A, Strongylus vulgaris; B, Trichonema tetracanthum; C, Stephanurus dentatus; D, Oesophagostomum bifurcum; E,* fourth-stage larva of *Oesophagostomum; F, Ternidens deminutus.* (*A, B, C,* and *F* after Yorke and Maplestone, *Nematode Parasites of Vertebrates,* 1926. *D* after Travassos and Vogelsang, Mem. Inst. Oswaldo Cruz, 1932; *E* after Mönnig, *Veterinary Helminthology and Entomology,* 1949.)

crown of leaflets (corona radiata) guarding its entrance (Fig. 129). The family contains a number of species which are injurious to domestic animals and a few which are more or less frequent parasites of man. Among the more important forms are species of *Oesophagostomum* in pigs, ruminants, and primates; *Strongylus* and members of the subfamily Cyathostominae (also called Trichoneminae or Cylicostominae) in horses; *Chabertia* in sheep and goats; and *Stephanurus* (kidney worm) in pigs.

The esophagostomes of pigs and ruminants and the strongyles of horses, as well as many of the trichostrongylids (discussed on pp. 443 to 445), are very susceptible to treatment with phenothiazine, but this drug is relatively ineffective for worms of the hookworm type.

All the members of the Strongylidae have a life cycle similar to that of the hookworms except that the infective larvae of most genera do not penetrate the skin but are ingested with vegetation. The larvae have long pointed tails and are protected from desiccation by their sheaths. They are frequently found in an apparently dry state, but viable, curled up on the under side of grass or leaves. Many species retreat to the upper layers of the soil during the heat of the day. They are susceptible to excessive heat or direct sunlight but not to freezing. Some of the species, e.g., *Strongylus* spp. (Fig. 129*A*), go on a roundabout tour through the body before growing to maturity in the intestine, but most of them spend their apprenticeship in nodules in the walls of the gut.

The Cyathostominae or "small strongyles" (Fig. 129*B*) of horses also pass the first part of their parasitic life as larvae in nodules in the walls of the large intestine and cecum, but the nodules are smaller. One species of *Strongylus* (*S. vulgaris*) frequently causes aneurisms of the mesenteric arteries in horses and is a cause of colic; the larvae migrate through the intestinal wall and between the layers of the mesentery, enter the mesenteric artery where they develop in blood clots, and finally migrate back to the colon wall and into the lumen of the intestine.

The esophagostomes or nodular worms are common and injurious parasites of pigs, sheep, goats, cattle, apes, and monkeys, and are occasional parasites of man. They are about the size of hookworms but have a shallow mouth capsule with a corona, and a groove behind the head on the ventral side (Fig. 129*D*). When the infective larvae are eaten and liberated from their sheaths they do not at once establish themselves in the lumen of the intestine but first burrow into the lining of the large intestine where the host forms a tumorlike nodule around them (Fig. 130).

In young nonimmune animals the worms return to the lumen of the intestine in 5 to 8 days and grow to maturity, but, as immunity develops, the tissue reaction becomes greater and many of the worms remain imprisoned in the nodules, even for months. Older animals may have the large intestine covered with nodules but have very few adult worms. The nodules are ½ to 1 in. in diameter and contain a greenish puslike substance surrounding the immature worm. Progressive weakness and loss of weight are characteristic, with diarrhea in earlier stages.

Several species of these worms, normally inhabiting apes and monkeys, have been found on rare occasions in man, but, since the eggs are

Fig. 130. Tumors or nodules of *Oesophagostomum bifurcum* in large intestine of an African. ¾ natural size. (After Brumpt, *Précis de parasitologie,* 1949.)

indistinguishable from those of hookworms, human cases may be much commoner than is suspected.

In ruminants and apes the infection commonly produces severe emaciation and prolonged dysentery, and sometimes fatal peritonitis. In the encysted stage the worms are unaffected by anthelmintics, and they are difficult to dislodge even when free on account of their location in the large intestine.

A related worm, *Ternidens deminutus,* is a common parasite in natives of parts of southern East Africa; Sandground (1931) found light infections in 50 to 65% of natives examined in two villages in southern Rhodesia. The eggs measure about 84 μ by 51 μ, and are usually in the eight-celled stage when passed. In various monkeys this parasite occurs all the way from the Atlantic coast of Africa to southeast Asia and Celebes. The worms, which are bloodsuckers, superficially resemble hookworms but have a deep goblet-shaped buccal capsule with three teeth in the depths (Fig. 129*F*). Its eggs may easily be mistaken for those of hookworms, but unlike hookworms *Ternidens* forms nodules in the intestine and is not expelled by the usual hookworm remedies, so it may be much commoner in countries inhabited by its reservoir hosts than the records indicate.

A strongylid which causes much damage in the colon of sheep is *Chabertia ovina,* a rather large worm, the females measuring up to 20 mm. long, and having a globular mouth capsule with no teeth but with two crowns of extremely fine leaflets. It is a northern parasite; its larvae are capable of development at very low temperatures. In sheep it is 9 to 10 weeks after infection before eggs appear.

Stephanurus dentatus, the kidney worm of pigs, has a globular capsule with a very feeble corona (Fig. 129*C*) and a poorly developed bursa that is subterminal. After developing to the infective stage outside the body the larvae enter the body either by mouth or through the skin and go by way of the blood stream to the liver, where they live and grow for a few months, eventually making their way to the kidneys. Here they become embedded, the eggs reaching the ureters and being excreted with the urine. They are rather injurious to pigs, and entail considerable economic loss from condemned liver and kidneys.

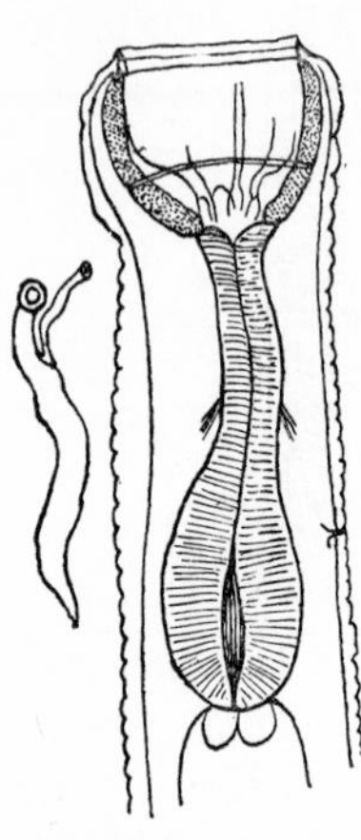

Fig. 131. *Right,* head of ♀ *Syngamus kingi* (after Leiper, *J. Soc. Trop. Med. Hyg.,* 1913). *Left,* a pair of worms in copula.

Family Syngamidae. This family (see p. 418) includes bloodsucking worms that live in the trachea and bronchi of birds and mammals. The worms are called gapeworms or forked worms because the male remains permanently attached to the vulva of the female by its bursa, giving a forked appearance (Fig. 131); the females are 15 to 20 mm. long and red in color. The eggs measure about 85 by 50 μ and are in early stages of segmentation when oviposited. From the air passages they are coughed up and swallowed, passing out with the feces. Although direct infection with embryonated eggs is possible, the eggs are frequently eaten by various invertebrates, in which the embryos hatch and become encapsulated, the invertebrates thus becoming "transport" hosts. Earthworms are particularly important hosts and may harbor the infection for years, but slugs, springtails, maggots, and others are also involved. When swallowed, the infective larvae penetrate the mucous membranes and are carried to the lungs by the blood stream.

One species, *Syngamus trachea,* is an injurious parasite of turkeys, young chickens, and pheasants, and a related worm affects geese. Many passerine birds such as robins, blackbirds, crows, and starlings also harbor gapeworms, although Goble and Kutz in 1945 showed that several species are involved; these birds can serve as carriers of the poultry parasite. A number of instances of human infection with gapeworms have been recorded, all but one of them in tropical America; in most of these cases the species concerned was S. *laryngeus* of cattle. The worms attack the pharynx, trachea, or neighboring air spaces in the head or throat, often causing nodules at the point of attachment. Coughing and gaping are the usual symptoms, and chicks may die from obstruction of the trachea. Immunity develops quickly, but chicks

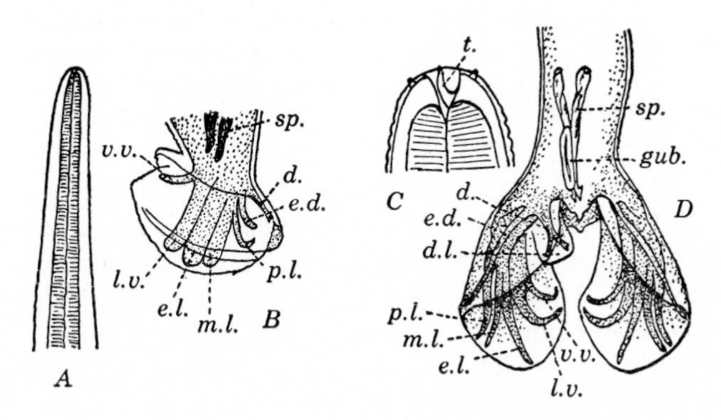

Fig. 132. *A* and *B*, head and bursa of *Trichostrongylus colubriformis; C* and *D*, head (greatly enlarged) and bursa of *Haemonchus contortus; t.*, buccal tooth; *d.l.*, dorsal lobe; *gub.*, gubernaculum; other abbreviations as in Fig. 121. (*B*, after Looss; *C*, after Yorke and Maplestone, *Nematode Parasites of Vertebrates*, 1926; *D*, after Ransom.)

lose their worms much more rapidly than turkeys. Barium antimonyl tartrate, inhaled as a dust, is recommended for treatment. Addition of 4% phenothiazine to the mash of chickens controls the infection, but it only kills the young worms when they hatch in the intestine, so it must be used continuously.

Family Trichostrongylidae. These worms, recognizable by the finely drawn-out head without a large buccal capsule, together with a well-developed bursa in the male (Fig. 132), are very important parasites of domestic animals. Sheep, goats, and cattle suffer severely from the stomach worm, *Haemonchus contortus*, and to a less extent from species of *Trichostrongylus, Cooperia, Nematodirus*, and *Ostertagia. Trichostrongylus axei* also lives in the stomach of horses. Pigs are infected with *Hyostrongylus*. These genera have the following characters:

1. ***Haemonchus:*** length, ♀ 20–30 mm., ♂ 10–20 mm.; small buccal cavity with a lancet; ♀ with conspicuous vulvar flap; ♂ with short, stout spicules and small asymmetrical dorsal lobe on bursa. (Fig. 132*C*, *D*.)
2. ***Ostertagia:*** length, ♀ 8–9 mm., ♂ 6–8 mm.; head with very small buccal cavity; ♂ with short spicules and small accessory bursal membrane dorsally.
3. ***Cooperia:*** length, ♀ about 6–7 mm., ♂ about 5–6 mm.; head 25 μ in diameter; no cervical or prebursal papillae; ♂ with short spicules, branches of dorsal ray lyre-shaped.
4. ***Nematodirus:*** length, ♀ 15–20 mm., ♂ 10–15 mm.; extremely slender; ♀ tail truncated with spinelike process; ♂ with filiform spicules, dorsal ray split to base.
5. ***Trichostrongylus:*** length, ♀ 5–6 mm., ♂ 4–6 mm.; head 10 μ in diameter; ♂ with short spicules, dorsal ray split only near tip. (Fig. 132*A*, *B*.)
6. ***Hyostrongylus:*** length, ♀ 5–8 mm., ♂ 4–5 mm.; very similar to *Trichostrongylus* but found only in pigs.

The life cycles of all these worms are essentially the same; the eggs develop outside the host's body into long-tailed, sheathed larvae most of which are capable of withstanding considerable desiccation and live for a long time. They gain access to their hosts by being ingested with vegetation. They grow to maturity directly in the intestine, although some species burrow into the mucous membrane before becoming established in the lumen of the intestine, and some, e.g., *Ostertagia* and *Cooperia,* like esophagostomes, become enclosed in nodules in partly immune animals.

All these worms may cause a condition known as "verminous gastroenteritis" by veterinarians, but more often as "black rush," "black scours," etc., by sheep men. As with hookworms, severe infection is the result of poor nutrition or of overwhelming initial infections, for otherwise the animals soon build up an immunity resulting in "self-cure" by expulsion of the worms and resistance to reinfection. The occurrence of diarrhea, emaciation, failure to gain weight, anemia, poor wool production, and general weakness and unthriftiness in the presence of considerable numbers of trichostrongylids is *prima facie* evidence of faulty nutrition, resulting either from poor pasturage or, in nursing animals, from inability to get adequate milk from the mothers. *Haemonchus contortus* is the only important bloodsucker in this group, and in heavy infections in young animals may cause enough blood loss to produce severe anemia before immunity can develop, and thus be a primary pathogen, whereas the others are all secondary to malnutrition as Whitlock (1949, 1951) pointed out. Phenothiazine treatment is indicated in *Haemonchus* infections that are producing anemia, but is of much less importance than improved nutrition under other conditions.

Haemonchus contortus (stomachworm or wireworm) (Fig. 132*C, D*) lives in the stomach (abomasum) of sheep, goats, and cattle, and may play havoc with young animals. It is much larger than other tricho-strongylids, the females being about 1 in. long and the males about ½ in. It has a world-wide distribution. One human case has been reported from Brazil.

Ostertagia are also stomach parasites; they are brownish hairlike worms less than ½ in. long. *O. ostertagi* is commoner than *Haemonchus* in cattle in western United States.

Trichostrongylus (Fig. 132A, *B*) contains many species of minute reddish hairworms only about ¼ in. in length. *T. axei* lives in the stomach of ruminants and horses, and rarely in man. Sheep and goats suffer from a number of species, *T. colubriformis, vitrinus,* and *capricola* being the commonest. Andrews (1939) showed that in pure experimental infections *Trichostrongylus* produces a profuse, continuous diarrhea. Animals with very heavy infections die after several weeks.

Human *Trichostrongylus* infections are fairly common in parts of the Middle East, Far East, and the tropics, with incidences running as high as 80% in some localities, e.g., in parts of Japan. In Egypt and other places a high incidence of *Trichostrongylus* infections probably reflects the close association of man and animals, especially where they share the same house. Since the infections are usually very light they are of little consequence, but the eggs resemble those of hookworms enough to be mistaken for them by people unfamiilar with them; they are, however, larger, more slender, and more pointed at one end. The worms are not expelled by hookworm remedies. For the most part the species involved have not been determined.

Nematodirus and *Cooperia* also contain species parasitic in the duodenum of sheep and goats, and the latter also in cattle. *Nematodirus* is remarkable for its large eggs, up to 200 μ or more long, and for its larvae, which undergo two molts in the egg before they hatch and climb to a vantage point on grass. *Hyostrongylus,* the red stomachworm of pigs, is only ⅕ in. long and red in color. In poorly nourished pigs it produces effects similar to those of most other tricho-strongyles.

Mention should also be made of the subfamily Heligmosominae, which are parasites of rodents; they are tiny red worms which have a single ovary and uterus and which roll their bodies in spirals.

Metastrongylidae (lungworms). These slender worms inhabit various parts of the respiratory system of mammals; most of them inhabit the fine branches of the bronchial tubes of the lungs, but one species lives in the heart and pulmonary arteries of dogs, and others in the frontal sinuses of tigers, skunks, and mink.

Species of the genus *Dictyocaulus,* which inhabit the bronchial tubes of sheep, cattle, and horses, are threadlike worms several inches in length. In a number of ways they show affinity with trichostrongylids; they have short, robust spicules, a vulva near the middle of the body, well-developed ovejectors, and no intermediate hosts. The larvae hatch in the bronchi of the host and are either coughed out or swallowed and passed in the feces. They are peculiar in not feeding at all in the free-living phase; they molt twice, and both shed cuticles are retained for a time. They infect by being swallowed, and they reach the lungs via the lymph system. These worms may cause coughing and bronchitis, and in sheep may block off so much of the lungs as to be fatal.

Lungworms of the subfamily Metastrongylinae have medium or very long spicules, and the vulva is a short distance in front of the anus (Fig. 133*A*). In the genera *Metastrongylus* and *Choerostrongylus,* important parasites of pigs, the spicules are very long and the female has a fingerlike tail (Fig. 133*D*). These pig parasites produce thick-shelled embryonated eggs which hatch when ingested by certain species of earthworms, in which they develop to the infective stage. Members of the subfamily Protostrongylinae (*Protostrongylus* and *Muellerius* of ruminants, *Aelurostrongylus* of cats, and several genera in deer) develop in mollusks. The embryos hatch before leaving the body and burrow into the foot of various land snails and slugs, where they undergo two molts, become encapsulated, and remain infective as long as the snail remains alive. As Hobmaier remarked, the utilization of mollusks as intermediate hosts by these worms probably grew out of their habit of seeking protection from desiccation in the slime of the mollusks.

Only one species of lungworm, *Metastrongylus elongatus* (Fig. 133) of pigs, has been found in man, and this only three times.

Attempts at treatment of lungworms in animals have been made by tracheal injection of various substances and by inhalation of chloroform or fumes of tar, sulfur, etc., but with good care the animals resist the infection and soon lose their worms.

Of very great interest for helminthology in general is the demonstration by Shope (1939) that swine influenza is caused by a combination of certain influenza bacteria and a virus, and that the virus is harbored by the larvae of lungworms (*Metastrongylus*), which serve as vectors for it. The virus survives as long as three years in lungworms encapsulated in earthworms; it is thus perpetuated from one outbreak to another. Species of *Strongylus* have likewise been found to harbor the virus of swamp fever of horses, and *Trichinella* has been shown to be capable of carrying the viruses of lymphocytic chorio-

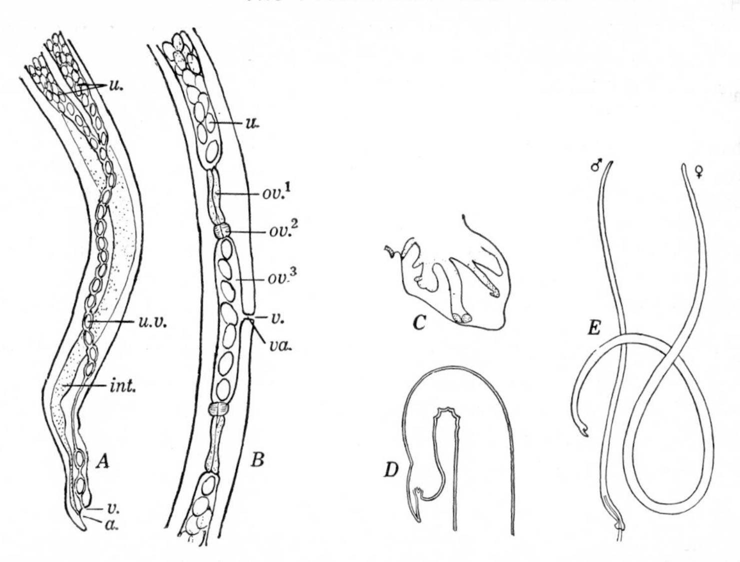

Fig. 133. *A* and *B*, comparison of female reproductive system of a metastrongylid (*A*) and a trichostrongylid (*B*); *a.*, anus; *int.*, intestine; *ov.*[1], ejector of ovejector; *ov.*[2], sphincter of ovejector; *ov.*[3], chamber of ovejector; *u.*, uterus; *u.v.*, uterine vagina; *v.*, vulva; *va.*, vagina. *C, D,* and *E, Metastrongylus elongatus; C,* bursa of male; *D,* posterior end of female; *E,* male and female worms. ×3. (Adapted from various authors.)

meningitis and poliomyelitis into the nervous system. The fluke *Nanophyetus salmincola* is a carrier of the rickettsia of salmon-poisoning, and *Heterakis* (p. 459) for the flagellate, *Histomonas* (p. 103). It has been suggested that *Dientamoeba* enters the host in the egg of *Enterobius*. The role of helminths as vectors for viruses and other disease agents is, however, still an almost virgin field.

REFERENCES

Hookworms

Andrews, J. 1942. New methods of hookworm disease investigation and control. *Am. J. Public Health,* 32: 282–288.

Ashford, B. K., Payne, G. C., and Payne, F. 1933. The larval phase of unicinariasis. *Puerto Rico J. Public Health Trop. Med.,* 9: 97–134.

Beaver, P. C. 1953. Persistence of hookworm larvae in soil. *Am. J. Trop. Med. Hyg.,* 2: 102–108.

Beaver, P. C. 1956. Larva migrans. *Exp. Parasitol.,* 6: 587–621.

Biocca, E. 1951. On *Ancylostoma braziliense* (de Faria, 1910) and its morphological differentiation from *A. ceylanicum* (Looss, 1911). *J. Helminthol.,* 25: 1–10.

Carr, H. P., Pichardo Sarda, M. E., and Nunez, N. A. 1954. Anthelmintic treatment of unicinariasis. *Am. J. Trop. Med. Hyg.,* 3: 495–503.

Chandler, A. C. 1929. *Hookworm Disease.* Macmillan, New York.

Copp, F. C., et al. 1958. A new series of anthelmintics. *Nature,* 181: 183.

Cort, W. W. 1925. Investigations on the control of hookworm disease. XXXIV. General summary of results. *Am. J. Hyg.,* 5: 49–89.

Cort, W. W., and Otto, G. F. 1940. Immunity in hookworm disease. *Rev. Gastro-enterol.,* 7: 2–11.

Cruz, W. O. 1948. Hookworm anemia—a deficiency disease. *Proc. 4th Intern. Congr. Trop. Med. Malaria,* 2, Sect. VI: 1045–1054.

Cruz, W. O., and de Mello, R. P. 1945. Profilaxia da anemia ancilostomotica-sindrome de carencia. *Mem. inst. Oswaldo Cruz,* 42: 401–448.

Dove, W. E. 1932. Further studies on *Ancylostoma braziliense* and the etiology of creeping eruption. *Am. J. Hyg.,* 15: 664–711.

Foster, A. O., and Cort, W. W. 1932. The relation of diet to the susceptibility of dogs to *Ancylostoma caninum. Am. J. Hyg.,* 16: 582–601.

Foster, A. O., and Landsberg, J. W. 1934. The nature and cause of hookworm anemia. *Am. J. Hyg.,* 20: 259–290.

Hill, A. W., and Andrews, J. 1942. Relation of the hookworm burden to physical status in Georgia. *Am. J. Trop. Med.,* 22: 499–506. Intern. Health Div. Rockefeller Fdn., Annual Reports.

Keller, A. E., Leathers, W. S., and Densen, P. M. 1940. The results of recent studies on hookworm in eight southern states. *Am. J. Trop. Med.,* 20: 493–509.

Lambert, S. M. 1936. A resurvey of hookworm disease in Fiji in 1935, ten years after mass treatment. *J. Trop. Med. Hyg.,* 39: 19–21.

Landsberg, J. W. 1939. Hookworm disease in dogs. *J. Am. Vet. Med. Assoc.,* 94: 389–397.

Mayhew, R. L. 1948. The life cycle of the hookworm, *Bunostomum phlebotomum* in the calf. *Am. J. Vet. Research,* 9: 35–39.

Millspaugh, J. A., and Sompayrac, L. M. 1942. Creeping eruption. Infestation with *Anklyostoma braziliense* larvae. *U. S. Naval Med. Bull.,* 40: 393–396.

Wells, R. S. 1931. Observations on the blood-sucking activities of the hookworm, *Ancylostoma caninum. J. Parasitol.,* 17: 167–182.

Young, M. D., et al. 1958. Biphenium, a new drug active against human hookworm. *J. Parasitol.,* 44: 611–612.

Young, M. D., Jeffrey, G., Freed, J., and Morehouse, W. 1958. The effectiveness of Dithiazanine against worm infections in mental patients. *A.M.A. Arch. Neurol. and Psychiatry,* 80: 785–787.

Other Strongylata

Alicata, J. E. 1954. A new method for the control of swine kidney worms. *J. Am. Vet. Med. Assoc.,* 124: 36–39.

Amberson, J. M., and Schwarz, E. 1952. *Ternidens deminutus* Railliet and Henry, a nematode parasite of man and primates. *Ann. Trop. Med. and Parasitol.,* 46: 227–237.

Cameron, T. W. M. 1934. *The Internal Parasites of Domestic Animals.* A. C. Black, London.

Dougherty, E. C. 1951. A further revision in the classification of the family Metastrongylidae Leiper (1909) (Phylum Nematoda). *Parasitology,* 41: 91–96.

Foster, A. O. 1936. A quantitative study of the nematodes from a selected group of equines in Panama. *J. Parasitol.,* 22: 479–510.

Hobmaier, M. 1934. Lungenwurmlarven in Mollusken. *Z. Parasitenk.,* 6: 642–648.

Kates, K. C. 1950. Survival on pasture of free-living stages of some common gastrointestinal nematodes of sheep. *Proc. Helminthol. Soc. Wash., D. C.,* 17: 39–58.

Mayhew, R. L. 1940–1948. Studies on bovine gastrointestinal parasites. II. Immunity to hookworm and nodular worm infection in calves. *J. Parasitol.,* 26: 345–357; B. Immunity to the stomach worm, *Haemonchus contortus. Ibid.,* 26: Suppl. 17; IX. The effects of nematode infections during the larval period. *Cornell Vet.,* 34: 299–307; X, The effects of nodular worm (*Oesophagostomum radiatum*) on calves during the prepatent period. *Am. J. Vet. Research,* 9: 30–34.

Pavlov, P. 1935. Recherches sur le cycle évolutif de *Metastrongylus elongatus* et de *Dictyocaulus filaria. Ann. parasitol. humaine et comparée,* 13: 430–434.

Porter, D. A. 1942. Incidence of gastro-intestinal nematodes of cattle in the southeastern United States. *Am. J. Vet. Research,* 3: 304–307.

Sandground, J. H. 1931. Studies on the life history of *Ternidens deminutus* with observations on its incidence in certain regions of southern Africa. *Ann. Trop. Med. Parasitol.,* 25: 147–184.

Schwartz, B. 1934. Controlling kidney worms in swine in the southern States. *U. S. Dept. Agr. Leaflet* 108.

Shope, R. E. 1940–1942. The swine lungworm as a reservoir and intermediate host for swine influenza virus I-IV. *J. Exptl. Med.,* 74: 41–68; 77: 111–126; 127–138.

Stoll, N. R. 1958. The induction of self-cure and protection, with special reference to experimental vaccination against *Haemonchus. Rice Inst. Pamphlet,* 45 (1): 184–208.

Taylor, E. L. 1935. *Syngamus trachea. J. Pathol. Therapy,* 48: 149–156.

Travassos, L., and Vogelsang, E. 1932. Contribuçao as conhecimento does especies de *Oesophagostomum* does primatos. *Mem. inst. Oswaldo Cruz,* 26: 251–328.

U. S. Department of Agriculture. 1942. Keeping Livestock Healthy. *Yearbook.* Washington.

Watson, J. M. 1946. The differential diagnosis of hookworm, *Strongyloides,* and *Trichostrongylus,* with special reference to mixed infections. *J. Trop. Med. Hyg.,* 49: 94–98.

Wehr, E. E., and Olivier, L. 1943. The efficiency of barium antimonyl tartrate for the removal of gapeworms from pheasants. *Proc. Helminthol. Soc. Wash., D. C.,* 10: 87–89.

Whitlock, J. H. 1950. The relationship of nutrition to the development of the trichostrongylidoses. *Cornell Vet.,* 39: 146–182.

1951. The relationship of the available natural milk supply to the production of the trichostrongylidoses in sheep. *Ibid.,* 41: 299–311.

Chapter 20

OTHER INTESTINAL NEMATODES

ASCARIDATA. I. ASCARIDOIDEA

As noted on p. 397, the suborder Ascaridata consists of two superfamilies, the Ascaridoidea and the Oxyuroidea. The former contains for the most part relatively large, opaque, *Ascaris*-like worms of which there are numerous species parasitic in all kinds of vertebrates, whereas the latter contains smaller, transparent, *Oxyuris*-like worms which are parasitic mostly in the cecum and colon of vertebrates and also of insects.

Ascaris lumbricoides

General Account. *Ascaris lumbricoides* has undoubtedly been one of man's most faithful and constant companions from time immemorial, probably since he began domesticating pigs and by his habits made possible the development of a special strain particularly adapted for residence in his own intestine. This worm has clung to mankind successfully through the stone, copper, and iron ages, but plumbing threatens eventually to dissolve the partnership if children can be "yard-broken" early enough. Wherever soil pollution prevails, if only by toddlers in the dooryards, and wherever there is warmth and moisture, *Ascaris* infections are common.

Although *Ascaris* is one of the longest-known human parasites, it is a remarkable fact that important details of its life cycle were unknown before 1916, and the factors influencing its epidemiology were not elucidated until after 1930. One reason for this is that *Ascaris* infections have in general not been taken very seriously and their injurious effects have been minimized, whereas the effects of hookworm have often been exaggerated. Early in the present century *Ascaris* came into the limelight as an injurious and sometimes dangerous parasite. When a parasite steps into prominence nowadays it has little more chance to keep any details of its life and habits under cover than has a candidate for public office.

Morphology. *Ascaris lumbricoides* is a large nematode; the females commonly reach a length of 8 to 14 in. or even more, and are 4 to 6 mm. in diameter. The males are 6 to 12 in. long but distinctly more slender than the adult females; they are always distinguishable by the curled tail, whereas the females have a blunt tail (Fig. 134). Both sexes are more slender at the head end.

In common with other members of the Ascaridoidea, *A. lumbricoides* has the mouth guarded by three lips, one dorsal and two latero-ventral, each with minute papillae (Fig. 134*H*). The esophagus is nearly cylindrical and is followed by a flattened, ribbonlike intestine. The vulva is situated about one-third the distance from head to tail. The coiled tail of the male is short and provided with a characteristic number and arrangement of papillae but no alae. This worm is a favorite object for the study of nematode anatomy, since it is always easily obtainable and is easily dissected. An *Ascaris* morphologically indistinguishable from the human species is a very common parasite of pigs, but the two worms are physiologically distinct; eggs from one host species do not readily infect the other, so pigs are negligible as reservoirs. Both pigs and dogs may, however, be important in the dissemination of eggs which they have ingested with human feces.

Life Cycle. The adult *Ascaris* normally lives in the small intestine, where it is supposed to feed on the semidigested food of the host, but there is evidence that it commonly bites the mucous membranes with its lips and sucks blood and tissue juices to some extent. Reid (1945) showed that a related worm, *Ascaridia galli* of chickens, is highly susceptible to host starvation for 48 hours, just as are tapeworms, and that many are expelled when their stored glycogen supply is depleted.

The egg production of *Ascaris* is astounding. Cram (1925) estimated the number of eggs contained in a mature female worm to be as high as 27,000,000, and the eggs per gram of feces for each female worm may be in excess of 2000. This would indicate a daily production of something like 200,000 eggs! Evidently the chances against the offspring of an *Ascaris* reaching a comfortable maternity ward in a human intestine are many millions to one.

The eggs (Fig. 55*U*, *V*) have a thick, clear, inner shell covered over by a warty, albuminous coat which is stained yellow or brown in the intestine; they usually measure about 60 to 70 μ by 40 to 50 μ. Unfertilized eggs (Fig. 55*W*) are more difficult for a beginner to identify, since they are more elongate and less regularly oval in shape and have amorphous contents instead of the well-defined round cell of the fer-

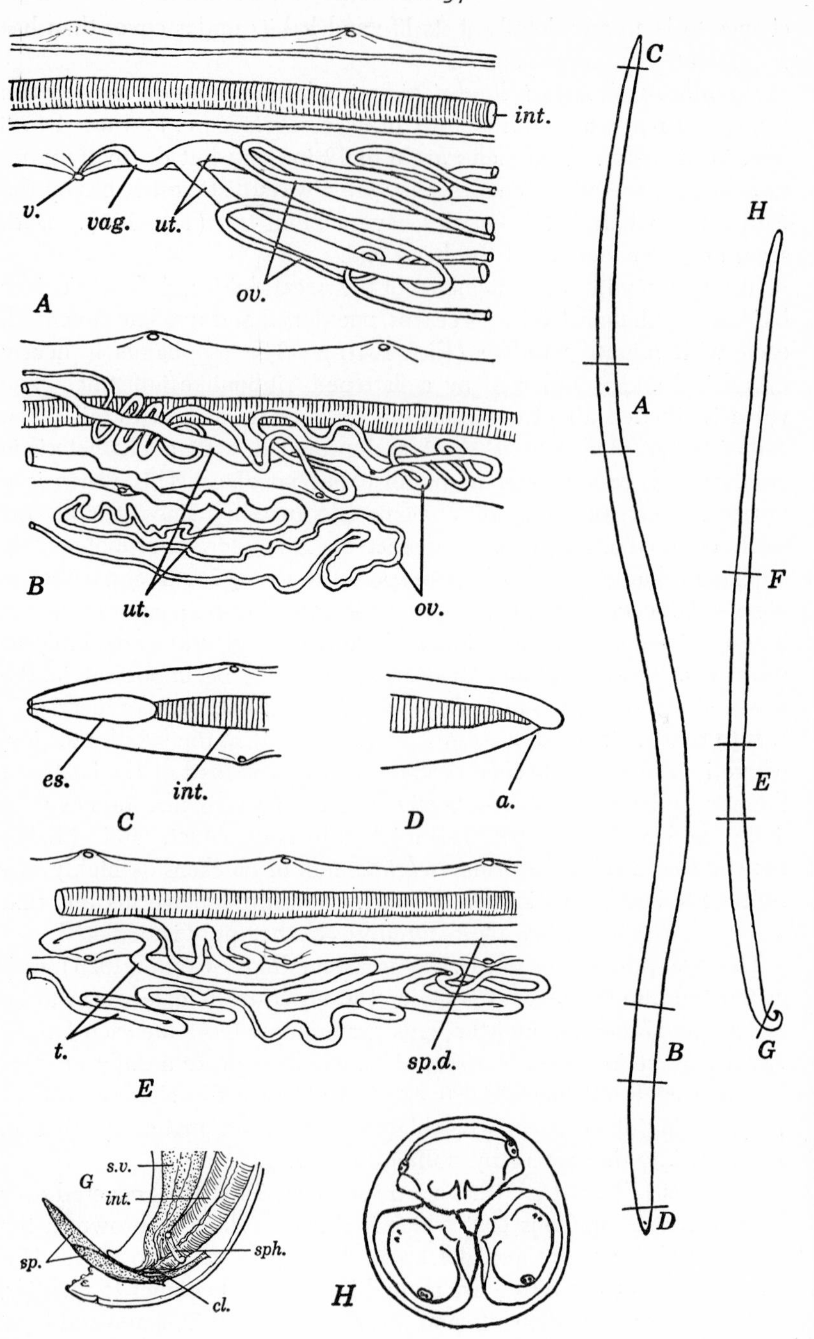

int.
v.
vag.
ut.
ov.
A
B
ut.
ov.
es.
int.
C
a.
D
t.
sp.d.
E
s.v.
G
int.
sp.
sph.
cl.
H
C
A
B
D
H
F
E
G

tilized eggs. The warty, albuminous coat dissolves off in sodium hydroxide, so in feces examined by Stoll's egg-count method the eggs may have only the thick inner shell.

The eggs are unsegmented when they leave the host. In order to develop they require a temperature lower than that of the human body, at least a trace of moisture, and oxygen. They are very resistant to chemical substances and will develop readily in weak formalin solutions or in sea water, but they can be killed by methyl bromide (see p. 437). They gradually degenerate at temperatures above 100°F. and cease development below about 60°F.; about 85°F. is the most favorable temperature. Complete drying is lethal, but in moist soil they remain viable for years. An enterprising German researcher seeded a plot of soil with *Ascaris* eggs; two persons ate unwashed strawberries raised on the plot each year for 6 years, and each year acquired a few *Ascaris.*

Under favorable conditions of temperature, moisture, and air the eggs develop active embryos within them in 10 to 14 days, but the embryos are not infective until they have molted inside the egg, becoming second-stage larvae; this requires an extra week.

When the eggs are swallowed the larvae hatch in the small intestine. They penetrate the mucous membranes and go on a sort of home-seeker's trip through the body, being carried by the blood stream to the liver, then the heart, and then the lungs. Here they burrow out and make their way through the trachea, throat, and esophagus back to the intestine, meanwhile having benefited from the trip by a growth from 200 to 300 μ to about ten times this length. The migration through the lungs takes place readily in rats, mice, guinea pigs, and other rodents as well as in the natural hosts, but after the return to the intestine the worms pass right on through in unnatural hosts and are voided in the feces.

The larvae of ascarids of pigs and horses perform the same migration through the liver and lungs and back to the intestine as does the human *Ascaris,* but work by Sprent and by Tiner (see Sprent, 1954) indicates

---

Fig. 134. *Ascaris lumbricoides.* Entire worms at right, ½ natural size. At left, dissected parts corresponding to letters on whole worms. *A,* region of vulva of ♀; *B,* region where uteri change to ovaries; *C,* anterior end; *D,* posterior end of ♀; *E,* region where sperm duct changes to testis; *F,* region where coils of testis end anteriorly (no dissection of this shown); *G,* posterior end of ♂; *H,* anterior view of lips. Abbreviations: *a.,* anus; *cl.,* cloaca; *es.,* esophagus; *int.,* intestine; *ov.,* ovary; *sp.d.,* sperm duct; *sp.,* spicules; *sph.,* sphincter; *s.v.,* seminal vesicle; *t.,* testis; *ut.,* uterus; *v.,* vulva; *vag.,* vagina.

that ascarids of carnivores behave differently (see p. 458). There is evidence that when eggs of *Toxocara canis* of dogs are swallowed by children the larvae may roam around in the liver and probably other viscera, a condition referred to as visceral larva migrans (see p. 459). If the larvae should invade the brain they might cause very serious effects as some ascarids do in mice (p. 458), and filariae in sheep and other animals (p. 495). Sprent (1955) has reviewed the knowledge concerning invasion of the nervous system by nematodes; a wide variety of species have been reported from the brain or spinal cord.

After reaching the human intestine young *Ascaris lumbricoides*, 2 to 3 mm. long, grow to maturity in 2 to 2½ months. The length of life in the host is short and averages only 9 months to a year.

Epidemiology. Since a combination of heat and dryness is injurious to them, *Ascaris* eggs in feces passed on sandy soil exposed to the sun in a hot climate die before the embryos can develop. *Ascaris* thrives best where there is abundant moisture and shade and where the soil is clayish rather than sandy, since in sandy soil the sorting action of rain drops concentrates the eggs on the surface, where they are exposed to sun and desiccation (Beaver, 1952).

Infection ordinarily results from swallowing embryonated eggs, which are more frequently conveyed to the mouth by fingers than by other methods. In some places in India heavy infection is directly correlated with polluted water supplies. Brown in 1927 observed that in Panama the infection is distinctly of household nature and is derived from contamination of hands and food by eggs developing in the soil on the floors and dooryards of huts polluted by young children. In Egyptian villages the mud floors of the houses are heavily seeded with *Ascaris* eggs, and, since the Egyptian fellaheen sit, lie, eat, and play on the floor, opportunities for infection are numerous. Raw leafy vegetables, contrary to popular belief, are relatively unimportant vectors of the eggs, even in China where night soil is used for fertilizer. Considerable *Ascaris* infection may occur in the riffraff living in crowded quarters on the edges of southern cities, when there are dense shade, abundant rain, and children who are careless in their defecation habits. The playing of children on polluted ground near their homes, tracking of pollution into the houses, and eating with dirty hands are the most important factors in the epidemiology.

In the United States *Ascaris* infection is largely limited to the mountainous areas of the southeastern states and to southern Louisiana, and is concentrated in young children. Here shelter in the immediate vicinity of the dooryards leads to close-in pollution, and the clay soil is protective for the eggs.

Pathology. In heavy experimental infections the migration of the larvae through the lungs causes hemorrhages and sets up a severe pneumonia which may be fatal. The invasion may be accompanied by a fever, a temporary anemia, leucocytosis, and eosinophilia. Pigs frequently show lung symptoms known as "thumps," and similar conditions have been observed in human beings preceding an *Ascaris* infection; ordinarily in nature, however, not enough eggs are ingested at a time to cause serious pneumonia.

After reaching maturity in the intestine, *Ascaris* may or may not disturb the peace of the host, but vague abdominal discomfort and acute colic pains are frequently felt, sometimes with vomiting, diarrhea, and mild elevations of temperature. Light infections may be entirely unsuspected until the eggs are found in the feces. On the other hand, the parasite is not always so docile. In heavy infections, especially if made uncomfortable by some food or drug taken by the host, the worms are likely to tangle themselves in masses and completely block the intestine. One thousand to five thousand worms have been recorded in some cases, but even less than a hundred worms may cause a blockage that is fatal if not surgically removed. A number of cases of death after carbon tetrachloride treatment for hookworm are known, due to obstruction of the intestine by squirming masses of irritated *Ascaris*.

Sometimes irritation of the mucous membranes may cause dangerous spasmodic contractions or permanent nervous constrictions of the intestine. The worms sometimes cause appendicitis by blocking the appendix. Toxic products may cause effects resembling anaphylactic shock and such nervous symptoms as convulsions, delirium, general nervousness, and coma. Sang in 1938 demonstrated a substance excreted by *Ascaris* which combines with trypsin, and he believes that when numerous *Ascaris* are present, enough destruction of trypsin may occur to interfere with digestion of proteins and account for the loss of condition and stunting of growth often seen in infected animals. Japanese workers found that *Ascaris*-infected schoolchildren were shorter than uninfected ones and had less memory and thinking capacity. Venkatachalam and Patwardhan (1953) carried out experiments with poorly nourished children infected with *Ascaris* and showed that the worms interfere with protein utilization. Deworming produced an immediate improvement in protein nutrition. Since the heaviest *Ascaris* burdens are in those areas of the world suffering most from malnutrition, particularly protein deficiency, it seems probable that the most important effect of *Ascaris* on populations is one of grossly aggravating a multitude of other health problems.

The list of dangerous complications of *Ascaris* infection is greatly enlarged by the fact that the worms have a "wanderlust" and tend to explore ducts and cavities. They frequently invade bile or pancreatic ducts and may enter the gall bladder or even go on into the liver; when children too young to have gallstones have symptoms of disease of the biliary tract, a misplaced *Ascaris* may well be suspected. That this complication is not a rare one is indicated by the report from Chinese workers in 1957 that of 1685 biliary tract surgery cases in Shanghai hospitals almost 10% involved invasion of the bile ducts by *Ascaris*. Occasionally an *Ascaris* creeps forward through the stomach and is vomited or emerges through the nose of a horrified patient; it may even enter the trachea and cause suffocation. On at least one occasion a young ascarid has emerged from the corner of the eye. *Ascaris* sometimes passes through the intestinal wall and causes fatal peritonitis or may even come through the umbilicus or groin, or the worms may make their way into the pleural cavity, urinogenital organs, etc. It is evident, therefore, that these worms, so far from being the "guardian angels" of children, as they were once considered, are more like bulls in a china shop.

Treatment and prevention. *Ascaris*, as long as it stays in the intestine, is fairly easily expelled by some anthelmintics, but some, e.g., tetrachlorethylene, merely irritate the worms and cause intestinal blockage. Therefore, when different treatments for *Ascaris* and some other worm infection are indicated, the *Ascaris* treatment should usually be given first. Brown in 1946, however, found no irritating action from gentian violet.

Oil of chenopodium and Santonin are efficient drugs for ascariasis, but both are very toxic. However, a mixture of oil of chenopodium and tetrachlorethylene, as pointed out on p. 434, is usually successful. Hexylresorcinol crystals in gelatin capsules (crystoids), with fasting for 12 hours before treatment and for 4 hours afterwards, followed by sodium sulfate to expel the dead worms, is one of the most effective treatments. At a dose rate of 1 gram for adults and 0.5 gram for children, it eliminates a high percentage of the worms, and makes a clean sweep in 40 to 80% of cases. Some workers recommend larger doses of 2 to 2.5 grams. Hetrazan (see p. 483), administered in a syrup, is also highly effective and nontoxic, and is particularly useful for babies and young children who cannot, or will not, swallow capsules. Hoekenga in Honduras reported 80% cures with a similar dosage. Loughlin et al. (1951) recommended a single daily dose of 13 mg. per kilogram for 4 days, administered in syrup. Extension of treatment to 5 or 6 days might increase the therapeutic efficacy without toxic effects.

In the past 6 years numerous workers have reported a high degree of success in the treatment of *Ascaris* infections with piperazine hydrate or citrate given in a syrup, with rare undesirable side reactions. A regime of 70 mg. or less per pound per day in divided doses, for 5 days, with a maximum daily dosage of 3 grams, is highly efficient. It has the advantage that neither purging nor hospital observation is required. The new broad spectrum anthelmintic, Dithiazanine, is effective against *Ascaris* as well as against several other nematodes.

In endemic localities treatment without sanitary improvement does little good, for a treated population usually gets back to the pretreatment level of infection within a year. On the other hand, when reinfection is stopped the worms are lost in about 9 to 12 months even without treatment. Prevention must depend mainly upon doing away with soil pollution near homes, even by very young children, and teaching children early in life to wash their hands before eating. The installation of privies is not always as successful as anticipated, because of only partial use of them. Careful washing of vegetables grown in polluted or night-soil-treated ground is desirable, for, although not as important as soil-to-mouth infection by children's dirty hands, such vegetables may cause infection in more fastidious adults.

Other Ascaridoidea

Ascaris lumbricoides var. *suum* is a very common parasite of pigs; about 75% of pigs in the United States and Canada harbor it before they are 6 months old. The principal effect is stunting of growth. Spindler found that pigs infected with 20 or more worms at 8 weeks of age failed in gain of weight, in proportion to the number of worms. One pig with 109 worms gained no weight at all, whereas uninfected pigs gained an average of 100 lb. Loss results also from condemnation of carcasses for jaundice owing to blockage of bile ducts. Sodium fluoride is highly efficient for removal of *Ascaris* and stomachworms of pigs (see p. 498) when 1% is added to 1 lb. of dry ground feed for one day for a 25-lb. pig; for heavier animals additional medicated feed up to a total of 4 lb. may be given at 12- to 24-hour intervals.

Other ascarids found in domestic animals are *Parascaris equorum* (=*megalocephala*) in horses, and *Neoascaris vitulorum* in calves. Sodium fluoride at 2.5 grams per 100 lb. is effective against ascarids in horses, but may have toxic effects. Dogs and cats harbor smaller ascarids, 3 to 5 in. long, belonging to the genera *Toxocara* and *Toxascaris.* They have cervical alae which give the anterior end an arrowhead shape (Fig. 135*A*). *Toxocara* males have a fingerlike process at the end of the tail (Fig. 135*B*), lacking in *Toxascaris* (Fig. 135*C*), and *Toxocara* eggs are delicately pitted whereas those of *Toxascaris* are smooth. Piperazine hydrate (100 mg./kg. daily for 10 days) eliminates these worms.

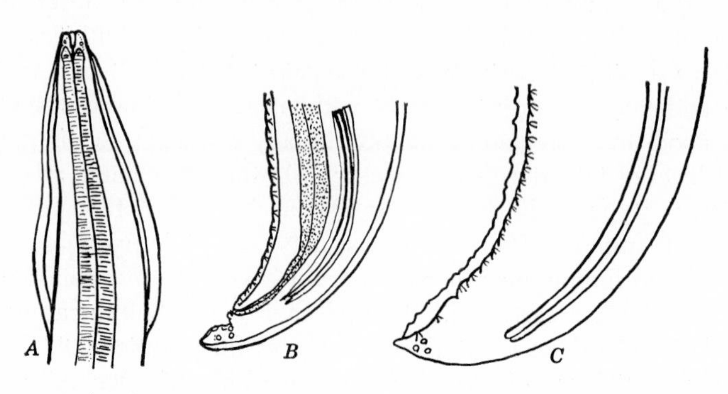

Fig. 135. *A,* head of *Toxocara canis; B,* tail of male of same; *C,* tail of male of *Toxascaris leonina.* (After Yorke and Maplestone, *Nematode Parasites of Vertebrates,* 1926.)

Toxocara apparently depends largely on prenatal infection as a means of spreading. Eggs swallowed by mother dogs migrate to the fetuses and establish themselves in their intestines; they may become encapsulated in the mother's tissues, but immunity prevents them from developing in the parental intestine. When fed to mice the larvae become encapsulated mainly in the intestinal walls. *Ascaris columnaris* of raccoons and skunks probably uses rodents as true intermediate hosts; the larvae become encapsulated in the viscera and frequently enter the brain, where even one larva may be fatal. Tiner (1953) believes brain damage may render intermediate hosts easier prey for the final hosts, and thus be of value to the parasites (see review by Sprent, 1955).

An ascarid rarely found in man is *Lagochilascaris minor,* normally found in the cloudy leopard. In several cases in Trinidad and Guiana sexually mature specimens have been found in subcutaneous or tonsillar abscesses about the head. The adults are about the size of hookworms and are identifiable by their lips and a keel-like expansion of the cuticle extending the whole length on each side.

Poultry are subject to two common types of ascarids, *Ascaridia galli* in the small intestine, and species of *Heterakis* or cecal worms in the ceca. The former, a worm about 2 to 4 in. long with a muscular preanal sucker on the male, causes retarded growth and droopiness in heavy infections in young chickens. Older birds develop a marked age immunity due to an increase in number of mucin-producing goblet cells (see p. 26). Phenyl mercuric compounds (50 mg.) plus 0.5 gram phenothiazine removes most of these worms. *Heterakis gallinarum* (=*gallinae*) is 7 to 15 mm. long with a chitin-rimmed preanal sucker and conspicuous caudal alae in the males. It seems to be harmless, even when the ends of the ceca contain swarming masses of them, except for its role as a carrier of the protozoan, *Histomonas meleagridis*, that causes "blackhead" in turkeys (see p. 103). Chickens become infected with ascarids by swallowing embryonated eggs; there is no migration through the body, but *Ascaridia* temporarily bury themselves in the intestinal wall.

Ascarids in fish-eating mamals, birds, and fish have more complicated life cycles involving first and second intermediate hosts, which are aquatic invertebrates and small aquatic vertebrates, respectively.

Visceral larva migrans. Beaver et al. (1952) proposed this term for the prolonged migration of larval nematodes in tissues other than skin. Dog and cat ascarids seem to be most often involved in cases showing the characteristic symptoms. The disease is seen most often in young children who typically have a marked eosinophilia and an enlarged liver. Lung inflammation also commonly occurs. Medical attention is sometimes sought when children have irregular fever, loss of appetite, fail to gain weight, chronically cough, or have muscle-joint or abdominal pains. Occasionally, *Toxocara* larva may wander into such locations as the eye where they produce serious effects. Wilder (1950) reported on a series of children who had suffered the loss of an eye due to nematode infection of this organ. Nichols (1956) identified some of these as *Toxocara* and some as *Ancylostoma caninum* larvae (see p. 420). Since the recognition of this disease a few years ago, cases have been reported from all over the United States and in England. Children apparently become infected by ingesting eggs which are promiscuously distributed by man's best friend, the dog. Cats are incriminated to a lesser extent, presumably because these animals are somewhat more fastidious in defecation habits. It becomes more and more evident that man pays a price for maintaining dogs in the household, since these animals serve as reservoirs for at least a dozen serious human diseases.

At present the diagnosis of visceral larva migrans must be based on clinical symptoms or on finding the larvae in biopsy specimens of liver tissue. Several groups of research workers in the United States are currently attempting to develop serological methods that will allow the ready diagnosis of larva migrans. It seems probable that such methods will be available shortly and should rapidly replace the unpleasant procedure of liver biopsy.

At present there is no treatment for this disease and it usually runs its course with chronic symptoms for 6 to 18 months.

ASCARIDATA. II. OXYUROIDEA

As already noted, the Oxyuroidea are almost exclusively parasites of the cecum or colon of their hosts, not only of vertebrates but also of insects. Only a single oxyurid, *Enterobius vermicularis,* occurs commonly in man.

Enterobius vermicularis

Most members of the Caucasian race, even in highly sanitated countries, fail to get through life without affording food and shelter for oxyuris, also popularly called the pinworm, seatworm, or threadworm (*Enterobius vermicularis*). It is found all over the world but unlike most helminthic infections is relatively rare in the tropics. Its great stronghold is in Europe and North America, but according to Neghme 60% of schoolboys in Chile are infected. As Stoll (1947) remarked, there seem to be factors in our modern way of living which are very favorable for the spread of *Enterobius* in high as well as low social levels. Sample surveys of white children in cities in the United States and Canada give incidences of 30 to 60%. Colored races are far less susceptible. In Negro children in Washington, D. C., the incidence is 16% as compared with 40% in Whites, and in Honolulu 40% in Caucasians and 21% in Orientals. *Enterobius vermicularis* is strictly a human parasite, although closely related species occur in apes and monkeys.

Morphology. The adult worms (Fig. 136) live in the cecum, appendix, and neighboring parts of the intestine, from which the gravid females migrate to the rectum. These are little white worms, often seen wriggling actively in stools passed after a purge or enema. Through the semitransparent cuticle can be seen the esophagus with a bulb at its posterior end, and the uteri and coiled ovaries. The head has three small lips and is set off by lateral expansions of the cuticle. The females, 8 to 13 mm. long, taper at both ends, but the tail is drawn out into a long, fine point. The minute males, only 2 to 5 mm. long, are less numerous than the females and are seldom noticed. The tail is curled and has a small bursalike expansion; there is only one spicule (Fig. 136).

Life Cycle. As the uteri of the females fill with eggs, the worms migrate down to the anus; according to MacArthur (1930), they may make regular nightly trips, deposit eggs in the perianal region, and

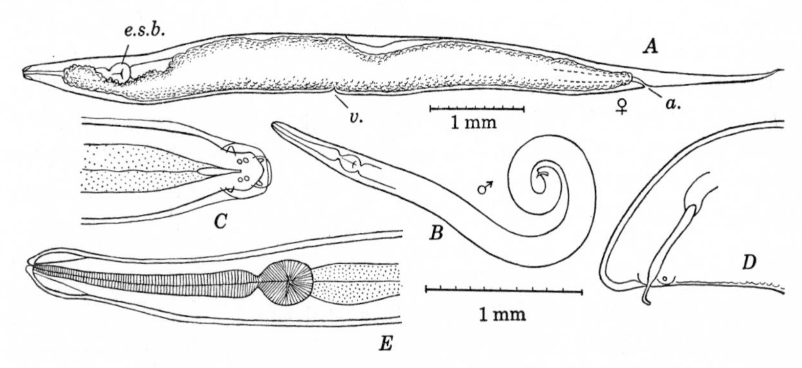

Fig. 136. *Enterobius vermicularis.* A, adult gravid ♀ (*a., anus; e.s.b.,* esophageal bulb; *v.,* vulva). *B,* adult ♂; *C,* posterior end of ♂, ventral view; *D,* same, much enlarged, showing form of single spicule; *E,* anterior end. (*C* and *E* after Yorke and Maplestone, *Nematode Parasites of Vertebrates,* 1926.)

retreat into the rectum, but many worms creep out of the anus, and others are passed in the feces. Their movements cause intense itching. Contact with air stimulates the worms to deposit eggs, and a trail of these is left behind as the worms crawl. Eggs are seldom found in the feces before the worms have disintegrated but can be obtained from scrapings from about the anus or lower part of the rectum. The worms eventually dry and explode, liberating all the remaining eggs in showers. The eggs when first laid contain partially developed embryos in the "tadpole" stage, but they develop to the infective stage in as little as 6 hours. The eggs (Fig. 55Z) are clear and unstained, measuring about 55 by 30 μ, and are flattened on one side. Reardon estimated the average number of eggs in a female oxyuris to be about 11,000. After being swallowed, the larvae hatch and temporarily burrow into the mucous membranes in the region of the cecum before growing to maturity in the lumen.

Mode of infection and epidemiology. The eggs regain access to the same or another person in various ways, but are probably most often air-borne or conveyed by the hands. The itching caused by the emigration of the worms from the anus results in scratching, and the eggs lodged under the fingernails may eventually reach the mouth in children or others who are careless in their habits. The eggs are easily liberated into the air when sheets, clothing, etc., contaminated with them are shaken or rubbed, and may be inhaled or may settle as dust which may be inhaled later.

The extent to which the eggs become scattered in infected households is almost incredible. Not only are they present on the hands, clothing, bed linen, towels, washcloths, and soap, but also on floor, upholstery, and furniture, often in every room of houses occupied by heavily infected children. The junior author observed the build up of very heavy infections in a university housing unit for students' families. Eggs were being disseminated in washing machines in which the water was tepid because of inadequate water heaters. Schüffner (1944) found that the smaller the enclosed space the greater the number of eggs; in 1 sq. ft. in a large dining hall he found 119 eggs, in a smaller classroom 305, in a toilet 5000. He pointed out that half the life of an infected child is spent in a still smaller enclosed space—between bed sheets, where the eggs are disseminated by movements of the sleeper. He believes, however, that very heavy infections result only from transfer of eggs by fingers after scratching. Schüffner believes that chronic adult infections may be due to "retrofection," i.e., re-entrance of larvae that sometimes hatch from eggs on the perianal skin.

Cram has called attention to the familial nature of pinworm infections, and numerous observations point to its ready spread in schools and institutions. The eggs survive longest (2 to 6 days) under cool humid conditions, but their life span in dry air above 25°C. is greatly shortened, few surviving as long as 12 hours. In dry air at 36 to 37°C. less than 10% survive for 3 hours and none for 16 hours. Since the worms have a life span of only 37 to 53 days, the infection would die out in this period if reinfection could be stopped; the periodic appearance of increased numbers of worms often observed at 4- or 5-week intervals is due to the maturation of new generations of worms from reinfections.

Diagnosis. No dependence can be placed upon examinations of the feces for the eggs of *Enterobius.* Direct fecal smears show less than 1% of the actual infections, and flotation methods less than 25%; even heavy infections often fail to be detected.

Far better results are obtained by scraping the perianal region. Of the several devices proposed for this, Beaver (1949) found the widely used Scotch tape method the easiest and also most efficient. A piece of Scotch tape is held by thumb and forefinger over the end of a tongue depressor, sticky side out, applied to the right and left perianal folds, and then flattened on a slide for examination. The NIH swab and wet pestle methods pick up fewer eggs and are less "foolproof." The success of these methods is affected by bathing, personal cleanliness, and irregular periodicity in the migration of the worms, so the number of eggs found has no relation to the size of the infection and one negative examination cannot be considered conclusive.

Pathology. The itching caused by migration of the worms in the anal region and by allergic irritation of the skin may be intense, causing loss of sleep, restlessness, nervousness, and even sexual disorders. In girls the worms may cause vaginitis by entering the vulva, and they may even wander into the Fallopian tubes or to the peritoneal cavity, where they become encysted.

Immature burrowing worms may cause inflammation in the cecal region, with some abdominal pain and digestive disturbances. Since the males and young females are often found in removed appendices they are often accused of causing appendicitis, but there is very little to support this view, since they are about equally common in healthy and inflamed appendices.

Treatment and prevention. Many of the nematode group of anthelmintics, including tetrachlorethylene, hexylresorcinol, and gentian violet, remove some of the worms. Other drugs which have been useful are Terramycin, lindane, Egressin, Diphenan, and Dithiazanine. In the past few years piperazine (Antepar) has proven to be the drug of choice. It may be given as the hydrate, citrate or adipate in a flavored syrup at the rate of 50 to 100 mg. per kilogram per day for 7 days followed by 7 days of rest and a second round of treatment in the third week. The rest period allows stray eggs in the environment to either get into the host or die, and the second period of treatment removes the new worms from the host before they have time to produce a new crop of progeny.

If reinfection could be stopped the infection would disappear without treatment in a few weeks, but, even with the most meticulous care in cleanliness, complete prevention of reinfection without treatment usually fails. It requires closed pajamas of nonporous material, daily changing and sterilization of bedclothes, towels, and underwear, use of anal bandages, and disinfecting ointments, frequent washing of hands, close-clipped fingernails, a dustless house, and unrelaxing parental vigilance. Treatment is easier! Schüffner (1944) believes that 100% of the children in Holland are infected in spite of the proverbial Dutch cleanliness. He thinks that efforts to eliminate the infection completely may lead to a "pinworm neurosis" that is worse than a mild pinworm infection. The junior author has seen American families in the grip of a neurotic frenzy far out of proportion to the seriousness of the pinworm infection. Many physicians have remained pitifully behind the times in treating enterobiasis and may contribute to psychological difficulties.

Other Oxyuroidea

The only domestic animal that suffers from oxyuris infection is the horse, which harbors *Oxyuris equi.* Rodents harbor numerous species, and one of these, *Syphacia obvelata* of mice and rats, was found once in a child in the Philippines. Common oxyurids for class study can nearly always be found in large cockroaches.

RHABDITATA

The suborder Rhabditata is of particular interest from an evolutionary standpoint since it contains nematodes showing every imaginable gradation from free-living, saprophagous forms to strict parasites. It presents a sort of pageant of parasites in the making. The genus *Rhabditis* alone contains many species which appear to be experimenting with parasitism. Some species have been found breeding in the feces-soiled hair of the perianal region of dogs; the larvae of the common soil nematode, *R. strongyloides,* have been found repeatedly in itching pustules in the skin of dogs and other animals after lying on soiled straw bedding. Members of a related genus, *Longibucca,* have been found breeding in the stomach and intestine of snakes and bats. Another member of the same family, *Diploscapter coronata* (see p. 469), is an opportunist which is capable of establishing itself in the human stomach or female urinogenital system when abnormal conditions make these environments favorable.

Members of the families Strongyloididae and Rhabdiasidae have bridged the gap between free-living and parasitic existence by a method peculiar to themselves—a true alteration of generations. There is a free-living generation consisting of males and females which are hardly distinguishable from *Rhabditis,* and a parasitic generation of parthenogenetic females which have a markedly different appearance. The eggs produced by one generation give rise to worms of the alternate generation. This routine is, however, short-circuited by many of the individual worms by omission of the free-living bisexual generation entirely, in spite of the fact that this is unquestionably the ancestral type. By this process we arrive at a form which is as truly parasitic as a hookworm.

The Strongyloididae pass the parasitic phase of their lives in the intestine of mammals, whereas the Rhabdiasidae pass theirs in the lungs of amphibians and reptiles. *Strongyloides stercoralis* is the only common and important human parasite in the Rhabditata, but *Rhabditis* (see p. 470) is frequently found in human stools, to the confusion of technicians examining them.

Strongyloides stercoralis

General account. This, the smallest nematode parasitic in the human body except the male *Trichinella,* is a very common human parasite in moist tropical or subtropical climates, having much the same distribution as hookworms. Faust found it in 20% of hospital and village populations in Panama and in 4% of cases examined in New Orleans hospitals and clinics. It is a common parasite in soldiers returning from the South Pacific. Statistics based on ordinary stool examinations do not give a correct idea of the prevalence of this parasite.

The parasitic females of *Strongyloides* are parthenogenetic. Kreis (1932) and Faust (1933) found a few male worms of the free-living type in the lungs of dogs, which were undoubtedly precociously developed free-living males. There is one species of Strongyloididae, *Parastrongyloides winchesi,* which *is* bisexual in the parasitic generation, but the males are filariform like the females.

The parasitic females (Fig. 137, *1*) are extremely slender worms 2 to 2.5 mm. long by only 40 to 50 μ in diameter, with a bulbless esophagus about one-fourth the length of the body. The uteri diverge from the vulva in the posterior third of the body; each contains a few developing eggs in single file.

Life cycle. (Fig. 137). The adult females burrow in the mucous membranes of the intestine anywhere from just behind the stomach to the rectum, although the upper part of the small intestine is their favorite spot. A few mature even in the bronchial tubes. The eggs, measuring about 50 by 32 μ, are deposited in the mucous membranes where they undergo development and hatch, the larvae then making their way into the lumen of the intestine, to be voided with the feces. The egg output per worm is relatively small, not more than 50 per day.

The passed larvae are rhabditiform (see p. 423) and have usually grown to a length of 300 to 800 μ. They resemble hookworm larvae but can be distinguished by the very short mouth cavity (Fig. 126, *9*).

The course of development of these larvae may follow either one of two lines: (1) direct or "homogonic," or (2) indirect or "heterogonic." In the *indirect course* of development the rhabditiform larvae develop, in 36 hours or more and after four molts, into free-living males and females (Fig. 137, *2*) which closely resemble soil nematodes of the

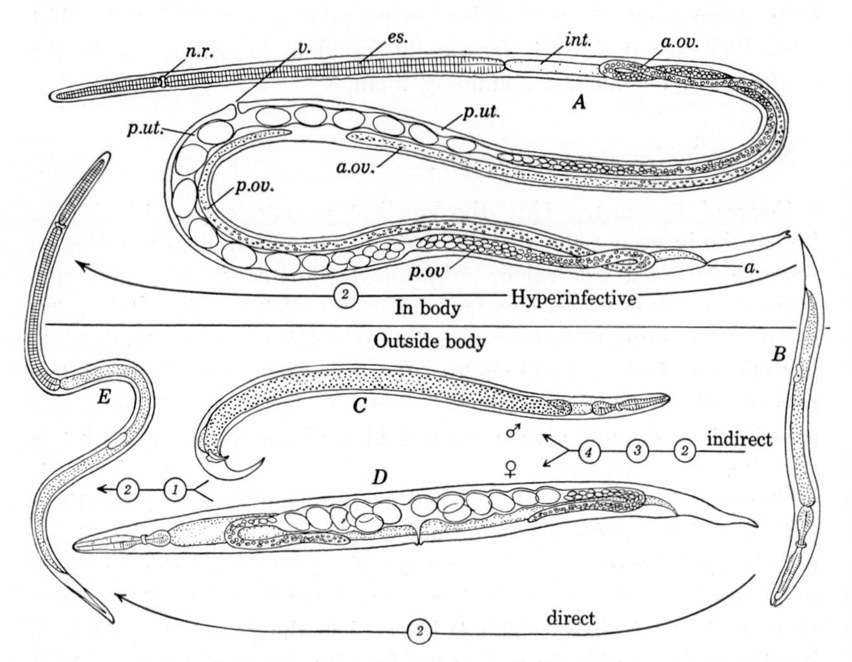

Fig. 137. *Strongyloides stercoralis,* life cycle. Direct (bottom arrow); indirect (middle arrows); hyperinfective (upper arrow). Circles enclosing numbers represent rhabditiform larval stages from 1 to 4 not drawn on diagram, and similar in appearance to *B*. Thus the direct cycle includes first and second rhabditiform and filariform larval stages outside body; indirect cycle includes four rhabditiform larval stages before becoming adult ♂ and ♀, then two rhabditiform stages (offspring of these) and one filariform stage before re-entering the body; hyperinfective cycle includes two rhabditiform larval stages and a filariform larval stage, all inside the body. Abbreviations: *a.*, anus; *a.ov.*, anterior ovary; *es.*, esophagus; *int.*, intestine; *n.r.*, nerve ring; *p.ov.*, posterior ovary; *p.ut.*, posterior uterus; *v.*, vulva. (Adapted from various authors.)

genus *Rhabditis;* they are about 1 mm. in length and 40 to 60 μ broad. These adults produce eggs which hatch into rhabditiform larvae very similar to the offspring of the parasitic females, which then ordinarily transform after two molts into slender filariform larvae characterized by a very long, slender esophagus and a long tail notched at the tip (Fig. 137*E*). The small, oval genital primordium is midway between the end of the esophagus and the anus. These larvae, 600 to 700 μ long, remain, like infective hookworm larvae, ensheathed by the molted cuticles of the rhabditiform larvae, and are now in the infective stage. They may appear in less than 48 hours, and they become numerous in 5 or 6 days. They infect by penetrating the skin or mucous membranes as do hookworm larvae. Occasionally, according to Beach (1936), more than one free-living generation may develop.

In the *direct* course of development the rhabditiform larvae produced by the parasitic females, usually after a brief period of feeding and growth, metamorphose directly into infective filariform larvae at the second molt. These penetrate the skin as do those produced indirectly.

A third possible course of development, called the *hyperinfective* method, occurs in exceptional cases when the larvae of the parasitic females rapidly undergo two molts inside the intestine without feeding or growing, transforming into filariform larvae which then burrow through the mucous membranes or perianal skin, causing reinfection without any outside existence (see Faust and de Groat, 1940).

The larvae of *Strongyloides* are easily destroyed by cold, desiccation, or direct sunlight, and are rather short-lived even under the most favorable conditions. This probably accounts for the infrequence of *Strongyloides* infections outside warm, moist climates.

After penetration some larvae remain in the skin for a long time, but they appear in the lungs from the third day onward. The larvae undergo development to adolescence in the lungs, and then migrate to the alimentary canal via the trachea and throat, although a few mature and reproduce in the lungs and bronchioles. Larvae begin to appear in the feces about 17 days or more after infection in man, but in dogs the prepatent period is only 12 days and in rats 6 days. The numbers rise rapidly but decrease again after some months, when immunity begins to develop.

Biology of direct and indirect development. The apparently willy-nilly appearance of the direct and indirect modes of development of *Strongyloides* has been very puzzling. Attempts have been made to explain it on the basis of environmental effects inside and outside the host, age of worms, fertilization by supposed parasitic males, and biologically different strains. Graham (1936–1939) started two pure lines of *S. ratti* in rats from original single-larva infections of the homogonic and heterogonic types, respectively, and found marked inherent differences between them. In each line over 85% of the total progeny were of its own type, with an extreme difference in the number of free-living males produced. Meanwhile Beach (1935,

1936) showed conclusively that the course of development can be influenced by nutritional conditions; as these become less favorable more and more of the rhabditiform larvae undergo direct transformation into filariform larvae instead of becoming males and females. The conclusion seems warranted, therefore, that the course of development is dependent upon nutrition or other environmental influences and not on genetic constitution, but that there are genetic differences in the extent to which different strains are influenced toward homogony by given degrees of unfavorableness in the environment.

Diagnosis. The infection is diagnosed by the finding and identification of the larvae in the stools; they can be found in simple fecal smears and can be floated satisfactorily in zinc sulfate solution (see p. 253), but they shrink badly in saturated sodium chloride. If scanty they can be found readily by culturing the stool mixed with an equal part of charcoal or sterilized earth. The rhabditiform larvae, as already noted, can be distinguished from those of hookworms by the very short mouth cavity, but are difficult to distinguish from coprophagic *Rhabditis* larvae in stale or contaminated stools unless cultured for 2 to 5 days, and the filariform larvae found by extraction into warm water. Embryonated eggs are occasionally found in cases of severe diarrhea. In examining stale stools there may be confusion with hookworm infections, but an excess of larvae over eggs in uncultured stools is indicative of *Strongyloides.* The eggs, if present, are decidedly smaller and always embryonated.

Pathology. Skin penetration by the larvae often causes redness and intense itching, a transient cutaneous larva migrans (see p. 430). Invasion of the lungs sometimes causes acute inflammation. The adults burrowing in the intestinal mucosa cause a catarrhal inflammation with so much erosion in severe cases as to give the appearance of raw beefsteak. In very light infections there may be no demonstrable symptoms; in moderate and chronic cases there are usually intermittent diarrhea and epigastric pain; in severe cases there may be uncontrollable diarrhea with blood and undigested food in the liquid stools. The loss of food and continued drain of liquids cause severe emaciation. In the tropics there is often evidence of allergic effects as well. de Langen described cases in Java with high eosinophilia, leucocytosis, anemia, slight fever, edema, and bronchial pneumonia in addition to the intestinal symptoms, and Faust called attention to the frequency of nervous symptoms in chronic infections. These are probably due to toxic allergic effects of disintegration of numerous larvae invading the bodies of people who have been repeatedly exposed.

Treatment and prevention. Gentian violet is moderately effective for *Strongyloides* infections. It stains the intestinal mucosa and kills some, but usually not all, of the adult worms buried in it. The standard treatment for adults is two ½-grain enteric-coated tablets (to open in 1½ hr.) with meals three times a day for 16 days; for children the dosage is 9 mg. per day per year of age. The difficulty is to get the tablets to open where needed in the duodenum but not in the stomach. Administration of 25 cc. of a 1% aqueous solution of gentian violet by duodenal tube may be effective in refractory cases, or even 25 cc. of a 0.5% solution intravenously. Hexylresorcinol is very toxic to *Strongyloides* in vitro but may not be effective in vivo. Hetrazan by stomach tube was reported to have given favorable results. Swartzwelder et al. (1958) have recently reported that a halogenated cyanine dye, Dithiazanine, is highly effective against *Strongyloides*. This drug also acts against *Ascaris, Trichuris, Enterobius,* and hookworms and may well become the treatment of choice.

Control is much the same as in hookworm infections, except that the delicacy of the *Strongyloides* larvae should make it easier.

***Strongyloides* in animals.** *Strongyloides stercoralis* is infective for dogs and cats as well as man but usually dies out in a number of weeks. In India, however, the writer found a high percentage of cats naturally infected with a *Strongyloides* which was very similar to, if not identical with, the human species. Other species occur in monkeys, sheep, rodents, pigs, and other animals. One human infection with *S. fülleborni* of monkeys has been reported. Most of the species in herbivorous animals differ from those in man and carnivores in that the eggs usually do not hatch until after they have left the body of the host.

It has been suggested that a *Strongyloides* of sheep may be involved in the transmission of foot-rot, a bacterial disease of sheep, presumably by carrying the bacteria through the barrier presented by the skin.

Diploscapter coronata

Some nematodes found in the aspirated stomach contents of nine patients who were suffering from complete or almost complete lack of hydrochloric acid were examined by Chandler (1938) and found to be *Diploscapter coronata* (Fig. 138). This nematode was previously known only as an inhabitant of soil or sewage beds; a related species is parasitic on living roots of plants. The worms from the stomach were abundant in some cases and scanty in others; they were in all stages of development, but no males were found. This corresponds with most previous observations on this worm; apparently, like *Strongyloides* and some species of *Rhabditis,* it can get along very well without the presence of the male sex. Adult females are about 420 μ long.

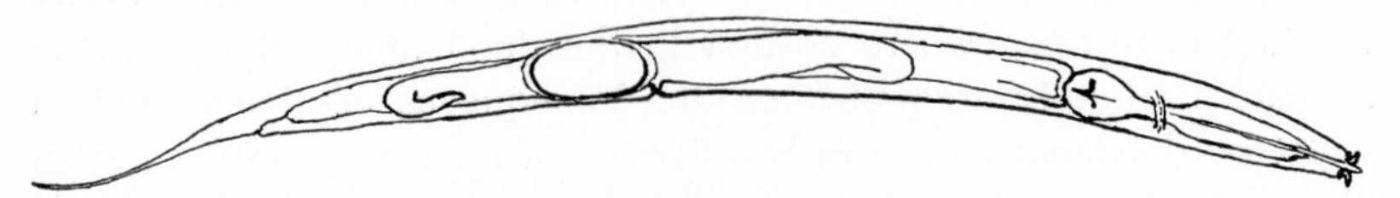

Fig. 138. *Diploscapter coronata,* adult female from human stomach. (After Chandler, *Parasitology,* 1938.)

All the cases were discovered in a Houston clinic, and similar cases are reported as having been seen frequently before, but incorrectly diagnosed as *Strongyloides.* In one case a re-examination 4 days later showed the worms still present, so they were undoubtedly established in the stomach. Oddly enough, only a single prior case of similar nature has been recorded in the literature. The same worm has, however, been found in the urine of women, once in Japan and three times in Israel.

Rhabditis

The genus *Rhabditis* contains numerous species of nematodes normally found in soil, organic matter, or water, and frequently in feces of man or animals. They closely resemble the free-living generation of *Strongyloides* but have no alternation of generations.

Rhabditis pellio is a species which has on a few occasions been found living in the human vagina, the larvae escaping in the urine. *R. hominis* and other species have been recorded from stools of man and animals. In most of these cases there was suspicion of their being true parasites, but the worms have not been found on re-examination, and in some cases clear evidence of contamination with soil or water was obtained. There is as yet no conclusive evidence that any of these species are more than coprophagous. Other pseudo-infections with *Rhabditis* were mentioned on p. 464. Their only importance is their possible confusion with *Strongyloides.*

REFERENCES

Ascaridoidea

Ackert, J. E. 1931. The morphology and life history of the fowl nematode *Ascaridia lineata* (Schneider). *Parasitology,* 23: 360–379.

Beaver, P. C. 1952. Observations on the epidemiology of ascariasis in a region of high hookworm endemicity. *J. Parasitol.*, 38: 445–453.

Beaver, P. C., et al. 1952. Chronic eosinophilia due to visceral larva migrans. *Pediatrics*, 9: 7–19.

Beaver, P. C. 1956. Larva migrans. *Exp. Parasitol.*, 6: 587–621.

Chandler, A. C. 1954. A comparison of helminthic and protozoan infections in two Egyptian villages two years after the installation of sanitary improvements in one of them. *Am. J. Trop. Med. Hyg.*, 3: 59–73.

Cram, E. B. 1926. Ascariasis in preventive medicine. *Am. J. Trop. Med.*, 6: 91–114.

Headlee, W. H. 1936. The epidemiology of human ascariasis in the metropolitan area of New Orleans, La. *Am. J. Hyg.*, 24: 479–521.

Lane, C. 1934. The prevention of *Ascaris* infection: A critical review. *Trop. Diseases Bull.*, 31: 605–615.

Nichols, R. L. 1956. The etiology of visceral larva migrans. I and II. *J. Parasitol.*, 42: 349–399.

Otto, G. F., and Cort, W. W. 1934. The distribution and epidemiology of human ascariasis in the United States. *Am. J. Hyg.*, 19: 657–712.

Scott, J. A. 1939. Observations on infection with the common roundworm, *Ascaris lumbricoides* in Egypt. *Am. J. Hyg.*, 30 D: 83–116.

Smith, M. H. D., and Beaver, P. 1955. Visceral larva migrans due to infection with dog and cat ascarids. *Ped. Clin. North Am.*, 2: 163–168.

Sprent, J. F. A. 1954. The life cycles of nematodes in the family Ascarididae Blanchard, 1896. *J. Parasitol.*, 40: 608–617.

Sprent, J. F. A. 1955. On the invasion of the central nervous system by nematodes. I. The incidence and pathological significance of nematodes in the central nervous system. *Parasitology*, 45: 31–40. *Ibid.* II. Invasion of the nervous system in ascariasis. *Parasitology*, 45: 41–55.

Venkatachalam, P. S., and Patwardhan, V. N. 1953. The role of *Ascaris lumbricoides* in the nutrition of the host. Effect of ascariasis on digestion of proteins. *Trans. Roy. Soc. Trop. Med. Hyg.*, 47: 169–175.

Wilder, H. C. 1950. Nematode endophthalmitis. *Trans. Am. Acad. Ophthalmol.*, Nov.-Dec., 99–109.

Winckel, W. E. F., and Treurniet, A. E. 1956. Infestation with *Lagochilascaris minor* (Leiper) in man. *Documenta Med. Geograph. et Trop.*, 8: 23–28.

Winfield, G. F., et al. 1937. Studies on the control of fecal-borne diseases in North China, II, IV. *China Med. J.*, 51: 502–518, 643–658, 919–926.

Oxyuroidea

Beaver, P. C. 1949. Methods of pinworm diagnosis. *Am. J. Trop. Med.*, 29: 577–587.

Gordon, H. 1933. Appendical oxyuriasis and appendicitis based on a study of 26,051 appendices. *Arch. Pathol.*, 16: 177–194.

Kessell, J. F., and Markell, E. K. 1957. Special therapeutics (helminthic diseases). *Ann. Rev. Med.*, 8: 415–426.

Lentze, F. A. 1935. Zur Biologie des *Oxyuris vermicularis*. *Zentr. Bakteriol. Parasitenk.* I Abt., Orig., 135: 156–159.

MacArthur, W. P. 1930. Threadworms and pruritis ani. *J. Roy. Army Med. Corps*, 55: 214–216.

Most, H. 1943. Studies on the effectiveness of phenothiazine in human nematode infections. *Am. J. Trop. Med.*, 23: 459–464.

National Institutes of Health, Studies on Oxyuriasis, I-XXVIII (Papers by Hall, Wright, Bozicevich, Cram, Jones, Reardon, Nolan, Brady, et al.). 1937–1943. XXVIII, Summary and conclusions (By Cram, E. B., *Am. J. Diseases Children,* 65: 46–59, 1943) contains references to entire series.

Schüffner, W. 1944–1946. Die Bedeutung der Staubinfektion für die Oxyuriasis, Richtlinien der Therapie und Prophylaxe. *Münch. med. Wochenschr.,* 44: 411–414; Review in *Trop. Diseases Bull.,* 43: 233–236.

Stoll, N. R. 1947. This wormy world. *J. Parasitol.,* 33: 1–18.

Rhabditata

Beach, T. D. 1936. Experimental studies on human and primate species of *Strongyloides.* V. The free-living phase of the life cycle. *Am. J. Hyg.,* 23: 243–277.

Chandler, A. C. 1938. *Diploscapter coronata* as a facultative parasite of man, with a general review of vertebrate parasitism by rhabditoid worms. *Parasitology,* 30: 44–55.

Faust, E. C. 1932. The symptomatology, diagnosis and treatment of *Stronglyoides* infection. *J. Am. Med. Assoc.,* 98: 2276–2277.

1933–1935. Experimental studies on human and primate species of *Strongyloides.* II. *Am. J. Hyg.,* 18: 114–132. III and IV. *Arch. Pathol.,* 18: 605–625; 19: 769–806.

Faust, E. C., and de Groat, A. 1940. Internal autoinfection in human stronglyoidiasis. *Am. J. Trop. Med.,* 20: 359–375.

Graham, G. L. 1936–1939. Studies on *Strongyloides* I, II. *Am. J. Hyg.,* 24: 71–87; 27: 221–234; III, *J. Parasitol.,* 24: 233–243; IV, *Am. J. Hyg.,* 30, D: 15–27; V, *J. Parasitol.,* 25: 365–375.

Napier, L. E. 1949. *Strongyloides stercoralis* infection, Parts 1 and 2. *J. Trop. Med. Hyg.,* 52: 25–30, 46–48.

Premvati. 1958. Studies on *Strongyloides* of primates. *Canad. J. Zool.,* 36: I: 65–77; II, 185–195; III, 447–452; IV, 623–628.

Sandground, J. H. 1925. Observations on *Rhabditis hominis* Kobayashi in the United States. *J. Parasitol.,* 11: 140–148.

1926. Biological studies on the life cycle in the genus *Strongyloides* Grassi, 1879. *Am. J. Hyg.,* 6: 337–383.

Swartzwelder, J. C., et al. 1958. Therapy of strongyloidiasis with Dithiazanine. *Arch. Internal Med.,* 101: 658–661.

Chapter 21

FILARIAE, SPIRUROIDS, AND GUINEA WORM

SUBORDER SPIRURATA. I. FILARIAE (SUPERFAMILY FILARIOIDEA)

The filariae, constituting the superfamily Filarioidea of the suborder Spirurata (see p. 398) are slender thread-like worms which inhabit some part of the blood or lymphatic system, connective tissues, body cavities, eye sockets, nasal cavities, etc. They have simple mouths without lips and rarely a vestibule; the females nearly always have the vulva forward near the mouth, and the relatively small males have spirally coiled tails, with or without alae but always with papillae. Many of them—all those that concern us here except *Parafilaria* in horses—produce embryos that live in the blood or skin, whence they are sucked out by bloodsucking arthropods which serve as intermediate hosts; they gain access to a new host through the skin when these arthropods bite.

The classification into families and subfamilies is still controversial. We shall follow Wehr's (1935) arrangement, as modified by Chabaud and Choquet (1953), who divided one of the families, Dipetalonematidae, into six instead of two subfamilies. Wehr recognized four families; of these three, Stephanofilariidae, Filariidae, and Dipetalonematidae, contain parasites of medical or veterinary interest. The first contains a single genus, *Stephanofilaria*, a skin parasite of horses, cattle, etc., with a row of small spines around the mouth. The other two, Filariidae and Dipetalonematidae, are distinguished mainly by the first-stage larvae, which are usually short and stout with spiny anterior ends in the Filariidae and long and slender with no spines in the Dipetalonematidae. Except *Parafilaria* and *Setaria* in horses and cattle, all the species parasitic in man or domestic animals belong to the Dipetalonematidae, distributed in three of the six subfamilies. The Dirofilariinae, with short tail, well-developed caudal alae in the males, and esophagus externally divided into separate muscular and glandular parts, contains *Loa* and *Dirofilaria;* the Dipetalonematinae, with long tail, very narrow caudal alae, if any, in the males, and no external division of the esophagus, contains *Wuchereria, Dipetalonema,* and *Mansonella;* and the Onchocercinae, with short tail, inconspicuous alae, and undivided esophagus, contains *Onchocerca.*

Guinea worms, *Dracunculus,* were formerly classed with filariae but are now placed in an entirely distinct suborder (see p. 398). At one time all the filariae were placed in the single genus *Filaria* and are sometimes still so referred to in medical and veterinary books.

Microfilariae. Many filarial infections are practically impossible to diagnose except by the embryos or "microfilariae," and it is therefore important to be able to distinguish these. When living they are colorless and transparent and may or may not be enclosed in "sheaths." In order to identify them it is usually necessary to stain them. The body will then be found to contain a column of nuclei, broken in definite places which serve as landmarks (Fig. 139). The principal ones are a nerve ring anteriorly, an excretory pore or "V" spot, an excretory cell somewhat farther back, a few genital cells posteriorly, and an anal pore or "tail spot." The spacing of these landmarks is fairly constant in different species. The presence or absence and arrangement of nuclei in the head and tail ends and the shape of the tail are also useful identification marks. The following table shows the outstanding characters of the microfilariae found in human blood or skin (see also Fig. 139).

Sheathed forms

Mf. bancrofti: about 225 to 300 μ by 10 μ; sheath stains red with dilute Giemsa stain; tail end tapers evenly; no nuclei in tail; does not stain with 1 : 1000 methylene blue when alive; lies in graceful coils when dried; nocturnal or non-periodic; in blood or urine.

Mf. loa: same size; sheath unstained in Giemsa; tail short and recurved, with nuclei to tip; stains with methylene blue when alive; lies in kinky scrawls when dried; diurnal; in blood.

Mf. malayi: about 160 to 230 μ by 5 to 6 μ; tail sharp-pointed, with a single nucleus at its tip and another 10 μ in front of it; nocturnal.

Unsheathed forms

Mf. perstans: about 200 μ by 4 μ; tail ends bluntly, with nuclei to its tip; stains with methylene blue when alive; no periodicity; in blood.

Mf. streptocerca: about 215 μ by 3 μ; tail ends in a crook and terminates bluntly with nuclei to tip; does not stain with methylene blue when alive; no periodicity; in skin.

Mf. ozzardi: about 200 μ by 5 μ; tail sharply pointed, with no nuclei at its tip; stains with methylene blue when alive; no periodicity; in blood.

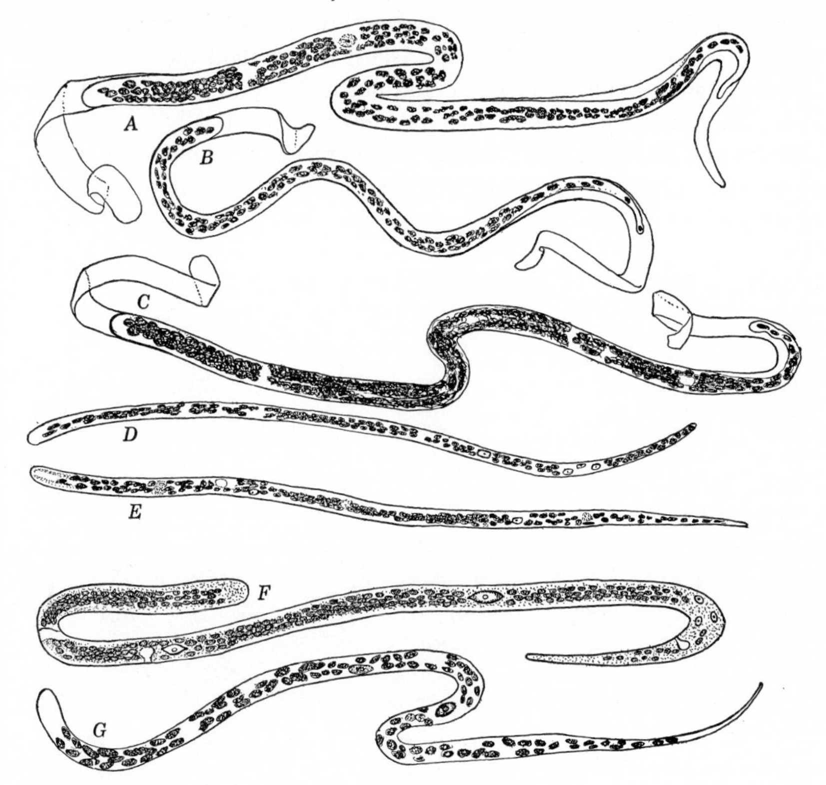

Fig. 139. Various species of microfilariae drawn to scale.

A, Wuchereria bancrofti; sheathed, no nuclei in tip of tail, 270 × 8.5 μ.
B, W. malayi; sheathed, 2 nuclei in tail, 200 × 6 μ.
C, Loa loa; sheathed, nuclei to tip of tail, 275 × 7 μ.
D, Dipetalonema perstans; no sheath, tail blunt with nuclei to tip, 200 to 4.5 μ.
E, Mansonella ozzardi; no sheath, pointed tail without nuclei at tip, 205 × 5 μ.
F, Onchocerca volvulus; no sheath, no nuclei in end of tail, 320 × 7.5 μ.
G, Dirofilaria immitis; no sheath, sharp tail without nuclei in end, 300 × 6 μ.

Mf. volvulus: about 300 to 350 μ by 5 to 8 μ; tail sharply pointed, with no nuclei at its tip; no periodicity; in skin.

Wuchereria (=Filaria) bancrofti

Distribution. This worm is a very widespread and important human parasite in warm countries but is not evenly distributed or uniformly prevalent throughout any country. As Augustine (1945) pointed out, it occurs almost entirely in coastal areas and islands where there is a fairly long hot season with high humidity. In Africa it is found on the Mediterranean and east and west coastal areas but not in the interior of Central Africa. In Asia it is prevalent on the coasts of Arabia, India, Malaya, Formosa, and north to China and the southern parts of Korea and Japan. It is prevalent in practically all the East Indian and South Pacific islands. It was formerly common in coastal Queensland but has almost disappeared in recent years.

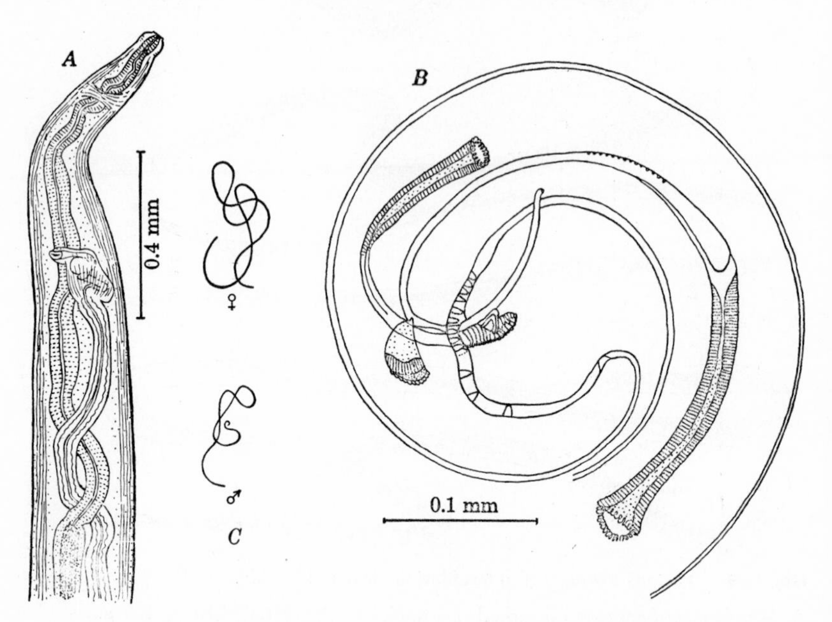

Fig. 140. *Wuchereria bancrofti.* *A*, anterior end of ♀; *B*, posterior end of ♂; *C*, adult ♀ and ♂, natural size. (*A*, after Vogel, *Arch. Schiffs- u. Trop.-Hyg.*, 32, 1928; *B*, after Fain, *Ann. parasitol. hum. et comp.*, 26, 1951.)

In the Western Hemisphere, where it was almost certainly introduced by Whites or Negroes, it is prevalent throughout the West Indies and on the northern coast of South America from northern Brazil to Colombia, but it is strangely scarce or absent on the Caribbean shores of Central and North America. In the United States it was once endemic in Charleston, South Carolina, but failed to become established elsewhere and has apparently died out there. Throughout this area it is almost completely restricted to towns and sometimes even to parts of them, but in some places it affects 80% or more of the local population.

Morphology. The adult worms (Fig. 140) live in the lymph glands or ducts, often in inextricable tangles. The females are 65 to 100 mm. long and only 0.25 mm. in diameter—about the caliber of coarse sewing thread; the males are about 40 mm. long and 0.1 mm. in diameter. The body tapers to a fine head slightly swollen at the end, with a simple pore as a mouth. The esophagus is partly muscular and partly glandular, with the vulva opening a little behind its middle. The males have the tail coiled like the tendril of a vine, with numerous pairs of papillae; there is one long and one short spicule.

Life cycle. The female worms give birth to microfilariae which are surrounded by delicate membranes or sheaths. Other characters are listed on p. 474. These microfilariae appear in the peripheral circulation chiefly between 10 P.M. and 4 A.M., except in the nonperiodic variety *pacifica* (see below).

There has been much speculation and experiment to determine the reason for this periodicity. One theory was that the larvae are concentrated in internal organs during the day, and circulate in the blood at night to keep a sort of tryst with their night-biting mosquito transmitters, chiefly *Culex quinquefasciatus* (=*fatigans*) or *C. pipiens* and certain species of *Anopheles.* Another theory was that the embryos are born at a certain time each day, and are then destroyed in the host within the next 24 hours. Recently Hawking and Thurston reviewed this matter and gave convincing evidence that, at least in the case of some filarial infections in monkeys and dogs, the microfilariae are concentrated mainly in the capillaries and other blood vesels of the lungs when not present in the peripheral circulation. The stimulus which induces the microfilariae to enter the general circulation is connected with periods of activity of the host, for it is gradually reversed in people who reverse their sleeping and working hours, or go halfway around the world.

Throughout the greater part of the range of *Wuchereria bancrofti* the microfilariae have nocturnal periodicity, i.e., appear in numbers in the peripheral circulation only at night, but in the Polynesian Islands (Samoa, Fiji, Tonga, and Cook Islands) there is a nonperiodic variety, with different transmitters and epidemiology, and differing somewhat in its pathogenic effects. In the Philippines this form coexists with the usual "periodic" form. Although the nonperiodic form is not morphologically distinguishable except for a slightly greater length of the adults, this form may properly be recognized as a variety or subspecies *pacifica;* Manson-Bahr and Muggleton (1952) think it deserves to be considered a distinct species.

The further development of the microfilariae depends on their being sucked with blood by certain species of mosquitoes which serve as intermediate hosts (see p. 756). Unlike malaria and yellow fever, *Wuchereria bancrofti* is not limited to transmission by species of one genus or group of mosquitoes; it is transmitted by certain species of *Culex, Aedes, Anopheles,* and others. *Culex quinquefasciatus* and in some places the closely related *C. pipiens* play a leading role as vectors of the periodic form in most parts of the world, but these are replaced by certain *Anopheles* in some places. *Aedes polynesiensis* is the principal vector of the nonperiodic type except in Fiji. *C. quinquefasciatus* is highly refractory as a transmitter of the nonperiodic form. For further details on transmitters see pp. 756–757.

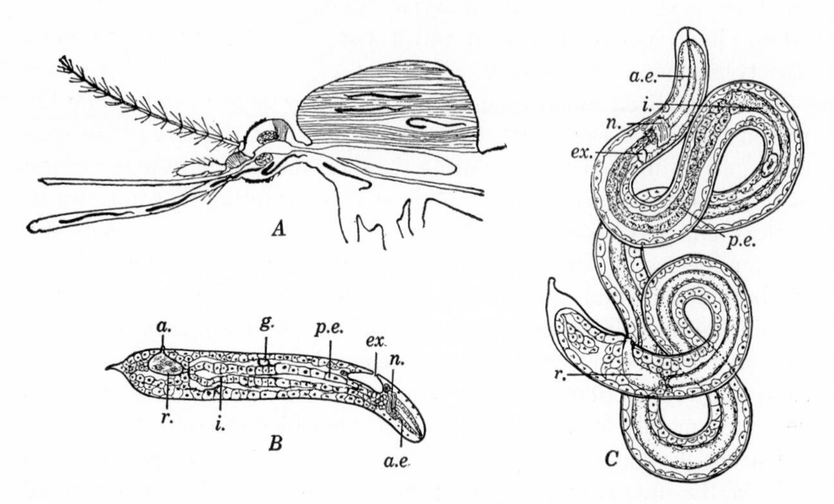

Fig. 141. *Wuchereria bancrofti.* *A,* mature larvae in thoracic muscles and proboscis of a mosquito (adapted from Castellani and Chalmers). *B,* "sausage" stage of development of larva in thoracic muscles; *C,* infective larva from mosquito proboscis (adapted from Looss). Abbreviations: *a.,* anus; *a.e.,* anterior portion of esophagus; *ex.,* excretory pore; *g.,* genital rudiment; *i.,* intestine; *n.,* nerve ring; *p.e.,* posterior portion of esophagus; *r.,* rectum.

In order to infect mosquitoes there must be about 15 or more microfilariae per drop of blood (20 cu. mm.); a high concentration of 100 or more per drop is fatal to the mosquitoes. Sometimes the blood contains up to 600 in a drop.

Shortly after being ingested by a mosquito the embryos penetrate through the stomach wall and migrate to the breast muscles, where they lie lengthwise between the muscle fibers (Fig. 141*A*). Here the body shortens to half its original length but grows several times as thick, thus changing into a sausage-shaped creature (Fig. 141*B*). Then the digestive tract differentiates, and the worms begin to grow in length as well as girth, eventually measuring about 1.5 to 2 mm. by 20 to 30 μ (Fig. 141*C*). During this time there have been two molts and the larvae have reached the infective stage. They now leave the thoracic muscles to make their way towards the head of the mosquito and down into the proboscis in the interior of the labium, although some get lost and end up in other parts of the body.

This development to the infective stage takes a minimum of 8 to 10 days but more frequently 2 weeks or more. The optimum conditions are 80°F. and 90% humidity. At best only a small percentage of the microfilariae ingested develop into infective larvae.

When the mosquito bites a warm moist skin the larvae break free from the labium where the labellum is joined, creep out on the skin of the host, and penetrate through the mosquito bite or other abrasions. This happens successfully only in warm, moist weather, for cold makes the larvae inert and dryness destroys them. Gradually, as the mosquitoes bite, the larvae escape from the proboscis until after about 3 weeks all are gone.

Nothing is known of the course pursued by the larvae after they enter the skin and very little as to the time required for sexual maturity to be reached. The large heart filaria of the dog, *Dirofilaria immitis,* matures 9 months after infection, and it is unlikely that the human filaria takes longer. The fact that in India children seldom show microfilariae in their blood under 5 years of age and Europeans only after many years of residence in an infected locality is due either to the scarcity of the embryos in the blood or to failure of the males and females to meet each other in the same glands or lymph ducts. Probably the adults live at least 4 or 5 years.

Pathology. Filarial symptoms are caused by the adult worms; the microfilariae usually produce no symptoms. The so-called signs and symptoms are due either to inflammatory reactions or to lymphatic obstruction.

It is very likely that the inflammatory effects are due largely to allergic reactions in sensitized tissues. They consist primarily in inflammation of lymph glands (lymphadenitis) and lymph channels (lymphangitis), particularly of the male genital organs (scrotum, spermatic cords, epididymis, and testes), and of the arms and legs. The attacks are usually recurrent, often being precipitated by exercise, and may be accompanied by chills, fever, aches, and general malaise. It is believed that the allergic irritation may be due either to fluid in which the embryos of the worms are discharged, or other metabolic products, or to proteins liberated from dead and phagocytized worms. Some workers, e.g., Grace, believe that hypersensitiveness to accompanying chronic *Streptococcus* infections is largely responsible for the symptoms. Failure of penicillin and sulfonamides to affect filarial lymphangitis is against this theory.

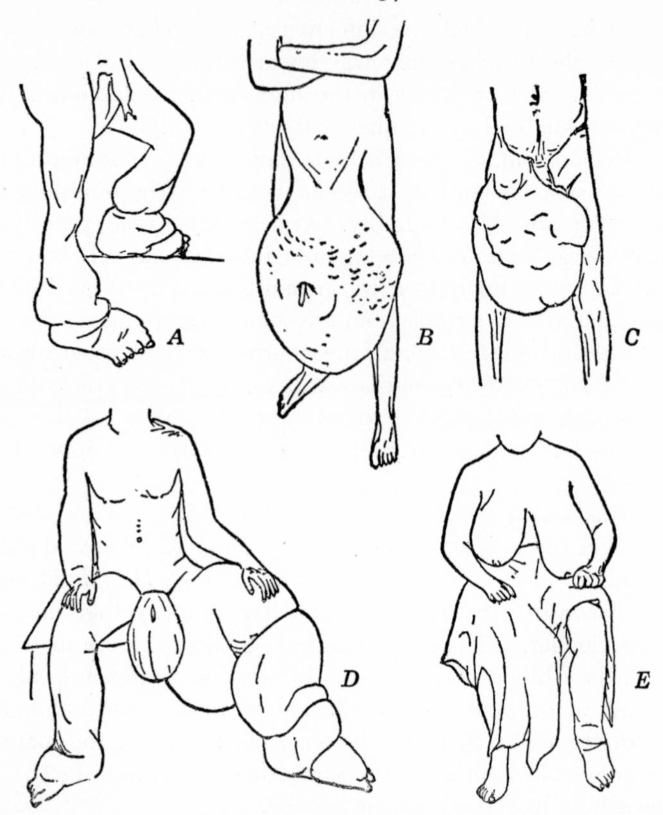

Fig. 142. Some extreme cases of elephantiasis. *A*, of legs and feet; *B*, of scrotum; *C*, varicose groin gland; *D*, scrotum and legs; *E*, of mammary glands. (*A* and *B* sketched from photographs from Castellani and Chalmers; *C*, *D*, and *E* from Manson.)

Obstruction of lymph channels may play a prominent part in the symptoms, especially in old infections. The dramatic end result of this is elephantiasis (Fig. 142), which, as Brown (1945) says, is popularly but mistakenly believed to be the inevitable final termination of every filarial infection. This belief caused a tremendous amount of unnecessary mental anguish and psychoneurosis during World War II among infected American troops in the South Pacific, who had visions of themselves ending up with anything from sterility to being attached to a 200-pound scrotum or leg.

The earliest obstructive effects are varicose lymph or chyle vessels behind places where lymph glands or channels are blocked by inflammatory tissue reactions. Such varices may burst and divert large amounts of lymph or chyle into the scrotum, bladder, kidney, or peritoneum, or even into the intestine. When obstruction occurs in the smaller lymph channels in the subcutaneous system and skin, especially in scrotum, limbs, breast, or vulva, the tissues become swollen and "blubbery." Eventually fibrous tissue increases and the skin becomes dense, hard, and dry, since the sweat glands also degenerate. This process may gradually increase until true elephantiasis appears, when certain parts of the body develop to monstrous proportions. It is a characteristic feature of obstructive forms of filariasis that microfilariae are commonly absent from the blood, either because they are dammed up in the lymph system or because the parental worms have died. There is a positive correlation between incidence of filarial disease and microfilaria rate in a community but a negative correlation between elephantiasis and blood microfilariae in an individual.

As Brown pointed out, any disease that may run its course for a period as long as 50 years is likely to vary greatly in its clinical course in different human hosts. Such factors as number of worms, speed with which they are acquired, and allergic sensitivity of the individual must be taken into account. Many cases never show any obvious symptoms. In a large group in the Virgin Islands, 20% had microfilariae in the blood, yet practically all were unaware of infection. One had 23,240 microfilariae per cubic centimeter of blood, yet had no signs or symptoms of filarial infection except a slight general glandular enlargement.

It has been common experience in India and other parts of the world that filarial symptoms are slow in appearing; in India, Europeans seldom show symptoms until they have resided in endemic localities for 10 to 15 years, and even native children seldom show symptoms until half grown. Sometimes, however, elephantiasis, once started, may develop rapidly. Brown saw a patient whose scrotum grew from normal size to a weight of 14 lb. in a year.

In contrast to all prior experience with filariasis, American troops exposed to the nonperiodic strain in Samoa and other South Pacific islands during World War II developed filarial symptoms in as short a time as 3½ to 6 months and in an average of 9 months (Dickson, Huntington, and Eichold, 1943). This disease, called by the native name "mu-mu," was characterized by lymphangitis, enlarged glands, swelling, and redness, most frequently in the genitals or arms and less often in the legs. Headache, backache, fatigue, and nausea were common, but fever and malaise were unusual; physical and mental depression was very pronounced. Microfilariae appeared in the blood in very few cases.

It seems probable that the differences between this rapidly developing disease and the slow-developing filariasis of other parts of the world was due to intensity of infection and consequent early development of strong allergic reaction. In most places in the tropics Europeans are segregated from infected natives at nights and protect themselves from mosquitoes sufficiently to escape heavy infections. In the Pacific islands, where the abundant day-biting *Aedes polynesiensis* is the transmitter, men working or fighting in or near native villages may get as many infective bites in a month as they would get in India in years. A very interesting and possibly significant fact is that in islands where only the nocturnal strain exists, e.g., North Guinea, few or no cases developed among white troops.

Little work has been done on immunity to filarial infections, but work by Scott and McDonald (1953) on *Litomosoides* in cotton rats indicates that as in other worm infections immunity is directed against metabolic products of the worms rather than against the body proteins (see p. 28).

Diagnosis. If microfilariae are present they can usually be demonstrated (in night blood in the nocturnal strain) by examination of a fresh drop of blood for squirming microfilariae, or of a dehemoglobinized and stained thick smear (see p. 186). A more accurate method in case the embryos are scanty is to take 1 cc. of blood in 10 cc. of 2% formalin, centrifuge, and examine the sediment. For specific identification the embryos should be stained by Giemsa or Wright methods.

Since microfilariae are frequently absent, especially in elephantiasis cases, clinical signs and symptoms must be relied on to a considerable extent. Skin tests with antigen prepared from *Dirofilaria immitis* or other filariae, since there is very little specificity, are very helpful. Injection of 0.01 cc. of a 1 : 8000 dilution gives positive reactions in most cases and a minimum of false positives, though many of the positives are not clinically active cases. False positives are probably due, as Augustine and Lherisson (1946) pointed out, to sensitization of man by larvae of nonhuman filarial worms, to which he must often be exposed. Negative skin reactions are helpful in ruling out filarial infections, though they sometimes occur in active cases, probably due to desensitization.

Treatment and prevention. Filariasis apparently balked all efforts to treat it until World War II, when several American workers (Brown, 1944, and Culbertson et al., 1945, 1947), following up successful experiments on *Dirofilaria* in dogs and on *Litomosoides* in cotton rats, found that a number of antimony and arsenic compounds given over a period of 2 weeks or more greatly reduced or completely eliminated the infections. The antimony compounds (especially Neostibosan) quickly kill the microfilariae but have slower effects on the adults; the latter have their reproductive organs injured, resulting in eventual sterility as in the case of antimony-treated schistosomes, but the response in some cases is disappointing. The arsenic drugs, of which arsenamide is most promising, have a slower effect on the microfilariae, but kill or sterilize the adult worms rather quickly. Arsenamide has been found very effective against both the periodic form of *W. bancrofti* (in Virgin Islands) and the nonperiodic form (in Samoa) (Otto, Brown et al., 1952; Otto et al., 1953), but has the disadvantage of having to be given intravenously daily for 15 days. The minimum curative dose is probably 0.6 mg. per kilogram daily for at least 10 to 12 days. The diarrhea and nausea which it sometimes causes can be relieved by giving ascorbic acid.

Hetrazan, a piperazine derivative given by mouth, very rapidly kills the microfilariae, and probably has a slow and gradual effect on the adult worms. The microfilariae disappear completely or nearly completely in a few hours and often fail to reappear for many months, whereupon they gradually return. In Samoa, however, Otto et al. found a number of treated cases showing few microfilariae, sometimes none, even after 2 years. Evidently in these cases the adult worms were either destroyed or sterilized. The dosage used is 3 mg. per kilogram body weight daily for 7 to 14 days. The principal disadvantage is that allergic symptoms always develop as the result of the sudden destruction of the microfilariae and liberation of their proteins in a sensitized body. The symptoms vary from brief chills and fever to almost complete prostration with severe headache, muscular aches, dizziness, sweating, etc. Suramin (Bayer 205 or Antrypol) also has some effect on *Wuchereria* infections, but is too toxic for routine use.

Some workers believe that elephantiasis is brought on by dead filariae, so the wisdom of killing the adult worms by chemotherapy has been questioned; but no evidence of elephantiasis has appeared in cured patients. In some cases of elephantiasis Knott (1938) has obtained good results from pressure bandaging. In some cases surgery can be used to advantage, as Auchincloss showed in 1930. Cortisone (100 mg. daily for several weeks) is helpful. Kessel has described the *joie de vivre* of patients who have experienced relief through cortisone therapy.

Control is largely a matter of mosquito control, and this varies with the local transmitters. The predominantly urban *Culex quinquefasciatus* can be controlled by local elimination of breeding places and DDT spraying; the latter is also effective against some of the *Anopheles* transmitters. The *Aedes* of the *scutellaris* group that transmit the variety *pacifica,* on the other hand, seldom enter houses and almost never rest in them, so they cannot be controlled by DDT. Here wholesale use of Hetrazan to eliminate most of the microfilariae may be the best solution.

Wuchereria malayi

Although *Wuchereria bancrofti* was long thought to be the only filaria commonly responsible for lymphangitis and elephantiasis, it has been found that in many places this species plays second fiddle to another species that was long known only by the embryo, *Microfilaria malayi* (see p. 475 and Fig. 139). The adult, which resembles *W. bancrofti* closely, was first found in 1940 by Rao and Maplestone in India.

W. malayi is common in many places in India and in southeastern Asia and the East Indies, sometimes along with *bancrofti,* sometimes alone; it may affect up to 50% of the rural population.

Recently various wild and domestic animals in Malaya, India (Orissa), and the Kenya coast of Africa have been found with *malayi*-like infections. In Malaya, it has been shown that *W. malayi,* and a closely related second species *W. pahangi,* occur in monkeys, slow loris, dogs, cats, tigers and other wild felines, and ant-eaters. Edeson and Wharton (1958) showed that *W. malayi* is readily transmissible from man to cats and various other animals and that 34% of the cats in one Malayan area were naturally infected. It is apparent that animal reservoirs are important in maintaining *W. malayi,* a facet of this disease which has been appreciated only in the past 5 years. In 1958 Buckley and his associates described *Wuchereria patei,* a *W. malayi*-like form, from Pate Island, Kenya, in cats and dogs. *W. malayi* has not been found in man in this region, although *W. bancrofti* infections are not uncommon.

The transmitting mosquitoes are mainly species of *Mansonia* (see p. 756) and sometimes *Anopheles.* Since *Mansonia* lives in swamps, with the larvae and pupae attached to the roots of water plants, the disease is for the most part rural. *W. bancrofti* infections increase toward the center of towns, *malayi* infections peripherally. In India *Mansonia annulifera* is the principal vector, and so the disease can be controlled by the delightfully simple method of removing *Pistia* (water lettuce) on which this mosquito lives almost exclusively (Iyengar, 1938). In Malaya, however, the chief vector is *M. longipalpis,* which pierces the fine roots of swamp-loving trees, and so only extensive drainage is effective. In 1958, Wilson et al. reported that the strain of

W. malayi transmitted by *Mansonia* showed a number of differences from the strain transmitted by *Anopheles*. The *Anopheles*-transmitted strain shows, among other features, marked microfilarial periodicity and develops poorly in cats, whereas *Mansonia*-transmitted strain is essentially nonperiodic and is established readily in cats. Further study is required to determine the significance of these observations.

The pathogenic effects of *malayi* infection are similar in most respects to those of *bancrofti* infection, but the elephantiasis is more frequently in the legs, and the genital organs are rarely affected. In Travancore, microfilariae were occasionally found in children only 2 years old, and elephantiasis was seen in a child of 6. In miltary personnel repatriated from Indo-China, a condition of "tropical eosinophilia," with swelling and inflammation of lymph glands and bronchial and pneumonic symptoms, has been observed.

The African Eye Worm, *Loa loa*

This worm is a common parasite of man in west and central Africa. In 1936 Sandground examined adult worms collected from monkeys in Belgian Congo and believed them to be *Loa loa*. Some years later, Gordon (1955) showed that *L. loa* of human origin can be transmitted to monkeys. Duke and Wijers (1958) concluded that the *Loa* of naturally infected monkeys in the Cameroons is probably of little significance as a reservoir of human infection since the vectors involved in monkey transmission live in the forest canopy and rarely come in contact with man. The adults live in the subcutaneous tissue of man and make excursions from place to place under the skin, causing itching and a creeping sensation; they show a special preference for creeping in and about the eyes (Fig. 143*D*), and are responsive to warmth. In a person sitting before a fire the worms become active and move to exposed parts; they have been observed to travel at the rate of about an inch in 2 minutes.

The adult worms resemble pieces of surgical catgut, the female varying from about 20 to 70 mm. in length, whereas the males measure about 20 to 35 mm. The general anatomy is not unlike that of *Wuchereria bancrofti*, but the cuticle is provided with numerous little dewdrop-like warts along the lateral lines (Fig. 143).

Loa loa produces sheathed embryos (see p. 474 and Fig. 139*C*) which make their way to the blood stream. They have a diurnal periodicity, swarming in the blood in the daytime and disappearing at night. The intermediate hosts are certain species of *Chrysops* (*C. dimidiata, C. silacea,* and possibly others), known as mango flies (see

p. 690). The larvae develop in the fly's abdomen, sometimes by hundreds, and invade the proboscis after development to the infective stage, which takes 10 to 12 days. When the fly bites, the larvae file out of the proboscis and enter the skin through the bite.

Loa worms seem especially active in their youth, later showing a tendency to retire to deeper parts of the body. In the eye they are painful, but it is here that they can most easily be extracted. The extraction, however, has to be done expeditiously, before the disturbed worm flees to hiding places deeper in the body.

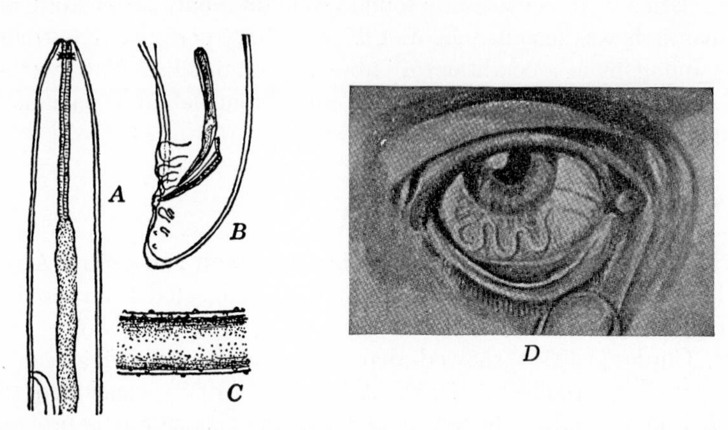

Fig. 143. *Loa loa: A,* anterior end, showing muscular and glandular parts of esophagus and position of vulva (×20); *B,* tail of male, showing spicules, narrow alae, and papillae (×100); *C,* portion of body showing dewdrop-like warts on cuticle along lateral lines; *D, Loa loa* in eye. (*A, B,* and *C* adapted from various authors; *D,* after Fülleborn in Kolle u. Wassermann, *Handbuch der path. Mikro-org.,* Vol. 6, 1929.)

Loa infections are usually accompanied by painless though sometimes itchy edematous swellings, commonly as large as pigeon eggs, which appear suddenly, last a few days, and then disappear to reappear later somewhere else. These "Calabar swellings" are often more troublesome a few months after removal to a cold climate than they are in West Africa, so much so that they take the joy out of leave trips home for some Europeans. The swellings are undoubtedly allergic reactions to metabolic products of the worms or to proteins liberated from injured or expired worms. Chandler, Milliken, and Schuhardt (1930) produced a large swelling by injection of a minute amount of *Dirofilaria* antigen into the skin of a patient. Kivits (1953) reported *Loa* microfilariae in the cerebrospinal fluid in four cases of fatal encephalitis. Possibly the microfilariae cause serious effects when they penetrate into the brain or spinal cord, as do *Setaria* larvae (p. 495), or perhaps they open the door for neurotropic viruses or even carry them in.

Hetrazan, even in small doses, kills the microfilariae of *Loa,* and relieves symptoms, but affects the adults slowly if at all.

Dipetalonema (=*Acanthocheilonema*) *perstans* and *D. streptocerca*

The genus *Dipetalonema* (including the old genus *Acanthocheilonema*), contains many species found in monkeys and small animals. Newton and Wright (1956) reported for the first time that a species of *Dipetalonema,* perhaps *D. reconditum,* is a relatively common parasite of dogs in the United States. The same authors (1957) showed that dog fleas serve as intermediate host for this worm. As yet,

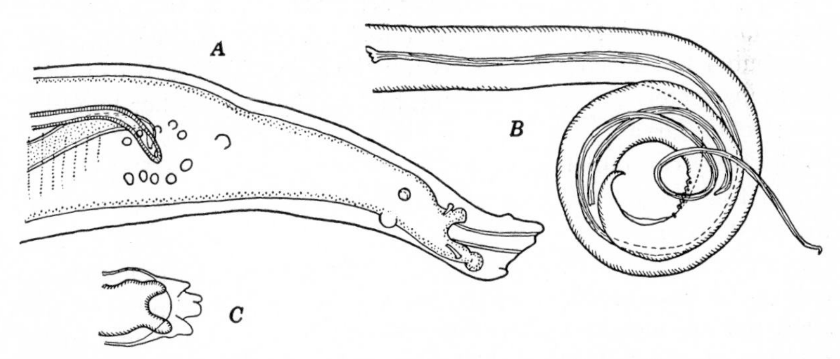

Fig. 144. *D. perstans: A,* posterior end of ♂, ventral view (large spicule cut off at level of cloaca); *B,* same, lateral view; *C,* tip of tail of ♀. (After Chabaud, *Ann. parasitol. hum. et comp.,* 27, 1952.)

there is no information concerning its pathogenicity in the dog, but it seems rather clear that microfilariae of this worm must have been mistakenly diagnosed as the pathogenic *Dirofilaria immitis* on literally hundreds of occasions. *D. perstans* (Fig. 144) is a very common parasite of man and apes in rain forests of west and central Africa. The incidence of infection increases with age, and in some areas in Congo, Uganda, and Cameroons practically all elderly people show microfilariae in the blood. It has also become established in northern South America and northern Argentina. The microfilariae (Fig. 139*D*) exhibit no periodicity. The threadlike adults live in deep connective tissue; the males are about 35 to 45 mm. long and only 60 μ in diameter, and the females 70 to 80 mm. and 120 μ in diameter.

The infection seems to produce no evident symptoms, at least in the majority of cases, but Enzer observed cases of persistent headache and drowsiness in individuals whose blood was teeming with the embryos and in whom there was no other evident cause for the symptoms. Duke in 1957 reported symptoms suggesting an allergic reaction, with a skin rash and eosinophilia. Others have observed continuous fever. Sharp (1928) showed that the intermediate hosts in the Cameroons are minute nocturnal midges, *Culicoides austeni* and *C. grahami* (see p. 676). Chardome and Peel (1949) found that *Culicoides grahami* readily ingested skin-inhabiting microfilariae of *Dipetalonema streptocerca* (see below) but not those of *D. perstans* and questioned *Culicoides* being a vector of *perstans*, but later work in British Cameroons confirmed Sharp's work (see p. 678). Development takes place in the breast muscles of the flies, and infective larvae invade the head in about 8 or 9 days. Since the vectors are very numerous but only harbor a few infective larvae, there is a high incidence of light infections, whereas in *Loa* infections (see above) unevenly distributed but heavier infections are the rule. This is because the *Chrysops* vectors are less abundant, but one fly may harbor hundreds of infective larvae.

There are contradictory reports on the effectiveness of Hetrazan in eliminating the microfilariae of *D. perstans;* possibly this varies with the location of the adult worms.

Another species of *Dipetalonema, D. streptocerca,* was long known only by the microfilariae (see p. 474), which resemble those of *D. perstans* but are usually longer and more slender and are found in the skin like those of *Onchocerca.* It occurs in 2 to 100% of natives in some parts of west and central Africa, and also occurs in apes. Only a few adults of this species have been found. No symptoms can definitely be ascribed to it. Similar larvae were found in six of eleven chimpanzees in Belgium Congo; the adult worms were located in connective tissue. The intermediate host was shown by Chardome and Peel (1949) to be *Culicoides grahami.*

Mansonella ozzardi

This worm, related to *Dipetalonema perstans,* is common in parts of the West Indies, Yucatan, Panama, and neighboring coasts of South America; it is also present in 25 to 30% of the people in northern Argentina. The adults, found in the mesenteries or visceral fat, are about the size of *Wuchereria bancrofti;* the females are characterized by a pair of flaplike processes with fleshy cores at either side of the tail. Only one single incomplete male has ever been found. The microfilariae (see p. 474 and Fig. 139*E*) are much like those of *D. perstans* but differ in having pointed tails without nuclei. There is no evidence that the worm is pathogenic although infected persons are reported to have an increase in blood eosinophiles. Buckley (1934) showed that the intermediate host in St. Vincent, W. I., is *Culicoides furens* (see p. 676), the development being similar to that of *D. perstans;* it is completed in about 7 or 8 days.

Onchocerca

The members of the genus *Onchocerca* are long, threadlike filarial worms which live in the subcutaneous and connective tissues of their hosts, where they are usually imprisoned in tough fibrous cysts or nodules. The females are so extremely long and hopelessly tangled that it is very difficult to get entire specimens. In man the females sometimes reach a length of 500 to 700 mm. (over 2 ft.), and in cattle twice this length or even more, with the diameter of a coarse sewing thread (about 0.3 to 0.4 mm.). The males are very small by comparison, about 20 to 50 mm. long, with a diameter of 0.2 mm. The microfilariae (see p. 474 and Fig. 139*F*) are sharp-tailed and unsheathed, and differ in that they do not enter the blood stream but localize in the skin and eye tissues.

A number of species have been described from horses, cattle, antelopes, and man, but they are very difficult to distinguish, and some have been differentiated mainly on the basis of the usual location of the nodules in the host's body. All the species are recognizable by the presence of thickened ridgelike rings on the cuticle (Fig. 145*A*), much more conspicuous in females than in males. The male has a coiled tail bluntly rounded at the tip and provided with papillae but no alae (Fig. 145*B*); there are two unequal spicules. The females have the vulva near the end of the esophagus and have a bluntly rounded tail.

Some species, especially *Onchocerca gibsoni,* injure the hides and carcasses of cattle by the hard nodules that form. Another, *O. reticulata* (=*cervicalis*), inhabits the neck ligament of horses, causing "poll ill" and fistulous withers; the microfilariae cause papular, itching skin sores. This species occurs in the United States. The intermediate hosts of these species of *Onchocerca* of horses and cattle are species of *Culicoides* (see p. 676). This may also be true of *O. armillatus,* which causes aneurysms in the aorta of cattle in Africa.

Onchocerca volvulus

Human onchocerciasis is caused by *O. volvulus,* which occurs in southern Mexico, Guatemala, Salvador, and northwest Venezuela in the Western Hemisphere, and in central Africa and Yemen in the Old World. It was probably originally an African infection introduced rather recently into Central America, where it was not discovered until 1915. In some localities 80 to 100% of the people harbor this worm, and 5% of them lose their sight.

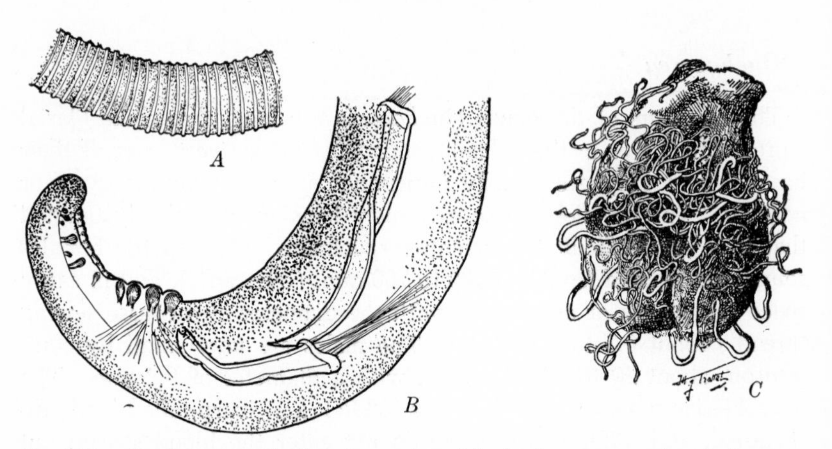

Fig. 145. *Onchocerca volvulus: A,* portion of body showing annular thickenings; *B,* tail of male, showing spicules and papillae; *C,* an opened *Onchocerca* nodule showing tangled worms inside. ×2. (*A* and *B* adapted from Fülleborn, in Kolle u. Wassermann, *Handbuch der path. Mikro-org.,* Vol. 6, 1929. *C,* after Brumpt, *Précis de parasitologie,* 1949.)

Life cycle. The developing worms creep about in the subcutaneous tissue, but when they come to rest there is an inflammatory reaction which results in the formation of the characteristic fibrous cysts; in one instance a nodule was found in a child 2 months old, but usually a somewhat longer time is required for them to appear. They may grow to a diameter of 1 cm. in a year, but usually the growth is slower. Strong et al. in Guatemala usually found 3 or 4 worms in a nodule, but in Africa there are composite nodules containing more than 100 worms. The worms lie in tangles in the cysts (Fig. 145*C*), which vary from the size of a pea or smaller to that of a pigeon's egg; usually a swarm of microfilariae is present also. In most localities infected people have only one to half a dozen nodules, but in some places in Africa 25 to 100 nodules are commonly seen, most of them only a few millimeters in diameter. There is ample evidence, however, that not all of the adult worms become encapsulated. The adults are long-lived; in Kenya microfilariae were still present in the skin seven years after reinfection was stopped by eradication of the intermediate host.

The microfilariae, 250 to 360 μ long and unsheathed (Fig. 139*F*), escape readily from the prisons which enclose their parents and make their way, not into the blood stream, but into the connective tissue just under the skin, where they accumulate in large numbers. Sometimes they emerge by hundreds when a bit of excised skin is placed in a saline solution for two or three hours.

The intermediate host in Africa was shown by Blacklock in 1926 to be a species of blackfly, *Simulium damnosum* (see p. 684); *S. neavei* is a vector in much of Congo and, before its recent eradication, transmitted the disease in Kenya. In Guatemala and Mexico *S. ochraceum* seems to be the only really anthropophilic species, and onchocerciasis is common only where this species is present; but *S. metallicum* also bites man, and *S. callidum* does so to a less extent. Several other species may be locally important. The microfilariae of *Onchocerca* are seldom found below the knee, where *S. metallicum* and some other species usually bite. It has been suggested that the *Onchocerca* infections often found in wild-caught specimens of *metallicum* may represent *O. gutturosa* of cattle or *O. reticulata* of horses, which are also present. However, as far as is known, the latter species, and perhaps some of the cattle species also, are transmitted by *Culicoides* (see p. 676).

When the fly is biting, its salivary secretions attract the microfilariae from adjacent areas of skin so that even 100 to 200 may be ingested in a single meal. Rapid development takes place in the thoracic muscles; infective larvae are produced in 6 to 7 days, according to Wanson (1950).

Epidemiology. In central Africa the infection has a wide distribution, largely coinciding with that of *Simulium damnosum* and *S. neavei*, but in America it is mostly limited to a narrow strip on the Pacific slope in Guatemala and southwestern Mexico, between about 2000 and 4500 ft. elevation, where coffee is extensively grown, and where there are numerous small, shaded, trickling streams arising from springs, in which *S. ochraceum* breeds.

In Mexico 20,000 people are affected in Chiapas and 11,000 in Oaxaca, but apparently the infection has not yet spread over all the areas where it could thrive. Dampf (1942) called attention to the danger of its spread along the Pan-American Highway, which passes through foci in both Mexico and Guatemala. The danger is greatest to natives since, as in the case of *Wuchereria bancrofti*, harmful effects develop only after continued exposure to infection.

Pathology. As already noted, the outstanding feature of onchocerciasis is the development of fibrous nodules enclosing the worms (Fig. 144*C*). In parts of Africa the nodules are largely confined to the trunk, especially just over the hips and on the knees, elbows, ribs, etc., but in some regions of the Belgian Congo, and especially in Central America, they are commonly found on the head. In Guatemala about 95% are on the head, especially about the ears. The location of the nodules seems to be influenced by pressure on the skin, either by bones or by hats or clothing, which might temporarily make the going hard for the migrating worms and impede them long enough for the tissues to start the imprisoning process. The site of the bites of the intermediate hosts is certainly not the determining factor.

Ordinarily the nodules are not painful, and seldom suppurate, so usually give very little trouble. The microfilariae, however, which creep in the skin, not necessarily in the immediate vicinity of the nodules, cause other disturbances. In Africa, onchocerciasis is commonly associated with a peculiar thickened, scaly, lizardlike skin, especially around the middle part of the body, but not on the head, whereas in Central America there are more likely to be erysipeloid rashes on the head (coastal erysipelis). A more marked difference between the infection in Africa and America is the degree of lymphoid involvement. In Africa enlarged lymph glands containing microfilariae, lymph scrotum, and elephantiasis of the scrotum and legs are frequently associated with *Onchocerca* infections, even where *Wuchereria bancrofti* is absent. The reasons for these differences have not been explained. It has been suggested that perhaps many of the adults, particularly in the African strain, do not form nodules and remain undetected in the body.

The most serious complication of the disease is interference with the eyes, often ending in blindness. This is very prevalent in the endemic zones of Guatemala and Mexico, where in some localities 10 to 25% of the population suffer from partial or total blindness. This is also true in some places in Africa, although in some localities where the nodules are mainly on the trunk, eye disturbances are much less frequent. As recently as 1949, Brumpt and Chabaud concluded that the ocular complications do not occur in Europeans, and the general impression has been that eye disease only appears after long years of infection. However, Woodruff and Murray in 1958 described eye involvement in 34% of 72 European patients, some of whom had been in the tropics a relatively short time. In contrast, there is a very low incidence of ocular onchocerciasis although almost 100% of the population is infected.

Strong (1934) carefully investigated this condition in Guatemala and found that the embryos escaping from nodules on the head had a tendency to invade the tissues of the eye—conjunctiva, cornea, iris, and other parts, sometimes even the optic nerve. Eye disturbances usually occur among adults with a history of nodules extending over 4 or 5 years or more. The lesions are chronic and progressive, beginning with injection of the conjunctiva, inflammation of conjunctiva and cornea, and development of opaque spots which run together. These lesions are due partly to irritation set up by the continual passage of numerous embryos through the eye tissues, and probably in part to allergic irritation. After the corneal tissues of the eye have become opaque, complete restoration of sight is not possible, but there is often some degree of improvement, or at least arrest of further harm, after removal of nodules on the head or destruction of microfilariae by Hetrazan treatment.

Diagnosis. Diagnosis can usually be made by puncturing and aspirating a nodule, by excising a small piece of skin with a razor (preferably not deep enough to draw blood), or by applying a cover-slip to the blood-stained exudate pressed out after four or five superficial scratches with a sharp instrument. Skin from the shoulder region, around the umbilicus, or in the vicinity of nodules is likely to provide the largest number of microfilariae, regardless of the situation of the nodules. Examination of fed blackflies (xenodiagnosis) may be an even better method when feasible. Precipitin and skin reactions to filarial antigens are unreliable, but a skin reaction following a single dose of Hetrazan is of diagnostic value.

Treatment. The most effective treatment is excision of the nodules, which is usually possible. By systematically doing this, the amount of infection has been markedly reduced in Mexico.

Only two drugs have thus far proved useful in treatment. One, Hetrazan (see p. 483), has spectacular effects on the microfilariae, completely destroying them in a few hours, but it has a slow and unreliable effect on the adult worms, so the microfilariae eventually return. The chief disadvantage is that the rapid destruction of the microfilariae commonly brings on severe allergic reactions—fever, joint pains, inflamed and itchy skin, enlarged lymph nodes, and irritation of the eyes. These symptoms reach their height in 12 to 15 hours, and may be severe enough so that the treatment is considered worse than the disease. Use of small doses to begin with, together with antihistaminic drugs may alleviate this trouble. The recommended dosage is 2 mg. per kilogram three times a day for 2 or 3 weeks, or 10 mg. per kilogram once daily for 1 week, repeated every 6 months. The other drug is Suramin (Bayer 205), 1 gram each week for 5 weeks. This kills the adult worms; the microfilariae then gradually disappear in the course of several months, and the nodules shrink. But this drug is very toxic, produces severe reactions, and requires intravenous injection. Present indications are that Suramin treatment following Hetrazan may be a good procedure. Systematic "denodulization" is still the safest and best method of treatment.

Prevention. Among natives exposure to bites of blackflies is unavoidable. Systematically destroying the parasites or nodules in human beings in more or less circumscribed foci, as in Mexico and Guatemala, might be possible, but it is thought to be impracticable in Africa. Animals are not believed to constitute important reservoirs, since the parasites, though morphologically indistinguishable, seem to be biologically distinct.

A better alternative is elimination of breeding places of blackflies or treatment of them with larvicides (see p. 686), which has been done with remarkable success in Mexico and Guatemala, and also in areas in Africa.

Other Filariae in Man

Scattered cases of a number of other adult or immature filariae that are of doubtful nature or unknown affinities are on record. Most of these are tentatively referred to as *Dirofilaria conjunctivae* by Faust (1957) who has reviewed the human infections. *D. conjunctivae* has been found in cystlike tumors of the eye, nose, arm, and mesentery in Europe, India, U. S. S. R., and Thailand. Three cases have been reported from Florida. Some or all of the Old World cases may be *D. repens,* a parasite of dogs, but it seems more likely that the Florida cases are another species, perhaps normally in wild animals, e.g., *D. tenuis* in raccoons or *D. scapiceps* in rabbits. There is a record from Brazil of a filaria from the heart, *D. magalhaesi,* and Faust recorded a single male *Dirofilaria* from the inferior vena cava of an elderly woman in New Orleans; these may be identical with *D. immitis* which lives in the heart of dogs.

In many tropical areas, people frequently suffer from a disease variously known as tropical eosinophilia or pulmonary eosinophilia. A persistent cough, difficulty in breathing, and a blood eosinophilia are usual symptoms. In 1956 Danaraj showed that the administration of filaricidal drugs resulted in a dramatic disappearance of symptoms and he suggested that the disease was due to filariae of animal origin which cannot develop satisfactorily in man. Buckley (1958) and Danaraj (1959) furnished further evidence that tropical eosinophilia is a form of larva migrans (see p. 459).

The possibility of larval filariae sometimes making their way to the human central nervous system and causing nervous disturbances has been pointed out by Innes and Shoho (1953), who showed that lesions in the brain, spinal cord, or eye of horses, sheep, and goats are due to invasion by larvae of species of *Setaria* (see below) in unnatural hosts; these lesions result in lumbar paralysis or, when in the eye, a disease called kumri. Whitlock (1952) found immature filariae of a new species, *Neurofilaria cornellensis,* in the central nervous system of sheep in New York State suffering from a similar disease. Sprent (1955) reviewed the pathology associated with invasion of the nervous system by nematodes.

Filariae in Domestic Animals

Except for *Onchocerca* infections in cattle and horses (see pp. 489–490) and occasional injury to horses from *Onchocerca* infections in the neck ligament, larger domestic animals suffer relatively little from filarial infections. *Setaria equina* and *S. labiato-papillosa* are often found in the peritoneal cavities of horses and cattle, respectively, but do no appreciable damage except when, during an early period of wandering through the tissues in abnormal hosts, they enter the eye or central nervous system (see preceding paragraph). Innes (1953) quotes a Korean report which states that mosquitoes (*Anopheles sinensis, Armigeres obturbans,* and *Aedes togoi*) are transmitters, but *Stomoxys* is said to transmit the cattle species. In the central and western states another species, *Stephanofilaria stilesi,* causes skin sores in cattle and sometimes in goats and pigs; in India *S. assamensis* causes "hump sore" in cattle, and sometimes ulcers in the ears. It is a small worm, the females only 6 to 8 mm. and the males 2 to 3 mm. long, with cuticular spines behind the mouth.

In the Old World horses are afflicted by *Parafilaria multipapillosa,* and cattle by *P. bovicola;* the females are 40 to 70 mm. long and the males about 30 mm. As in the genus *Filaria* the vulva opens just beside the mouth. These worms live in subcutaneous tissue and pierce the skin to deposit their embryonated eggs, causing "summer bleeding" from small nodules, and injuring the hides. Muscoid flies feeding on the blood suck up the eggs and serve as intermediate hosts. Sheep suffer from sores on the head caused by *Elaeophora schneideri,* the adult of which lives in the internal maxillary arteries.

Heartworm of dogs (*Dirofilaria immitis*). Dogs suffer severely from this worm, which usually inhabits the right ventricle of the heart and the adjacent parts of the pulmonary arteries. The females are 20 to 30 cm. long, the males about 12 to 18 cm. The microfilariae are unsheathed and show a partial periodicity. The infection is found in all warm climates and has been reported frequently in southern United States, heavy infections being limited to coastal areas. There

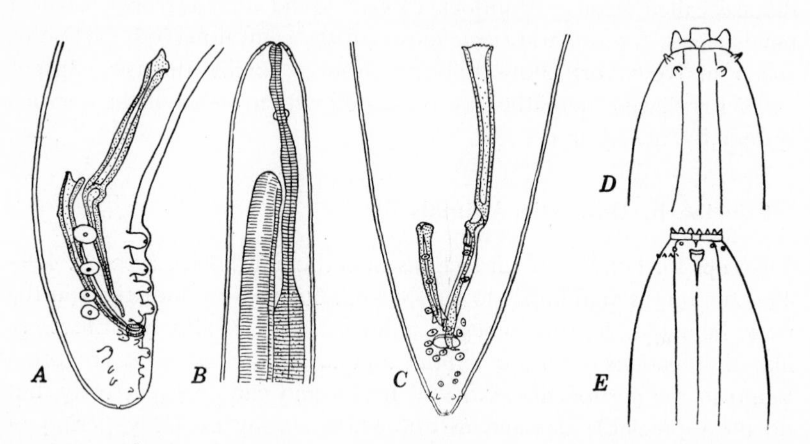

Fig. 146. Filariae of animals. *A, Dirofilaria immitis* of dogs, posterior end of ♂; *B*, same, anterior end of ♀; *C, Setaria equina*, posterior end of ♂; *D*, same, anterior end; *E, Stephanofilaria*, anterior end. (*A–C*, after Mönnig, *Veterinary Helminthology and Entomology*, Williams and Wilkins. *D* and *E* after Whitlock, *Practical Identification of Endoparasites for Veterinarians*, Burgess.)

is recent evidence that, in the past, it may have been confused so frequently with a *Dipetalonema* in dogs that information on incidence in United States dogs may be unreliable (see p. 487). Various mosquitoes, especially certain species of *Aedes*, are intermediate hosts and *Anopheles quadrimaculatus* has been shown experimentally to be an efficient host for *D. immitis.* Five other species of *Dirofilaria* have been shown to develop in mosquitoes. The heavy infections near the coast are possibly due to salt-marsh mosquitoes serving as vectors. Development takes place in the Malpighian tubules. Fleas have been reported to serve as intermediate hosts, but the studies of Newton and Wright (1957) cast considerable doubt on this.

The adult worms usually remain in the right ventricle, but may spread into the pulmonary arteries when excessive numbers are present. Pulmonary circulation is interfered with, and dogs with fifty or more worms cough and quickly show respiratory difficulties on exercise, or may collapse entirely. Treatment with antimony compounds, especially Fuadin, kills the microfilariae and eventually kills or sterilizes the adults. Arsenamide kills the worms but has little or no effect on the microfilariae. Hetrazan quickly kills the microfilariae, but its effect on adult worms is questionable. Since dead worms tend to enter and clog the pulmonary arteries, exercise should be reduced to a minimum during and for two months after treatment.

SUBORDER SPIRURATA
II. SPIRUROIDS (SUPERFAMILY SPIRUROIDEA)

Morphology. The superfamily Spiruroidea contains a large number of worms that are parasitic in all kinds of vertebrates. They vary enormously in form and include slender, filarialike worms such as *Thelazia* and *Gongylonema;* large heavy-bodied forms superficially resembling ascarids, such as *Physaloptera;* short, thick forms such as *Gnathostoma;* and forms with bizarre females nearly spherical in shape, such as *Tetrameres.* Some have the head or body armed with spines or other cuticular embellishments. The mouth opens into a chitinized vestibule; in some, e.g., *Thelazia* (Fig. 147*B*), it has no lips, but in the majority there is either a single pair of lateral lips, e.g., *Physaloptera* and *Gnathostoma* (Fig. 147*C, D*), or a pair of dorsoventral lips in addition to the lateral pair, but never three or six lips. The vulva usually opens in the middle region of the body, but near the anus in *Gongylonema.* In the males the tail is spirally coiled; it usually has broad alae often ornamented with cuticular markings and provided with pedunculated papillae.

Important species. The table on page 498 gives a list of the forms that are of interest as parasites of domestic animals, including those that are accidental parasites of man. It will be seen that, though some of them live in the alimentary canal, most of them live in its walls or in more distant parts of the body.

Life cycles. Except for *Thelazia* and specimens of *Spirocerca* that get misplaced in aortic cysts, the eggs of all these spiruroids get access to the alimentary canal and are voided with the feces. In all cases in which the life cycles have been worked out, except *Thelazia,* the thick-shelled, embryonated eggs are swallowed by arthropods either in soil

SPIRUROIDS OF INTEREST AS PARASITES OF DOMESTIC ANIMALS. ACCIDENTAL PARASITES OF MAN MARKED "*"

Name of Parasite	Definitive Hosts	Habitat	Intermediate Hosts
Ascarops (*Arduenna*) and *Physocephalus*	Pigs	Stomach	Dung beetles
Cheilospirura spp.*	Chickens and turkeys	Walls of gizzard	Grasshoppers, sow bugs (also beetles and sandhoppers)
Echinuria spp.	Ducks and geese	Stomach and small intestine	*Daphnia,* amphipods
*Gnathostoma spinigerum**	Fish-eating carnivores	Stomach tumors	First host: *Cyclops;* second: fish, frogs, or snakes
Gongylonema spp.*	Ruminants, pigs, horses, rodents, fowls	Walls of esophagus or rumen	Dung beetles or roaches
Habronema microstoma, muscae and *macrostoma*	Horses	Mucosa or lumen of stomach	Maggots of *Stomoxys* or *Musca* (escape from proboscis of adults)
Hartertia gallinarum	Chicken (Africa)	Small intestine	Workers of termites
Physaloptera spp.*	Insectivorous and carnivorous mammals, birds, reptiles (common sp. in opossum)	Stomach or intestine	Cockroaches, earwigs, beetles, and crickets
Protospirura spp.	Rodents, monkeys, etc.	Esophagus and stomach	Roaches, fleas?
Spirocerca sanguinolenta	Dogs	Tumors on esophagus, stomach, or aorta	Dung beetles
Tetrameres spp.	Poultry	Glands of proventriculus	Grasshoppers, roaches, amphipods, *Daphnia*
*Thelazia callipaeda**, *rhodesi* and *californiensis**	Ruminants, dogs and man	Eye	*Musca* spp.
Oxyspirura mansoni	Chicken	Eye	Roaches

or in water, and in these the larvae develop. In most cases infection of the final host results from the swallowing of the intermediate host, but in at least some of the species accessory methods of transfer have been evolved. Although the species of *Habronema* of horses, which develop in maggots of stableflies or houseflies, may infect their hosts through the swallowing of infected adult flies, the larvae, after development in the Malpighian tubules or fat bodies, make their way to the head and voluntarily escape from the labium on warm wet surfaces as do filariae. They thus reach the lips, nose, or wounds, and finally infect via the mouth when licked off and swallowed. It has been suggested that this type of life cycle is an evolutionary step toward the filarial type, but Anderson (1957) has argued quite persuasively that *Habronema* has a highly specialized life cycle which evolved quite independent of the filarial life pattern. *Thelazia* larvae also escape from the proboscis of flies feeding around the eyes, thus reaching their destination directly.

Infective spiruroid larvae if eaten by abnormal hosts may burrow into the tissues and become re-encapsulated. The senior writer found armadillos from hog lots with hundreds of cysts containing dead larval stomach-worms of pigs (*Ascarops* and *Physocephalus*) obtained from eating infected grubs before the pigs got them. For *Gnathostoma* this seems to be routine procedure, for whereas cats are easily infected by feeding them gnathostome larvae encysted in second intermediate hosts (fish, frogs, snakes), attempts to infect them by feeding infected *Cyclops* have so far failed.

Since a considerable number of spiruroids are capable of partial and sometimes complete development in human beings, it is obvious that we owe our relative immunity to spiruroid infections to the fact that we are not for the most part voluntarily insectivorous. In the following paragraphs are considered briefly the principal forms recorded from man.

Gongylonema. These slender, filarialike worms (Fig. 147A) live in the walls of the esophagus or mouth cavity. The females reach a length of 15 cm. and the males 6 cm., but the diameter is only 0.2 to 0.5 mm. Eight rows of wartlike bosses on the anterior end are a characteristic feature; the vulva of the female is not far from the anus, and the male has very unequal spicules and a coiled tail with asymmetrical alae.

A number of human infections have been recorded, all of them with immature worms; although given the name *Gongylonema hominis*, they are probably identical with *G. pulchrum* of pigs and ruminants. Several cases occurred in southern United States. All the patients were aware of the active migrations of the worms under the lips or cheeks and were much annoyed by them; the worms move so rapidly that considerable dexterity is required to remove them. Two of the patients

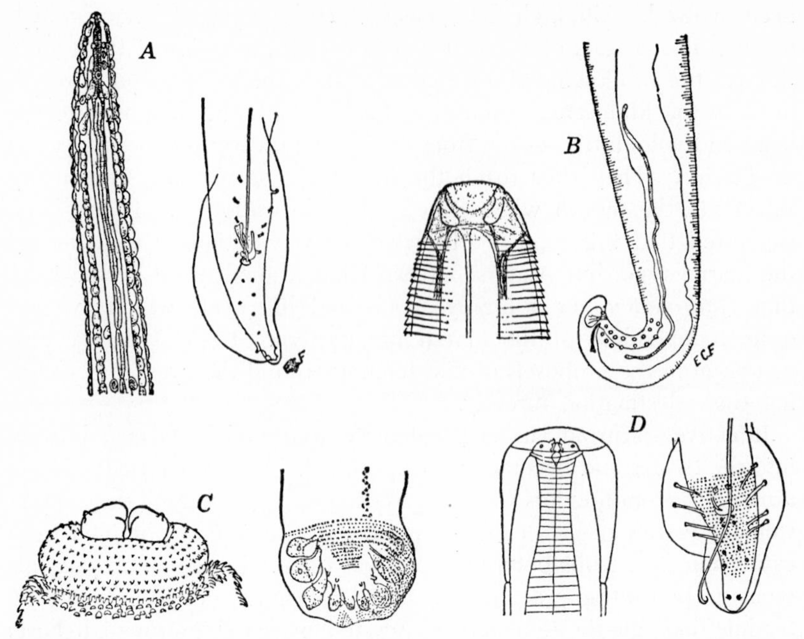

Fig. 147. Heads and tails of male spiruroid worms found in man. *A, Gongylonema pulchrum;* head, ×45; tail, ×48. *B, Thelazia callipaeda;* head, ×210; tail, ×33. *C, Gnathostoma spinigerum;* head and tail, ×39. *D, Physaloptera caucasica;* head, ×22; tail, ×16. (*A, B,* and *D* after various authors, adapted from *Human Helminthology,* by Ernest Carroll Faust, Lea and Febiger, Philadelphia. *C* from Yorke and Maplestone, *Nematode Parasites of Vertebrates.*)

also had nervous disorders which disappeared after they got rid of their parasites. Since dung beetles and roaches are the intermediate hosts it is obvious that human infection could not be common, for our appetites tend in other directions.

It is of interest to note that two species in rats, *G. neoplasticum* and *G. orientale,* frequently stimulate cancerous growths, but there is no evidence that other species do so.

Physaloptera. The genus *Physaloptera* contains numerous species parasitic in all sorts of carnivorous and insectivorous land vertebrates. They are large worms (Fig. 147*D*), superficially resembling ascarids; they live most frequently in the stomach but may also live in the intestine and occasionally even the liver; they bury their heads in the mucous membranes and cause sores and ulcerations. The females are usually 3 to 10 cm. long by 1.2 to 2.8 mm. in diameter; the males about half this size. A characteristic feature is a collarette surrounding the head end and a pair of trilobed lips. The vulva is anterior in position. The male has a coiled tail with broad asymmetrical alae which meet in front of the anus and have very long papillae; the spicules are very unequal. One species, *P. caucasica,* normally parasitic in African monkeys, is said by Leiper (1911) to be fairly common in natives of tropical Africa; one case was found in the Caucasus in Europe.

Protospirura muricola. Though not yet recorded from man, this rodent parasite has been reported by Foster (1938) as causing an injurious and often fatal infection of captive monkeys. In general appearance it resembles a small *Physaloptera* but lacks the collarette. The parasites block the esophagus and irritate the stomach wall, sometimes perforating it. Cockroaches serve as intermediate hosts. It becomes more and more evident that eating roaches is a very bad habit for the animals that habitually indulge in it.

Gnathostoma spinigerum. This is a very robust worm, 25 to 50 mm. long, with a globular swelling at the head end which is armed with eight or more rows of thornlike hooks (Fig. 147*C*). The mouth is bounded by a pair of fleshy lateral lips. Behind the swollen head the body is clothed with overlapping rows of toothed scales, which gradually dwindle away near the middle of the body.

The natural hosts of this species are wild and domestic cats and less frequently dogs. The adults inhabit large tumors, sometimes an inch in diameter, in the stomach wall, which open into the stomach by one or more pores. Other species in the Orient occur in the stomach of pigs, and one, *G. nipponicum,* in esophageal tumors in a very high percentage of mink in Japan. In the United States *G. spinigerum* has been reported rarely from mink; other species occur in raccoons and opossums.

The stomach tumors may cause fatal peritonitis when, as sometimes happens, they open into the body cavity. The seasonal occurrence of the parasites in cats, as seen by the senior writer in Calcutta, suggests that they may very commonly be fatal, for it seems impossible that the tumors could disappear completely soon after the worms had left.

In 1925 Chandler found a high percentage of snakes near Calcutta to harbor larvae of *Gnathostoma,* which, when fed to cats, developed first in the liver and subsequently invaded the stomach wall. Later Prommas and Daengsvang (1933) showed that *Cyclops* served as first intermediate hosts, and a few years later it was shown that when infected *Cyclops* are swallowed by fishes, amphibians, or snakes the larvae escape, invade the tissues, and become re-encysted in the flesh of these second intermediate hosts. Carnivores become infected when they eat these hosts, but not when fed *Cyclops.* In Thailand 92% of frogs, 80% of eels, and 30 to 37% of certain other food fishes in the markets were found to harbor larval gnathostomes. In southern Japan Miyazaki (1954) reported finding larvae in about 60 to 100% of a fresh-water fish, *Ophicephalus argus,* in some parts; this fish is commonly eaten raw by the Japanese.

When ingested by man, *Gnathostoma spinigerum* larvae develop to morphologically mature worms, but remain sexually immature. Like many other helminths in a strange host, they fail to find their way to their proper destination, in this case the stomach wall. Instead they wander aimlessly in the body, usually in or under the skin, but sometimes in the mucous membranes or viscera. Occasionally they blunder into the eye or even the brain. During their wanderings they most commonly cause migrating but intermittent swellings or edema, but sometimes a creeping eruption. Eventually they usually become encapsulated or escape through an abscess.

Human gnathostome infections have long been known to be of frequent occurrence in southeast Asia, especially in Thailand. The senior author found eggs of the worm in presumably human feces on two occasions in Burma, so it is possible that the worm does occasionally mature in the human stomach. Since the end of World War II edema or creeping eruption caused by *G. spinigerum* was found by Miyazaki (1954) to be very common in southern Japan. In one report over one-third of 3900 patients examined were found infected. A single infection with a gnathostome of pigs, probably *G. doloresi,* has been reported from a man in Tokyo. A third species in Japan has apparently not found its way into humans.

***Thelazia* spp.** These slender little worms, possibly more nearly related to the filariae than to the spiruroids, inhabit the conjunctival sac and lachrymal ducts of animals and occasionally man. At times they creep out over the eyeball, later returning to their nest in the inner corner of the eye. *T. callipaeda,* primarily a parasite of dogs in India, Burma, and China, has been reported from man four times in China. *T. californiensis,* reported by Stewart from sheep, deer, and dogs in brushy, mountainous places in California, has been found in man twice. Other species are important parasites of the eyes of cattle and horses in some places; altogether nineteen species have been described from various mammals and birds.

The female worms are 7 to 19 mm. long, the males somewhat smaller. The cuticle is pleated into well-defined striations with sharp edges; there are no lips, but there is a short vestibule (Fig. 147*B*). The vulva is anterior as in filariae, and the male has no caudal alae. Krastin (1950) showed that certain flies of the genus *Musca* which cluster around the eyes of cattle serve as intermediate hosts for *T. rhodesi,* which is harbored by over 90% of cattle in late summer in parts of eastern Siberia.

By their movements the worms irritate the eye considerably, causing a free flow of tears and injection of blood vessels, and sometimes severe pain and nervous symptoms. At first the eye is not seriously affected, but Faust (1928) observed that in the course of time the repeated scratching of the surface of the eyeball by the serrated cuticle of the worm causes the formation of scar tissue, and the eye gradually develops a cloudiness, progressing outward from the worm nest, which ultimately reduces the vision. Cattle are sometimes blinded by a *Thelazia* in Africa and Asia.

After the eye is desensitized with 1% cocaine, the worms are easily removed with a forceps or swab if seen, but several examinations are usually necessary in order to get a complete catch.

***Cheilospirurua* sp.** Africa and Garcia (1936) found a specimen belonging to this genus in a nodule on the conjunctiva of a Filipino. Members of this genus, so far as known, normally live under the lining of the gizzard of birds. It is another example of abnormal behavior in an abnormal host.

SUBORDER CAMALLANATA. GUINEA WORMS (SUPERFAMILY DRACUNCULOIDEA)

The superfamily Dracunculoidea, placed by Chitwood in the suborder Camallanata, was formerly included with the filarial worms. It contains several genera of worms that are peculiar in the relatively enormous length of the female worms as compared with the midget males, and in the fact that during the course of their development the alimentary canal and vulva atrophy, leaving the body of the adult almost entirely occupied by the embryo-filled uterus. The embryos are liberated by the bursting of a loop of the uterus prolapsed through the mouth or through a rupture of the anterior end of the body. One genus, *Philometra,* contains parasites of the body cavity of fishes; the others, *Dracunculus, Avioserpens,* and *Micropleura,* contain parasites of the connective tissues of mesenteries of reptiles, birds, and mammals. The guinea worm, *Dracunculus medinensis,* is a common human parasite in parts of Asia and Africa. Another species, *D. insignis,* is a parasite of raccoons in America, and occurs sporadically in mink, dogs, etc. The senior writer (1942) found *D. insignis* to be very common in the hind feet of raccoons in eastern Texas. The females closely resemble the human guinea worm of the Old World except for their smaller size (up to 16 in. long), but there are minor differences in the males. Price has recently found other species in various wild mammals in Maryland. In the Old World guinea worm infections only rarely occur in animals, although dogs and other carnivores are susceptible.

Dracunculus medinensis

Occurrence and distribution. The guinea worm, *Dracunculus medinensis* (meaning the little dragon of Medina), has been known since remote antiquity, for one of its main strongholds is in the region of western Asia which cradled civilization. The "fiery serpents" which molested the Israelites by the Red Sea were probably guinea worms. It is still, as it was in ancient times, one of the important scourges of life from central India to Arabia, and it is locally important in the East Indies, Egypt, and central Africa. Stoll (1947) estimated that there are 48,000,000 human guinea-worm infections in the world. The disease is commonly associated with dry climates because of the concentration of water supplies in step-wells or reservoirs and the greater opportunity for *Cyclops* in the drinking water supply to become contaminated from human skin. In innumerable villages in central and western India up to 25% or more of the population suffer annually from guinea-worm infections. The human guinea worm became established in a few localities in tropical America but seems to have died out.

Morphology. The gravid female worm, long the only form known, lives in the deeper layers of the subcutaneous tissues, where she usually can be seen lying in loose coils, like a small varicose vein, under the skin. Sometimes she is more easily felt than seen until she produces a skin ulcer through which she gives birth to myriads of embryos. She reaches a length of 2.5 to 4 ft. with a diameter of 1 to 1.5 mm. The head end is bluntly rounded, and commonly ruptured in worms which have begun expelling embryos. The tail is attenuated and sharply hooked.

The males were practically unknown until Moorthy and Sweet (1936) obtained them in experimentally infected dogs. Mature specimens measured 20 to 29 mm. in length and were found 15 to 20 weeks after infection, but were not found when the gravid females were found in the skin at the end of 15 months. Males and young females of similar size (Fig. 148) have the simple mouth surrounded by papillae and have an esophagus about 10 mm. long; in the females the vulva is a little anterior to the middle of the body. The males have a spirally coiled tail with four pairs of preanal and six of postanal papillae, but no alae, and two nearly equal spicules 0.5 to 0.7 mm. long.

Life cycle. When ready to bring forth her young, the guinea worm is instinctively attracted to the skin, especially to such parts as are likely to, or frequently do, come in contact with cold water, such as the arms of women who wash clothes at a river's brink or the legs and

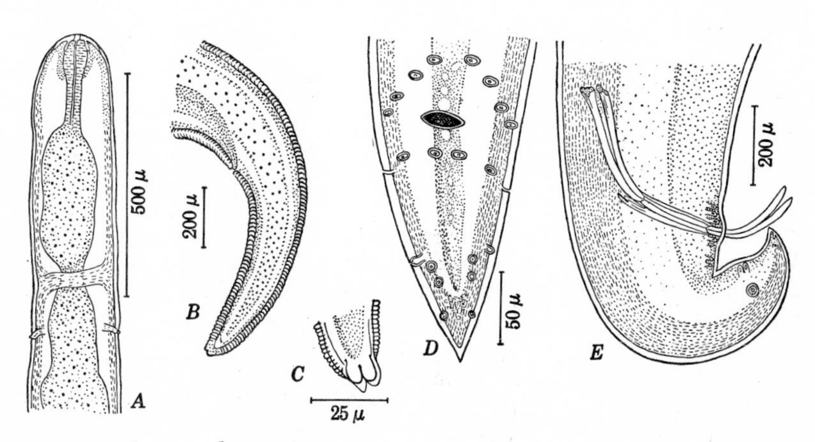

Fig. 148. *Dracunculus medinensis*, guinea worm. *A*, anterior end of ♂; *B*, posterior end of immature ♀; *C*, tip of tail of immature ♀, showing four processes (mucrones); *D*, tail of ♂, ventral view; *E*, tail of ♂, lateral view. (After Moorthy, *Parasitol.*, 23, 1937.)

backs of water carriers. The worm pierces the lower layers of the skin with the front end of her body and excretes a toxic substance that irritates the tissues and causes a blister to form over the injured spot (Fig. 149). The blister eventually breaks, revealing a shallow ulcer, about as large as a dime, with a tiny hole in the center. When the ulcer is douched with water a milky fluid is exuded directly from the hole or from a very delicate, transparent projected structure which is a portion of the worm's uterus. This fluid is found to contain hordes of tiny coiled larvae with a length of about 600 μ, one-third of which is occupied by the long filamentous tail (Fig. 150).

An hour or so later a new washing with cold water will bring forth a fresh ejection of larvae, and so on until the supply is exhausted, a little more of the uterus being extruded each time. After each ejection of the larvae the protruded portion of the uterus dries up, thus sealing in the unborn larvae and saving them for the next douching. This procedure, of course, increases the chances of some of the larvae finding *Cyclops*-inhabited water. The whole process is one of the neatest adaptations in behavior in all the realm of biology,

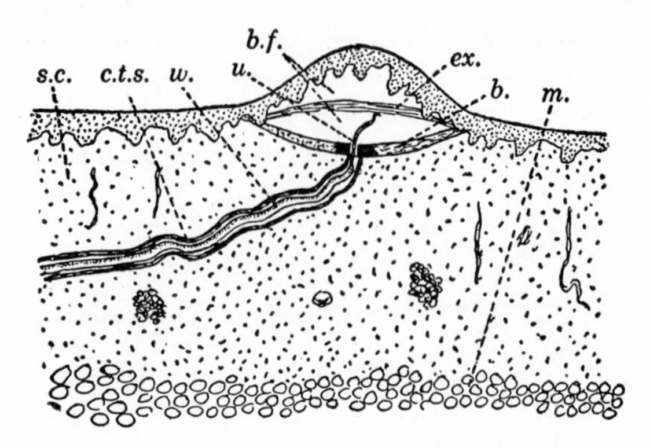

Fig. 149. Diagram of guinea worm in the skin at the time of blister formation: *b.*, base of ulcer; *b.f.*, blister fluid; *c.t.s.*, connective tissue sheath of worm; *ex.*, layer of exudate; *m.*, muscle layer; *s.c.*, subcutaneous tissue; *u.*, extruded uterus of worm; *w.*, worm. (Adapted from Fairley.)

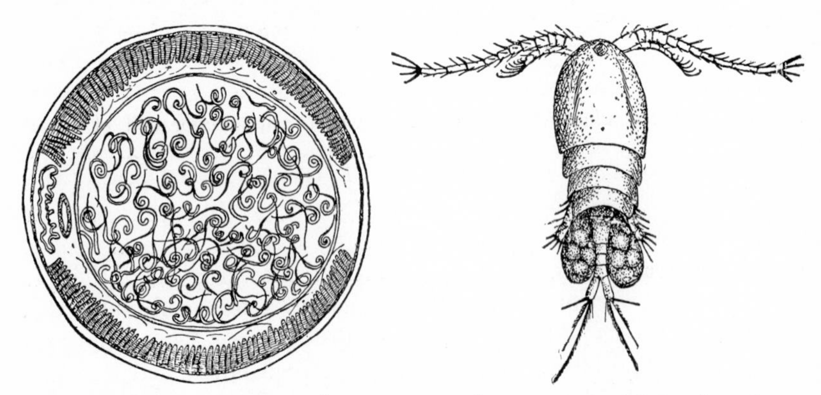

Fig. 150. *Left,* cross-section of guinea worm showing uterus filled with embryos, about ×30 (after Leuckart). *Right,* a *Cyclops,* some species of which serve as intermediate hosts of guinea worm, about ×25.

enabling a blind, unmeditative, burrowing worm to give her aquatic *Cyclops*-inhabiting offspring a fair chance in life even on a desert. She turns what would seem to be a hopeless handicap into an actual advantage.

When all the young have been deposited under the stimulus of contact with water the parent worm shrivels and dies and is soon absorbed by the tissues.

The embryo worms, safely deposited in water, unroll themselves and begin to swim about. They remain alive for several days but eventually perish unless swallowed by a *Cyclops* (Fig. 150). When this happens they burrow into the body cavity of the surprised *Cyclops,* reaching that destination in 1 to 6 hours. Moorthy never found a *Cyclops* with more than one larva in nature; when infected with more than four to five in the laboratory, development is interfered with. The senior writer never found more than two larvae of *Dracunculus insignis* to develop in experimentally infected *Cyclops.*

The larvae molt twice in the body cavity and reach the infective stage within 3 weeks. At this time they vary from 240 to 600 μ in length, the tail now being short. After feeding infected *Cyclops* to dogs the first specimens were found by Moorthy after about 10 weeks; they were deep in the connective tissues and only 12 to 24 mm. long, although the vaginas of females 24 mm. long already contained a mucoid plug, indicating that they had already been fertilized. The worms appear to come to maturity in about 11 to 12 months after infection.

Epidemiology. In western India the infection is always associated with step-wells which, instead of being provided with buckets and ropes, are approached by steps, the people standing foot- or knee-deep in the water while filling containers. During this time the parent worm ejects her offspring, and at the same time previously infected *Cyclops* are withdrawn with the water. In African villages, ponds function in a similar manner.

In an epidemiological study in the Deccan, India, almost no infection was found in children under 4, but after that the incidence increased steadily to 85% in the 30- to 35-year age group, then gradually fell off again. There may be one to fifty worms per person but in most instances only one in a year. Few people suffer from infections for more than 4 years, after which immunity usually develops. The worms form their ulcers on the legs in about 90% of cases.

Pathology. The first symptoms appear simultaneously with the beginning of the blister formation, and consist of urticaria, nausea and vomiting, diarrhea, asthma, giddiness, and fainting; some or all of these symptoms may be present. Fairley (1925) believes they are due to absorption of the toxin employed by the worm to form the blister. The symptoms strongly suggest an allergic reaction; injection of adrenalin brings about rapid improvement. Eosinophilia is marked.

Later symptoms result from secondary invasion of the ulcer by bacteria. The worms are usually mechanically extracted and, being elastic, are likely to break. The broken end of the worm draws back, carrying with it into its connective-tissue sheath various bacteria which produce abscesses. These may cause such severe infection as to necessitate amputation or may even lead to fatal blood poisoning. Joints are frequently involved, leading to permanent deformities. These occur with deplorable frequency in villages of the Deccan in India. There is some evidence that reinfections do not occur while an adult worm is still in the body. Most victims are incapacitated for several weeks; fortunately only a minority suffer permanent deformities or more serious consequences.

Treatment. Most drugs used against guinea worms have proved to be of little or no value, often, in fact, harmful, since local applications by natives after the ulcer has formed succeed only in causing secondary infections. Elliott (1942) reported excellent results from intramuscular injection of Phenothiazine emulsified in olive oil into a number of places close to the worm. Two to four grams of Phenothiazine can be injected at a sitting, with repetitions at weekly intervals; more than two courses are rarely needed. It takes 5 to 7 days for the drug to act; if a worm is being or is to be extracted, it is better to wait this long after injections.

Extraction of the worm by winding it out on a stick is a time-honored method which, with a few scientific refinements, is still widely used. Native medicine men extract the worm through the ulcer by repeatedly dousing its head with water and then winding it out a little at a time. Care must be taken not to pull hard enough to rupture the worm; a safe extraction takes 10 to 14 days. If the ulcer is carefully treated with antiseptics, the worm can be withdrawn a little faster by exposing a loop and pulling it from both ends. Natives in India apply to the wound a green powder made of neem leaves, together with a choice assortment of contaminating bacteria, and the unfortunate patient has to fight his battle with the bacteria instead of the relatively innocent worm. If he loses his leg or his life it is the will of the gods and no fault of the doctor. Another native method is to apply a cone-shaped piece of metal over the exposed part of the worm and suck it vigorously until a negative pressure is created sufficient to draw the tissue up into the cylinder. The tongue is then applied and the finger quickly substituted, and after a few minutes the worm may be found in the tube.

By the use of local anesthetics and aseptic precautions, mechanical extraction is usually successful, and complete healing may follow in less than a week, as contrasted with the usual month.

Prevention. Prevention of the infection would be extremely simple if it were not for the scruples of the natives, often of religious nature, as to where and how they obtain and use their water. In India wherever step-wells are replaced by other types which keep the legs or arms out of the water, guinea worm disappears. If the water were strained through muslin to remove *Cyclops*, guinea worm would disappear, but even this is objected to. However, education and governmental pressure eventually bring results, and many areas in India that have suffered from guinea worm for centuries have been freed in recent years by altering the wells. Moorthy has had some success in destroying *Cyclops* by treating wells with dilute copper sulfate and "perchloron," and he reports that a fish, *Barbus puckelli,* feeds on them voraciously, but he emphasizes that abolition of step-wells is the only permanent and foolproof method of control.

REFERENCES

Filariae in General and Wuchereria

Augustine, D. L. 1945. Filariasis. *N. Y. State J. of Med.,* 45: 495–499.

Augustine, D. L., and Lherisson, C. 1946. Studies on the specificity of intradermal tests in the diagnosis of filariasis. *Am. J. Hyg.,* 43: 38–40.

Basu, C. C., and Rao, S. S. 1939. Studies on filariasis transmission. *Ind. J. Med. Res.,* 27: 233–249.

Beye, H. K., et al. 1952. Preliminary observations on the prevalence, clinical manifestations and control of filariasis in the Society Islands. *Am. J. Trop. Med. Hyg.,* 1: 637–661.

Brown, H. W. 1945. Current problems in filariasis. *Am. J. Public Health,* 35: 607–613.

Byrd, E. E., St. Amant, L., and Bromberg, L. 1945. Studies on filariasis in the Samoan Area. *U. S. Naval Med. Bull.,* 44: 1–20.

Chabaud, A. G., and Choquet, M. T. 1953. Nouvel essai de classification des filaires (superfamille des Filarioidea). *Ann. parasitol. humaine et comparée,* 28: 172–192.

Dikmans, G. 1948. Skin lesions of domestic animals in the United States due to nematode infections. *Cornell Vet.,* 38: 3–23.

Edeson, J. F. B., and Wharton, R. H. 1958. The experimental transmission of *Wuchereria malayi* from man to various animals in Malaya. *Trans. Roy. Soc. Trop. Med. Hyg.,* 52: 25–38.

Edeson, J. F. B., and Buckley, J. J. C. 1959. Studies on filariasis in Malaya. On the migration and rate of growth of *Wuchereria mayali* in experimentally infected cats. *Parasitology,* 53: 191–209.

Lavoipierre, M. M. J. 1958. Studies on the host-parasite relationships of filarial nematodes and their arthropod hosts. II. The arthropod as a host to the nematode: A brief appraisal of our present knowledge based on a study of the more important literature from 1878 to 1957. *Ann. Trop. Med. Parasitol.,* 52: 326–345.

Fülleborn, F. 1929. Filariosen des Menschen, in *Handb. der path. Mikro-org.* Kolle u. Wassermann, 6, No. 28, 1043–1224.

Hawking, F. 1950. Some recent work on filariasis. With addendum. *Trans. Roy. Soc. Trop. Med. Hyg.*, 44: 153–192.

Hawking, F., and Thurston, J. P. 1951. The periodicity of microfilariae. *Trans. Roy. Soc. Trop. Med. Hyg.*, 45: 307–340.

Hodgkin, E. P. 1939. The transmission of *Microfilaria malayi* in Malaya. *J. Malaya Branch Brit. Med. Assoc.*, 3: 8–11.

Huntington, R. W., Jr., Fogel, R. H., Eichold, A., and Dickson, J. G. 1944. Filariasis among American troops in a South Pacific Island group. *Yale J. Biol. Med.*, 16: 529–537.

Iyengar, M. O. T. 1938. Studies on the epidemiology of filariasis in Travancore. *Indian Med. Res. Mem.*, 30, 179 pp.

Knott, J. 1938. The treatment of filarial elephantiasis of the leg by bandaging. *Trans. Roy. Soc. Trop. Med. Hyg.*, 32: 243–252.

McDonald, E. M., and Scott, J. A. 1953. Experiments on immunity in the cotton rat to the filarial worm, *Litosomoides carinii*. *Exp. Parasitol.*, 2: 174–184.

Manson-Bahr, P. 1952. The clinical manifestations and ecology of Pacific filariasis. *Documenta Med. Geograph. et Trop.*, 4: 193–204.

Manson-Bahr, P., and Muggleton, W. J. 1952. Further research on filariasis in Fiji. *Trans. Roy. Soc. Trop. Med. Hyg.*, 46: 301–326.

Napier, L. E. 1944. Filariasis due to *Wuchereria bancrofti*. *Medicine,* 23: 149–179.

O'Connor, F. W., and Hulse, C. R. 1932. Some pathological changes associated with *W. bancrofti* infection. *Trans. Roy. Soc. Trop. Med. Hyg.*, 25: 445–452.

Otto, G. F., Brown, H. W., et al. 1952. Arsenamide in the treatment of infections with the periodic form of the filaria, *Wuchereria bancrofti*. *Am. J. Trop. Med. Hyg.*, 1: 470–473.

Otto, G. F., et al. 1953. Filariasis in American Samoa III. Studies on chemotherapy against the non-periodic form of *Wuchereria bancrofti*. *Am. J. Trop. Med. Hyg.*, 2: 495–516.

Raghavan, N. G. S., et al. 1957. Filariasis. Epidemiology-pathogenesis, chemotherapy, vectors—control. *Bull. World Health Organization,* 16: 553–564.

Rozeboom, L. E., and Cabrera, B. D. 1956. Filariasis in the Philippine Islands. *Am. J. Hyg.*, 63: 140–149.

Wartman, W. B. 1947. Filariasis in American armed forces in World War II. *Medicine,* 26: 333–394.

Wehr, E. E. 1935. A revised classification of the nematode superfamily Filaroidea. *Proc. Helminthol. Soc. Wash.*, 2: 84–88.

Onchocerca

Adams, A. R. D., et al. 1958. Symposium on onchocerciasis. *Trans. Roy. Soc. Trop. Med. Hyg.*, 52: 95–134.

Burch, T. A., and Ashburn, L. L. 1951. Experimental therapy of onchocerciasis with duramin and hetrazan; results of a three-year study. *Am. J. Trop. Med.*, 31: 617–623.

Lewis, D. J. 1953. *Simulium damnosum* and its relation to onchocerciasis in the Anglo-Egyptian Sudan. *Bull. Entomol. Res.*, 43: 597–644.

Mazzotti, L. 1948. Oncocercosis en Mexico. *Proc. 4th Intern. Congr. Trop. Med. Malaria,* 2: 948–956.

Moignoux, J. B. 1952. Les onchocerques des equidés. *Acta Trop.*, 9: 125–150.
Ortlepp, R. J. 1937. The biology of *Onchocerca* in man and animals. *J. S. African Vet. Med. Assoc.*, 8: 1–6.
Steward, J. S. 1937. The occurrence of *Onchocerca gutturosa* Neumann in cattle in England with an account of its life history and development in *Simulium ornatum* Mg., *Parasitology*, 29: 212–218.
Strong, R. P., Hissette, J., Sandground, J. H., and Bequaert, J. C. 1938. Onchocercosis in Africa and Central America. *Am. J. Trop. Med.*, 18, No. 1, Suppl.
Strong, R. P., Sandground, J. H., Bequaert, J. C., and Ochol, M. M. 1934. *Onchocercosis.* Harvard University Press, Boston.
Wanson, M. 1950. Contribution à l'étude de l'onchocercose Africaine humaine. *Ann. soc. belge de méd. trop.*, 30: 667–863.
World Health Organization. 1954. Expert committee on onchocerciasis. First Report. *Tech. Report Ser.*, 87.

Other Filariae

Anderson, R. C. 1957. The life cycles of dipetalonematid nematodes (Filarioidea, Dipetalonematidae): The problem of their evolution. *J. Helminthol.*, 31: 203–224.
Buckley, J. J. C. 1934. On the development in *Culicoides furens* Poey of *Filaria* (*Mansonella*) *ozzardi*. *J. Helminthol.*, 12: 99–118.
Buckley, J. J. C. 1958. Tropical pulmonary eosinophilia in relation to filarial infections (*Wuchereria* pp.) of animals. Preliminary note. *Trans. Roy. Soc. Trop. Med. Hyg.*, 52: 335–336.
Chabaud, A. G. 1952. Le genre *Dipetalonema* Diesing, 1861, essai de classification. *Ann. parasitol. humaine et comparée*, 27: 250–285.
Chandler, A. C., Milliken, G., and Schuhardt, V. T. 1930. The production of a typical calabar swelling in a Loa patient by injection of a Dirofilaria antigen. *Am. J. Trop. Med.*, 10: 345–351.
Connal, A., and Connal, S. L. M. 1922–1923. The development of *Loa loa* in *Chrysops silacea* and in *Chrysops dimidiata*. *Trans. Roy. Soc. Trop. Med. Hyg.*, 16: 64–89, 437.
Danaraj, T. J., Da Silva, L. S., and Schacher, J. F. 1959. The serological diagnosis of eosinophilic lung (tropical eosinophilia) and its etiological implications. *Am. J. Trop. Med.*, 8: 151–159.
Desportes, C. 1939–1940. *Filaria conjunctivae* Addario, 1885, parasite accidentel de l'homme, est un *Dirofilaria*. *Ann. parasitol. humaine et comparée*, 17: 380–404.
Drudge, J. H. 1952. Arsenamide in the treatment of canine filariasis. *Am. J. Vet. Res.*, 13: 220–235.
Duke, B. O. L., and Wijers, O. J. B. 1958. Studies on loiasis in monkeys. I. The relationship between human and simian *Loa* in the rain-forest zone of the British Cameroons. *Ann. Trop. Med. Parasitol.*, 52: 158–175.
Fain, A. 1947. Répartition et étude anatomo-clinique des filarioses humaines dans le territoire de Banningville (Congo Belge). *Ann. soc. belge méd. trop.*, 27: 25–63.
Faust, E. C. 1937. Mammalian heart worms of the genus *Dirofilaria*. *Festschrift Nocht.* 131–139. Hamburg.
Faust, E. C. 1957. Human infection with species of *Dirofilaria*. *Z. Tropenmed. Parasitol.*, 8: 59–68.

Gordon, R. M. 1955. A brief review of recent advances in our knowledge of loiasis and of some of the still outstanding problems. *Trans. Roy. Soc. Trop. Med. Hyg.*, 49: 98–105.

Gordon, R. M., et al. 1950. The problem of loiasis in West Africa. *Trans. Roy. Soc. Trop. Med. Hyg.*, 44: 11–41.

Innes, J. R. M., and Shoho, C. 1953. Cerebrospinal nematodiasis: focal encephalomyelomalacia of animals caused by nematodes (*Setaria digitata*); a disease which may occur in man. *Arch. Neurol. Psychiat.*, 70: 325–349.

Kershaw, W. E. 1955. The epidemiology of infections with *Loa loa*. *Trans. Roy. Soc. Trop. Med. Hyg.*, 49: 143–150.

Kivits, M. 1952. Quatre cas d'encéphalite mortelle avec invasion du liquide cephalorhachidien por *Microfilaria loa*. *Ann. soc. belge méd. trop.*, 32: 235–242.

Newton, W. L., and Wright, W. H. 1956. The occurrence of a dog filariid other than *Dirofilaria immitis* in the United States. *J. Parasitol.*, 42: 246–258.

1957. A reevaluation of the canine filariasis problem in the United States. *Vet. Med.*, 52: 75–78.

Sharp, N. A. D. 1928. *Filaria perstans;* its development in *Culicoides austeni*. *Trans. Roy. Soc. Trop. Med. Hyg.*, 21: 371–396.

Sprent, J. F. A. 1955. On the invasion of the central nervous system by nematodes. I. The incidence and pathological significance of nematodes in the central nervous system. *Parasitology*, 45: 31–40.

Steuben, E. B. 1954. Larval development of *Dirofilaria immitis* (Leidy) in fleas. *J. Parasitol.*, 40: 580–589.

Summers, W. A. 1943. Experimental studies on the larval development of *Dirofilaria immitis* in certain insects. *Am. J. Hyg.*, 37: 173–178.

Vogel, H. 1928. Zur Anatomie der *Microfilaria perstans*. *Arch. Schiffs- u. Tropen-Hyg.*, 32: 291–306.

Whitlock, J. H. 1952. Neurofilariosis, a paralytic disease of sheep: II. *Neurofilaria cornellensis*, n.g., n.sp. (Nematoda, Filarioidea), a new nematode parasite from the spinal cord of sheep. *Cornell Vet.*, 42: 125–132.

Spiruoidea

Africa, C. M., and Garcia, E. Y. 1936. A new nematode parasite (*Cheilospirura* sp.) of the eye of man in the Philippines. *J. Philippine Isl. Med. Assoc.*, 16: 603–607.

Chandler, A. C. 1925*a*. A contribution to the life history of a gnathostome. *Parasitology*, 17: 237–244.

1925*b*. (Helminthic parasites of cats.) *Indian J. Med. Res.*, 13: 213–227.

Daengsvang, S. 1949. Human gnathostomiasis in Siam with reference to the method of prevention. *J. Parasitol.*, 35: 116–121.

Faust, E. C. 1928. Studies on *Thelazia callipaeda*. *J. Parasitol.*, 15: 75–86.

Foster, A. O., and Johnson, C. M. 1938. Protospiruriasis, a new nematode disease of captive monkeys. *J. Parasitol.*, 24: No. 6, Suppl., Abst. 75.

Hiyeda, K., and Faust, E. C. 1929. Aortic lesions in dogs caused by infection with *Spirocerca sanguinolenta*. *Arch. Pathol.*, 7: 253–272.

Hosford, G. N., Stewart, M. A., and Sugarman, E. I. 1942. Eye worm (*Thelazia californiensis*) infection in man. *Arch. Ophthalmol.* (Chicago), 27: 1165–1170.

Krastin, N. I. 1950. Elucidation of the life cycle of *Thelazia rhodesii* (Desmarest,

1827), parasitic in eyes of cattle, and epizoology of thelaziasis in cattle. Abstracts in *Helminthol. Abstr.*, 18: 84–85, 118–119.

Leiper, R. T. 1911. On the frequent occurrence of *Physaloptera mordens* as an internal parasite of man in tropical Africa. *J. Trop. Med.*, 14: 209–211.

Lucker, J. T. 1932. Some cross-transmission experiments with *Gongylonema* of ruminant origin. *J. Parasitol.*, 19: 134–141.

Miyazaki, I. 1954. Studies on *Gnathostoma* occurring in Japan (Nematoda, Gnathostomidae) I. Human gnathostomiasis and imagines of *Gnathostoma. Kyushu Mem. Med. Sci.*, 5: 13–27.

Prommas, C., and Daengsvang, S. 1933–1937. (Life cycle of *Gnathostoma spinigerum*), *J. Parasitol.*, 19: 287–292; 22: 180–186; 23: 115–116.

Stewart, M. A. 1940. Ovine thelaziasis. *J. Am. Vet. Med. Assoc.*, 96: 486–489.

Stiles, C. W. 1921. *Gongylonema hominis* in man. *Health News, U. S. Public Health Service,* June.

Witenberg, G., et al. 1950. A case of ocular gnathostomiasis, *Ophthalmologica* 119: 114–122.

Guinea Worms

Chandler, A. C. 1942. The guinea worm, *Dracunculus insignis* (Leidy, 1858) a common parasite of raccoons in east Texas. *Am. J. Trop. Med.*, 22: 153–157.

Chitwood, B. G. 1933. Does the guinea worm occur in North America? *J. Am. Med. Assoc.*, 100: 802–804.

Elliott, M. 1942. A new treatment for dracontiasis. *Trans. Roy. Soc. Trop. Med. Hyg.*, 35: 291–301.

Fairley, N. H., and Liston, W. G. 1925. Studies on guinea worm disease. Collected papers from *Indian J. Med. Res.* and *Indian Med. Gaz.*, Calcutta.

Moorthy, V. N. 1932. An epidemiological and experimental study of dracontiasis in the Chitaldrug District, India. *Indian Med. Gaz.*, 67: 498–504.

1937. A redescription of *Dracunculus medinensis. J. Parasitol.*, 23: 220–224.

1938. Observations on development of *Dracunculus medinensis* larvae in *Cyclops. Am. J. Hyg.*, 27: 437–460.

Moorthy, V. N., and Sweet, W. C. 1936. Experimental infection of dogs with dracontiasis. *Indian Med. Gaz.*, 71: 437–442; Biological methods of control. *Ibid.*: 565–568; natural infection of *Cyclops. Ibid.*: 568–570.

Onabamiro, S. D. 1956. The early stages of the development of *Dracunculus medinensis* (Linnaeus) in the mammalian host. *Ann. Trop. Med. Parasitol.*, 50: 157–166.

Part III

ARTHROPODS

Chapter 22

INTRODUCTION TO ARTHROPODS

To the average person it is astonishing to learn that the insects and their allies, constituting the phylum Arthropoda, include probably more than four times as many species as all other animals combined. It is even more startling for egotistical humanity to realize that this is not the age of man but the age of insects, and that man is only beginning to dispute with insects for first place in the procession of animal life in the world.

The Arthropoda are the most highly organized of invertebrate animals. Their nearest allies are the segmented worms or annelids, i.e., earthworms and leeches, but most of them show a great advance over their lowly cousins. Like the annelids they have a segmented type of body, though in some types, such as the mites, all the segments become secondarily confluent. Like the annelids, also, the arthropods are protected by an external skeleton, but this usually consists of a series of sclerotized rings encircling the body. The most obvious distinguishing characteristic of the arthropods is the presence of jointed appendages in the form of legs, mouth parts, and antennae. Internally they are distinguished from other invertebrates in that the body cavity, so conspicuous in the annelids, has been entirely usurped by a great expansion and running together of blood vessels, so that a large blood-filled space called a hemocoele occupies the space of the usual body cavity or coelom. Within this space are blood vessels and a so-called heart, which retained their individuality while the other vessels fused. These vessels are not closed, however, but open into the hemocoele at each end.

Classification

The phylum Arthropoda is divided by Comstock into thirteen classes, but only four of these concern us as human parasites or disease transmitters, namely, the Crustacea, the Arachnoidea, the Pentastomida, and the Insecta or Hexapoda.

Crustacea. The Crustacea rival insects in their diversity of form and habits; included are well-known large forms such as crabs and crayfish; minute plankton forms such as copepods or water fleas; sessile barnacles; and terrestrial isopods (sowbugs). Primarily they are gill-breathing arthropods of water. They are geologically of great antiquity, and among them are the most primitive of the typical arthropods. Their appendages are usually numerous and, taking the group as a whole, show a wonderful range of modifications for nearly every possible function.

The Crustacea include many parasites of aquatic animals, among them some very highly modified and bizarre forms. Most of the so-called fish lice belong to the subclass Copepoda, the North American forms of which were dealt with by Wilson (1902–1922). The Cirripedia (barnacles) also include some parasites, among them the remarkable *Sacculina*, a parasite of crabs, which ends up as an external reproductive sac with a network of roots that ramify through the entire body of the host. A number of isopods and at least one amphipod (whale louse) have also become parasites. The biology of these crustacean parasites is interestingly discussed by Baer (1952). Small crustaceans serve as intermediate hosts of several worms parasitic in man and animals, namely, certain species of *Cyclops* for the guinea worm and for *Gnathostoma spinigerum;* certain species of *Cyclops* and *Diaptomus* for the tapeworms *Dibothriocephalus* and *Spirometra;* and *Diaptomus* for *Hymenolepis lanceolata* of ducks. Crabs and crayfish (members of the order Decapoda) serve as second intermediate hosts for the lung flukes, *Paragonimus.*

Arachnoidea. The Arachnoidea, including spiders, scorpions, ticks, and mites (all belonging to the subclass Arachnida) represent the terminus of a separate line of evolution. They probably had a common origin with the Crustacea, but they have become adapted to terrestrial life. The members of this class have four pairs of legs as adults, two pairs of mouth parts, and no antennae. The head and thorax are grown together, forming a cephalothorax, and in the ticks and many mites not even the abdomen remains as a distinct section. The Arachnida breathe by means of "book lungs," or may have a system of tracheae similar to those found in the insects and myriapods. Some of the small mites, however, lack both book lungs and tracheae and respire through the cuticle. Only one of the eight orders of Arachnida, the Acarina (mites and ticks), contains parasitic species; many of these are important disease vectors. Some of the Arachnida are very poisonous, including some scorpions and centipedes and certain spiders, especially the black widows, *Latrodectes,* and the skin-destroying *Loxosceles laeta* of Chile, Argentina, and Uruguay.

Pentastomida. The Pentastomida are degenerate wormlike creatures which in the adult stage have no appendages except two pairs of hooks near the mouth. If it were not for the larval forms, which have two pairs of short legs, their affinities with the arthropods might be doubted. They were formerly included with the mites for want of a better way of disposing of them. They have no circulatory or respiratory organs. Like many of the parasitic worms, they undergo their larval development in intermediate hosts.

Insecta. The insects represent the zenith of invertebrate life. They are primarily terrestrial arthropods which breathe by tracheae. Their appendages, however, are reduced to one pair of antennae (except in Protura, which have none), three pairs of mouth parts (one pair more or less fused together), three pairs of legs, and usually two pairs of wings if not secondarily lost. Nearly all adult insects are readily divisible into three parts, the head, the thorax, and the abdomen.

Insect morphology and anatomy

The cuticle. The cuticle of insects (Fig. 151) serves as an external supporting skeleton and as a place of attachment for muscles. It is a very complicated structure made up of an extremely thin outer layer of lipoid material, the epicuticle (Fig. 151*B*), under which is a more or less thick layer composed of chitin and protein; in the outer portion of this, the exocuticle, the protein is tanned and hard, whereas in the deeper portion, the endocuticle, the chitin and untanned protein are arranged in horizontal lamellae. Under the endocuticle is a single layer of epidermal cells. The exo- and endocuticle are provided with countless minute "pore canals," sometimes over 1,000,000 per square millimeter, which are believed to be filamentous processes of the epidermal cells around which the cuticle was secreted.

The thin waxy layer of the epicuticle is responsible for the relative impermeability of the cuticle to water so that even insects with very soft skins can survive in dry places. Since the cuticle covers the fore- and hindgut and the larger tracheae as well as the body surface, only the midgut is not plated with this moisture-conserving material. The insect cuticle is freely permeable by lipoid substances, which is of great importance in connection with such lipoid-soluble insecticides as the chlorinated hydrocarbons (DDT, etc., see p. 533) and their solvents.

Movement and expansion are allowed for by thin, lightly sclerotized areas between rings or plates, except in the head and frequently the thorax. In the abdomen each ring has a dorsal plate or *tergite* and a ventral one or *sternite* (Figs. 194, 226). The thorax may also have lateral plates or *pleurites*. Since chitin is unaffected by alkalies, all the soft parts of insects may be dissolved away by treatment with potassium hydroxide, leaving all the cuticular characters, which are principally used in identifications, more easily examinable.

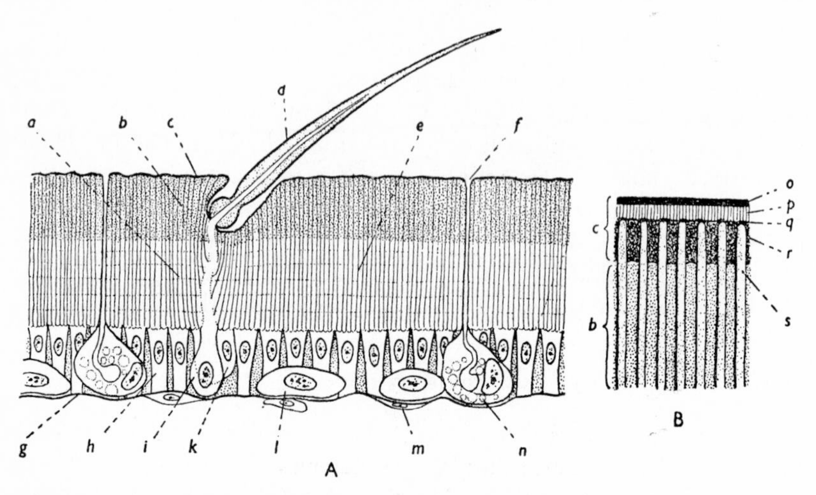

Fig. 151. *A*, section of typical insect cuticle; *B*, detail of epicuticle, schematic; *a*, laminated endocuticle; *b*, exocuticle; *c*, epicuticle; *d*, bristle; *e*, pore canals; *f*, duct of dermal gland; *g*, basement membrane; *h*, epidermal cell; *i*, trichogen cell; *k*, tormogen cell; *l*, oenocyte; *m*, haemocyte adherent to basement membrane; *n*, dermal gland; *o*, cement layer of epicuticle; *p*, wax layer; *q*, polyphenol layer; *s*, pore canal. (After Wigglesworth, *Biol. Revs.*, 23, 1948.)

As arthropods grow they gradually become too large for their cuticles. The underlying hypodermis then lays down a new, thin, elastic cuticle under the old one. Certain cells produce a molting fluid which partially dissolves the old cuticle, making it easier to shed after a split has been formed in it. After the molt the new cuticle hardens and then gradually thickens again by formation of more chitin.

Mouth parts of insects. Incredible as it may seem, the mouth parts of all kinds of insects, from the simple chewing organs of a grasshopper to the highly modified piercing organs of mosquitoes and the coiled sucking tube of butterflies and moths, are modifications of a single fundamental type which is represented in its simplest form in the chewing or biting type, as found in grasshoppers and beetles

(Fig. 152). The mouth parts in these insects consist of an upper lip or *labrum;* a lower lip or *labium* bearing segmented *labial palpi;* a pair of *mandibles* or jaws; a pair of *maxillae* lying ventral to the mandibles, bearing segmented *maxillary palpi,* and with two distal processes, the *galea* (lower in Fig. 152) and the *lacinia;* and the *hypopharynx* on the floor of the mouth, through which the ducts of the salivary glands

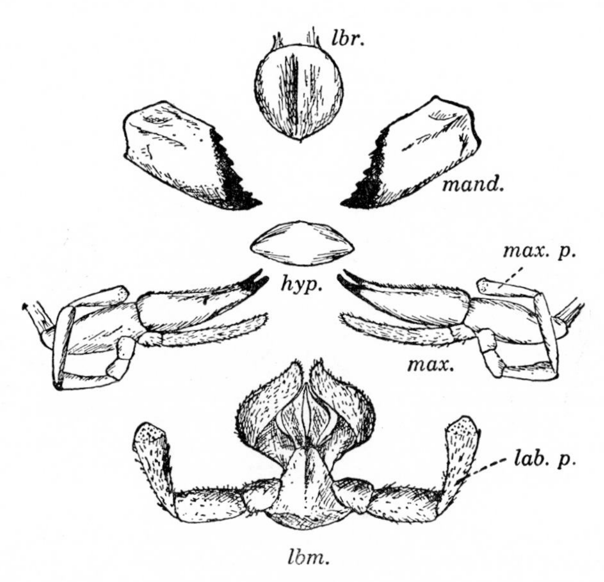

Fig. 152. Primitive mouth parts of a chewing insect: *lbr.,* labrum; *mand.,* mandible; *hyp.,* hypopharynx; *max.,* maxilla, consisting of a basal segment, the *stipes,* to which the maxillary palpus (*max.p.*) is articulated, and two distal parts, lateral *galea* and a mandible-like inner or medial lobe, the *lacinia* (black-tipped in figure); *lbm.,* labium, really a second pair of maxillae fused together, and bearing the labial palpi (*lab.p.*).

open. In addition the roof of the pharynx, under the labrum, has a sclerotized *epipharynx;* this is often combined with the labrum to form a *labrum-epipharynx.*

Legs. The legs of insects (Fig. 153*A*) consist of five parts: the coxa, trochanter, femur, tibia, and tarsus. The *coxa* articulates the leg with the body and sometimes appears more like a portion of the body than a segment of the leg. The *trochanter* is a very short inconspicuous segment and sometimes appears like a portion of the femur. The *femur* and *tibia* are long segments. The *tarsus,* or foot, consists of a series of segments, most commonly five; often the first segment is much the longest. Usually the tarsus is terminated by a pair of claws but sometimes only one. Often there are padlike structures, *pulvilli,* which have glandular hairs or pores through which an adhesive substance is excreted, permitting the insects to walk on the under side of objects. Sometimes there is a pulvillus at the base of each claw and also a similar median structure between them, called an *empodium* (Fig. 153*B*).

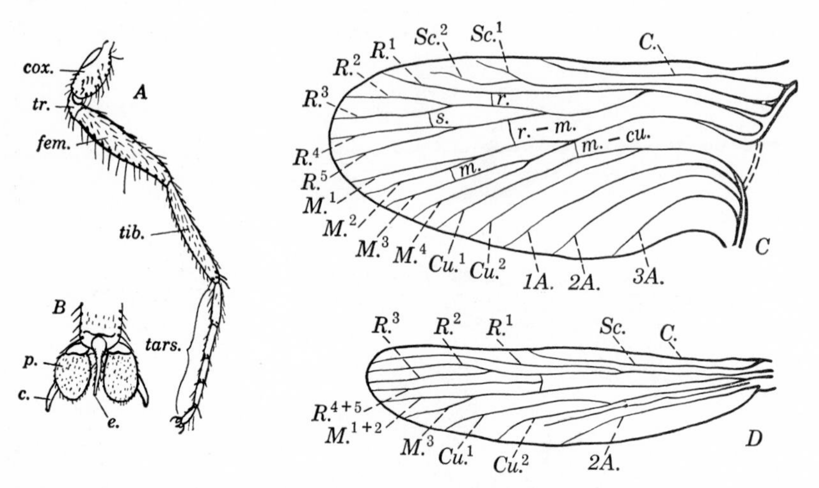

Fig. 153. Leg and wings of insects. *A*, leg; *cox.*, coxa; *fem.*, femur; *tars.*, tarsus; *tib.*, tibia; *tr.*, trochanter. *B*, foot, enlarged; *c.*, claw; *e.*, empodium; *p.*, pulvillus. *C*, diagram of primitive tracheae of wing from which the veins are derived. *D*, venation of a mosquito wing showing a comparatively simple modification; *c.*, costa; $Sc.^{1 \text{ and } 2}$, subcosta, branches 1 and 2; $R.^{1 \text{ to } 5}$, radius, branches 1 to 5; $M.^{1 \text{ to } 4}$, media, branches 1 to 4; $Cu.^{1 \text{ and } 2}$, cubitus, branches 1 and 2; *1A.*, *2A.*, and *3A.*, first to third anal; *r.*, radial cross vein; *s.*, sectorial cross vein; *r.–m.*, radio-medial cross vein; *m.*, medial cross vein; *m.–cu.*, medio-cubital cross vein (inadvertently omitted in mosquito wing). (*A* and *B* after Matheson, *Medical Entomology*, Comstock. *C* after Comstock, *Introduction to Entomology*, Comstock.)

Wings and venation. The structure of the wings of insects is often of great use in classification and identification. Only in a few primitive orders are the wings primarily absent, although in many forms, especially parasitic ones, e.g., lice and fleas, they are secondarily lost. Typically there are two pairs of wings, borne by the second and third segments of the thorax.

The wings originate as saclike folds of the body wall, but the upper and lower surfaces become applied to each other and thus they appear as simple membranes. Where they flatten down against the tracheae hollow supports or *veins* are formed. In most insects the majority of the veins are longitudinal, but there are usually a few cross-veins, which in some kinds of insects are very numerous. Figure 153*C* shows the hypothetical primitive arrangement of the venation of an insect wing, but in many insects the modifications brought about by coalescence, anastomosis, atrophy, and addition of extra branches and cross-veins often make it as difficult as a Chinese puzzle to determine the true homologies of the resulting veins. The spaces between the veins, called *cells*, are named after the longitudinal veins behind which they occur. Figure 153*D* shows the wing of a mosquito as an example of a comparatively simple modification.

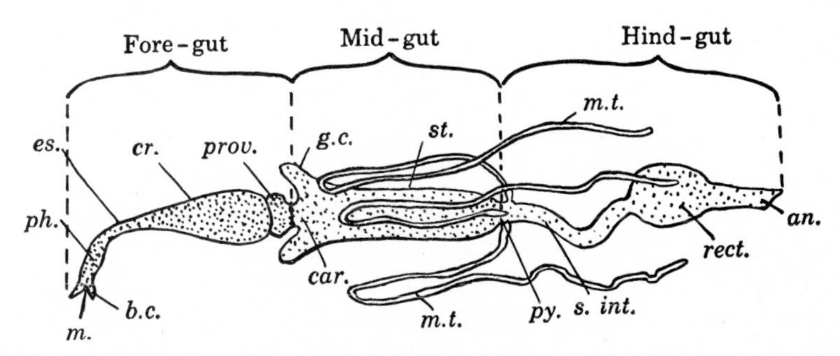

Fig. 154. Diagram of alimentary canal of an insect, showing portions pertaining to foregut, midgut, and hindgut, respectively: *an.*, anus; *b.c.*, buccal cavity; *car.*, cardium; *cr.*, crop; *es.*, esophagus; *g.c.*, gastric ceca; *m.*, mouth; *m.t.*, Malpighian tubules; *ph.*, pharynx; *prov.*, proventriculus; *py.*, pylorus; *rect.*, rectum; *s.int.*, small intestine; *st.*, stomach. (Adapted from Snodgrass, *Principles of Insect Morphology*, McGraw-Hill.)

Internal anatomy. The alimentary canal of insects (Fig. 154) has three primary divisions which may be of very unequal extent, namely, (1) the *foregut*, including pharynx, esophagus, crop, and proventriculus; (2) the *midgut*, including the stomach and sometimes a midintestine; and (3) the *hindgut*, including the small intestine and rectum. The foregut and hindgut epithelium is of ectodermal origin and these regions are lined by cuticle; the midgut lining is endodermal. The junction of the midgut and hindgut is marked by the entrance of the *Malpighian tubules* (see second paragraph following).

The *pharynx* in bloodsucking insects is muscular and acts like a suction pump. The ducts of the *salivary glands*, which themselves lie in the thorax, may open into the floor of the pharynx or in bloodsucking forms may unite and continue to the tip of an elongated hypopharynx. The pharynx is followed by an *esophagus*, which in some insects is expanded into a capacious *crop* and in some into a muscular *proven-*

triculus provided with chitinous teeth and serving the same function as the gizzard of a bird; in the mosquitoes three pouchlike *food reservoirs* are connected with the esophagus.

The true *stomach* follows the proventriculus and is sometimes provided with ceca that produce digestive juices; the stomach may constitute the entire midgut or it may be narrowed behind into a *mid-intestine*. The exposed cells of the midgut are protected in many insects by a delicate tubular membrane of cuticular material, the peritrophic membrane, secreted by cells in its anterior portion, but since this membrane lacks the epicuticular layer it is freely permeable. Enzymes may be liberated into the lumen either in vacuoles eliminated by the cells, or by disintegration of cells. The latter is important in the liberation of intracellular rickettsias into the lumen in lice. A number of slender Malpighian tubules enter at the posterior end of the midgut. These function as excretory organs, corresponding to the kidneys of vertebrate animals; they extract waste products from the blood, convert them into less soluble form, and pass them into the hindgut, to be voided through the anus along with the feces.

The *hindgut* in some insects has a distinct *small intestine* followed by a more expanded *rectum*, but in others there is only a rectum, which is lined by chitin. Some insects have an expanded *anal pouch* at the posterior end.

The *tracheae* of insects constitute a ventilation system of air tubes ramifying all through the body even to the tips of the antennae and legs. They open by a series of pores along the sides known as *spiracles*.

The *nervous system* of insects is very highly developed. In some species the instincts simulate careful and accurate reasoning, and it is difficult not to fall into the error of looking upon them as animals endowed with a high degree of intelligence.

Sense organs. Nearly all the spines and hairs on the body surface of insects are sensory end organs (*sensilla*). These structures assume a variety of forms and serve many different senses. The simplest are articulated setae which are stimulated by movement in the socket and thus serve as tactile sensilla. Some tactile sensilla are modified so that they respond to pressure, body movement, or even sound vibrations. These may be exceedingly complex, as in mosquitoes, and involved in the control of equilibrium in flight and in sound detection.

Sensilla connected with a cuticular drum or *tympanum* are specifically associated with the detection of sound by insects. They may be located on the legs or on the abdomen. These organs are frequently sensitive to sound frequencies above those detected by the human ear.

Sensilla of several kinds seem to be sensitive to chemical stimulation. Organs of taste have been demonstrated in association with the mouth in many insects but also occur on the feet and antennae. In some insects contact chemical receptors occur in the ovipositor. Dethier (1957) has summarized elegant experiments on chemoreception in insects.

The *compound eyes* are the main visual organs of insects. They are made up of a number of elongated light-refracting units, each having a light-sensitive structure beneath it. Each unit receives a fraction of the external light corresponding to the limited area at the eye surface. Vision is thus a mosaic of bundles of stimuli received by the units of the eye. Form and color can be perceived, but it is not feasible to compare this directly with human vision. Many insects also have simple eyes or *ocelli* which seem to be designed for light perception rather than image perception. In addition, some insects have light receptors on the body, but these have not been specifically identified.

Reproductive organs. The reproductive organs consist of paired ovaries or testes with their respective oviducts or sperm ducts opening near the posterior end of the abdomen, usually on the ventral side. Most female insects have a *spermatheca* or storage sac for sperms, for most kinds of insects mate only once, whereas the egg-laying period may extend over a long time. In some insects, e.g., the fleas, the shape of the chitinized spermatheca is a good identification character. The eggs are fully formed with shells (chorion) before fertilization; one or more minute pores, the micropyles, permit entry of sperms, usually during passage through the vagina. Many insects have an *ovipositor* which may simulate a miniature saw, borer, or piercing organ for depositing the eggs; this versatile organ may be, in the Hymenoptera, modified into a *sting*. In the male the sperm ducts unite to form an *ejaculatory duct*, the terminal part of which may be sclerotized and evaginated as an intromittent organ. Many male insects have highly developed *external genitalia* in the form of *claspers* and accessory parts, details of which often provide valuable identification marks. Some species of *Culex* cannot be differentiated with certainty in any other way. (See Figs. 206 and 229.)

Life History

Most insects hatch from eggs deposited by the mother, but in some instances free young are born; in the hippoboscids and tsetse flies the eggs hatch before birth and the young are retained and nourished in the body of the mother until they are ready for pupation. In these cases only a few young are produced, but most insects lay large numbers of eggs, some all at once, some in batches at intervals, and others individually at short intervals.

Three principal types of life history can be recognized among insects. In the primitive subclass Apterygota alone there occurs *direct development,* in which the newly hatched insect is almost a miniature of its parent and merely increases in size, continuing to molt periodically after becoming an adult. Among the higher insects, Pterygota, which are winged or secondarily wingless, the two common types of development are by *incomplete* and *complete metamorphosis.* Insects with an incomplete metamorphosis may differ more or less from their parents when hatched, but gradually assume the parental form with successive molts. There is no quiescent phase. The young or *nymphs* of such insects invariably lack wings and often have other characteristics different from their parents.

Insects with a complete metamorphosis are usually totally different from the parents when newly hatched and do not assume the parental form gradually. The early stages of such insects, usually wormlike, are called *larvae* in distinction from nymphs of insects with an incomplete metamorphosis. Upon completion of larval growth and development they go into a quiescent, more or less inactive stage, and are then known as *pupae.* The pupa may have no special protection, as in mosquitoes and midges; it may retain the last larval skin as a protecting case called the *puparium,* as in muscid flies; or it may be encased in a cocoon of silk thread spun by the larva as a protection from the hostile world before going into its mummylike pupal state, as in fleas and many moths.

Although apparently inactive, the pupal stage is frequently one of feverish activity from a physiological standpoint, for the entire body has to be practically made over. This transformation necessitates the degeneration of almost every organized structure in the body and a reformation of new organs out of a few undifferentiated cells left in the wreckage. The time required for this wonderful reorganization is sometimes amazingly short. Many maggots transform into adult flies in less than a week, and some mosquito larvae transform into perfect mosquitoes in less than 24 hours.

The length of life of insects in the larval and adult stages varies greatly. The larval stage may occupy a small portion of the life, as in the case of many mosquitoes and flies, or it may constitute the greater part of it. Some mayflies, for instance, live the greater part of two years as nymphs, but they exist as adults not more than a few hours. As a rule male insects are shorter-lived than females. The length of life of the female is determined by the laying of the eggs—when all the eggs have been laid the female insect has performed her duty in life and is eliminated by nature as a useless being. The result is the paradoxical fact that ideal environmental conditions *shorten* the life of these insects, since they facilitate the early deposition of the eggs.

Classification of Insects

The identification of insects is based mainly on three characteristics: the type of development, the modification of the mouth parts, and the number, texture, and venation of the wings. All bloodsucking insects have mouth parts adapted in some way for piercing and sucking, but the types vary greatly in different groups. Many of the more thoroughly parasitic insects, e.g., lice, bedbugs, and "sheep ticks," have secondarily lost their wings entirely or have them in a rudimentary condition. In the whole order of Diptera the second pair of wings is reduced to inconspicuous club-shaped appendages known as *halteres*.

Not all entomologists agree on the division of insects into subclasses and orders. According to Ross (1956) there are 28 orders of which 5, including the proturans, springtails and "silver fish," are primitively wingless and are placed in a subclass Apterygota; all the others belong to the subclass Pterygota which have wings except when these are secondarily lost, as in such parasitic forms as lice and fleas.

Many of the orders of insects comprise small or little-known groups. There are six "big" orders, members of which are known to everybody. These are (1) Orthoptera, the grasshoppers, crickets, etc.; (2) Hemiptera, the true bugs, aphids, etc.; (3) Coleoptera, the beetles; (4) Lepidoptera, the moths and butterflies; (5) Diptera, the flies, mosquitoes, etc.; (6) Hymenoptera, the bees, wasps, and ants. Important parasites and disease vectors of man and animals are found in two of these big orders, the Hemiptera and the Diptera, but others are found in small orders, the Siphonaptera or fleas, the Anoplura or sucking lice, and the Mallophaga or chewing lice (bird lice). These two groups of lice are considered by many entomologists as suborders of one order, Phthiraptera. The characteristics of these five orders containing parasites are briefly as follows:

Hemiptera: metamorphosis incomplete; mouth parts fitted for piercing and sucking, the piercing organs being ensheathed in the jointed labium and folded under the head: in the suborder Heteroptera; first pair of wings, unless reduced, leathery at base and membranous at tip; second pair of wings, when present, membranous with relatively few veins. Parasites: bedbugs, conenoses, kissing bugs.

Anoplura: metamorphosis incomplete; wings secondarily lost; body flattened, the thoracic segments more or less fused; legs short, the tarsi with only one or two segments, adapted for clinging to hairs of the host; mouth parts highly modified, adapted for piercing and sucking. All parasitic on mammals: sucking lice.

Mallophaga: similar to Anoplura except mouth parts fitted for chewing, being reduced to a pair of mandibles, and thorax as narrow as or narrower than the head, with not all the segments fused as much as in Anoplura. Parasitic on birds and mammals: biting or "bird" lice.

Siphonaptera: metamorphosis complete; mouth parts fitted for piercing and sucking, the piercing organs being ensheathed in the labial palpi and part of the maxillae modified as holding organs; wings secondarily lost. Parasites: fleas, chiggers.

Diptera: metamorphosis complete; mouth parts fitted for piercing and sucking, for sucking alone, or rudimentary; first pair of wings (absent in a few species) membranous with few veins; second pair of wings represented only by a pair of club-shaped organs, the halteres. Bloodsuckers: sandflies, mosquitoes, blackflies, tabanids, hornflies, stableflies, tsetse flies; parasites: Pupipara, maggots.

Arthropods as Parasites and Bloodsuckers

Degrees of parasitism. All gradations exist between arthropods that are strictly parasitic throughout their lives, e.g., itch mites, hair follicle mites, and lice, and species that are purely predatory, existing entirely apart from their living restaurants except when actually feeding, e.g., mosquitoes, tabanids. Close to the strict parasite end of the series are the ixodid ticks and some adult fleas, which only intermittently leave their hosts. Somewhat farther removed are the bedbugs, triatomids, argasid ticks, and other adult fleas, which not only attend to their reproductive functions off the host but also leave to take their after-dinner naps; these forms, however, are normally inhabitants of the nests or habitations of their hosts and may be looked upon as parasites of the homes. Fleas are parasitic only as adults; certain mites and flies, only as larvae.

Effects of bites. The effects produced by arthropod bites are brought about mainly by direct or indirect reactions to the salivary secretions. Many insects in biting create a subcutaneous pool of blood from which they suck instead of directly from a capillary. This results in most parasites that are transmitted being locally deposited in the tissues, and not directly in the circulation (Gordon and Crewe, 1948). People commonly show immediate and delayed reactions to bites; the former reaction is due to prior sensitization and eventually may disappear as immunity develops, but the delayed reaction is due to slow-acting toxic substances in the injected saliva and usually disappears after repeated exposures before the immediate reaction does. The usual effects are local redness and swelling, with varying degrees of itching and sometimes pain, resulting in restlessness and loss of sleep.

Allergy and immunity. The susceptibility of different individuals to the toxic effects of insect bites varies widely and is certainly dependent to a considerable extent upon sensitization. The reaction to the bites of particular arthropods before sensitization has developed may be very painful immediately (e.g., those of tabanids and reduviids), or may slowly become extremely irritating over a period of hours or even days (e.g., those of redbugs and blackflies), or may be practically unnoticed (e.g., those of itch mites and lice) until after sensitization has developed (see pp. 29 and 30). Sometimes, as in bites of sandflies and fleas, a papule appears after a week or two, probably due to slow-acting toxic substances in the saliva. After subsequent bites the papules appear sooner and larger, are more inflamed and irritating, and eventually the whole area around the bites may become inflamed and swollen, and old bites are reactivated. Eventually, with increasing sensitization an immediate reaction appears in the form of a wheal, and at long last, maybe after many years, desensitization develops and the bites again pass unnoticed. During the period of sensitization such generalized allergic symptoms may appear as urticaria, fever, fatigue, restlessness, and a rotten disposition! Sometimes, however, the unthriftiness commonly seen in animals heavily infested with ectoparasites is more the cause of the heavy ectoparasite burden than the result of it (see p. 30). Some parasites produce *special* toxic effects, such as the paralysis produced by certain ticks (see p. 585) and the blue spots caused by crab lice.

Immunity is an important factor. The senior writer has seen innumerable newcomers to Texas, including himself, who suffered intolerably from redbug bites during the first season or two of exposure to them, but who gradually became more and more immune to them. In New Jersey it is a common experience for people from inland to suffer far more severely from salt-marsh mosquitoes while vacationing on the coast than do the residents, whereas people from the coast react similarly to the inland species of mosquitoes. Cherney, Wheeler, and Reed (1939) called attention to the fact that California fleas do not usually encroach on the comforts of the local population but are a source of great misery to newcomers for several months to several years.

Trager (1939) found that guinea pigs previously exposed to bites of larval or nymphal ticks developed so much immunity that larval ticks were incapable of feeding on them at all, and nymphs were unable to feed to repletion, owing to such rapid cellular reaction to the bites that the parasites were cut off from their food. Persons susceptible to flea bites react positively to injection of flea extract, whereas immunes react negatively; most susceptible persons immunized by injections of the flea extract either become oblivious to fleas or are much less annoyed by them. Some retain their immunity several years, others for only a few weeks. Immunized persons are not actually ignored by the fleas, but are unaware of their bites.

Although acquired immunity plays a large part, there appear to be some true instances of distastefulness of individuals to insects, based on some difference in skin metabolism which is not yet understood. One instance has been recorded in which a man's skin was highly toxic to ticks. Riley and Johannsen report a case of two brothers who volunteered to act as feeders for some experimental stock lice; the lice fed greedily on one but absolutely refused to feed on the other, even when hungry. This matter of natural repellance to arthropods needs further investigation.

The effect of diet on ectoparasites is a disputed question. Most observers have seldom seen healthy, well-fed animals swarming with fleas, lice, or mites, though this is often seen in sick, poorly nourished animals. Kartman (1949), however, obtained some evidence indicating *loss* of parasites in animals on inadequate diets. There is probably an interplay of several factors—need of the parasites for vitamins, ability of the host to develop immunity, and activity in picking or nipping the parasites.

Arthropods as Disease Transmitters

Important as arthropods sometimes are as parasites or bloodsuckers, it is in their capacity as carriers of germs or as intermediate hosts of other parasites that they have to be reckoned with as among the foremost of human foes. Since the beginning of the twentieth century many of the most important human and animal diseases have been shown not only to be transmitted by arthropods but also to be *exclusively* transmitted by particular genera or species. In addition, there are a number of other diseases, such as yaws, pinkeye, Q fever, tularemia, and anthrax, in which arthropods play an important but not exclusive role, and still others, such as many bacterial, protozoan, and helminthic infections of the digestive tract, in which they play a minor but not a negligible part.

Mechanical transmission. The simplest method of disease transmission by arthropods is *indirect mechanical transmission,* in which the arthropods function as passive carriers of disease agents, picking them up on the bodies or in the excretions of man or animals and depositing them on food. The importance of any particular species of arthropod depends on the degree to which its structure and habits facilitate such transportation. Prominent among these indirect mechanical transmitters are houseflies and roaches.

Slightly more specialized is *direct mechanical transmission,* in which the insects pick up the germs from the body of a diseased individual and directly inoculate them into the skin sores, wounds, or blood of other animals. Biting flies, such as tsetse flies, *Stomoxys,* and tabanids, transmit blood diseases in this manner, e.g., anthrax, fowl pox, and some animal trypanosomiases. Flies that feed on sores or wounds, such as eye flies and many muscids, transmit skin or eye diseases, e.g., yaws, trachoma, and Oriental sore. In most of these cases the organisms do not live for more than a few minutes to a few days in the vectors.

Biological transmission. When an arthropod plays some further part in the life of the parasite or germ than merely allowing it to hitchhike, and multiplication or cyclical changes or both take place within its body, the process is called *biological transmission.* Huff (1931) proposed a classification of different types of biological transmission as follows:

1. *Propagative:* the organisms undergo no cyclical changes but they multiply as in culture tubes. Examples: plague, yellow fever.
2. *Cyclopropagative:* the organisms undergo cyclical changes and multiply in the process. Examples: malaria, trypanosomes.
3. *Cyclodevelopmental:* the organisms undergo developmental changes but do not multiply. Examples: filariae, guinea worm.

To these we add:

4. *Paratenic* or *Transporting* (including vertebrates as well as arthropods): the organisms invade and often become encysted in some host, specific or nonspecific, after developing elsewhere, and are transported by this host to the final host. Example: *Syngamus.*

Transovarial transmission. Transmission of disease agents to offspring by invasion of the ovary and infection of the eggs, often loosely called "hereditary" transmission, is characteristic of arthropod infections where a state of almost perfect adaptation of parasite and host has been reached. It is especially frequent in mites and ticks, which transovarially transmit some Protozoa, tularemia bacilli, rickettsias, relapsing fever spirochetes, and some viruses. In Texas fever (see p. 595) and scrub typhus (see p. 553), transovarial transmission is a necessary part of the mechanism of transmission. Not many insect-borne infections are thus passed on from generation to generation, though some viruses may be.

Airplane dissemination of arthropods. Many domestic arthropods, or parasites on man, rats, or domestic animals, succeeded in making this "One World" for themselves in bygone days of slow boat travel. The airplane has made this a possibility for many more. Even if stowaway arthropods do not become established in their new surroundings, they may live long enough to pass diseases they may carry to local vectors or reservoirs.

Two unpleasant possibilities exist: (1) the introduction of more efficient vectors for diseases already in existence, e.g., *A. gambiae* to Brazil; (2) the introduction of new diseases with vectors or reservoir hosts. Ticks shipped from South America have arrived in this country harboring three different diseases not now known in America, and mosquitoes and ticks from Russia successfully carried with them an encephalitis virus. One shudders to think of the consequences if a yellow fever-infected mosquito were landed in India or China.

The only protection is very strict regulation of fumigation of airplanes from foreign countries. If the world is to be made safe from arthropod as well as human invaders in the future, these precautions will have to be used with the utmost care, for international air traffic is constantly expanding.

Insecticides and Repellents

Insecticides. Some insecticides, such as hydrocyanic acid and methyl bromide, are as toxic to vertebrates as to arthropods, which limits their uses, whereas others, such as pyrethrins and rotenone, are practically harmless in the amounts used. The chlorinated hydrocarbons (ClHC) and organic phosphorus compounds (OPhC), developed during and after World War II, are relatively much more toxic to arthropods than to vertebrates, but vary somewhat in this respect. Some of these newer chemicals, because of their slow deterioration, have long-lasting residual effect which may make surfaces treated by them lethal for weeks, months, or even a year. They have revolutionized insect control by chemicals. Disadvantages are their toxicity to man if carelessly handled, tendency to accumulate in milk, meat, or fat when applied to animals, and *most* unfortunately the tendency of arthropods to develop resistance to them.

Of the older inorganic insecticides, the deadly hydrocyanic acid and sulfur dioxide are still used as fumigants in emergencies, e.g., to kill both rats and fleas in a plague outbreak. Nicotine sulfate painted on roosts (1 pint per 200 ft.) and sodium fluoride or fluoro-silicate in dips (1 ounce per gallon) or as dusts applied as pinches are still used against lice and mites on poultry, though they have been largely superseded by rotenone or ClHC or OPhC dusts. Arsenic (As_2O_3) in 0.175 to 0.19% solution is still useful in dips to kill ticks on cattle, especially when the ticks have developed resistance to other insecticides.

Rotenone and pyrethrins are the only important insecticides derived from plants. Rotenone, a resin, is the active principle (5%) in powdered roots of *Derris* or *Lonchocarpus* (cubé), first known as fish poisons. Pyrethrins are oil-soluble esters extracted from the powdered flower heads of certain species of *Chrysanthemum*. They are nonvolatile, but are unstable and so have no residual effect. The addition of certain synergists, especially piperonyl butoxide or sulfoxide at a rate of 10:1 enhances the toxicity of pyrethrins and slows down their deterioration. Pyrethrins are absorbed only through the cuticle. Even very small amounts paralyze insects and thus give a quick knockdown, but large amounts are required to kill them. They are practically nontoxic to warm-blooded vertebrates. There is a related synthetic compound, allethrin, but this is less effective than synergized pyrethrins.

Of the ClHC the first to come into use and the one still used most extensively is DDT, but there are numerous others. These include several analogues of DDT, especially methoxychlor, which has a special advantage for use on dairy cattle because it does not accumulate in milk. Others widely used are dieldrin, benzene hexachloride (BHC), chlordane, and toxaphene. Dieldrin is many times more toxic to most insects than is DDT, and also has less tendency to repel them. BHC and chlordane have special value, e.g., when used on absorbent surfaces of mud houses in the tropics, because they are slowly volatile and so have continued fumigant action from under the surface. Commercial grades of most of the ClHC (and OPhC also) contain at least 80 to 85% active ingredients, but BHC only about 13%, since only the gamma isomer is actively insecticidal. This isomer is obtainable in practically pure form and marketed as "lindane" or "gammexane." All the ClHC act primarily on the central nervous system, causing slow paralysis, but the exact mechanism of action is unknown.

The OPhC, at first used mainly for agricultural pests, are in general more toxic to vertebrates than the ClHC, but are coming into more and more extensive use as larvicides in water or in manure, as space sprays, in baits, impregnated cords, etc., used against adult houseflies.

The most extensively used OPhC are malathion, diazinon, dypterex, and chlorthion, and the more toxic parathion for agricultural pests. These compounds act primarily as inhibitors of cholinesterase, permitting acetylcholine to accumulate.

Both the ClHC and OPhC are absorbed through the stomach as well as the cuticle, and are commonly ingested when the insects groom themselves to remove the clinging crystals or powder. Different insects vary in their susceptibility to these insecticides either in general or to particular ones, and also in their ability to develop resistance (see below). Mosquitoes, of the *Culex pipiens* group, for instance, are considerably more resistant to most insecticides than are *Aedes* or *Anopheles*, and triatomids and many ticks (e.g., *Ornithodoros moubata*) are more resistant to DDT than to the other ClHC.

Formulations and dosages. All the organic insecticides are soluble in oils but only slightly so in water. They are prepared for use as 5% (or less) solutions in oil; as 20% emulsifiable concentrates (usually in xylene), or as 50% wettable powders (25% for some OPhC) for aqueous emulsions or suspensions; or as dusts. Application may be as residual sprays, dips, dusts, granules, volatile fumigants, or pellets, aerosols, fogs or smokes, or internally by injection or in food. There are also specialized methods for particular purposes, such as additions of concentrates to running water for blackfly larvae, oil films for mosquito larvae, wet or dry sugar baits or impregnated cords for houseflies, and automatic "do-it-yourself" back rubbers and treadle sprayers for biting flies on cattle.

In general the concentrations employed for dusts are about ten times those for sprays, and for dips one-half to one-tenth as much. The usual concentration of the ClHC and OPhC sprays on animals is 0.5%, but dieldrin can be used in about one-fourth this concentration, and lindane one-tenth. Pyrethrin sprays, having no residual effect, are used mainly on animals, usually 0.1% plus 1% synergist, but one-fourth of this suffices for lice. Enough to wet the coat (1 to 2 quarts per cow) is required. For residual sprays on surfaces of houses, barns, etc., application at a rate just short of run-off is best. At the usual rate of about 1 gallon per 1000 sq. ft. the recommended dosages of 200 mg. per sq. ft. of DDT, 40 to 50 of dieldrin, and 20 of lindane are obtained by using 5%, 1%, and 0.5% sprays. However, ticks and triatomids require more. In some situations, local applications by brushing on may be better than general spraying. For aquatic larvae about 0.05 to 0.1 lb. per acre of DDT or 0.025 of dieldrin is desirable, applied as a solution in 1 gallon of fuel oil, but where fish need not be considered 1 lb. per acre may give control for as much as a year.

Suspensions made from wettable powders with a particle size of 10 μ or less are most satisfactory for general use. The solutions and emulsions have disadvantages in greater absorption and in forming large recumbent crystals when the solvent evaporates instead of fine erect crystals that are more readily picked up by insects. Much depends on the nature of the surface, particle size, accompanying substances, etc. (see p. 534). Also the situation is different with BHC which is *more* effective if absorbed (not too deeply) because of the longer-lasting fumigant effect. For dips either suspensions or emulsions can be used.

For dusts, ClHC and OPhC and sometimes rotenone are diluted with pyrophyelite, talc, or other inert substances. The inert particles tend to stick to and abrade the cuticle, and the insecticides dissolve in the lipoids of the epicuticle (p. 519), thus facilitating outward passage of water and inward penetration of the insecticide. Dusts are used for many outdoor applications, e.g., preflood treatment of rice fields; underclothing for lice; in ratruns and bait boxes to kill fleas; on animals; in poultry houses; as pinches applied to chickens, etc.

To kill aquatic larvae (mosquitoes, *Culicoides*, tabanids, etc.) in swamps, marshes, rice fields, etc., where there is heavy vegetation and the water is shallow, granules (30 to 60 mesh), composed of bentonite, clays, or tobacco by-products, with insecticides incorporated, are more effective and longer lasting than sprays. Two lb. per acre of 5% dieldrin granules is sufficient in rice fields. Pellets (10 grams) of sand and cement (5:1) impregnated with 16% by weight of dieldrin or lindane are excellent against mosquitoes in some situations (see p. 760).

Space sprays indoors are designed to give immediate kills with little or no residual effect; these are used extensively where residual insecticides are undesirable, as in food-processing plants, restaurants, ships, planes, etc. Combinations of ClHC (aq. 3% DDT) and pyrethrins (0.4%) are commonly used; dispensed as kerosene solutions in hand sprayers, or preferably oil solutions dissolved (15%) in freon (85%) under pressure and released from aerosol "bombs." The freon immediately evaporates liberating the insecticide in minute particles 2 to 10 μ in diameter, as compared with droplets 5 to 150 μ in diameter from atomizers. The fine particles float in the air and permeate every crevice. In vaporizers lindane is used almost exclusively.

For outdoor use there are various devices for producing fogs, thermal aerosols, smokes, etc., usually dispensed from ground vehicles or with the exhaust from planes (see Davidson, 1955). They may be used for protection of outdoor gatherings, or in barrier strips a few 100 ft. wide to protect camps, etc.

Attempts have been made with some success to get action of insecticides from within by adding them to feed or injecting them under the skin (see Lindquist and Knipling, 1957). It may be useful during outbreaks of anthrax, etc., against biting flies. Phenothiazine in salt, also used for intestinal nematodes, reduces warble infestation.

Resistance. A serious drawback to the use of chemical insecticides is the ability of arthropods to develop resistance to them, sometimes to the extent of complete immunity, as micro-organisms do to chemotherapeutics. From the standpoint of resistance, the ClHC fall into two groups—DDT and its analogues on the one hand, and dieldrin, BHC, chlordane, toxaphene, etc., on the other. Resistance to one group does not entail resistance to the other, though this may develop more quickly. The OPhC form a third group to which houseflies, mosquitoes, and a few other insects have thus far developed resistance. No appreciable resistance has developed to the pyrethrins or allethrin, and none to the synergist, piperonyl butoxide.

Some insects, and particularly houseflies, are far more prone to develop resistance than are others, but some degree of resistance to some of the ClHC has been developed by nearly all arthropods of medical and veterinary importance, but usually in localized areas in various parts of the world.

Much research has been done on the nature of insecticide resistance. The resistance is not due to adaptation by exposure to the chemicals, but to natural selection and reproduction of resistant individuals when the susceptible ones are killed. The resistance is sometimes controlled by a single pair of recessive genes that are already there, though resistance to DDT is controlled by a different pair of genes than that to the other ClHC. Resistance to DDT is largely due to presence of an enzyme that converts it into nontoxic DDE inside the body. This conversion is strongly inhibited by piperonyl cyclonone, which therefore acts as a synergist enhancing the effectiveness of DDT. When a chemical to which resistance has developed is withdrawn from use, susceptibility gradually returns. Behavioristic resistance also has been reported, when insects refuse to alight on treated surfaces.

Repellents. Thousands of chemicals have been tested to make skin or clothing obnoxious to biting arthropods, but only a few pass the criteria of being effective and long-lasting without being toxic, irritating, or sensitizing; injurious to clothing; or objectionable to the nose. Some of these, either alone or in mixtures, give protection as skin repellents for several hours and on clothing for several days or weeks even, in some cases after several washings. No repellent is equally effective against all kinds of arthropods, so mixtures are commonly made to try to get as wide a coverage as possible. Some of the best individual chemicals are diethyltoluamide, diethylbenzamide, indalone ethyl hexanediol (Rutgers 612), dimethyl and dibutyl phthalate, dimethyl carbamate, and benzyl benzoate. The last four are toxicants as well as repellents against mites and ticks. For use on the skin the repellents are dispensed as liquids, or pressurized sprays, commonly diluted 50% or more with alcohol. They can be sprayed or rubbed on clothing, or outer clothing can be impregnated by wetting with the repellent dissolved in dry-cleaning fluid; soaking in an emulsion (5 tablespoons to 3 pints water and 1½ tablespoons of soap or Tween 80), wringing and drying; or (for redbugs, ticks, and fleas) by the barrier method—merely applying solutions to socks or stockings and to trouser cuffs (and other openings if you sit or lie down). All the repellents affect paints, varnishes, and some plastics, and are injurious to rayon but not cotton, wool, or nylon clothing, but "612" and diethyltoluamide less so than the others.

Diethyltoluamide is the best known repellent for mosquitoes, biting flies, and fleas, both for skin and clothing applications; it is effective for the longest time on the skin, and is most resistant to wiping off, on the skin, and to rinsing, on the clothing. According to Traub it gives solid protection against mosquitoes in Malaya for at least 4 hours even when the attack rate is extremely high. For redbugs it lasts longer but is more easily washed out than the standard army repellent, M1960, and it is a little less effective against ticks (*Amblyomma americanum* at least). All the repellents have their period of effectiveness cut in half by profuse sweating.

REFERENCES

Andrews, J. M., and Simmons, S. W. 1948. Developments in the use of the newer organic insecticides of public health importance. *Am. J. Public Health,* 38: 613–631.

Baer, J. G. 1952. *Ecology of Animal Parasites.* University of Illinois Press, Urbana, Ill.

Barlow, F., and Hadaway, A. B. 1952. Some factors affecting the availability of contact insecticides. *Bull. Entomol. Research,* 43: 91–100.

Brown, A. W. A. 1951. *Insect Control by Chemicals.* Wiley, New York.

Busvine, J. R. 1951. *Insects and Hygiene.* Methuen, London.

Busvine, J. R. 1957. Insecticide-resistant strains of public health importance. *Trans. Roy. Soc. Trop. Med. Hyg.,* 51: 11–31.

Busvine, J. R., and Nash, R. 1953. The potency and persistence of some new synthetic insecticides. *Bull. Entomol. Research,* 44: 371–376.

Buxton, P. A. (Arranger). 1952. Symposium on Insecticides. *Trans. Roy. Soc. Trop. Med. Hyg.*, 46: 213–274.

Cherney, L. S., Wheeler, C. M., and Reed, A. C. 1939. Flea antigen in prevention of flea bites. *Am. J. Trop. Med.*, 19: 327–332.

Comstock, J. H. 1940. *An Introduction to Entomology.* Comstock, Ithaca, New York.

Davidson, G. 1955. The principles and practice of the rise of residual contact insecticides for the control of insects of medical importance. *J. Trop. Med. Hyg.*, 58: 49–56, 71–80.

Day, M. F. 1955. Mechanisms of transmission of viruses by arthropods. *Exp. Parasitol.*, 4: 387–418.

Day, M. F., and Bennetts, M. J. 1954. A review of problems of specificity in arthropod vectors of plant and animal viruses. *Commonwealth Sci. and Ind. Res. Org., Australia.*

Dethier, V. G. 1957. The sensory physiology of blood-sucking arthropods. *Exp. Parasitol.*, 6: 68–122.

Essig, E. O. 1942. *College Entomology,* Macmillan, New York.

Ewing, H. E. 1929. *A Manual of External Parasites.* Thomas, Springfield, Ill.

Fay, R. W., and Kilpatrick, J. W. 1958. Insecticides for control of adult Diptera. *Ann. Rev. Entomol.*, 3: 401–420.

Gilbert, I. H., Gouck, H. K., and Smith, C. N. 1957. New insect repellent. *Soap and Chem. Specialties,* May 1957 (Reprint).

Gordon, R. M., and Crewe, W. 1948. The mechanisms by which mosquitoes and tsetse-flies obtain their blood meal, the histology of the lesions produced, and the subsequent reactions of the mammalian host; together with some observations on the feeding of *Chrysops* and *Cimex.* *Ann. Trop. Med. Parasitol.*, 42: 334–356.

Graham-Smith, G. S. 1913. *Flies in Relation to Disease* (*Non-Bloodsucking Flies*). Cambridge University Press, Cambridge.

Granett, P., and Haynes, H. L. 1945. Insect-repellent properties of 2-ethylhexanediol-1,3 (Rutgers 612). *J. Econ. Entomol.*, 38: 671–675.

Green, H. L., and Lane, W. R. *Particulate Clouds: Dusts, Smokes, and Mists.* E. and F. N. Spon, London.

Hadaway, A. B., and Barlow, F. 1952. Studies on aqueous suspensions of insecticides, Pt. III. *Bull. Entomol. Research,* 43: 281–311.

Hall, M. C. 1929. Arthropods as intermediate hosts of helminths. *Smithsonian Misc. Collections,* 81: No. 15.

Herms, W. B. 1950. *Medical Entomology.* 4th ed. Macmillan, New York.

Hindle, E. 1914. *Flies and Disease* (*Bloodsucking Flies*). Cambridge University Press, London.

Huff, C. G. 1931. A proposed classification of disease transmission by arthropods. *Science,* 74: 456–457.

Imms, A. D. 1951. *A General Textbook of Entomology,* 8th ed. Methuen, London.

Kartman, L. 1949. Preliminary observations on the relation of nutrition to pediculosis of rats and chickens. *J. Parasitol.*, 35: 367–374.

Kearns, C. W. 1956. The mode of action of insecticides. *Ann. Rev. Entomol.*, 1: 123–166.

King, W. V. 1951. Repellents and insecticides available for use against insects of medical importance. *J. Econ. Entomol.*, 44: 338–343.

Knipling, E. F., et al. 1948. Evaluation of selected insecticides and drugs as chemotherapeutic agents against external blood-sucking parasites. *J. Parasitol.,* 34: 55–70.

Lindquist, A. W., and Knipling, E. F. 1957. Recent advances in veterinary entomology. *Ann. Rev. Entomol.,* 2: 181–202.

Matheson, R. 1950. *Medical Entomology,* 2nd ed. Comstock, Ithaca, N. Y.

Metcalf, C. L., Flint, W. P., and Metcalf, R. L. 1951. *Destructive and Useful Insects: Their Habits and Control.* 3rd ed. McGraw-Hill, New York.

Metcalf, R. L. 1955. Physiological bases for insect resistance to insecticides. *Physiol. Rev.,* 35: 197–232.

Metcalf, R. L. (Ed.) 1957. *Advances in Pest Control Research.* Vol. 1. Interscience Publishers, New York.

Mönnig, H. O. 1949. *Veterinary Helminthology and Entomology,* 3rd ed. Williams and Wilkins, Baltimore.

Neveu-Lemaire, M. 1938. *Traité de zoologie médicale et vétérinaire, II Entomologie.* Vigot Frères, Paris.

Patton, W. S., and Cragg, F. W. 1913. *A Textbook of Medical Entomology.* Christian Lit. Soc. for India, London.

Patton, W. S., and Evans, A. M. 1929. *Insects, Ticks, Mites, and Venomous Animals of Medical and Veterinary Importance.* Pt. I, Medical; Pt. II, Public Health. Grubb, London.

Roeder, K. D. (Editor), 1953. *Insect Physiology.* Wiley, New York.

Ross, H. H. 1956. *A Textbook of Entomology.* Wiley, New York.

Shepard, H. H. 1951. *The Chemistry and Action of Insecticides.* McGraw-Hill, New York.

Smart, J. 1956. *A Handbook for the Identification of Arthropods of Medical Importance.* 3rd ed. Brit. Mus. Natl. History, London.

Smith, C. N., Gilbert, I. H., and Gouck, H. K. 1958. Use of insect repellents. U. S. Dept. Agr., Agr. Research Serv. Leaflet 33–26.

Stage, H. H. 1947. DDT to control insects affecting man and animals in a tropical village. *J. Econ. Entomol.,* 40: 759–762.

Steinhaus, E. A. 1946. *Insect Microbiology.* Comstock, Ithaca, New York.

Trager, W. 1939. Acquired immunity to ticks. *J. Parasitol.,* 25: 57–81, 137–139.

U. S. Department of Agriculture. 1952. Insects. *Yearbook of Agriculture,* Washington.

Webb, J. E., and Green, R. A. 1945. On the penetration of insecticides through the insect cuticle. *J. Exptl. Biol.,* 22: 8–20.

Wigglesworth, V. B. 1948. The insect cuticle. *Biol. Rev.,* 23: 408–451.

Wilson, C. B. 1902–1922. (A series of 16 papers on North American parasitic copepods, Proc. U. S. Natl. Museum. Vols. 25, 27, 28, 31, 32, 33, 35, 39, 42, 47, 53, 55, and 60.)

U. S. Department of Agriculture. 1957. Insecticide recommendations. *Agr. Handbook,* 120.

World Health Organization. 1956. Specifications for pesticides, insecticides, rodenticides, molluscicides and spraying and dusting apparatus. World Health Organization, Geneva.

Chapter 23

THE ACARINA (MITES) AND PENTASTOMIDA

ACARINA (EXCEPT TICKS)

Acarina in general. The order Acarina of the class Arachnida includes a large number of species commonly known as mites or ticks. Most of them are very small, some barely visible to the naked eye, but some of the ticks reach half an inch or more in length. The ticks, although constituting only one of the five suborders of Acarina, are distinct and important enough to warrant special consideration in a separate chapter (24).

There is a great variety of body form in the Acarina, some appearing quite grotesque. The majority are more or less round or oval, without division into head, thorax, or abdomen, but some have a suture dividing the body into anterior and posterior divisions (Fig. 159). Many have dorsal or ventral plates to reinforce the cuticle. There is no true head, but the mouth parts are borne on an anterior part that is usually rather distinctly set apart, called a gnathosoma or capitulum. There are two pairs of mouth parts, the chelicerae and the pedipalps or palpi (Figs. 160, 168). The chelicerae usually end in little pincers made up of a movable and an immovable digit, but they may be modified into needlelike structures, as in *Dermanyssus*, or in other ways. The palps usually consist of four to six segments, sometimes modified as a thumb-and-claw. In the ticks the ventral wall of the gnathosoma is elongated between the palpi and armed with recurved teeth (Fig. 168). In the first or larval stage there are only three pairs of legs, but after the first molt when the mite or tick becomes a nymph, a fourth pair of legs is acquired. The legs typically consist of six or seven segments called the coxa, trochanter, femur (sometimes divided), genu, tibia, and tarsus. The tarsi usually bear claws and often a suckerlike caruncle, sometimes on long stalks (Fig. 155).

Many Acarina have pouches on the midgut which give them great food capacity; a well-fed female tick gets so distended that she looks more like a bean than an arthropod. For breathing, many mites have tracheae which open by one to four pairs of stigmata or spiracles. In the Mesostigmata and ticks the adults have only one pair, situated near the third coxae (Fig. 170), and in the Mesostigmata there is a sclerotized plate or *peritreme* leading forward from each spiracle. In the Prostigmata the spiracles are on or near the gnathosoma, whereas in the Sarcoptiformes there are either no spiracles (or tracheae) or many inconspicuous ones.

The anus opens on the ventral surface of the abdomen; in most mites, but not all, the genital openings are anterior to it and are closed by specialized plates. Usually there are morphological differences between the sexes, but sometimes the sexes are not easily distinguished.

Many mites are free-living and prey upon decaying matter, vegetation, stored foods, and the like; some are predaceous and feed upon smaller animals; some are aquatic, even marine; and many are parasitic on other animals during all or part of their life cycle. Some of these rank among the most important disease vectors, and some function as intermediate hosts of protozoans or helminths.

Life history. There are usually four stages in the development of mites and ticks: the egg, the larva, the nymph, and the adult (see Figs. 176, 177). The eggs are usually laid under the surface of the soil or in crevices or, in some parasites, under the skin of the host, but some species are ovoviviparous. After a varying period of incubation the larva hatches in the form of a six-legged creature, often quite unlike the parent. After a single good feed, or in many species without one, the larva molts and becomes an eight-legged nymph. In most parasitic Acarina except ticks (see Chapter 24) this protonymph, after a blood meal, molts and becomes a deutonymph, which may or may not feed again before becoming an adult. No molting occurs after the adult stage is attained. There may be additional intermediate stages, and some mites have specialized devices for dispersal; the larvae may parasitize hosts that transport them, e.g., trombiculids on vertebrates (p. 550) and trombidiids, and some water mites on insects, or they may merely hitchhike on insects, e.g., the grain mites (p. 559).

Some mites have become adapted to live as internal parasites in the lungs and air sacs of snakes, birds, and mammals, and there are records of mites which are not normally parasitic at all living and multiplying in the human urinary bladder; but all the species normally infesting man are either external or subcutaneous in their operations.

Classification. Baker and Wharton (1952) in a modification from the classification of Vitzthum (1940–1942), classify the Acarina as follows:

Suborder **Onchopalpida.** Palpi with claws as on legs; 2 or more pairs of stigmata on body. Contains no parasitic forms, although a species of *Holothyrus* in Mauritius is said to secrete a poison that may cause death of ducks and illness of children.

Suborder **Mesostigmata.** Body well-chitinized, with dorsal and ventral plates; gnathosome small, anterior; 1 pair of lateral stigmata near third coxae, each with a sinuous, chitinous peritreme leading forward; tarsi usually with claws and caruncles. Contains several families of interest:

- **Halarachnidae,** including parasites of respiratory passages of seals (*Halarachne*), and monkeys (*Pneumonyssus*).
- **Entonyssidae,** lung parasites of snakes (*Entonyssus*).
- **Rhinonyssidae,** nasal parasites of birds.
- **Dermanyssidae,** bloodsucking "red mites" of birds and mammals (*Dermanyssus, Ornithonyssus, Allodermanyssus*).
- **Laelaptidae,** bloodsucking mites of rodents, etc.

Suborder **Ixodides.** Ticks; size large; body leathery with or without plates; 1 pair of lateral spiracles near fourth coxae but no tubular peritreme; hypostome present, armed with recurved teeth. Includes 2 families:

- **Argasidae,** "soft" ticks; no dorsal shield; body tuberculated; gnathosome ventral.
- **Ixodidae,** "hard" ticks; dorsal shield present; body not tuberculated; gnathosome anterior.

Suborder **Trombidiformes.** One pair of stigmata on or near gnathosome; palpi free, highly developed; chelicerae developed for piercing. Contains several families of interest:

- **Demodicidae,** hair follicle mites; wormlike, with very short legs.
- **Pyemotidae(= Pediculoididae),** louse mites; a club-shaped organ between coxae 1 and 2; legs all similar; larvae hatch and develop to maturity in saclike abdomen of female.
- **Tarsonematidae,** plant pests; occasionally found in human lungs or intestine.
- **Trombiculidae,** harvest mites with parasitic larvae (redbugs or chiggers); tarsus of palpus forms thumb closing against tibia; body covered with feathered hairs; larvae with dorsal shield with hairs and a pair of pseudostigmatic organs; body of adult divided into two sections.
- **Cheyletidae,** *Psorergates* (mange mites of sheep and mice), and mites that prowl in foods, fur, or feathers hunting other mites.
- **Myobiidae,** soft-skinned mites attacking feathers or skin of birds, and skin of small mammals. This suborder also contains water mites, some of which have larvae parasitic on insects.

Suborder **Sarcoptiformes.** No well-developed stigmata or conspicuous tracheae. Chelicerae usually scissorslike for chewing; palpi simple; oral suckers often present.

Group **Acaridiae.** Soft-skinned; without stigmata or prominent club-shaped pseudostigmatic organs; tarsi with caruncles. Includes following families of interest:

Acaridae (= Tyroglyphidae), Glycyphagidae and related families, usually called tyroglyphids, infesting foods and causing grocer's itch.

Sarcoptidae, itch mites; soft, unsegmented body; skin with fine striations interrupted by scaly areas; legs short; includes *Sarcoptes, Notoedres,* and *Knemidokoptes.*

Psoroptidae, mange mites; dorsal shield present; bell-shaped caruncles on stalks; abdomen of male bilobed posteriorly; includes *Psoroptes, Chorioptes,* and *Otodectes.*

Group **Oribatei.** Heavily chitinized; prominent clublike pseudostigmatic organs; no caruncles. Free-living "beetle mites." Includes numerous families, of which at least seven contain species that serve as intermediate hosts of anoplocephalid tapeworms.

In the present chapter we shall consider all the Acarina except the ticks, which will be dealt with separately in Chapter 24.

Itch and Mange Mites (Sarcoptidae and Psoroptidae)

Species. The minute rounded or oval, short-legged, flattened mites of the family Sarcoptidae are the cause of scabies or "itch" in man; similar mites belonging to the related family Psorotidae (see p. 544) are the cause of mange or scab in many kinds of animals. The species that attacks man, *Sarcoptes scabiei,* is so similar to forms of *Sarcoptes* causing mange in many other animals—dogs, foxes, cats, rabbits, ruminants, horses, and pigs—that all of these are considered mere biological varieties of one species. These varieties are so adapted to the hosts in which they have been living that it is difficult to transfer them to other hosts. Other genera of these two families attack various domestic animals.

Following is a key to the most important genera:

1*a*. Posterior pairs of legs nearly or quite concealed under abdomen 2
1*b*. At least third pair of legs projecting 3
2*a*. Dorsum with spines and pointed scales (Fig. 155) ***Sarcoptes***
2*b*. Dorsum with spines and rounded scales ***Notoedres***
2*c*. No dorsal spines or scales (Fig. 156) ***Knemidokoptes***
3*a*. Pedicles of tarsal suckers very long ***Psoroptes***
3*b*. Pedicles short 4
4*a*. ♀ with suckers on legs 1, 2, 4; with posterior abdominal lobes ***Chorioptes***
4*b*. ♀ with suckers on legs 1, 2; abdomen without lobes ***Otodectes***
4*c*. ♀ with suckers on all legs; abdomen without lobes ***Dermatophagoides***

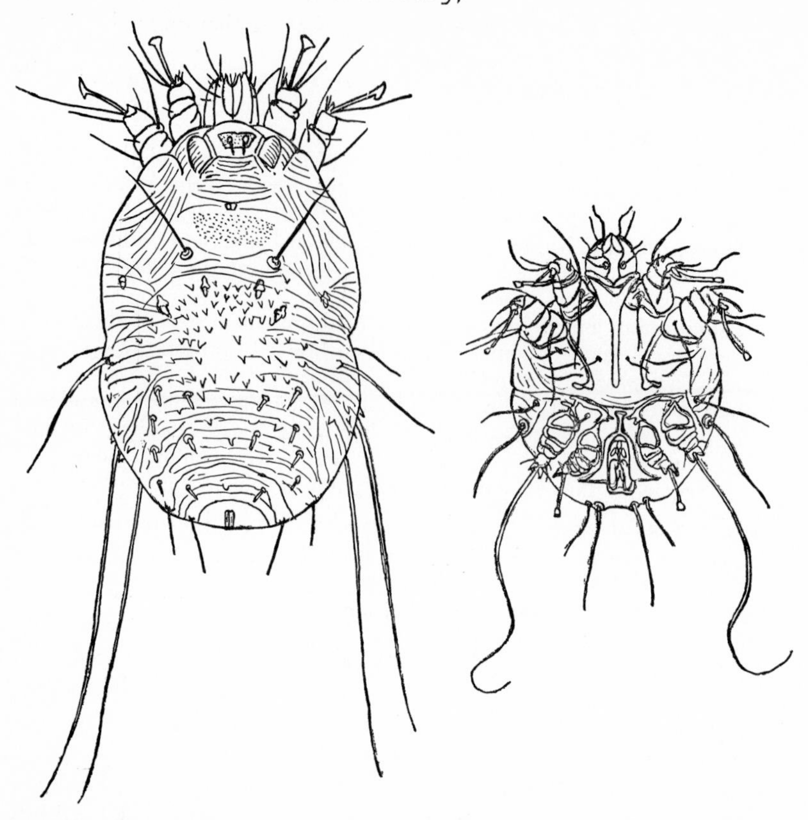

Fig. 155. *Sarcoptes scabiei,* itch mite. *Left,* ♀; *right,* ♂; ×150. (♀ adapted from Buxton, *Parasitology,* 13, 1921. ♂ after Bedford from Mönnig, *Veterinary Helminthology and Entomology,* 1949.)

Notoedres cati causes a very severe and sometimes fatal mange in cats; it temporarily infests man but soon dies out. *Psoroptes* does not burrow under the skin but causes "scab" in ruminants and horses; *Chorioptes* causes foot scab in horses, and *Otodectes* ear mange in carnivores. *Knemidokoptes* has several species causing scaly-leg in birds, one *K. mutans,* a pest of poultry and another, *K. pilae,* a pest of parrakeets; another species, *K. laevis* var. *gallinae,* is the "depluming mite" of poultry. *Dermatophagoides* is usually found on the skin of birds but has been reported to produce a dermatitis in man (Traver, 1951). *Psorergates ovis,* less than half the size of *Sarcoptes,* causes mange of sheep and injury to the wool, but this mite belongs to an entirely different group (see p. 542).

Sarcoptes scabiei. The itch mites, *Sarcoptes scabiei* (Fig. 155), are minute whitish creatures, scarcely visible to the naked eye. They are nearly round, and the cuticle is delicately sculptured with numerous wavy parallel lines, pierced here and there by stiff projecting bristles or hairs. They have no eyes or tracheae. The mouth parts, consisting of a pair of minute chelicerae and a pair of three-jointed triangular pedipalps, are attached to a capitulum or gnathosome in the front of the body. The legs are short and stumpy and are provided with sucker-like organs at the tips of long unjointed pedicels in the first two pairs of legs in the females, and in the first, second, and fourth pairs in the

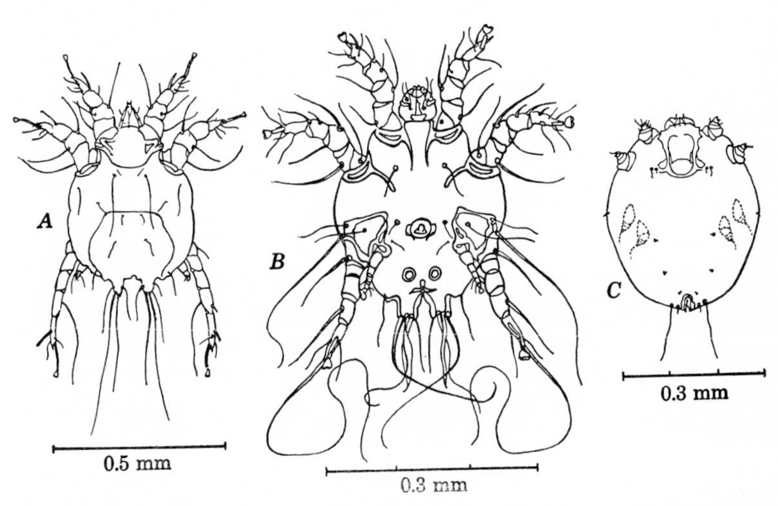

Fig. 156. *A, Psoroptes* ♂, dorsal view; *B, Chorioptes* ♂, ventral view; *C, Knemidokoptes* ♀, dorsal view. (After Whitlock, *Practical Identification of Endoparasites for Veterinarians*, Burgess.)

males; the other legs terminate in long bristles. The number of legs with pediceled suckers, and whether or not the pedicels are jointed, are important characters in differentiating other genera. In the human itch mite the male is less than 0.25 mm. in length and the female about 0.3 to 0.4 mm. in length.

The impregnated females excavate thin tortuous tunnels in the epidermis (Fig. 157), especially where the skin is delicate and thin. The tunnels measure a few millimeters to over an inch in length and are usually gray from the eggs and excrement deposited by the female as she burrows; under a lens they look like a chain of minute grayish specks punctuated at intervals by a tiny, hard, yellow blister. The daily excavations of a mite amount to 2 or 3 mm.

Life cycle. The eggs, about 160 μ long, are laid in the burrows at the rate of 2 or occasionally 3 a day until a total of about 35 to

50 have been laid, after which the female dies, usually at the end of a single tortuous burrow. The eggs hatch in a few days into larvae which resemble the adults except in minor details and in the absence of the fourth pair of legs. The larvae transform in 2 or 3 days into nymphs. The nymphs commonly build burrows for themselves and molt twice, the second time becoming adult male and female mites. The duration of the two nymphal periods is 3½ to 6 days, the entire development of the mites therefore requiring 8 to 14 days. The adults live about 4 weeks.

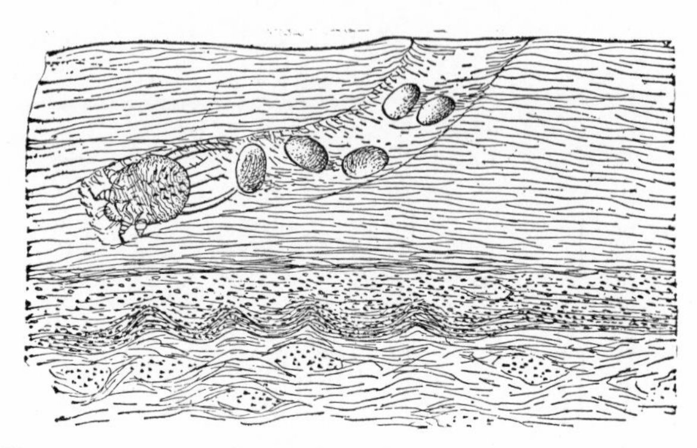

Fig. 157. Diagrammatic tunnel of itch mite in human skin, showing ♀ depositing eggs. About ×30. (Adapted from Riley and Johannsen, *Medical Entomology,* 1938.)

The mites are not necessarily nocturnal as was formerly supposed, but wander about on the surface of the skin when it is warm, most frequently when the host is in bed. The males are usually stated to be short-lived and to remain on the surface of the skin, but Munro in 1919 questioned this. The males are not, however, very commonly found. The young impregnated females make fresh excavations of their own. Since there is a new generation about every 3 weeks, the rate of increase is potentially enormous, yet according to Mellanby (1943) the average number of adult females in an infested person is less than 12, and not one person in ten has over 30.

THE DISEASE. The "itch" (scabies) is a disease which has been known much longer than it has been understood; it was formerly attributed to "bad blood." In the past, itch swept over armies and populations in great epidemics, but it has decreased with civilization and cleanliness.

As shown by Mellanby (1943, 1944), the intense itching that characterizes the disease does not begin until a month or so after an initial infection, when the skin has become sensitized; prior to this there is very little discomfort. After 6 weeks there is enough irritation to disturb sleep, and after about 100 days the irritation may be continuous and unbearable. In previously uninfected persons the mites reach a peak population of 50 to 500 in 7 to 16 weeks, after which the number declines sharply to 10 or less. The lesions, however, get worse and often appear where there are no longer any mites, and secondary infections, such as impetigo, develop. In reinfections intense local irritation, redness, and edema begin in 24 hours, often causing the parasites to be removed by the fingernails or to leave voluntarily an environment that is unfavorable for them because of edema or septic infections. The itching may persist for days or weeks after the mites are removed or killed. In reinfections the average number of mites present is only 3 or 4. A few mites may, however, persist for a very long time.

The mites invade the skin of the hands and wrists most frequently. Mellanby found them there in 85% of cases, but they also attack the groin and external genitals, breasts, feet, or other parts. The head is rarely attacked, although a severe "crusted" form of the disease called Norwegian itch occurs in Europe and attacks the head as well as other parts.

Although the burrows of the mites are often sufficiently characteristic to make a diagnosis possible, it should usually be confirmed by finding the mites, which are not in the vesicles but usually near them at the ends of the burrows. Scrapings from the blind ends of the burrows should be examined microscopically for adults or larvae; the latter are only about 0.15 mm. long.

EPIDEMIOLOGY. Infection can result only from the passage of male and female mites or of an impregnated female from an infected to a healthy individual. Normally this takes place by actual contact, rarely in the daytime on account of the secretive habits of the mites, but commonly at night, especially from one bedfellow to another. Mellanby found that transmission through bedding or clothing is relatively rare, except after contact with the small percentage of cases having a large number of mites. He was uniformly successful in establishing infections in previously uninfected volunteers by transfer of young impregnated mites but never younger stages. The adults can live apart from a host for 2 or 3 days under favorable conditions. It is possible for infection to be derived from mangy animals, though the mites, once adapted for several generations to a given host, do not often survive a transfer to a different species of host for more than a few days.

TREATMENT AND PREVENTION. For many years the standard remedy for scabies was sulfur ointment (½ ounce sulfur in 16 ounces lard or lanolin) applied after softening the skin with soap and warm water. Now a single painting of the body from the neck down with 25% benzyl benzoate emulsion, even without sterilization of bedding or clothing (but with simultaneous application to bed mates) is usually successful. Lindane (1% in ointments or vanishing cream), left on for 4 days, is also effective, and Tetmosol in bath soap has both curative and prophylactic value (Baker et al., 1956).

Prevention of this annoying infection consists merely in avoiding contact with infected individuals and of shunning public towels or soiled bed linen. When introduced among groups of previously uninfected individuals, scabies may cause extensive epidemics.

Treatment of mange in animals. Most forms of animal mange and scab respond to dips or sprays containing lindane (about 0.05%) repeated after 10 to 14 days, or, in dogs and pigs, lindane in kerosene can be applied to affected areas. An emulsion of malathion with a wetting agent on animals and on bedding and walls has also been recommended. Lime-sulfur and nicotine solutions can also be used for dips. For scaly-leg a 0.1% emulsion of lindane can be applied, or 15% lime-sulfur in lard, after loosening the scales by soaking in warm water.

Hair Follicle Mites (*Demodex*)

The hair follicle or face mite, *Demodex folliculorum* (Fig. 158), of the family Demodicidae, is a wormlike creature, very unmitelike in general appearance, which lives in the hair follicles and sebaceous glands of various mammals. In man it occurs especially on the face and has been found in the ear wax. In dogs it has been also found in lymph glands. Numerous forms from various animals have been described as different species, but they are all strikingly alike and the extent of their specificity is questionable.

These mites have a short, broad head, four pairs of short, stumpy, three-jointed legs, and a very elongated abdomen with fine transverse striations. The females are 0.35 to 0.4 mm. long, the males a little smaller.

The multiplication of these mites is slow. The eggs hatch into tiny six-legged larvae in which the legs are mere tubercles. It requires four molts to bring the larvae to sexual maturity.

The occurrence of these parasites in the hair follicles of man, particularly about the nose, is extremely common; in Germany Gmeiner

(1908) found them in 97 of 100 random examinations of individuals with healthy skins. In man they rarely cause any symptoms whatever, although there are occasional reports of redness, irritation, or other symptoms. When discovered in cases of acne, blackheads, and other skin conditions, it is natural to suspect them of being the cause, but Gmeiner found them much less frequently in cases of acne and blackheads than he did in healthy skins and thought that the altered contents of the diseased follicles and skin glands was unfavorable for their development. In dogs, on the other hand, *Demodex* causes a severe

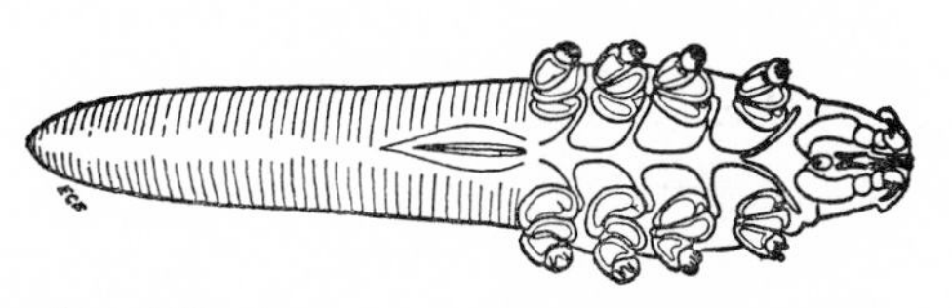

Fig. 158. *Demodex folliculorum,* hair follicle mite, ×200. (After Mégnin, from Faust in Brenneman, *Practice of Pediatrics,* Prior.)

and sometimes fatal form of mange. Some authors think the infection is extremely common in dogs, as it is in man, but that it produces symptoms only under conditions of poor health, vitamin deficiencies, etc. There is a scaly form of the disease in which the skin becomes red, wrinkled, and scaly, and loses its hair, and a pustular or abscessed form in which the skin is invaded by staphylococci, to which dogs are usually resistant. In cattle, pigs, goats, and horses *Demodex* causes pustules to develop, sometimes as large as walnuts. Potentially *Demodex* infections may persist for life, but spontaneous disappearance may occur in animals that are well fed and healthy.

Transmission is usually thought to be by direct contact, or by towels, etc., but internal systemic migration has been suggested. Infected dogs sometimes associate with other dogs for a long time without infecting them. Schiller et al. (1954) failed to infect any of eleven other dogs by mechanical transfer of deep skin scrapings from a *very* mangy mongrel. Experiments with transmission from dog to man have invariably failed. There is a high natural resistance; animals in poor condition from malnutrition or debilitating diseases are usually the ones that suffer.

No entirely satisfactory treatment is known although practically every insecticide, singly or in combination, and in all forms of application, have been tried with variable results (Baker et al., 1956). Schiller et al. (1954), however, got exceptionally good results in dogs with an ointment proposed for human cases by Ayres and Anderson in 1932

(Beta-naphthol, 4 grams; sublimed sulfur, 8 grams; balsam of Peru, 30 cc.; and petrolatum, 30 cc.).

Redbugs or "Chiggers" (Trombiculidae)

There is probably no creature on earth that can cause more torment for its size than a redbug, but in the Far East even this distinction is not enough. In that area some species add injury to insult by trans-

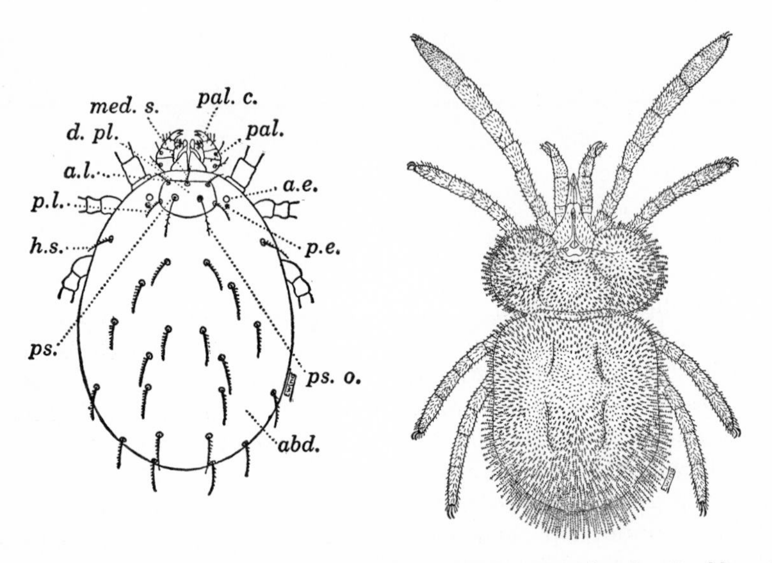

Fig. 159. *Left,* common American redbug or chigger, *Trombicula alfreddugesi,* about ×160. *Right,* adult ♀ of same. (After Ewing: *left, J. Wash. Acad. Sci.,* 28, 1938; *right, Proc. Biol. Soc. Wash.,* 38, 1925.)

mitting a disease, scrub typhus, which during World War II caused more trouble in the Pacific area than any other insect-borne disease except malaria. These mites are also suspected on epidemiological grounds of transmitting epidemic hemorrhagic fever (Traub et al., 1954). This is a virus disease in Siberia, Manchuria, and Korea which causes fever, dilation and increased permeability of capillaries, kidney damage, etc., and is fatal in about 5% of cases (Brown, 1954). A field mouse, *Apodemus agrarius,* is believed to be a reservoir host.

The redbugs (Fig. 159) are the six-legged larvae of mites of the family Trombiculidae, formerly considered a subfamily of Trombidiidae. The larvae of the latter are parasitic on insects, whereas trombiculid larvae are always parasitic on vertebrates. The nymphs and adults are velvety, scarlet-red mites that are free-living; their principal food seems to be insect eggs or minute insect larvae.

The parasitic larvae, called redbugs, rougets, chiggers, harvest mites, scrub mites, or various local names, are minute reddish or orange creatures barely visible to the naked eye (about 0.2 by 0.15 mm.) when unfed. Just behind the capitulum is a small dorsal scutum

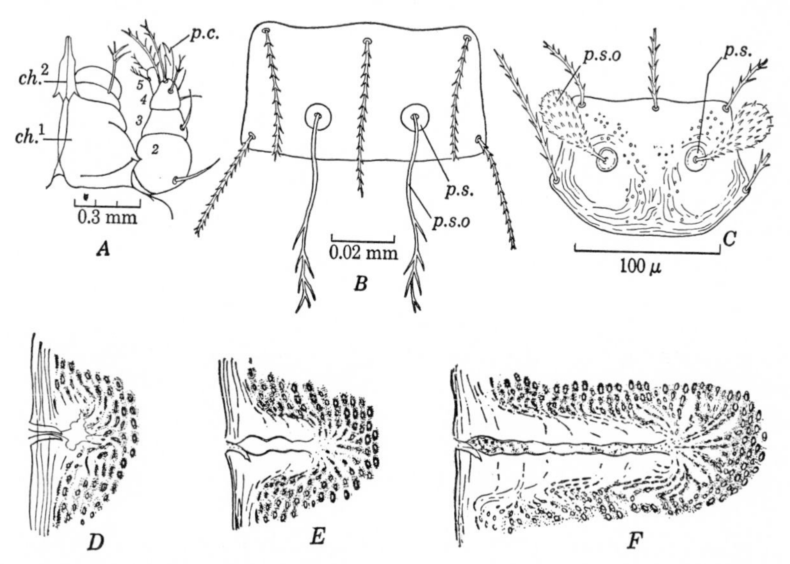

Fig. 160. *A*, Gnathosoma or capitulum of *Trombicula akamushi; ch.*[1], basal segment of chelicera; *ch.*[2], distal segment of chelicera; *p.c.*, palpal claw; numbers indicate second to fifth segments of palpus. *B*, scutum of *Trombicula akamushi. C*, same of *Neoschongastia; p.s.*, pseudostigma; *p.s.o.*, pseudostigmatic organ. *D–E*, stages in formation of tissue canal or stylostome by a trombiculid larva alternately injecting saliva and withdrawing tissue nutrients. *D*, initial injection of saliva. *E*, tissue canal forming. *F*, flow of nutrient along well-formed tissue canal. (*A* and *B* adapted from Wharton, *Proc. Ent. Soc. Wash.*, 48, 1946. *C*, from Brennan, *J. Parasitol.*, 37, 1951. *D–F*, after Jones, *Parasitology*, 40, 1950.)

ornamented with five (in some species six) feathered hairs and a pair of pseudostigmatic organs (Fig. 160*B, C*) from which arise sensory hairs, long and slender in the human species, club-shaped in certain others. There are also feathered hairs on other parts of the body and on the palpi and legs. Genera and species are distinguished by details of the dorsal scutum and of the hairs on the palpi and legs.

Life history. The life cycle has been worked out completely in only a few species but is probably similar for all; it is peculiar in that extra cuticular coverings are produced between the usual stages, and that the fleshy parts of the legs are resorbed and totally new legs are formed in each successive stage. The eggs, laid singly or in small groups, are deposited on the ground. After the general body form is laid down in the egg a cystlike membrane develops around the embryo, which is exposed by the splitting of the egg shell. This stage, called a deutovum, develops in about 6 days. Six days later the fully developed larva hatches, attaches itself to a vertebrate host at the first opportunity, and remains attached for a few days to a month. Williams found that the larvae may pass the winter comfortably holed up in the ears of rabbits or squirrels. After engorgement the larva drops off and molts; meanwhile the tissues of the appendages undergo lysis, and a thin chitinized shell is laid down under the old larval skin; this stage is called a protonymph or nymphochrysalis. In a few days an eight-legged nymph develops inside and emerges. After feeding and growing, the nymph changes to a preadult or imagochrysalis, from which an adult male or female emerges in about 6 days or more. The nymphs and adults (Fig. 159) are similar in appearance, the nymphs being about 0.5 mm. and the adults about 0.75 to 1 mm. long. They probably feed principally on insect eggs or minute larvae, and may do away with appreciable numbers of eggs of *Aedes* and *Psorophora* mosquitoes which are laid on dry ground in their hunting grounds. The minimum time from egg to eggs is about 7 weeks.

Important species and their habits. Redbugs are anchored to the skin surface by a tissue reaction to the irritating saliva that is superficially injected. The redbugs then drool into the skin; the saliva dissolves the skin tissue as it penetrates, forming a tubular structure in the skin called a stylostome, nearly as long as the body of the mite and filled with semidigested tissue debris on which the mite feeds (Fig. 160, *D–F*). The mites do not feed on blood, although their red color gives that impression.

Some redbugs show marked host preferences, different species normally confining themselves to such hosts as rodents, bats, birds, or reptiles, respectively, but a few do not show much discrimination and are content to drool into the skin of almost anything they can get access to, whether it be a turtle, snake, robin, rabbit, mouse, or human being. Most species on mammals have a tendency to get into the ears. On man they run over the skin or through the meshes of clothing, most commonly coming to rest about the garters or belt.

Fortunately relatively few of the many hundreds of species spread over all parts of the world engage in sucking blood from man or domestic animals, and only one very small group of closely related species of the genus *Trombicula* are concerned with transmission of scrub typhus, at least to man.

The species that commonly attack man in North America belong to the genus *Trombicula,* subgenus *Eutrombicula;* these are *T. alfred-dugesi,* which ranges from Canada to South America and the West Indies, but is most frequent in the southeastern states and the Mississippi Valley; *T. splendens* in wet localities in the southeastern states, and *T. batatas,* a grassland pest from the Gulf states and California to northern South America. In Texas and the southeastern states a species of *Neoschongastia* is a pest of chickens and birds but leaves man alone. In Europe *T.* (*Neotrombicula*) *autumnalis* is an annoying pest of man and domestic animals. In the Far East and Australia there are about a dozen species, in several genera, that can make life miserable for man and domestic animals. One in Australia, *T.* (*E.*) *sarcina,* causes nasty irritating sores on the legs of sheep.

The common pest redbug of the United States, *T. alfreddugesi,* attacks principally turtles, snakes, ground birds, and rabbits and is content to feed on man and domestic animals, but unlike many species it does not often attack rodents. Both this species and *T. splendens* are partial to reptiles; on snakes and lizards they may be so abundant as to suggest a new color pattern, characterized by rusty red patches between the scales! *T. batatas* favors birds as a source of food. *T. alfreddugesi* is particularly abundant near thickets, under blackberry bushes, around the base of trees, etc., but seldom in hardwood forests (Williams, 1946). Before attachment they run about actively on or near the ground, eager to climb on a host, but they do not climb upon grass or brush, and ordinarily do not travel far.

The irritation caused by redbugs, as in other arthropod attacks, is largely due to sensitization to the saliva injected. The reaction reaches its height of itching in 12 to 24 hours, when the stylostome is well developed. Eventually the reaction becomes so rapid that very little saliva gets into the skin and the mites are unable to engorge; in some individuals almost complete immunity develops, except for a few reactive bites early in the season which act like a booster shot of a vaccine to revive immunity. Alcohol or camphor helps to allay the itching, and a bath with baking soda or ammonia in the water gives some relief, especially if taken soon after exposure. Dusting sulfur inside the stockings and on the legs is undoubtedly a helpful prophylactic if better repellents are not available (see below).

Transmission of scrub typhus. A few closely related species of the genus *Trombicula,* subgenus *Leptotrombidium,* in the Far East are responsible for the transmission of scrub typhus, caused by *Rick-*

ettsia tsutsugamushi (see p. 215). *T. akamushi* and *T. deliensis* (possibly only a subspecies) inhabiting a wide area from India to Japan, southeastern Asia, East Indies, New Guinea and Australia, are the only proved transmitters to man, but another, *T. scutellaris,* is believed to transmit a mild strain of the disease (Sasa, 1954), and *T. tosa,* a virulent strain, in some parts of Japan. These are all primarily parasites of rats, field mice (*Microtus* and *Apodemus*), shrews, and rabbits, all of which have been found naturally infected with scrub typhus. They may serve as transient reservoirs, although the mites themselves are probably the principal reservoirs, since they pass the organisms transovarially to their offspring generation after generation. Since the mites normally attack only one host in the larval stage and are not parasitic at all in their later stages, transovarial transmission is a *necessary* feature in the epidemiology of this disease.

These mites feed also on other mammals and on birds, which may disperse them. Other species of trombiculids may spread the infection among rodents, keeping up a jungle cycle analogous to jungle yellow fever. Such mites might *occasionally* cause human infections. One rat mite, in southeast Asia, *Euschöngastia andyi,* has been reported to harbor not only *Rickettsia tsutsugamushi,* but also *R. typhi* of murine typhus (see p. 653).

Scrub typhus usually begins with a black sore or "eschar" at the site of the infective bite. Fever, insomnia, generalized inflammation of lymph glands, aches, and neuritis are the usual symptoms; often there is a rash also. The mortality varies from 3 to over 50% in different places. The disease is differentiated from "shop typhus" (endemic or murine typhus) by the OXK Weil Felix reaction. Treatment of this and other rickettsial diseases is discussed on p. 217.

Control and protection. Elimination of vegetation, intensive cultivation, etc., greatly reduces the number of redbugs. Spraying the ground with dieldrin emulsion gives effective control for at least a year. Dusting with chlorinated hydrocarbons (2 lb. per acre) is also effective.

For personal protection, clothing should be sprayed or wiped, or preferably impregnated, with dimethyl or dibutyl phthalate, benzyl benzoate, benzil, and diethyltoluamide (see Baker, 1956). At a dose of 2 grams per square foot of clothing, diethyltoluamide was still effective after 6 weeks as compared with 3-weeks' duration for the standard army repellent M2020 (see p. 537), but it is more easily washed out.

Bloodsucking Mites (Dermanyssidae)

The suborder Mesostigmata (see p. 542) contains several families of bloodsucking mites. The families Dermanyssidae and Laelaptidae contain many parasites of rodents and birds. Some live in nests or burrows, feeding on the host when it is "at home" or sitting on nests, but some stay on the host all their lives. In most species the larvae

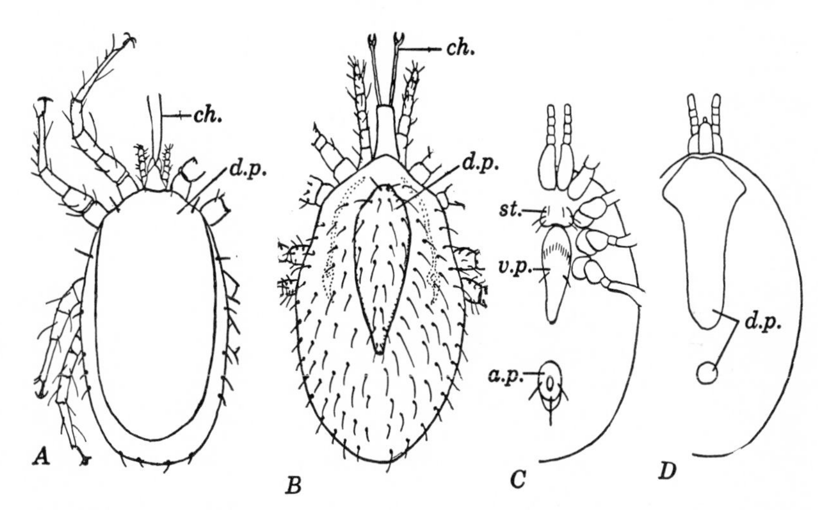

Fig. 161. Dermanyssid mites. A, *Dermanyssus gallinae;* B, *Ornithonyssus bacoti;* C, D, *Allodermanyssus sanguineus,* ventral and dorsal. Abbreviations: *ch.,* chelicerae, needle-like in *Dermanyssus,* pincer-like in *Ornithonyssus; d.p.,* dorsal plate, large in *Dermanyssus,* narrow in *Ornithonyssus,* and divided in *Allodermanyssus; a.p.,* anal plate; *v.p.,* ventral plate; *st.,* sternal plate. (Adapted from various authors.)

do not feed, and in some the deutonymphs do without food also. A number of these bloodsucking mites are important as "conservators" or vectors of rickettsias or viruses. Members of the related family Haemogamasidae transmit tularemia and probably other infections among rodents.

Dermanyssus gallinae. This, the "red mite" of poultry (Fig. 161A), along with one or two other species of poultry mites belonging to the family Dermanyssidae, often causes irritation and annoyance to people who work in chicken houses, live poultry markets, etc. Mites of this genus have a large dorsal shield rounded posteriorly (Fig. 161) and chelicerae that are needlelike in females but have pincerlike tips in males. The mites live and lay their eggs in cracks and crevices, nests, etc., feeding on the chickens mainly at night and sometimes doing much damage to them; they may actually bleed them to death. Although only able to live and multiply on birds, they may remain on human skin for a day or two, causing annoying bites. There is a report of a London hospital which was literally dusted with these mites originating from pigeon nests in the roof. Infested chicken houses, dovecotes, live poultry markets, etc., may remain infested for weeks or months after the birds have been removed.

The unfed larvae molt to become protonymphs; after one blood meal these molt and become deutonymphs and after one more meal, adults. Thereafter the female lays batches of eggs after successive feedings.

This mite has been found to harbor and transovarially transmit the encephalomyelitis viruses of the St. Louis and western equine types. Many chickens develop antibodies, and the mites were suspected of being important reservoirs of these viruses in winter and between epidemics. It was further shown that *Ornithonyssus sylviarum* (see below) of wild birds sometimes harbors these viruses, and an interesting epidemiological situation was visualized—the virus harbored by mites which transovarially transmit them, and infect their bird hosts; the bird hosts bitten in summer by mosquitoes, which then pass the viruses on to horses, whence other mosquitoes transmit them to other horses and to man. Later work, however, casts some doubt on this matter (see Eklund, 1954).

Use of chlorinated hydrocarbons and organic phosphorus compounds, especially lindane and malathion, in sprays have replaced the time-honored use of carbolineum and creosote, which had the disadvantage of giving eggs flavors that did not come from the chickens. The sprays on walls and roosts, plus dusts applied to the litter, nests, etc., gives good control of both this mite and *Ornithonyssus sylviarum* (see below).

Ornithonyssus. Members of this genus (formerly *Liponyssus* or *Bdellonyssus*) are important parasites of birds and rodents and are concerned in transmission of certain rickettsial and virus diseases. These mites have relatively narrow dorsal shields (Fig. 161) and chelicerae that end in pincers in both sexes.

O. sylviarum, in temperate climates, and *O. buarsa,* in the tropics, are common parasites of chickens and many wild birds. *O. sylviarum* spends its whole life on the feathers of the birds. *O. buarsa,* on the other hand, may be found either on the bodies of the birds or in their nests. Usually these mites, as noted above, can be controlled by dusting litter and nests with malathion or nicotine sulfate, or a mixture.

O. bacoti (Fig. 161*B*), the tropical rat mite, is a common mite of commensal rats and other rodents in southern United States and all warm parts of the world. It temporarily becomes an annoying human pest in rat-infected buildings when orphaned by its normal hosts being killed or driven off. This mite plays a minor role in transmission among reservoir hosts, and occasionally to man, of endemic typhus, rickettsialpox, Q fever, tularemia, plague and certain viruses, and it also serves as the intermediate host of the filaria of cotton rats, (*Litomosoides* (see p. 482). Rats and other rodents harbor other blood-sucking mites of this family and of the families Laelaptidae and Haemogamasidse. *Echinolaelaps echidninus* is a common rat parasite all over the world, and transmits *Hepatozoon muris* (see p. 204) when swallowed.

Mites and rickettsialpox. This disease caused by a *Rickettsia, R. akari,* related to that of spotted fever, was first recognized in an outbreak in a new housing development in New York City in 1946. It was traced to house mice as reservoir hosts, from which a dermanyssid mite, *Allodermanyssus sanguineus,* found in the United States and Africa, transmitted it to the residents. Subsequently this disease was reported in other eastern cities, and serological tests suggest its presence elsewhere in the world. This mite has a divided dorsal plate (Fig. 161*C, D*). Rickettsialpox begins like scrub typhus with a black eschar at the site of the bite, followed a week later by a sudden fever and rash resembling chickenpox.

Mites in the lungs, intestine, urinary passages, etc. Lung mites. In addition to the lung mites of the families mentioned on p. 541, the airsac mite, *Cytodites nudus,* related to the mange mites, lives in the airsacs, lungs, and linings of other internal cavities of chickens, canaries, and other birds, causing serious, sometimes fatal, disease. No true lung mites have been found in man. A number of cases of mites in the lungs or sputum accompanied by bronchial asthma and eosinophilia, have been reported. These have usually been species that are normally free-living or plant-parasitic, so were presumably accidentally inhaled, or possibly taken into the mouth with food.

Mites in the intestine. There have been numerous reports of intestinal irritation by grain or cheese mites (see below) ingested with infected foods. The eggs and dead mites in all stages may be found in the feces of man, or dogs, etc. This pseudoparasitism with mites sometimes fills unsuspecting technicians with wonder and excitement when they discover the large eggs in the feces. However, the ingestion of the mites in large numbers may cause gastrointestinal symptoms, e.g., eating the famous German cheese, "altenburger mulbankäse," that owes its piquant flavor to the presence of myriads of acarid mites with which it is inoculated, and which, with their feces, produce a moving grayish powder on its surface. However, people who are accustomed to eating it suffer no ill effects. In many cases the intestinal disturbances are probably allergic in nature, like grocer's itch (see below).

Urinary infections. Urinary infections with mites have frequently been reported, but in most of these cases it seems likely that the mites observed are really contaminations from containers or other sources. However, there are a number of apparently *bona fide* infestations of the urinary tract in which no sources of contamination could be found. Mackenzie and Mekie in 1926 reported finding mites in the urine of patients with uncontrollable nocturnal enuresis; the urine contained abundant epithelial cells, parts of mites, and a black deposit. The mites concerned were a tarsonematid, *Tarsonemus floricolus,* and acarids (see below). How the mites in such cases find their way into the urinary passages and bladder is unknown.

In rare instances acarid mites establish residence in the canal of the outer ear and occasionally even penetrate to the middle ear and mastoid.

There have been reports of acarid mites being found in cancers and other situations in the tissues of the body, but it is practically certain that these are cases of contamination.

Grocer's Itch and Allied Forms of Mite Dermatitis

Grain mites. Many mites, most of them belonging to the families Acaridae (=Tyroglyphidae) and Glycyphagidae, are common pests of human dwellings, stores, and warehouses, where they attack all sorts of food materials, stored seeds, stuffing of furniture, etc. When conditions are favorable they multiply until the infested materials are literally alive with them. They are especially commonly found in animal feeds, hay, grain, flour, sugar, dried fruits, copra, cottonseed, and cheese.

People who come into close association with infested goods develop symptoms which Hase (1929) thinks are of allergic nature. Though allergy undoubtedly plays a part, there is evidence that the bodies or excretions of the mites are toxic. When Hase fed mice with mite dust (dead mites, feces, cast, skins, etc.), 7 of 12 died with dysenteric symptoms. Sometimes dermatitis appears in whole groups of people working with infested materials, e.g., copra workers, or living where exposed to dust from mite-infested grain. Contact with living mites is unnecessary. Symptoms are produced as readily, if not more so, by infested materials rubbed on the skin, or dust blown on the skin or inhaled.

Both dermal and respiratory symptoms occur. The skin develops a typical itching urticaria, sometimes with large hives, sometimes eczematous. Asthma is common, and other frequent symptoms are quickened pulse, general aches, fever, and sometimes nausea, vomiting, and diarrhea. It is obvious that these symptoms are by no means peculiar to mite infections but are frequently observed in severe arthropod infections of other kinds. Few arthropods, except mites, develop in sufficiently prodigious numbers to produce symptoms by their dust; in almost all other cases, the production of symptoms depends upon the inoculation of saliva at the time of biting.

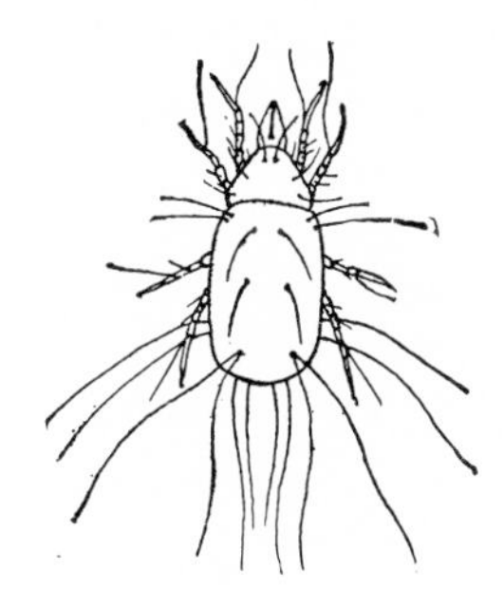

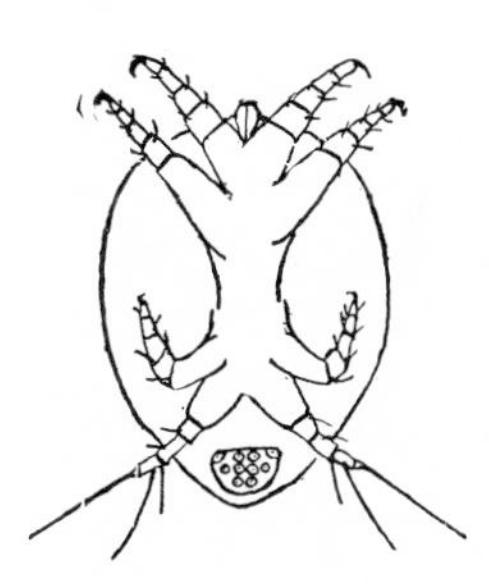

Fig. 162. *Left,* grain mite, *Tyrophagus putrescentiae* (=*Tyroglyphus longior*), ×30. (After Fumonze and Robin.) *Right,* hypopus or traveling stage, ventral view. Much enlarged. (After Banks, *U. S. Dept. Agric. Repts.,* 108.)

Some of the well-known examples of dermatitis from mites are grocer's or baker's itch, known all over the world; "copra itch" in copra mills in Ceylon; "miller's itch," familiar in most grain-raising countries; "vanillism" in handlers of vanilla pods; "cottonseed itch"; "barley itch"; etc. The affliction is not confined to man, for horses sometimes get dermatitis when provided with mite-infested hay.

Grain mites are small white or yellowish, soft-bodied mites. They have prominent pincerlike chelicerae which are entirely unsuited for piercing the skin. The three commonest genera are *Acarus* (=*Tyroglyphus*), *Tyrophagus* (Fig. 162), and *Glycyphagus*. The first two have elongate bodies, with a suture separating the body into anterior and posterior parts, and with a few long simple hairs; *Glycyphagus,* and also the sugar mite, *Carpoglyphus,* have no body suture.

The life cycle of many species is remarkable in that, after reaching a nymphal stage in orthodox mite style, the mites change into a form called a hypopus (Fig. 162, *right*), which is a special adaptation for hitchhiking. There are no mouth parts, the legs are short and stumpy, and there are ventral suckers on the abdomen. In some species (*Glyciphagus*) an encysted type of hypopus is produced to withstand desiccation. Thus equipped for travel, the mites attach themselves to insects or other objects and are transported to new localities. They have frequently been mistaken for parasites, but they are no more

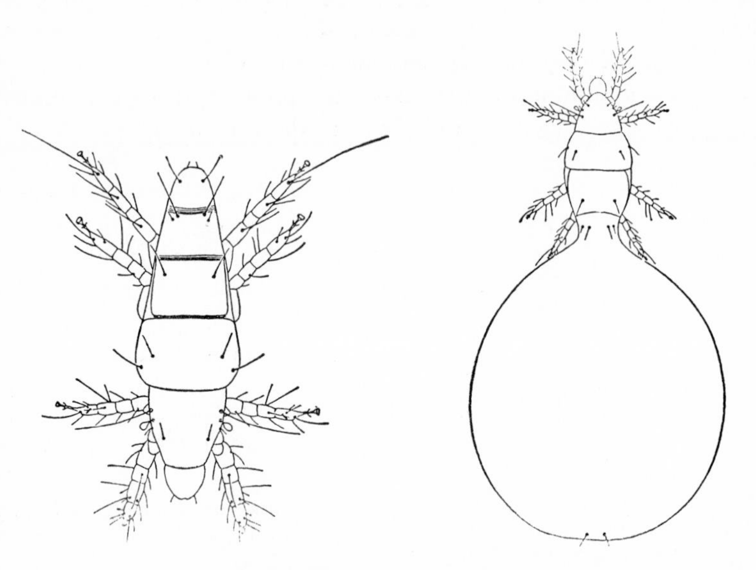

Fig. 163. *Pyemotes* (=*Pediculoides*) *ventricosus.* *Left,* virgin ♀. *Right,* partially gravid ♀. In fully gravid females the abdomen swells to several times the size shown here. (From Baker and Wharton, *Acarology,* Macmillan, 1952.)

parasitic than a man on horseback. After dropping from their animated conveyances they molt into eight-legged nymphs, which after feeding become adults.

Infested substances are best burned or otherwise disposed of, and the containers, rooms, etc., then fumigated, preferably with methyl bromide (see p. 535) or other fumigants. Carbon dioxide snow added to grain or feed in containers is helpful in keeping down mites as well as mealworms and weavils. Modern packaging gives protection if the contents are not already infested.

Pyemotes ventricosus. This mite (Fig. 163), formerly called *Pediculoides ventricosus,* belongs to the family Pyemotidae (see p. 542). The males and unencumbered females are only about 0.2 mm. long, barely visible to the naked eye. The pregnant females, however, retain their eggs and young in the abdomen until they are fully developed, the abdomen becoming a grotesquely enlarged brood sac, which may reach a diameter of 1.5 mm. The females retain their slender virginal form very briefly, for it may be only 6 days from the time they leave their distorted mother until they have a sacful of young of their own.

These mites are normally parasitic on grain-moth caterpillars and other insects in straw, grains, cottonseed, etc. In stored products the transformation and escape of their insect prey leave them with their normal food supply cut off, and the hungry and thirsty mites then attack any flesh that comes their way. Serious infestations occur among grain thrashers, millers, etc., and sometimes new straw mattresses turn out to be veritable beds of fire.

An itching rash begins about 12 to 16 hours or sooner after exposure to the mites. The bites, at first red and inflamed, itch unbearably. Little blisters form and, when scratched and ruptured, develop into pustules or scabs; in bad attacks the usual constitutional symptoms of severe arthropod infestation develop—fever, rapid pulse, headache, nausea, etc. One case has been reported in which dust from a mite-infested grain elevator blew into the cottages in the neighborhood and produced dermatitis in all the inhabitants. Since the mites cannot thrive on human blood they soon withdraw, disillusioned, to try some other source of food, and consequently the symptoms usually subside within a week unless fresh detachments of mites are constantly being acquired. This can be prevented by fumigation with sulfur or other fumigants, or pyrethrum sprays (see p. 532).

The itching can be alleviated by alkaline baths or application of soda and soothing ointments. People exposed to infestation can get protection from the bites by application of acaricides or repellents (see p. 536) and a change of clothing, but those who develop dust dermatitis will get no relief from these measures.

Other dermatitis-producing mites. A number of other mites may occasionally produce dermatitis. Brief mention should be made of members of the family Cheyletidae, belonging to the Trombidiformes, which prey on truly parasitic mites in the fur or plumage of animals. One species, *Cheyletiella parasitivorax*, often found on rabbits and cats, occasionally attacks the mammalian host. This mite has been found responsible in a few cases of human eczema from handling cats.

Oribatids

The oribatid mites (see Fig. 108) are free-living mites living in soil, moss, etc., feeding on molds and organic debris. They are very numerous and probably important, as are earthworms, in connection with soil fertility. The life cycle involves several nymphal stages, and may occupy a year or more. They are of interest to parasitologists because many species, belonging to a number of different families, serve as intermediate hosts of tapeworms of the families Anoplocephalidae and Catenotaeniidae (see p. 372). Estimated populations on pastures may run to several millions per acre, of which several hundred thousands may harbor larvae of cattle or sheep tapeworms. These mites creep out of the soil when the dew is on the grass, and are eaten by herbivorous animals with the vegetation. The principal factors determining their importance as tapeworm vectors are their abundance where the hosts of the tapeworms feed and their ability to swallow the tapeworm eggs.

THE PENTASTOMIDA

Tongue Worms and Their Allies

At one time this aberrant group of arthropods was classified with the Arachnida and was thought to be related to the mites, but it is now usually considered a separate class. The animals have become so modified by parasitic life that their affinity with the arthropods would be difficult to recognize if it were not for the form of the larvae, which are more or less mitelike and have either two or three pairs of legs. Even in life cycle they resemble parasitic worms in that they pass the immature stages in an intermediate host.

The adults have elongate bodies which are either flattened or cylindrical and divided into a series of unusually conspicuous rings which are not, however, true segments. There is no distinct division into head, thorax, or abdomen. On either side of the mouth at the anterior end there are two pairs of hollow, fanglike hooks, in some forms situated on fingerlike parapodia, which can be retracted into grooves like the claws of a cat (Fig. 164). These are believed to be vestiges of some of the appendages. At the bases of the retractile hooks there open a number of large glands, the secretion of which is believed to be hemolytic. The Pentastomida have a simple nervous system, a usually straight digestive tract, and a reproductive system. The anus is at the posterior end of the body. The females, which are larger than the males, have the genital opening either near the anterior or near the posterior end of the abdomen, but that of the males is anterior.

The life cycle involves two hosts. The adults usually live in the lungs or air passages of their hosts; the larvae live free or encysted in the viscera of some other host.

Classification. The classification of the Pentastomida according to Heymons and Vitzthum (1936) is as follows:

Order 1. **Cephalobaenida.** Hooks situated on fingerlike processes or at least swellings of the body behind mouth; genital opening anterior in both sexes.

Family 1. **Cephalobaenidae.** In lungs of lizards and snakes.

Family 2. **Reighardiidae.** In air sacs of gulls and terns.

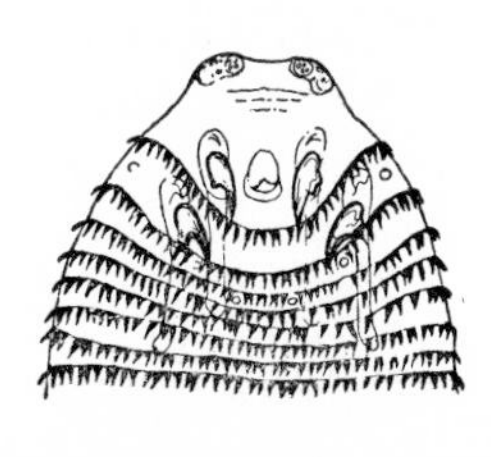

Fig. 164. *Left,* head of *Armillifer armillatus,* ×3 (after Sambon, *J. Trop. Med. Hyg.,* 25, 1922). *Right,* head of nymph of *Linguatula serrata,* ×25 (after Faust, *Am. J. Trop. Med.,* 7, 1927).

Order 2. **Porocephalida.** Hooks not on prominences, arranged trapezelike or in a curved line on either side of mouth; ♀ genital opening posterior.

Family 1. **Porocephalidae.** Body cylindrical. Adults in lungs of reptiles, young in a great variety of vertebrates; young of the genus *Armillifer* usually in mammals, including man.

Family 2. **Linguatulidae.** Body flattened. Adults in nasal passages of dog and cat family, except one in crocodiles; young in all sorts of mammals, including man.

There is a single well-authenticated instance of human infection with an adult *Linguatula serrata* in the nasal passages, but visceral infection with immature stages of this species and of several species of Porocephalidae is surprisingly common.

Linguatula serrata. The adult worms are nearly colorless; the females are 100 to 130 mm. long with a maximum width of about 10 mm.; the males are only about 20 mm. long and 3 to 4 mm. wide. They occur in the nasal passages and frontal sinuses principally of dogs (Fig. 165*A*, *F*) and occasionally other animals, where they suck blood. They sometimes cause severe catarrh, bleeding, and suppuration and may cause much sneezing and difficulty in breathing when they obstruct the nasal passages, but often they produce no symptoms at all.

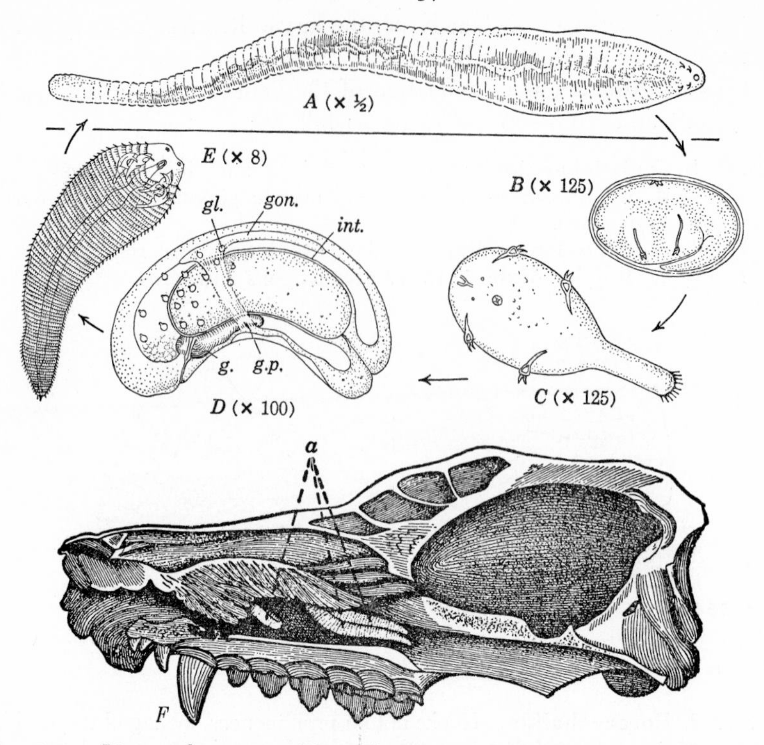

Fig. 165. *Linguatula serrata,* life cycle. *A,* adult ♀ from nasal passage of a dog; *B,* egg containing embryo; *C,* first-stage larva from viscera of sheep, man, etc.; *D,* third-stage larva, ninth week; *E,* nymph, from liver of sheep; *F,* head of dog split open to show three tongue worms, *Linguatula serrata,* (*a*) in nasal cavity. (*A* adapted from Brumpt, *Précis de parasitologie,* 1949. *B–D,* from Leuckart, *Bau und Entwicklungsgeschichte der Pentastomen,* 1860. *E,* from Railliet, *Recueil méd. vét.,* Alfort, 1884. *F,* after Colin, *Recueil méd. vét.,* 1863.)

The eggs (Fig. 165*B*), containing embryos with four rudimentary legs, are expelled either with nasal mucus or in the feces. They can be found in feces of dogs usually only after treatment with KOH to free them from viscid surface material.

The eggs are resistant and live for a long time outside the body. When ingested by an intermediate host, e.g., cattle, sheep, rabbits, rats, man, etc., the embryos (Fig. 165*C*), 75 μ long, migrate to the mesenteric nodes and various other viscera and there become encapsulated (Fig. 165*D*). They molt twice and assume a pupalike stage in which they are devoid of mouth parts, hooks, or segmentation, and are 0.25 to 0.5 mm. long. A number of other molts follow, and after 5 or 6 months a nymphal stage is attained in which the animal possesses two pairs of hooks and has its body, 4 to 6 mm. in length, divided into 80 to 90 rings, each bordered posteriorly by a row of closely set spines (Figs. 164, *right,* 165*E*). These are shed when the nymph transforms into an adult. For a long time this nymph was looked upon as a distinct species. The nymphs may remain alive in the intermediate host for at least 2 to 3 years, but their capsules become thick so that they are not easily liberated. This undoubtedly interferes with successful infection of a final host.

According to the Hobmaiers, contrary to the generally accepted belief, the nymphs do not leave their cysts during the life of the host but quickly liberate themselves after its death. Nor do swallowed nymphs succeed in migrating back to the pharnyx from the stomach. To cause infection the nymphs must cling to the mucous membrane of the mouth before being swallowed or when vomited. The worms begin laying eggs about 6 months after infection and seem to live for about 2 years.

L. serrata is nowhere abundant, even in its normal hosts, though it has a wide geographic distribution. In parts of Europe 10% of dogs may harbor the adults, and in some series of autopsies 10% of human beings may harbor the nymphs, but usually they are dead and calcified and of no pathological significance. In North Ireland the adults are frequent in foxes, and the nymphs in lymph nodes of cattle. Undoubtedly human infections result from too intimate contact with dogs. In the single human infection with the adult stage a frequent bleeding of the nose which had persisted for seven years ceased when an adult *Linguatula* was expelled in a violent fit of sneezing. Dogs may be treated by contact insecticidal aerosols.

Armillifer. Man is frequently parasitized by the nymphs of at least two species of *Armillifer* (*Nettorhynchus,* according to Dollfus, 1950), the adults of which live in the lungs of pythons and other snakes. The intermediate hosts include many kinds of mammals but particularly monkeys, which are important in the diet of pythons. Human infections with the encysted larvae of *A. armillatus* are common in Africa; Broden and Rodhain found 30 cases in 133 postmortems of natives in the Congo. Since some African natives esteem python for dinner, infection may result from handling them, as well as from contaminated water or vegetables. In the Oriental region this species is replaced by a closely related one, *A. moniliformis;* only a few human infections with this species have been seen—in Manila, Sumatra, and China.

These species of *Armillifer* have bright lemon-yellow cylindrical bodies, marked by braceletlike annulations which give them a screwlike appearance (Fig. 166*A*, *B*). The females are 90 to 130 mm. long, the males about 30 to 45 mm. In the intermediate hosts the nymphs lie coiled up in cysts either embedded in or attached to the liver or other organs; they resemble miniatures of the adults. When ingested

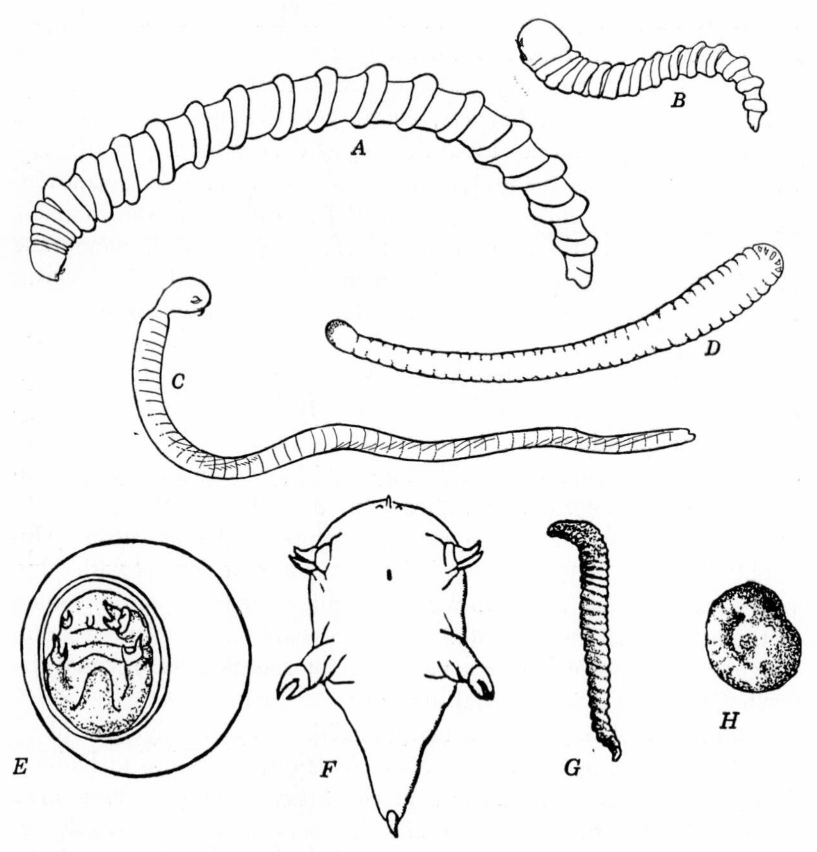

Fig. 166. *A, Armillifer armillatus* ♀; *B*, same, ♂; *C, Kiricephalus coarctatus*, common in American colubrine snakes; *D, Porocephalus crotali*, in American rattlesnakes. (*A–C, E, F*, adapted from Sambon, *J. Trop. Med. Hyg.*, 25, 1922. *D*, from Self and McMurray, *J. Parasitol.*, 34, 1948. *G* and *H*, from Fülleborn, *Arch. Schiffs- u. Tropen-Hyg.*, 23, 1919.)

by pythons they are said to reach the lungs by burrowing through the stomach wall, but this may be incorrect, since both *Porocephalus crotali* and *Linguatula* reach the lungs via the throat and trachea (Penn, 1942). In the intermediate host development is very slow, the nymphs requiring 1½ to 2 years to reach a length of 16 to 22 mm.

Two American cases of infection with porocephalid worms have been recorded. Since no American species of *Armillifer* are known, these worms may have been the young of *Porocephalus crotali* of rattlesnakes (Fig. 166*D*), or of *Kiricephalus coarctatus* of Colubridae (Fig. 166*C*). Both these genera although having annulated bodies lack the conspicuous rings of *Armillifer*. *P. crotali* nymphs are common in muskrats and other mammals.

Heavy experimental infections with immature worms produce injurious or even fatal effects, but there is no evidence that the light infections usually seen in man are pathogenic. One heavily loaded case reported by Cannon (1942) suffered from partial obstruction of the colon due to thickening of its parasite-studded walls. Since there are no characteristic symptoms, infections are recognized only at autopsies.

REFERENCES

Allred, D. M. 1954. Mites as intermediate hosts of tapeworms. *Proc. Utah Acad. Sci., Arts and Letters,* 31: 44–51.

Baker, E. W., Evans, T. M., Gould, D. J., Hull, W. B., and Keegan, H. L. 1956. A manual of parasitic mites of medical and of economic importance. *Natl. Pest Control Assoc. Tech. Publ.,* New York.

Baker, E. W., and Wharton, G. W. 1952. *An Introduction to Acarology.* Macmillan, New York.

Banks, N. 1915. The Acarina or mites. *U. S. Dept. Agr. Repts.* 108.

Booth, B. H., and Jones, R. W. 1952. Epidemiological and clinical study of grain itch. *J. Am. Med. Assoc.,* 150: 1575–1579.

Camin, J. H., and Rogoff, W. M. 1952. Mites affecting domesticated mammals. S. *Dakota State Coll., Agr. Exp. Sta. Tech. Bull.* 10.

Cannon, D. A. 1942. Linguatulid infestation of man. *Ann. Trop. Med. Parasitol.,* 36: 160–167.

Chamberlain, R. W., and Sikes, R. K. 1955. Laboratory investigations on the role of bird mites in the transmission of eastern and western equine encephalitis. *Am. J. Trop. Med. Hyg.,* 4: 106–118.

Davis, J. W. 1954. Studies of the sheepmite, *Psorergates ovis. Am. J. Vet. Res.,* 15: 255–257.

Dove, W. E., and Shelmire, B. 1932. Some observations on tropical rat mites and endemic typhus. *J. Parasitol.,* 18: 159–168.

Eklund, C. M. 1954. Mosquito-transmitted encephalitis viruses. A review of their insect and vertebrate hosts and the mechanisms for survival and dispersal. *Exp. Parasitol.,* 3: 285–305.

Evans, G. O., and Browning, E. 1955. Some British mites of economic importance. *Brit. Mus.* (*Natl. Hist.*) *Econ. Series,* No. 17, X + 46 pp.

Ewing, H. E. 1944. The trombiculid mites (chigger mites) and their relation to disease. *J. Parasitol.,* 30: 339–365.

Finnegan, S. 1945. Acari as agents transmitting typhus in India, Australia, and the Far East. *Brit. Mus. (Natl. Hist.) Econ. Series,* No. 16.

Fonseca, F. 1948. A monograph of the genera and species of Macronyssidae Oudemans, 1936 (synn.: Liponyssidae Vitzthum, 1931). *Proc. Zool. Soc. London,* 118: 249–334.

Fülleborn, F. 1919. Über die Entwicklung von Porozephalus und dessen pathogenen Bedeutung. *Arch. Schiffs- u. Tropen-Hyg.,* 23: 5–36.

Fuller, H. S. 1954. Studies of rickettsialpox III. Life cycle of the mite vector, *Allodermanyssus sanguineus. Am. J. Hyg.,* 59: 236–239.

Furman, D. P. 1954. A revision of the genus *Pneumonyssus. J. Parasitol.,* 40: 31–42.

Gmeiner, F. 1908. *Demodex folliculorum* des Menschen und der Tiere. *Arch. Dermatol. Syphilol.,* 92: 25–96.

Greenberg, M., Pellitteri, O. J., and Jellison, W. L. 1947. Rickettsialpox—a newly recognized disease, III. Epidemiology. *Am. J. Public Health,* 37: 860–868.

Hase, A. 1929. Zur path.-parasit. und epid.-hyg. Bedeutung der Milben, insbesondere der Tyroglyphinae. . . . *Z. Parasitenk.,* 1: 765–821.

Heymons, R., and Vitzthum, H. G. 1936. Beiträge zur Systematik der Pentastomiden. *Z. Parasitenk.,* 8: 1–103.

Hill, H. R. 1948. Annotated bibliography of the Linguatulida. *Bull. S. Calif. Acad. Sci.,* 47: 56–73.

Hobmaier, A., and Hobmaier, M. 1940. On the life cycle of *Linguatala rhinaria. Am. J. Trop. Med.,* 20: 199–210.

Hoffman, R. A. 1956. Control of the northern fowl mite and two species of lice on poultry. *J. Econ. Entomol.,* 49: 347–349.

Hughes, A. M. 1948. The mites associated with stored food products. 168 pp. H.M. Stationery Office, London.

Jenkins, D. W. 1948–1949. Trombiculid mites affecting man, I–IV. *Am. J. Hyg.,* 48: 22–35, 36–44; *J. Parasitol.,* 35: 201–204; *Ann. Entomol. Soc. Amer.,* 42: 289–319.

Johnson, C. G., and Mellanby, K. 1942. The parasitology of human scabies. *Parasitology,* 34: 285–290.

Jones, B. M. 1950. The penetration of the host tissue by the harvest mite, *Trombicula autumnalis* Shaw. *Parasitology,* 40: 247–260.

Kemper, H. E., and Peterson, H. O. 1953. Cattle scab and methods of control and eradication. *Farmer's Bull. 1017, U. S. Dept. Agr.*

Kohls, G. M. 1947. Vectors of rickettsial diseases. *Ann. International Med.,* 26: 713–719.

Lombardini, G. 1942. Contributo alla conoscenza della morphologia dei Demodicidae. Chiave analitica de genere *Demodex* Owen. *Redia,* 28: 89–102.

Mekie, E. C. 1926. Parasitic infection of the urinary tract. *Edinburgh Med. J.,* 33: 708–719.

Mellanby, K. 1943. *Scabies.* Oxford University Press, London.

1944. The development of symptoms, parasitic infection, and immunity in human scabies. *Parasitology,* 35: 197–206.

Michener, C. D. 1946. Observations on the habits and life history of a chigger mite, *Eutrombicula batatus. Ann. Entomol. Soc. Amer.,* 49: 101–118.

Nesbitt, H. H. J. 1945. A revision of the family Acaridae (Tyroglyphidae). order Acari, based on comparative morphological studies. *Can. J. Research, D* 23: 139–188.

Nichols, E., Rindge, M. E., and Russell, J. G. 1953. The relationship of the habits of the house mouse and the mouse mite (*Allodermanyssus sanguineus*) to the spread of rickettsialpox. *Ann. Internal Med.*, 39: 92–102.

Penn, G. H., Jr. 1942. The life history of *Porocephalus crotali*, a parasite of the Louisiana muskrat. *J. Parasitol.*, 28: 277–283.

Philip, C. B. 1948. Tsutsugamushi disease (scrub typhus) in World War II. *J. Parasitol.*, 34: 169–191.

Radford, C. D. 1950. The mites (Acarina) parasitic on mammals, birds, and reptiles. *Parasitology*, 40: 366–394.

Rogers, G. K. 1943. Grain itch. *J. Am. Med. Assoc.*, 123: 887–889.

Sasa, M., and Jameson, E. W. 1952. The trombiculid mites of Japan. *Proc. Calif. Acad. Sci.*, 28: 247–321.

Schiller, E. L., McIntyre, J. C., and Shirbroun, R. E. 1954. Preliminary observations on an experimental treatment for demodectic mange in dogs. *J. Parasitol.*, 40: 704–706.

Seddon, H. R. 1951. Diseases of domestic animals in Australia. Part 3, Tick and mite infestations. *Service Publ. No. 7, Dept. of Health, Div. Vet. Hyg.*, Canberra.

Sengbusch, H. G. 1954. Studies on the life history of three oribatoid mites, with observations on other species. *Ann. Entomol. Soc. Am.*, 47: 646–667.

Skaling, P., and Hayes, W. J. 1949. The biology of *Liponyssus bacoti* Hirst, 1913. *Am. J. Trop. Med.*, 29: 759–772.

Traub, R., Hertig, M., Lawrence, W. H., and Harriss, T. T. 1954. Potential vectors and reservoirs of hemorrhagic fever in Korea. *Am. J. Hyg.*, 59: 291–305.

Traver, J. R. 1951. Unusual scalp dermatitis in humans caused by the mite *Dermatophagoides*. *Proc. Entomol. Soc. Wash.*, 53: 1–25.

Vitzthum, H. G. 1940–1942. Acarina, in *Bronn's Klassen u. Ordnungen des Thier-reichs* 5, Sect. 4, Book 5. Akad. Verlagsgesellsch., Leipzig.

Wharton, G. W. 1946. The vectors of tsutsugamushi disease. *Proc. Entomol. Soc. Wash.*, 48: 171–178.

Wharton, G. W., and Fuller, H. S. 1952. A manual of the chiggers. *Mem. Entomol. Soc. Wash.*, 4.

Wharton, G. W., et al. 1951. The terminology and classification of trombiculid mites (Acarina: Trombiculidae). *J. Parasitol.*, 37: 13–31.

Williams, R. W. 1946. A contribution to our knowledge of the bionomics of the common North American chigger, *Eutrombicula alfreddugési* Oudemans, with a description of a rapid collecting method. *Am. J. Trop. Med.*, 26: 243–250.

Willman, C. 1952. Parasitische Milbenan Kleinsangern. *Z. f. Parasitenk.*, 15: 292–428.

Wolfenbarger, K. A. 1952. Systematic and biological studies on North American chiggers of the genus *Trombicula*, subgenus *Eutrombicula*. *Ann. Entomol. Soc. Amer.*, 45: 645–677.

Womersley, H. 1952. The scrub-typhus and scrub-itch mites of the Asiatic-Pacific region, Parts 1 and 2. *Records Australian Museum*, 10: 1–673.

Chapter 24

TICKS

Although the ticks constitute only one of the suborders of the order Acarina, they are popularly regarded as a quite distinct group because they are large and easy to recognize. They are not merely annoying pests but surpass all other arthropods in the number and variety of disease agents for which they are carriers. As carriers of human disease they rank next to mosquitoes, but as carriers of animal diseases they are preeminent.

General anatomy. The ticks are classed in two families, Argasidae or "soft" ticks, and Ixodidae or "hard" ticks, which differ considerably both in their structure and life cycle, as will be seen in the following pages. Structurally the Argasidae (Fig. 167*E, F*) are distinguished by having the body covered by a leathery cuticle marked by numerous tubercles or granulations, and sometimes small circular discs, also, but no plates or shields. The Ixodidae, on the other hand (Fig. 167*A–D*) have a dorsal shield or scutum that almost completely covers the back in males, but only the anterior portion of it in females. In some genera (*Dermacentor* and *Amblyomma*) the dorsal shield or scutum is ornamented with silvery markings and is then said to be "ornate" (Figs. 175, 176). In several genera the dorsal shield of the male is marked with "festoons" on the posterior border (Fig. 167*A, B*). Another character distinguishing these two families is the ventral position of the mouth parts in the Argasidae and their anterior position in the Ixodidae, where they fit into a groove or *camerostome* at the anterior end of the body. The females of both families when unfed are flat, but after their gluttonous meals they become grotesquely engorged and resemble beans or nuts (Fig. 169). The dorsal shield or scutum of engorged female ixodids becomes quite inconspicuous.

The mouth parts, as in other Acarina, are borne on a movable *capitulum* or *gnathosoma,* which is not a true head although popularly so called; it consists of a base (*basis capituli*) and the mouth parts (Fig. 168*A, B*). The latter consist of a *hypostome,* a pair of *palpi,*

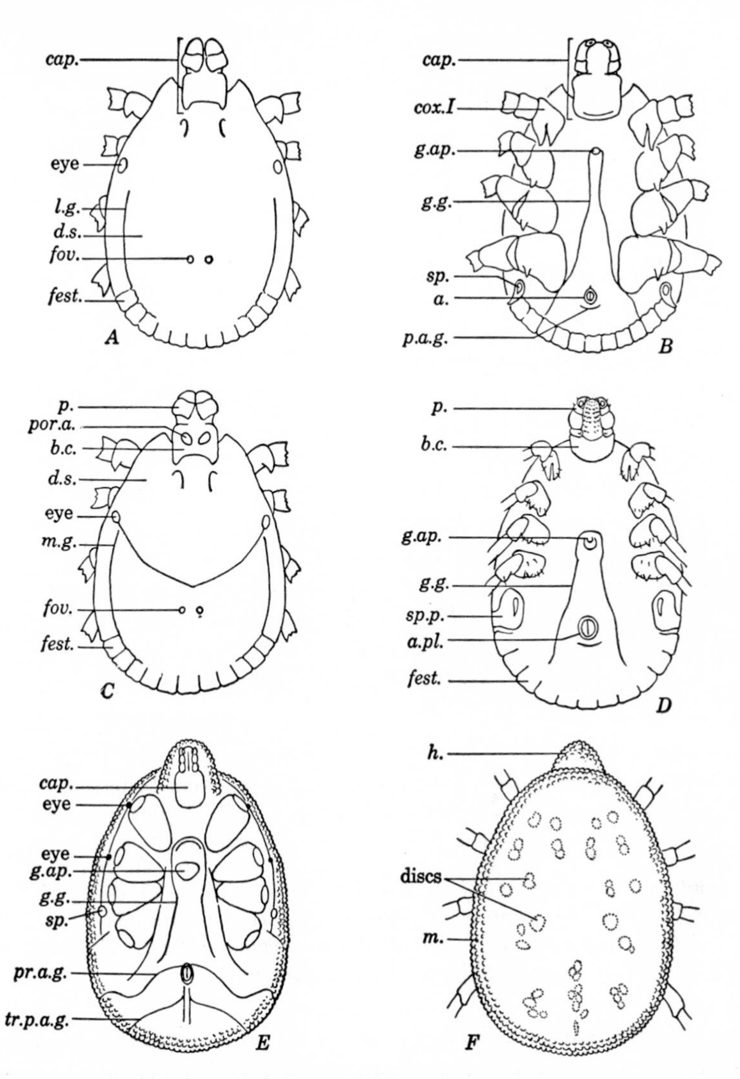

Fig. 167. Dorsal and ventral views of ixodid and argasid ticks. *A* and *B* of ♂ ixodid (*Dermacentor*); *C* and *D* of ♀ ixodid (*Dermacentor*); *E* and *F* of argasid (*Ornithodoros*). Abbreviations: *a.*, anus; *a.pl.*, anal plate; *b.c.*, basis capituli; *cap.*, capitulum; *cox.I*, first coxa; *d.s.*, dorsal scutum or shield; *fest.*, festoons; *fov.*, fovea; *g.ap.*, genital aperture; *g.g.*, genital groove; *h.*, hood; *l.g.*, lateral groove; *m.*, mammillae; *mg.*, marginal groove; *p.*, palpus; *p.a.g.*, post-anal groove; *por.a.*, porose area; *pr.a.g.*, pre-anal groove; *sp.*, spiracle; *sp.p.*, spiracular plate. (Adapted from Cooley, *N.I.H. Bull.* 171, and Cooley and Kohls, *Am. Midland Nat., Monogr.* 1, 1944.)

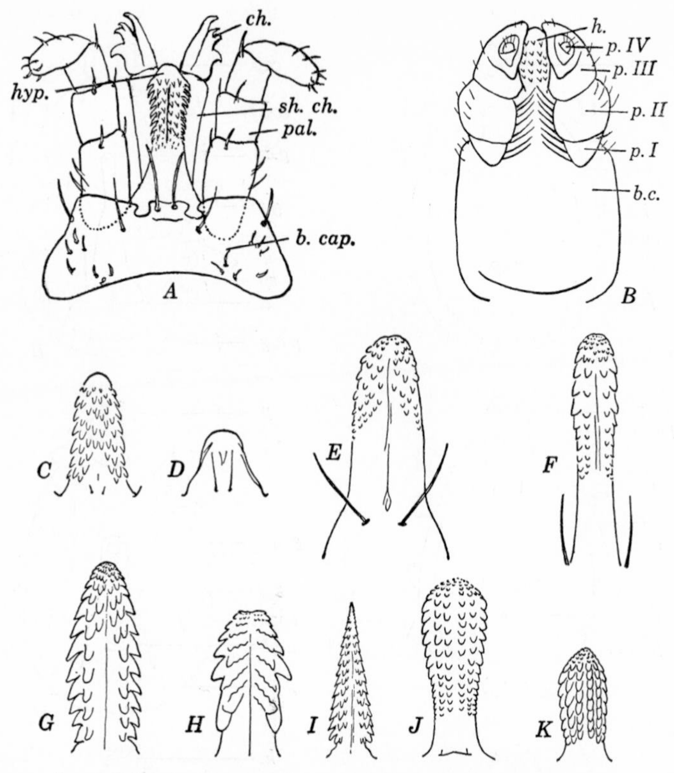

Fig. 168. *A*, capitulum of argasid tick, ventral view (from Matheson, *Medical Entomology*, Comstock). *B*, capitulum of ixodid tick (*Dermacentor*), ventral view (after Cooley, *N.I.H. Bull.* 171); *b.c.*, *b. cap.*, basis capituli; *ch.*, chelicera; *h.*, *hyp.*, hypostome; *pal.*, palpus; *p.I–IV*, palpal segments; *sh. ch.*, sheath of chelicera. *C–K*, hypostomes of: *C*, *Otobius mégnini*, nymph; *D*, same, adult; *E*, *Ornithodoros savignyi*, adult; *F*, *O. turicata* ♀; *G*, *Ixodes scapularis* ♀; *H*, same, ♂; *I*, *Ixodes mexicanus* ♀; *J*, *Rhipicephalus sanguineus* ♀; *K*, *Haemaphysalis leporispalustris* ♀. (Sketched from various authors.)

and a pair of *chelicerae*. The hypostome is a prolongation of the ventral wall of the capitulum. It is a formidable piercing organ beset with row after row of recurved teeth (Fig. 168*C–K*); these cause it to hold so firmly in the flesh into which it is inserted that forcible removal of the tick is liable to tear the body away from the capitulum, which remains embedded in the skin. The chelicerae are elongate, slender structures lying above the hypostome; at the tip they have a movable articulated digit armed with teeth. The palpi are limber, leglike structures in the Argasidae, but rigid and closely associated with the hypostome in the Ixodidae; in the latter the fourth segment is embedded in a pit on the ventral side of the third segment (Fig. 168*A*, *B*). The single pair of spiracles are situated on the sides of the body near the fourth coxae. The legs of all four pairs (three pairs in the larvae) are much alike and are terminated by a pair of claws on a stalk. The legs are long and conspicuous when the body is empty but are hardly noticeable after engorgement.

The genital aperture is situated on the ventral side between the first or second pair of legs, the anus about halfway between the fourth pair of legs and the hind margin of the body. The presence and position of grooves on the ventral side of both sexes and the presence or absence of ventral shields in male ixodids are characters of taxonomic value.

Habits and life history. All ticks are parasitic during some part of their lives. The majority of them infest mammals, though many species attack birds and some are found on cold-blooded animals. A decided host preference is shown by some species, whereas others appear to be content with practically any bird or mammal that comes their way. Many ticks tend to attack birds or small mammals as larvae and nymphs, and larger mammals as adults. This is important in connection with disease transmission since rodents and birds are important reservoir hosts of disease organisms, particularly rickettsias and viruses. As is true of mosquitoes, some species of ticks, including such important vectors as *Ornithodoros moubata* and *Rhipicephalus sanguineus,* have geographical strains or varieties that differ in habits and food preferences.

According to Philip (1953) ticks are attracted by animal smells up to distances of at least 50 ft., and tend to collect along game trails. Ticks can be collected by dragging a muslin cloth over infested vegetation or ground.

The life histories of argasid and ixodid ticks differ principally in that the Argasidae, except *Otobius* (see p. 579), feed repeatedly as nymphs and adults, and lay their eggs in batches at intervals of weeks or months. Ixodid adult females, on the other hand, take a single enormous meal, after which they drop off the host and lay all their eggs at once, from several hundred for some species to upwards of 18,000 for others, piling them up in elongate masses in front of them (Fig. 169, *right*). The process of converting the engorged blood into eggs and depositing them occupies several days. The eggs require from 2 or 3 weeks to several months to develop. Eggs deposited in the fall do not hatch until the following spring. Newly hatched ticks are called larvae or "seed ticks" and are recognizable by having only six legs (Fig. 177*B*).

In the Ixodidae the seed ticks assume a policy of watchful waiting until some suitable host passes within reach; often they crawl up on a blade of grass or a twig to reach a strategic position. Seed ticks must be imbued with almost unlimited patience, since in many if not in the majority of cases long delays must fall to their lot before a suit-

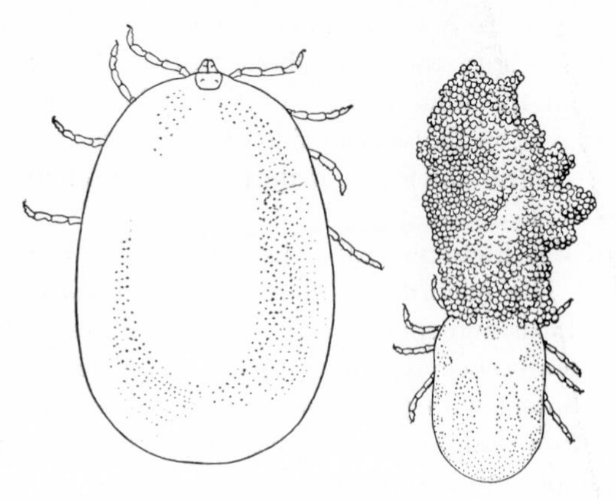

Fig. 169. *Left,* engorged tick (*Dermacentor*). *Right,* cattle tick, *Boophilus annulatus,* laying eggs (adapted from Graybill, *Farmer's Bull.,* 1912.)

able host comes their way, like a rescue ship to a stranded mariner. The larvae of some species survive unfed for a year or longer. The jarring of a footstep or rustle of bushes causes the ticks instantly to stretch out to full length, feeling with their clawed front legs, eager with the excitement of a life or death chance to be saved from starvation.

If success rewards their patience, even though it may be after many days or weeks, they feed for only a few days, becoming distended with blood and then dropping to the ground again. Retiring to a concealed place they rest for a week or more while they undergo internal reorganization. Finally they shed their skins and emerge as eight-legged but sexually immature ticks known as nymphs (Fig. 176*C*), distinguishable from females by absence of the genital aperture and of porose areas on the basis capituli (Figs. 167*C*, 170). The nymphs climb up on bushes or weeds and again there is a period of patient waiting, resulting either in starvation or a second period of feasting. Once more the ticks drop to the ground to digest the meal, transform, and molt, this time becoming fully adult and sexually mature. In this condition a host is awaited for a third and last time, and if again successful the females search for mates, copulate, and begin on their final engorgement, which results in distending them out of all proportion. Some ixodid females commence feeding before mating, but fill up more rapidly after mating. Some species of *Ixodes* which live on hosts with fixed lairs copulate before finding a host, and in such species the male is often not parasitic at all and may differ markedly

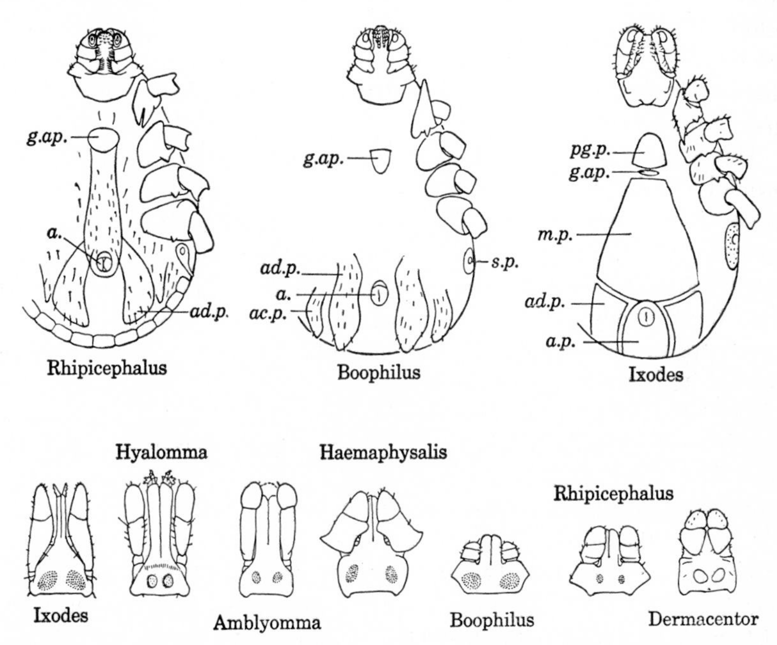

Fig. 170. *Upper figures,* ventral views of ♂, showing ventral plates; *a.,* anus; *ac.p.,* accessory plate; *ad.p.,* adanal plate; *g.ap.,* genital aperture; *m.p.,* median plate; *pg.p.,* pregenital plate; *s.p.,* spiracle. *Lower figures,* capituli characteristic of various genera; note porose areas on bases capituli. (Adapted from various authors.)

from the female in the reduced structure of its hypostome. The males usually die shortly after copulation. Most ticks, it will be seen, spend more time off their hosts than on; *Ixodes ricinus* spends only about 3 weeks of its 3 years of life on its hosts. Although some ticks can reproduce parthenogenetically, males appear not to be entirely *de trop.*

This, in general, is the life history of ixodid ticks, but it is subject to considerable variation in different species. In many species there are two nymphal periods instead of one. In some species one or both molts take place directly on the host, thus doing away with the great risk of the tick's being unable to find a new host after each successive molt, but the majority have not yet discovered the tremendous advantage of molting on the host. According to the number of times ticks risk their future by leaving their host to molt and then seeking new hosts, they are called one-host, two-host, or three-host ticks. The most important asset of ticks to counterbalance the disadvantage of having to find new hosts is their extraordinary longevity. Larvae of ticks, as noted above, may live a year or more without food, and adults have been kept alive in corked vials for 5 years. Although ixodid ticks are little troubled by inhospitable actions on the part of their hosts, in Africa tickbirds or oxpeckers (*Bucephalus*) dispose of great numbers of them.

The Argasidae differ in that they inhabit the homes of their hosts instead of the hosts themselves, and are seldom dropped in the inhospitable outside world. The eggs laid in batches number hundreds instead of thousands—a safe condition since the young argasids, reared in the home of the host, are in a much more advantageous position than the progeny of Ixodidae, which drop off and deposit their eggs anywhere in the wanderings of their host. The Argasidae lead more regular and less precarious lives.

The larvae of *Ornithodoros moubata* and *O. savignyi* of Africa molt and become nymphs a few hours after hatching, before partaking of their first meal. This is not true of the American species, however, which start looking for a place to drill almost at once. In some species, e.g., *talaje,* the larvae remain attached to a host for several days, but in others, e.g., *turicata,* they stay on only 10 minutes to a few hours; this is always true with the nymphs and adults, which can become distended like berries in this time. The nymphs may feed once or several times between molts, and may molt two to five times. The adult females, except *Otobius,* indulge in one or more gluttonous feeds between the layings of batches of eggs. *Otobius* does its last engorging as a second-stage nymph; the adults are not parasitic and have a toothless hypostome (Fig. 168*D*). During and just after feeding most species exude fluid from a pair of coxal glands opening just behind the first coxae, enough to bathe the ventral surface of the tick and contaminate the bite with organisms contained in it. Egg laying begins a week to several months after mating and feeding. The minimum time to reach the adult stage varies from 3 to 12 months. Adults may survive 5 to 12 years, including several years of starvation.

Classification, and Important Species

Classification. Both Argasidae and Ixodidae contain numerous disease transmitters and many others that are troublesome on account of the painfulness or subsequent effects of their bites. The family Argasidae contains four genera, *Argas, Ornithodoros, Otobius,* and *Antricola,* whereas in the Ixodidae there are about a dozen genera and about 500 species. The following table gives the principal distinguishing characters of the genera which are of interest as parasites of man or domestic animals.

Argasidae. No dorsal shield; capitulum ventral; segments of palpi movable (see Figs. 167, 168 and 171)

1. Dorsal and ventral surfaces demarcated by a marginal line; cuticle with small circular discs but without distinct protuberances (Fig. 171*A, C*) ***Argas***
2. Margins of body not clearly demarcated or differentiated; cuticle warty, with ridges or other types of distinct protuberances (Figs. 171*D*, 172, 174) ***Ornithodoros***

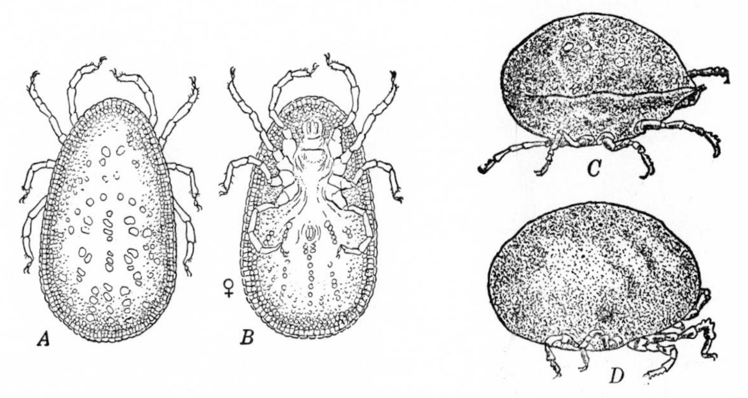

Fig. 171. *A* and *B*, *Argas persicus,* fowl tick, dorsal and ventral views of ♀. *C* and *D,* side views of engorged *Argas* and *Ornithodoros,* respectively, showing demarcation between dorsal and ventral parts of body (sutural line) in former, absent in latter.

3. Same as *Ornithodoros* but body of nymphs spiny; adults not parasitic (Fig. 171*D*) ***Otobius***

Ixodidae. Dorsal shield present; capitulum anterior; palpi rigid (see Fig. 167, 170)

1*a*. Anal groove in front of anus, horseshoelike; scutum inornate; abdomen not festooned; long mouth parts; no eyes; male with many ventral plates ***Ixodes***
1*b*. Anal groove behind anus or absent 2
2*a*. Palpi longer than width of capitulum 3
2*b*. Palpi shorter than width of capitulum 4
3*a*. Scutum inornate; 2nd and 3rd segments of palpi almost equal; male with 2 pairs of ventral plates; festoons present ***Hyalomma***

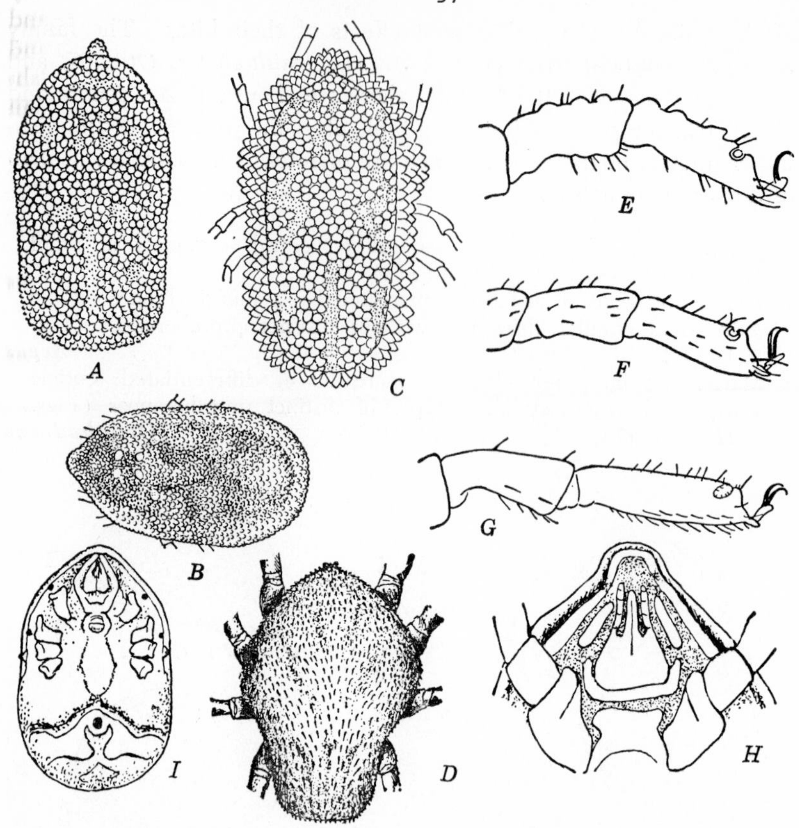

Fig. 172. Species and details of *Ornithodoros* and *Otobius: A, O. turicata; B, O. rudis; C, O. talaje; D, Otobius megnini,* nymph; *E, F,* and *G,* first leg of ♀ of *O. turicata, O. rudis,* and *O. talaje,* respectively; *H,* anterior end of *O. talaje; I,* ventral view of *O. coriaceus.* (*B* and *C* adapted from Brumpt, *Précis de parasitologie,* Masson, 1949. *D,* after Marx from Banks, *U. S. Bur. Ent. Tech. Bull.* 15. *E, F,* and *G* after Cooley and Kohls, *Am. Midland Naturalist Mon.* 1, 1944.)

3*b*. Scutum ornate; 2nd segment of palpi elongated; males without ventral plates; festoons present; only distal half of hypostome toothed **_Amblyomma_**
4*a*. Anal groove absent or very indistinct; no festoons; male with 2 pairs of ventral shields 5
4*b*. Anal groove distinct; festoons present 6
5*a*. Palpi ridged; legs of male normal **_Boophilus_**
5*b*. Palpi not ridged; segments of male legs beadlike **_Margaropus_**
6*a*. Ornate; first coxa deeply cleft (Fig. 173); basis capituli rectangular; second joint of palpi longer than third; male without ventral shields, but with festoons **_Dermacentor_**

6*b*. Inornate 7
7*a*. Palpi conical, second joint flaring at base; basis capituli rectangular; first coxa not deeply cleft; male without ventral shields ***Haemaphysalis***
7*b*. Palpi not conical, 2nd and 3rd segments of palpi about equal; basis capituli pointed at sides; first coxa deeply cleft, male with one pair of ventral shields ***Rhipicephalus***

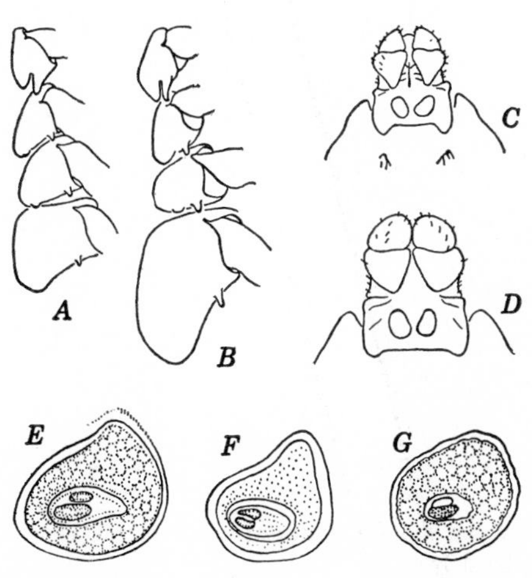

Fig. 173. Details of species of *Dermacentor* to illustrate key. *A*, coxa of *D. parumapterus; B*, same of *D. andersoni; C*, capitulum of *D. andersoni; D*, same of *D. occidentalis; E*, spiracular plate of *D. andersoni; F*, same of *D. variabilis; G*, same of *D. albipictus.* (Adapted from Cooley, *N.I.H. Bull.* 171.)

Important species and genera of Argasidae. The genus *Argas* is primarily parasitic on birds and bats. *A. persicus* (Fig. 171) is a serious pest of poultry in warm, dry parts of the world, and transmits fowl relapsing fever, "range paralysis," and fowl piroplasmosis (*Aegyptianella*). *A. reflexus* is the pigeon tick. People may suffer "grave molestation" by this tick in buildings formerly shared with pigeons.

The genus *Otobius* contains the spinose eartick, *O. megnini,* of southwestern United States and Mexico, now in South Africa also. This tick has nonconformist habits; the spiny nymphs remain attached to the ears of domestic animals, and sometimes children, for months, and then drop off, molt, mate, and lay eggs without feeding again.

The genus *Ornithodoros* contains about 50 species of thick podlike ticks, oval or elongated, and usually somewhat pointed anteriorly. (Figs. 171*D*, 172, 174.) The cuticle is leathery, mud-colored, and covered with tubercles, ridges, or variously-shaped protuberances.

These ticks attack mammals primarily. Some species live largely on rodents and other small mammals, some attacking man and domestic animals much more readily than others. About a dozen species confine their attentions to bats. *O. moubata* (Fig. 174) in Africa habitually lives in human habitations, but a number of others enter with rats or other animals, and then may bite the human inhabitants.

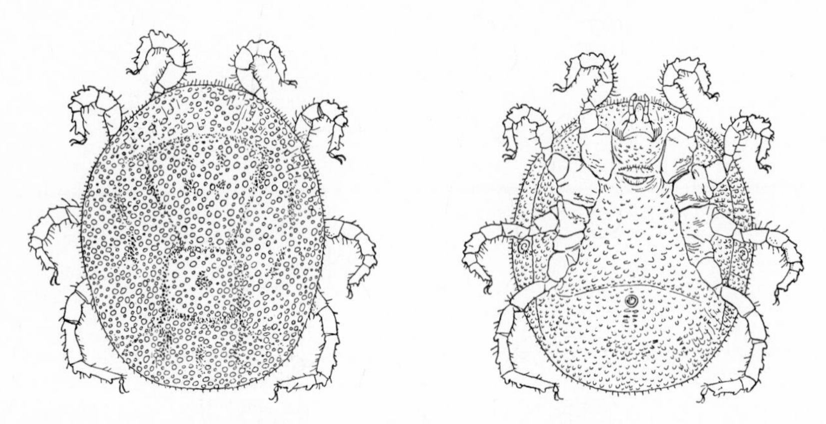

Fig. 174. *Ornithodoros moubata,* dorsal and ventral views. (From drawings supplied by Hoogstraal.)

Some species of *Ornithodoros,* e.g., *O. coriaceus* of California, cause painful and serious bites, but these ticks are particularly important as transmitters of relapsing fever (see p. 221). *Ornithodoros* ticks can also harbor and transmit a number of other diseases, e.g., leptospirosis, various rickettsial diseases, Q fever, tularemia, and virus encephalitides. The species of importance in connection with relapsing fever are discussed further under "Ticks and Relapsing Fever." Since these ticks are so important as potential transmitters of relapsing fever all over western North America, from Mexico to British Columbia, a key for the identification of important American species is given here.

Key to Important American Species of *Ornithodoros*

1*a*. Two pairs of eyes present; first coxa distinctly separated from others (Fig. 172*I*, 172*H*); large irregular depressed areas on back lacking tubercles; ♀ up to 9 mm. long; Southern California and Mexico *coriaceus*

1*b*. No eyes; first coxa barely, if at all, separated from others; small disc-like or irregular areas without tubercles; length of ♀ 5 to 7 mm. 2

2*a*. A pair of movable cheeks (Fig. 172*H*) at sides of camerostome (*talaje* group) 3

2*b*. No movable cheeks at sides of camerostome — 4

3*a*. Tubercles coarse; numerous irregular areas without tubercles (Fig. 172*C*); no marked distal hump on tarsus 1; Mexico and Central and South America, sporadic all over the United States — *talaje*

3*b*. Tubercles small; a few small disclike areas without tubercles (Fig. 172*B*); a distal hump on tarsus 1; Panama and northern South America; principal relapsing fever vector in those sections — *rudis*

4*a*. No cheeks; hood of capitulum projects beyond anterior end of body (Fig. 172*A*) (*turicata* group) — 5

4*b*. Cheeks are rounded nonmovable flaps at sides of camerostome; hypostome very small (145 μ long); tubercles very fine. High mountains west of Continental Divide, Arizona to Idaho, also in eastern Colorado; vector of relapsing fever — *hermsi*

5*a*. Tubercles in mid-dorsal region about 10 per linear mm.; hypostome over 600 μ long. Southwestern U. S. and Mexico north to Kansas, and Florida; vector of relapsing fever — *turicata*

5*b*. Tubercles in mid-dorsal region about 18 per linear mm.; hypostome 400 μ long or less; Wyoming and Washington; probably vector of relapsing fever — *parkeri*

Important genera and species of Ixodidae. Numerous species of the family Ixodidae occasionally attack man, but few habitually do so. Species belonging to a number of different genera are concerned with transmission of many important human and animal diseases and with the causation of tick paralysis. Some of the commoner and more important pests of man and animals consist of the following:

Ixodes. Although most species are parasites of small mammals, *I. ricinus,* the castor bean tick of Europe and Asia feeds almost indiscriminately, but in all its stages is a pest of cattle and sheep. Each stage feeds for a few days once a year, usually in the spring; the rest of its 3-year life cycle it just rests and digests, until at the end it has to convert its last vast meal into several thousand eggs. Other important species are *I. persulcatus* in northern Asia and *I. holocyclus* in Australia.

Amblyomma. The numerous species have a world-wide distribution. *A. americanus,* the lone star tick, so-called because of the single white spot on the scutum of the female (Fig. 175) is the commonest tick infesting man in the south central states. It also attacks rabbits, dogs, and large mammals in all its stages, and less frequently other mammals and large birds. *A. cajennense* is a common pest of domestic animals and man from southern Texas to Argentina. *A. maculatum,* the Gulf Coast tick, like the majority in this genus, attacks birds or rodents in its early stages, but large mammals, including man occasionally, as an adult. It is especially common on the ears of cattle where its bites often attract screwworms (see p. 776). Of the many species in the Old World, *A. hebraeum* of South Africa and *A. variegatus* of tropical Africa are particularly common pests and disease transmitters of large mammals.

Hyalomma. The tough, hardy species of this genus, confined to the Old World, can survive in hot arid regions where most ticks, sturdy as they are, find life too difficult. As adults they live primarily on large mammals, but in their early stages they feed on birds, rodents, and hares as well, from which they acquire many kinds of rickettsias and viruses for which these birds and mammals are reservoirs. When man or animals stop nearby, young adults, freshly molted from nymphs, come rushing from beneath every shrub. Hoogstral (1956) thinks these ticks, because of their nasty bites and efficiency as disease vectors, may rank among the most important ectoparasites of domestic animals in the entire world.

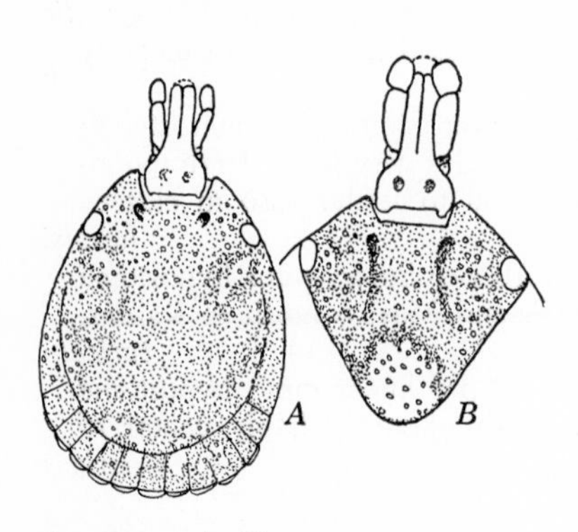

Fig. 175. *Amblyomma americanum,* lone-star tick. *A,* male; *B,* dorsal shield and capitulum of female. (Adapted from Cooley and Kohls, *J. Parasitol.,* 30, 1944.)

Rhipicephalus. Most of the 46 species of the "brown ticks" are African, but a few occur in southern Europe and Asia and one, *R. sanguineus,* distributed on its dog host, is almost cosmopolitan. Most species have rather wide host ranges, mostly on mammals, but there are a few exceptions, e.g., *R. distinctus* which confines its attentions to hyraxes. *R. sanguineus* feeds most frequently on carnivores and rabbits, largely neglecting ruminants and man. It is strangely attracted to habitations shared by man and dogs, whether it be an African hut or a Texas manor. However, except in the Mediterranean region where it transmits boutonneuse fever, it rarely bites man. This is fortunate since this species vies with *Dermacentor andersoni* in the number of disease agents it can harbor or transmit. In Africa *R. appendiculatus* is a particularly important pest of cattle, not only being the chief vector of deadly East Coast fever (see p. 198) but also predisposing its hosts' ears to bacterial and screwworm infection. *R. bursa* is one of the more important pests of domestic animals around the Mediterranean and in Asia.

Haemaphysalis. These small ticks especially abundant in Asia and Madagascar are mostly parasites of small mammals and birds. They are important as disease vectors among reservoir hosts; e.g., in North

America *H. leporispalustris,* keeps spotted fever going among rabbits, in Africa *H. leachii* plays a similar role for tick typhus (see p. 589), and in Australia *H. humerosa* distributes Q fever among bandicoots (see p. 590). *H. concinna* is reported as a tick typhus vector in eastern Siberia.

Boophilus. The three species in this genus are one-host ticks that feed primarily on cattle, less frequently on other large herbivores. The device of staying on one host from larva to adult, oddly, has been discovered by only a few other ticks. In spite of its tremendous biological value, this habit, plus close host specificity, boomeranged for *B. annulatus,* for it made possible the extermination of this species, and this species alone, in the United States (see p. 595). *B. annulatus* still survives in the West Indies and Mexico and in the Mediterranean region, where it was introduced. *B. microplus* is found in tropical areas in America, Africa, Asia, and Australia, and *B. decoloratus* in Central and South Africa.

Margaropus. *M. winthemi,* the beady-legged winter horse tick is a pest of horses, and less frequently other large herbivores, in South Africa.

Dermacentor. This genus contains species of prime importance to man in the United States. *D. andersoni* and *D. variabilis* are important transmitters of spotted fever. *D. andersoni* has been referred to as a "veritable Pandora's box" of disease-producing agents, among which, besides spotted fever, are anaplasmosis, tularemia, brucellosis, *Salmonella enteritidis,* a bacterial "moose disease," Q fever, Colorado tick fever, and several forms of virus encephalomyelitis. Many of these can be transmitted by *D. variabilis* also, and both species can cause tick paralysis (see p. 585). *D. sylvarium* of Siberia transmits a rickettsial disease and a virus encephalitis, and *D. marginatus* of Europe and Siberia transmits piroplasmosis of dogs and horses. (Figs. 176, 177.)

The species are largely confined to North America, Europe, and Asia. Most species, including the "wood ticks," *D. andersoni, D. variabilis,* and *D. occidentalis* in the United States, attack rabbits and rodents in the larval and nymphal stages, and rabbits and large mammals as adults, but *D. albipictus* is a one-host tick of large mammals. The following is a key to North American adult Dermacentors according to Cooley (1938).

Key to Important Species of *Dermacentor* in North America

1*a*. Spurs on coxa I widely divergent (Fig. 173A); southwestern United States, mainly on rabbits; a possible transmitter of spotted fever among rabbits ... *parumapterus*

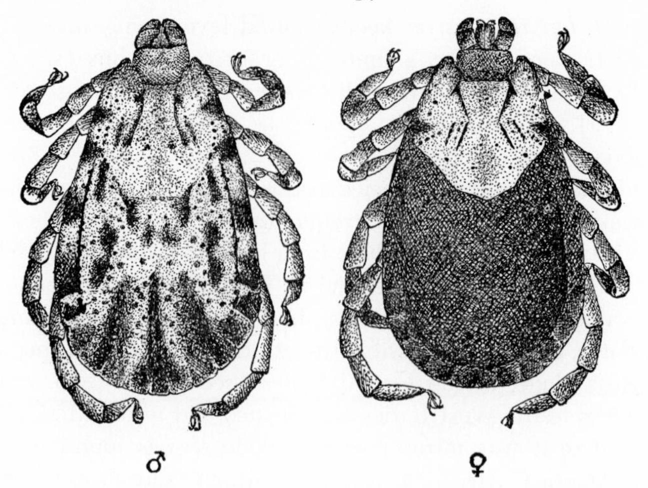

Fig. 176. Spotted fever tick, *Dermacentor andersoni,* ♂ and ♀, ×12.

1*b*. Spurs on coxa I with proximal edges parallel or a little divergent (Fig. 173*B*) 2

2*a*. Spiracular plate oval, without dorsal prolongation and with goblets (bead-like structures under spiracular plate) few and large (Fig. 173*G*); widely distributed in North America, a one-host tick, mainly on deer, etc.; probably not concerned with spotted fever, though an experimental vector *albipictus*

2*b*. Spiracular plate oval, with dorsal prolongation (Fig. 173*E, F*), and with goblets many or of moderate numbers 3

3*a*. Caudal projections from postero-lateral angles of dorsal side of basis capituli (cornua) long (Fig. 173*C*); west coast, southern Oregon;

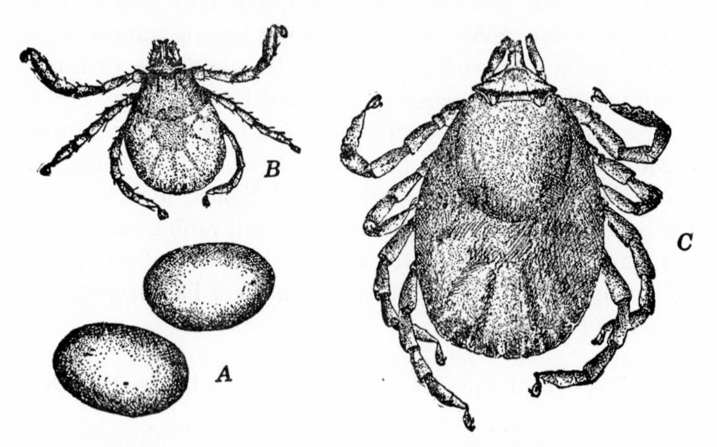

Fig. 177. Development of spotted fever tick, *Dermacentor andersoni; A,* eggs; *B,* larva; *C,* nymph. ×30.

larvae on rodents, adults on horse, deer, sheep, cow, dog, and man; a known carrier of tularemia and a suspected one of spotted fever *occidentalis*

3*b*. Cornua short or of moderate length (Fig. 173*D*) 4

4*a*. Spiracular plate with goblets very numerous and small (Fig. 173*F*); eastern N. A., west to eastern Montana and central Texas, also western California; larvae on rodents, adults on many large animals but principally dogs; a transmitter of spotted fever and tularemia, experimental vector of anaplasmosis *variabilis*

4*b*. Spiracular plate with goblets moderate in size and number (Fig. 173*E*); northwestern North America, south to northern New Mexico and Arizona, west to Sierras and Cascades, east to western Dakotas and western Nebraska, scattered records on west coast; larvae on small rodents, adults on all sorts of large mammals; vector of Rocky Mountain spotted fever and other diseases (see p. 591) *andersoni*

Injury from Bites

The wounds made by ticks, especially if the capitulum is torn off in a forcible removal, are very likely to become infected and result in inflamed sores or extensive ulcers, not infrequently ending in blood poisoning. Some species seem more prone to do this than others. The senior writer was once nearly "done in" by the bite of a tick in California, probably *Dermacentor occidentalis,* which has a bad reputation.

Ticks may also be the cause of a serious or even fatal anemia when present in large numbers. Such anemias have been observed in horses, moose, sheep, and rabbits. Jellison and Kohls in 1938 found that 60 to 80 or more female *D. andersoni* feeding on rabbits would kill them in 5 to 7 days. According to Schuhardt in 1940, rats exposed to *Ornithodorus turicata* in his "ticktorium" die after 3 hours' exposure. Development of immunity to tick bites is discussed on p. 529.

Ticks can usually be removed successfully by gentle pullng, although sometimes the mouth parts of species of *Ixodes* and *Amblyomma,* which have long hypostomes with ugly barbs, may break off in the flesh. If the tick is *jerked* off, the whole capitulum may tear off. Most ticks will not let go even if touched by chemicals that kill them. Repellents and acaricides for ticks are discussed on p. 597. Application of a disinfectant should follow removal of ticks.

Tick paralysis. More serious than the painful wounds made by ticks is a peculiar paralyzing effect of tick bites, known as tick paralysis. This effect is produced only by rapidly engorging female ticks, especially when attached on the back of the neck or at the base of the skull. In one case a male tick was suspected, but the evidence is not convincing. There is no evidence of any infective organism being involved. The cause of the paralysis is still obscure, but several investigators have obtained evidence that the eggs of ticks contain a highly toxic substance or that such a substance is formed during their development; it evidently makes its way to the salivary glands, since it is transmitted by the bites. Not all ticks produce the effect, but it is not limited to any one genus, nor does it extend to all the members of any one genus. In North America, *Dermacentor andersoni* and *D. variabilis* are responsible; in Australia, *Ixodes holocyclus;* in Crete, *Ixodes ricinus* and *Haemaphysalis punctata* (suspected); in Somaliland *Rhipicephalus simus,* and in South Africa *Ixodes rubicundus, Hyalomma transiens,* and *R. simus.* In Russia *Ornithodoros lahorensis* causes paralysis, but only when present in large numbers. *Rhipicephalus sanguineus* in Yugoslavia was found to contain the toxin. In one case in British Columbia, *Haemaphysalis cinnabarina* was incriminated.

Since the paralysis is not invariably produced even by ticks situated at the base of the neck, it is possible that the bite must pierce or come in contact with a nerve or nerve ending. The paralysis usually begins in the legs and may result in complete loss of their use; it gradually ascends during the course of 2 or 3 days, affecting the arms and finally the thorax and throat. Unless the heart and respiration are affected, recovery follows in 1 to 6 or 8 days after removal of the engorging female ticks, even though other ticks remain. If the engorging ticks are not removed, the affection may result in death from failure of respiration or in spontaneous recovery after a few days or a week. The disease as observed in Australia differs from the North American type in that improvement is less immediate after removal of the offending tick. Paralysis of man and animals, particularly cattle, sheep, dogs, and cats, is frequent in northwestern United States and Canada. In South Africa sheep are paralyzed but human cases are doubtful. It is by no means certain that all cases reported in animals are true tick paralysis, since symptoms that might be confused may be caused by tick-borne infections—Babesiidae, Anaplasma, rickettsias, viruses, or bacteria. A "moose disease" in northern Minnesota and Ontario suspected of being tick paraylsis was seemingly due to a paralysis-causing bacillus, *Klebsiella paralytica,* harbored by the tick, *Dermacentor albipictus* (see Wallace, Cahn, and Thomas, 1933). Most human cases are in children and are most frequent in girls, whose long hair conceals attached ticks. Some of the cases are fatal.

Ticks as Vectors of Disease

Ticks play an extremely important role as transmitters of disease to domestic animals and, fortunately to a somewhat less extent, to man. They are of outstanding importance in the transmission of organisms of six principal types: (1) spirochetes of relapsing fever; (2) rickettsias of spotted fever and related diseases of man and animals; (3) Babesiidae, causing many diseases of prime importance to domestic animals; (4) *Pasteurella tularensis,* the bacterium of tularemia; (5) *Anaplasma;* and (6) filtrable viruses of several types, including some causing encephalomyelitis. The special relation of ticks to the diseases caused by these six types of disease agents is considered in separate sections below. In addition, *Rhipicephalus sanguineus* and probably others are intermediate hosts for *Hepatozoon canis* (see p. 204), causing infection when swallowed. Many ticks, both ixodids and argasids, harbor and transmit a bacilli of the genus *Salmonella* which cause paratyphoid-like disease in rodents, and sometimes give trouble in experimental animals. Species of *Ornithodoros* can also harbor and transmit leptospiras (see p. 224).

Ticks and Relapsing Fever

Many, but not all, species of argasid ticks harbor and transmit spirochetes of the genus *Borrelia* that cause relapsing fever in birds and mammals (see p. 220). Those of birds are transmitted by *Argas,* whereas those of mammals are always transmitted by species of *Ornithodoros,* except one strain that has become adapted to transmission by lice and has become independent of ticks (see p. 630). The various strains of *Borrelia,* all alike morphologically, differ immunologically and in their pathogenicity for particular hosts, and show remarkable vector specificity. With rare exceptions each vector harbors and transmits its own strain of spirochetes and fails to transmit spirochetes from other species. Even strains or geographically separated clones of one tick species may differ in their ability to transmit a certain strain of spirochete. This goes so far that sometimes closely related species of ticks, e.g., *O. turicata* and *O. parkeri* in North America, are more readily distinguished by the spirochetes they harbor than by their morphological characters. On the other hand, it was shown by Nicolle and Anderson in 1927 that *O. moubata,* the transmitter of the severe *Borrelia duttonii* strain in Africa, is also capable of harboring and transmitting *B. hispanica* and a rodent strain normally carried by *O. erraticus.* The spirochetes, if acquired by a tick feeding on an infected host, quickly disappear from the digestive tract and invade the hemocoele and all the tissues, where they may persist for life. That may be a long time, for these ticks can live for at least 7 years with food and for 5 years without it, and meanwhile may pass the spirochetes on to many hundreds of offspring.

In most species, at least, the spirochetes are transovarially transmitted, so they may persist entirely independent of vertebrate hosts, although in some strains the percentage of transovarial transmission is low. Transmission to birds or mammals usually takes place directly by bite, but *O. moubata* adults commonly transmit the spirochetes via the coxal fluid, which it exudes copiously during feeding. In most species this fluid is not infective and in some it is absent. In some species transmission by feces may also be possible.

The spirochetes of relapsing fever, in contrast to their extreme finickiness about their tick hosts, are surprisingly indiscriminate about their vertebrate hosts, except that the bird strains do not infect mammals and vice versa. Probably most of the strains are capable of infecting man. The pathogenic effects of particular strains on particular species of hosts, including man, vary from inapparent to very severe. Relasing fever is not a serious disease of any large mammals except man. Many species of birds are susceptible to infection with *B. anserina,* transmitted principally by *Argas persicus* and *A. reflexus;* severe outbreaks occur among chickens, ducks, pigeons, and other domestic birds.

Species involved. *O. moubata,* the eyeless tampan of tropical Africa and Madagascar (Fig. 174), is the only species of its genus which has, in some places, become domestic, living in floors, crevices, thatch, etc., of native huts and rest houses along routes of travel. There appear to be at least three biological races of this tick (Walton, 1957), one limited to cool, wet areas, feeding primarily on man and sometimes rats; one, with a wide range, feeding primarily on fowls; and one found in burrows of wart hogs, which has not been found infected with spirochetes. The spirochete transmitted by this tick, *B. duttonii,* causes a particularly severe form of relapsing fever in man, with numerous relapses; this strain is also virulent for rats and mice, but hardly at all for guinea pigs. The closely related eyed tampan, *O. savignyi,* ranging from India to South Africa, has not been found naturally infected. This outdoor tick is common under trees where camels or other animals rest, and sometimes buries itself in the dust of native bazaars.

In North Africa from Morocco to Tunis, and in Spain, a large race of *O. erraticus* transmits a spirochete, *B. hispanica,* which causes sporadic human infections. The more widely distributed small race of this tick harbors a whole complex of spirochete strains which are rarely if at all infectious for man or adult guinea pigs, but very infective for small rodents and shrews. *O. erraticus* commonly lives in rodent burrows and sometimes in pig pens and fox dens; in the burrows it feeds on anything from toads and lizards to burrowing owls and porcupines. Ticks of this species taken from one locality may show all degrees of refractoriness or susceptibility to spirochetes from the same species in other localities, or sometimes even in other burrows (Baltazard, 1954).

In the Middle East and western and central Asia *O. tholozani*, primarily a parasite of bovids in all its stages, transmits another strain (*B. persica*) to man. The common and widely distributed sheep tick *O. lahorensis* is not a vector to man.

In the New World, *O. rudis* (Fig. 172*B*) enters human habitations with rats in Panama and northern South America, and transmits *B. venezuelensis*. *O. talaje* (Fig. 172*C*), widely distributed in the United States, Mexico, and Central America and often confused with *O. rudis*, harbors *B. mazzottii* in Mexico. It thrives on the blood of rats and pigs, but shows a definite aversion to that of man, so if it enters houses it seldom makes its presence known. In California *O. hermesi* inhabits the nest of chipmunks. These move into vacated mountain cabins during the winter and leave some of their spirochete-infected ticks behind when they move out as the summer residents return; the latter may come down with relapsing fever and have their vacations ruined. *O. turicata*, carrier of a strain, *B. turicatae*, which causes severe human cases, is distributed from Kansas to Mexico. It is an indiscriminate feeder, taking its blood where it finds it. In Mexico it has become semidomestic and frequents pigsties, abbatoirs, and thatched huts, hence often infects man. In Texas it haunts caves, where it promptly transmits relapsing fever to anyone venturing into them, and in Kansas it inhabits rodent burrows.

With the few exceptions in which the ticks invade human residences, human relapsing fever results from intrusion upon the natural wild habitats. *O. parkeri* causes sporadic cases in the northwest woods. The vicious-biting *O. coriaceus* of southern California is *not* a vector. A key to the important species of Ornithodoros in North America is given on pp. 580–581.

Ticks and Spotted Fever and Other Rickettsial Diseases

General considerations. Tick-borne rickettsial diseases (see p. 214), or tick typhus, as Megaw calls them collectively, occur in many parts of both the Old and New World. A number of different strains, or species according to some, are recognized by their immunological reactions and sometimes by their pathogenicity. Unlike the rickettsias of true typhus, the tick-borne varieties invade the nuclei as well as cytoplasm of the cells and so are placed in a separate subgenus, *Dermacentroxenus.* However both argasid and ixodid ticks can also transmit epidemic and endemic typhus; apparently bovid animals and their ticks serve as a reservoir for *R. prowazekii* in the tropics (Reiss-Gutfreund, 1956), and may constitute the original source of the louse-borne and flea-borne strains of typhus rickettsias (p. 216). Another rickettsia, *R. pavlovskyi,* which causes severe kidney damage (nephrosonephritis) has been reported in several areas in the Soviet Union. It has been recovered from patients, rodents, and a number of ectoparasites, including trombiculid and blood-sucking (gamasid) mites (see p. 215), and fleas as well as several species of ixodid ticks. It is said to be transovarially transmitted in both the ticks and the gamasid mites. Various atypical rickettsialike organisms (see p. 216) such as *Ehrlichia canis* and *E. bovis,* affecting dogs and bovids respectively; *Coxiella burnetii* of Q fever (p. 217); *Cowdria ruminantium* of South African heartwater fever (p. 217); and various other diseases involving rickettsialike organisms of uncertain relationships have been found, e.g., Bullis fever in Texas (p. 218).

The principal recognized forms of tick typhus, due to species or strains of *Dermacentroxenus,* are (1) spotted fever in North and South America, and probably Siberia, caused by *R. rickettsii;* boutonneuse fever around the Mediterranean and the same or a very closely related disease, called tick typhus or tick-bite fever, in tropical and South Africa and south and central Asia caused by *R. conorii;* North Queensland tick typhus, caused by *R. australis.* These diseases are characterized by a severe rash or blotching of the skin, including face, palms, and soles; headache; body pains; fever; and a positive Weil-Felix reaction, i.e., agglutination of OX19 or OX2 strains of *Proteus* by the serum of infected persons or animals. In the Old World forms there is usually a buttonlike black ulcer (eschar) at the site of the infective bite, hence the name boutonneuse, French for buttonlike.

Spotted fever and its vectors. This disease has long been known as a common and dangerous infection in the Rocky Mountain region of northwestern United States and Canada, especially Montana and Idaho. Cases occur more sporadically in other parts of the United States, particularly on the Middle Atlantic coast. In Maryland alone nearly 1000 cases were reported in the 20 years from 1931 to 1950, with 10% mortality. An immunologically indistinguishable disease occurs in northern Mexico, Colombia, and Brazil.

The disease is transmitted by the bites of ticks, but it takes about 2 hours of attachment before transmission is successful. The rickettsias in this as in other forms of tick typhus invade all the tissues of the ticks, and can be transovarially transmitted. Only a minority of the eggs are infected, all producing infected ticks. To keep the infection going in nature, however, transmission from tick to tick via infected mammals appears to be necessary as well. The infection is inoculable into rabbits and numerous rodents, but it is a curious fact that it has never been isolated from wild hosts, and only a very small percentage of ticks is found naturally infected. The distribution of spotted fever in northwestern United States corresponds closely with that of the cottontail rabbit, *Sylvilagus nuttalli,* and cottontails occur in other spotted fever areas, and since rabbits are the only hosts fed upon by the vector to man in the northwest, *Dermacentor andersoni* in all stages of its life cycle, suspicion attaches to them as important reservoirs of the disease. In eastern United States, however, Price (1954) considers the meadow mouse (*Microtus*) the most important reservoir host. Dogs may also be involved since they harbor inapparent infections, as they do of boutonneuse fever. Strains of *R. rickettsii* vary from very low to very high virulence; apparently this depends on the vertebrate rather than the tick hosts from which the organisms are derived; repeated passage through rabbits seems to lower virulence.

The only vector to man in the northwest is *Dermacentor andersoni,* but this infection is kept going among rabbits by several ticks, including the two rabbit ticks, *Haemaphysalis leporispalustris* and *Dermacentor parumapterus,* and also *Ornithodoros parkeri.* In eastern and southern states the principal vector is *Dermacentor variabilis;* in Texas and Oklahoma, *Amblyomma americanum,* and perhaps *Rhipicephalus sanguineus;* in Mexico probably *Rhipicephalus sanguineus* and *Ornithodoros nicollei;* and in South America, *Amblyomma cajennense* and *A. striatum.* On the west coast of the United States a few cases of spotted fever have been found within the domain of *Dermacentor occidentalis.*

***Dermacentor andersoni* and *D. variabilis*.** These are excellent transmitters of spotted fever because in their larval and nymphal stages they feed on rodents and rabbits, but as adults attack large mammals, including man. *D. andersoni* (Fig. 176) is a handsome reddish-brown tick with the dorsal shields conspicuously marked with silver. The six-legged larvae (Fig. 177*B*), of which there are about 5000 in a brood, although feeding on many small mammals, show some preference for squirrels, attaching themselves about the head and ears. After a few days the larvae drop, leisurely transform into nymphs (Fig. 177*C*), and again attack rodent or rabbit hosts. After dropping off these and transforming into adults they lose interest in the small mammals; but now seek large mammals such as horses, cattle, deer, etc., and man and dogs as well. The entire life cycle from egg to eggs takes 2 to 2½ years. The winter is passed in either the nymphal or adult stage.

D. variabilis and *D. occidentalis* have similar life cycles. *D. variabilis* exhibits a preference for meadow mice (*Microtus*) and white-footed mice (*Peromyscus*) in its immature stages, and for dogs in its adult stage. A key for the differentiation of the important North American species of *Dermacentor*, with geographic distributions, is given on pp. 583–584.

***Amblyomma americanum* and *A. cajennense*.** These ticks (see p. 588 and Fig. 175), unlike the dermacentors, bite man and larger animals in all their stages. This probably accounts for the multiple family infections that sometimes occur where these ticks are involved, for the nymphs may be present in immense concentrations in small areas, including many infected ones. In contrast, by the time an infected brood of dermacentors gets around to biting man as adults they are scattered far and wide, and most of them are dead.

Boutonneuse fever and "tick typhus." Tick-borne typhuslike diseases, caused by *Rickettsia conorii* and perhaps other species or strains of *Rickettsia*, occur in many parts of southern Europe, Asia, and Africa. These diseases are transmitted readily by *Rhipicephalus sanguineus* and certain other ticks but not, experimentally, by *Dermacentor andersoni*. In central Asia and Siberia "tick typhus" is transmitted by dermacentors (*D. nuttalli* and *D. sylvarum*), with marmots and *Microtus* suspected as reservoirs, this strain of the disease may be closer to spotted fever than to boutonnense fever. The rickettsia involved has been named *R. sibiricus*. In eastern Siberia *Haemaphysalis concinna*, also, is said to be a vector for this infection. In Africa south of the Sahara there are potential transmitters of tick typhus in practically all ecological situations. The dog ticks, *Rhipicephalus sanguineus* and *Haemaphysalis leachii*, both of which attack wild rodents in their early stages, play leading roles since rodents, especially ground squirrels, are probably the prime reservoirs in nature, and dogs, with inapparent infections, constitute a link between the rodents and man. *R. appendiculatus* and *R. simus*, both with wide host ranges, and *Amblyomma hebraeum* in its immature stages, also participate as vectors.

Tick typhus of Queensland, caused by *Rickettsia australis*, is transmitted by *Ixodes holocyclus; Ehrlichia canis* (see p. 216) of dogs, common in some parts of Africa and Asia, is transmitted by *R. sanguineus.*

Another *Rickettsia,* producing a typhuslike disease (maculatum fever) in guinea pigs, has been isolated from *Amblyomma maculatum* and *A. americanum* collected from cattle in Texas and Mississippi, and a related one from *Dermacentor parumapterus* from rabbits in the northwest. The organisms from *A. maculatum* are mild, those from *A. americanum* virulent. The interrelations of these various rickettsial infections are still very obscure, and much is still to be learned about the extent to which they are modified by passage through different vectors or development in relatively insusceptible hosts such as dogs, pigs, and cattle.

Q fever. This rickettsial disease, caused by *Coxiella burnetii,* was first discovered in 1937 in Australia where it is transmitted among bandicoots by *Haemaphysalis humerosa* and from these animals to cattle and possibly occasionally to man by *Ixodes holocyclus.* A year later this infection was discovered in *Dermacentor andersoni* in Montana. Subsequently it has been found to be widely distributed over the world, but strikingly absent from some countries. Small mammals and birds may serve as reservoirs, as well as large domestic animals and poultry. In such animals it may be a very common though frequently overlooked infection, since it produces clinical illness only in man.

Many ticks, of both families, may serve as vectors and transovarial transmission has been demonstrated experimentally. However, unlike tick typhus, Q fever is not primarily dependent upon ticks for transmission. It is commonly transmitted among bovids through the placenta or by milk, and to man by raw milk, raw eggs, dust in laboratories or cattle sheds, handling carcasses, or even by contaminated laundry. There is no direct transmission from person to person. Outbreaks may occur among stock handlers, or workers in dairies or slaughterhouses.

Q fever is an influenzalike disease without a rash, causing fever, chest pains, and cough; in Europe it was called Balkan grippe or atypical pneumonia. Cattle and horses acquire light or inapparent infections, as do dogs and cats, but sheep and goats develop respiratory symptoms similar to those in man. Natural infection in the latter animals has been demonstrated by inoculation of milk into guinea pigs.

Other diseases caused by rickettsialike organisms. A previously unknown rickettsial disease, first observed in a few cases at Camp Bullis, Texas, in 1941, seems now to be permanently established in central Texas; it is known as Bullis fever. Rickettsias, said not to be related to those of either spotted fever or Q fever, were isolated from cases, and also from a naturally infected *Amblyomma americanum* by Anigstein and Bader (1943).

Heartwater fever is another disease caused by an organism related to *Rickettsia, Cowdria ruminantium* (see p. 215). It severely affects ruminants in South Africa, causing large accumulations of fluid around the heart. The principal transmitters are *Amblyomma hebraeum* and *A. variegatus.* A disease resembling heartwater fever has been reported from goats in Dalmatia. Two species of the related genus *Ehrlichia,* one in dogs and one in cattle, are also tick transmitted.

Tularemia and Other Bacterial Diseases

Tularemia or rabbit fever is another disease of rabbits and rodents transmissible to man and many other animals, and widely distributed over the world. It is characterized by a local ulcer at the site of inoculation, enlarged and painful lymph glands in the vicinity, and such generalized symptoms as fever, prostration, general aches and localized pains, but it is rarely fatal. Symptoms may last for several months, and a lasting immunity develops. Diagnosis is made by inoculating material from the ulcer or inflamed glands into laboratory animals or by an agglutination test. Streptomycin is useful in treatment.

The disease is caused by a bacillus, *Pasteurella tularensis* (see p. 226), closely related to the plague bacillus. It infects many mammals and even some birds; rabbits, ground squirrels, and small rodents are important reservoirs. Severe outbreaks have been reported among sheep, jack rabbits, and beavers. In Arkansas hunting dogs are commonly infected by *Amblyomma americanum.* Several foci have been reported in the Soviet Union where floods force voles above ground, and the disease is then transmitted to man both by ticks and biting Diptera (see Pavlovsky et al., 1955; ref. on p. 34). Experimentally tularemia can be transmitted by many kinds of arthropods including lice, fleas, deer flies, and ticks, but the last are probably the primary transmitters among natural hosts. *Haemaphysalis leporispalustris* and species of *Dermacentor* are the commonest transmitters among rabbits in this country. The ticks transmit the disease either by their bites or by fecal contamination of skin abrasions; mere handling of an infected tick and subsequent rubbing of the eye may cause infection. The disease is also transmitted by contact and is frequently acquired by handling diseased animals, especially rabbits. In an epizoötic among beavers and muskrats in Montana and Wyoming, there were thirty-eight known human cases from handling these animals. A remarkable feature, still not adequately explained, was the pollution of all the water and mud in rivers and tributaries over a large area for a period of 16 months. In spite of this, human infections from drinking the water or swimming in it are few. In 1934 a water-borne epidemic was reported in Turkey.

Another bacterial disease that Tovar in 1947 found ticks able to transmit is undulant fever or brucellosis, a disease commonly acquired from the milk or meat of infected animals. Ticks of several ixodid genera on cattle, and also *Ornithodoros lahorensis,* have been found naturally infected and the disease is said by Russian parasitologists to be conserved in these ticks between outbreaks. Small rodents may harbor this infection as well as tularemia.

Ticks are not known to transmit other bacterial diseases to man, but some species, especially argasids, may harbor salmonellas, which they can transmit to laboratory animals, and perhaps among domestic animals (Floyd and Hoogstraal, 1956). These bacteria cause paratyphoidlike disease in rodents, and are a cause of food poisoning in man. *D. albipictus* has been found to harbor *Klebsiella paralytica,* associated with a "moose disease" (see p. 586). However, Anigstein and his colleagues have shown that ticks contain a substance that inhibits the growth of many kinds of bacteria.

Piroplasmosis

The small blood protozoans belonging to the family Babesiidae (see p. 196) are the cause of numerous important diseases in domestic animals. To these the name "piroplasmosis" is generally applied, since the organisms were once named *Piroplasma.* Man is peculiarly exempt.

Texas fever and other piroplasmoses. Texas fever or "redwater fever" (briefly described on p. 198) is of enormous economic importance in cattle-raising countries. It was formerly prevalent in southern United States but has now been wiped out. Smith and Kilbourne in 1893 set a milepost in history when they discovered its transmission by cattle ticks, *Boophilus annulatus.* Since this is a one-host species, it is obvious that a tick becoming infected on one animal would have no opportunity to infect another. It could not do so even if transplanted from one animal to another, for the organisms (*Babesia bigemina*) invade the eggs of the tick, and cyclical development takes place in the embryonic tissues of the developing offspring (see p. 198). Many other species of *Babesia* are known to infect ruminants, horses, pigs, dogs, and even poultry. The life cycles of the various species are probably similar, since all of them are hereditarily transmitted in their tick vectors.

All the species of *Boophilus* seem to be able to serve as intermediate hosts for the *Babesia* of cattle, sheep, etc. In Europe *Ixodes ricinus* is an important transmitter of *B. bovis,* and species of *Hyalomma* and *Rhipicephalus* have been implicated in North Africa and elsewhere. *B. canis,* causing piroplasmosis in dogs, is transmitted by *R. sanguineus* in the tropics, by *Dermacentor marginatus* (=*reticulatus*) in Europe (this species also transmits a *Babesia* of horses), and by *Haemaphysalis leachii* in South Africa. According to Shortt, *B. canis* can be transmitted by subsequent stages of an infected tick as well as by its offspring. A related parasite of horses, *Nuttallia equi* (see p. 198), is transmitted by *Dermacentor nuttalli* and also by a one-host tick, *Hyalomma detritum* in the Old World.

The Protozoa of the related genus *Theileria* (see p. 198), one of which causes the deadly East Coast fever of cattle in Africa, differ in their life cycles since they are transmitted only by two-host or three-host ticks and never transovarially. Species of *Rhipicephalus* and *Hyalomma savignyi* are the principal transmitters of East Coast fever, whereas various species of *Hyalomma* and *Boophilus annulatus* are reported as vectors of a milder infection in North Africa and Turkey.

Aegyptianella pullorum, a protozoan inhabiting the blood corpuscles of chickens, ducks, and geese and believed to belong to the Babesiidae, occurs in southern Europe and Africa and is transmitted by *Argas persicus.* A similar organism has been reported from fowls in New York and Philadelphia.

Ticks and Virus Diseases

Ticks have been shown to harbor and in some cases to transmit a number of viruses infective for man and animals, but as Hoogstraal (1956) pointed out, their propensities in virus transmission constitute a vast, largely unexplored field for research. In the United States *Dermacentor andersoni* can transmit western equine encephalomyelitis, and *D. variabilis* the St. Louis strain. In the Far East, *Ixodes persulcatus* is reported to be a carrier of Japanese B virus (see p. 229).

Colorado tick fever, which causes a disease strikingly like dengue (see p. 750), characterized by a diphasic fever, severe aches, and a marked leucopenia is widespread in western United States and Canada and has been reported from Long Island. *Dermacentor andersoni,* in which the virus is transovarially transmitted, is thus far the only proved vector to man, but this virus has been isolated from several species of rabbit-feeding ticks.

In northern Britain and Russia *Ixodes ricinus* transmits "louping ill," a virus disease which attacks the central nervous system of sheep and other ruminants, and rarely man. In the Soviet Far East, a closely related virus causing "spring-summer encephalitis" is commonly transmitted by *Ixodes persulcatus.* The immature stages of this tick, like those of the dermacentors, feed on rodents and birds, which serve as reservoir hosts for this virus. Human diseases caused by the same or another closely related virus, transmitted by *I. ricinus,* have been reported from western Russia and Czechoslovakia.

In various places in the Soviet Union human outbreaks of acute "hemorrhagic fever" occur where, under suitable environmental and ecological conditions, man is infected from reservoir hosts by particular vectors—e.g., in Crimea from hares by *Hyalomma marginatum;* in Omsk from *Microtus* by *Dermacentor pictus;* in Uzbekistan from rodents by *Hyalomma anatolicum;* etc. (Gajdusek, 1953). These are examples of Pavlovsky's "natural nidality" (see p. 23). Various "meningo-encephalitis" viruses have been reported from various parts of eastern Europe, transmitted by *I. ricinus* and probably other ticks.

In Kenya, the virus of Rift Valley fever (see p. 228) survives in ticks (*R. appendiculatus*) for a week after an infective feed. This tick is also a vector of Nairobi sheep disease, characterized by fever and severe diarrhea.

Anaplasmosis. This frequently fatal disease, which causes fever, jaundice, and a very severe destruction of blood corpuscles in cattle and other animals, is characterized by dotlike bodies in the blood corpuscles, called *Anaplasma* (see p. 218). The disease is very commonly associated with *Babesia* or *Theileria* infections, since the tick vectors often have double infections. It can be transmitted by at least seventeen species of ticks belonging to several different genera and also by the intermittent feeding of biting flies. It can also be transmitted by ticks to their offspring.

Control of Ticks

Argasids. The domestic *O. moubata* in Africa, like triatomids in South America, can be eliminated by improved housing and cleanliness, but in most places this is still a dream of the distant future. Meanwhile spraying floors and walls (to a height of a foot or two) with a suspension of BHC may eliminate it for a year or more. Dusting with BHC or other chlorinated hydrocarbons or organic phosphorus compounds is also effective. Since this tick is very much a "stay-at-home," reinfection of cleared areas does not readily occur. Control of *O. rudis* involves rat elimination, and getting rid of *O. hermsi* demands exclusion of chipmunks. Dusting cracks in stone walls eliminates *O. lahorensis*, sometimes for over 2 years. Most other species of *Ornithodoros* need only to be left alone in their natural habitats.

Argas can be eliminated from poultry houses by spraying walls, roosts, and litter with chlorinated hydrocarbons, or organic phosphorus compounds, or spraying creosote, turpentine, or crude oil into their hiding places.

Against the eartick, *Otobius megnini*, Kemper (1953) recommends pine oil containing 1% lindane or 5% chlordane, 2 : 1 mixture of pine oil and cottonseed oil, or a "pyridine-adhesive" remedy. These would probably be useful for other earticks as well.

Ixodids. Ticks on domestic animals were formerly controlled by sodium arsenite dips (about 0.16%), which are still used to some extent, and by pasture rotation. By these methods and by federal quarantine the one-host cattle tick, *Boophilus annulatus*, and with it Texas or redwater fever of cattle, was eliminated from the United States, although a few of these ticks persist on deer in a few places in Florida, and sometimes cross the Rio Grande from Mexico on cattle or deer.

On animals sprays or dips of emulsions of chlorinated hydrocarbon emulsions, or dusts, can be used, but may be toxic on repeated applications and should never be repeated in less than 30 days. Methoxychlor is best for dairy cows, since it does not get into the milk. Unfortunately, ticks resistant to arsenic or BHC (not both) have appeared in some parts of the world; for these sprays of diazinon or malathion are useful, although these chemicals are too unstable for dips. The effects of different chemicals vary somewhat not only with the species of tick but also with rainy and dry seasons. Sprays or dusts containing pyrethrins plus a synergist (see p. 533) are safe and effective, as they are for lice (see p. 632).

Houses, kennels, and barns can be freed of ticks by spot-spraying harboring places with 5% DDT, 3% chlorinated hydrocarbons, or organic phosphorus compounds.

On pastures, yards, etc., fairly good control effective for 6 to 8 weeks can be obtained by spraying or dusting heavily with insecticides. For large areas, helicopters or planes are used. Dieldrin seems to be the most effective chemical for this purpose.

Most repellents are somewhat less effective against ticks than against insects; indalone and diethyltoluamide are the best skin repellents, and M 1960 best for treatment of clothing (see p. 537).

Important current progress is being made in the use of silica aerogels.

REFERENCES

Amer. Assoc. Adv. Science. 1941. *A Symposium on Relapsing Fever in the Americas.* Publ. 18: 130 pp. Science Press, Lancaster, Pa.

Anigstein, L., and Bader, M. N. 1943. Investigations on rickettsial diseases in Texas, 1 and 2. *Tex. Repts. Biol. Med.*, 1: 105–116, 117–140, 298, 389–409.

Banks, N. A. 1908. A revision of the Ixodoidea or ticks of the United States. *U. S. Bur. Entomol. Tech. Bull.* 15.

Bequaert, J. C. 1945. The ticks or Ixodoidea of the northeastern United States and Canada. *Entomologia Americana*, 25: 73–120, 121–184, 185–232.

Bishopp, F. C., and Wood, H. P. 1913. The biology of some North American ticks of the genus *Dermacentor. Parasitology*, 6: 153–187.

Burgdorfer, W. 1956. The possible role of ticks as vectors of leptospirae. I. *Exp. Parasitol.*, 5: 571–579.

Burgdorfer, W., and Owen, C. R. 1956. Experimental studies on argasid ticks as possible vectors of tularemia. *J. Infectious Diseases*, 98: 67–74.

Burroughs, A. L., et al. 1945. A field study of latent tularemia in rodents with a list of all known naturally infected vertebrates. *J. Infectious Diseases*, 76: 115–119.

Cooley, R. A. 1938. The genera *Dermacentor* and *Otocentor* in the United States. *Natl. Insts. Health Bull.* 171.

1946. The genera *Boophilus, Rhipicephalus* and *Haemaphysalis* (Ixodidae) of the New World. *Natl. Insts. Health Bull.*, 187. 54 pp.

Cooley, R. A., and Kohls, G. M. 1944. The genus *Amblyomma* (Ixodidae) in the United States. *J. Parasitol.*, 30: 77–111.

1944*b*. The Argasidae of North America, Central America, and Cuba. *Am. Midland Naturalist, Monogr.*, 1: 152 pp.

1945. The genus Ixodes in North America. *Natl. Insts. Health Bull.*, 184.

Davis, G. E. 1940. Ticks and relapsing fever in the United States. *Public Health Repts.*, 55: 2347–2351.

Eklund, C. M. 1955. Distribution of Colorado tick fever and virus-carrying ticks. *J. Am. Med. Assoc.*, 157: 335–337.

Florio, L., and Miller, M. S. 1948. Epidemiology of Colorado tick fever. *Am. J. Public Health*, 38: 211–213.

Floyd, T. M., and Hoogstraal, H. 1956. Isolation of *Salmonella* from ticks in Egypt. *Am. J. Trop. Med.*, 5: 388–389.

Fuller, H. S. 1956. Veterinary and medical acarology. *Ann. Rev. Entomol.*, 1: 347–366.

Gajdusek, D. C. 1953. Acute infectious hemorrhagic fevers and mycotoxicoses in the U.S.S.R. *Army Med. Serv. Grad. School, Med. Sci. Publ.* 2 Washington.

Gregson, J. D. 1956. *The Ixodoidea of Canada. Dept. Agr. Canada Publ.* 930.

Hoogstraal, H. 1956. *African Ixodoidea,* Vol. 1, 1101 pp. U. S. Dept. Navy Research Rept. NM 005–050. 29.07.

Hooker, W. A., Bishopp, F. C., and Wood, H. P. 1912. The life history and bionomics of some North American ticks. *U. S. Bur. Entomol. Bull.* 106.

Huebner, R. J., et al. 1949. Q Fever—a review of current knowledge. *Ann. Intern. Med.,* 30: 495–509.

Jellison, W. L., and Gregson, J. D. 1950. Tick paralysis in northwestern United States and British Columbia. *Rocky Mt. Med. J.,* Jan. 1950.

Kemper, H. E., and Peterson, H. O. 1953. The spinose ear tick. *U. S. Dept. Agr. Farmer's Bull.* 980.

Nuttall, G. H. F., Warburton, C., and Robinson, L. E. 1908–1926. *Ticks, A Monograph of the Ixodoidea.* Pts. 1–4. Cambridge University Press, Cambridge.

Parker, R. R., Kohls, G. M., and Steinhaus, E. A. 1943. Rocky Mountain spotted fever: spontaneous infection in the tick *Amblyomma americanum. Public Health Repts.,* 58: 721–729.

Parker, R. R., Philip, C. B., Davis, G. E., and Cooley, R. A. 1937. Ticks of the U. S. in relation to disease in man. *J. Econ. Entomol.,* 30: 51–69.

Parker, R. R., and Steinhaus, E. A. 1943. *Salmonella enteritidis:* Experimental transmission by the Rocky Mountain wood tick, *Dermacentor andersoni* Stiles. *Public Health Repts.,* 58: 1010–1012.

Parker, R. R., et al. 1952. The recovery of strains of Rocky Mountain spotted fever and tularemia from ticks of the Eastern United States. *J. Infectious Diseases,* 91: 231–237.

Philip, C. B. 1952. Tick transmissions of Indian tick typhus and some related rickettsioses. *Exptl. Parasitol.,* 1: 129–142.

1953. Tick Talk. *Sci. Monthly,* 76: 77–84.

Price, W. H. 1954. The epidemiology of Rocky Mountain spotted fever. II. *Am. J. Hyg.,* 60: 292–319.

Reiss-Gutfreund, R. J. 1956. Isolement de souches de *Rickettsia prowazeki* a parter du sang des animaux domestiques d l'Ethiopie et de leurs tiques. *Bull. soc. pathol. exotique,* 48: 602–606.

Stanbury, J. B., and Huyck, J. H. 1945. Tick Paralysis: A critical review. *Medicine,* 24: 219–242.

Stoker, M. G. P., and Marmion, B. P. 1955. The spread of Q fever from animals to man. *Bull. World Health Organization,* 13: 781–806.

Wallace, G. J., Cahn, A. R., and Thomas, L. J. 1933. *Klebsiella paralytica,* a new pathogenic bacterium from "Moose disease." *J. Infectious Diseases,* 53: 386–414.

Walton, G. A. 1957. Observations on biological variation in *Ornithodoros moubata* (Murr.) *East African Bull. Entomol. Research,* 48: 669–710.

Chapter 25

BEDBUGS AND OTHER HETEROPTERA

The order Hemiptera; suborder Heteroptera. The suborder Heteroptera, comprising the true bugs, contains numerous species, most of which are predaceous or feed on plant juices, but some of which habitually or occasionally suck blood. The most important of these are the bedbugs, which probably first became acquainted with man when he shared caves with bats and swallows during the Ice Age, and which have since become fully domestic, to the disgust of good housekeepers all over the world. Also important are the conenoses (Triatominae); these are large, fierce bloodsuckers, some species of which have become habitual residents in human habitations, and in tropical America are the transmitters of Chagas' disease. In addition, not only the wild bloodsuckers but also many forms which are predaceous on insects may inflict painful and even dangerous bites.

The Heteroptera have an incomplete metamorphosis, the adult condition being attained gradually by successive molts of the nymphs (see p. 526). The mouth parts (Fig. 180, *left*) are fitted for piercing and sucking. There is a short labrum covering the bases of the mouth parts. The labium is in the form of a three- or four-jointed beak bent back under the head and thorax and grooved on the dorsal surface (ventral when bent under the head) to contain the styletlike mandibles and maxillae. The maxillae are usually coarser and fit together to form two grooves—a large food channel and a small salivary duct. The hypopharynx and palpi are absent.

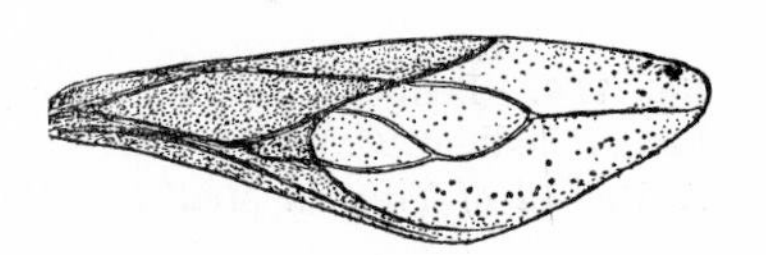

Fig. 178. A heteropteran wing (reduviid).

The wings, except in those forms, like the bedbugs, in which they are vestigial, are very characteristic; the first pair, called hemelytra, have the basal portion thickened and leathery while the terminal portion, which is sharply demarcated, is membranous (Fig. 178). The second pair of wings are membranous and fold under the others when at rest. Many bugs have "stink glands" between the bases of the hind legs in adults or on the abdomen in nymphs which secrete a clear volatile fluid by means of which they emit a strong offensive odor.

BEDBUGS (*CIMEX*)

General account. The bedbugs belong to the family Cimicidae. They have broad, flat, reddish-brown bodies and are devoid of wings, except for a pair of bristly pads which represent the first pair of wings

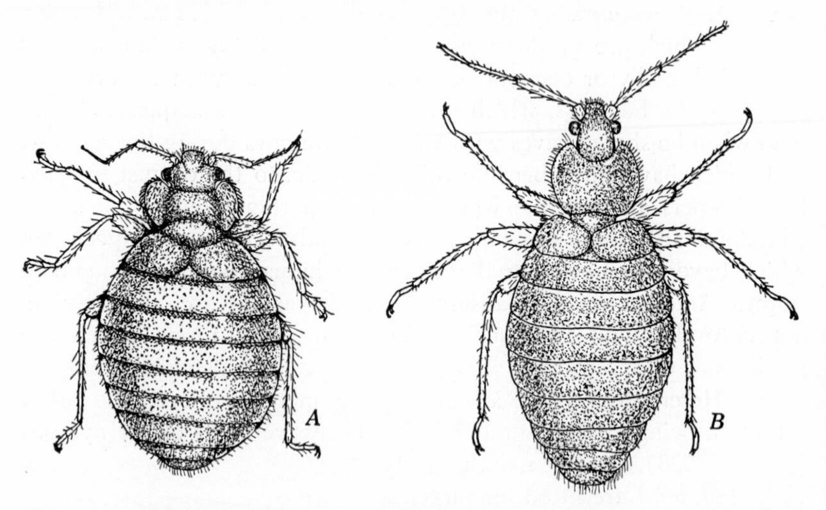

Fig. 179. Bedbugs. *A*, common bedbug, *Cimex lectularius; B*, Oriental or Indian bedbug, *C. hemipterus.* ×8. (Adapted from Castellani and Chalmers, *A Manual of Tropical Medicine,* 1920.)

(Fig. 179). The eyes project prominently at the sides of the head, the antennae are four-jointed, and the beak is three-jointed (Fig. 180). The legs have the usual segments, the tarsi being three-jointed. The prothorax is large, indented in front for the head, and has flat lateral expansions. The mesonotum is small and triangular, bearing the wing pads, which nearly cover the metanotum. The abdomen is flat, its contour an almost perfect circle in unfed bugs, but elongated in full ones; it has eight visible segments, the first two (of nine) being fused. In males the abdomen is pointed at the tip, whereas in females it is evenly rounded. The greater part of the body is covered with bristles set in little cup-shaped depressions.

Bedbugs have a peculiar pungent odor known to all who have had to contend with these pests; the adults have the stink glands situated in the last segment of the thorax, opening through a pair of ducts between the coxae of the hind legs. In the first four nymphal stages these glands are not present but are preceded by glands situated on the dorsal side of three of the anterior abdominal segments. The nasty odor of bedbugs has evidently inspired some faith in their medicinal value. Seven bugs ground up in water was said by Pliny to arouse one from a fainting spell, and one a day would render hens immune

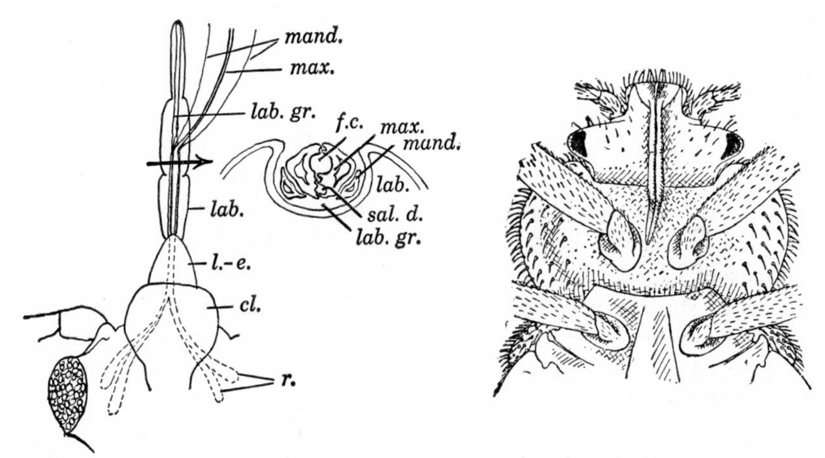

Fig. 180. *Left,* mouth parts of bedbug, with cross-section; *cl.,* clypeus; *f.c.,* food channel; *lab.,* labium; *lab.gr.,* labial groove; *l.-e.,* labrum-epipharynx; *mand.,* mandibles; *max.,* maxillae; *r.,* roots of mandibles and maxillae; *sal.d.,* salivary duct. (After Matheson, *Medical Entomology,* Comstock, 1950.)

to snake bites. Even at the present time there are places in civilized countries where bedbugs are given as an antidote for fever and ague.

Species. The true bedbugs belong to the genus *Cimex,* but not all the species are human parasites; some confine their attentions, ordinarily at least, to birds and others, to bats. There are two widely distributed species that attack man; one is the common bedbug, *C. lectularius,* found in all temperate climates and sometimes also in tropical ones; the other is the tropical or Indian bedbug, *C. hemipterus* (formerly *rotundatus*), which is the prevalent species in the tropical parts of the world. It is distinguished by having less marked lateral expansions of the prothorax (Fig. 179). In west Africa another species, *Leptocimex boueti,* attacks man; it is a silky-haired, long-legged bug with small rectangular thorax.

Some other Cimicidae may become nuisances under special conditions. The proverbially superclean housekeepers of Holland villages, for instance, are sometimes greatly chagrined to find bugs in their spotless houses, the bugs being *C. columbarius* derived from pigeons nesting in the roofs. Similar temporary invasions by *C. pilosellus* of bats sometimes occur in dwelling houses, especially when the bats are driven away or migrate. The silky-haired bugs of the closely related genus *Oeciacus*, which live in swallows' nests, also occasionally invade houses. The Mexican poultry bug, *Haematosiphon inodora,* is a related species resembling a bedbug but having longer legs, no odor, and a very long beak that reaches to the hind coxae. It is often a serious poultry pest in the dry parts of the southwest and sometimes invades houses and torments man. The Mexicans sometimes abandon or burn their huts to escape from them.

The interhostal traffic in bedbugs is not by any means a one-way affair, for the nests of sparrows and starlings, the burrows of rats, the attic roosts of bats, and also chicken houses and pigeon cotes are often invaded by hungry bugs that have been abandoned by their human sources of blood. The sparrows and starlings may frequently be a means of starting new colonies in other houses, but bats, contrary to popular opinion, probably rarely do, since they are not much given to visiting. Bugs found under bark and moss out of doors are not bedbugs but immature stages of other bugs that superficially resemble them.

Following is a key for the differentiation of the commoner Cimicidae likely to invade houses in America:

1*a*. Beak short, not extending behind first coxae (Fig. 180) 2
1*b*. Beak long, extending to hind coxae; legs long; no odor; on poultry in southwestern United States *Haematosiphon inodora*
2*a*. Body hairs short, set in sockets only on dorsal side; pronotum deeply concave in front; third and fourth joints of antennae markedly slender (*Cimex*) 3
2*b*. Body hairs long and silky, set in sockets on ventral side also; pronotum not deeply concave; terminal joints of antennae only slightly more slender than basal joints; in nests of swallows; L. 4 mm. *Oeciacus vicarius*
3*a*. Second and third joints of antennae about equal, third longer than fourth; L. 4 mm.; on bats *Cimex pilosellus* and *C. adjunctus*
3*b*. Second joint of antennae shorter than third 4
4*a*. Lateral parts of pronotum with flat lateral expansions (Fig. 179A); in human houses in temperate climates; L. 5 to 6 mm. *C. lectularius*
4*b*. Similar but smaller, rounder; shorter and coarser antennae; prothorax less concave in front; in pigeon houses (doubtful if a distinct species) *C. columbarius*

4*c*. Pronotum rounded to margin on dorsal side (Fig. 179*B*); in human houses in tropics; L. 5 to 6 mm. *C. hemipterus*

Habits. Bedbugs are normally night prowlers and exhibit a considerable degree of cleverness in hiding away in cracks and crevices during the daytime. When hungry they will frequently come forth in a lighted room at night and may even feed in broad daylight. Favorite hiding places are in old-fashioned wooden bedsteads, in crevices between boards, under wallpaper, etc., into which they can squeeze their flat bodies. They sometimes go considerable distances to hide in the daytime and show remarkable resourcefulness in reaching sleepers at night. In the tropics newcomers often wonder why their heads itch under their sun helmets until they discover a thriving colony of bugs in the ventilator at the top.

When a bug is about to drill for blood the beak is bent forward and the piercing organs sunk into the flesh. Bugs seldom cling to the skin while sucking, preferring to remain on sheets or clothing. Since a fresh meal apparently acts as a stimulus for emptying the contents of the rectum, the adherence to the clothing is a fortunate circumstance, inasmuch as it precludes to some extent the danger of bedbugs infecting their wounds with excrement, as do ticks.

In the course of 10 or 15 minutes a full meal is obtained and the distended bug retreats to its hiding place. According to Cragg, *Cimex hemipterus* does not entirely assimilate a full meal for at least a week, although the bug is ready to feed again in a day or two, thus having parts of several meals in the stomach at once. Most bloodsucking insects completely digest one meal before another is sought.

Bedbugs are able to endure long fasts; they have been kept alive without any food whatever for over a year. Sometimes, however, bugs migrate from an empty house in search of an inhabited area. In cold weather they hibernate in a semitorpid condition and do not feed, but in warm climates they are active the year around. *C. lectularius*, according to Marlatt, succumbs at temperatures above 96° to 100°F. if the humidity is high. According to Bacot, unfed newly hatched bugs are able to withstand cold between 28° and 32°F. for as long as 18 days, though they are destroyed by exposure to damp cold after a full meal.

Hosts. Although man is undoubtedly the normal and preferred source of blood for *Cimex lectularius* and *C. hemipterus*, the bugs manage to get along surprisingly well on other kinds of blood. Johnson in 1937 found experimentally that bedbugs thrive even better on mice than on men and about equally well on chickens. They sometimes multiply in great numbers in chicken houses, dovecotes, and white-rat cages. Dogs and cats are frequently bitten also.

Life history. The eggs of bedbugs (Fig. 181*A*) are pearly white oval objects, furnished at one end with a little cap which is bent to one side. The eggs are relatively large, about 1 mm. in length, and are therefore laid singly or in small batches, averaging about two a day. The total number of eggs laid is about 100 to 250. The bugs frequently return to the same places to ovipost until sometimes as many as 40 eggs have been accumulated.

Fig. 181. Egg (*A*) and newly hatched larva (*B*) of bedbug. ×20. (After Marlatt, *Farmer's Bull.*, 754, 1925.)

The eggs hatch in 6 to 10 days during warm weather but are retarded in their development by cold. A week of freezing temperature reduces the hatching to 25%. The freshly hatched bugs (Fig. 181*B*) are very small, delicate, and pale in color. The skin is normally molted five times at intervals of about 8 days before the final adult stage is reached, at least one gluttonous feed being necessary before each molt in order to insure normal development and reproduction. However, the bug may gorge itself several times between molts. The several nymphal stages of the insect resemble each other quite closely except in the constantly increasing size and deepening color. The wing pads appear only after the last molt. The total time required for development to maturity under favorable conditions is about 7 to 10 weeks, but starvation, low temperature, etc., may drag out the period much longer. There may, however, be three or four generations in a year. The adults under ordinary conditions usually live for several months to a year or more.

Effects of bites. The degree of irritation caused by bites of bedbugs undoubtedly depends to a large degree on development of hypersensitivity or eventual immunity (see p. 529). The irritation is produced by the salivary secretion, but when an engorging bug is undisturbed much of the secretion is redrawn with the blood meal and the irritation is lessened. Continued excessive biting by bugs may cause anemia, nervousness, insomnia, and general debility. Titschack found that the bites of 50 adults were enough to produce influenzalike symptoms, whereas over 100 caused palpitation of the heart, headache, and eye disturbances. Hase (1938) considered these eye disturbances to be a fairly common phenomenon when numerous bugs bite day after day.

Bedbugs and Disease

General considerations. The bedbug would appear at first sight to be eminently adapted for human disease transmission. Like an ex-criminal, it is under constant suspicion. With or without reason it has been on trial in connection with kala-azar and other forms of leishmaniasis (see p. 115), Chagas' disease, relapsing fever, leptospirosis, tularemia, plague, typhoid, leprosy, typhus, yellow fever, poliomeylitis, and even malaria, beri-beri, and some filaria infections.

In spite of all this suspicion and some circumstantial evidence there is still to be produced any single instance in which the bedbug has been shown to be more than a relatively unimportant accessory in the transmission of any human disease.

Of the numerous pathogenic organisms mentioned above, none except leptospirosis (from guinea pig to guinea pig) has been shown to be transmitted by the bites except occasionally when a bug begins a meal on one host and finishes it on another, which is no more significant than the equally successful experiment of infecting an animal by pricking it with an injection needle that has just been inserted into an infected animal. Nevertheless bedbugs may conserve pathogenic microorganisms in their bodies for days or weeks, e.g., spirochetes, *Schizotrypanum cruzi,* rickettsias, viruses, and a few bacteria (those of tularemia, plague, brucellosis, and typhoid), and infections may occasionally result when bites or scratches are contaminated by the feces or crushed bodies of the bugs. However, the evidence is not sufficient to warrant much excitement over the danger, although bedbugs will probably continue to be suspected as disease transmitters when other obvious transmitting agents are not discovered. Epidemiologists would do well to read more detective stories and learn that the most obvious explanation is often not the correct one.

One factor that limits the effectiveness of bedbugs as disease transmitters is their tendency to stay at home. Although frequently carried about in clothing, they normally live in the homes and not on the persons of their hosts, and would therefore usually be limited in the spread of a disease except in theaters, cheap hotels or motels, sleeping cars, etc.

Remedies and Prevention

Prevention of "bugginess" consists chiefly in good housekeeping, but occasional temporary infestations are likely to occur in almost any inhabited building; often the bugs gain entrance with second-hand furniture, laundry, visitor's luggage, etc. Bedbugs are easily controlled by spraying or dusting the hiding places in walls, bed frames, mattresses, etc., with chlorinated hydrocarbons or organic phosphorus compounds. The residual effect is good for several months. For theaters, hotels, etc., power sprayers are needed. Sometimes it is desirable to fumigate furniture with HCN or methyl bromide in special fumigation vans.

Triatominae

Most of the other true bugs which may be looked upon as normally human parasites belong to the subfamily Triatominae, which comprises members of the large family of predaceous bugs, Reduviidae, which have become addicted to sucking blood from vertebrates instead of juices from captured insects. Morphologically they differ from the predaceous reduviids in having a slender straight beak instead of a coarse curved one, and in having the antennae inserted on the sides of the head somewhere between the eyes and the end of the "snout" instead of on top of the head (see Fig. 182).

These bugs, of which there are nearly 100 species, are large and often brightly colored, and are active runners and good fliers. They are especially numerous in the warmer parts of the New World, from the southern half of the United States to Argentina; but one species, *Triatoma rubrofasciata*, is cosmopolitan, and there are a few Old World species. Most of the species live in the nests or burrows of rodents or other animals on which they feed, but a few species have become household pests and feed primarily on man and his household pets, or on the pigs in his yard.

The head of triatomids is small and narrow. The slender beak, bent straight back under the head, is three-jointed, and the long filamentous antennae are four-jointed. The pronotum flares posteriorly and is usually fairly distinctly divided into an anterior and posterior lobe; the posterior angles of the latter may be round or pointed (see Fig. 182,

7 and 8). Behind the pronotum is a triangular scutellum which may have a posterior spine. The abdomen has flattened lateral margins,

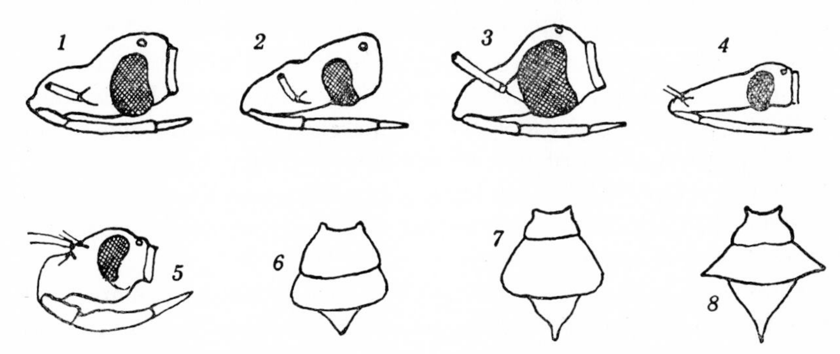

Fig. 182. Heads and thoraces of various genera of triatomids and reduviids. *1* and *2*, *Triatoma* (*T. protracta* and *T. rubida*); *3*, *Panstrongylus* (*P. megista*); *4*, *Rhodnius* (*R. prolixus*); *5* and *6*, head and thorax of *Melanolestes*, a reduviid; *7*, thorax of *Triatoma*; *8*, thorax of *Eratyrus*. (*1–4*, *7*, *8* adapted from Pinto, *Bol. Biológico*, 19, 1931.)

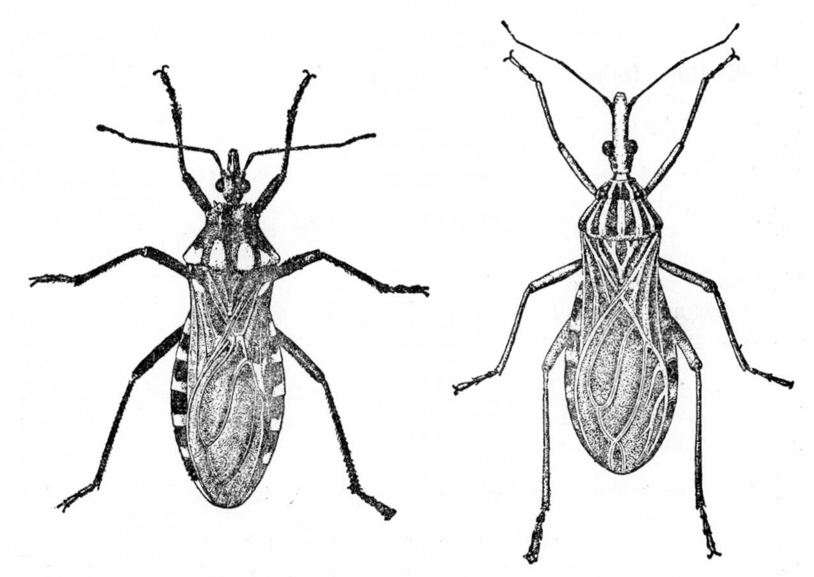

Fig. 183. *Left, Panstrongylus megistus; right, Rhodnius prolixus.* ×2. (After Brumpt, *Précis de parasitologie*, 1949.)

the connexiva, not covered by the wings (Fig. 183). This and the leathery basal portion of the wing (corium) are usually marked with red or yellow, as is the pronotum.

The bloodsucking habit of the Triatominae is probably a recent experiment on the part of these bugs, evolutionarily speaking, for some of them still sometimes pursue bedbugs and are cannibals on each other, obtaining their blood second-hand by sucking it out of their brothers and sisters. The bites of triatomids are usually much less painful than those of many nonbloodsucking Heteroptera, but in some individuals they may cause severe allergic symptoms. Sometimes there is little or no irritation at the time of biting, but redness and itching develop several days later. The triatomids are of great importance because they are the natural and only important transmitters of *Schizotrypanum cruzi.*

Genera and species. According to Usinger (1943) there are nineteen genera of triatomids in the Americas, but some of these contain species which are of no interest here. Used in separating genera are the place of insertion of the antennae, the shape and other characters of the prothorax and scutellum, and hairiness of the body.

The important genera of American Triatominae, and of other Reduviidae which occasionally cause painful bites, can be distinguished by the following key:

Reduviidae (other than Triatominae): Beak thick and curved; antennae inserted just in front of eyes on top of head; head relatively short (Fig. 182, *5*).

1*a*. A cogwheel-like ridge on pronotum (wheelbug) ***Arilus***
1*b*. No cogwheel ridge 2
2*a*. Thorax constricted at or anterior to middle; color nearly uniformly dark brown ***Reduvius***
2*b*. Thorax constricted behind middle (Fig. 182, *6*) 3
3*a*. Wings and body all black or dark brown ***Melanolestes***
3*b*. A large yellow spot on wings ***Rasahus***

Triatominae: Beak slender and straight; antennae inserted on sides of head; head elongated in front of eyes; thorax constricted anterior to middle; black or brown with red or yellow markings.

1*a*. Antennae inserted near apex of head, which is somewhat widened apically (Fig. 182, *4*, 197*B*) ***Rhodnius***
1*b*. Antennae inserted just in front of eyes (Fig. 182, *3*) 2
1*c*. Antennae inserted about midway between eyes and apex of head (Fig. 182, *2*) 3
2*a*. Body clothed in long curved hairs ***Parastrongylus***
2*b*. Body nearly naked ***Panstrongylus***
3*a*. Scutellum with a long pointed spine posteriorly and pronotum with pointed posterior angles (Fig. 182, *8*) ***Eratyrus***
3*b*. Scutellum without long spine; angles of pronotum rounded (Fig. 182, *7*) body ½ to 1 in. long, distinctly colored ***Triatoma***

Most triatomids will feed on a great variety of hosts, although some are especially associated with certain animals, particularly pack or wood rats (*Neotoma* spp.), armadillos, and opossums. They will feed to repletion on lizards also, and lizards frequently eat the bugs. A few species, discussed below, habitually live in human habitations, in the daytime hiding like bedbugs in the cracks of walls, debris on the floor, etc., issuing forth at night to feed on their sleeping hosts. Most species are so active and hide so rapidly when a light is produced that they are hard to catch. It is probable that any of them would accept a human meal if the opportunity presented itself, but their habits and habitats render some species much more frequent human biters than others.

Life cycle. The life cycle of triatomids is similar in general to that of bedbugs. The eggs are white oval objects when first laid, in some species turning yellowish or pinkish later; they are laid singly or in small batches by most species, but *Rhodnius prolixus* lays them in a mass joined together by a secretion. The total number laid by a female (not a *single* female as some writers say—they have to mate!) is usually about 100 to 300. The eggs require 2 to 3 weeks to hatch, the time depending on temperature. The wingless nymphs are light in color when they first hatch but soon darken. They molt a total of five times, always after a full blood meal, which is six to twelve times their own body weight! The later nymphal instars have the wings represented as rounded lobes (Fig. 184*B*). The whole development from egg to adult requires about a year in some species and 2 years in others.

Important species. Comparatively few species invade houses and become human pests. In South America three species are of outstanding importance. *Panstrongylus megistus* (Fig. 183A), the "barbeiro" of Brazil, a large, handsome red-trimmed black insect widely distributed in Brazil and neighboring countries, is one of the principal vectors of Chagas' disease. It is thoroughly domestic in its habits and normally lives in the huts of natives. *Triatoma infestans,* known in Argentina as the "vinchuca," replaces this species farther south and west as a house-infesting pest. This is the species whose habits were vividly described by Darwin in his *Voyage of a Naturalist.* In northern South America *Rhodnius prolixus* (Fig. 183*B*) is the most annoying domestic species. This bug is garbed in brown and yellow. Although a common pest in human habitations, it also inhabits burrows of armadillos and pacas.

Other species that breed in houses and are partially "domestic" are *T. braziliensis* and *T. sordida* in South America, *T. dimidiata* in Ecuador, Central America and Mexico; and *Panstrongylus geniculatus,* a species usually partial to armadillos, in Panama.

The cosmopolitan *T. rubrofasciata* is semidomestic in habits. Many other species frequently enter human habitations to bite if not to breed. Most of the species in the United States, ten of which have been found to harbor *Schizotrypanum cruzi* (see p. 150), are com-

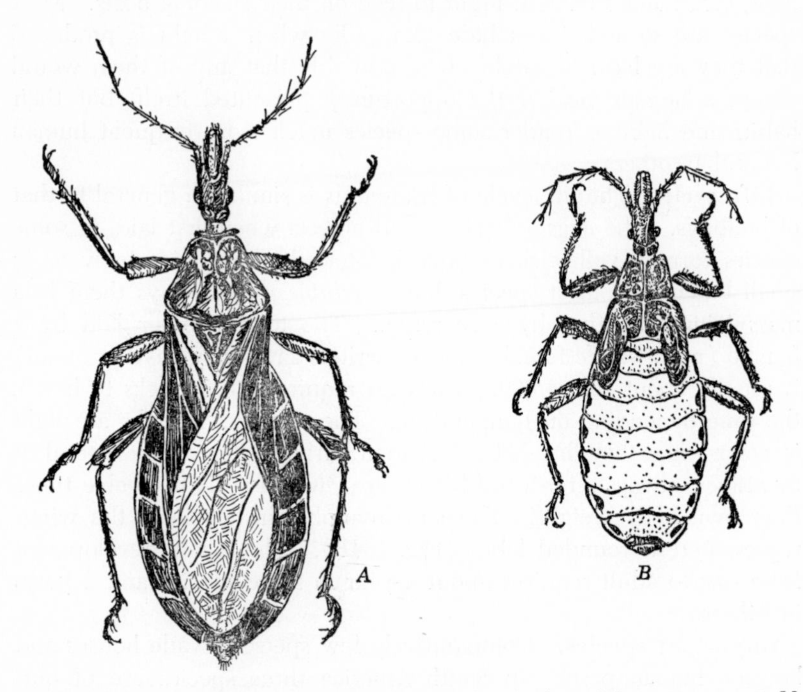

Fig. 184. *Triatoma protracta,* adult and nymph. (Adapted from Usinger, *Publ. Health Bull.* 288, 1944.)

monly found associated with wood rats (*Neotoma*) or in nests of opossums or burrows of armadillos. The adults particularly are likely to invade barns, tents, or houses or to prowl about in the beds of outdoor sleepers. They are found throughout the southern half of the United States but are especially common in Texas, New Mexico, Arizona, and southern California. The species most often found invading houses are *T. gerstaeckeri* in Texas, which sometimes invades in great numbers; *T. rubida* (= *uhleri*), *T. protracta* (Fig. 184), and *T. recurva* (= *longipes*), found farther west and in Mexico; and *T. sanguisuga,* throughout the South.

Other Hemiptera

Several reduviid bugs, which are not normally bloodsuckers, often bite man when provoked, and their bites, as noted previously, are far more painful and toxic than those of the true bloodsuckers. In North and Central America there are a number of species of "kissing bugs" and "corsairs" of the genera *Melanolestes, Reduvius,* and *Rasahus.* The common kissing bug or black corsair, *Melanolestes picipes,* became very abundant in the United States at one time and gave opportunity for many startling newspaper stories. The wheelbug, *Arilus cristatus,* is another vicious biter.

All Hemiptera have piercing mouth parts and many are capable of inflicting painful wounds, so it is safest not to handle any of them with the bare hands. Although the majority of the bad biters belong to the Reduviidae, the malodorous pito bug, *Dysodius lunatus,* of South America is worthy of mention. It is a broad, flat bug belonging to the family Aradidae; it frequents houses and bites severely. Among other species that bite when handled are various kinds of water bugs. The large "electric light bugs" are venomous enough to kill fish and even birds and to cause in man severe pain lasting for several days.

Triatominae and Disease

At least thirty-six of the hundred or more species of Triatominae have been found to be capable of acting as intermediate hosts of *Schizotrypanum cruzi* or a species which is morphologically indistinguishable from it. Species of the genus *Rhodnius* are vectors of *Trypanosoma rangeli* also (see p. 155). *Panstrongylus megistus* seems to be the species most frequently involved in human transmission in the greater part of Brazil. In southern and western Brazil, Uruguay, Argentina, Paraguay, Bolivia, and Chile, *Triatoma infestans* plays the leading role, whereas in northern South America and Central America *Rhodnius prolixus* is the principal vector. All three of these species have been shown by Dias to have a tendency to defecate while filling up with blood, a process that usually takes about 15 to 20 minutes; this results in frequent contamination of bites by the feces, or rubbing of the feces with the eye, which are the methods by which transmission of *S. cruzi* occurs (see p. 151). The rare occurrence of Chagas' disease in the United States is probably due in part to failure of species that are not habitually domestic, such as *T. protracta* to defecate while feeding (see Fig. 184).

In the United States 10 species of *Triatoma* have been found naturally infected—*gerstaekeri, protracta, protracta woodi, rubida, recurva, sanguisuga, neotomae, uhleri, lecticularius,* and *ambigua.* Packchanian in 1939 reported the heaviest infections—92 out of 100 *gerstaeckeri* collected in a house and barn in southern Texas and 65% of *lecticularius* collected at Temple, Texas.

Rhodnius spp. are transmitters of *Trypanosoma rangeli* (= *T. ariarii*) also (see p. 155). Unlike *S. cruzi,* this trypanosome is transmitted by the bite as well as by fecal contamination. *Rhodnius* often harbors mixed infections of *T. rangeli* and *S. cruzi.* Trypanosomes of the African type (*T. equinum* and *T. gambiense*) survive in triatomids for only a short time.

There are certain habits of Triatominae that are of interest in connection with the transmission of trypanosomes. Cannibalism is a common habit among many of them. Young bugs, especially, often suck blood from the distended bodies of nest mates, and they sometimes suck blood from bedbugs. The robbed bugs seem quite untroubled and unharmed by it. Perhaps like the Romans of old they are glad to be rid of their meals in order to enjoy the ingestion of more. *Rhodnius* is said to feed upon the excreta of fellow bugs. In these ways and perhaps by contamination from feces, *Schizotrypanum cruzi* may occasionally spread from bug to bug, but transovarial transmission does not occur. The percentage of infected bugs steadily increases with age up to the adult stage. Infection is usually acquired from infected mammalian hosts.

In India and also in South America *Triatoma rubrofasciata* is a vector of a trypanosome of rodents, *T. conorrhinae.* In Mauritius, *T. rubrofasciata* has been found to harbor rickettsialike bodies which were transmissible to laboratory animals. Spirochetes of relapsing fever persist in triatomids for some time, and in Kansas the virus of western equine encephalomyelitis has been isolated from *T. sanguisuga.*

Control

Spraying houses with BHC or Dieldrin is very effective in controlling the house-infesting species. Spraying with 5% emulsions, suspensions, or kerosene solutions controls triatomid infestations even up to a year. The town of Bambui, State of Minas Gerais, Brazil, was entirely freed of triatomids by repeated applications of insecticides in the houses, and there were no new cases of Chagas' disease for over 5 years. Pinotti found that when the walls of native huts are plastered with a mixture containing cowdung, which prevents cracks from forming, the bugs moved out and stayed out.

REFERENCES

Busvine, J. R. 1957. Recent progress in the eradication of bedbugs. *Sanitarian,* 65: 365–369.

Dunn, L. H. 1924. Life history of the tropical bedbug (*Cimex rotundatus*) in Panama. *Am. J. Trop. Med.,* 4: 77–83.

Hase, A. 1938. Zur hygienischen Bedeutung der Parasitären Haus- und Vogelwanzen. *Z. Parasitenk.,* 10: 1–30.

Horvath, G. 1912. Revision of the American Cimicidae. *Ann. Mus. Hungary,* 10: 257–262.

Johnson, C. G. 1942. The ecology of the bedbug, *Cimex lectularius* L. in Britain. *J. Hyg.,* 41: 345–461.

McKenny-Hughes, A. W., and Johnson, C. G. 1942. The bedbug. Its habits and life history and how to deal with it. *Brit. Mus. Nat. Hist. Econ. Ser.,* 5: London.

Mellanby, K. 1935. A comparison of the physiology of the two species of bedbugs attacking man. *Parasitology,* 27: 111–122.

Reports of the Committee on Bedbug Infestation, 1935–1940. *Med. Research Council Special Repts. Ser.,* 245. London.

U. S. Dept. of Agriculture. 1953. Bedbugs. How to control them. Leaflet 337.

Wright, W. H. 1944. The bedbug—its habits and life history and methods of control. *Public Health Repts. Suppl.* 175.

Triatominae

Buxton, P. A. 1930. The biology of a bloodsucking bug, *Rhodnius prolixus. Trans. Entomol. Soc. London,* 78 Pt. 2, 227–236.

Dias, E. 1956. Observações sôbre eliminação de dejeções e tempo de sucção em alguns triatomíneos sul-americanos. *Mem. inst. Oswaldo Cruz,* 54: 115–124.

Dias, E., and Chandler, A. C. 1949. Human diseases transmitted by parasitic bugs. *Mem. inst. Oswaldo Cruz,* 47: 403–422 (Portuguese), 423–441 (English).

Herbig-Sandreuter, A. 1955. Experimentelle Untersuchungen über den Cyclus von *Trypanosoma rangeli* Tejera, 1920, in Warmblüter and in *Rhodnius prolixus. Acta Trop.,* 12: 261–264.

Hoare, C. A. 1934. The transmission of Chagas' Disease. A critical review. *Trop. Diseases Bull.,* 31: 757–762.

Packchanian, A. 1942. Reservoir hosts of Chagas' disease in the State of Texas. *Am. J. Trop. Med.,* 22: 623–631.

Readio, P. A. 1927. Studies on the biology of the Reduviidae of America north of Mexico. *Univ. Kansas Sci. Bull.,* 17: 5–291.

Sullivan, T. D., et al. 1949. Incidence of *Trypanosoma cruzi* Chagas in *Triatoma* (Hemiptera, Reduviidae) in Texas. *Am. J. Trop. Med.,* 29: 453–458.

Usinger, R. L. 1944. The Triatominae of North and Central America and the West Indies and their public health significance. *Public Health Bull.*, 288, 83 pp.

Wood, S. F. 1941. Notes on the distribution and habits of reduviid vectors of Chagas' disease in the southwestern United States. *Pan-Amer. Entomol.*, 17: 85–94.

1942–1943. Observations on vectors of Chagas' disease in the United States. I, Calif. *Bull. Southern Calif. Acad. Sci.*, 41: 61–69. II, Ariz. *Am. J. Trop. Med.*, 23: 315–320.

1951. Importance of feeding and defecation times of insect vectors in transmission of Chagas' disease. *J. Econ. Entomol.*, 44: 52–54.

See also References, Chapter 8.

Chapter 26

LICE (ANOPLURA AND MALLOPHAGA)

Opinion is divided as to whether the sucking lice of mammals and the chewing lice of birds and mammals should be placed together in one order Anoplura, the former in a suborder Siphunculata and the latter in a suborder Mallophaga, or should be placed in separate orders Anoplura and Mallophaga, respectively. They have many features in common, but differ in the radically different structure of the mouth parts. Both are small, wingless, flattened insects, probably derived from the free-living book lice or bark lice, Corrodentia. Although probably they came from the same original stock, they form clearly defined groups with perhaps a single form, *Rhynchophthirina,* which can be considered intermediate. It thus is convenient to consider them as separate orders.

The Mallophaga are mainly bird parasites, although many of them also make themselves at home in the fur of mammals. They have nipperlike mandibles fitted for nibbling on feathers, hair, and epidermal debris, although some of them quite regularly dine on blood by piercing the pulp of young growing feathers or even gnawing through the skin of the host, but many are apparently content with parts of feathers or epidermal debris. The Anoplura, on the other hand, have piercing mouth parts consisting of three "piercers" which when not in use are drawn back into a special pouch under the pharynx. These feed entirely on blood. Mallophaga annoy their hosts principally by irritating the skin surface, but those that also suck blood (members of the suborder Amblycera) are also potential disease transmitters. The sucking lice cause trouble by sucking considerable amounts of blood, by causing allergic reactions by their bites or feces, and by transmitting disease.

The lice of these two orders are readily distinguishable by the presence in the Mallophaga and absence in the Anoplura of heavily sclerotized mandibles which are brown or blackish. They can also be distinguished at once by the fact that the Mallophaga have very broad heads, always at least as broad as the thorax (Fig. 185), whereas in the Anoplura the head is always narrower than the thorax (Fig. 188). None of the Mallophaga are human parasites, but some are annoying to domestic animals, particularly poultry. They are briefly considered on pp. 634–638.

ANOPLURA

Morphology and physiology. The Anoplura have the body clearly divided into a narrow and often elongate head, a broad thorax, the segments of which are fused in nearly all species, and an abdomen that is more or less distinctly divided into segments (Fig. 188). There are primitively nine segments, but one or two of the anterior ones are fused or lost, so usually seven or eight are recognizable. In the crab

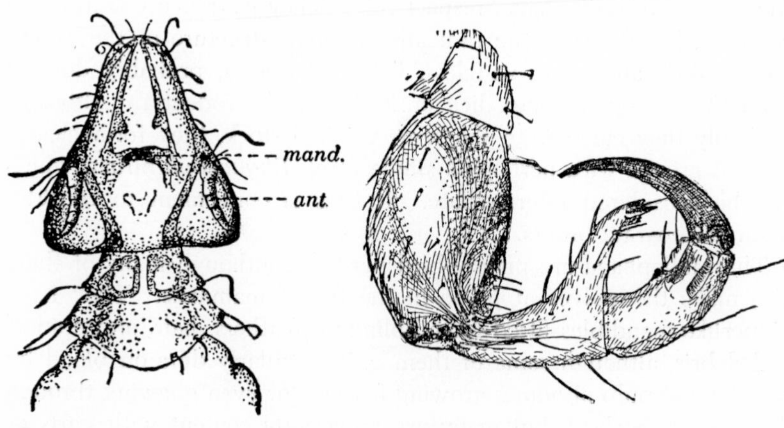

Fig. 185. Head of a species of Mallophaga (*Uchida* sp.) from golden eagle. Note breadth of head compared with thorax, and pair of black-tipped mandibles; *ant.*, antenna; *mand.*, mandibles.

Fig. 186. Front leg of ♂ body louse, *Pediculus humanus*. Note huge claw and thumb-like opposing process of tibia. ×100.

louse, *Phthirus*, segments III to V are fused. Ordinarily segments III to VIII bear spiracles, and there is one pair of spiracles on the thorax also. The abdomen of lice is poorly sclerotized except for the pleural plates at the sides in some genera, e.g., *Pediculus*. In the females the terminal segment of the abdomen is indented, whereas in the males it is rounded, with the large spikelike vaginal dilator often projecting from the sex opening just ventral to the anus.

The head has short four- or five-segmented antennae; eyes are absent in most species, but the human lice have small but prominent ones. The legs, except in the separate suborder Rhynchophthirina on elephants, have each tarsus, consisting of one segment, armed with a large curved claw, quite grotesque in appearance in some species, which closes back like a finger against a thumblike projection of the tibia (Fig. 186). There are not even rudiments of wings.

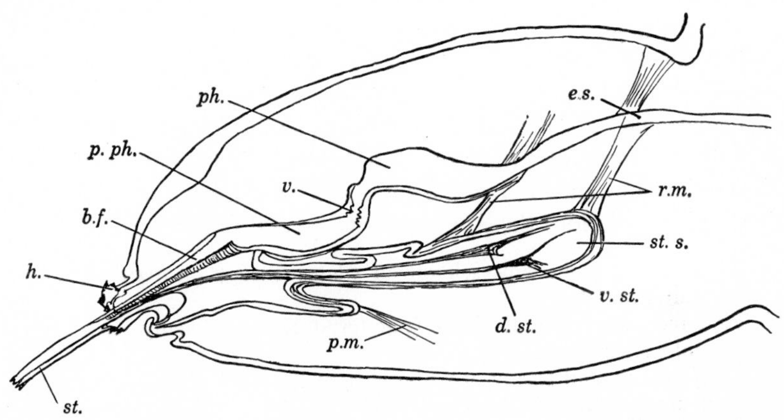

Fig. 187. Mouth parts of *Pediculus humanus: b.f.*, buccal funnel; *d.st.*, dorsal stylet; *es.*, esophagus; *h.*, haustellum (everted); *p.ph.*, prepharynx; *ph.*, pharynx; *p.m.*, protractor muscle; *r.m.*, retractor muscles; *st.*, stylets; *st.s.*, stylet sac; *v.*, valve; *v.st.*, ventral stylet. (From Sikora, *Centr. Bakt.*, 1, Orig., 76, 1915.)

The mouth parts (Fig. 187), fitted for piercing and sucking, are so highly modified that their homology is in doubt. There is a short tubular haustellum (*h.*) armed with teeth which can be everted so that the teeth are on the outside, as shown in the figure. This appressed to the skin makes a tight seal. Through this can be protruded three slender stylets which, when not in use, are retracted into a blind sac lying under the pharynx. The dorsal and ventral stylets are forked at their proximal ends in the stylet sac. The ventral stylet is believed by Ferris (1951) to represent the terminal part of the labium, the more basal part forming part of the floor of the stylet sac. This stylet is stouter than the others and toothed at the tip, and used for piercing the skin. In a groove on its dorsal surface lies the extremely slender middle stylet, a hypostome, at the tip of which the salivary duct opens. The dorsal stylet, partly embraced by the ventral one, is believed by Ferris to represent fused maxillae. It has a groove which Buxton (1947) and others believe acts as a food channel, although Ferris denies this.

The stomach is provided with lateral pouches in order to increase the food capacity. Normally lice do not fill themselves to repletion and then wait until this food is digested before feeding again but enjoy more moderate meals several times a day.

All except the human body louse glue their eggs to hairs near the base. The eggs hatch in a few days, and the young nymphs closely resemble the adults except for their size and paler color. There are three nymphal instars; the third molt brings them to the adult stage in 10 to 20 days.

Most species of lice are quite closely limited to a single host, and sometimes even genera are thus limited. Kellogg has suggested that the evolutionary affinities of different birds and mammals may be demonstrated by the kinds of lice that infest them. Only about 230 species of Anoplura are known, but the species of Mallophaga are numerous. Ferris (1951) divides the Anoplura into several families as follows:

1*a*. Abdomen without any sclerotized plates except on terminal and genital segments — 2
1*b*. Abdomen with pleural plates (paratergites), and usually tergal and sternal plates also — 4
2*a*. Body robust, and bristly or scaly; legs, at least last 2 pairs, with very stout undivided tibiotarsus; on seals — **Echinophthiriidae**
2*b*. Body with setae but not scales; 1st pair of legs smaller than 2nd or 3rd — 3
3*a*. Abdominal spiracles several; on Artiodactyla and hyraxes — **Linognathidae**
3*b*. Only 1 abdominal spiracle, on 8th segment, on shrews — **Neolinognathidae**
4*a*. Pleural plates (paratergites) with their apical parts projecting free from body; legs all about equal, or 1st pair smaller; majority on rodents, a few on insectivores and primates, one on Equidae — **Hoplopleuridae**
4*b*. Pleural plates without free apices — 5
5*a*. Eyes reduced or absent, legs and claws approximately alike on all 3 pairs; on Artiodactyla and Equidae — **Haematopinidae**
5*b*. Eyes with well-developed lenses and pigment; legs all about equal, or 1st pair smaller; on Primates, including man — **Pediculidae**

Human Lice

In former times human lice were looked upon with less disgust and loathing than they are in most civilized countries now. In days when a bath on Saturday nights (except in winter) was considered adequate and most laundering was done on river banks, lice intruded themselves in everyone's company, from the royal family to the lowest peasant. Even today, however, body lice are distressingly common in camps, jails, trenches, etc., where association with careless people cannot be avoided and facilities for cleanliness are not all that could be desired, especially in time of war. In some parts of the world lice are believed to be indicative of robust health and fertility. Head lice are even more prevalent among the poorer classes in cities, particularly among children. A survey in England in 1939–1940 showed that in industrial cities 50% of preschool children and of school girls harbored them. In women the incidence never fell under 5 to 10%, and in young women in industrial areas it was 30 to 50%. Permanent waves tend to bring lice more peace and quiet in the heads of young women. The incidence was only 2% in men, and relatively rare (under 5%) in rural areas, where lousiness is considered a disgrace.

The lice infesting man belong to two genera, (1) *Pediculus*, including the head and body lice, and a few closely related species on monkeys, and (2) *Phthirus*, with only one species, the crab louse.

No doubt both the head louse and body louse are the descendants of a species that roamed the hairy bodies of our forefathers in the days when we fought our struggle for existence with mammoths and cave bears instead of tax collectors and strike organizers. With the developing hairlessness of the host, the hunting grounds of human lice became more and more restricted. The crab louse adapted itself to the coarse residual body hair, but *P. humanus* solved the problem in two different ways: some retired to the fine hair of the head and became head lice, whereas others, more resourceful, adapted themselves to living on the clothing next to the skin and became body lice. The body louse is larger than the head louse, with more slender forelegs in the males, but few if any strains can boast a pure heritage. With the differentiation of the principal races of man some slight differentiation of the head lice may also have occurred; Ewing recognized four varieties, found on Caucasians, Negroes, Chinese, and American Indians, respectively, but other louse specialists do not recognize them. Even if such varieties once existed it would be almost impossible to find pure strains now, for the lice interbreed as do their hosts. Studies of ethnology of lice and of man are beset with the same difficulties.

Head and Body Lice (Pediculus humanus)

Morphology. The general appearance of *Pediculus humanus* can be seen from Fig. 188. The head has eyes and five-jointed antennae, the latter distinctly longer in the body than head lice. The thorax has a single pair of spiracles, situated between the first and second pairs of legs. The abdomen is composed of seven segments (III to IX), of which the first six bear spiracles. Pleural plates are well developed, but more so in head than in body lice, resulting in a more festooned appearance in the head lice. Females, somewhat larger than males, are usually 2.4 to 3.3 mm. long in head lice and 2.4 to 3.8 mm. in body lice. They can readily be distinguished from males by the indented posterior end of the abdomen and by the slenderer anterior legs with smaller grasping apparatus. In males the anus and sex openings are dorsal in position.

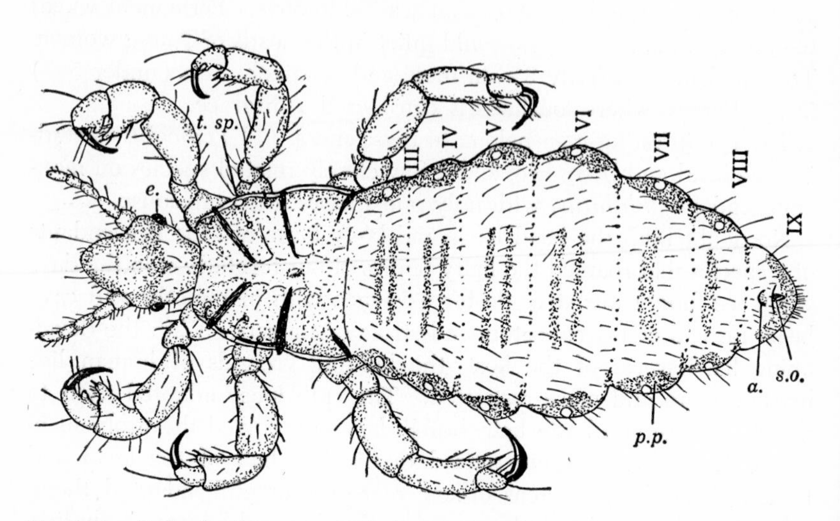

Fig. 188. *Pediculus humanus,* body louse, ×40; *a.,* anus; *e.,* eye; *p.p.,* pleural plate; *s.o.,* sex opening; *t.sp.,* thoracic spiracle; III–IX, abdominal segments. (Adapted from Keilin and Nuttall, *Parasitology,* 22, 1930.)

Biology. Head lice prefer to live in the fine hair of the head, though they sometimes wander to other parts of the body. They occur on all races of man in every part of the world. Buxton found that the majority of infested persons, however, harbor very few lice. In England a high percentage have only 1 to 10, and less than 10% have over 100, usually in girls 5 to 8 years old. Apparently brushing and combing keep them at a low level.

The body louse, on the other hand, lives on the clothing instead of the hair of its host; the German name "Kleiderlaus" is a very appropriate one. Possibly this louse developed independently from the ancestral ape-man louse, shifting its position from the waning hair to the clothes as nudism temporarily went out of style, but more likely it developed from the head louse. It is a true radical in its habits, for of all the lice in the world it alone lives elsewhere than in the hair of its host. A person infested with hundreds of body lice may remove his clothing and find not a single specimen on his body. An examination of the underwear will reveal them adhering to the inside surfaces. Here they live and lay their eggs, reaching across to the body to suck blood, holding to the clothing by their hind legs.

Life cycle. The eggs of lice, commonly called "nits," are oval, whitish objects fitted with a little lid at the larger end which is provided with air cells pierced by pores through which air can enter the egg. The eggs of head lice are slightly less than 1 mm. in length, and are glued to the hairs by means of a cementlike excretion (Fig. 189*A*). The favorite "nests" are in the vicinity of the ears. The average number laid by each female, according to Bacot, is 80 to 100.

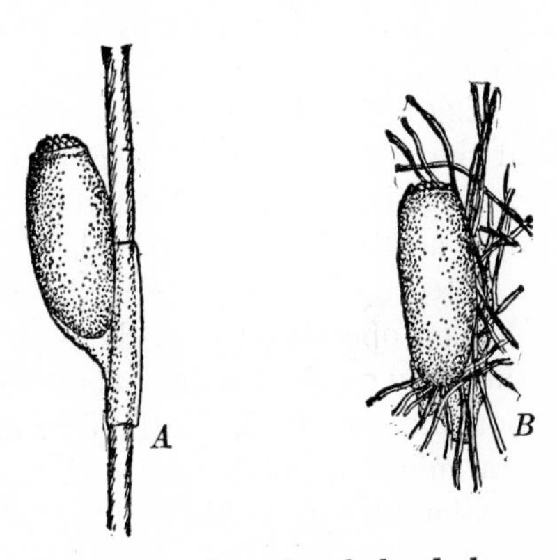

Fig. 189. *A*, egg of head louse attached to a hair; *B*, egg of body louse attached to fibers of clothing. ×25. (After Cholodkowsky.)

Body lice lay slightly larger eggs and glue them to the fibers of the clothing (Fig. 189*B*), especially along the seams or creases. Under experimental conditions the body louse will sometimes lay eggs on hairs, but it nearly always selects the crossing point of two hairs and shows less skill than the head louse in attaching the eggs. The body louse shows a marked "homing" instinct in laying her eggs, and tends to cluster them until 50 or 75 have been collected. The total number of eggs may be 200 to 300. The female begins production slowly, but after a week of practice she reaches an output of about 8 or 10 per day, although there is never more than one developed egg in the body at a time. Unfertilized eggs are sometimes laid, but they do not develop. Egg-laying ceases at temperatures below 77°F., and a daily exposure to a temperature of 60°F. for only 2 or 3 hours causes a marked falling off in egg production.

At 80° to 85°F., the eggs hatch in about 8 to 10 days; at higher temperatures many die. No hatching occurs below 70°F., and eggs held below 60°F. for 7 to 9 days do not hatch even if warmed. Either excessive humidity or complete drying is fatal to the eggs. It is evident that in winter the laying off of the clothing at night in a cold room or the leaving of mattresses or bedclothes in the daytime is sufficient to prevent the laying or hatching of eggs.

The young lice have an interesting way of escaping from the eggs. They suck air into the body and expel it from the anus until a cushion of compressed air is formed sufficient to pop open the lid of the egg. The newly hatched lice are almost perfect miniatures of the adults except that they have three-segmented antennae. They are ready to feed almost as soon as they emerge from the egg and will usually die in less than 24 hours if not allowed to feed. At a temperature of 95°F. and with as many daily feeds as would willingly be taken, namely six, the lice pass through the first molt in 3 days, the second in 5 or 6 days, and the third, which brings them to maturity, in 8 or 9 days. With fewer feeds or lower temperatures the development is slower.

Egg-laying begins 1 to 4 days after the final molt and continues at the rate previously described until the death of the insect. The average length of life for the females is about 35 or 40 days and probably a little less for the males.

Feeding habits. Lice show less tendency to vacillate in their drilling operations than do fleas. They make a single puncture and then rely on the salivary secretion to dilate the capillaries by its irritation and thus facilitate the flow of blood; this sometimes requires several minutes. They may suck to repletion in a few minutes but often continue to pump blood into their stomachs intermittently for several hours, meanwhile voiding feces containing undigested blood corpuscles.

There seems to be some degree of specificity in the salivary secretion, since it is more efficient in aiding the lice to suck blood from man than from other animals. Rat blood seems to disagree with human lice, although they can be adapted to feed on rabbits.

Lice do not have the remarkable resistance to starvation displayed by ticks and bugs. At high temperatures they succumb in 2 or 3 days but at about 40°F. can live for 8 or 10 days without food. Adult lice stand exposure to moderate cold very well but are killed in a few hours at temperatures of 10°F. or below. They are highly susceptible to heat, especially when the humidity is high, and die in a few minutes at 122° to 126°F.

The maximum favorable temperature for the development and reproduction of lice is about 95°F. The absence of lice from hot countries—observable in Mexico, for instance, where they are abundant on the central plateau above 5000 to 6000 ft., but absent from the hot coastal strips—is apparently not due to the high temperature but probably to the disastrous effect of profuse perspiration and consequent excessive humidity between the clothes and skin. Head lice are found in hotter countries than body lice, especially on bareheaded people.

Effect of bites. The effect of louse bites varies greatly with individuals and with the degree of sensitivity to them, for according to Peck, Wright, and Gant (1943) the principal symptoms appear to be allergic in nature. When persons previously unexposed to lice are experimentally bitten there is at first only a slight sting and little or no itching or redness. After about a week such a person may become sensitized, and then the bites cause considerable irritation and inflamed red spots. When such persons are bitten by large numbers of lice there may be a general skin eruption, mild fever, and a marked feeling of tiredness and irritability. The bites themselves are only partially responsible; much of the reaction is due to contact of the bites with the feces of the lice, to which infested individuals also become sensitized.

Eventually the increased sensitivity gives way to immunity as with other insect bites (see p. 529), and people long infested become oblivious to them. During the stage of increased sensitivity the irritation leads to scratching, and sometimes the scratched bites become secondarily infected, causing pustules to form. Often areas around the bites turn brown, giving the skin a mottled appearance. In very negligent individuals badly infested with head lice the hair may become matted and form a sort of filthy carapace under which fungus growths develop, and the head may exude a fetid odor.

The interesting observation that white rats that have remained free from lice for years promptly become lousy (probably from wild rats or from a very few they previously harbored) when placed on a diet deficient in riboflavin (one of the vitamin B complex) may possibly have some bearing on the fact that some individuals seem to be attacked more readily than others, and that lice become especially prevalent under conditions of hardship and starvation.

Crab Louse

The crab louse, *Phthirus pubis* (Fig. 190), is quite distinct from the other two species of human lice. It has a very broad short body with long, clawed legs, presenting the general appearance of a tiny crab, from which it derives its name. The first pair of legs are smaller than the others. The thorax is very broad with all the segments fused, and the abdomen is greatly foreshortened. Its first three segments (III to V) are fused into one, which bears three pairs of spiracles. The last four segments bear wartlike processes on the sides, the last pair of which is particularly large. This louse is grayish in color, with slightly reddish legs. The females are about 1.5 to 2 mm. in length, the males somewhat smaller. The favorite haunts are the pubic regions and other parts of the body where coarse hair grows, as in the armpits and in the beard, eyebrows, and eyelashes. Occasionally they infest almost the entire body except the head; one of the writers once

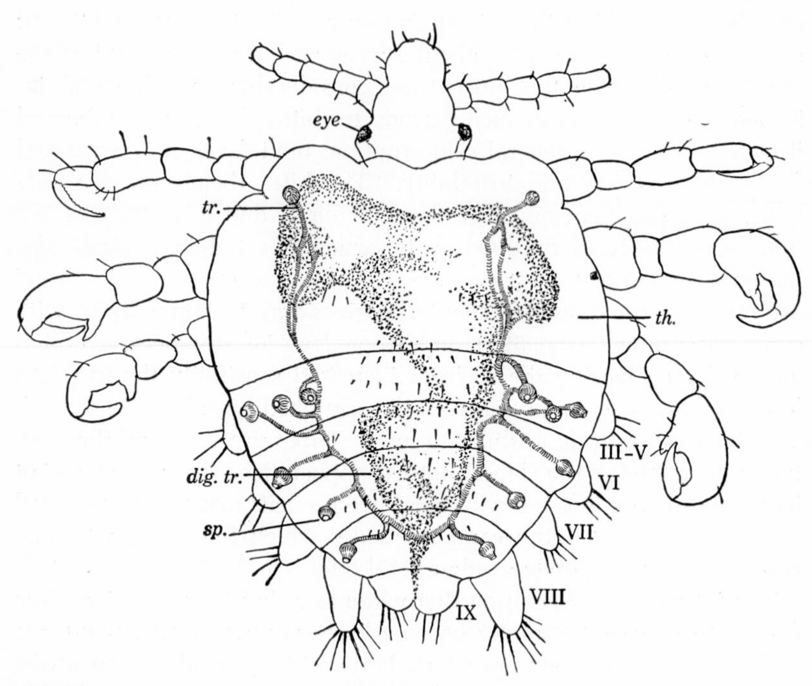

Fig. 190. *Phthirus pubis*, ♀, ×35; *dig.tr.*, digestive tract; *sp.*, spiracle; *th.*, thorax; *tr.*, trachea; III–IX, abdominal segments.

saw a hairy individual who was covered with them from eyebrows to ankles. Unlike the other human lice this species is almost exclusively confined to the Caucasian race.

The females produce 25 or more eggs and glue them, one at a time, to the coarse hairs among which they live. A number of eggs may be glued to a single hair and often at some distance from the skin. The eggs hatch in 6 or 7 days, and the young become sexually mature in about 2 to 3 weeks. This species, even under favorable conditions, can live apart from its host only 10 or 12 hours. The eggs are said not to develop except at temperatures between 68° and 86°F., which are approximately the temperatures to which eggs attached to hairs beneath the clothing would be exposed in cool climates. The adults suck blood intermittently for hours at a time, and often cause severe itching. The secretion of one of the pairs of salivary glands has a peculiar effect on hemoglobin, causing it to turn a violet color in the absence of air; in the body this often causes the formation of telltale pale blue spots from ⅛ to 1 in. in diameter.

Crab lice move about much less than head or body lice, but usually shift their positions daily. The adults probably do not live more than a few weeks. The infestation rate is usually rather low, even when other lice are common. Since it is very often, though by no means always, venereally transmitted, the French call it the "papillon d'amour."

Lice and Disease

Louse-borne diseases have in the past ranked high among the minor horrors of war—higher, certainly, than they ever will again. In peacetime large louse-borne epidemics do not regularly occur since people and communities are rather sharply divided into those who tolerate lousiness and those who do not. The former are exposed to bites of lice from birth; they suffer from louse-borne diseases early in life, become immune, and are kept immune by repeated reinfection. The louse-free population suffers only sporadic cases since there is not sufficient louse traffic between the two components of a community to cause an epidemic except under disrupted conditions brought on by disaster, famine, or local conditions where facilities for keeping body and clothes clean are inadequate, as in construction camps, prisons, etc. Furthermore, there is ordinarily no opportunity for building up the virulence of the organisms concerned by a rapid sequence of human cases, as happens at the beginning of epidemics.

There are three diseases for which lice are the primary transmitters —epidemic typhus, trench fever, and relapsing fever. The first is caused by *Rickettsia prowazekii* (see p. 215), the second is believed to be caused by another rickettsia, *R. quintana*, and the third is caused by a spirochete, *Borrelia recurrentis* (see p. 220). Lice are certainly not the primary transmitters of either typhus or relapsing fever. *Rickettsia prowazekii* of epidemic typhus may have been derived from *R. typhi* of flea-borne endemic typhus (see p. 653) or, perhaps more likely, both forms may have originally been tropical tick-borne diseases, as *R. prowazekii* still is in Africa, where it has a tick-bovid-tick cycle independent of the louse-man-louse cycle (see pp. 216 and 592).

Typhus organisms are invariably fatal to lice, though unfortunately not quickly enough to prevent transmission, whereas fleas suffer no evident ill effects, and the rickettsias of mites and ticks unlike those of lice or fleas, are even transovarially transmitted. The spirochete of louse-borne relapsing fever has undoubtedly been derived from tick-borne strains, quite likely in North Africa (see p. 587), it has become better adapted to lice than the rickettsias, but is not transovarially transmitted.

Epidemic typhus. Louse-borne typhus, known in Europe for centuries, seldom makes itself evident except in extensive outbreaks during wars or other conditions when, as already noted, lice have an opportunity to extend their acquaintanceships widely and to transmit the organisms to many people who had no chance to develop immunity early in life. Zinsser says that typhus has killed more human beings than any other disease. It was typhus that caused Napoleon to withdraw his armies from Russia. It is estimated that this disease killed 3,000,000 Russians during World War I. In former days the disease broke out on sailing ships or in prisons so frequently that it was sometimes called ship fever or jail fever. Hence the name epidemic typhus. In New York and Boston, however, and probably elsewhere, the disease exists in endemic form and is called Brill's disease. Some workers believe these cases to be due to relapses in individuals harboring the rickettsias for many years, perhaps for life, but von Bormann (1952) has made out a strong case for louse transmission. The majority of the cases occur among people who handle worn clothes, such as tailors and cleaners.

Lice are also able to transmit the endemic or murine strain of typhus, *Rickettsia typhi,* which is normally flea-borne (see p. 653); such louse-borne outbreaks have been reported in Mexico, Spain, Manchuria, and Africa. Also, the rat louse, *Polyplax spinulosa,* is capable of transmitting *R. typhi* of murine typhus among rats. However, ordinarily at least, *R. typhi,* even when transmitted by lice, retains its destructive characteristic of causing scrotal swelling and other scrotal reactions in guinea pigs, which *R. prowazekii* fails to do. Ordinarily, at least, *R. typhi* retains its distinctive characteristics even when transmitted by lice.

Epidemic typhus is practically absent from the tropics, e.g., India and tropical parts of Africa and South America, even when lice are abundant. Body lice are the principal disease transmitters, perhaps because head lice and crab lice are less frequently passed around, for these lice experimentally are good hosts for *Rickettsia prowazekii* and *R. quintana* also. Epidemics occur most frequently in winter when people are closely huddled together and the lice nightly migrate from one pile of clothing to another. Lice become infected from typhus patients from early in the disease to about the tenth day or later, even after the temperature has fallen, although by no means all lice become infected. The rickettsias multiply in the epithelial cells, which become greatly distended, and in the lumen of the midgut, and are voided in the feces after a few days. From these time considerations, as Buxton (1947) pointed out, lice from convalescing or dead typhus patients are more dangerous than those from cases at the height of the disease. Transmission results from contamination of the bites or scatches by the feces, or even from inhalation of dried feces; the latter may remain infective for over two months, long after the living lice are gone. A person who wears the boots or clothing of typhus victims may also become a victim.

In man epidemic typhus under epidemic conditions is much more severe than endemic typhus, sometimes causing the death of 70% of its victims. This is doubtless due in part to exaltation of virulence by rapid passage through numbers of susceptible people and in part to lowered resistance under wartime conditions. Poor diet and undernourishment are big factors. The disease is marked by severe headache, prostration, high fever, and a rash caused by small skin hemorrhages. It can be prevented by vaccines and is susceptible to treatment with antibiotics, particularly chloramphenicol and Terramycin. The evidence indicates that these drugs suppress the multiplication of the rickettsias for about a week, by which time the patient will have initiated his own defensive mechanism. With DDT or chlordane to destroy lice, control of typhus is now a relatively simple matter, at least until lice develop resistance to these chemicals, as they have already done in Korea and Egypt. It is unlikely that typhus will ever again cause devastating epidemics in civilized countries.

Trench fever. It is generally believed that this disease is also caused by a rickettsia, *R. quintana* (see p. 215), which, like the nonpathogenic *R. pediculi,* lives in the lumen of the intestines of lice but not in the epithelial cells. Although unknown previously, this disease became so common during World War I that it caused more sickness than any other disease except scabies. After the war it fell into obscurity again, but an outbreak of what was probably the same disease occurred among louse-feeders in a typhus-vaccine laboratory in Warsaw just prior to World War II, and some small outbreaks occurred in the Balkans during that war. *R. quintana* has also been reported from Mexico City.

The best explanation for this "here again, gone again" disease is, in our opinion, that the supposedly nonpathogenic *R. pediculi,* when lice are restricted to their home-folks, produces mild or inapparent human infections early in life, resulting in immunity, but when the lice get access to numerous previously unexposed and susceptible people, and are passed through a series of these, the rickettsias have their virulence exalted and cause epidemics which subside when normal relations between lice and their perennially infested home-folks are re-established.

The infection is transmitted, like typhus, by louse feces, but since urine and feces of human cases are infective, it may not be spread exclusively by lice. The symptoms are headache, body pains, a double rise of fever, albumin in the urine, and usually a rash.

Relapsing fever. When lice ingest the blood of a relapsing-fever patient, most of the spirochetes die in the alimentary canal within a very short time, but a few survive by penetrating into the body cavity. Although very sparse for a few days they become abundant after about a week. Neither the bites nor the feces of the lice are infective, and transmission occurs only by breaking or crushing lice and inoculating bites with the infective body fluid. Although louse-borne strains were undoubtedly derived originally from tick-borne infections, lice seem not to be susceptible to some tick-borne strains although they are to others.

Outbreaks of louse-borne relapsing fever have been suffered in many parts of Europe, Africa, and Asia; in the past they occurred in the Western Hemisphere also, but not now. After World War II a great epidemic spread across Africa south of the Sahara; its southward extension was limited by the nudity of the Central African natives. In temperate climates epidemics of louse-borne relapsing fever, like louse-borne typhus, rage fiercest in winter, and have accompanied wars in the past.

Borrelia recurrentis, although undoubtedly originally derived from argasid ticks, is now so completely estranged from its ancestral hosts that it will no longer infect them, nor will it infect other mammals than man, except monkeys and new-born rabbits. However, since this strain of spirochete has not become sufficiently adapted to its louse host to be transovarially transmitted by it, it is dependent on human infections for survival.

Lice and other diseases. Lice may also serve as mechanical transmitters of certain other diseases. The bacilli of bubonic plague, *Pasteurella pestis,* have been found alive in both body lice and head lice taken from victims of the disease, and both species have been experimentally proved able to transmit plague from rodent to rodent in Java. Lice do not transmit plague by their bites, but they may transmit it when crushed, and possibly with their feces. Natives in Java kill lice by mashing them against the head of the host, which should make infection through the scratched sores on the head very easy. In Ecuador and Peru natives are said to kill lice by crushing them between the teeth; there is much more danger involved when man bites louse than when louse bites man. Lice can also harbor and transmit (via their feces) the related bacillus, *Pasteurella tularensis,* that causes tularemia (see p. 226).

Most other bacteria are quickly destroyed by antibacterial substances in the alimentary canal of lice, as they are in most insects, but salmonellas, species of which cause typhoid and other diseases in man or animals, multiply rapidly in human lice, and kill them in 24 to 48 hours. The lice do not transmit them directly by bite, but do by contamination of the bites with their feces, and the bacilli may remain alive in dead lice and their feces for a year or more.

Prevention and Remedies

Methods of dispersal. The prevention of lousiness consists primarily in personal cleanliness. However, no amount of personal hygiene and cleanliness will prevent temporary lousiness if there is association with unclean and careless companions. Lousiness and human wretchedness and degradation have always been companions, but this does not imply that lice have any inherent abhorrence of a clean body if they can get access to it. From the nature of their habitats the common modes of infection of the three different species of human lice vary somewhat. The head louse depends for distribution largely on crowded cloakrooms, on promiscuous use of combs and brushes or borrowed hats and caps, and on the free-for-all trying on of headgear in haberdasheries and millinery shops. The body louse is dispersed by clothing and bed linen, usually at night, and finds fresh hunting grounds by nocturnal migrations from one pile of clothes to another. The crab louse sometimes utilizes public toilets for dissemination and is commonly spread by promiscuous sexual intercourse.

Where men are crowded together in prisons or war camps lousiness is almost sure to develop unless particularly guarded against, since some unclean persons are nearly always in the aggregation, and conditions are such that the infestation is given every opportunity to spread. Even under normal conditions there are many opportunities for dispersal. Buxton (1947) recalls seeing some men sitting on a wall in Iran, catching lice and dropping them into the street, with typhus and relapsing fever epidemic at the time. Of importance in connection with the spread of these diseases is the fact that lice desert a febrile patient and seek a new host. In Europe during World War II lice were mostly well controlled in military forces, but they spread extensively among civilians, especially in crowded bomb shelters and in crowded homes for evacuated children.

Elimination of lice. Few preventive measures have given more spectacularly good results than the use of DDT in destroying body lice and protecting against them. During the latter part of World War II almost every American soldier and sailor in the European theater was provided with a dusting powder consisting of 10% DDT in pyrophyllite and was free of cooties for the first time in any war. The dust was applied by blowguns to the head, skin, and clothing, without undressing. However, body lice are eliminated by dusting the inner surface of the underwear, socks, and seams of outer garments, or dusting under the clothing when it is not feasible to remove it, using a sifter-top can. One ounce per treatment is enough. The eggs are not killed, but since they hatch within the period of effectiveness of the DDT, one application suffices.

Unfortunately lice have developed resistance to DDT in many localities, particularly in Korea, Japan, and the Middle East; in these places 1% lindane powder can be substituted, but a second application in 7 to 10 days is advocated. Synergized pyrethrum powders are also satisfactory for individual use, but have less residual action, so repeated treatments are needed. Lice resistant to DDT are also very susceptible to 0.5 to 1% malathion dusts, but these have not yet been cleared toxicologically for human use. For head or crab lice a highly effective treatment is NBIN emulsion (68% benzyl benzoate, 6% DDT, 12% benzocaine, and 14% Tween 80 (a detergent), one part diluted with 5 parts of water for application by spraying or sponging. Left on for 24 hours, this kills the eggs as well as the lice. DDT or lindane powders can also be used, but should be reapplied at weekly intervals if the hair is washed between treatments.

For disinfestation of clothing, hot water, dry heat, pressing with a hot iron, live steam, or laundering in various disinfectant solutions are all effective. Dipping in 5% DDT solution or emulsion renders clothing protective against lice through several washings. Fumigation with methyl bromide has been used at United States ports of disembarkation for eliminating lice from clothing of prisoners of war.

Anoplura on Domestic Animals

The Anoplura that torment domestic animals belong to the families Linognathidae and Haematopinidae. The latter have well-developed pleural plates, vestigial eyes if any, and all the legs about equal. There is only one genus, *Haematopinus,* on domestic animals, including *H. suis,* common on pigs; *H. asini* of horses; and *H. eurysternus* (Fig. 191*B*), the short-nosed louse of cattle, of which *H. quadripertusus,* the tail louse of cattle, may be only a variety.

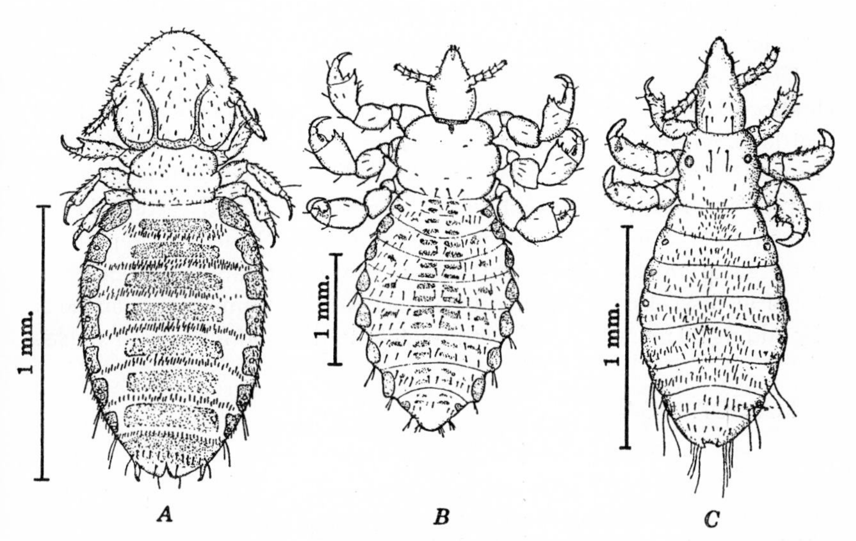

Fig. 191. Cattle lice: *A, Damalinia* (=*Bovicola*) *bovis; B, Haematopinus eurysternus; C, Linognathus vituli.* (After Mönnig, *Veterinary Helminthology and Entomology,* Williams and Wilkins, 1949.)

The Linognathidae, which have no sclerotized abdominal plates, include three genera with important species on domestic animals: (1) *Solenopotes,* with only one row of bristles on each abdominal segment, mostly parasitic on deer but with one species, *S. capillatus,* on cattle; (2) *Linognathus,* with two or more rows of bristles on the abdominal segments and with no eyes, including *L. vituli* (Fig. 191*C*), the long-nosed or blue louse of cattle; *L. stenopsis* of goats; *L. pedalis,* the foot louse of sheep; and *L. setosus* of dogs and foxes; and (3) *Microthoracius,* also with numerous bristles, but with very long heads, and eyes present, containing *M. cameli* of camels, and three species on llamas. Two species of Hoplopleuridae (see p. 620), *Polyplax spinu-*

losa and *Hoplopleura oenomydis,* are common rat lice in the United States, and one, *Haemodipsus ventricosus,* is common on rabbits.

Like the Mallophaga (see p. 637) these lice stick closely to their hosts and are usually transmitted only by body contact, though horse lice are spread by blankets, saddles, etc.

The hosts are much more annoyed by these lice than they are by the Mallophaga. Their fur and skin are affected, and they become restless, lose their appetites, and become susceptible to other diseases. Constant licking by cattle produces hair balls in the stomach, and the foot louse of sheep may cause lameness.

Treatment consists of sprays, dips, or dusts containing chlorinated hydrocarbons, organic phosphorus compounds, or rotenone. For dairy cattle synergized pyrethrum or methoxychlor are recommended because they do not accumulate in the milk; malathion, also, does not appear in appreciable amounts in the milk of sprayed cattle. Dusts are preferable in the winter to avoid chilling. Sometimes repetition after 14 to 18 days is needed, but usually one treatment is effective for 6 to 8 weeks. Lice have been successfully eliminated from hogs by adding 30 mg. per kg. of methoxychlor to the food for several weeks.

MALLOPHAGA

Morphology. Two characters make it easy to identify Mallophaga: the head as broad as, or broader than, the thorax; and the pair of strongly sclerotized, pincerlike mandibles on the ventral side of the head. In most species the second to seventh abdominal segments bear spiracles. The legs are short and all much alike; in the suborder Ischnocera they are fitted for clasping as in the Anoplura, but in the Amblycera, which roam over the body, the tarsi are longer and modified for clinging to smooth surfaces. The Amblycera usually have short, four-segmented maxillary palpi, but these are absent in the Ischnocera. These two orders differ also in the antennae; in Amblycera these are short and tucked into grooves as in fleas, whereas in Ischnocera they are longer and free (Fig. 192). Following is a brief key to some of the families and genera common on domestic animals:

Suborder **Amblycera.** Antennae club-shaped and mostly concealed in grooves; maxillary palpi present (Fig. 192*B*). Only family important on domestic animals is Menoponidae, infesting poultry.

A. All tarsi with 2 claws, antennae 4-segmented; head evenly expanded behind and broadly triangular. **Menoponidae** 1

1*a.* Thorax with 3 distinct segments (Fig. 193*A*) ***Trinoton***
2 or 3 spp. on geese, ducks, and swans.

1*b.* Thorax apparently with 2 segments (Figs. 192*A*, 193*B*) 2

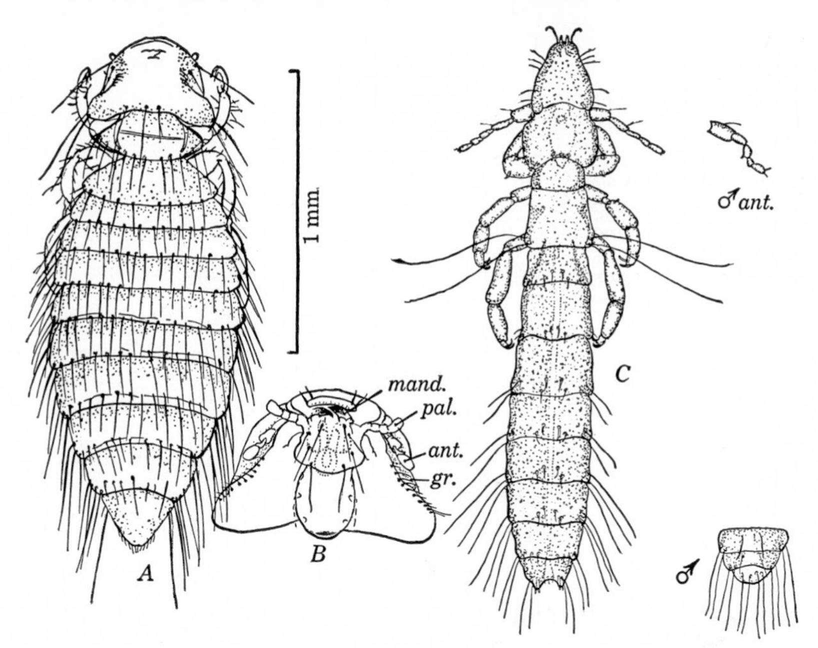

Fig. 192. Examples of two suborders of Mallophaga. *A, Menopon gallinae* of chickens, dorsal view; *B,* same, ventral view of head (adapted from Ferris, *Parasitology,* 16, 1924). *C, Columbicola columbae* of pigeons, ♀, and antenna and posterior end of abdomen of ♂ (after Martin, *Canad. Ent.,* 66, 1934).

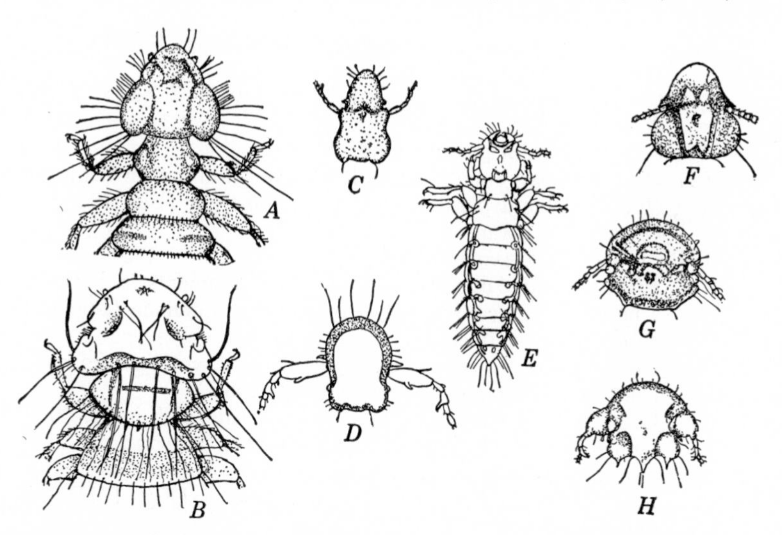

Fig. 193. Details of various Mallophaga to illustrate key. *A, Trinoton querquedulae* of ducks; *B, Colpocephalum pectiniventre* of geese; *C, Esthiopterum crassicorne* of ducks; *D, Lipeuris caponis* of chickens; *E, Ornithobius cygni* of swans; *F, Philopterus dentatus* of ducks; *G, Goniocotes gigas* of chickens; *H, Goniodes pavonis* of peacock. (Adapted from various authors.)

2*a*. Abdominal segments with 2 rows of bristles ***Menacanthus***
M. stramineus (body louse) of chickens.
2*b*. Abdominal segments with 1 row of bristles 3
3*a*. Eye lodged in a shallow sinus (Fig. 192*A*) ***Menopon***
M. gallinae (shaft louse) of chickens.
3*b*. Eye lodged in a deep sinus (Fig. 193*B*) ***Colpocephalum***
C. turbinatum of pigeons and *C. pectiniventre* of geese.

Suborder **Ischnocera.** Antennae filiform and exposed; no maxillary palpi (Fig. 191*C*).
1. Antennae 5-segmented; tarsi with 2 claws; infesting birds (Fig. 193*C–H*) **Philopteridae**
2. Antennae 3-segmented; tarsi with 1 claw (Fig. 191*A*); infesting mammals **Trichodectidae**

A summary of the principal genera of these two families, infesting birds and mammals respectively, is given below, with mention of the commoner and more important parasites of domestic animals in North America:

Ischnocera of Birds **(Philopteridae)**
A. Body long and slender.
(1) ***Lipeurus:*** Head nearly hemispherical in front of antennae; in ♂ 1st segment of antenna much enlarged, third segment with prong (Fig. 193*D*).
L. gallipavonis on turkey; *L. caponis* on chickens.
(2) ***Esthiopterum:*** Head elongated in front of antennae; antennae as in (1); clypeus without dorsal spines (Fig. 193*C*).
E. crassicorne on ducks.
(3) ***Columbicola:*** Like (2) but clypeus with 2 pairs of spines dorsally.
C. columbae on pigeons (Fig. 192*C*).
(4) ***Ornithobius:*** ♂ antenna without prongs; abdomen with second chitinized band paralleling chitinized margin (Fig. 193*E*).
O. cygni on swans.
B. Body broad, abdomen rounded.
(1) ***Philopterus:*** Antennae with 5 similar segments in both sexes; a horn-like process in front of insertion of antennae (Fig. 193*F*).
P. dentatus on ducks.
(2) ***Goniocotes:*** No prongs on segments of ♂ antenna; no spine in front of insertion of antennae (Fig. 193*G*).
G. gigas (3–4 mm. long) and *G. hologaster,* fluff louse, (0.8–1.3 mm. long) on chickens; *G. bidentatus* on pigeons.
(3) ***Goniodes:*** Antenna of ♂ with prong on 3rd segment at least; no spine in front of insertion of antennae (Fig. 193*H*); head angular behind.
G. dissimilis on chickens; *G. damicornis* (2 mm. long) and *G. minor* (less than 2 mm. long) on pigeon; *G. parviceps* (2 mm. long) and *G. pavonis* (3 mm. long) on peacock; *G. meleagridis* on turkey.

(4) ***Cuclotogaster:*** Antennae of ♂ with prong on 3rd segment, head rounded behind, without spines.
C. heterographus, head louse of chickens.

Ischnocera of Mammals **(Trichodectidae)**

A. ***Damalinia:*** With pleural plates; antennae alike in both sexes (Fig. 191A).
D. bovis on cattle, *D. caprae* on goats, *D. equi* on horses, *D. ovis* on sheep.

B. ***Trichodectes:*** With pleural plates; antennae of ♂ with first segment of antenna enlarged.
T. canis on dogs.

C. ***Felicola:*** Without pleural plates; antennae alike in both sexes.
T. subrostrata on cats.

Life cycle, habits, etc. The Mallophaga live their entire lives, generation after generation, on their hosts. The Ischnocera, to which the mammalian parasites belong, are relatively sedentary and feed exclusively, or nearly so, on feathers or epidermal scales. Many stout-bodied species live only on the head where they cannot be picked, other slender forms live on the back or wings. The Amblycera roam about more freely and escape by good broken-field running; most of these commonly vary their diet with blood obtained by nipping live feather papillae or nibbling through the skin. In sick birds that do not peck actively, lice may become excessively abundant; one hen is reported to have harbored over 8000 body lice (*Menacanthus stramineus*).

The eggs are glued to feathers or hairs, sometimes in selected places, usually one a day. According to Martin (1933) eggs of the pigeon louse, *Columbicola columbae* hatch in 4 days and the nymphs pass through their three molts and reach maturity in about 3 weeks.

Except for occasional hitchhiking on the bodies of hippoboscid flies, Mallophaga move from host to host only by direct body contact. For this reason they are usually kept in the family and pass from generation to generation of the same species. Thus isolated zoologically, many species and even genera have arisen which are closely confined to particular kinds of birds. They tend to change more slowly than their hosts, so even after their hosts have become widely separated and isolated, and have developed into new species or genera, they may harbor the same or closely related lice as did their remote ancestors. Strong host specificity keeps lice from transferring their attentions to unrelated species even when there is opportunity, e.g., from a goldfinch to a cuckoo, or from a quail to a falcon.

Control. Control of Mallophaga on larger animals is the same as for Anoplura (see p. 632). On chickens, lice can be eliminated by putting pinches of sodium fluoride dust or 5% DDT dust on head, back, vent region, and under the wings, or by painting the roosts with 1% Lindane emulsion or 40% nicotine sulfate. If the birds are given an opportunity for self-treatment by dusting in litter, etc., that has had 1% malathion dust added to it at the rate of about 1 lb. per 75 sq. ft. (or per 25 birds), lice may be completely eliminated in about 10 weeks.

REFERENCES

Bacot, A. W. 1917. Contribution to the bionomics of *Pediculus humanus* (*vestimenti*) and *Pediculus capitis*. *Parasitology*, 9: 229–258.

Busvine, J. R. 1957. Insecticide-resistant strains of public health importance. *Trans. Roy. Soc. Trop. Med. Hyg.*, 51: 11–31

Buxton, P. A. 1936–1941. Studies on populations of head lice (*P. humanus capitis*). *Parasitology*, 28: 92–97; 30: 85–110; 32: 296–302; 33: 224–242.

1947. *The Louse, An Account of the Lice Which Infest Man, Their Medical Importance and Control.* 2nd ed. Arnold, London.

Chung, H., and Wei, Y. 1938. Studies on the transmission of relapsing fever in North China, II. *Am. J. Trop. Med.*, 18: 661–674.

Dyer, R. E., Rumreich, A., and Badger, L. F. 1931. Typhus fever. *Public Health Repts.*, 46: 334–338.

Ewing, H. E. 1936. The taxonomy of the mallophagan family Trichodectidae, with special reference to the New World fauna. *J. Parasitol.*, 22: 233–246.

Ferris, G. F. 1951. The Sucking Lice. *Mem. Pacific Coast Entomol. Soc.*, 1: ix + 320 pp., San Francisco.

Fuller, H. S., et al. 1949. Studies of the human body lice; P.h. corp. *Public Health Repts.*, 64: 1287–1292.

Grinnell, M. C., and Hawes, I. L. 1943. Bibliography on lice and man. *U. S. Dept. Agr. Bibliogr. Bull.* 1.

Hopkins, J. H. E. 1949. The host-associations of the lice of mammals. *Proc. Zool. Soc. London*, 119: 387–604.

Keilin, D., and Nuttall, G. H. F. 1930. Iconographic studies on *Pediculus humanus*. *Parasitology*, 22: 1–10.

Kemper, H. E., and Peterson, H. O. 1953. Cattle lice and how to eradicate them. *U. S. Dept. Agr. Farmer's Bull.* 909.

Mellanby, K. 1942. Natural population of the head louse (*Pediculus humanus capitis:* Anoplura) on infected children in England. *Parasitology*, 34: 180–184.

Milner, K. C., Jellison, W. L., and Smith, B. 1957. The role of lice in transmission of *Salmonella*. *J. Infectious Diseases*, 101: 181–192.

Nuttall, G. H. F. 1917–1919. Biology of *Pediculus humanus*. *Parasitology*, 10: 1–42, 80–185, 411–590; 11: 201–220, 329–346. Biology of *Phthirus pubis*, *ibid.* 10: 375–382, 383–405; 11: 329–346.

Peck, S. M., Wright, W. H., and Gant, J. Q. 1943. Cutaneous reactions due to the body louse. *J. Am. Med. Assoc.*, 123: 821–825.

Rothschild, M., and Clay, T. 1952. *Fleas, Flukes and Cuckoos,* Chap. 8. Philosophical Library, New York.

Sikora, H. 1915. Beiträge zur Biologie von *Pediculus vestimenti. Centr. Bakt.,* 1, Orig., 76, 523–537.

Soper, F. L., Davis, W. A., Markham, F. S., and Riehl, L. A. 1947. Typhus fever in Italy, 1943–1945, and its control with louse powder. *Am. J. Hyg.,* 45: 305–334.

U. S. Department of Agriculture. 1952. Insects. *Yearbook.*

Zinsser, H. 1935. *Rats, Lice, and History.* Little, Brown, Boston.

See also References of Chapters 4, 10, and 22.

Chapter 27

FLEAS (SIPHONAPTERA)

David Harum says, "A reasonable amount of fleas is good for a dog. They keep him from broodin' on bein' a dog." A goodly supply of fleas might likewise keep man from brooding over anything deeper than the presence of these fleas, but in some cases this in itself is a rather serious thing to brood over. Not only are fleas very annoying pests and a common cause of insomnia, but they may also serve as the disseminators of two important human diseases, bubonic plague and endemic typhus.

General structure. Fleas, constituting the order Siphonaptera, are believed by most entomologists to be more or less distantly related to the Diptera. Their bodies are much compressed to facilitate gliding between the hairs or feathers of their hosts. The head is broadly joined to the thorax, which is relatively small. The abdomen consists of ten segments, the last three of which are modified for sexual purposes, particularly in the male (Figs. 194, 195). In both sexes the tergum (dorsal plate) of the tenth segment has a pitted area covered with little bristles, called the *pygidium* or *sensilium*. The seventh tergum in most fleas bears one to four pairs of long *antepygidial bristles.*

All parts of the body are furnished with backward-projecting bristles and spines which aid the flea in forcing its way between dense hairs and prevent it from slipping backward. The efficiency of these spines is apparent when one attempts to hold a flea between his fingers. Many fleas have specially developed, thick, heavy spines arranged in rows suggestive of the teeth of combs and therefore known as *ctenidia* or "combs" (Fig. 198). There may be a *genal ctenidium* along the ventral margin of the head, or a *pronotal ctenidium* on the hind margin of the pronotum (the dorsal plate covering the first segment of the thorax), or in both places; a few fleas also have abdominal ctenidia. The presence or absence of these combs and the number of teeth in them are of considerable use in identification of species.

The legs of fleas are very long and powerful and at first glance seem to possess one more segment than do the legs of other insects (Fig. 194). They really consist of the usual number of segments with five-segmented tarsi, but are peculiar in the enormous development of the coxae, which in most insects are quite insignificant. The shape of the

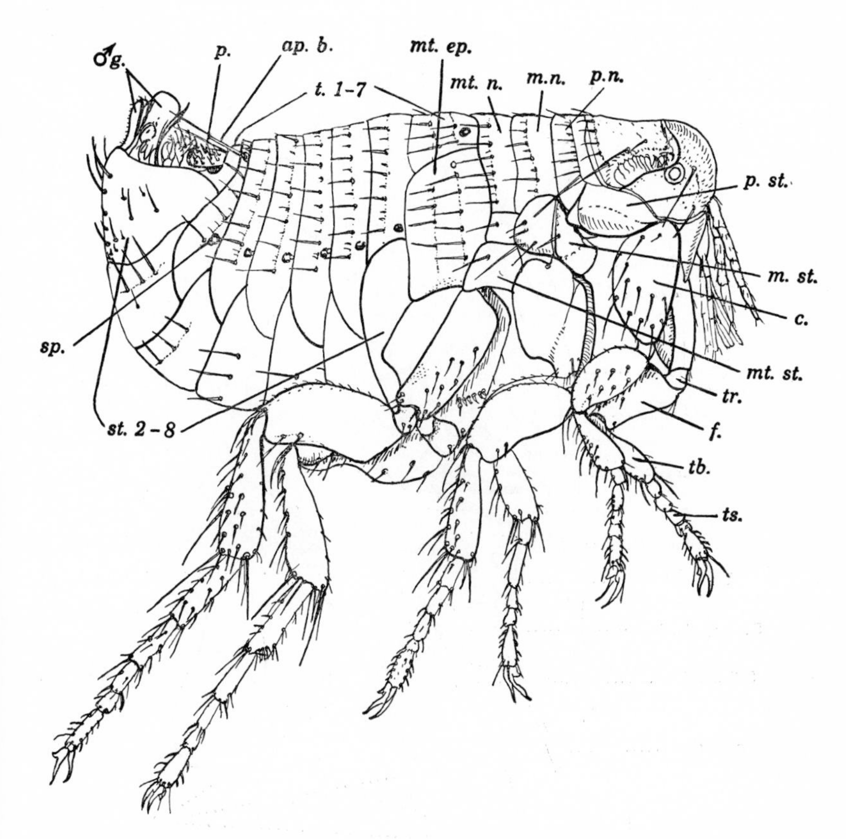

Fig. 194. *Xenopsylla cheopis,* rat flea: *ap.b.,* antepygidial bristles; *c.,* coxa; *f.,* femur; *m.n,* mesonotum; *m.st.,* mesosternum; *mt.ep.,* metepimeron; *mt.n.,* metanotum; *mt.st.,* metasternum; *p.,* pygidium; *p.n.,* pronotum; *p.st.,* prosternum; *sp.,* spiracle; *st. 2–8,* abdominal sternites 2–8; *t. 1–7,* abdominal tergites 1–7; *tb.,* tibia; *tr.,* trochanter; *ts.,* tarsus; *♂g.,* male genitalia. (Modified from Jordan and Rothschild, *Parasitology,* 1, 1908.)

sternal plate to which the coxae are attached is suggestive of still another segment. The great development of the coxae as well as of the other segments of the leg gives unusual springiness and consequently enormous jumping power. The so-called human flea, *Pulex irritans,* has been observed to jump 13 in. horizontally and as much

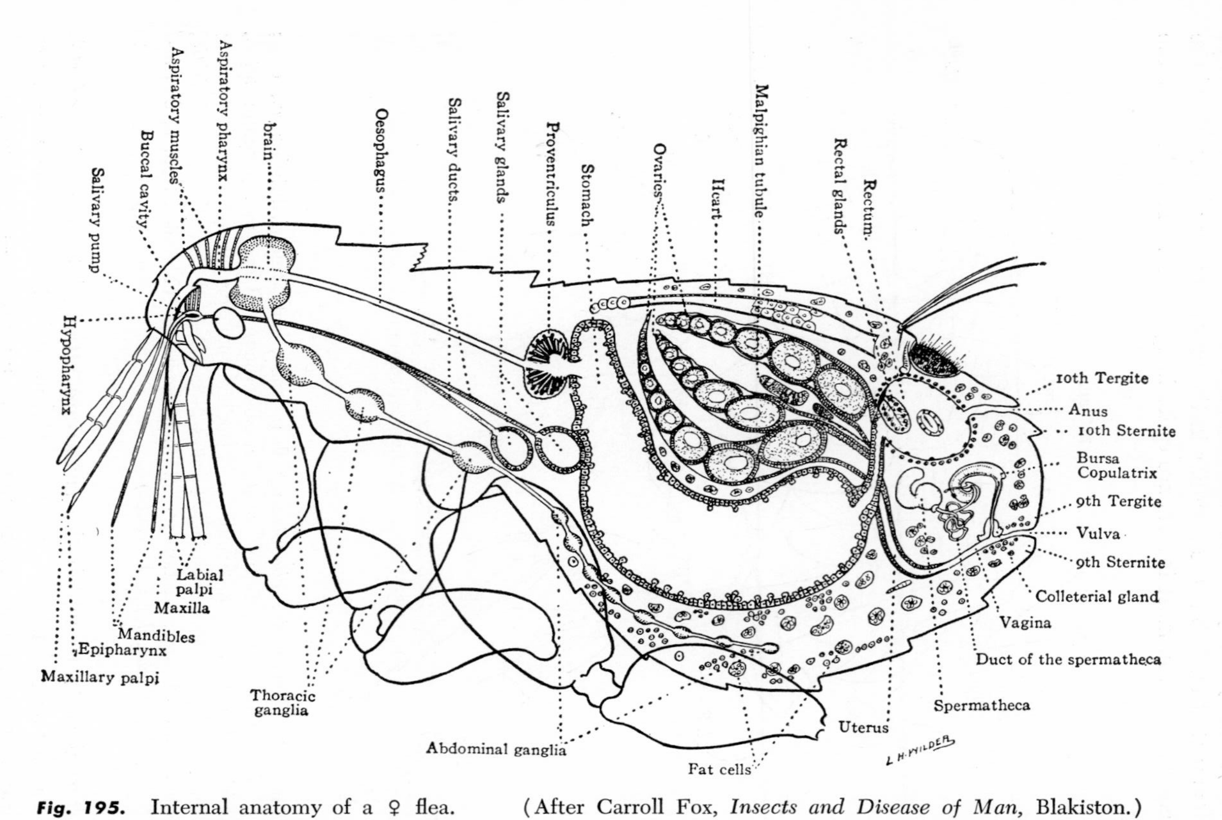

Fig. 195. Internal anatomy of a ♀ flea. (After Carroll Fox, *Insects and Disease of Man*, Blakiston.)

as 7¾ in. vertically; equivalent jumps for a man of average height would be a broad jump of 450 ft. and a high jump of 275 ft. All the legs are furnished with rows of stout spines and are armed at the tip with a pair of large claws.

Simple eyes are present in some species of fleas but not in others. The antennae are short and club-shaped and consist of three segments, two small, the third large and laminated. When not in use they are

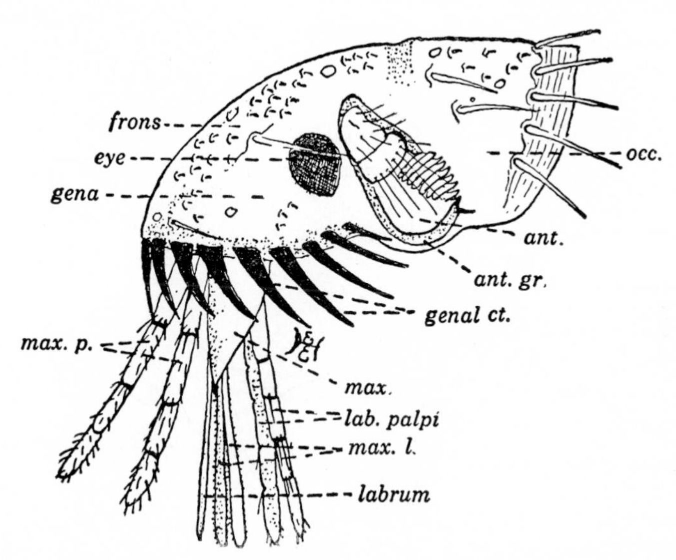

Fig. 196. Head and mouth parts of a flea, *Ctenocephalides felis* ♀; *ant.*, antenna; *ant. gr.*, antennal groove; *genal ct.*, genal ctenidium; *lab. palpi,* labial palpi; *max. l.*, maxillary laciniae or stylets; *max.*, maxillary plate; *max. p.*, maxillary palpi; *occ.*, occiput. (In Fig. 195 the labrum is erroneously labeled "epipharynx," and the maxillary laciniae, "mandibles.") (Adapted from Ewing and Fox, *U. S. Dept. Agric. Misc. Publ.*, 500, 1943.)

folded back into special grooves for them on the sides of the head (Fig. 196, *ant. gr.*), but they can be rotated out or up over the head like a pair of horns. In many fleas there is a frontal notch or tubercle on the front of the head. The region behind the antennal groove is called the *occiput;* the region in front of it is more or less divided into a forward *frons* and a lower *gena* (see Fig. 196).

The mouth parts (Fig. 196) are fitted for piercing and sucking. The labial palpi, usually with three to five segments, are elongated, grooved structures that fit together to form a sheath for the piercing organs. These consist of a pair of slender maxillary stylets (laciniae), which are serrated at the tip like little saws, and a bristlelike epipharynx grooved on its posterior surface to form a food channel in conjunction with the maxillary stylets. The stylets also form a canal that serves as a salivary duct. The basal parts of the maxillae (maxillary plates) are large pyramidal structures with a spine at the tip, used to hold

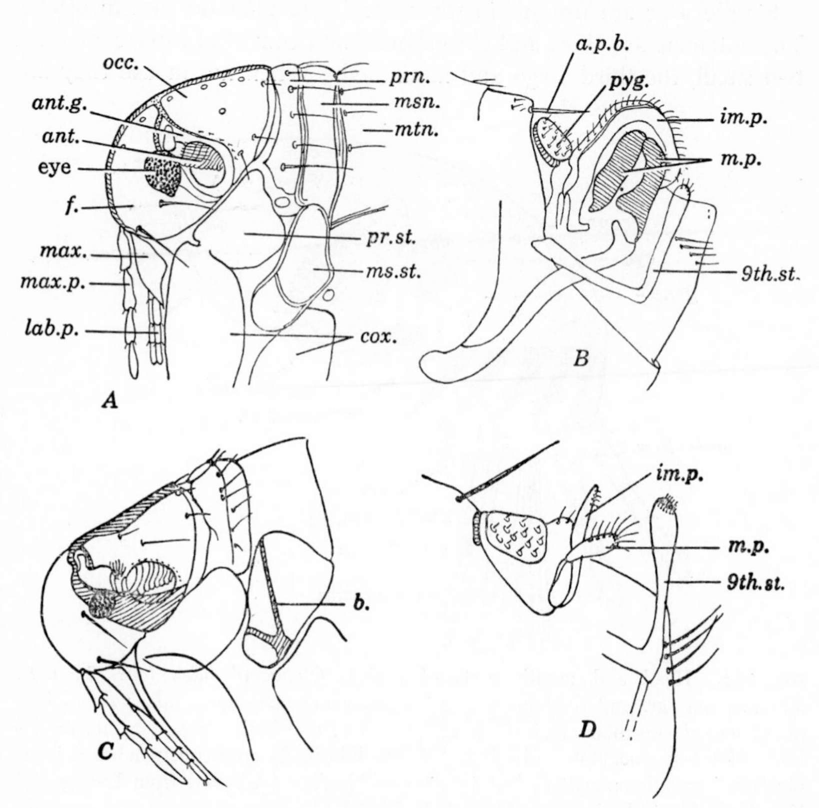

Fig. 197. *A* and *B*, head of posterior end of ♂ of *Pulex irritans; C* and *D*, same of *Xenopsylla cheopis; ant.*, antenna; *ant.g.*, antennal groove; *a.p.b.*, antepygidial bristle; *b.*, sclerotized bar of mesosternum of *Xenopsylla; cox.*, coxae; *f.*, frons; *im.p.*, immovable process of clasper; *lab.p.*, labial palpi; *max.*, maxilla; *max.p.*, maxillary palpi; *m.p.*, movable process of clasper (double in *Pulex*); *msn.*, mesonotum; *ms.st.*, mesosternum; *mtn.*, metanotum; *occ.*, occiput; *prn.*, pronotum; *pr.st.*, prosternum; *pyg.*, pygidium; *9th st.*, 9th sternite. (Adapted from Fox, *Fleas of Eastern United States*, Iowa State College Press.)

the flea in position while feeding; they are provided with large four-segmented maxillary palpi which might be mistake for antennae.

Male fleas are easily distinguished by the rakish upward tilt of the abdomen, which in females is rounded (Figs. 194, 195). The terminal segments are modified into complicated claspers for holding the female, and a remarkable intromittent organ which is so complex that one morphologist concluded that "truly, the thing does not make sense." The claspers have a broad immovable lobe or process and a movable *finger.* The details of these genital organs are of great taxonomic value, but difficult to use unless there are illustrations to go by (Fig. 197). In the females the terminal segments of the abdomen are reduced and inconspicuous. In this sex a taxonomic character of great importance is the form of the spermatheca (Figs. 195, 201), which is sclerotized and easily seen inside the abdomen in cleared specimens.

Classification. Over a thousand species of fleas have been described. The differentiating characters are sometimes so subtle that only an experienced siphonapterologist can do the job, and then perhaps only if he has a particular sex under his microscope. Also, importance attached to taxonomic characters changes. Major groups were once based on whether or not the head was "broken" dorsally between the antennae, and on the "telescoping" of the thorax. In the latest and well-received classification of Jordan (in Smart's *Insects of Medical Importance,* 1956) these characters are reduced to subfamily value, and fleas with narrow thorax fall into two different families (*Tunga* in Tungidae and *Echidnophaga* in Pulicidae).

Jordan divided the fleas into two superfamilies, Pulicoidea and Ceratophylloidea, based not on any single character but on a combination of them, of which perhaps the most reliable for identifying the Pulicoidea are (1) the absence of an internal vertical ridge on the wall of the midcoxa, (2) metepimeron extending far upward with its spiracle well out of line with the other abdominal spiracles, and (3) at most one row of bristles on the abdominal segments. The Pulicoidea contain the families Tungidae and Pulicidae; the former contains the genus *Tunga* (chiggers) and the latter the majority of the important annoying and disease-carrying fleas—*Pulex, Echidnophaga, Ctenocephalides,* and *Xenopsylla.* The Tungidae have a "telescoped" thorax, a pygidium with only 8 pits on each side, and no antepygidial bristles, whereas the Pulicidae have broad thoracic segments (except *Echidnophaga*), 14 pits on each side of the pygidium, and 1 or more antepygidial bristles.

The Ceratophylloidea are divided into 15 families, but only 3 of these contain species that are important to man or his domestic animals: Ceratophyllidae (see below), Leptopsyllidae, and Hystrichopsyllidae. The family Ceratophyllidae constitutes what was once a single genus *Ceratophyllus,* but it is now split into about 20 genera. Most of them are parasites of rodents or rabbits, but the genus *Ceratophyllus* (in the strict sense) contains parasites of birds, two of which are pests of poultry (see p. 658). These have 24 or more spines in the pronotal comb. Another ceratophyllid, *Nosopsyllus fasciatus* (Fig. 198*C*), is the commonest flea of domestic rats in temperate climates; it has 18 or 20 teeth in the pronotal comb. There are many potential transmitters

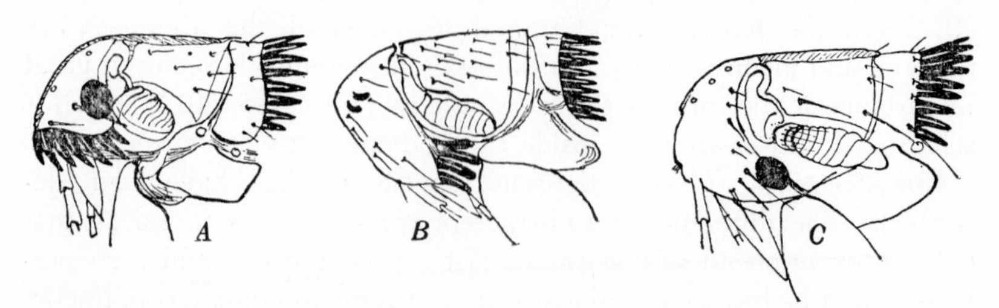

Fig. 198. Heads of fleas, showing arrangement of ctenidia. *A*, *Ctenocephalides canis*; *B*, *Leptopsylla segnis*; *C*, *Nosopsyllus fasciatus*. (After Carroll Fox, *Insects and Disease of Man*, Blakiston.)

of sylvatic plague in this family. Identification of the genera and species is beyond the scope of this book, and students are referred to Ewing and Fox (1943) and Hubbard (1947).

The following is a purely artificial key to the common fleas in or around human habitations, or important as vectors of plague or typhus:

1*a*. No ctenidia — 2
1*b*. At least one ctenidium present — 5
2*a*. Dorsal plates of thorax very narrow, appearing telescoped — 3
2*b*. Metanotum as long as or longer than 1st abdominal tergite — 4
3*a*. Pygidium with only 8 pits on each side; inner side of hind coxa without spinelike bristles; no antepygidial bristles; female with abdominal spiracles II–IV very small or absent, V–VII very large — ***Tunga***
Important species: *T. penetrans*, the jigger.
3*b*. Pygidium with 14 pits on each side; inner side of hind coxa with spine-like bristles; 1 or more antepygidial bristles present — ***Echidnophaga***
Important species: *E. gallinaceum*, sticktight flea of poultry, rats, etc.
4*a*. Mesothorax without a vertical rod; frons smoothly rounded — ***Pulex***
Important species: *P. irritans*, human flea.
4*b*. Mesothorax with a vertical rod; suture between antennal grooves feebly sclerotized — ***Xenopsylla***
Important species: *X. cheopis* (most important plague and typhus vector), *astia, braziliensis, philoxera, vexabilis, hawaiiensis.*
5*a*. Genal and pronotal ctenidia present — 6
5*b*. Only pronotal ctenidium present — 7
6*a*. Genal ctenidium horizontal, in the common species on cats and dogs with 8 or 9 teeth on each side — ***Ctenocephalides***
Important species: *C. canis* and *C. felis*, cat and dog fleas.

6*b*. Genal ctenidium of 3 spines with a broad genal lobe above and posterior to the spines ***Ctenophthalmus***
Important species: *C. argyrtes* of rats and mice in Europe.
6*c*. Genal ctenidium vertical, with 4 spines ***Leptopsylla***
Important species: *L. segnis* of rats and mice.
7*a*. One row of bristles on each abdominal segment ***Hoplopsyllus***
7*b*. Two rows of bristles on each abdominal segment, metanotum with short apical spines ***Ceratophyllidae*** **(= Dolichopsyllidae)**
Important species: *C. gallinae* and *C. niger* on poultry; *Nosopsyllus fasciatus* on rats.

Habits. Most fleas are neither as strictly host parasites as lice nor as strictly nest parasites as bedbugs. The nests or lairs of the host are the normal breeding places and are the homes of the eggs, larvae, and pupae; frequently adults are found in them too. Many rodent and bird fleas are more frequently found in the nests than on the bodies of the hosts; Wayson (1947) found over 1000 fleas in each of 13 rodent nests in an area where the average number on each of 500 trapped animals was 3. The relative numbers on hosts and in the nests are markedly influenced by climate; fleas tend to stay in out of cold or rain. Fleas move in and out of rodent burrows on warm nights and may then be picked up by other animals. A few species, e.g., the cat and dog fleas (*Ctenocephalides felis* and *C. canis*), lay their eggs in the fur of the host, whence they drop off when the animal shakes himself and prepares to sleep, but most species leave the host to deposit the eggs in nests or burrows. It is significant that mammals that have no permanent habitations, such as monkeys and deer, are nearly free of fleas, although they seldom lack lice.

Most fleas are not so closely limited as are lice to particular kinds of hosts. A few, especially the so-called human flea, *Pulex irritans,* probably primarily a pig parasite, are remarkably indiscriminate. Most fleas of rodents and other small mammals limit themselves largely to particular species or genera of preferred hosts, e.g., prairie dogs, ground squirrels, voles, rabbits, shrews, etc., but usually can survive and even reproduce on one or more "secondary" hosts. When these are closely associated with a "preferred" host, random collections may give false ideas about true host-flea relationships. Accidental, temporary transfers are frequent and may be important, e.g., when a cat brings home a flea from a plague-sick mouse that it was mauling, or a vulture transports one for miles after acquiring it from a dead prairie dog. Most fleas when hungry suck blood wherever they can find it, and it is for this reason that rodent fleas are of importance to man as transmitters of rodent diseases. Some species require much less provocation to bite man than others—*X. cheopis* does this with little reluctance, whereas *L. segnis* rarely ever does it.

The fleas found in human houses are mainly of three types: (1) so-called human fleas (*Pulex*) (Fig. 197); (2) cat and dog fleas (*Ctenocephalides*) (Fig. 197A); and (3) rat and mouse fleas. These are discussed further on pp. 655–657.

Life cycle. The eggs of fleas are oval, pearly-white objects of relatively large size, sometimes one-third the length of the parent flea. Except for the chiggers they are laid several at a time, the total number laid by one female being 300 to 400 or more. The eggs hatch in from 2 to 3 days to over 2 weeks, depending on temperature. Eggs of *N. fasciatus* will hatch at temperatures as low as 41°F., but those of

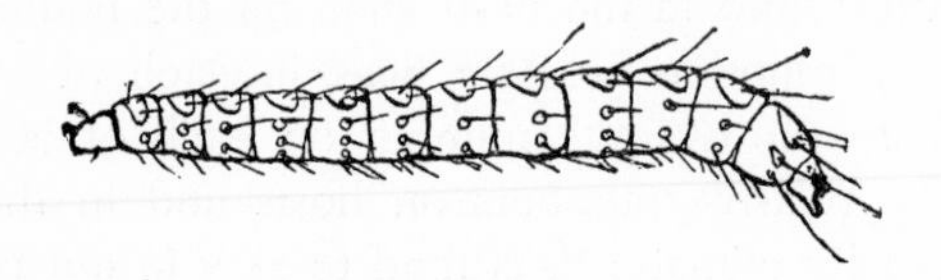

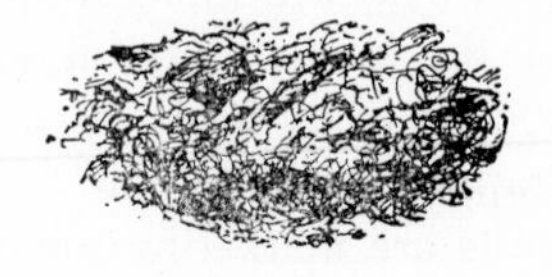

Fig. 199. Developmental stages of fleas. *Left,* larva of *Xenopsylla cheopis.* (After Bacot and Ridewood, *Parasitology,* 7, 1914.) *Right,* dust-covered cocoon of *Pulex irritans.* ×12.

Pulex and *Xenopsylla cheopis* require higher temperatures. The most favorable conditions for the development of most species are temperatures between 65° and 80° and a humidity of 70% or more. The higher the temperature the greater the humidity required. In the nests and holes where fleas breed, however, the "microclimate" may be favorable even when conditions in the open are highly unfavorable.

The larvae (Fig. 199) are tiny cylindrical maggotlike creatures with neither legs nor eyes, although they are very sensitive to light. They have small brown heads and whitish bodies composed of 13 visible segments and a hidden terminal one, all provided with rather sparse bristly hairs to aid in crawling. The last segment is terminated by a pair of tiny hooks.

The larvae avoid light and feed upon what bits of organic matter they can find, such as mouse "pills," crumbs, hairs, epidermal scales, their own shed skins, and feces of the adult fleas. This contains semidigested blood on which the larvae thrive; some, e.g., *Nosopsyllus fasciatus,* are said to require it, but others, e.g., *Xenopsylla cheopis,* can develop on flour alone. The duration of the larval stage varies with climatic conditions and food, and to some degree also with the species, and may be from a week to several months.

When ready to pupate the larvae spin little silken cocoons that are viscid, so that particles of sand, dust, or lint readily adhere to them (Fig. 199). Under favorable conditions the adults emerge in from 1 to 2 weeks to a month, but at low temperatures or in dry weather the insects may remain dormant in their cocoons for several months and thus tide over unfavorable seasons such as northern winters or dry seasons in the tropics. The complete life cycle takes a minimum of 3 weeks in the tropics and from 4 to 6 weeks in temperate climates.

The adult fleas do not become sexually mature or copulate for some days after they escape from the cocoon. Soon after copulation egg laying begins, but no breeding takes place without blood meals.

The length of life of adult fleas depends largely on food supply, temperature, and humidity. At low temperatures (60°F.) well-fed fleas may live for several years. In the absence of a host they have less endurance than ticks or even bugs, but may survive for several months, even for a year or more at low temperatures. The optimum climatic conditions and normal length of life probably vary a great deal with different species. Most of the *Ceratophyllus* group and also *Pulex irritans* are fleas of temperate or cold climates, whereas the species of *Xenopsylla* are characteristic of hot climates.

Unlike most bloodsucking insects, fleas usually feed at frequent intervals, generally at least once a day and sometimes oftener. Fleas frequently feed even when the digestive tract is already well filled, and may pass practically unaltered blood in their feces to be utilized, second-hand, by the larvae.

The susceptibility of different individuals to flea bites is variable. Some people are apparently entirely immune to flea bites and feel no pain from them. The senior writer on his first visit to California, warned to expect trouble from fleas (*Pulex irritans*), was pleasantly surprised to feel no discomfort other than tickling as the fleas promenaded, while a roommate spent many sleepless hours in pursuit of the wily fleas and in violent massaging of painful wounds. Some people suffer immediate reactions, others delayed reactions in which the irritation develops from 1 to 3 days later. In allergic persons a course of events like that described for sandfly bites on p. 529 takes place. Immunity can be artificially induced in most people by injection of antigens made from pulverized fleas (see p. 530).

Fleas and Disease

Fleas are the prime transmitters, both among their reservoir hosts and to man, of two human diseases of outstanding importance—plague and murine or endemic typhus. They are also important transmitters of tularemia among rodents, play a part in the spread of myxomatosis among rabbits, and serve as intermediate hosts of certain tapeworms.

FLEAS AND PLAGUE

Plague (see comprehensive review by Pollitzer, 1954), like smallpox and cholera, once ranked among the great scourges of mankind. In a fourteenth century epidemic in Europe it is estimated that one-fourth of the population of that continent, some 25 million people, died of the disease, and superstition and unreasoning terror led to horrible persecution and torture. Only early in the twentieth century was the role of commensal rats as reservoir hosts in cities clearly established. In 1914 an Indian Plague Commission (see Bacot and Martin, 1914) discovered the usual method of transmission—the bacilli that cause the disease, *Pasteurella pestis* (see p. 226), when acquired from an infected host by a flea that is a good vector, such as the Oriental rat flea, *Xenopsylla cheopis,* multiply so prodigiously that the flea's digestive tract becomes solidly blocked by masses of the organisms (Fig. 200); such "blocked" fleas are unable to ingest more blood and in attempting to do so they regurgitate plague germs into their victims.

For many years this domestic epidemiology involving rat epizootics followed by human epidemics, brought on by the biting of man by infected rat fleas orphaned when plague killed their natural hosts, was the only form of the disease known, just as was the case with the epidemic form of yellow fever. Gradually during the last fifty years it has become clear that in many parts of the world plague exists in a sylvatic form among wild rodents, constituting a vast, often unrecognized, reservoir of infection comparable with the jungle form of yellow fever. Among some of these sylvatic hosts, e.g., prairie dogs, ground squirrels, and marmots, as among domestic rats, plague may cause extensive die-offs, just as yellow fever does among monkeys, but some wild hosts in this country, particularly voles (*Microtus*), and in the Old World, gerbils, are susceptible but seldom die. These are probably of prime importance in keeping plague alive.

As with yellow fever, the classic, epidemic disease is fast dying out; in 1956 less than 700 cases of human plague were reported in the entire world—a contrast to the half-million cases a year formerly occurring in India alone. Sylvatic plague causes extremely few human cases since the probability of transmission in the wilds is very small, but there is a constant threat that the smoldering sylvatic disease may burst into epidemic flame by transfer to rats and thence eventually to man. Few fleas are strictly host-specific, so where wild rodents and domestic rats coexist in suburbs, recreation areas, dumps, etc., transfer of fleas can occur, and with them the disease, after which *X. cheopis* takes over and trouble lies ahead. Having sylvatic plague in your suburbs is like having a Typhoid Mary in your kitchen or a rattlesnake on your golf course. It is an unpleasant threat.

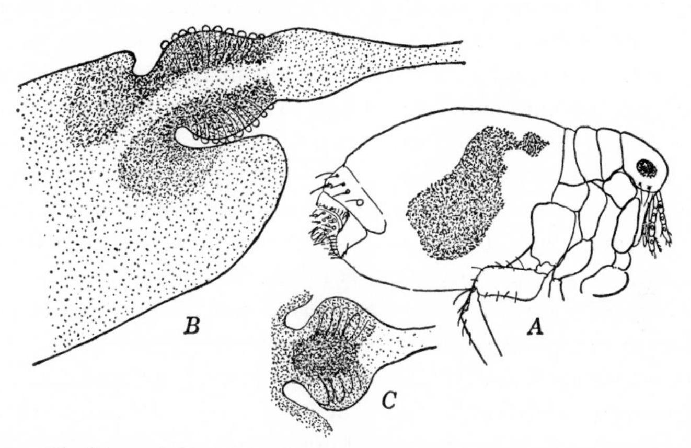

Fig. 200. Blocking of fleas by plague bacilli. *A*, position of proventriculus and stomach of flea as they appear when full of blood. *B*, proventriculus and fore part of stomach partially blocked (heavily shaded area). *C*, proventriculus completely blocked. (*B* and *C* adapted from Esky and Haas, *Publ. Health Bull.* 254, 1940.)

Plague is a highly fatal disease. Infection by bites of blocked fleas causes the "bubonic" form; after a short incubation period there is a sudden high fever, mental disturbance, and severe prostration, with characteristic large swollen lymph glands called buboes. Direct passage from man to man may occur via the respiratory tract, causing even more fatal pneumonic plague (mortality around 90%), but this is limited to small local outbreaks following a bubonic case.

The importance of particular kinds of fleas as vectors of plague depends on many factors, among which are their abundance; the percentage that become infected when feeding on infected hosts; the frequency with which these become blocked; the time required for blocking; how long unblocked fleas live before they either lose their infections (up to 15 days in cool weather); and how long they survive after blocking—factors which for any particular flea may be greatly influenced by climatic conditions. All these factors affect their efficiency as vectors among their preferred hosts. As far as man is concerned, another important factor is the freedom with which the species will transfer its attention from its preferred hosts, before or after these hosts die off, either directly to man, or indirectly via rats (see Pollitzer, 1954, Kartman et al., 1958).

Human epidemics in urban areas are always preceded by epizootics among commensal rats, the disease being transmitted to man by hungry blocked fleas orphaned by the dying off of the rats. Members of the genus *Xenopsylla,* widely distributed in warm parts of the world, are the best transmitters. *X. cheopis* is an excellent vector on all counts, needs less provocation to feed on man than do most rodent fleas, and is the dominant flea on rats in all warm parts of the world. In the United States this flea has established itself locally even in places like Denver, St. Louis, and Washington, but it is prevalent throughout the southeastern states, Texas, and California seaports. *A. braziliensis* is also a good rat-to-man transmitter in parts of Africa, but *X. astia* in India and Ceylon is relatively inefficient.

Just as yellow fever may be transported and temporarily established far outside the usual range of *Aedes aegypti,* so may plague be dispersed by infected rats, or by fleas abandoned by them—such fleas might be able to start an outbreak even up to 2 weeks. However, unless *X. cheopis* is present in abundance, such outbreaks soon die out since the other rat fleas, mainly *Nosopsyllus fasciatus* and *Leptopsylla segnis,* are relatively inefficient vectors to man. *N. fasciatus* tends to stay in rat nests, and *L. segnis* is very reluctant to bite man.

Sylvatic plague occurs in many parts of the world, and involves very many vertebrate hosts and many different fleas. Macchiavello (1954) listed 344 mammals and 2 birds that are at least potential harborers of plague, the majority of them rodents. Although blocked fleas undoubtedly play an important part in sylvatic plague as well as in human plague, other mechanisms are also involved, e.g., nipping fleas, infection of bites by flea feces, and mechanical transmission by infected mouth parts when fleas and rodents are present "en masse." Therefore even fleas like *Malaraeus telchinum* on *Microtus,* although they do not become blocked, may be concerned in maintaining sylvatic plague. In addition, rodent cannibalism is a factor.

Plague is generally believed to have been introduced into San Francisco by seafaring rats in 1900, whence it spread to ground squirrels and other rodents, but it now seems possible that it could have had a sylvatic origin. In 1920 it entered ports on the Gulf Coast of Texas, but died out without establishing itself in sylvatic form. Sylvatic plague is now present throughout the western half of North

America, and has been found in most species of rodents and numerous fleas. The importance of either a vertebrate or a flea varies not only with the species, but varies under different environmental conditions. Much more needs to be learned about the ecology of plague, as Kartman et al. (1958) have ably demonstrated. As far as human infection is concerned but less so conservation in nature, the large rodents that die off extensively, e.g., prairie dogs, ground squirrels, and marmots, are important. Ranchers aid in distribution of plague by capturing sick prairie dogs and setting them free on their own ranches to transmit the infection to new colonies and thus exterminate them. This is reminiscent of a friend who unwittingly kept a blackwidow spider as a pet in her house because it killed flies. One enterprising boy in Colorado was found catching sick prairie dogs and selling them to unsuspecting tourists as souvenirs!

Other important known foci of sylvatic plague exist in Central Asia and southeast Russia; in Manchuria, where 60,000 people hunting marmots for their skins fell victim in 1910–1911; in South Africa where the semidomestic multimammate mouse comes into houses and passes the disease on to domestic rats harboring *X. braziliensis;* and in several parts of South America. Now that epidemic plague is dying out in India, it is becoming apparent that even there the basic reservoir of the infection is in wild rodents in foothills not under full cultivation.

Control of plague will be considered along with that of endemic typhus on p. 660.

Fleas and Typhus

In 1926 Maxcy pointed out that the epidemiology of sporadic cases of typhus occurring in southern United States was not suggestive of louse transmission but had the earmarks of a disease transmitted only occasionally to man from a rodent reservoir. Subsequent work by Dyer and his colleagues of the U. S. Public Health Service showed that this was true and that this endemic type of typhus (see p. 216) existed among rats and was transmitted principally by fleas. It is associated with rat-infested localities such as granaries, markets, and restaurants, and is now known to occur in nearly all temperate and tropical parts of the world, where it is variously known as endemic, murine, or shop typhus. It exists in wild rodents and other small mammals as well as in house rats. A number of species of fleas, as well as occasionally rat lice, bloodsucking mites, and even ticks are capable of transmission among reservoir hosts, but *Xenopsylla cheopis* is believed to be the only species of importance in maintaining the disease among domestic rats and man, at least in the United States. However, after initial infection by a rodent flea, the disease may be spread from person to person by lice (see p. 627). In fleas the rickettsia (*Rickettsia typhi*) multiply in the cells lining the gut and are voided with the feces, where they may remain viable for weeks, just as *R. prowazekii* does in lice. Unlike lice, fleas are not killed by the infection and may remain infected for life and for some time after death, but there is no transovarial transmission.

The disease in man is much milder than the epidemic louse-borne type and also differs from it in causing scrotal swellings in guinea pigs. In rats and mice it is usually symptomless; since the rats do not die from it their fleas are not forced by hunger to seek human blood. Besides, there is no mechanism like the blocking in plague to enable the fleas to transmit the disease by their bites; only their feces and crushed bodies are infective. Since this is much less likely to lead to human infection, human cases are sporadic, although local outbreaks may follow rat-eliminating campaigns. Occasionally cases may result from contamination of food by urine of infected rats, or from handling rats themselves.

In the United States typhus is endemic principally in the southern states. Up to 1930 few cases were reported, but after that time the disease became commoner year by year at an accelerating rate, until in 1945 over 5000 cases were reported, the largest numbers being in Texas and Georgia. In a survey in one Texas county 94% of urban and 80% of rural buildings harbored rats which by blood test were positive for typhus; and of 213 pools of the three commonest rat fleas (*Xenopsylla cheopis, Nosopsyllus fasciatus,* and *Leptopsylla segnis*), 53 harbored typhus. Some of these were found on kittens, puppies, and opossums.

The disease has repeatedly been introduced into more northern cities along with *X. cheopis,* but in such places this flea remains very localized or fails to establish itself permanently. Otherwise the gradual encroachment of this infection from the south and of plague from the west would be an unpleasant outlook for midwestern cities. Since 1945 the number of typhus cases has steadily decreased until in 1952 and 1953 there were only a few over 200 cases each year, a 96% reduction. By 1959, the number dropped even more. This has been due in part, no doubt, to vigorous anti-typhus campaigns including DDT dusting of rat runs (see p. 660), poisoning of rats, and rat-proofing of buildings, but a similar decrease took place in Puerto Rico *without* human effort. Possibly when a high percentage of local rats develop antibodies against typhus, sucking immune blood interferes with development of the rickettsias in fleas.

Fleas and Other Diseases

Fleas easily become infected with tularemia (see p. 226), and may remain carriers for a month, but they are probably not important transmitters. They also transmit Whitmore's bacillus, causing a glanders-like disease in rodents and man in southeastern Asia. Tovar in 1947 reported that fleas, as well as bedbugs and ticks, could transmit brucellosis (see p. 595). Fleas serve as intermediate hosts for the non-pathogenic rat trypanosome, *Trypanosoma lewisi*, but *Schizotrypanum cruzi*, although it develops in such different arthropods as ticks and bugs, undergoes rapid degeneration in fleas. This is also true of relapsing-fever spirochetes. Fleas were long suspected of transmitting infantile and canine leishmaniasis (see p. 116) in the Mediterranean region; the occurrence of a natural *Leptomonas* in fleas provided considerable circumstantial evidence against them, but they were finally acquitted.

Fleas serve as intermediate hosts for certain tapeworms, among them *Dipylidium caninum* (see p. 371) of dogs and cats, which is occasional in children, and *Hymenolepis diminuta* and *H. nana* of rats, mice, and man (see pp. 367 and 370). The eggs are ingested by larvae, but the cysticercoids finish their development in the adult fleas.

Notes on Important Species of Fleas

Human Flea. The only species of flea that is known to be a frequent parasite of man, with the exception of the jigger, is the appropriately named *Pulex irritans* (Fig. 197*A*, *B*), though in many places man is annoyed more by certain other species that are primarily parasites of his domestic animals. This flea is also a pest of pigs, and it has been found on many other animals. It probably originated in Europe, whence it has been introduced to all parts of the world, but is relatively rare in the tropics. This flea is the species that has made northern California as famous for its fleas as New Jersey is for its mosquitoes. A cool, humid summer climate combined with a mild, wet winter is ideal for this pest. It is not an important plague transmitter because it seldom becomes "blocked," but it may cause some infections by its feces.

Dog and cat fleas. Next in frequence to the human flea as parasites of man are the dog and cat fleas, *Ctenocephalides canis* and *C. felis*. In the southeastern United States where the flea scourge competes very well with that of California, these are the species usually met with. During the moist, hot summers they become exceedingly abundant. Although primarily parasites of dogs and cats they willingly include man in their bill of fare when the preferred hosts are not readily available. A case once came to the senior writer's attention in which the residents of a house were unmercifully bitten after disposing of a badly infested dog and two or three cats, although there had previously been no annoyance. Cat and dog fleas readily go from one of these hosts to the other. These fleas can easily be distinguishable from any other common species by the presence of *two* well-developed ctenidia (Fig. 198*A*), each with numerous teeth.

The eggs are usually laid loosely in the fur, whence they fall out when the host shakes himself or is settling himself for a nap. They develop in the dust and dirt of kennels, woodsheds, house floors, or other places where infested animals are likely to go. Patton and Cragg found the inside of a hat in which a kitten had slept overnight so full of flea eggs that it looked as if it had had sugar sprinkled in it.

Dog and cat fleas, from their habits, are the species most frequently implicated in the transmission of the dog tapeworm (*Dipylidium*) to children. These species are even less frequently concerned with plague transmission than is *Pulex irritans.*

Rodent fleas. The various species of fleas that infest rats, ground squirrels, and sometimes other rodents are only accidental parasites of man. If it were not for their importance in the spread of plague and typhus they would need no special consideration. For identification of the commoner species refer to the key on p. 646.

The combless fleas of the genus *Xenopsylla* (Fig. 194), which are of the greatest importance in transmission of epidemic plague among domestic rats and man, and of murine typhus to man, are primarily residents of Asia and Africa, but *X. cheopis* (see p. 649) has accompanied rats to all important seaports in the world and thence inland in many places. This species attacks man more readily than most rodent fleas when deprived of its normal hosts. *X. cheopis* and *X. astia* (Figs. 194, 197, 201) are both common in India; *cheopis,* however, has accompanied its rat hosts to all warm parts of the world and is the most important transmitter to man of both plague and murine typhus. It attacks man more readily than most rodent species when deprived of its normal hosts. In temperate climates this flea is most abundant in summer. *X. braziliensis* is the most prevalent species on domestic rats in Africa, and *X. philoxera* in South Africa, and *X. vexabilis* and *hawaiiensis* are common on field rats.

Nosopsyllus fasciatus (Fig. 198*C*) is the most prevalent flea on domestic rats in temperate climates, and *Leptopsylla segnis* (Fig. 197*B*) is very common on rats and mice in southern United States. These two fleas are the prevalent species on rats in winter and spring. All three of the commonest rat fleas, as well as the domestic rats (*Rattus* spp.) and the house mouse, are imported from the Old World. The sticktight flea (see following section) is another potential plague vector found on rats.

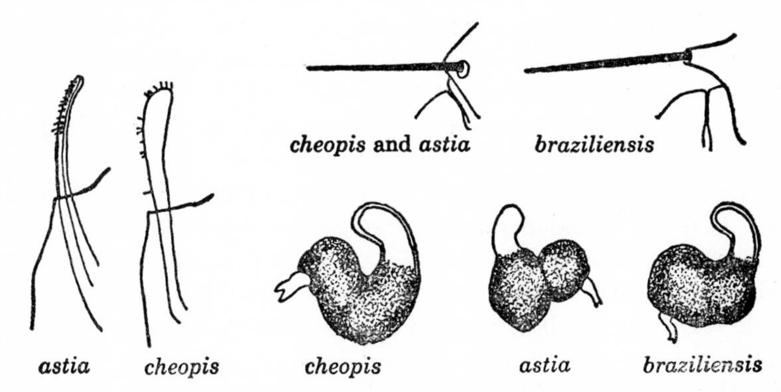

Fig. 201. Differential characters between common species of *Xenopsylla*. *Left,* 9th sternites; *top right,* antepygidial bristles; *lower right,* spermathecae. (From *Trop. Diseases Bull.,* 20, 1923.)

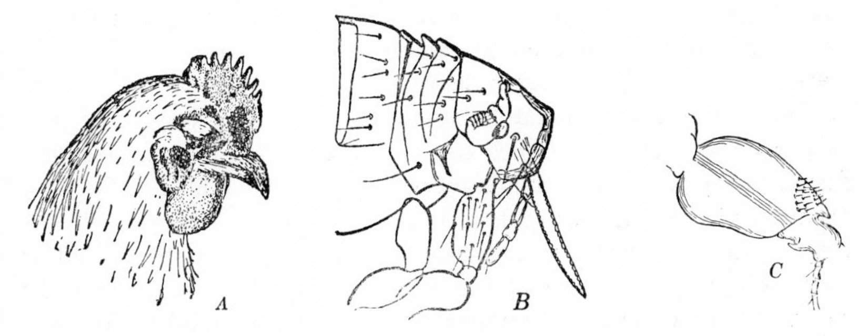

Fig. 202. *Echidnophaga gallinacea,* sticktight flea. *A,* clusters on head of a chicken; *B,* head of flea; *C,* inner aspect of hind coxa. (*A,* adapted from Bishopp; *B* and *C* from Fox, *Fleas of Eastern North America.*)

Bird fleas. There are three important fleas of poultry. One, the small sticktight flea, *Echidnophaga gallinacea* (Fig. 202), belonging to the family Pulicidae (see p. 645), has a world-wide distribution in warm countries, and is the most important flea pest of poultry in the United States. The female of this flea sticks tight to its host in one place; it appears to be gregarious, collecting in clusters on the heads of poultry, in the ears of mammals, etc. It burrows to some extent, causing considerable irritation. Besides poultry, it attacks dogs, cats, rabbits, rats, and other mammals and birds, and not infrequently children. The eggs are scattered on hen-house floors, dog yards, etc. Since this flea is susceptible to plague and attacks both birds and rodents, it may carry infection from wild to domestic rodents, and when attached to such birds as vultures, hawks, or pheasants may be a means of carrying it to distant places.

The other two important bird fleas are *Ceratophyllus gallinae* in Europe and eastern United States and *C. niger* in western United States. These have habits like those of rodent fleas, living primarily in the nests.

Jiggers. The jigger, chigoe, or sand flea, *Tunga penetrans* (Fig. 203), is the pest which inspired the sailor's oath, "I'll be jiggered." Originally found in tropical America, it was introduced to west Africa with some ballast sand in 1872. It spread rapidly over nearly the whole of Africa but has failed to establish itself in Europe or India. The jigger is a small flea, only about 1 mm. in length, of the family Tungidae (see key p. 646). The males and virgin females are similar to other fleas in habits, except that they attack a wider range of hosts. Man and pigs seem to be the principal hosts of this pest, but cats, dogs, and rats are also attacked.

The jigger breeds especially in regions with sandy soil shaded by heavy underbrush or in the earth floors of native houses. After emergence the fleas lie in waiting in debris on the ground and attack mainly the feet of animals or human beings which come their way. The particular importance of this flea lies in the fact that the impregnated females have the aggravating habit of becoming imbedded in the skin, especially in such tender spots as under the toenails. They retain the eggs in the abdomen, causing it to swell into a great round ball as large as a pea. The head and legs appear as inconspicuous appendages (Fig. 203*B*). Only the two posterior segments of the abdomen do not enlarge; these project from the swollen inflamed skin in which the flea is embedded. The eggs, up to a hundred in number, mature in about a week and are then expelled by the female through the protruding end of the abdomen. Sometimes the entire gravid female is expelled.

The eggs fall to the ground, where they undergo development in the orthodox siphonapteran manner; the larvae (Fig. 203*C*) feed on organic debris, grow to maturity, pupate in a cocoon, and finally emerge as adults after 17 days or more. Faust and Maxwell, however, have observed an unusual case in which the thriving larvae were found in various stages of development in skin scrapings from the inguinal and pubic regions of a man who had been attacked in these parts by adult jiggers while sitting on some bales of sisal imported from Yucatan.

The wounds made by the burrowing female in the skin become much inflamed and very painful, especially if the distended abdomen of a flea is crushed and the eggs released in the wound. After the eggs are laid and the flea expelled, secondary infections may cause loss of toes or limbs, or death from tetanus or gangrene.

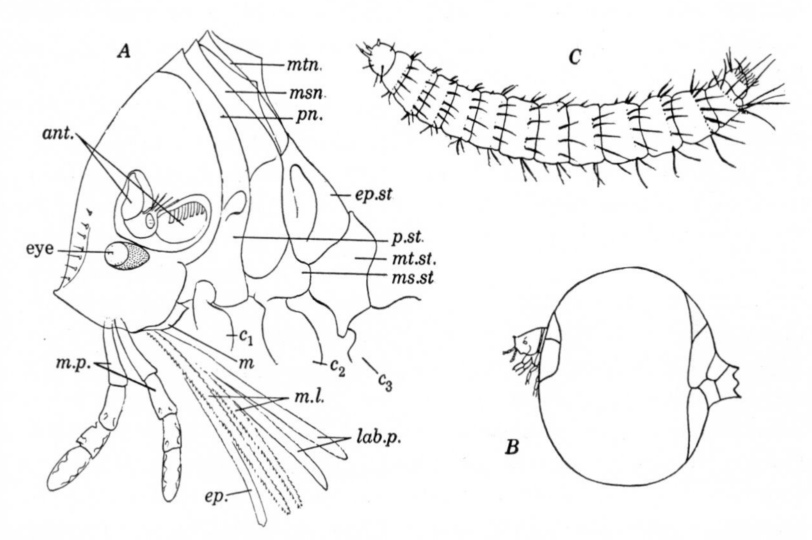

Fig. 203. *Tunga penetrans,* jigger. *A,* head and thorax (after Hubbard, *Fleas of Western North America,* 1947). *B,* gravid ♀ showing legs disintegrating (after Patton and Evans, *Insects, Ticks, Mites and Venomous Animals of Medical and Veterinary Importance, I, Medical,* Grubb). *C,* larva (after Faust and Maxwell, *Arch. Dermatol. Syphilol.,* 22, 1930). Abbreviations approximately as in Figs. 194 and 197A, except *m,* base of maxilla, and *m.l.,* maxillary laciniae.

Although usually only a few jiggers are present at a time, there are cases in which hundreds infest a person at once, literally honeycombing the skin and making the feet or other parts of the body so sore that the victim is rendered a complete invalid.

The best treatment is to enlarge the opening around the flea with a sterile needle and remove the parasite entire, then carefully dress the wound and protect it until healed.

Houses, yards, etc., in jigger regions should be kept as free as possible of dust, dirt, and debris. In Central America Quiros recommended a prohibition against driving hogs affected with jiggers through the streets, along with regulations for treating affected hogs where they are raised. Spraying premises with a 2.5% DDT suspension quickly eliminates jiggers. Dusting the feet and socks with DDT is the best means of personal protection.

Control of Fleas and Flea-Borne Diseases

Except in a few instances where minor resistance has developed, fleas are very susceptible to DDT and other modern insecticides. In homes sprays applied to floors and lower walls, and light mists to rugs, furniture, etc., are recommended, and in animal quarters or yards, sprays, dusts or silica aerogels. On cats or dogs use powders containing pyrethrum plus a synergist, or rotenone; on older dogs and on pigs, goats, etc., the hydrocarbons or phosphorus compounds can be used. Cats lick off DDT and are sickened by it, but lightly dusting the sleeping places soon controls the situation. When pyrethrum is used the fleas begin to leave the animal, and though temporarily paralyzed, sometimes recover, so the animals should be treated outdoors, or where the fleas can be swept up and destroyed.

Control of plague and endemic typhus. In controlling plague in the past reliance has been placed almost entirely on rat control, but this had its dangers, since the fleas, if not killed with the rats by fumigation with HCN, methyl bromide, or burning sulfur, were likely to turn their attention to man. This was especially true with typhus, since this disease does not kill the rats, and so only occasional human cases occur until the fleas are orphaned by poisoning, trapping, or driving off of their rat hosts. Often a rat campaign precipitates a number of human typhus cases. There are situations where immediate destruction of both rats and fleas by fumigation is indicated.

Since the advent of DDT, it has been possible to control these diseases without this danger from orphaned fleas. The procedure is first to dust harborage places and rat runs heavily with DDT or Chlordane and to blow it into the openings of burrows. Rats pick up enough DDT on their feet, tails, and fur to kill fleas not only on their own bodies but also in their nests. Depending on how thoroughly the dusting is done, the flea index on rats is reduced 80 to 99% almost at once, and 75 to 80% control can be expected after 4 to 6 months. In practice it takes about 2 to 3 lb. of 10% dust per premise. Good control of fleas is also obtained by applying 5% DDT emulsion to runways, harborage places, etc., in residences or other places where the dust is undesirable. Two or three days after dusting, rats may be destroyed by poisoning with some of the new rodenticides, such as Warfarin, "1080" (sodium fluoroacetate), or Antu. These methods are rapidly making epidemic plague a thing of the past. Additional protection, e.g., for military forces, is possible by the use of vaccines, which are available for both typhus and plague.

Both DDT dusting and rat poisoning are temporary measures. Permanent good comes only from ratproofing buildings where this is possible, as it is in American cities. The cost is low compared with loss from rats, even without considering their relation to disease, since rats usually outnumber the human population three to two, and the average cost of upkeep per rat for food and goods destroyed is estimated at one-half cent per day. That amounts to $1.80 per rat per year, or a total of over $250,000 for a city of 100,000 population.

Control of sylvatic plague is feasible only as an emergency measure, e.g., near villages, suburbs, or garbage dumps where domestic and wild rodents mingle, or when epizootics occur near military camps, in national parks, etc. Dusting burrows of larger rodents and using insecticidal "bait boxes" for small ones (Kartman, 1958) makes flea control an effective do-it-yourself proposition for these wild rodents. In the long run, however, control of domestic rats is the only permanent safeguard.

REFERENCES

Bacot, A. W., and Martin, J. C. 1914. Observations on the mechanism of the transmission of plague by fleas. *J. Hyg.*, 14. *Plague Suppl.* III, 423–439.

Burroughs, A. L. 1947. Sylvatic plague studies. *J. Hyg.*, 45: 371–396.

Dyer, R. E. 1942. Endemic typhus in the United States. *Proc. 6th Pacific Sci., Congr., V:* 731.

Esky, C. R., and Haas, V. H. 1940. Plague in the western part of the United States. *Public Health Bull.*, 254.

Ewing, H. E., and Fox, L. 1943. The fleas of North America. *U. S. Dept. Agr. Misc. Publ.*, 500. Washington.

Fox, I. 1956. Murine typhus fever and rat ectoparasites in Puerto Rico. *Am. J. Trop. Med. Hyg.*, 5: 893–900.

Garnham, P. C. C. 1949. Distribution of wild-rodent plague. *Bull. World Health Organization*, 2: 271–278.

Hirst, L. F. 1953. *The Conquest of Plague. A Study of the Evolution of Epidemiology.* Oxford University Press, London.

Hopkins, G. H. E., and Rothschild, M. 1953, 1956. *An illustrated catalogue of the Rothschild collection of fleas (Siphonaptera) in the British Museum. I. Tungidae and Pulicidae, II. (Five Small Families).* British Museum, London.

Hubbard, C. A. 1947. *Fleas of Western North America.* Iowa State College Press, Ames, Iowa.

Jellison, W. L., and Good, N. E. 1942. Index to literature of Siphonaptera of North America. *Natl. Insts. Health Bull.* 178.

Jellison, W. L., Locker, B., and Bacon, R. 1953. A synopsis of North American fleas, north of Mexico, and notice of a supplementary index. *J. Parasitol.*, 39: 610–618.

Kartman, L., et al. 1958. New knowledge of the ecology of sylvatic plague. In *Animal Disease and Human Health, Ann. N. Y. Acad. Sci.*, 70, Art. 3, 668–711.

Link, V. B. 1955. A history of plague in the United States of America. *Public Health Monogr.* 26. Government Printing Office, Washington.

Macchiavello, A. 1954. Reservoirs and vectors of plague. *J. Trop. Med. Hyg.*, 57: 3–8, 45–48, 65–69, 87–94, 116–121, 191–197, 220–224, 238–243, 275–279, 294–298.

Mohr, C. O. 1951. Entomological background of the distribution of murine typhus and murine plague in the United States. *Am. J. Trop. Med.*, 31: 355–372.

Morlan, H. B., et al. 1952. Domestic rats, rat ectoparasites and typhus control, Parts I–III. Public Health Serv. Publ., 209, *Public Health Monogr.*, 5: 1–37.

Pollitzer, R. 1954. Plague. *World Health Organization Monogr.* No. 22, Geneva.

Public Health Monograph. 1952. Plague in Colorado and Texas. *Public Health Monogr.*, 6.

Simmons, S. W., and Haynes, W. J. 1948. Fleas and disease. *Proc. 4th Intern. Congr. Trop. Med. Malaria,* 2: 1678–1688.

U. S. Department of Agriculture. 1955. Fleas. How to control them. *Leaflet* 392.

U. S. Public Health Service, 1949. Rat-borne disease prevention and control, XIV + 293 pp. *Communicable Disease Center,* Atlanta, Ga.

Wayson, N. E. 1947. Plague—field surveys in western United States during 10 years. (1936–1945). *Public Health Repts.*, 62: 780–791.

Chapter 28

DIPTERA
I. BLOODSUCKING AND DISEASE-CARRYING FLIES OTHER THAN MOSQUITOES

Importance. Although outstripped by mites and ticks as vectors of veterinary diseases, the Diptera are as important as all other arthropods combined as far as human disease is concerned, for in this order are included the normal transmitters of malaria, African trypanosomiasis, leishmaniasis, Oroya fever, various types of filariasis, yellow fever, dengue, sandfly fever, and numerous encephalitis and other virus infections. Without their dipteran transmitters these diseases would probably entirely disappear. Other diseases, such as anthrax, yaws, and pinkeye, are mechanically conveyed by flies, and the housefly and other nonbiting flies are involved in the mechanical transmission of all kinds of filth diseases. Besides all this, the Diptera include nearly all the insects which infect wounds, skin, nasal passages, or digestive tract in the larval (maggot) stage.

General structure of Diptera. To understand the relations of these numerous important insects and their classification, we must make a brief survey of the characteristics and classification of the order Diptera. The whole order can usually be distinguished readily from other insects by the fact that there is only one pair of membranous wings, the second pair of wings being represented only by an insignificant pair of knobbed rodlike appendages known as *halteres* (see Fig. 226). Even in those forms in which the wings are secondarily absent the halteres are usually present. These are vibrated with great speed during flight (300 times per second) and act as balancers. In the Cyclorrhapha (houseflies, etc.) there is a membranous lobe at the base of the wing on the posterior side, called the *alula;* behind this, in many Cyclorrhapha are one or two additional lobes called squamae or calypters (Fig. 246).

The legs consist of the usual segments (see p. 521), generally with long coxae. The tarsi are usually terminated by two claws with pad-like "pulvilli" under them, and often a third appendage, the "empodium," between them, either bristlelike or resembling a third pulvillus. The head is joined to the thorax by a very narrow, flexible neck. The thorax has its three component parts fused. The abdomen usually

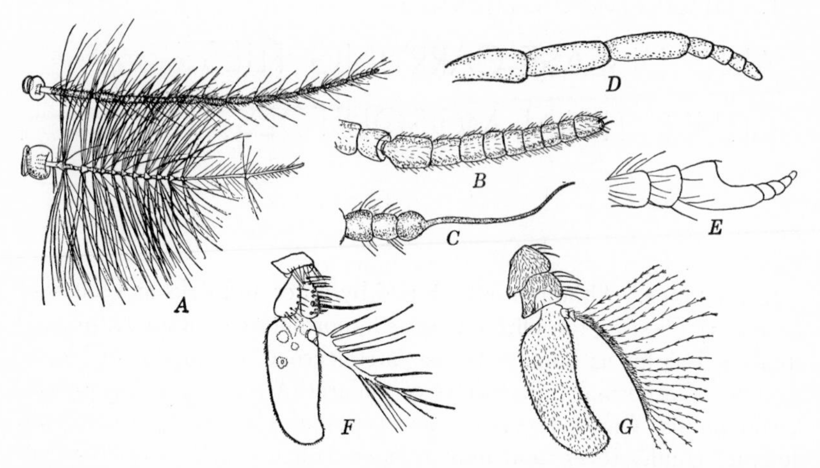

Fig. 204. Types of antennae of Diptera: *A*, ♀ and ♂ mosquito (*Culex*); *B*, *Simulium*, blackfly; *C*, *Chrysopila*, a leptid; *D*, *Chrysops dissimilis*; *E*, *Tabanus*; *F*, *Musca*, housefly; *G*, *Glossina*, tsetse fly.

consists of four to nine visible segments and is terminated by egg-laying organs in the female and by the clasping or copulatory organs in the male.

The antennae and also the palpi are of considerable use in classification; the extent of the variations in the antennae may be gathered from Fig. 204. In the more generalized families, e.g., the Nematocera, the antennae consist of many segments which, except the basal two, are similar in form (Fig. 204*A*, *B*) and often bear whorls of hairs which in the males give a plumose effect, e.g., in male mosquitoes. In more specialized families the terminal segments tend to coalesce more or less (Fig. 204*C*, *D*, *E*), as in the tabanids and other Brachycera, or the segments beyond the third are reduced to a simple or plumose bristle or arista which appears as an appendage of the enlarged third segment (Fig. 204*F*, *G*), as in nearly all the Cyclorrhapha.

The mouth parts are profoundly modified in accordance with the habits of the flies. In the botflies, in which the adults live only long enough to reproduce their kind, the mouth parts and even the mouth are much degenerated; in the nonbloodsucking forms, such as the common housefly, the mouth parts are developed as a fleshy proboscis which is used for lapping up dissolved foods; in the bloodsuckers, which are the forms that particularly interest us here, the mouth parts are developed into an efficient sucking and piercing apparatus. In the suborder Orthorrhapha (mosquitoes, sandflies, blackflies, and horse-flies) the labium acts as a sheath for the other parts which are fitted for piercing and sucking; in the Cyclorrhapha (*Stomoxys*, hornfly, and the tsetse flies), the labium itself forms a piercing organ, and the epipharynx and hypopharynx form a sucking tube, the mandibles and maxillae being absent. The evolution of this type of proboscis is discussed and illustrated on p. 692.

Life histories. All Diptera have a complete metamorphosis, but beyond that fact the life history varies within wide limits. Most flies lay eggs, which require from minutes to days to hatch; others, e.g., the sheep nasal fly, *Oestrus ovis*, deposit newly hatched larvae, and still others, e.g., the tsetse flies and the Hippoboscidae, do not deposit their offspring until they have undergone their whole larval development and are ready to pupate.

The larvae of Diptera may be simple maggots without distinct heads or appendages and capable of only limited squirming movements, e.g., the screwworms, or they may be active creatures with well-developed heads, e.g., the larvae of mosquitoes and midges. Many are aquatic, many others terrestrial. Usually the eggs are laid where the larvae will find conditions suitable for their development, and the flies often show such highly developed instincts in this respect that it is hard not to credit them with actual forethought.

The pupae of Diptera also vary widely. In the suborder Orthor-rhapha the pupa is protected only by its own hardened cuticle, or, as in the blackflies, a spun cocoon, and is often capable of considerable activity; from this "obtected" type of pupa (Fig. 205*B*) the adult insect emerges through a transverse or T-shaped slit, usually near the anterior end. In the suborder, Cyclorrhapha the pupa retains the hardened skin of the larva as a protective covering or "puparium" and is usually capable of very slight movement; from this "coarctate" type of puparium (Fig. 205*A*) the adult escapes by a circular opening made by pushing off the anterior end of the puparium, hence the name Cyclorrhapha, meaning "circular slit." Except in a few families (group Aschiza) this is done by means of a hernialike outgrowth on the front of the head. This outgrowth, called the "ptilinium" (Fig.

205*C*), shrinks after the fly has emerged, but leaves a permanent crescent-shaped mark on the head known as the "frontal lunule" (Fig. 205*D*), which embraces the bases of the antennae.

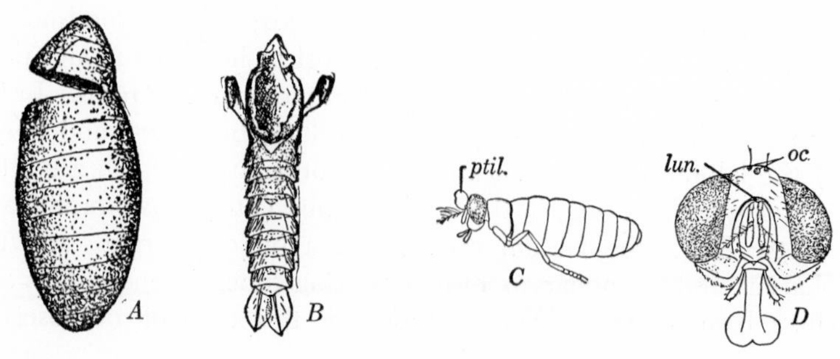

Fig. 205. *A* and *B*, types of pupal cases, showing manner of emergence of adults; *A*, empty puparium of blowfly, typical coarctate pupa of Cyclorrhapha; *B*, empty case of mosquito, typical obtected pupa of Orthorrhapha; *C*, newly emerged fly showing bladder-like ptilinium (*ptil.*) by which the end of the pupal case is pushed off; *D*, face of fly showing crescent-shaped scar of lunule (*lun.*) left by drying up of ptilinium; *oc.*, ocelli. (After Alcock, *Entomology for Medical Officers*, 1920.)

The classification of the Diptera into major divisions and the characteristics which distinguish the forms that are of medical or veterinary importance are shown in the following key:

Suborder **Orthorrhapha.** Pupa not encased in old larval skin and often active; adults emerge through straight or T-shaped dorsal slit (Fig. 205*B*); larvae usually with well-developed or somewhat reduced head; wing venation usually fairly simple.

1*a*. Antennae of at least 6 similar joints, and usually long (Fig. 204*A*, *B*); larvae with well-developed head; series **Nematocera** 2

1*b*. Antennae short, with 3 segments, of which the third may show some annulation (Fig. 204*E*); series **Brachycera** 5

2*a*. Antennae much longer than head, with distinct whorls of hairs at joints, often plumose in males 3

2*b*. Antennae not much longer than head, with no long hairs (Fig. 204*B*); body stout and "humped"; wings broad, with only anterior veins well developed (Fig. 212); (blackflies) **Simuliidae**

3*a*. Body clothed with scales; scales on wing veins and fringe (Fig. 238); (mosquitoes) **Culicidae**

3*b*. Body and wings without scales 4

4*a*. Wings with 9 to 11 long, parallel veins, with no cross-veins except at base (Fig. 206); body hairy and mothlike **Psychodidae**

4*b*. Wing veins not all nearly parallel; body not very hairy; small and short; broad wings fold flat over abdomen, often mottled, anterior veins thickened (Fig. 210) **Heleidae (=Ceratopogonidae)**

5*a*. Third antennal segment annulated, never with a bristle (Fig. 204*D*,

E); a forked vein near tip of wing (Fig. 214*A*); mouth parts fitted for piercing; wings held apart when at rest; large robust flies **Tabanidae**

5*b*. Antennae short, with a bristle or style (Fig. 204*C*); abdomen long and tapering; snipe flies **Rhagionidae**

Suborder **Cyclorrhapha.** Pupa encased in old larval skin (puparium) and usually inactive; adults escape through circular split at one end (Fig. 205*A*); larvae maggotlike without distinct head; wing venation highly modified.

1*a*. No frontal lunule (Fig. 205*D*) series **Aschiza**
No bloodsuckers, but includes a few myiasis-producers, *Eristalis* of family Syrphidae, and *Aphiochaeta* of family Phoridae.

1*b*. Frontal lunule present; series **Schizophora** 2

2*a*. Opposite legs of each pair close together; abdomen distinctly segmented (Fig. 258); larval development outside uterus except in tsetse flies 4

2*b*. Opposite legs of each pair widely separated (Fig. 224); body flattened, leathery; wings sometimes absent, if present, with stronger

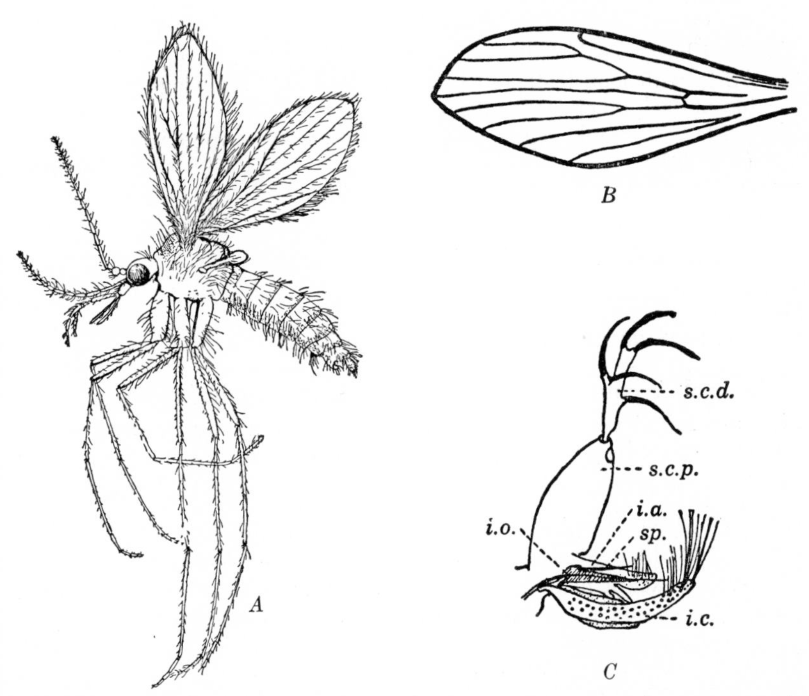

Fig. 206. *Phlebotomus argentipes:* A, adult ♀; *B*, venation of wing; *C*, male genitalia; *s.c.d.*, distal segment of superior clasper; *s.c.p.*, proximal segment of superior clasper; *i.a.*, intermediate appendage, with spine (*sp.*); *i.o.*, intromittent organ; *i.c.*, inferior clasper. (Adapted from Sinton, *Trans. Far Eastern Assoc. Trop. Med.*, 7th Congr., 1928.)

veins crowded along costal margin; larval development in uterus (pupiparous); palpi forming sheath for proboscis; parasitic; **Hippoboscidae** 3

3*a*. Wingless, abdomen unsegmented; pupae glued to wool; on sheep (sheep tick, Fig. 225*C*) **Melophagus**

3*b*. Winged (Fig. 225*A*); pupae develop off host; many genera and species on birds; on horses, camels, dogs **Hippobosca**

4*a*. Squamae (or calypteres) rudimentary or absent 5

4*b*. Squamae well developed 6

5*a*. Mouth parts vestigial; horse botflies **Gasterophilidae**

5*b*. Mouth parts normal; wing without subcostal or anal veins, and no closed cells (Fig. 224) (eye flies) **Chloropidae**

6*a*. Proboscis well developed; wings about as in Fig. 218*A* (Muscoidea) 7

6*b*. Mouth parts vestigial; no palpi; hairy, beelike flies (botflies); (for key to genera, see p. 771) **Oestridae**

7*a*. Proboscis fleshy, fitted for lapping or at most for scratching (Fig. 216, *A* and *B*): arista with lateral hairs, or, rarely, bare; houseflies, fleshflies, blowflies, screwworm flies. See key on p. 768.

7*b*. Proboscis fitted for piercing 8

8*a*. Palpi short, not forming sheath for proboscis (Fig. 217, *C*) (stableflies) ***Stomoxys***

8*b*. Palpi long, capable of sheathing proboscis (Fig. 217, *D*) 9

9*a*. Arista with long feathered hairs on upper surface; wings folded on top of each other (tsetse flies) ***Glossina***

9*b*. Arista with simple hairs on upper surface; wings diverge when folded (hornfly) ***Siphona***

SUBORDER ORTHORRHAPHA

Phlebotomus or Sandflies (Psychodidae)

General Account. Phlebotomus flies, commonly known as sandflies, are minute, hairy midges found in nearly all warm and tropical climates of the world. They belong to the family Psychodidae, most of which resemble tiny moths on account of their very hairy bodies and mothlike pose. They are easily recognized by the characteristic wing venation, with a series of more or less parallel veins and with no cross-veins except near the base (Fig. 206*B*.) Phlebotomus flies hold their wings erect over the body when resting, and are less mothlike than other psychodids.

Theodor (1948) divided the Psychodidae into four subfamilies, of which only the Phlebotominae are bloodsuckers. All of these are usually included in a single genus *Phlebotomus;* but Theodor recognized four genera, and at least one has been added since. *Phlebotomus,* however, contains nearly all the species that feed on man and other mammals and all that are implicated in the transmission of disease except some South America species, e.g., *Lutzomyia longipalpis.*

Morphology. The sandflies (Fig. 206A) are small, dull-colored insects, usually yellowish or buff, slender in build, with hairy body, very long and lanky legs, and narrow hairy-veined wings. They have long slender antennae, long maxillary palpi, and a proboscis longer than the head. The proboscis consists of a fleshy labium containing daggerlike mandibles and maxillae, both with sawlike teeth at the tips; a bladelike hypopharynx containing the salivary duct; and a flat daggerlike labrum-epipharynx which is provided with sensory hairs and spines and is probably not used as a piercing organ. The piercing organs project beyond the tip of the labium when at rest, and the labium does not bow back when the other parts are in action, as it does in mosquitoes. The male genitalia are highly complex and of great value in identification. Other characters useful in classification are details of the palpi, form of the spermatheca, and nature of the pharyngeal teeth. Identification, especially of females, is difficult.

Habits. The females are bloodsuckers whose bites cause irritation quite out of proportion to their size, due to allergic reactions. As described by Theodor (1935) there is little or no immediate reaction to first bites except a needlelike prick, but a week or two later a persisting papule appears. With development of sensitization after repeated exposures the reaction to the bites follows the course described on p. 529. Some species refeed only after laying a batch of eggs and others may refeed several times without relation to egg laying, but some, e.g., *P. argentipes,* commonly die after depositing one batch of 50 or 60 eggs; preferred hosts vary; some species feed almost exclusively on lizards, some on rodents or rabbits; the majority are zoophilic, feeding on a variety of mammals such as dogs, horses, and cattle. Some of these bite man readily, others rarely do; and some species vary in their habits in different localities, as do mosquitoes. Only *P. papatasii* (see p. 672) seems to *favor* human blood; it is the commonest domestic species in the Old World. *P. argentipes* in India feeds by preference on cattle, but attacks man when cattle are not readily available. The preferences are an important factor in determining the importance of various species as disease vectors, though by no means the only one.

The males are not bloodsuckers, and some appear not to feed at all. Sandflies are very short-lived and seldom survive more than a fortnight in the laboratory. Most are nocturnal; in some places they seem to forage for only an hour or so after sundown. During the day they hide in cool, damp places, and are much more abundant in masonry or thick mud-walled houses than in bamboo-and-plaster huts. They favor dark rooms on the sheltered side of the ground floor of houses away from breezes, coming forth on still, warm nights to seek food. Ceiling fans usually keep them away, and they seldom rise to a second floor. Sandbags used for raising the sides of tents supply good hiding places for *P. papatasii.* When disturbed on walls they usually fly only a few inches, appearing to hop rather than fly. Their

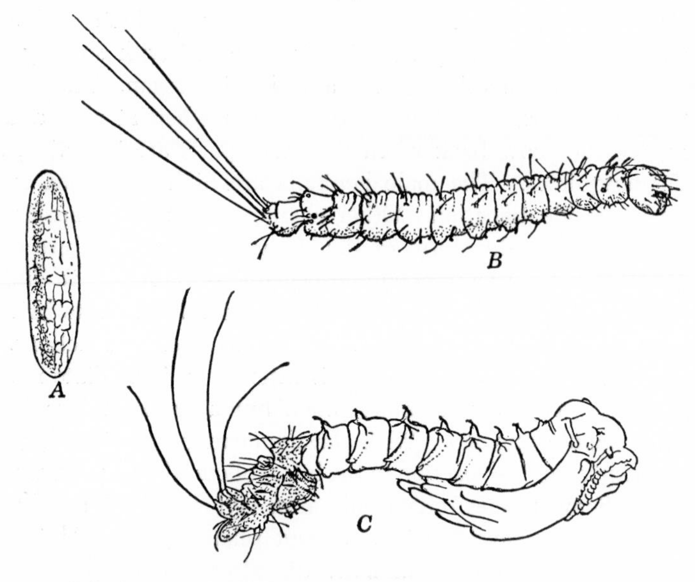

Fig. 207. Life history of *Phlebotomus: A*, egg; *B*, larva; *C*, pupa. (Adapted from Patton and Evans, *Insects, Ticks, Mites and Venomous Animals of Medical and Veterinary Importance, 1, Medical,* Grubb.)

flight range is very limited, rarely over 50 yards, but they sometimes steal rides in public conveyances. Usually their breeding places are within a few hundred feet of their feeding places.

Life history (Fig. 207). The species of *Phlebotomus* whose life histories have been studied lay their eggs in crevices in rocks, masonry, or crumbling buildings, between boards in privies or cesspools, in rubbish, or in the burrows of animals or in deep soil cracks, where there are moderate temperatures, high humidity, and darkness. The females usually deposit their eggs within a few days after emergence and a blood meal, but only in places where the microclimatic relative humidity is practically 100%. The eggs vary from 40 to 60 in number; some individuals lay second batches of similar size after a subsequent feed. When deposited, the eggs are literally shot out by the female to a distance several times the length of the abdomen. The eggs are viscid and adhere to the surfaces with which they come in contact; it would seem that the peculiar method of ejecting the eggs is a protective adaptation, facilitating their deposition in the farthest reach of a

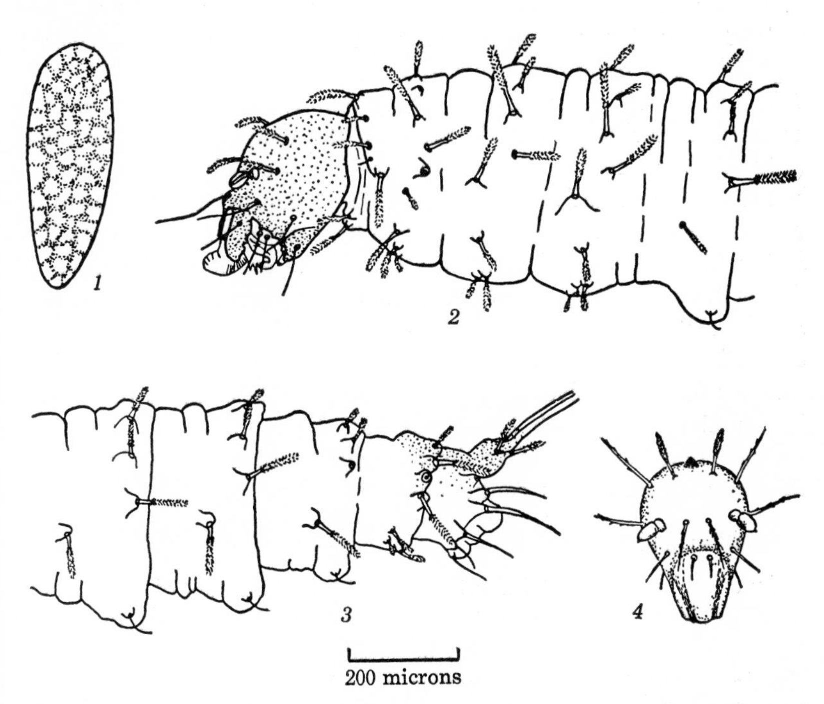

Fig. 208. Egg and larva of *Phlebotomus anthophorus: 1,* egg; *2,* anterior end of larva; *3,* posterior end of larva (only bases of caudal bristles shown); *4,* front view of head of *P. minutus.* (*1–3* after Addis, *J. Parasitol.,* 31, 1945.)

crevice where even the tiny insect itself could not penetrate. The eggs are elongate and of a dark, shiny brown color, with fine surface markings which vary in different species (Fig. 207A).

The incubation of the common Old World *Phlebotomus papatasii* requires 6 to 9 days under favorable conditions, but the eggs are very susceptible to external conditions and die quickly if exposed to sunlight or if not kept damp. *P. argentipes* eggs may hatch in 4 days. The larvae (Fig. 207*B*) are tiny caterpillarlike creatures with a relatively large head and heavy jaws (Fig. 208) and with two pairs of long bristles on the last segment of the abdomen which are held erect and spread out fanwise; in the newly hatched larvae there is only one pair of bristles. The body is provided with numerous toothed spines which give it a rough appearance (Fig. 208). These spines differ in different species and, together with the relative length of the caudal bristles, form good identification marks. The larva of *P. papatasii* when full grown is less than 5 mm. long and it is therefore not so large as an ordinary rice grain. The larvae feed on decaying vegetable matter, fecal particles, and other organic debris. For all species a high degree of humidity is required.

The full development of the larvae requires 2 weeks to 2 months or more, depending almost entirely on the temperature. The pupa (Fig. 207*C*) is characterized by a very rough cuticle over the thorax but can be identified best by the last larval skin, which adheres to its posterior end. It is colored much like its surroundings and looks like a tiny bit of amorphous matter. The pupae are less susceptible to drying than the larvae. In warm weather the adult insect emerges after 6 to 10 days, but this is much delayed by low temperatures. The entire life cycle seldom occupies less than 2 months and in cool weather may take several months. In temperate climates the winter is passed in the fourth larval stage so that, after pupation in the spring, great numbers of adults may appear quite suddenly, but in the tropics breeding is continuous.

Sandflies and Disease

Sandflies are of great importance as the transmitters of the various types of leishmaniasis, of a filtrable virus disease called three-day fever, or more commonly sandfly or papatasi fever, and of Oroya fever. (See review by Adler and Theodor, 1957.)

Sandfly fever. This relatively mild virus disease (see p. 229) in many respects resembles dengue and may be confused with influenza. It comes on suddenly with fever, headache, pain in the eyes, stiffness of neck and back, and rheumatic pains. As in dengue, a reduction in white blood cells (leucopenia) is a prominent feature. It is often followed by a prolonged period of malaise and depression. Sabin et al. (1944) isolated two different strains of the virus, from Sicily and the Middle East, respectively.

It was experimentally shown by Doerr in 1908 to be transmitted by *Phlebotomus papatasii*. The insects become infective about 6 or 7 days after feeding on a patient in the first or second day of the fever. Since sandflies are so short-lived and frequently suck blood only once, and since the disease appears as soon as the adults emerge in May in the Mediterranean area, evidently having passed the winter in the larvae, transovarial transmission had long been suspected. This was demonstrated in 1922 (Whittingham and Rook, 1923) and later confirmed in Russia, though Sabin et al. (1944), however, got negative results. Since only man is known to be susceptible, transovarial transmission seems necessary for survival of the virus from one season to another. Thus far *P. papatasii* is the only known transmitter throughout the definite range of the disease from Italy to central India, but fevers which *may* be sandfly fever are reported from the western Mediterranean, tropical Africa, China, and Colombia, and other sandfly vectors are suspected.

P. papatasii is of medium size, reaching about 2.5 mm. in length. It is pale yellowish gray with a dull red-brown stripe down the middle of the thorax and a spot of the same color at either side. The disease is of little consequence to local populations since immunity is acquired in early childhood, but it may cause serious epidemics in military forces or other nonimmune newcomers.

Leishmaniasis. The baffling problem of transmission of the various forms of leishmaniasis (see pp. 111–129) was not finally solved until 1941, but there is no longer any doubt that sandflies are the principal transmitters of all forms of the disease, and as Adler and Theodor (1957) pointed out, there is no evidence of endemic leishmaniasis in the absence of sandflies. When the parasites are ingested by the flies in the leishmania form they develop into flagellated forms and multiply in the stomach, gradually moving forward to the pharynx and even the proboscis. Although massive growths occur, Adler and Theodor believe that the mechanism of transmission is not by blocking as in fleas (see p. 116), but by deposit in the wound when the flagellates descend to the tip of the proboscis.

The species involved in the different forms of leishmaniasis are not the same. *P. sergenti* and *P. papatasii* are the principal vectors of cutaneous leishmaniasis or Oriental sore in the Old World. Quite likely *P. caucasicus,* which inhabits burrows of gerbils and ground squirrels in Turkestan, but has become domestic in Iran, is the primitive transmitter among the rodent reservoir hosts, whence it became adapted to *P. sergenti* in western Asia and to *P. papatasii* in the Mediterranean area, and through these to dogs and man (see p. 125). Visceral leishmaniasis or kala-azar, on the other hand, is transmitted almost exclusively by members of the *major* group—*argentipes* in India (where man is the only known host), by *chinensis* in China, and by *major, perniciosus,* and *longicuspis* in the Mediterranean area where dogs are the reservoirs. In Sudan the vectors are unknown.

In tropical America the vectors of leishmaniasis are not known with certainty, but there is considerable evidence against *P. intermedius,* and some against *migonei, pessoai,* and others, as transmitters of cutaneous forms of the disease in Brazil, whereas the most probable vector of "chiclero ulcer" in Yucatan is *P. cruciatus* (Biagi, 1953). In Brazil the strongest evidence is against *P. longipalpis* as a vector of kala-azar (Deane, 1956). Very little is known about the bionomics of any of these species.

Oroya fever. Oroya fever or Carrion's disease is an acute febrile disease caused by a very minute organism, *Bartonella bacilliformis* (see p. 218), occurring in valleys on the slopes of the Andes in Peru, Chile, Ecuador, and Bolivia. In 1938 it appeared in epidemic form in the province of Nariño in Colombia, causing 1800 deaths in 9 months in a population of perhaps 200,000. The acute stage of the disease is characterized by high fever, severe anemia, aches, and albuminuria, and is often fatal. In more chronic cases it is followed by an eruption of nodules called verruga peruviana. In mild cases the eruption may be the only manifestation, and it is considered a good sign when it appears. Some cases seem to be symptomless.

The probable relation of sandflies to this disease was pointed out by Townsend in 1913. In the Peruvian Andes the disease is limited to a comparatively small zone and is contracted exclusively at night. *Phlebotomus verrucarum* is the principal, probably the sole, transmitter. It is a strictly nocturnal species which enters houses readily and bites man freely. In Colombia the transmitter is believed to be *P. colombianus,* possibly merely a variety of *P. verrucarum;* the females of these two species are indistinguishable. The *Bartonella* organisms have been found in the distal part of the proboscis and are presumably deposited in wounds as are leishmanias.

P. verrucarum is found in the deep-cut canyons of the west slope of the Peruvian Andes at elevation between 2800 and 8000 ft. Within this zone it would be extremely dangerous to be caught at night, for no ordinary screening would afford adequate protection. At lower elevations the valley is too arid for sandflies to breed, and at higher elevations the cold nights inhibit their activity. In the verruga zone there is a favorable combination of cool summers, warm sunny winters, moderate rainfall, and mild nights—a zone of perpetual spring.

Control

Because of their short flight range, which is seldom over 75 to 100 yards, and their tendency to stay close to the ground, sandflies are very easily controlled locally by destruction of *adults* by application of insecticidal sprays, particularly DDT, to the inside of buildings, and in some places also to rock walls or other outdoor breeding places.

The long life cycle and sedentary habits result in a very slow comeback of a depleted population; often one annual preseason house-spraying is enough. Remarkable results in sandfly control have been reported by Hertig and Fairchild (1948) in Peru and by Hertig (1949) in Greece and Italy. Residual spraying for malaria control eliminates sandflies too.

When venturing beyond protected areas after sundown it is necessary to use protective clothing and repellents, of which dimethyl phthalate is probably best.

Biting Midges or No-see-ums—Heleidae (=Ceratopogonidae)

Morphology. The tiny flies belonging to this family (Fig. 209) go by various names: midges, gnats, punkies, or sandflies, and, in the west, "no-see-ums" because of their minute size, which is seldom over 1 to 2 mm. in length. They can usually be distinguished from allied insects by the peculiar venation of the wings, the first two veins being very heavy, whereas the others are more or less indistinct. In most but not all species the wings are mottled with white spots on a grayish background (Fig. 209*A*). The proboscis is short and essentially similar in structure to that of *Simulium* (Fig. 212*B*); one marvels at the irritation which can be inflicted by such a small insect with such a small organ. The maxillary palpi have 5 segments, the

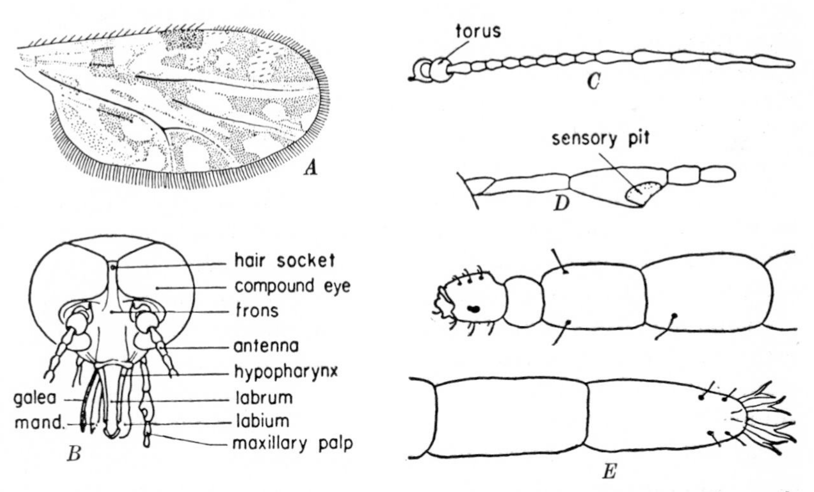

Fig. 209. *Culicoides.* *A,* wing of *C. furens; B,* head; *C,* antenna; *D,* maxillary palp; *E* anterior and posterior ends of larva of *C. furens.* (*A–D,* after Foote and Pratt, *Publ. Hlth. Mon.* 18, 1954; *E,* after Painter, Med. Dept. United Fruit Co., 15th Rept., 1927.)

third enlarged with a characteristically placed sensory pit (Fig. 209*D*). The antennae are long, with 13 segments in the flagellum; in the males they are more plumose with the last 3 or 4 segments lengthened; in females they are filamentous with the last 5 segments differing in character from the others. In most species the wings are mottled in characteristic patterns (Figs. 209*A* and 211). The male genitalia are useful in identification of species.

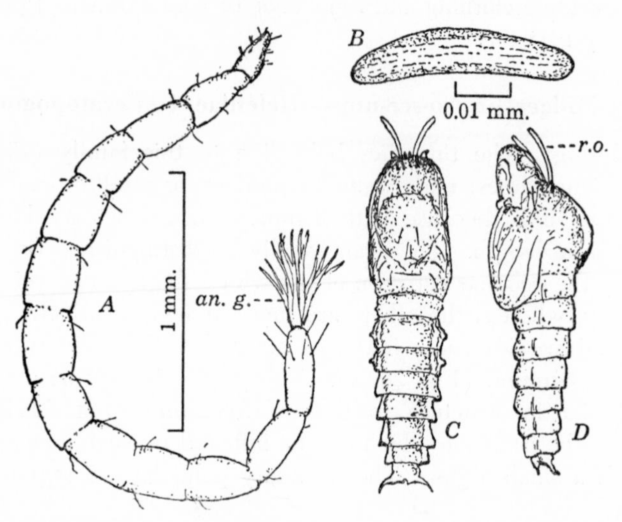

Fig. 210. Immature stages of *Culicoides*. *A*, larva (the anal gills (*an.g.*) are retractile, there is no proleg, and the anal segment has three pairs of hairs); *B*, egg, much enlarged; *C* and *D*, dorsal and lateral views of a pupa; note slender respiratory organs (*r.o.*) on thorax. (After Dove, Hall and Hull, *Ann. Ent. Soc. Amer.*, 25, 1932.)

The family includes many genera (see Johannsen, 1943), but the majority of species that attack man and animals belong to the genus *Culicoides*. There are, however, some annoying pests in the genera *Leptoconops*, *Helea* (=*Ceratopogon*), *Forcipomyia*, and a few others.

Habits and life cycle. Only the females are bloodsuckers. They become active at dusk, but if disturbed many of them will bite in the shade, even on bright days. Both sexes are attracted by lights. They are much more active fliers than *Phlebotomus*, and are said to go as far as half a mile in search of a host, but only when the air is still. Most of the species appear to prefer cattle, camels, or other animals as food, but some species readily attack man.

The eggs of *Culicoides*, several hundreds in number, are deposited in gelatinous masses like miniature masses of frog eggs and are usually moored to some object under water in swamps, ponds, or tree holes; some species breed in rotting vegetation. Relatively cool shaded places are preferred by most species, but the pestiferous *C. furens* of Central America and the West Indies breeds only where exposed to sun; it is a salt-water breeder that thrives in open marshes regularly washed by tides.

After a few days slender larvae (Fig. 210*A*) hatch. They burrow in decaying vegetation or mud either in or out of water, according to the species. When swimming, their movements suggest giant spirochetes. *Culicoides* larvae, unlike the related chironomid larvae, do not have pseudopods on the first or last segments of the abdomen. At the posterior end there are gill-like structures that can be protruded; the larvae do not need air as do mosquito larvae. The food of most species consists of microscopic plant and animal life or organic debris. The pupa (Fig. 210*C*, *D*) rather resembles that of a mosquito, except that the abdomen is kept extended instead of curled under and the pupa hangs from the surface in a vertical position, breathing through a pair of trumpetlike tubes as do mosquitoes. Both larvae and pupae are hard to find, and the presence of a breeding place is more frequently discovered by finding the floating pupal cases from which the adults have emerged.

Dove et al. (1932) believe that the larvae of *C. furens* (which they called *dovei*), a pestiferous salt marsh species and filarial vector (see below), may live for 6 months to a year. *C. milnei* of Africa, however, has an egg-to-adult cycle of 25 to 28 days. In contrast, *C. tristriatulus* of Alaska has one brood a year, the larvae living through the winter buried in soil.

Annoyance. The bites of *Culicoides* produce nettlelike pricks which are sometimes followed by burning sensations and intolerable itching; in many sensitized individuals the bites last longer and are more painful than mosquito bites, but eventually most people develop some degree of immunity to them. The ability of these insects to go through ordinary mosquito screens makes them particularly obnoxious. Some species are so minute as to be individually overlooked, but they may be so abundant as to give a gray tinge to the skin and to cause extreme irritation. Dove et al. (1932) think the flies are attracted by the heat emanating from warm bodies, but the senior writer believes they are attracted largely by animal odors, since he has been attacked by great swarms upon opening a rabbit carcass. The flies usually attack exposed parts of the body but will bite through thin clothing. Salt-marsh breeders (*C. furens* and others) are intolerable pests along the southern Atlantic seaboard of the United States, so much so as to have retarded development of certain areas. They seriously interfere with romance on moonlit beaches in the West Indies and Central America.

Midges as Disease Carriers

Culicoides serve as intermediate hosts for three filarial worms of man, *Dipetalonema perstans, D. streptocerca,* and *Mansonella ozzardi,* and at least two filariae of animals, *Onchocerca reticulata* of horses and *O. gibsoni* of cattle. In addition, *Culicoides* have been accused of transmitting several virus diseases: blue tongue of sheep and "horse

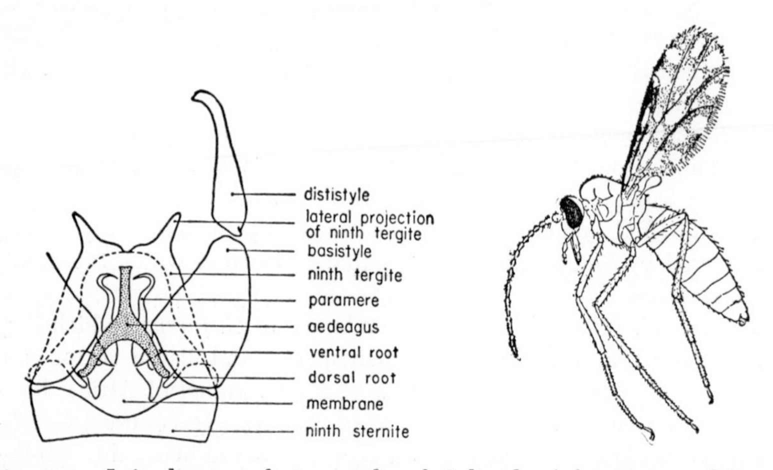

Fig. 211. *Left,* diagram of ♂ genitalia of *Culicoides* (after Foote and Pratt, *Publ. Health Mon.* 18, 1954). *Right, Culicoides milnei,* vector of *Dipetalonema perstans,* ×20 (after Sharp, *Trans. Roy. Soc. Trop. Med.,* 19, 1928).

sickness" in South Africa, and fowlpox of poultry in Japan. Two land-breeding species, *Forcipomyia utae* and *F. townsendi,* were believed by Townsend to act as transmitters of uta, a form of leishmaniasis in Peru, but this work needs confirmation.

***Culicoides* and filarial infections.** Sharp (1928) proved two species of *Culicoides, C. milnei* (=*austeni*) (Fig. 211) and *C. grahami,* to be intermediate hosts of the filarial worm, *Dipetalonema* (=*Acanthocheilonema*) *perstans* (see p. 488) in British Cameroons, where over 90% of the natives are infected in some areas. This was later confirmed by Hopkins and Nicholas (1952), although work by Chardome and Peel (1949) in Belgian Congo threw doubt on it (see p. 488). The latter workers showed that *C. grahami* was an intermediate host for

Dipetalonema streptocerca in that country. *D. perstans* can also develop in *C. inornatipennis,* a rainforest fly.

C. milnei is a night-biting fly which is abundant in villages in or near rain forests, readily entering houses to bite in darkness, but often leaving again during the night. A blood meal is necessary before eggs are laid. Breeding is common in banana stumps in early stages of decay, which tends to increase the density of the insects near villages; however, they also breed in damp soil containing decaying vegetation. The flies appear to disperse not over 400 yards from their breeding places in clearings, but their dispersal may be different in the forest canopy. *C. grahami,* a smaller species (1 mm. long as compared with 2 mm. for *milnei*), bites in early morning and evening, or all day on dark, rainy days. It also breeds freely in banana stumps and frequents villages, but it is commoner than *milnei* in forest reserves, grasslands, and transition areas. It is a less efficient vector than *milnei,* but may nevertheless be important under some circumstances.

Buckley (1934), working on the island of St. Vincent, W. I., showed that a *Culicoides, C. furens* (Fig. 209), serves as an intermediate host for another filarial worm, *Mansonella ozzardi* (see p. 488). The development in *Culicoides* is similar to that of *Dipetalonema perstans. C. furens,* a notorious biter, is easily recognized by the speckled appearance of the mesonotum. It is widely distributed on the Atlantic and Gulf coasts from Massachusetts to Texas, Mexico, West Indies, and Brazil, and is the most troublesome species in Florida.

Steward in 1933 showed that *Onchocerca reticulata,* the cause of fistula of the withers or head (poll-evil) in horses, is transmitted by *C. nubeculosis* in England. This species breeds in liquid manure and other foul stagnant water. Buckley (1938) showed that in Malaya *Onchocerca gibsoni* of cattle can undergo development to the infective stage in several species of *Culicoides;* although less than 0.5% of the flies were successful hosts, the flies were so numerous that Buckley calculated that a cow would be bitten by at least one infective fly every day. Circumstantial evidence points to species of *Culicoides* as vectors of *O. gibsoni* in Australia and South Africa. Attempts to get development of *O. volvulus* of man and *O. gutturosa* of cattle have failed.

Control

The control of *Culicoides* is often difficult, and the methods must depend on knowledge of the breeding places of the species to be controlled. Marsh-breeding species, such as *C. furens,* may sometimes be reduced 90% by draining, ditching, or diking, and also, for periods of

3 to 6 months, by dieldrin applied as sprays of emulsion or in granular form (see p. 534). DDT, however, is quickly detoxicated by mud so is less effective than it is against mosquitoes. In England a BHC emulsion, applied at 100 mg. per square foot, gave good results in small breeding areas, especially when rain caused deep penetration; preseason spraying was suggested. Spraying or painting DDT solutions on screens and on resting places near habitations is helpful. Thermal aerosols are useful for temporary emergency treatments. Repellents (see p. 536) are effective for shorter periods than for most insects. In Africa it seems that considerable protection against the *C. austeni* and *grahami* might be obtained by elimination or treatment of banana stumps near villages.

Blackflies or Buffalo Gnats (Simuliidae)

The blackflies, as annoyers of domestic animals and man, are among the most important of insect pests, since they often appear in overwhelming hordes. They kill animals and even children not only by their irritation and loss of blood but also by filling up their bronchial tubes and literally suffocating them. In bad years *Simulium colombaschense*, which breeds in the Danube River, kills over 10,000 domestic animals in the Balkans, and in the United States and Canada the animal losses may run to many hundreds in a year. Blackflies have a world-wide distribution, being found from the arctic to the tropics, and to the perpetual snow on mountains, wherever there is running water. Unlike most small biting flies, they are strictly diurnal. All the blackflies were once included in the single genus *Simulium*, but they are now split into half a dozen genera, largely on the basis of wing venation (Smart, 1945). The genus *Simulium* still contains the great majority of the species, including all the important disease transmitters.

Morphology. Unlike the usual slender, midgelike flies with long legs and long antennae in the group Nematocerca, the blackflies are small, robust, hump-backed creatures with short legs, broad wings, and short, eleven-segmented antennae without whorls of hairs at the joints (Fig. 212*A*). The proboscis in the female is short but heavy and powerful; in the males, which are not bloodsuckers, it is poorly developed. The mouth parts (Fig. 212*B*), consisting of toothed daggerlike mandibles and maxillae, and also a hypopharynx and labrum-epipharynx, resemble in general those of *Phlebotomus*.

Most of the northern species are black, whence their name, but some of the species are reddish brown or yellowish, and they may be variously striped and marked. The wings are either clear or of a grayish or yellowish color, with the few heavy veins near the anterior margin often distinctively colored. Some of the species are not more than 1 mm. in length, and the largest of them scarcely exceed 4 mm.

Life history and habits. Unlike the mosquitoes and midges, blackflies breed exclusively in running water, but the species vary greatly in the kinds of streams they select and in the speed of current they prefer. Some breed in lazy rivers flowing 2 or 3 miles an hour, some

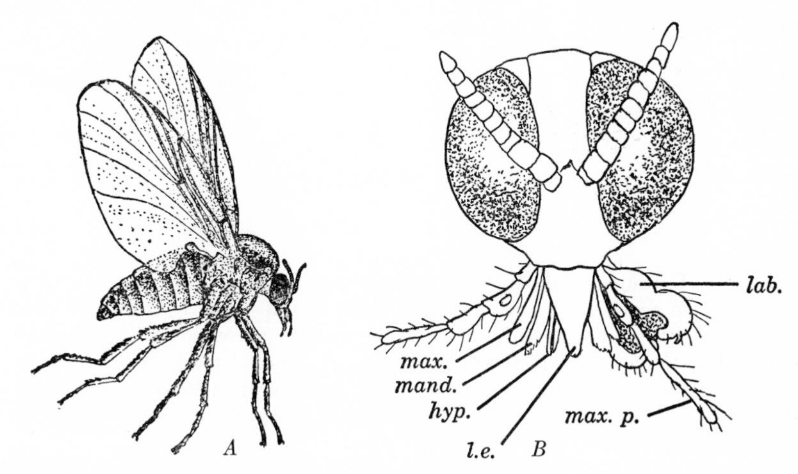

Fig. 212. *A*, Blackfly (buffalo gnat), *Cnephia pecuarum*, ×7 (after Riley). *B*, Mouth parts of *Simulium: hyp.*, hypopharynx; *lab.*, labium; *l.e.*, labrum-epipharynx; *mand.*, mandible; *max.*, maxilla; *max.p.*, maxillary palpus (after Alcock, *Entomology for Medical Officers*, 1920).

in bubbling brooks or trickling rivulets; S. *pictipes* in North America breeds only in or at the foot of falls. The eggs, provided with a viscid coat, are usually deposited on vegetation, stones, etc., licked by water or partly submerged, but S. *arcticum*, a pest of the northwest, avoids getting her feet cold by dropping them while flying over water. S. *ochraceum* in Guatemala drops its eggs while hovering over emergent vegetation. S. *damnosum* (see p. 684) has been observed to make a communal project out of egg-laying, swarms of flies depositing them on water-licked leaves—up to 2000 to 3000 per square centimeter. Usually the eggs hatch in a few days, but those of S. *arcticum* sink and hatch next spring.

The larvae hold on by means of circles of minute hooklets at the blunt posterior end of the body. They spin glutinous silken threads from modified salivary glands which extend to the posterior end of

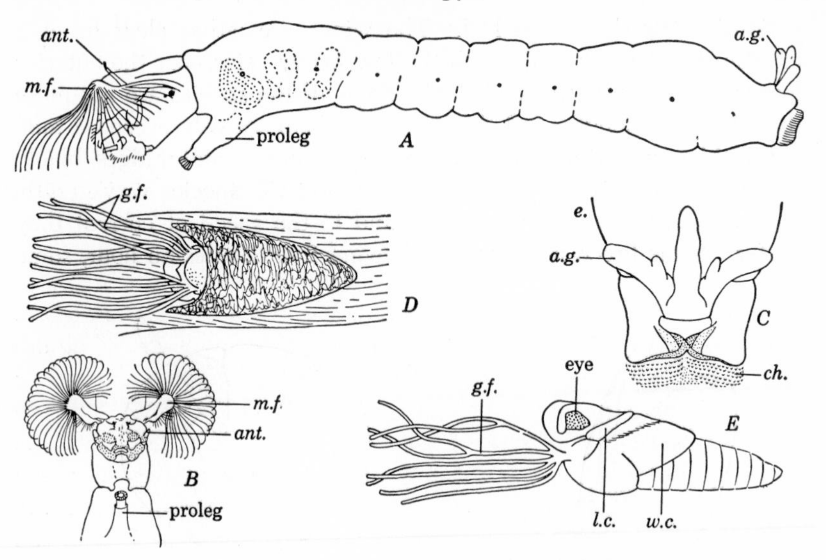

Fig. 213. Developmental stages of *Simulium.* *A,* larva; *B,* anterior end of larva, ventral view; *C,* posterior end of larva, dorsal view; *D,* wallpocket-like pupal case, with gill filaments of pupa protruding; *E,* pupa; *a.g.,* anal gills; *ant.,* antenna; *ch.,* caudal circlets of hooks; *g.f.,* gill filaments; *l.c.,* leg case; *m.f.,* mouth fan; *w.c.,* wing case. (*A* and *B,* adapted from Peterson, *Larvae of Insects,* pt. 2, 1951. *C* from Brumpt, *Précis de parasitologie,* 1940. *D* and *E* from Jobbins-Pomeroy, *U. S. Dept. Agric. Bull.* 329, 1916.)

the body; these threads serve for anchoring and as life lines, and are also used to spin cocoons. The larvae have a stumpy proleg, also provided with hooks, and by using this and the posterior circlet of hooks they loop along like "measuring worms." There are a pair of "mouth fans" by means of which microscopic organisms are swept into the mouth. The larvae breathe by means of tiny gills that can be projected through the anal slit in the last segment of the abdomen.

The larvae live in communities, attached to submerged vegetation, rocks, boards, etc., sometimes completely coating them, but they avoid hairy plants or objects covered by algae or slime. The larvae of *S. neavei* (see page 685) attach themselves to the shells of fresh-water crabs; those of *S. colombaschense,* the scourge of the Balkans, live to a depth of 190 ft. in the Danube River.

When ready to pupate the larvae spin for themselves a partial cocoon which is variously shaped like a jelly glass, slipper, wall pocket, etc., open at the anterior end for the extrusion of the branching gill filaments which are used as breathing organs (Fig. 213*D*). Some species simply spin a snarl of threads, the work of a whole community, in the meshes of which the pupae exist in a fair state of protection. The general form of the pupae can be seen in Fig. 213*E*. The breathing filaments vary greatly in different species and may have four to sixty branches. The adults emerge in 3 days to a week or more and are carried to the surface by a bubble of air which has been collected inside the old pupal skin. The adults are short-lived and lay their eggs soon after emergence.

The far northern species have only one or possibly two broods a year, most of them coming out in devastating hordes for a few weeks in late May or June after overwintering as larvae or in some cases, e.g., *S. arcticum*, as eggs. In warmer climates there are several broods and in the tropics many, the generation time being only about 3 or 4 weeks. Here survival of a dry season is the problem; some evidence indicates that adults seek shelter in rock crevices in dry stream beds, but it is also possible that resistant eggs may be laid. In the tropics the eggs may hatch in 2 days, the larvae go through 6 molts and pupate in 5 to 7 days, the adults emerge from pupae in 4 days, and the adults lay eggs in 4 days more (Wanson, 1950). In some and possibly all species a blood meal is necessary before development of the eggs.

The adults, at least in Guatemala, are not found resting in grass or bushes, but high up in trees. Nearly all species travel for considerable distances. *S. damnosum* in Africa may migrate up to nearly 50 miles from its breeding area, and it commonly goes 6 to 12 miles. In Guatemala recaptured dyed flies had flown an average of about 7½ miles. In the Balkans swarms of the Columbacz fly rise high in the air and are carried by air currents 100 miles or more, after which they come down and begin an active migration covering 3 to 6 miles per day, killing many unprotected animals on their way. Blackflies feed only in fairly bright light, and so do not bite in dark habitations. In poor light, white-skinned people may be bitten when dark-skinned ones are not.

Blackflies, having very short mouth parts, have to rasp a hole through the skin, from which blood trickles down; it takes several minutes for the flies to engorge. The bites are usually painless at first, but after several hours begin to swell and become more agonizingly itchy for about 3 days, when the annoyance is almost intolerable. A feeling of malaise and despondence, with some fever, may also develop. Apparently the salivary secretion is particularly toxic. Fortunately repeated attacks convey some degree of immunity but, as with other immunities of this sort, it is much more effective in some individuals than in others.

Some blackflies confine their attentions largely to birds or cold-blooded animals, and some to large mammals. Fortunately comparatively few habitually attack man. In North America common tormentors of man and animals are *Cnephia pecuarum,* the buffalo gnat, in the south central states; *Simulium venustum* in the north and east; S. *arcticum* in the northwest; and *Prosimulium hirtipes,* a springtime pest in the northeast.

Comparatively few species habitually attack man. Bequaert in 1938 said that of 57 species known in Africa only 5 have been reported as biting people. In Mexico and Guatemala S. *ochraceum* seems to be the only truly anthropophilic species, but a number of others (*metallicum, callidum, exiguum, haematopotum,* and *veracruzanum*) have no marked aversion to human blood. In Africa S. *damnosum* and S. *neavei,* the two vectors of onchocerciasis, are the only species that really prefer human blood, and even S. *damnosum* avoids biting man in some localities. In a part of Sudan (Dongola) S. *griseicolle* is an important human pest even though only a small proportion of the flies molest man, for there are countless numbers of them.

Simulium and Disease

Blackflies are probably the sole vectors of human onchocerciasis and of *Leucocytozoon* infections of birds. They probably also collaborate with mosquitoes in spreading myxomatosis of rabbits.

Onchocerciasis. Blacklock (1926), working in Sierra Leone in Africa, first showed that the important filarial parasite of man, *Onchocerca volvulus,* was transmitted by a blackfly, *Simulium damnosum.* Subsequently S. *neavei* in Africa and several species in southern Mexico and Guatemala were found to be important vectors.

The flies appear to exert a chemotactic attraction on the microfilariae in the skin, so that as many as 100 to 200 may be ingested during engorgement. Most of these are held in the fly's gut by the peritrophic membrane that is secreted during feeding (see p. 524), so only a few (average three in *damnosum*) succeed in developing. Too many larvae kill the flies, and even a few developing in the thoracic muscles tend to retard flight and perhaps hasten death. Unlike infective filarial larvae in mosquitoes and *Chrysops, Onchocerca* larvae may not lie in wait in the proboscis but, stimulated by the warmth, migrate from the body cavity of the fly during the slow engorgement.

Food habits constitute the most important factor in determining which species are important vectors of onchocerciasis, but some species seem to be poor hosts for the worms, and others are unimportant because they nearly always bite on the legs.

Of the American transmitters, as noted on p. 491, S. *ochraceum* is most important, althought other man-biting species (*metallicum, callidum, veracruzanum*) may be locally important. Finding flies naturally infected with *Onchocerca* larvae may be misleading, for in some areas in the onchocerciasis zone 100% of cattle are infected with *Onchocerca gutturosa* and a high proportion of horses with *O. reticulata*. S. *ochraceum* is a small species 1.5 to 2 mm. long, with yellowish-red thorax, black legs, and yellow and black abdomen. It breeds only in "infant" or "young" streams—varying from mere trickles to brooks several feet in width, deeply shaded by emergent and overhanging vegetation (Dalmat, 1955). It never breeds in older streams that have beaches and are not overhung by vegetation. These "young" streams are largely limited to a zone between about 2000 and 5000 ft. elevation, but the flies, which may live for about 4 weeks, have a flight range up to about 6 miles. However, the entire onchocerciasis zone in Guatemala covers only about 600 sq. miles. The natural infection rate in blackflies is less than 1%. Zoophilic blackflies breed mainly in "mature" streams, primarily on rocks rather than vegetation.

In Africa only S. *damnosum* and S. *neavei* are vectors. The former is widely distributed south of the Sahara and breeds in large rivers as well as in smaller tributaries not choked by papyrus, preferring rapids or cataracts of broken water, but breeding also in stretches where the current may be only 2 or 3 miles per hour. The larvae attach to stones, vegetation, etc., at various depths. The adults are abundant where there is high grass or dense vegetation along the river banks, and also at distant places. S. *neavei*, confined to central Africa, lays its eggs in clusters on vegetation, but the larvae and pupae are found only on the shells of crabs of the genus *Potamonautes*, which live in sunny cascades and rocky falls in rivers, usually at elevations of 1400 to 2000 ft. The adults congregate in humid forests, whence they make short sorties to bite.

In Europe S. *ornatum* is a transmitter of *Onchocerca gutturosa* of cattle, but the transmitters of this and of other *Onchocerca* infections in animals elsewhere have not been determined.

***Leucocytozoon* Infections.** Blackflies are transmitters of *Leucocytozoon* infections in birds (see p. 165). *L. simondi*, causing disease in young ducks and geese in northern United States and Canada, is believed to be transmitted primarily by *Simulium rugglesi*, which crawls under the feathers (Shewell, 1955). *L. smithi*, injurious to turkeys in various parts of the United States, is transmitted by several species of *Simulium*, including S. *jenningsi* (= *nigroparvum*), S. *slossonae*, and S. *occidentale*.

Control

The extreme susceptibility of the larvae of blackflies to DDT and dieldrin, and also phosphorus compounds, renders their extermination feasible in localized areas. In laboratory experiments exposure to as little as 0.01 ppm of DDT or 0.001 ppm of dieldrin for one hour produces 100% mortality within 24 hours, and the efficacy is even greater in the field. In Canada a single minute dose of 0.13 ppm of dieldrin controlled blackfly breeding for 100 miles downstream. Apparently the insecticides are concentrated by absorption on fine suspended particles that the larvae eat.

Applications for 15 minutes of a 5 or 10% solution of DDT in fuel oil at the rate of 1 part in 10,000,000 of water eliminates them for many miles, and at this rate it is not even harmful to fish. Spectacular and long-lasting results have been obtained by Wanson et al. (1948) by airplane spraying against S. *damnosum* in Belgian Congo, and by McMahon et al. (1958) against S. *neavei* in Kenya, which has been practically freed of this fly. In some areas spraying or fogging of vegetation on river banks, brush clearing, etc., is feasible for destruction of adults since they rest in such locations for several days before laying eggs. In Guatemala Lea and Dalmat (1955) found that using 0.1 ppm for 3 minutes eliminated larvae completely from small streams flowing less than 5000 gallons per minute, and therefore effective in the numerous small rivulets where most S. *ochraceum* breeding occurs. Larger streams require either heavier concentrations or more prolonged applications. Sometimes DDT-saturated briquettes are also useful. In any attempt at extermination it is necessary, because of the long flight ranges, to cover an entire area simultaneously, not zone by zone. Effective use of airplanes has been reported in Alaska and Pennsylvania.

Protection against blackflies can be obtained by the use of repellents (see p. 536), particularly dimethyltoluamide. Few blackflies enter houses, and they do not bite in dark places. DDT thermal-generated aerosols (see p. 535) are also helpful. In camp life and for the protection of animals in pastures, smudges are indispensable. The flies will not tolerate the smoke, and domestic animals soon learn to take advantage of its protection. Cheap repellents made of emulsions of kerosene or various resinous oils with soap and water have also been recommended for spraying animals.

Horseflies (Tabanidae)

The tabanids are the only Brachycera (see p. 666) which suck blood except some species of the family Rhagionidae (=Leptidae), the snipe flies. One genus of these, *Symphoromyia,* contains vicious bloodsuckers in North America, but they are not known to transmit disease. These flies have a long tapering abdomen, and antennae with a terminal bristle (Fig. 204*C*). The tabanids, known as gadflies, deerflies, horseflies, etc., are mainly animal pests, but many species attack man

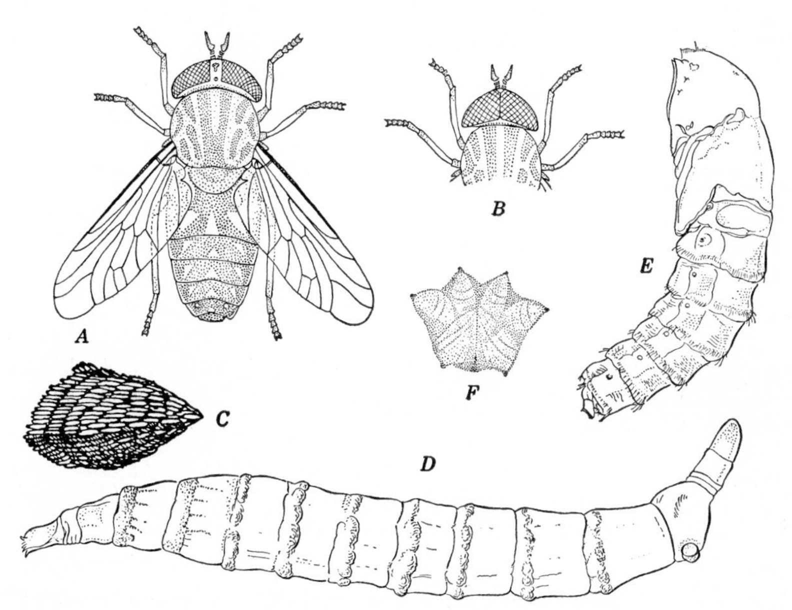

Fig. 214. Life cycle of tabanids. *A,* adult (*Tabanus fairchildi* ♀); *B,* head of ♂ (note spacing of eyes); *C,* egg mass; *D,* larva; *E,* pupa; *F,* "aster" at posterior end of pupa. (*A, B,* and *F* adapted from Hines, *U. S. Dept. Agric. Misc. Papers, Tech. Ser.,* 12, pt. 2, 1906. *D* and *E* from Patton and Evans, *Insects, Ticks, Mites and Venomous Animals of Medical and Veterinary Importance,* Grubb.)

also, inflicting painful bites. They are also implicated in the spread of certain diseases of man and animals. The females alone are bloodsuckers, the males living chiefly on plant juices; even the females in some genera feed on flowers. There are over 2500 species of worldwide distribution with species, but not individuals, most abundant in the tropics. (For a review, and revision of classifications, see Mackerras, 1954–1955.)

Morphology. The tabanids are of large size and heavy build Fig. 214*A*). They are often beautifully colored in black, brown, and orange tones, sometimes with brilliant green, or green-marked eyes, though in most species of temperate climates the huge eyes are brown or black. The head is large and in the male is almost entirely occupied by the eyes, which meet across the crown of the head; in the females a narrow space is left between them. The antennae are of characteristic shape (Fig. 204*D, E*), varying somewhat in the different genera. The mouth parts (Fig. 215, *left*) are almost exactly like those of the black-flies on a large scale. The stabbing and cutting parts are usually short, heavy, and powerful in biting species. The wings are usually held at a broad angle to the body, as shown in Fig. 214; they have a characteristic forked vein near the tip, and often have smoky markings which help in identification.

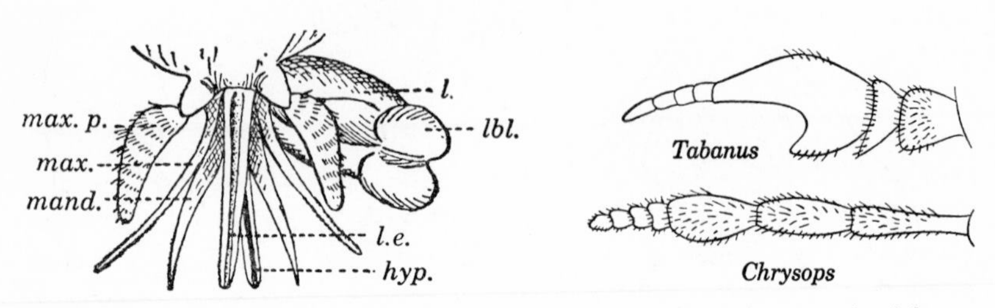

Fig. 215. *Left,* mouth parts of a tabanid; *hyp.,* hypopharynx; *l.,* labium; *lbl.,* labellum; *l.e.,* labrum-epipharynx; *mand.,* mandible; *max.,* maxilla; *max.p.,* maxillary palpus. *Right,* antennae of *Tabanus* and *Chrysops.*

Of the genera most important as human pests, *Tabanus* (Fig. 214) is large and has clear or smoky wings, with no spots or a few small scattered ones; *Chrysozona* (=*Haematopota*) is of moderate size and has wings with profuse scroll-like markings; and *Chrysops,* the species of which are often even smaller than a housefly, has conspicuous black bands and spots on the wing (Fig. 216).

Life history and habits. Many tabanids breed in wet or damp places or without a covering of water; some breed in rotting wood in treeholes or logs; others breed in relatively dry soil. The eggs (Fig. 214*C*), several hundred in number, are laid in definitely shaped masses on the leaves of marsh or water plants, on the leaves or twigs of trees overhanging breeding places, or in crevices of rocks along the sides of streams. When deposited the eggs are covered by a gluey waterproof secretion that binds them together. They are deposited during the summer or wet season and under favorable circumstances hatch in 4 to 7 days. Many are attacked by small hymenopteran parasites.

The larvae (Fig. 214*D*) are cylindrical legless creatures, tapering at each end. The body has eleven segments exclusive of the very small and often retracted head. Each segment has a row of wartlike processes provided with spines or hairs. The larvae are voracious feeders; most species prey upon soft-bodied animals such as earthworms and insect larvae and are not averse to cannibalism if food is scarce, but most species of *Chrysops* feed on dead organic matter. The larvae are active and grow rapidly during the summer, but in winter in the north they bury themselves several inches deep in soil or dead vegetation, which may freeze around them. In the spring the mature larvae migrate to drier ground and pupate. The pupae (Fig. 214*E*) are suggestive of the chryalids of butterflies. Many species are recognizable by the characteristics of the "aster" (Fig. 214*F*) at the posterior end. The pupal period lasts only 1 to 2 weeks. Some tabanid larvae may take 3 years to mature. In the tropics a buried resting stage during the dry season has been suggested. Most species have but one brood a year and are seasonally abundant.

Nearly all tabanids are diurnal, and often active in bright sunlight, though some species prefer shade; a few in central Africa are said to extend their activities into the night and some nonbiting species in North America are nocturnal. They are strong fliers, and may travel a mile or more from their breeding places. Most species are very deliberate in their feeding, and not easily disturbed after beginning a meal. They like to skim over the surface of water, a habit that can be taken advantage of to trap them in oil-covered pools.

Annoyance and damage. Tabanids cause serious injury to cattle by annoyance and loss of blood, the latter easily amounting to 100 to 200 cc. per day. This, together with loss from *Stomoxys* and hornflies (see pp. 693 and 694), is sufficient to seriously reduce weight gain and/or milk production. Animals may gain 20 to 25 lb. and increase their butterfat production by 15% within a month after being freed from annoyance by tabanids. Even human beings may suffer severely from bites of tabanids, particularly deerflies.

Tabanids and Disease

Trypanosomes. Tabanids are of importance in connection with the transmission of some of the trypanosomes of animals. *Trypanosoma evansi,* causing surra in horses, cattle, camels, dogs, etc., and the related *T. equinum,* causing mal-de-caderas of horses and other animals in South America, are commonly transmitted by tabanids and *Stomoxys* from one animal in a herd to another by a soiled proboscis. Even the tsetse-borne trypanosomes are frequently transmitted in this manner (see p. 156), especially *T. vivax*, which has become established outside of tsetse areas in Africa, and even in South America. Trypanosomiases of man are probably much less frequently transmitted by tabanids and *Stomoxys* than are those of animals. At least one species of tryanosome, the nonpathogenic *T. theileri* of cattle, undergoes cyclical development in tabanids.

Tularemia. In the western United States, particularly in Utah and Colorado, there is a form of tularemia known as "deerfly fever," which seems to be associated with the bites of a single species of deerfly, *Chrysops discalis* (Fig. 216, *left*), since it does not occur outside the range of that species. Why only this species should be a capable

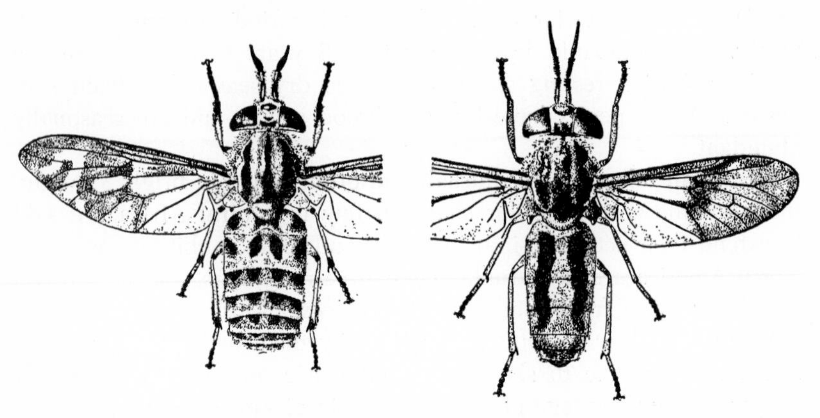

Fig. 216. *Left, Chrysops discalis,* the vector of deerfly fever. *Right, Chrysops dimidiata,* vector of *Loa loa.* (After Brumpt, *Précis de parasitologie,* Masson.)

vector of this disease is unknown; Jellison (1950) suggested that it might be a predilection of *C. discalis* for feeding on rabbits, which constitute an important reservoir of the disease (see pp. 594–595).

Transmission of other diseases by interrupted feeding. Tabanids may be of importance in transmitting a number of other infectious blood diseases by means of a soiled proboscis during interrupted feeding on herds of animals. Anaplasmosis (see p. 218) can be transmitted within 5 minutes after feeding on an infected animal. Anthrax can also be transmitted in this manner by either tabanids or *Stomoxys.* This is a very destructive bacterial disease affecting domestic animals and transmissible to man. It may enter the body through skin abrasions, aerial spores, or contaminated food. Since ticks, fleas, and bedbugs have been found capable of transmitting undulant fever or brucellosis, it will not be surprising if tabanids are incriminated also. Tabanids can also transmit swamp fever of horses.

Loa loa. Certain species of Chrysops, locally known as mangrove flies, are the intermediate hosts of a human filarial worm, *Loa loa* (see p. 485), in Africa. The diurnal microfilariae are ingested in very irregular numbers (relative to the numbers in the circulating blood), since *Chrysops* is a "pool" feeder (see p. 528). The filarial larvae develop in the abdomen of the fly, and the infective larvae invade the proboscis after 10 or 12 days to 3 or 4 weeks, according to temperature. Although all species of *Chrysops* are probably potential vectors, *C. silacea* and to a less degree *C. dimidiata* (Fig. 216, *right*) and *C. distinctipennis* are important vectors since they are numerous, good hosts, enter houses freely, and bite man readily. Oddly enough *C. silacea,* at least, is attracted by wood smoke, and is thus lured to human victims in clearings or houses. *C. silacea* and *C. dimidiata* are forest species that bite man in forest clearings, and monkeys in the canopy; possibly they derive their infections from these (see p. 485). Other species may be important vectors among monkeys. The flies breed along river banks in wet mud, covered by thick decaying vegetation in or out of shallow water where there is dense shade (Davey and O'Rourke, 1951).

Control

Cattle can be protected by spraying with emulsions containing pyrethrins with a synergist, applied as wet or mist sprays, or making them douse themselves by means of automatic treadle sprayers (Bruce and Decker, 1951). In some locations aerial spraying of insecticides in oil has been used successfully against larvae. Diethyltoluamide is a good repellent to make life more pleasant for fishermen, picnickers, etc. In Africa the *Loa* vectors may be controlled by clearing brush along stream banks and canalizing stagnant pools, or using larvicides where this is impractical. *But*—clearing brush might encourage *Anopheles gambiae,* and there would be malaria to contend with instead of *Loa!*

SUBORDER CYCLORRHAPHA

This suborder includes flies that are important to us (1) as blood suckers, some of which are important as intermediate hosts and/or direct inoculations of pathogenic organisms; (2) as mechanical transmitters of germs directly to wounds or eyes, or indirectly via food; and (3) as myiasis-producing larvae. The last will be considered separately in Chapter 30.

The majority of the Cyclorrhapha to be considered in this chapter belong to the group Muscoidea (6*a* in key on p. 668), which were once all included with the housefly in one family, Muscidae. They have a frontal lunule (see p. 666), have the opposite legs of each pair set close together, well-developed squamae, a well-developed proboscis, and have fairly smooth fleshy larvae that taper anteriorly (Fig. 218*C*).

The majority of the Muscoidea have fleshy proboscides fitted for lapping up liquid foods. Some have acquired the habit of devoting nearly all their time to the skins of animals, flitting from spot to spot in search of blood or exudations. Such, according to Patton (1932),

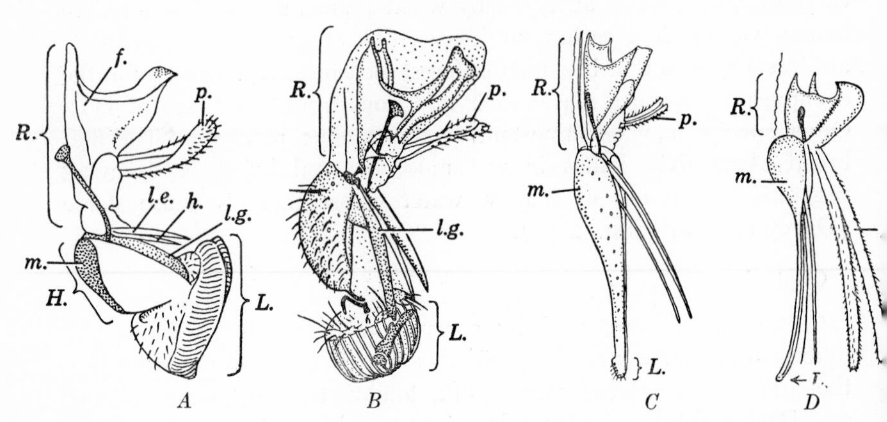

Fig. 217. Evolution of bloodsucking proboscides in muscoid flies. *A, Musca domestica* (for sucking only); *B, Philaematomyia insignis* (for scratching and tearing skin); *C, Stomoxys,* and *D, Glossina,* for piercing. Note progressive shortening of upper segment or rostrum (*R.*); increase in proportionate length of haustellum (*H.*), and increase in sclerotized area of it (*m.*); reduction of labellum; freeing of hypostome (*h.*) from labial gutter; development of prestomal armature of spines at tip of labellum; and finally, in *Glossina* (and also in *Pupipara*), elongation and modification of maxillary palpi to form sheath for labium, now the piercing organ. (Adapted from various authors.)

are *M. bezzii* and *M. lusoria*, which are therefore potential mechanical transmitters of blood infections. From these scavengers Patton has traced an interesting evolutionary development of the proboscis into a scratching and tearing and finally a piercing organ (see Fig. 217). *Philaematomyia insignis* (Fig. 216*B*) is able to rasp and tear a hole through the skin of cattle and suck the exuding fluid. A much higher development is reached in *Stomoxys* and *Siphona* (Fig. 216*C*), in which the elongated, strengthened, and styletlike proboscis acts as a piercing organ after a hole has been rasped and torn. The proboscis of the tsetse flies (Fig. 217*D*) and of the more distantly related Hippoboscidae is the culmination in this line of evolution. The tsetse flies and the Hippoboscidae, together with a few other flies once classed with the Hippoboscidae into a group Pupipara, are viviparous and deposit their third-stage larvae one at a time just before pupation. Of the blood-feeding species of *Musca*, *M. planiceps*, does this also, but other blood-feeding *Musca* deposit their larvae in the second stage. Strict dependence on blood as food and viviparity seem to go together.

All the blood-sucking members of the genus *Musca*, all the tsetse flies, and most of the hornflies and stableflies are Old World species. Our houseflies, stableflies, and hornflies are probably all importations from across the seas.

Hornflies and Stableflies

Hornfly (*Siphona irritans*). This small blackish fly, about half the size of a housefly, causes endless misery to cattle; no other pest, except possibly screwworms is, as inimical to the contentment of cows. The fly sometimes attacks other domestic animals but rarely man. On ranches in the southwest an average of 4000 flies per animal is frequent. They stay on the animals most of the time, night and day, stabbing them and sucking blood usually twice a day. The loss of blood from a herd of 500 cattle with 4000 flies apiece is estimated at 7 quarts a day. The irritated animals become restless, cease to graze, and lose vitality not only from loss of blood but also from loss of food. The resulting heavy loss of meat and milk is inevitable. In 1945 hornflies were estimated to have caused a loss of 86 million pounds of meat in the United States. Fortunately, since they stick fairly closely to one animal, hornflies do not transmit diseases as frequently as stableflies and tabanids.

Hornflies usually leave the animals only to lay eggs. They swarm to a fresh dropping, crawl under it and lay their eggs, and return to the animal in 5 to 10 minutes. The larvae develop in the dropping in about 10 to 12 days; they are about 7 mm. long and have large black stigmal plates (see p. 770) very close together. They pupate in soil under the dropping; the flat and wingless adults that emerge crawl off to rest for an hour, distending the abdomen and unfolding the wings. The flies live for 6 or 7 weeks and lay about 400 eggs.

Hornflies, because of their habit of staying on animals, are easily controlled by spraying animals with emulsions or suspensions of insecticides (see pp. 532–535) or, in small herds, hand dusting at about 3-week intervals. Cables wrapped with burlap soaked in oil solutions of insecticides can be rigged up as back-rubbers to enable cattle to do their own work of hornfly control. Space sprays in barns with such

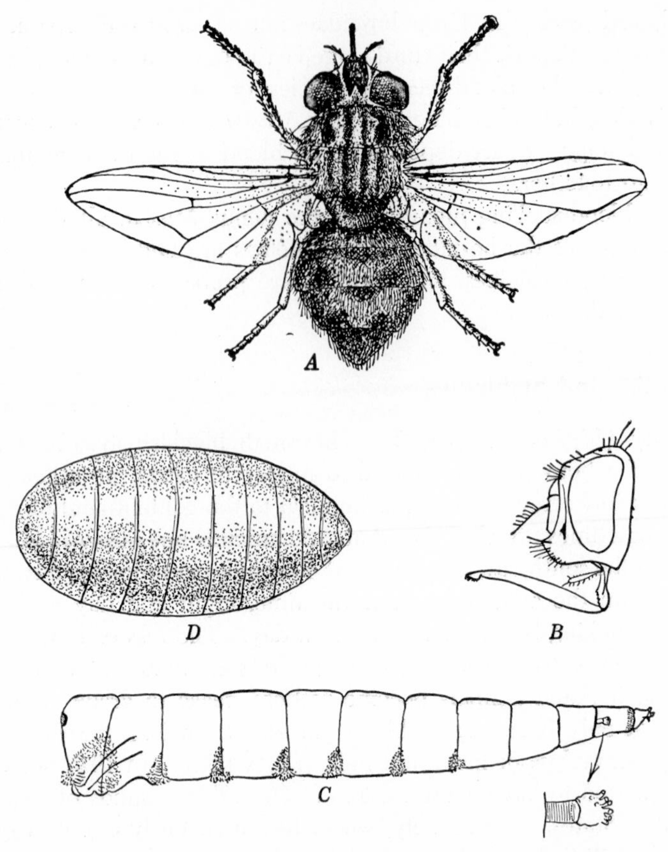

Fig. 218. Stablefly. *Stomoxys calcitrans. A,* adult; *B,* head, side view; *C,* larva, with enlargement of anterior spiracle; *D,* pupa. (Adult, ×5; larva and pupa, ×7.)

volatile insecticides as lindane or diazinon also kill many hornflies, and the eggs of helminths. Weight gains quickly follow treatments. Laake estimated that a pound of DDT caused a gain of 2360 lb. of animals. Complete extermination of hornflies might not be extremely difficult unless the flies develop immunity to the chemicals as do houseflies. Unfortunately, however, the day has not yet come when the animals could comfortably dispense with their tails.

Stableflies (*Stomoxys*). The genus *Stomoxys* contains a number of Old World species, but one, *S. calcitrans,* the stablefly or dogfly (Fig. 218), is an annoying pest of animals and man all over the world. In some outbreaks the stablefly may cause as much loss in beef and milk as the hornfly. The stablefly resembles the housefly so closely that it is often mistaken for it, whence the common belief that houseflies can bite. It is easily distinguished by its narrow, pointed, shiny black proboscis (Figs. 217*C*, and 218*B*).

Stableflies breed by preference in decaying straw or rotting vegetable matter or straw mixed with manure, particularly of horses. An unusual scourge of them, sufficient to harass cattle severely and to denude the beaches of bathers, occurs on parts of the Florida and New Jersey coasts where washed-up piles of seaweed, trapped in lakes behind sand dunes, afford ideal breeding places, especially in September. Later, in December, extensive breeding occurs in piles of peanut litter and celery strippings farther inland in Florida.

The eggs are deposited in small batches; in a few days they hatch into white, semitransparent, footless maggots (Fig. 218*C*) distinguishable by the form and position of the stigmal plates at the posterior end (see Fig. 247). The larvae mature in 10 days to a month or more and pupate in drier parts of the breeding material. The chestnut-colored pupae (Fig. 218*D*), 6 to 7 mm. long, hatch in a week or more in warm weather.

Stableflies frequently begin a meal on one animal and finish it on another, so they may mechanically transmit diseases such as trypanosomiasis, anthrax, vesicular stomatitis, etc., just as do tabanids (see p. 689). They are particularly important as transmitters of *Trypanosoma evansi*, the cause of surra. *Stomoxys* and *Musca* also serve as intermediate hosts for spiruroid nematodes of the genus *Habronema* (see p. 499), parasites in the stomach of horses. *Habronema* larvae may cause human conjunctivitis, when liberated by flies, in the sore eyes of children. *Stomoxys* is also an intermediate host for the filaria, *Setaria cervi*, of deer, and for a tapeworm of chickens, *Hymenolepis carioca*.

Where immunity has not developed, control can be obtained by means of sprays on animals or residual sprays in barns. However, elimination of breeding places is far more feasible than in the case of tabanids. Along the west Florida coast, where marine vegetation washed up on shore breeds enough of these flies to make life miserable, application of 2 gallons of 0.5% DDT emulsion per 100 sq. ft. gave excellent results; about 100 to 600 gallons per mile of shore was needed.

Housefly (*Musca domestica*)

The common housefly, *Musca domestica*, world-wide in distribution, is too well known to require detailed description. In some areas about

99% of flies in houses are usually of this species. In the United States the housefly spends *some* of its time creeping over the baby's toys or contemplating the view from a window pane, but in Egypt the variety *vicina* has dedicated itself with an incredible singleness of purpose to crawling over the skin of human beings and driving visitors, at least, frantic. A smaller but related species, *Musca sorbens*, has somewhat different habits; it clusters around the eyes and mouth or sores and sits by the hour, trading germs with its environment. It is also partial to meat in butcher shops, and human feces, which in the Near East may be found practically anywhere, but especially in the animal room of village houses. In many places in the Near East a fly brush for sweeping flies from the face is a most essential piece of equipment and is *one* souvenir a tourist does well to buy.

The housefly is a particularly important transmitter of filth germs, especially those affecting the eyes and the alimentary canal. In Egypt, when flies were practically eliminated from certain villages by chemicals for a few months, pussy eyes, due to gonococcal infections, were strikingly decreased. The relation to enteric diseases is due to the fact that the fly frequents privies and feces for egg laying, and dining tables and kitchens for food. It harbors vast numbers of germs on its sticky feet; vomits them from its food reservoirs with liquid to melt sugar or cake; and deposits germs in its feces (fly specks)—and a well-fed fly is said to produce such specks about every 5 minutes all day! It is probably the greatest single factor in the epidemiology of bacillary dysentery, and is an important one in other enteric diseases of both man and animals. In addition houseflies have been found to pick up and harbor the virus of poliomyelitis, though there is still no evidence that they are an important factor in causing the paralytic form of the disease; their possible role is discussed on pp. 227–228. Helminth eggs can be carried externally and also internally if under 50 μ in diameter, so houseflies as well as blowflies could be a factor in causing hydatid infections (see p. 362).

The housefly lays its eggs by preference in horse manures, but will breed in almost any kind of animal or vegetable wastes, in pen litter, or in slop-soaked soil. The eggs hatch in 12 to 24 hours, and the maggots, recognizable by their stigmal plates (see Fig. 247), are full grown in a few days. They then move to drier places, pupate, and emerge as flies in about 2 to 3 weeks. In her lifetime, averaging about 6 or 8 weeks, a female deposits about 2000 eggs.

The housefly suffered a serious setback when automobiles replaced horses in American cities, and the advent of DDT sprays was thought by many people to spell "finis" for this pest. In fact, *Science News*

Letter, in April 1948, carried a big headline "Flyless Age Now in Sight," and some cities and even states seriously set out on campaigns of extermination. Five years earlier this would have seemed a fantastic dream; five years later it is recognized to have been just that. The housefly, to a greater extent than any other insect, has adapted itself to living with DDT, just as it adapted itself to living without horses. After a few months or even weeks a fly population, by natural selection, may become entirely immune (see p. 536). When immunity to one chlorinated hydrocarbon is developed, immunity to others follows more quickly, so *mixtures* of these insecticides should not be used. Also, resistance develops even faster in the larval stage; any chemical used against the adults should not be used against the larvae. Residual treatments of the contents of privies *increase* the fly output.

A fly population that has become immune to DDT and related chemicals consists of superflies which are hardier in other respects as well, and may become more abundant than in untreated places. As yet, however, most populations are still susceptible to the organic phosphorus compounds. Excellent control of adults has been obtained by hanging 3/16-in. cotton cords impregnated with parathion or diazinon in dairies, dining halls, etc. (30 ft. of cord per 100 sq. ft.). Use of these substances mixed with sugar or syrup as sprays (1:2.5) or as dry or wet baits gives spectacular results. In one barn 50,000 flies were swept up about 2 hours after distribution of bait! Such baits painted on likely resting places may remain effective an entire summer. The organic phosphorus compounds are also very effective against maggots when sprayed on manure with plenty of water (0.25% in a total of 225 cc. per sq. ft. in several applications).

It is obvious that we must return to basic sanitation for fly control—sanitary disposal of feces, frequent thin spreading of manure on farms, burial or burning of garbage or other materials (e.g., at slaughter houses) where flies breed. We must also, for protection against fly-borne diseases, screen houses, protect food on trucks or in shops, etc. It is no easy road, but it is not blocked by impassable barriers as is the attractive-looking road of chemical control.

Blowflies

Although the housefly is the most important mechanical carrier of pathogenic organisms, many other flies may also serve in the same capacity. Attention should be called to the blowflies (Calliphoridae), some of which breed in dung but most of which breed in carcasses and garbage. They may carry not only germs of enteric diseases but also those of plague, anthrax, undulant fever, and tularemia, and may be a factor in transmitting hydatid disease. *Phaenicia sericata* (=*Lucilia*) is a very common domestic fly in warm parts of the world including the United States, and is sometimes more common than *Musca domestica.* An unsolved enigma, but one that may have significance, is the fact that a high percentage of blowflies (principally *Phaenicia sericata* and *Phormia regina*) collected in the vicinity of poliomyelitis cases were found to harbor the virus. They probably did not get it from human feces; where, then, did they acquire it? The answer might help to solve the problem of polio transmission.

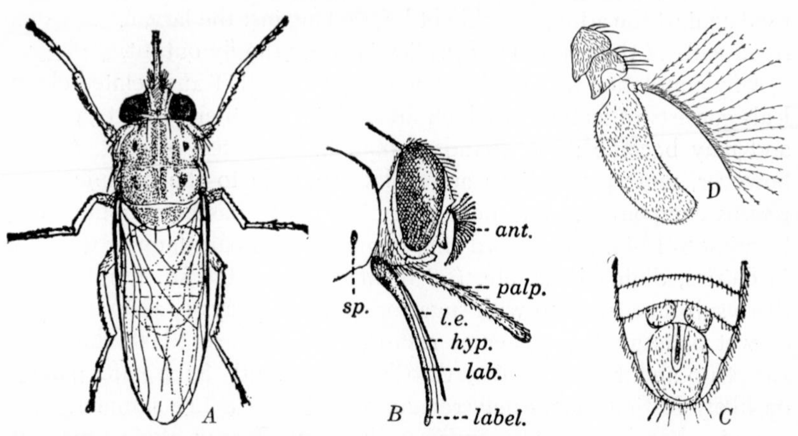

Fig. 219. A, Tsetse fly (*Glossina*) in resting position, ×4 (adapted from Austen and Hegh, *Tsetse Flies,* 1922). *B,* head and mouth parts; *ant.,* antenna; *hyp.,* hypopharynx; *lab.,* labium; *label.,* labellum; *l.e.,* labrum-epipharynx; *palp.,* labial palpus; *sp.,* spiracle (after Alcock, *Entomology for Medical Officers,* 1920). *C,* hypopygium of ♂, retracted. *D,* antenna.

Tsetse Flies (*Glossina*)

Tsetse flies are of paramount importance in Africa because of their role as vectors of the trypanosome infections of man and domestic animals, which have had a profound effect on the economy and development of that continent (see p. 131). There are about 20 species of tsetse flies, all belonging to the one genus, *Glossina,* of the family Glossinidae, sometimes considered a subfamily of Muscidae (Fig. 219). Except for one species that enters southwest Arabia, the tsetse flies are confined to Africa from south of the Sahara to the northern parts of the Union of South Africa (Fig. 220).

Morphology. The tsetse flies (Fig. 219) are elongated dark brown or yellowish-brown flies, some species no larger than ordinary houseflies, others larger than blowflies. When at rest they fold their wings flat over the back, one on top of the other, instead of spreading them as do most other flies. Even on the wing the darting manner of flight and buzzing sound make them fairly easily recognizable. The antennae (Fig. 219) have a conspicuous arista with long-feathered bristles on one side only. The proboscis, which projects horizontally in front of the head, consists of a piercing labium with a bulblike base and special structures at its tip for rasping and tearing. In its anterior groove are a delicate labrum-epipharynx forming a food trough and a fine hypopharynx containing the salivary duct. The elongated maxillary palpi, each grooved on its inner face, close together to form a sheath for the proboscis (see p. 692 and Fig. 217*D*).

The large quadrangular thorax and the tapered abdomen have characteristic patterns in the different species (cf. Figs. 222 and 223). In the riverine species (*palpalis* and *tachinoides*), vectors of *Trypanosoma gambiense,* the hind tarsi are all dark, whereas in the *morsitans* group, vectors of *T. rhodesiense* and the animal trypanosomiases, they are partly pale. Male tsetses have a large oval swelling on the underside of the last segment of the abdomen, the "hypopygium" (Fig. 219*C*), containing the genitalia.

Fig. 220. Approximate ranges of tsetse flies.

≡ . . . range of entire genus *Glossina*

⫽ . . . range of *Glossina palpalis*

⫽ . . . range of *Glossina morsitans*

Habits. Tsetses are diurnal, although *G. pallidipes* will bite on moonlight nights. *G. tachinoides* stays close to the ground and usually bites below the knee, whereas *palpalis* and others commonly fly 5 or 6 ft. above the ground and more frequently bite above the waist. The usual flight range is only about 500 yards to at most a mile or two, although the flies are often carried considerable distances by vehicles, animals, or bicyclists. Most species avoid either thick underbrush or parklike areas under a canopy, since they commonly rest on the underside of twigs near the ground, and deposit their young under low shade. *G. tachinoides* is entirely barred by thick underbrush, and even the *morsitans* group avoids dense unbroken thickets. Most species can stand considerable ranges of climate and environment. *G. palpalis* and *tachinoides* are found only close to water along rivers or lake shores except in wet rain-forests; *palpalis* seldom thrives where there is less than 45 in. of rain, but *tachinoides* may extend along rivers to a 15-in. zone. The *morsitans* group, on the other hand, is partial to savannah-woodland, frequenting clumps or stretches of certain types of vegetation surrounded or interspersed by open grassland, into which the flies make hunting forays to feed on game animals. Moving objects or even large quiet objects attract them, so small clearings for distances of only 100 yards or so along streams may increase rather than decrease fly-man contacts. Tsetses show marked preference for certain colors; the dark skin of Negroes is selected in preference to pale skin to such an extent that a white man is seldom troubled when accompanied by natives.

Both sexes of tsetse flies are bloodsuckers, but they also suck plant juices. Different species have different tastes. The *G. morsitans* group, including *pallidipes* and *swynnertoni,* feed mainly on large mammals (game animals), but they seem to have a preference for warthogs and pigs. These species are the most important transmitters of *T. rhodesiense* of man and of the animal trypanosomes—*brucei, congolense,* and *vivax,* and their relatives—which are so deadly to domestic animals that in the presence of these flies it is impossible to keep anything but poultry and a few fly-hardened goats. Small mammals are of little use to these or other tsetses because of their retiring and nocturnal habits. It is a case of game and tsetses or man and cattle—it cannot be both. The riverine species of tsetses (*G. palpalis* and *G. tachinoides*) which transmit *T. gambiense* to man, on the other hand, are able to thrive where there is no game and man is the only source of food, except for occasional blood meals supplied in a pinch by lizards, crocodiles, or birds.

Life history. Tsetse flies differ from most related flies in their remarkable manner of reproduction. They do not lay eggs; a single developing larva is retained within the body, being nourished by special "milk glands" on the walls of the uterus while lying with its stigmal plates, containing the spiracles, close to the genital opening of the mother. The larva passes through its molts and is full grown and ready to pupate before it is born, occupying practically all of the swollen abdomen of the mother. As soon as one is born another begins its development, and new larvae are born about every 10 or 12 days, provided that the temperature is around 75° to 85°F. and food is abundant. There are few data on the total number of young produced, but in one captive fly eight larvae were produced in 13 weeks, and only one egg was found left in the body. Pregnant flies often abort when disturbed, and cases are known in which the larvae pupated within the abdomen of the mother, to the destruction of both of them. Breeding occurs mainly in the dry season.

The full grown (third stage) larva (Fig. 221*C*) is a yellowish-white creature about 8 to 10 mm. in length, with a pair of dark knoblike protuberances at the posterior end, between which are the stigmal plates. Immediately after birth it hides itself at a depth of 1 to 2 cm. in loose soil or under dead leaves and transforms to a pupa (Fig.

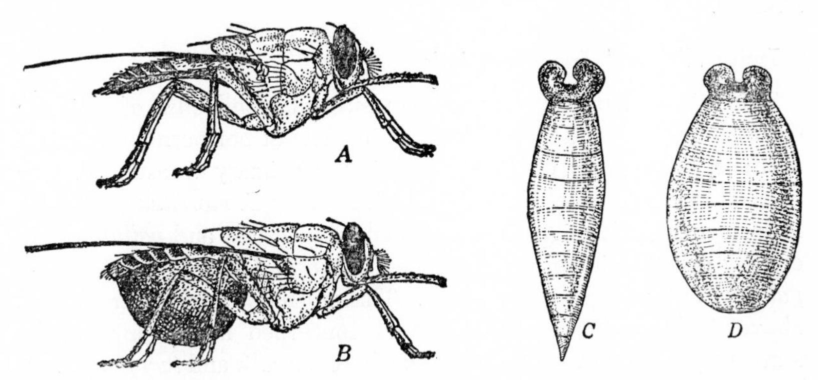

Fig. 221. *A* and *B*, *Glossina morsitans* before and after feeding, ×4 (adapted from Austen and Hegh, *Tsetse Flies,* 1922). *C*, newly born larva of *G. palpalis. D*, pupa of *G. palpalis.*

221*D*). This is olive shaped, and turns a mahogany color, with the blackish knobs still present at the posterior end; the shape and size of these knobs and of the notch between them are useful species characters. The duration of the pupal stage may be 17 days to nearly 3 months. Few adults emerge at temperatures below 70° or above 86°F.

All species select dry, loose soil in shaded, protected spots for deposition of the grown larvae. As maternity spots, *G. morsitans* and its allies select places under logs, hollows under trees, etc., whereas *palpalis,* less resistant to unfavorable temperature and humidity, gives birth to its young in sites protected by *low* as well as high shade; *tachinoides,* except in more austere parts of its range, needs only high shade.

Adult tsetses apparently live for only a few months in the wet season, and possibly for only about 3 weeks under dry conditions. In the drier parts of their range *palpalis* and *tachinoides* often concentrate in the vicinity of a few persisting water holes along the course of a stream. Since the human population does likewise, it is obvious that even though the flies may not be numerous, the contacts with man *are* numerous, and conditions are ideal for transmission of human trypanosomiasis (see p. 140). The infected people, frightened by outbreaks of the disease, or seeking new water holes, or migrating along trade routes, spread the disease further. There is one record of 30 of 43 people in a hamlet becoming infected, the source being a small water hole where four tsetse flies did the job. In wet coastal areas, even though the flies may be far more numerous, they are more dispersed, have more alternative sources of food, and there is much less sleeping sickness.

Tsetse Flies and Trypanosomiasis

The principal transmitters of *Trypanosoma gambiense* of man are *Glossina palpalis* and *G. tachinoides;* in parts of northern Nigeria and northern Cameroons *G. tachinoides* is the primary transmitter. The principal transmitter of *T. rhodesiense* of man is *G. morsitans,* although in some areas in East Africa the related *swynnertoni* and *pallidipes* are equally or more important; in an epidemic in Uganda in 1940–1943 *pallidipes* was the principal vector. This *morsitans* group, as remarked above, includes the principal species concerned in transmitting the animal trypanosomes, even in places where *palpalis* and *tachinoides* are commoner. In parts of West Africa *longipalpis* is another important vector to animals, and in East Africa *brevipalpis.* Other species, either because of their rarity or their food habits, are less important. The *morsitans* group is also able to transmit *T. gambiae* experimentally, but its dependence on game results in its being scarce except where there are few human beings (less than 40 per square mile). The importance of this group of flies in transmitting *rhodesiense* to man is undoubtedly due to this species of trypanosome utilizing wild game as a reservoir to a much greater extent than does *T. gambiense.* As noted in Chapter 8, even in many places where 20 or 30% of game harbor trypanosomes only a small percentage of the flies may be infected. Some strains of flies seem more refractory than others.

Glossina palpalis (Fig. 222) is a large dark species with blackish-brown abdomen and gray thorax with indistinct brown markings. Its wide distribution in west and central Africa, shown in Fig. 220, is nearly coincident with that of Gambian sleeping sickness. In the rainy season the flies may extend their range and retreat again as the water dries. This species is seldom found more than 30 yards from the edge of a river or lake where vegetation overhanging the water is abundant, although it follows man or animals for a few hundred yards from such positions. It is feared that *G. palpalis* may sometime bridge the short

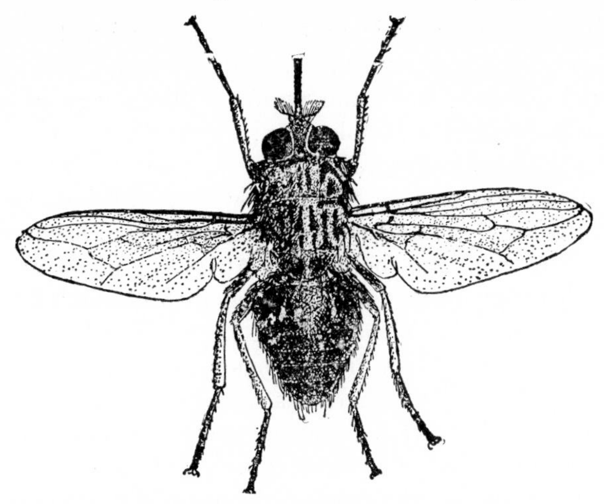

Fig. 222. *Glossina palpalis,* carrier of Gambian and Nigerian sleeping sickness. ×4. (After Austen and Hegh, *Tsetse Flies,* 1922.)

gap between the headwaters of the Congo and the Zambesi and become established along the latter river and its tributaries, carrying sleeping sickness with it.

G. tachinoides is one of the smallest tsetse flies, about the size of a housefly, occurring around the southern border of the Sahara. It has habitats similar to those of *palpalis,* but it is able to live in more

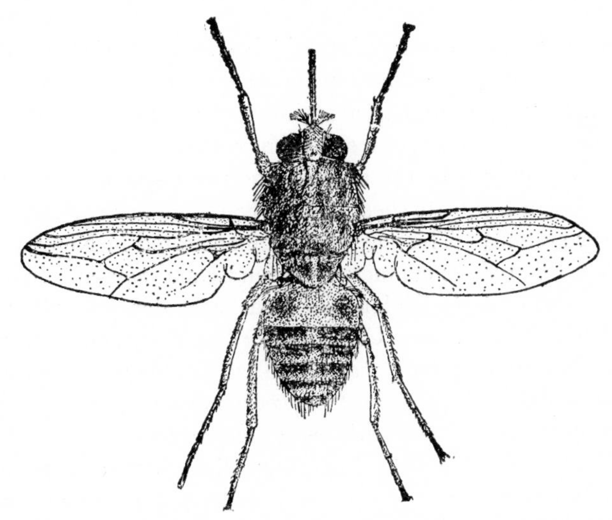

Fig. 223. *Glossina morsitans,* carrier of Rhodesian sleeping sickness. ×4. (After Austen and Hegh, *Tsetse Flies,* 1922.)

sparsely shaded places with less rainfall, and it is absent from the wet coastal areas. It feeds on animals primarily but is not entirely dependent on game. It is an important vector of sleeping sickness only in sacred groves, around water holes, etc., where there is little choice of hosts and contact with man is close.

G. morsitans (Fig. 223) is the most widely distributed species of tsetse fly (see Fig. 220) and is also the best known, having attracted the attention of big-game hunters in Africa for many years. It is slightly smaller and much lighter colored than *G. palpalis,* with distinctly banded abdomen. *G. morsitans* is not confined to the immediate vicinity of water. It prefers dry savannahs interspersed with patches of bush or woodland where there is a moderate amount of shade, but shows strong predilections for certain types of vegetation. It is not found in open grassland or in deep forest. A thin, deciduous woodland called "miombo" is an especially favored habitat, but within any environment this and other species exercise a very precise choice of places to feed, rest, or breed. With sufficient knowledge tsetses can sometimes be eliminated by surprisingly little "discriminative" clearing in local areas.

G. swynnertoni favors somewhat drier and sparser vegetation than does *morsitans,* whereas *pallidipes* and *brevipalpis* favor somewhat greater humidity and denser thickets. All these species are closely related and are good trypanosome transmitters.

Control

G. palpalis and *G. tachinoides,* which live along rivers and shores of lakes, are easier to control than the *morsitans* group, which is independent of water. Partial clearing along river banks by elimination of brush and low branches of trees, with "ruthless" (complete) clearing for a mile at the ends of the cleared areas to prevent reinfestation in wet seasons, will eliminate these species from sections of stream or even whole river systems. This has been demonstrated in Nigeria and the Gold Coast. To reduce contacts between fly and *man* and thus reduce human trypanosomiasis, "defensive" clearing is done on either side of villages, bridges, fords, etc. Clearings for less than 200 yards *increases* man-fly contacts by making man more visible. Ruthless clearings of 500 to 800 yards reduce FBH (flies per boy hour) 60%. Clearings of 800 to 1000 yards give the highest rate of freedom from flies that is economically practicable. For these flies game extermination or exclusion is useless.

Control of the *morsitans* group is more difficult. In earlier days the only solution when sleeping sickness outbreaks occurred was to relinquish the land to the tsetses and game, and move the human population to less dangerous places. This is sometimes still the best policy where the land is poor and population sparse. Now, however, in areas that are worth it, the land *can* be reclaimed for man and domestic animals and the tsetses banished. The methods used include (1) direct attacks on the flies by insecticides, traps, burning, etc.; (2) game destruction or exclusion; (3) clearing of certain types of vegetation; and (4) human settlement.

Treatment of water's edge vegetation with dieldrin emulsion controls *G. palpalis* and might eradicate it. The *G. morsitans* group can be greatly reduced by airplane spraying of savannah woodland, locally spraying cattle or bicyclists' clothes and then having them pass back and forth through an area, generating insecticide smokes, using traps, and burning breeding areas, but the necessary *complete* elimination can seldom be accomplished.

Destruction of game is a successful method on more or less isolated areas of a few hundred square miles, but one which is often looked upon with considerable repugnance. It is, however, a choice between preserving the game for hunters and naturalists or making land available for development by man. Game reserves in areas unsuitable for development—and there are plenty of them—will prevent extermination of the animals. Danger that with game destruction the tsetses will turn their attention to domestic animals is remote, for as Buxton (1948) pointed out there is no moment when numerous hungry flies could do this. Until all the game is gone, and the tsetses with it, there are not yet any domestic animals available. Fencing of various types has been tried, but the expense and difficulties involved rarely make it feasible. It was primarily by game destruction that over 6000 square miles in Southern Rhodesia have been reclaimed from tsetses for use of man and animals.

Reclamation by "discriminative" clearing of vegetation favored by flies has been successful in over 1300 square miles in Tanganyika. Methods include late burning, felling, and bulldozing, with help from cultivation and browsing goats. With increasing knowledge of the fly's predilections, successful clearing becomes less difficult, but much is yet to be learned.

Human settlement can hold land that has been cleared of tsetses and makes advances into a fly belt possible; the game and tsetses abandon more and more land instead of *man* doing this. However, there are problems. Most of the land to be reclaimed is marginal, and if enough settlers occupy it to drive off the game, the land is overstocked or overworked and starvation eventually replaces trypanosomiasis. There is clearly no easy road to the development of Africa unless the use of prophylactic drugs in animals and man may bring about a partial solution (see pp. 146–147, 158).

Hippoboscidae

These peculiar insects, sometimes called louseflies, were once included with two families of bat parasites ("bat ticks") and a family of bee parasites in a group "Pupipara," but Bequaert (1953–1957) has shown that this is an unnatural assemblage, and that the Hippoboscidae are most nearly related to the tsetse flies.

The hippoboscids are dorso-ventrally flattened, with a leathery, saclike abdomen without evidence of segmentation in most species, with the head directly forward in a horizontal position, and with the opposite

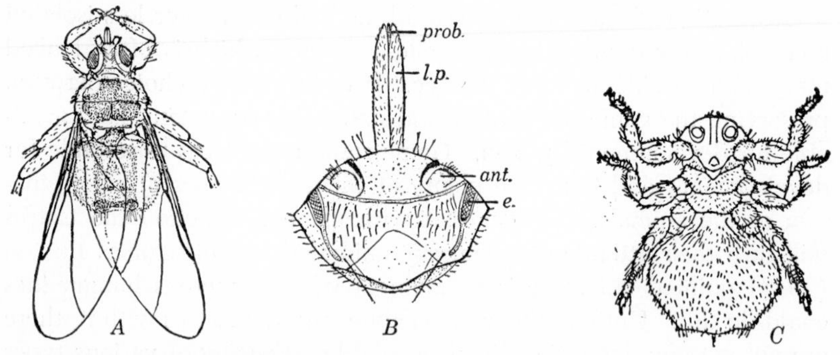

Fig. 224. *A, Pseudolynchia canariensis;* note large eyes and exposed antennae; *B* and *C, Melophagus ovinus,* the ked or "sheep tick." In *B,* note small eyes, antennae sunk in pits, and long palpi sheathing mouth parts, as in tsetse flies. (*A* and *B* adapted from Massonat, *Ann. Univ. Lyon,* 1909. *C* from Metcalf and Metcalf, *A Key to the Principal Orders and Families of Insects,* 1928.)

legs set wide apart. The mouth parts are similar to those of a tsetse fly but with the piercing proboscis partly retractile into a pouch in the ventral part of the head, the projecting part then sheathed by the labial palpi. In the winged forms the eyes are large and the antennae exposed, but in the wingless "sheep tick," *Melophagus ovinus,* the eyes are small and the reduced antennae sunken in pits on top of the head (Fig. 224*B, C*). They give birth, one at a time, to larvae which are fully mature and ready to pupate. In most species the larvae are deposited in dry soil or humus, in the nests of birds, or in other protected spots where they transform at once into pupae and turn a shiny black. In the subfamily Melophaginae, however, the motionless larvae are deposited in the wool or fur of the host; the wingless ked or "sheep tick," *Melophagus ovinus,* has the larval skin covered by a gluelike substance which sticks it to the wool, whereas the pupae of *Lipoptena,* parasitic on deer, mostly fall out of the fur in wallows when the deer shed their winter coat. These undergo a diapause (cessation of development) until fall, when the winged adults emerge and seek a new host, soon after which the wings break off.

The genus *Hippobosca* includes winged forms parasitic on horses, cattle, camels, dogs, etc., in the Old World. *Pseudolynchia,* also winged, attacks nestling birds; one species, *P. canariensis* (=*maura*) (Fig. 224*A*), is the intermediate host of the common pigeon parasite, *Haemoproteus columbae* (see p. 165). *Lynchia fusca* transmits *H. lophortyx* among California quail. The ked or "sheep tick," *Melophagus ovinus* (Fig. 224*B, C*), mentioned above, is annoying to sheep. The females give birth to a full-grown larva every 7 or 8 days until a total of 12 or 15 have been born. The pupal stage lasts for about 3 weeks, and the adult is ready to be a mother about 2 weeks after emergence. The ked serves as an intermediate host for the nonpathogenic *Trypanosoma melophagium* of sheep; apparently the blood of the sheep merely serves as a means of distribution of the trypanosomes for infection of other keds.

The hippoboscids only exceptionally bite man but produce painful bites when they do so.

Keds are easily controlled by DDT, rotenone, or other sprays applied under pressure, or by dips or even dusts. The pupae are not killed, but the residual effect lasts long enough to kill the adults when they emerge.

Eye Flies (Chloropidae)

The eye flies (Fig. 225) are small, nearly hairless flies, about 1.5 to 2.5 mm. long, of the family Chloropidae (=Oscinidae). The larvae of some species are pests of growing wheat, etc., but most of those annoying to man breed in excrement or decaying organic matter. *Hippelates collusor* (so-called by Sabrosky; usually called *H. pusio,* a common southeastern species) was found by Hall to breed in almost any substance in an advanced state of decay, but when artificially bred it did best in human excrement.

According to Hall the entire development from egg to adult required 11 days or more, averaging about 18 days, but Burgess (1951) found 28 days to be required and thought in nature the time might be extended to several months or nearly a year. Although the eye flies are not bloodsuckers in the ordinary sense, many of them are habitually attracted by the skin and natural orifices of man and animals, lapping up perspiration, excretions, exudations of sores and wounds, or blood from scratches or insect bites. Some species appear to be especially attracted to the eyes and lachrymal secretions of man, whence their name. The proboscis is fitted for lapping, as in the housefly, but is capable of being used as a rasping instrument to cause minute scarifications on the delicate conjunctival epithelium or on granulation tissue of sores, thus assisting pathogenic organisms in gain-

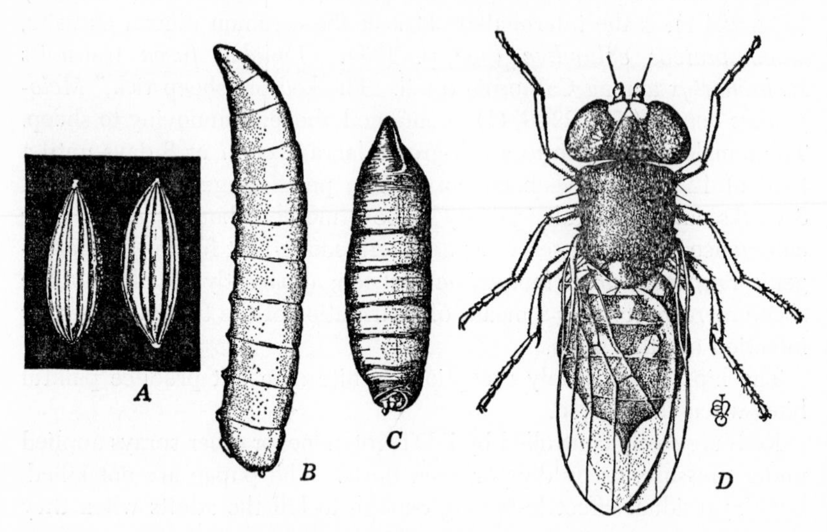

Fig. 225. Stages in life cycle of California eye fly, *Hippelates collusor*. *A*, eggs, in dorsal and lateral views; *B*, mature larva; *C*, pupa; *D*, adult ♀. *A*, about ×45; *B, C*, and *D*, about ×18. (After D. G. Hall, *Am. J. Hyg.*, 16, 1932.)

ing entrance. The habits of these insects, therefore, render them particularly dangerous mechanical carriers of eye infections and of various diseases of the skin and mucous membranes as well. Graham-Smith (1930) gives a review of the principal species concerned in disease transmission.

Relation to eye diseases. In the Coachella Valley of California *Hippelates collusor* (Fig. 225), and in parts of Florida *H. pusio*, are a sufficient nuisance to be limiting factors in the development of the country. They are small flies, 2 mm. in length, which are active throughout the day for 9 or 10 months of the year, but particularly in spring and fall. They persistently buzz around the heads of men and animals, frequently darting at the eyes or into the ears or feeding on sores or mucous membranes. In one high school 1500 children were reported as suffering from "pinkeye" in 1929; 50% of the young children of the region had some conjunctivitis, and 10% had chronic trachoma.

In India and the East Indies another member of the family, *Siphunculina funicola,* with similar habits, is responsible for spreading conjunctivitis and probably skin infecton also. In Egypt, where eye infections are particularly common, houseflies and related species are usually considered to be the principal transmitting agents, but chloropid flies are very common in some places and should be investigated in connection with the ophthalmia that is so prevalent.

Eye flies and yaws. Members of the family Chloropidae are an important, perhaps one of the most important, factors in the transmission of yaws (see pp. 222–223). Nichols in 1912 was convinced that eye flies, *Hippelates flavipes,* were responsible for the majority of cases of yaws in the West Indies, and similar views have been expressed by a number of other writers in the West Indies and Brazil. In Trinidad this species is called the "yaws fly."

Kumm and Turner (1936) found that the spirochetes of yaws remain motile in the pharynx and esophageal diverticula of *H. flavipes* for at least 7 hours but lose their motility in the midgut and do not undergo development in the fly as do the spirochetes of relapsing fever in lice. The flies commonly regurgitate drops of fluid after feeding; these drops contain viable spirochetes. Kumm and Turner experimentally transmitted the disease to rabbits both by bites of the flies and by inoculation of esophageal diverticula. There is a close correlation between the distribution of this fly and that of yaws in the West Indies. It was possible to catch over 1500 flies feeding on yaws sores within 15 minutes, so the opportunities for transmission are obvious.

In Assam, Fox in 1921 showed that epidemics of another spirochetal infection, Naga sore (see p. 221), are associated with plagues of *Siphunculina funicola,* which swarm on the sores and mechanically transmit the infective material.

Control. Control of these flies is difficult, but the suppression or treatment of decaying organic matter and improved sanitation to prevent breeding in human excrement would be of some value. Parman in 1932 advocated the use of box traps baited with odoriferous decaying infusions. DDT spraying of animals for other flies, and of buildings, etc., for houseflies, will certainly reduce the number of eye flies.

REFERENCES

General

Curran, C. H. 1934. *The Families and Genera of North American Diptera.* Ballou, New York.

Downes, J. A. 1957. The feeding habits of biting flies and their significance in classification. *Ann. Rev. Entomol.,* 2: 203–226.

Hayes, W. P. 1938–1939. A bibliography of keys for the identification of immature insects, Pt. I, Diptera. *Entomol. News,* 49: 246–251; 50: 5–10.

Lindsay, D. R., and Scudder, H. I. 1956. Non-biting flies and disease. *Ann. Rev. Entomol.,* 1: 323–346.

Snodgrass, R. E. 1943. The feeding apparatus of biting and disease-carrying flies. *Smithsonian Misc. Collections,* 104, No. 1.

Psychodidae (Phlebotomus, etc.)

Addis, C. J., Jr. 1945. Collection and preservation of sandflies (Phlebotomus) with keys to U. S. species. *Trans. Am. Microscop. Soc.,* 64: 328–332.

Adler, S., and Theodor, O. 1957. Transmission of disease agents by phlebotomine sandflies. *Ann. Rev. Entomol.,* 2: 203–226.

Biagi, F., and Biagi, A. M. 1953. Alganos flebotomus de area endemica de leishmaniasis tegumentaria Americana del Estado de Compeche (Mex.). *Medicina,* 33: 315–319.

Deane, L. M. 1956. Leishmaniosevisceral no Brasil. Estudos sobre reservatorios e transmissores realizados no estado do Ceara. *Serv. nac. educ. sanitaria,* Rio de Janeiro.

Hertig, M. 1942. Phlebotomus and Carrion's disease, I–IV. *Am. J. Trop. Med.,* 22, Suppl. 75.

1948. Sandflies of the genus *Phlebotomus*—a review of their habits, disease relationships, and control. *Proc. 4th Intern. Congr. Trop. Med. Malaria,* 2: 1609–1618.

1949. *Phlebotomus* and residual DDT in Greece and Italy. *Am. J. Trop. Med.,* 29: 773–809.

Hertig, M., and Fairchild, G. B. 1948. The control of *Phlebotomus* in Peru with DDT. *Am. J. Trop. Med.,* 28: 207–230.

Sabin, A. B., Philip, C. B., and Paul, J. R. 1944. Phlebotomus (pappataci or sandfly) fever, a disease of military importance. *J. Am. Med. Assoc.,* 125: 603–606, 693–699.

Theodor, O. 1948. Classification of the Old World species of the subfamily Phlebotominae. *Bull. Entomol. Research,* 39: 85–116.

Townsend, C. H. T. 1913. Progress in the study of verruga transmission by bloodsuckers. *Bull. Entomol. Research,* 4: 125–128.

Whittingham, H. E., and Rook, A. F. 1923. Observations on the life history and bionomics of *Phlebotomus papatasii. Brit. Med. J.,* 1144–1151.

Young, T. C. M., Richmond, A. E., and Brendish, G. R. 1925–1926. Sandflies and sandfly fever in the Peshawar district. *Indian J. Med. Research,* 13: 961–1021.

See also References, Chapter 7.

Heleidae (=Ceratopogonidae) (Culicoides, etc.)

Buckley, J. J. C. 1934. On the development in *Culicoides furens* Poey of *Filaria (Mansonella) ozzardi*. *J. Helminthol.*, 12: 99–118.

1938. On *Culicoides* as a vector of *Onchocerca gibsoni* (Cleland and Johnston, 1910). *J. Helminthol.*, 16: 121–158.

Chardome, M., and Peel, E. 1949. Le répartition des Filaires dans la région de Coquilhatville et la transmission de *Dipetalonema streptocerca* par *Culicoides grahami*. *Ann. soc. belge méd. trop.*, 29: 99–119.

Dove, W. E., Hall, D. G., and Hull, J. B. 1932. The salt marsh sandfly problem (*Culicoides*). *Ann. Entomol. Soc. Amer.*, 25: 505–522.

Foote, R. H., and Pratt, H. D. 1954. The *Culicoides* of the eastern United States (Diptera, Heleidae), a review. *Public Health Monogr.* 18.

Hopkins, C. A., and Nicholas, W. L. 1952. *Culicoides austeni*, the vector of *Acanthocheilonema perstans*. *Ann. Trop. Med. Parasitol.*, 46: 276–283.

Kershaw, W. E., et al. 1950–1953. Studies on the epidemiology of filariasis in West Africa, with special reference to the British Cameroons and the Niger Delta, I–IV. *Ann. Trop. Med. Parasitol.*, I, 44: 361–378; II, 45: 261–283; III, 47: 95–111; IV, 47: 406–425.

Nicholas, W. L. 1952. The bionomics of *Culicoides austeni*, vector of *Acanthocheilonema perstans* in the rain-forest of the British Cameroons, together with notes on *C. grahamii* and other species which may be vectors in the same area. *Ann. Trop. Med. Parasitol.*, 47: 187–206.

Root, F. M., and Hoffman, W. A. 1937. The North American species of *Culicoides*. *Am. J. Hyg.*, 25: 150–176.

Sharp, N. A. D. 1928. *Filaria perstans;* its development in *Culicoides austeni*. *Trans. Roy. Soc. Trop. Med. Hyg.*, 21: 371–396.

Williams, R. G. 1951. Observations on the bionomics of *Culicoides tristriatulus* Hoffman with notes on *C. alaskensis Wirth* and other species at Valdez. Alaska, Summer 1949. *Ann. Entomol. Soc. Amer.*, 44: 173–183.

Woke, P. A. 1954. Observations on Central American biting midges (Diptera, Heleidae). *Ann. Entomol. Soc. Amer.*, 47: 61–74.

Simulidae (Blackflies)

Blacklock, D. B. 1926. The development of *Onchocerca volvulus* in *Simulium damnosum*. *Ann. Trop. Med. Parasitol.*, 20: 1–48, 203–218.

Dalmat, H. T. 1954. The blackflies of Guatemala and their role as vectors of onchocerciasis. *Smithsonian Misc. Collections,* 125, No. 1 (Publ. 4173), 425 pp. Government Printing Office, Washington.

Fairchild, G. B., and Barreda, E. A. 1945. DDT as a larvicide against *Simulium*. *J. Econ. Entomol.*, 38: 694–699.

Garnham, P. C. C., and McMahon, J. P. 1947. The eradication of *Simulium neavei* Roubaud from an onchocerciasis area in Kenya Colony. *Bull. Entomol. Research*, 37: 619–628. (See also W. H. O. Rept. No. 5.)

Jobbins-Pomeroy, A. W. 1916. Notes on five North American buffalo gnats of the genus *Simulium*. *U. S. Dept. Agr. Bull.* 329.

Lea, A. O., and Dalmat, H. T. 1955. Field studies on the larval control of blackflies in Guatemala. *J. Econ. Entomol.*, 48: 274–278.

Muirhead-Thomson, R. C. 1957. Laboratory studies on the reactions of *Simulium* larvae to insecticides. I–III. *Am. J. Trop. Med. Hyg.*, 6: 920–934.

Nicholson, H. P., and Mickel, C. E. 1950. The black flies of Minnesota (Simuliidae). *Univ. Minn. Tech. Bull.* 192.

Shewell, G. E. 1955. Identity of the blackfly that attacks ducklings and goslings in Canada. *Can. Entomologist,* 87: 345–349.

Smart, J. 1945. The classification of the Simuliidae (Diptera). *Trans. Roy. Entomol. Soc. London,* 95: 463–532.

Twinn, C. R. 1936. The blackflies of eastern Canada. *Can. J. Research,* 14: 97–130.

Twinn, C. R., and Peterson, D. G. 1955. Control of blackflies in Canada. *Con. Dept. Agr. Publ.* 940.

Vargas, L., et al. 1946. Simulidos de Mexico. *Rev. del Inst. de Saludbridad y Enfermedad Trop.,* 7: 101–192.

Wanson, M., and Henrard, C. 1948. Habitat et comportement larvaire du *Simulium damnosum* Theobald. Review in *Rev. Appl. Entomol.* B, 36: 146–148.

World Health Organization. Report of expert committee on Onchocerciasis, Geneva, 1953. (Important paper by Bornley on control of S. *damnosum,* No. 18; Elishewitz on entomological investigations in Guatemala, No. 16; Freeman on S. *neavei,* No. 2; Lewis on Simuliidae in Anglo-Egyptian Sudan, No. 3; Garnham and McMahon, results of S. *neavei* control in Kenya, No. 5.)

See also references, Chapter 21.

Tabanidae

Bruce, W. N., and Decker, G. C. 1951. Tabanid control on dairy and beef cattle with synergized pyrethrins. *J. Econ. Entomol.,* 44: 154–159.

Cameron, A. E. 1926. Bionomics of the Tabanidae of the Canadian prairies. *Bull. Entomol. Res.,* 17: 1–42.

Davey, J. T., and O'Rourke, F. J. 1951. Observations on *Chrysops silacea* and *C. dimidiata* at Benin, southern Nigeria, I–III. *Ann. Trop. Med. Parasitol.,* 45: 30–37, 66–72, 101–109.

Duke, B. O. L. 1955. Studies on the biting habits of Chrysops, I–IV. *Ann. Trop. Med. Parasitol.,* 49: 193–202, 260–272.

Hays, K. L. 1956. A synopsis of the Tabanidae (Diptera) of Michigan. Misc. Publ. Museum Zool., Univ. Michigan, 98, 79 pp.

Jellison, W. L. 1950. Tularemia. Geographical distribution of "deerfly fever" and the biting fly, *Chrysops discalis Williston. Public Health Repts.,* 65: 1321–1329.

Kershaw, W. E. 1955. The epidemiology of infections with *Loa loa* (and discussion). *Trans. Roy. Soc. Trop. Med. Hyg.,* 49: 97–157.

Kershaw, W. E., et al. 1950–1955. Studies on the epidemiology of filariasis in West Africa, I–VII. *Ann. Trop. Med. Parasitol.,* 44: 361–378; 45: 261–283; 47: 95–111, 406–425; 48: 110–120; 49: 66–79, 455–460.

Mackerras, I. M. 1954–1955. The classification and distribution of Tabanidae. I–III. *Australian J. Zool.,* 2: 431–554; 3: 439–511, 583–633.

Miller, L. A. 1951. Observations on the bionomics of some northern species of Tabanidae (Diptera). *Can. J. Zool.,* 29: 240–263.

Mitzmain, M. B. 1913. Tabanids and surra. *Philippine J. Sci.* B., 8: 197–221, 223–229.

Philip, C. B. 1931. The Tabanidae (horseflies) of Minnesota. *Minn. Agr. Exp. Sta. Tech. Bull.,* 80, 32 pp.

Philip, C. B. 1947. A catalogue of the blood-sucking fly family Tabanidae of the nearctic region north of Mexico. *Am. Midland Naturalist,* 37: 257–324.

Schwadt, H. H. 1936. Horseflies of Arkansas. *Univ. Ark. Exptl. Sta. Bull.* 332.

Stone, A. 1930. The bionomics of some Tabanidae. *Ann. Entomol. Soc. Am.,* 23: 261–304.

1938. Horseflies or Tabanidae of the nearctic region. *U. S. Dept. Agr. Misc. Publ.* 305.

Muscoidea

Blakeslee, E. B. 1945. Surface sprays for control of stablefly breeding in shore deposits of marine grass. *J. Econ. Entomol.,* 38: 548–552.

Buxton, P. A. 1948. The problem of tsetse flies (*Glossina*). *Proc. 4th Intern. Congr. Trop. Med. Malaria,* 2: 1630–1637.

Buxton, P. A. 1955. *The Natural History of Tsetse Flies.* K. Lewis, London.

Decker, G. C., and Bruce, W. N. 1951. Where are we going with fly resistance? *Soap and Sanit. Chemicals,* 27: 139–145.

Haines, T. W. 1953, 1955. Breeding media of common flies. I and II. *Am. J. Trop. Med. Hyg.,* 2: 933–940; 4: 1125–1130.

Lindsay, D. R. 1951. The significance of house fly resistance to insecticides in fly control operations. *Communicable Disease Center Bull.,* November.

McLintock, J., and Depner. K. R. 1954. A review of the life history and habits of the hornfly, *Siphona irritans* (L.). *Can. Entomologist,* 86: 20–33.

Nash, T. A. M. 1937. Climate, the vital factor in the ecology of *Glossina. Bull. Entomol. Research,* 28: 75–127.

1948. *Tsetse flies in British West Africa.* Colonial Office, London.

Nash, T. A. M., and Steiner, J. D. 1957. The effect of obstructive clearing on *Glossina palpalis* (R.-D.). *Bull. Entomol. Research,* 48: 323–339.

Patton, W. S. 1932. A revision of the species of the genus *Musca,* etc. *Ann. Trop. Med. Parasitol.,* 26: 347–405.

Rogoff, W. M., and Moxon, A. L. 1952. Cable type back rubbers for hornfly control on cattle. *J. Econ. Entomol.,* 45: 329.

Schoof, H. F., and Kilpatrick, J. W. 1957. House fly control with parathion and diazinon impregnated cords in dairy barns and dining halls. *J. Econ. Entomol.,* 50: 24–27.

Séguy, E. 1937. Diptera, Fam. Muscidae. *Gen. Insectorum.* Fasc. 205, 1–600.

Simmons, S. W., and Dove, W. E. 1941–1942. Breeding places of the stable fly or "dog fly," *Stomoxys calcitrans* (L.) in northwestern Florida. *J. Econ. Entomol.,* 34: 457–462. See also *ibid.,* 35: 582–589, 589–592, 709–715.

Swynnerton, C. F. M. 1936. The tsetse flies of east Africa. *Trans. Roy. Entomol. Soc. London,* 84.

Toomey, J. A., Tokacs, W. S., and Tischer, L. A. 1941. Poliomyelitis virus from flies. *Proc. Soc. Exptl. Biol. Med.,* 48: 637–639.

U. S. Department of Agriculture. 1953. *Stable Flies, How to Control Them.* Leaflet 338. Washington.

U. S. Department of Agriculture. 1955. Hornflies on cattle–how to control them. Leaflet 388.

Weitz, B., and Glasgow, J. P. 1956. The natural hosts of some species of *Glossina* in East Africa. *Trans. Roy. Soc. Trop. Med. Hyg.,* 50: 593–612.

West, L. S. 1951. *The Housefly.* Comstock, Ithaca, N. Y.

See also References, Chapter 8.

Chloropidae

Burgess, R. W. 1951. The life history and breeding habits of the eye gnat, *Hippelates pusio* Loew, in the Coachella Valley, Riverside County, California. *Am. J. Hyg.*, 53: 164–177.

Graham-Smith, G. S. 1930. The Oscinidae (Diptera) as vectors of conjunctivitis and the anatomy of their mouth parts. *Parasitology*, 22: 457–467.

Kumm, H. W., and Turner, T. B. 1936. The transmission of yaws from man to rabbits by an insect vector, *Hippelates pallipes*. *Am. J. Trop. Med.*, 16: 245–271.

Sabrosky, C. W. 1951. Nomenclature of the eye gnats (*Hippelates* spp.). *Am. J. Trop. Med.*, 31: 257–258.

Hippoboscidae

Bequaert, J. 1942. A monograph of the Melophaginae; or ked-flies of sheep, goats, deer and antelopes (Diptera, Hippoboscidae). *Entomol. Americana*, 22: 1–64.

Bequaert, J. 1953–1957. The Hippoboscidae or louse-flies (Diptera) of mammals and birds, Parts I and II. *Entomol. Americana*, 32: 1–209; 33: 211–242; 34: 1–232; 35: 233–416; 36: 417–611.

Evans, G. O. 1950. Studies on the bionomies of the sheep ked, *Melophagus ovinus* L. in West Wales. *Bull. Entomol. Research* 40: 459–478.

Kemper, H. E., and Peterson, H. O. 1953. The sheep tick and its eradication. *U. S. Dept. Agr. Farmer's Bull.*, 2057. Washington.

Massonnat, E. 1909. Contribution à l'étude des Pupipares. *Ann. Univ. Lyon., n. ser. I., Sci. Med. Fasc.* 28.

Chapter 29

DIPTERA
II. MOSQUITOES

Importance. Of all existing insect pests mosquitoes are Public Enemy No. 1. The mere annoyance which the enormous numbers of them cause by their bites is sufficient to have made some parts of the world practically uninhabitable. Some of the choicest hunting and camping grounds have been transformed by countless millions of mosquitoes into an intolerable hell.

Unlike most insect pests the mosquitoes in the Far North are more abundant than they are in the tropics. In places in Alaska mosquitoes have been observed to land on the back of a woolen glove at the rate of 70 per minute, and spray equipment for outdoor insect control is as necessary to the civilized life of residents as are roads or a kitchen stove.

Fortunately these far northern mosquitoes are not disease transmitters, whereas many tropical mosquitoes have their spears poisoned with death-dealing disease germs. Unfortunately many of the tropical species do not annoy when biting and thus are not particularly noticed. No less than five important human diseases are normally transmitted by mosquitoes exclusively—malaria, yellow fever, dengue, filariasis, and various forms of encephalomyelitis (see p. 229). A South American fly, *Dermatobia,* usually depends on mosquitoes for transportation of its eggs to the skin of man or animals, where they hatch and cause myiasis. Many diseases of animals, including various virus infections, such as fowlpox of poultry, myxomatosis of rabbits, Rift Valley fever of sheep, and encephalitis of horses, as well as of birds, and heartworm of dogs, are also transmitted primarily or exclusively by mosquitoes.

General structure. Mosquitoes, comprising the family Culicidae (see key, p. 666), can easily be distinguished from all other Diptera, some of which superficially resemble them, by the presence of scales along the wing veins and a conspicuous fringe of scales along the hind margin of the wings. The venation (see Figs. 153 and 234) is very similar in all the species, but the coloration produced by the scales,

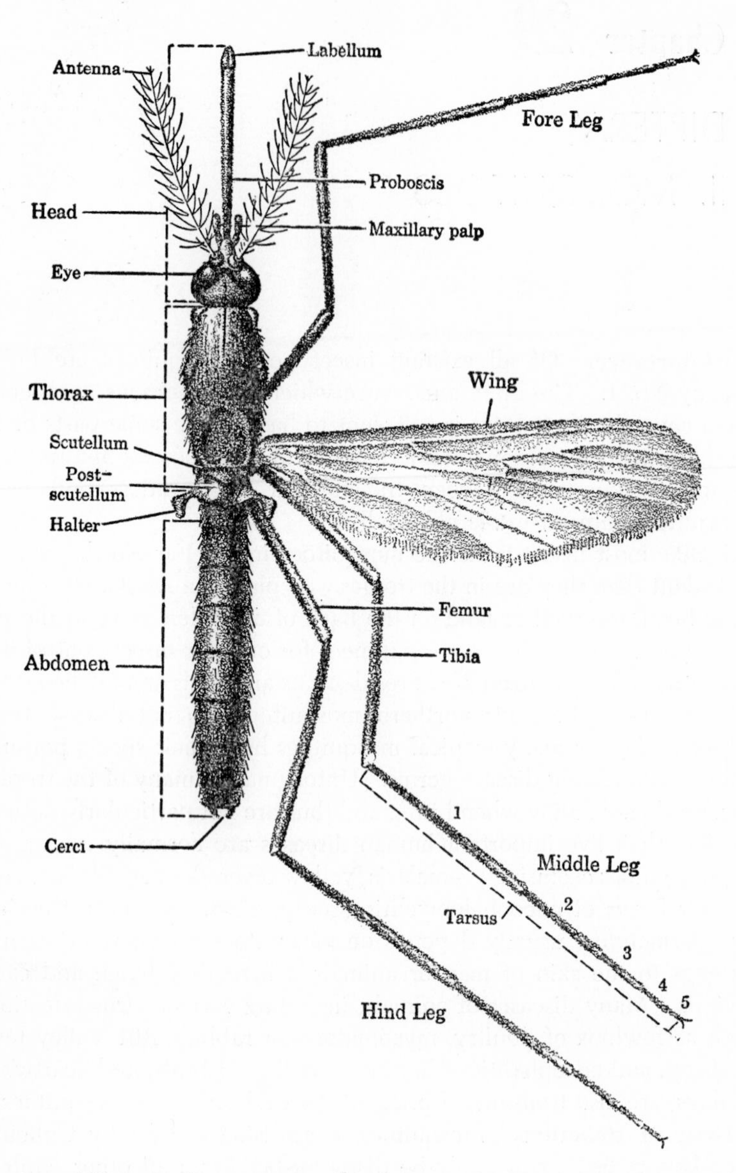

Fig. 226. Diagram of a mosquito, dorsal view. (After MacGregor, *Mosquito Surveys*, 1927.)

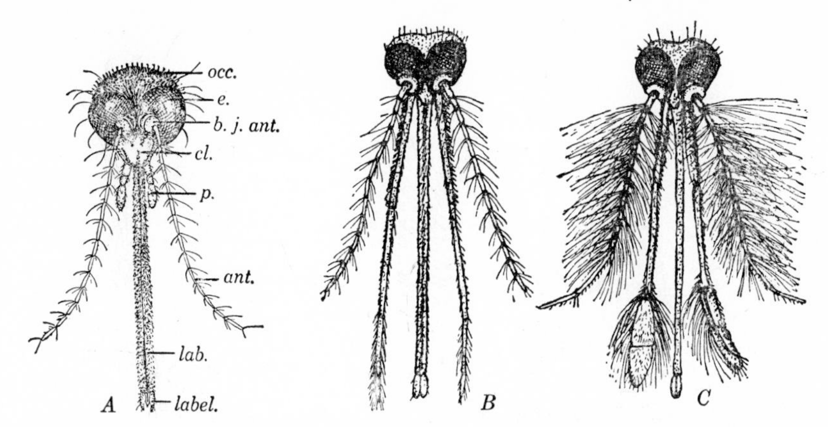

Fig. 227. Heads of mosquitoes. *A,* ♀ *Culex; B,* ♀ *Anopheles; C,* ♂ *Anopheles: ant.,* antenna; *b.j.ant.,* basal joint of antenna; *cl.,* clypeus; *e.,* eye; *lab.,* labium; *label.,* labellum; *p.,* palpus; *occ.,* occiput.

especially in *Anopheles,* is useful in identification of species in that genus. Most of the Culicidae have a long prominent proboscis containing needlelike organs for piercing and sucking, but in two subfamilies there is no long proboscis.

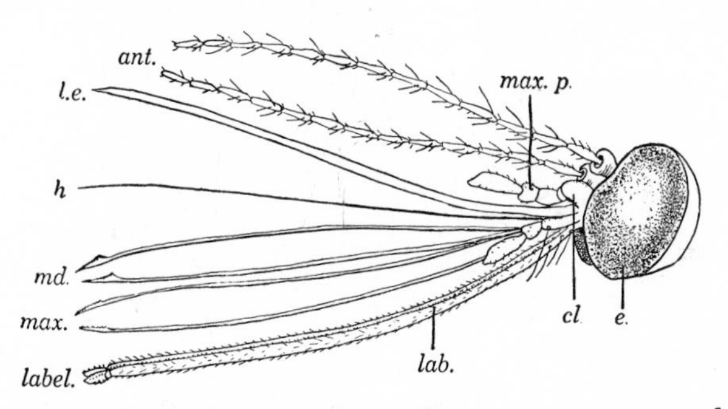

Fig. 228. Head and mouth parts of ♀ culicine: *ant.,* antenna; *cl.,* clypeus; *e.,* eye; *h.,* hypopharynx; *lab.,* labium; *label.,* labellum; *l.e.,* labrum-epipharynx; *max.,* maxilla; *max.p.,* maxillary palpus; *md.,* mandible. (After Matheson, *Medical Entomology,* Comstock.)

Figures 226, 227, 228, 229, and 234 illustrate the main features of a mosquito. The sexes can usually be distinguished most readily by the antennae (Fig. 226); in the female they are long and slender with a whorl of a few short hairs at each joint, whereas in the male they have a feathery appearance due to tufts of long and numerous hairs at the joints. In many mosquitoes the palpi also furnish a means of distin-

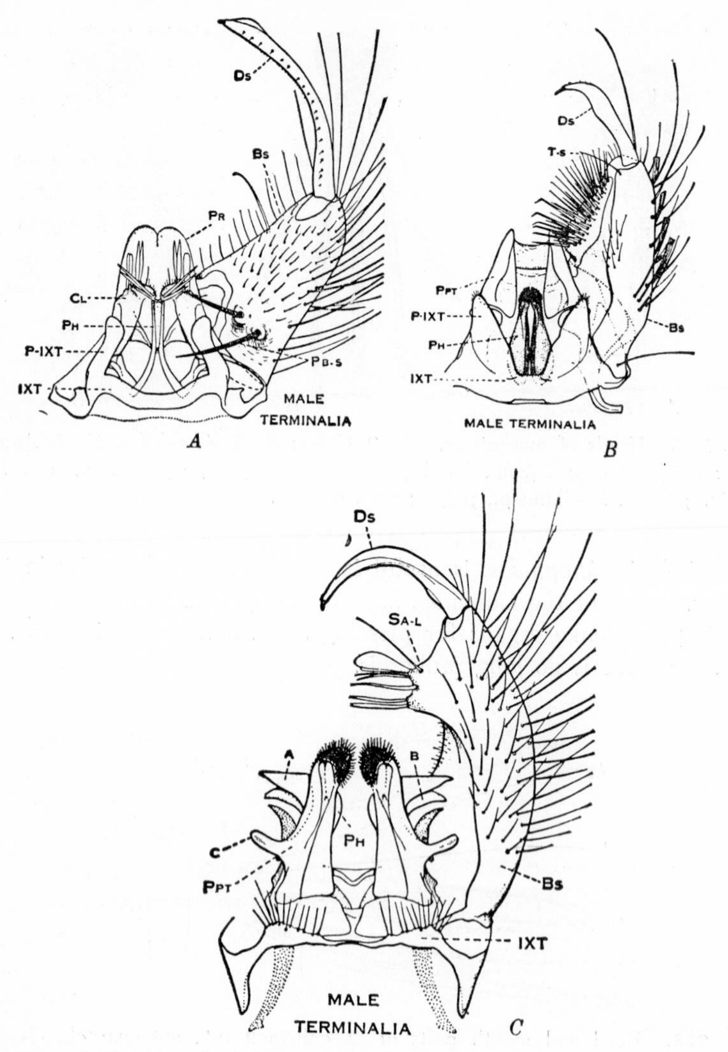

Fig. 229. Male genitalia of mosquitoes. *A, Anopheles quadrimaculatus; B, Aedes aegypti; C, Culex fatigans.* Abbreviations: *b.s.*, basistyle or side piece; *cl.*, claspette, with spines; *d.s.*, dististyle or clasper; *ph.*, phallosome or aedeagus; *pr.*, proctiger or anal lobe; *sa.l.*, subapical lobe (only in *Culex*). (After Ross and Roberts in *Mosquito Atlas,* I.)

guishing the sexes; they are usually long in the males but short in the females (Fig. 227), but in *Anopheles* they are long in both sexes, and in some mosquitoes, e.g., *Uranotaenia,* they are short in both.

The proboscis also differs in the sexes and fortunately is so constructed in the male that a mosquito of this sex could not pierce flesh if he wished. At first glance the proboscis appears to be a simple bristle, sometimes curved, but when dissected and examined with a microscope it is found to consist of a number of needlelike organs lying in a groove in the fleshy labium which was the only part visible before dissection. In the female mosquito there are six of these needlelike organs, the nature and names of which are shown in Fig. 228. The labrum-epipharynx and hypopharynx are the principal piercing organs and act together to form a tube for drawing up blood into the mouth. A tiny tube run down through the hypopharynx, opening at its tip, through which saliva is poured into the wound as through a hypodermic needle to prevent blood from coagulating. The maxillae and mandibles are thin, flat and flexible, the former recognizable by their sawtooth tips. The ensheathing labium bows back as the mosquito bites, the flexible tip or labellum acting as a guide for the piercing organs as they are sunk into the flesh. In male mosquitoes the maxillae and mandibles are much degenerated, only the remaining part of the apparatus being well developed.

Besides the variations of the parts mentioned already, adult mosquitoes vary in the form, distribution, and color of the scales that clothe much of the body and the edges and veins of the wings, the distribution of bristles on the thoracic sclerites (Fig. 234), the details of the male reproductive organs at the tip of the abdomen (Fig. 229), the details of the female hypopygium, and in other respects. Mosquitoes have three food reservoirs connected with the esophagus, in addition to a large stomach (Fig. 37). These reservoirs are used for storage of "aspirated" foods such as fruit juices, but not for blood, which passes directly to the stomach. This reduces the probability of *direct* transmission of disease germs immediately after an infective feed. Connected with the proboscis is a pair of salivary glands consisting of three lobes each lying in the anterior part of the thorax. It is in these that the malaria parasites collect after development; from here they are poured with the secretions of the glands into the wounds.

Life History and Habits

In a general way the life histories of all mosquitoes are much alike, but they differ in details.

Eggs. The eggs of mosquitoes (Fig. 230) are usually oval with various surface markings and, in *Anopheles,* with a peculiar "float" of air cells. The number of eggs laid by one female mosquito varies from 40 to 50 to several hundred. Species of *Aedes* and *Psorophora* lay their eggs singly out of water; *Anopheles* lays them singly in loose clusters on water (Fig. 231A); *Culex, Culiseta,* and *Uranotaenia* lay them in little boat-shaped rafts called egg-boats, the individual eggs standing upright (Fig. 231*B*); and *Mansonia* lays them in irregular clusters on the underside of floating leaves or in rafts.

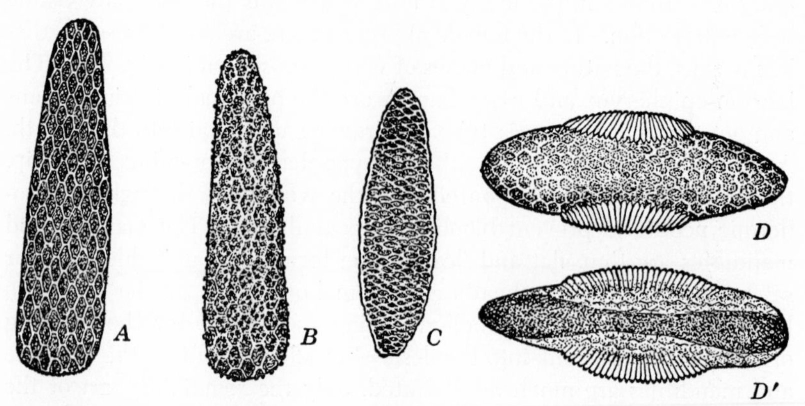

Fig. 230. Eggs of mosquitoes: *A*, *Culiseta inornata;* *B*, *Mansonia perturbans;* *C*, *Aedes aegypti;* *D*, *Anopheles punctipennis,* dorsal view; *D′*, same, ventral view. ×75. (After Howard, Dyar, and Knab, *Carnegie Inst. Wash. Publ.* 159, 1912–1917.)

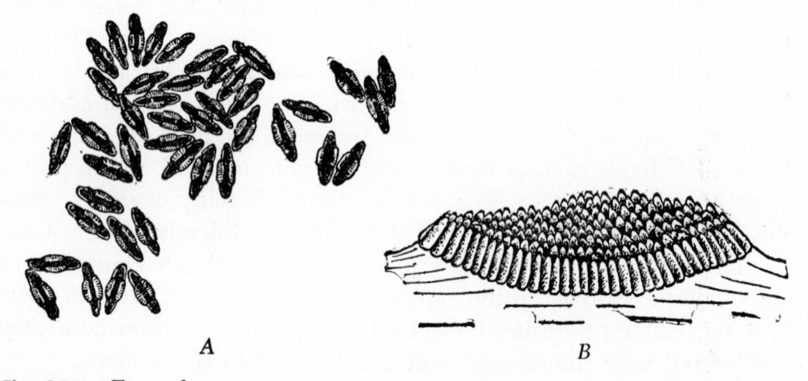

Fig. 231. Eggs of mosquitoes: *A*, eggs of *Anopheles* on surface of water, ×13 (after Howard, *Farmer's Bull.* 155, 1908). *B*, egg-boat of *Culex* floating on water, about ×6.

Species of *Anopheles, Culex, Culiseta,* and *Mansonia,* which are common in warm climates, lay their eggs on the open surface of water or attach them to some partially submerged object. Species of *Aedes* and *Psorophora,* on the other hand, lay their eggs *out* of water in places likely to be submerged later, e.g., in dry depressions in the North that will be filled with melted snow the following spring, in dry depressions in marshes or meadows that will be flooded after rains or high tides, or just above the water line in tree holes or containers. This is a useful biological adaptation to make an adequate supply of water likely when the eggs hatch.

When the eggs are laid on water they may hatch in a few days, or even within 24 hours, but those laid out of water lie unhatched until submerged, which may be weeks or months, and even then only a fraction of them usually hatch at the first immersion. The eggs of the far northern species of *Aedes* lie dormant until the following spring, solidly frozen through the long winter. Mosquitoes of dry, hot countries lay eggs that are highly resistant to desiccation and do not lose their vitality during months of dryness. Such species must almost "live while the rain falls" and must be prepared to utilize the most transitory pools for the completion of their aquatic immature stages. In such cases the embryo within the egg shell develops to the hatching point, so that it is ready to begin the larval existence almost with the first drop of rain. Eggs of some species of *Aedes* (*vexans* and *sticticus*) may remain viable in the soil for several years.

Such mosquitoes further fortify their race against the unkind environment by laying their eggs in a number of small batches instead of in a single mass, as is the habit of mosquitoes where water is plentiful. Just as a man runs less risk of ruin if he deposits his money in a number of insecure banks rather than in a single uncertain one, so it is with mosquitoes and the places where they deposit their eggs. The failure of all of a batch of dried eggs to hatch at any one immersion is a similar adaptation.

The vagaries of different species in selecting breeding places are discussed on the following pages. The only feature common to all is the fact that the eggs do not hatch except in the presence of water.

Larvae. The larvae, which are always aquatic, are well known as wrigglers or wiggle-tails (Fig. 232). When first hatched they are almost microscopic, but they grow rapidly to a length of 8 to 15 or 20 mm. The bunches of long bristly hairs on the body aid the larva in maintaining a position in the water. There is a rotary mouth brush of stiff hairs used to sweep small objects toward the mouth; in cannibalistic species that feed largely on other mosquito larvae, such as *Toxorhynchites* and the huge larvae of the "gallinippers" of the genus *Psorophora*, the mouth brushes are modified into rakelike structures or into strong grasping hooks for holding prey (Fig. 236, 8).

A trumpet-shaped siphon or breathing tube is present on all true mosquito larvae except *Anopheles*, in which it is undeveloped. It is used to pierce the surface film of the water, attach larva to surface film by means of special "valves," and to draw air into the tracheae, for mosquito larvae are air breathers and make frequent trips to the surface to replenish their air supply. The leaflike anal gills on the last segment of the abdomen differ from true gills in that air tubes or

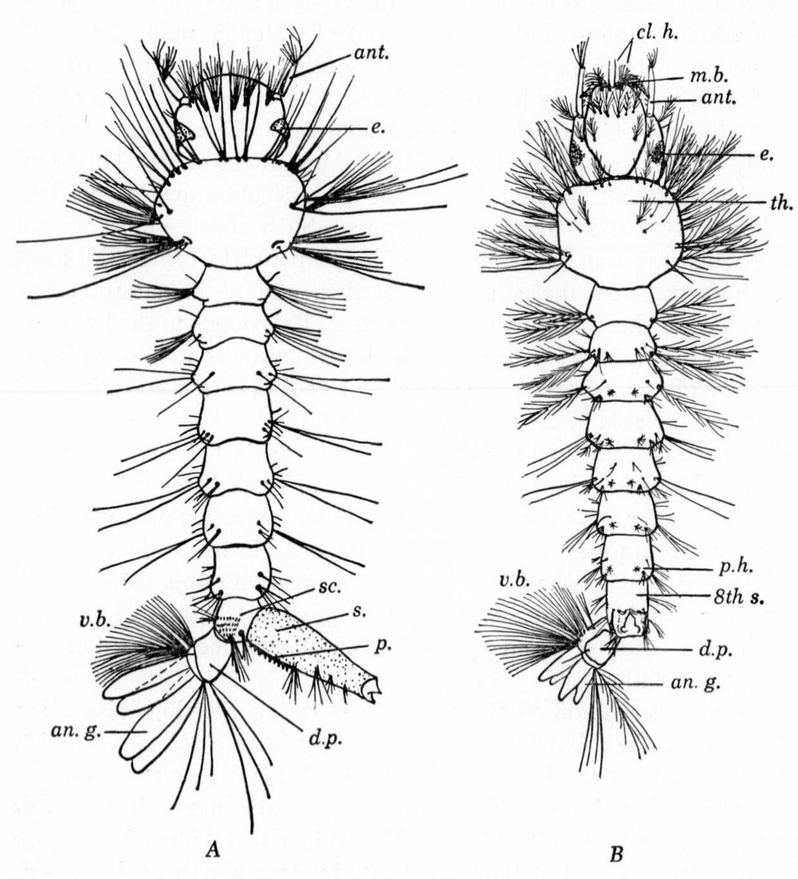

Fig. 232. *A*, Larva of *Culex quinquefasciatus* (after Soper); *B*, larva of *Anopheles punctipennis* (after Matheson); *an.g.*, anal gills; *ant.*, antenna; *cl.h.*, clypeal hairs; *d.p.*, dorsal plate of 9th segment; *8th s.*, 8th segment of abdomen; *e.*, eye; *m.b.*, mouth brush; *p.*, pecten; *p.h.*, palmate hair; *s.*, siphon or breathing tube; *sc.*, patch of scales on 8th segment; *th.*, thorax; *v.b.*, ventral brush.

tracheae instead of blood vesels ramify in them. They may function primarily as osmotic regulators rather than respiratory organs since they are always larger in mosquitoes living in saline water. In well-aerated water larvae can live for a long time, but they die within a few hours if shut in water without dissolved air. In one genus, *Man-*

sonia, the larvae absorb air from the air-carrying tissues in the roots of certain aquatic plants, piercing them with the apex of the breathing tube (Fig. 236, *11*) and thus avoiding the necessity of rising to the surface of the water.

Mosquito larvae, unless suspended from the surface film by means of the breathing tube, have a tendency to sink, and they rise again only by an active jerking of the abdomen, using it as a sculling organ. Some species are habitual bottom feeders; others feed at the surface; some live on microscopic organisms, others on dead organic matter; and still others attack and devour other aquatic animals, including young mosquito larvae of their own and other species. Soluble and colloidal substances in water can also be utilized.

The larvae shed their skins four times and then go into the pupal stage. Mosquitoes of temperate climates usually take 5 days to 2 weeks or longer to complete the larval existence, depending on the species and on temperature and abundance of food. In the mosquitoes adapted to take advantage of transitory rain pools the larvae may transform into pupae within 2 days. On the other hand, some mosquitoes habitually pass the winter as larvae. Larvae of most mosquitoes can live for several days on a wet surface but do not succeed in pupating except in water. Adults usually succeed in emerging from pupae stranded on mud.

A key for the identification of the larvae of different genera, and of important American species, is given on pp. 731–734.

Pupae. The general form of the pupa can be seen in Fig. 233; it resembles a tiny lobster deprived of appendages and carrying its tail bent. The pair of earlike breathing tubes on the cephalothorax takes the place of the trumpetlike tube of the larva and is used in the same manner except that there are no apical "valves." Unlike the larva, the pupa is lighter than water and requires muscular effort to sink instead of to rise. The pupae of *Mansonia,* like the larvae, do not come to the surface for air but pierce the air channels in the roots of aquatic plants with their pointed breathing tubes.

Adults. The transformation into the adult during the pupal stage may be a matter of a few hours in the case of the dry-climate mosquitoes, but in most species it requires 2 days to a week, depending on the temperature. The adult mosquitoes emerge head first through a longitudinal slit along the back of the cephalothorax. After their exit they rest a few moments on the old pupal skin, stretch and dry their wings, and then take flight. They may suck blood within 24 hours, but several days to a week or more elapse before they lay eggs.

In warm climates most mosquitoes have as many generations a year as the length of the life cycle and the climate permit, probably up to 15 or 20 for some species in the tropics where year-round breeding is possible. All the mosquitoes in the far north have a single brood (and what a brood!), but there are two types of life cycle: the *Aedes* type in which the females mate, feed, and deposit their eggs within a few weeks after emerging, the eggs then lying dormant until the following spring; and a type exemplified by *Culiseta impatiens* in which the females court and mate the first summer but have no lust for blood, then seek shelter and hibernate until the following spring when they engorge on blood, lay their eggs, and die (Frohne, 1956).

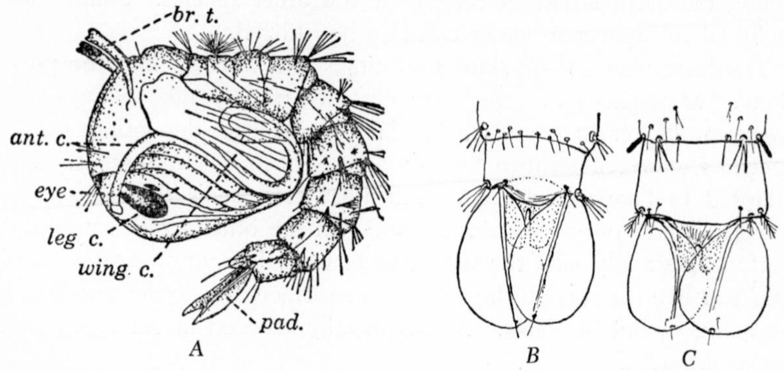

Fig. 233. *A*, pupa of *Culex pipiens; ant.c.*, antennal case; *br.t.*, breathing tubes; *leg c.*, leg cases; *pad.*, paddles; *wing c.*, wing case. ×10. (After Howard, Dyar, and Knab, *Carnegie Inst. Wash. Publ.*, 159, 1912–1917.) *B* and *C*, end of abdomen of pupae of *Aedes vexans* and of *Anopheles m. occidentalis*, respectively. *Anopheles* has peg-like spines at apical angles of all but last segment, and culicines have single tuft of branched fine hairs anterior to angle. (Adapted from Matheson, *Handbook of Mosquitoes of North America*, Comstock, 1944.)

Adult mosquitoes vary to a remarkable degree in habitats, feeding habits, mode of hibernation, choice of breeding grounds, and other habits. Knowledge of the habits and habitats of particular species is of great economic importance, since it does away with useless expenditure in combating harmless or relatively harmless species and aids in the fight against particularly noxious ones. The measures that would be required to control the breeding of the yellow fever mosquito, *Aedes aegypti*, would have little or no effect on *Anopheles quadrimaculatus*, and vice versa. Cutting down jungle to admit sunlight to water would eliminate malaria in some places but would be the best way to increase it in others. In dealing with mosquito-borne diseases or mosquito plagues it is obvious, therefore, that the particular species involved should be determined and their habits thoroughly understood.

Relations to man. Because of their variable breeding habits, choice of breeding places, tastes in blood, extent of travels, and willingness to come indoors to bite or to rest, different species of mosquitoes have very different relations to human beings. The majority that are annoying or dangerous to man can conveniently be divided into the following categories.

1. A few domestic species that breed in or near houses, readily enter them, and feed primarily on human blood. *Aedes aegypti* and members of the *Culex pipiens* complex stand out in this group. These species tend to be very local, seldom scattering more than a few hundred yards, so that people at 2100 Main Street may complain bitterly of mosquitoes while those at 2500 Main Street seldom see one. A number of species of *Anopheles* are also house pests, which have been a big factor in malaria control, e.g., *A. quadrimaculatus* in United States, *A. maculipennis* in Europe, *A. darlingi* in South America, and *A. gambiae* in Africa, and *A. stephensi* in India—the last is a truly urban species.
2. Species that regularly come in contact with man outdoors, and may come into houses to feed but not to sit. These breed in rice fields, in irrigation, seepage and overflow, in swamps or temporary rain pools or in tree holes, axils of banana leaves, coconut shells, etc., near villages or cultivated areas. Important in this group are *Culex tarsalis* in the United States, *Aedes albipictus* and the *A. scutellaris* group in the Pacific Islands, *Mansonia* spp. in Asia, and many species of *Anopheles* in various parts of the world. Many other species belonging to various genera make life miserable for picnickers, campers, hunters, fishermen, and porchsitters.
3. Species that breed in salt or fresh-water marshes or in temporary rain pools in meadows. These deposit their eggs in dry depressions, and emerge by billions about 5 days after rains, floods, or high tides. Sometimes marshes average up to 50 million larvae per acre. In the United States two species, *Aedes sollicitans* and *A. taeniorhynchus*, migrate up to 30 or 40 miles under favorable conditions, and towns in their path may be blanketed by mosquitoes, though fortunately for only a few days. A number of other species, especially in the genus *Psorophora*, contribute locally to these mass invasions, but are not such long-distance fliers. In this group we may also include the single-brooded mosquitoes of the far North, whose life cycles are outlined on p. 724.
4. Forest canopy mosquitoes that breed in tree holes or water-holding tracts of leaves, especially the bromeliads that grow in tree

tops. These come in contact with man extensively only under special circumstances as when the bromeliad-bearing trees are planted for shade in coffee groves or along city streets. To this group belong *Anopheles bellator* of Trinidad and *A. cruzi* of Brazil, both malaria-carriers, and the vectors of jungle yellow fever—in South America, especially the gay-colored species of *Haemagogus*.

Breeding places. The breeding places of mosquitoes include practically any kind of water except the open sea; some species show very little preference, whereas others seem to be unreasonably choosy. There are species which breed in reedy swamps, woodland pools, eddies of rivers, slow-flowing streams, holes in trees, pools of melted snow, salt marshes, tide pools, crab holes, pitcher plants, treetop bromeliads, broken bamboo stems, coconut shells, or artificial containers from tin cans to cisterns and flooded basements. The species of *Mansonia* breed only in pools in which certain water plants grow, especially water lettuce (*Pistia*) or water hyacinths.

Some species breed only in pure, clear water, others prefer filthy water; some breed only in sunlit water, others only in the shade; some demand quiet water, others breed only in flowing streams. Sometimes, of apparently similar pools, some will produce vast numbers of mosquitoes whereas others are left uninhabited; for the most part nobody knows why. Attempts have been made to correlate the preferences shown by mosquitoes with food, acidity or alkalinity, oxygen concentration, dissolved solids, etc., but with little success. The complexity of the problem is great, for odors may attract or repel the females searching for places to lay eggs; substances in the water may be directly injurious to the larvae; or, what is probably usually the dominant factor, the quantity or quality of the food may or may not be suitable. Most mosquito larvae, however, are able to use a considerable variety of foods, though living organisms are usually preferred. Barber successfully reared certain larvae on algae, bacteria, or ciliates alone. Biological control of mosquitoes may eventually be possible, but so far only the surface has been scratched.

Migration. As already noted, the distance mosquitoes travel from their breeding places varies greatly with the species; knowledge of this is of great importance in connection with control. *Aedes aegypti* in cities and *A. polynesiensis* in jungles are examples of species that seldom stray more than 100 yards from where they were bred, and members of the *Culex pipiens* complex not much farther. When mosquitoes are complained of in cities, a breeding place can nearly always be found within a block, and often right on the premises. Most *Anoph-*

eles make nightly migrations but seldom appear in appreciable numbers more than half a mile to a mile from their breeding places; a few species are known to migrate up to 10 miles. The yellow fever-carrying *Haemagogus* of South America travel at least 4 to 6 miles. *Aedes* on northern prairies are attracted by moving herds of animals and may follow them for many miles. Salt-marsh mosquitoes, however, are the only really migratory species, sometimes going 30 to 40 miles from home (see p. 725). *A. aegypti* is annually carried inland from the Gulf Coast by trains and busses and to northern ports on ships. Hawaii originally had no mosquitoes, but three species, *Culex quinquefasciatus, Aedes aegypti,* and *A. albopictus,* have been introduced with sailing vessels. In 1930 the deadly African malaria transmitter, *Anopheles gambiae,* was introduced and established in Brazil, as it was some years earlier in Mauritius and the Seychelles and later in Egypt.

By no means are all mosquitoes nocturnal. *Aedes aegypti* and most others of the subgenus *Stegomyia,* e.g., the *A. scutellaris* group, are diurnal, though most active in early morning and late afternoon, or in dense shade. Numerous forest-dwelling species, and northern species that live where the nights are too cold for them, are also diurnal. Most *Anopheles* and *Culex,* however, seek food at dusk or night. Here again a knowledge of its habits may aid in intelligent avoidance of a disease-carrying species.

Food habits. Heretical as it may sound, mosquitoes feed mainly on plant juices, honey, etc. Philip found flowers, e.g., goldenrod, to be good collecting places for all but *Anopheles* mosquitoes. All adult males are strictly vegetarians and some females are also, e.g., *Toxorhynchites* (=*Megarhinus*), which, however, is strictly cannibalistic on other mosquito larvae in its larval stage. Some females, although also feeding on nectar, etc., are bloodsuckers, and some require blood before they can lay fertile eggs. Dining on blood is not, however, alway a necessary prelude to maternity; a domestic "citified" variety of *Culex pipiens, C. molestus,* is autogenous, i.e., oviposits without the need of blood meals. Roubaud bred twenty generations of this species without ever allowing the adults to feed at all. Some species indiscriminately attack any warm-blooded or even cold-blooded animal, but others show strong preferences. *Aedes spencerii* of our northern prairies flies towards any large object which its instinct leads it to suspect as a source of food. The importance of various species of *Anopheles* as malaria transmitters (see pp. 734–742) depends largely on the extent to which they choose man as food. This is why *A. gambiae* is more dangerous than *quadrimaculatus,* and *quadrimaculatus* than *punctipennis. Anopheles gambiae* is one of the few species showing a strong preference for human blood. *Aedes aegypti* shows no distaste for man but readily bites other mammals, birds, and even cold-blooded animals.

Hibernation. The method employed by mosquitoes for passing the cold or dry seasons varies with the species. Many mosquitoes of temperate or tropical climates hibernate or pass the dry season as adults, the females stowing themselves away in hollows in trees, caves, crevices in rocks, cellars, barns, etc., to come forth and lay their eggs in the spring. A few species hibernate in the larval stage; *Wyeomyia smithii* larvae become enclosed in solid ice in the leaves of the pitcher plant in which they live. Most hibernating larvae retire to the bottom of their breeding pools during cold weather and do not survive freezing. Many temperate- and warm-climate mosquitoes and the northern *Aedes* pass the unfavorable season in the egg state, which may be looked upon as the *common* method of hibernation. Most *Anopheles* survive the cold season either as adults or as larvae but usually not as eggs.

Length of life. The length of life of mosquitoes varies with the species and the sex. Male mosquitoes seldom live more than 1 to 3 weeks; their duty in life is done when they have fertilized the females. Paradoxically, the more favorable the conditions the shorter the lives of the females. They die soon after all their eggs are laid; with plenty of blood meals and readily available breeding places this may be only 3 or 4 weeks, whereas under less favorable conditions it may be several months.

Classification

More than 2000 species of Culicidae have been described, the majority of which belong in the tropics, although the north is rich in individuals; over 120 species have been described in the United States and Canada. There are two subfamilies, the Corethrinae and Dixinae, with a short nonpiercing proboscis, and a third subfamily, the Culicinae, including all the true mosquitoes. The further divisions into tribes and genera are shown in the following key, which is provided for the identification of the genera found in North America and of the more common or important species:

Key to Adults of North American Genera of Mosquitoes, with Notes on Commonest Species

I. Tribe **Sabethini.** Postnotum (Fig. 234, *postn.*) with tuft of setae; tropical, nonbloodsucking species, breeding in water-holding plants; two wide-spread species in North America
Wyeomyia smithii and ***W. haynei***

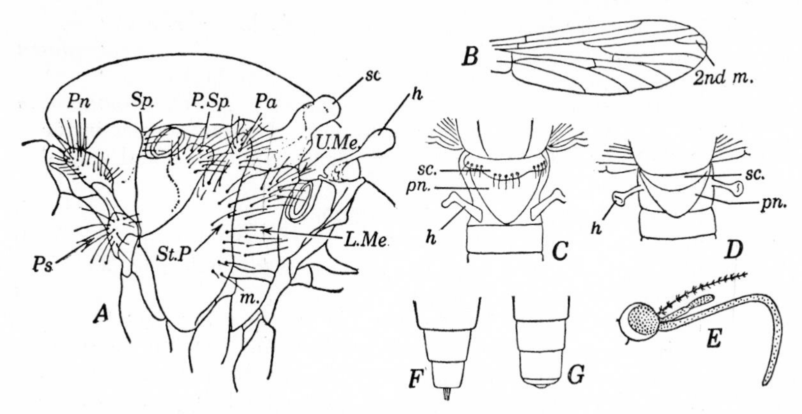

Fig. 234. Details of adult mosquitoes to illustrate key. *A*, thorax of *Psorophora* showing plates and groups of bristles used in distinguishing genera; *h.*, haltere; *l.me.* and *u.me.*, lower and upper mesepimeral bristles; *pa.*, prealar; *pn.*, pronotal; *ps.*, prosternal; *p.sp.*, postspiracular; *sc.*, scutellum; *st.p.*, sternopleural. *B*, wing of *Uranotaenia* with short 2nd marginal cell (*2nd m.*); *C*, trilobed scutellum (*sc.*) of culicines. *D*, scutellum of anophelines, without lateral lobes. *E*, head of *Toxorhynchites* (=*Megarhinus*) with curved proboscis. *F*, end of abdomen of *Aedes* ♀. *G*, end of abdomen of *Culex* ♀. (*A*, adapted from Matheson, *Handbook of the Mosquitoes of North America*, Comstock, 1944.)

(Palpi short in both sexes; very small; color metallic, with underside of abdomen, tips of middle legs, and spot on top of head silvery white; breed in pitcher plants)

II. Tribe **Anophelini.** Postnotum without setae; scutellum not lobed (Fig. 234*D*); palpi long in both sexes (Fig. 227); wings usually spotted or mottled; resting position usually not humpbacked; proboscis nearly parallel with axis of body (Fig. 238, *bottom*); only 1 genus ***Anopheles***

(Key to 11 North American species or varieties on p. 738)

III. Tribe **Culicini.** Postnotum without setae; scutellum usually trilobed (Fig. 234*C*); palpi short in female; wings rarely spotted, never mottled; resting position humpbacked (Fig. 237, *bottom*)

1*a*. Second marginal cell ("2nd m." in Fig. 234*B*) less than half as long as petiole; palpi short in both sexes ***Uranotaenia***
(Small, tropical, nonbiting, pool-breeding mosquitoes)

1*b*. Second marginal cell over half as long as petiole (Fig. 153*D*) 2

2*a*. Spiracular bristles present (Fig. 234*A*); size large; metallic colors present or absent 3

2*b*. Spiracular bristles absent; size medium or small; no metallic colors 4

3*a*. No postspiracular bristles (Fig. 234*A*); large; not metallic ***Culiseta***
(*C. melanura*, small, dark, clear-winged, sylvatic vector of EEE (see p. 752); *C. incidens*, dark spots on wings, unstriped legs, common on west coast; *C. impatiens*, an Alaskan pest)

3*b*. Postspiracular bristles present; size usualy large; metallic colors sometimes present **_Psorophora_**
(One species, *P. ciliata*, is largest mosquito in the United States)

4*a*. Fourth segment of front tarsi very short **_Orthopodomyia_**
(Tree-hole breeders, mostly in tropics. The one species in the United States has thorax with 6 narrow white lines and striped legs.)

4*b*. Fourth segment of front tarsi normal 5

5*a*. Female with tip of abdomen truncated or blunt (Fig. 234*G*) 6

5*b*. Abdomen of female pointed, with exserted cerci (Fig. 234*F*); eggs laid singly out of water **_Aedes_**
(Three main ecological groups: (1) tree-hole breeders (*Stegomyia*), from which *A. aegypti* was derived; (2) salt- or fresh-water marsh species with successive broods; (3) single-brood species breeding in spring pools from eggs laid previous summer.)

Common North American species:

(*a*) *aegypti;* black with white striped legs and abdomen; black proboscis; lyre-shaped mark on back of thorax.

(*b*) *Salt-marsh mosquitoes; sollicitans* and *taeniorhynchus* with white band on proboscis, *sollicitans* brown with median stripe on abdomen, *taeniorhynchus* darker, abdomen barred black and white; *cantator* with less prominent leg bands and no stripe on proboscis, confined to Atlantic coast.

(*c*) *vexans;* fresh-water marsh and floodwater breeder; thorax bronze, abdomen black with white stripes; tarsi with very narrow white rings; widespread in United States and Canada.

(*d*) *dorsalis;* in salt marshes and on western plains; tarsi white-ringed on both ends of joints; abdomen white-scaled, with two black patches on each segment; northern United States and Canada.

(*e*) *spring pool-breeders;* numerous species, without the combination of characters of above species
(*A. communis* (in woods) and *A. punctor* (on tundras) especially common in Far North, and *A. canadensis* in eastern United States and Canada)

6*a*. Female abdomen truncated; palpi of female usually one-fourth as long as proboscis or longer; wing scales large and broad **_Mansonia_**
(Vicious biters in New World tropics, especially *M. titillans*, which reaches southern Texas and Florida; 1 species, *M. perturbans*, widely distributed in North America, has white-striped abdomen and legs, yellow band on proboscis.)

6*b*. Female abdomen blunt (Fig. 234*G*); palpi less than one-fifth as long as proboscis; wing scales narrow **_Culex_**

(*a*) Brown species with proboscis and legs unstriped; abdomen with distinct white bands; common in and about houses; *pipiens* and *territans* in north, *quinquefasciatus* in south.

(*b*) Similar, with very narrow or no bands on abdomen; breeds in marshes; less domestic; *salinarius* in eastern United States.

(*c*) Proboscis with white band; abdomen and legs white-striped; breeds in ground pools, seldom enters houses; *tarsalis* in western United States.

IV. Tribe **Toxorhynchitini.** Postnotum without setae; scutellum not lobed; proboscis rigid, down-curved; large, brilliantly colored, non-bloodsucking, mostly tropical, only 1 or 2 species in United States ***Toxorhynchites***

Identification of Larvae of Common or Important North American Species

Anopheles Larvae. No breathing siphon (Fig. 232); lie parallel with surface. (See Fig. 235 for other anatomical details mentioned in key.)

1*a*. Both pairs of anterior clypeal bristles simple or slightly feathered 2

1*b*. Outer pair of anterior clypeal bristles profusely branched; leaflets of palmate hairs notched at tip 3

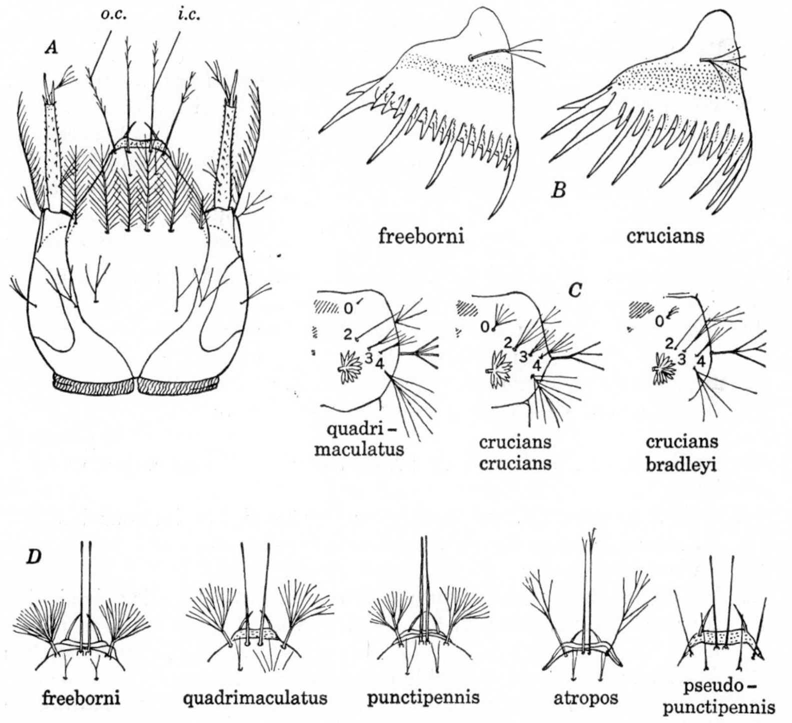

Fig. 235. Details of North American *Anopheles* larvae to illustrate key. *A*, head, showing hairs; *i.c.*, inner clypeal hairs; *o.c.*, outer clypeal hairs. *B*, pectens. *C*, fourth abdominal segments showing hairs *0*, *2*, *3*, and *4* (hair *1* is the palmate hair). *D*, clypeal region of various species. (Adapted from Ross and Roberts, *Mosquito Atlas I*.)

2*a*. Both pairs of anterior clypeal bristles slightly feathered; leaflets of palmate hairs narrow, pointed; palmate hairs on all of first 7 abdominal segments, those on first small **_albimanus_**
2*b*. Both pairs of anterior clypeal bristles simple; leaflets of palmate hairs narrowed to slender point at tip; palmate hairs rudimentary on first and second segments, all about same size on other segments **_pseudopunctipennis_**
3*a*. Inner clypeal hairs close together at base 4
3*b*. Inner clypeal hairs separated at base 9
4*a*. Inner clypeal hairs forked beyond middle; in northern United States and Canada, west to continental divide **_maculipennis earlei_**
4*b*. Inner clypeal hairs single 5
5*a*. Palmate hairs well developed on only 3 segments (4–6); southeastern United States **_crucians georgianus_**
5*b*. Palmate hairs well developed on 5 segments (3–7) 6
6*a*. Pecten with no more than 2 or 3 small teeth between long ones 7
6*b*. Pecten with 3 to 5 small teeth between long ones 8
7*a*. Hairs 0, 2, 3, and 4 on 4th abdominal segment all well-developed, branched **_crucians crucians_**
7*b*. Hair 0 very small, hair 2 usually single **_crucians bradleyi_**
8*a*. Posterior clypeal hairs double; antepalmate hair (2) double or triple **_punctipennis, maculipennis freeborni_**
8*b*. Posterior clypeal hairs multiple, antepalmate (2) of fourth abdominal segment usually single; on west coast **_occidentalis_**
9*a*. Antepalmate hair (2) of 4th abdominal segment usually single; southeastern United States **_quadrimaculatus_**
9*b*. Antepalmate hair (2) of 4th abdominal segment multiple; central valley of Mexico **_maculipennis aztecus_**

Culicine Larvae. Breathing siphon present (Fig. 232A). See Fig. 236 for other anatomical details mentioned in key.

1*a*. Breathing siphon spinelike at tip, used to pierce air channels in roots of aquatic plants; larvae do not come to surface **_Mansonia_**
1*b*. Breathing siphon normal with movable apical valves 2
2*a*. Anal segment with ventral brush composed of only 1 pair of hairs **_Wyeomyia_**
2*b*. Anal segment with ventral brush composed of several pairs of hairs 3
3*a*. Siphon without pecten; anal segment ringed by chitinous band; chitinous plate on eighth segment also 4
3*b*. Siphon with pecten 5
4*a*. Mouth brushes modified into coarse prehensile lamellae, hooked for seizing prey; plate on 8th segment with 2 spiny hairs; in tree holes **_Toxorhynchites_**
4*b*. Mouth brushes normal; double row of stout spines on 8th segment; in water-holding plants or tree holes **_Orthopodomyia_**
5*a*. Siphon with several pairs of ventral tufts; anal segment ringed; siphon usually at least 4 or 5 times as long as wide **_Culex_**
a′ Antenna with tuft at or before middle; all but one of ventral tufts of siphon represented by single hairs *territans*
a″ Antenna with tuft well beyond middle; several tufts on siphon *b*

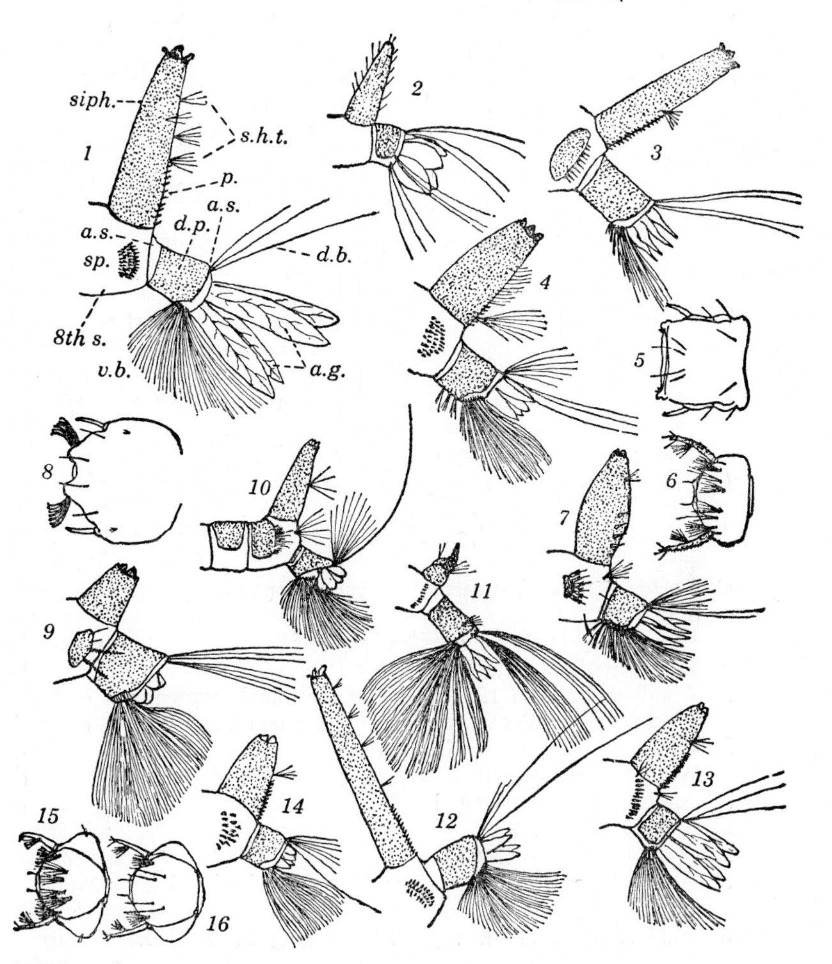

Fig. 236. Details of structure of culicine larvae to illustrate key. (Adapted from Dyar, *Carnegie Inst. Wash. Publ.* 387.) *1,* posterior portion of larva of *Culex quinquefasciatus; a.s.,* anal segment; *8th s.,* eighth segment; *a.g.,* anal gills; *d.b.,* dorsal brush; *d.p.,* dorsal plate (a complete ring in this and many other species); *p.,* pecten or comb; *siph.,* siphon; *sp.,* patch of spines on eighth segment; *s.h.t.,* siphonal hair tufts. *2, Wyeomyia smithii,* posterior end. *3, Uranotaenia sapphirina,* posterior end. *4, Culiseta incidens,* posterior end. *5, Psorophora ciliata,* head. *6, Psorophora (Grabhamia) confinnis,* head. *7, Psorophora (Grabhamia) confinnis,* posterior end. *8, Toxorhynchites (=Megarhinus) septentrionalis,* head. *9, Toxorhynchites (=Megarhinus) septentrionalis,* posterior end. *10, Orthopodomyia signifera,* posterior end. 11, *Mansonia perturbans,* posterior end. *12, Culex salinarius,* posterior end. *13, Aedes aegypti,* posterior end. *14, Aedes taeniorhynchus,* posterior end. *15, Aedes canadensis,* head. *16, Aedes dorsalis,* head.

b′ Siphon 7 × 1 *salinarius*
b″ Siphon 5 × 1 or less *c*
c′ Siphon with 5 hair tufts, all in line *tarsalis*
c″ Siphon with 4 hair tufts, third out of line *pipiens* and *quinquefasciatus*

5*b*. Siphon with a single pair of ventral tufts 6
6*a*. Spines of 8th segment attached to posterior margin of a chitinous plate ***Uranotaenia***
6*b*. Spines of 8th segment not attached to a chitinous plate 7
7*a*. Hair tuft at base of siphon; spines of distal part of pecten produced into long hairs ***Culiseta***
7*b*. Hair tuft near middle of siphon or beyond; pecten of short spines 8
8*a*. Anal segment ringed by chitinous plate, and ventral brush partly inserted into it; in temporary rain pools ***Psorophora***
(*a*) Mouth brushes prehensile; antennae not projecting anterior to head; predaceous; very large subgenus ***Psorophora***
In North America two spp., *ciliata* and *howardi.*
(*b*) Mouth brushes normal, and antennae large; size smaller; several common United States species subgenus ***Grabhamia***
8*b*. Anal segment not ringed, or if ringed, ventral brush posterior to it ***Aedes***
a′ Pecten with teeth detached outwardly; antenna spined all over; siphon 3 × 1 *vexans*
siphon 2½ × 1, on northern prairies *spencerii*
a″ Pecten without detached teeth *b*
b′ Comb scales 8 to 12, in a single row; anal segment not quite ringed; head hairs all single; anal segment short *aegypti*
b″ Comb scales in triangular patch *c*
c′ Anal segment ringed *d*
c″ Anal segment not ringed *e*
d′ Siphon about 2 × 1 *sollicitans*
d″ Siphon less than 2 × 1 *taeniorhynchus*
e′ Anal segment nearly twice as long as wide; head hairs single or double; anal gills large, *stimulans;* anal gills very small *dorsalis*
e″ Anal segment about 1½ times as long as wide; head hairs multiple *canadensis*

Mosquitoes and Malaria

Anopheles mosquitoes are the sole transmitters of human malaria, which at least until recently has ranked as the most important human disease in the world. Species of *Culex* and *Aedes,* and less frequently *Anopheles,* are vectors of malaria of birds and reptiles (see p. 165). The role of mosquitoes in transmitting human malaria has been suspected by various peoples as far back as any records go. The steps which led to the *proof* of it are briefly outlined on pp. 167–168.

The genus *Anopheles* contains numerous species, and they are divided into a number of subgenera. The majority of the species can be experimentally infected with malaria parasites, but some much more readily than others. There is also a difference in the facility with which certain species can be infected with different malarial species and strains (see p. 741). Mere experimental infection of an insect with a disease germ or even successful transmission by it under experimental conditions means very little with respect to its role in nature. Many other factors come into play which cannot be studied in the laboratory and the combined effect of which can be learned only by extensive and carefully studied epidemiological evidence. *A. punctipennis,* for example, though a proved transmitter of malaria in the laboratory, is eliminated in most areas in nature by its habits. It is usually a "wild" species which seldom enters occupied houses and which shows a strong preference for animal over human blood. The factors which determine the importance of particular species of *Anopheles* as malaria vectors are discussed on pp. 184–186.

Biological races of *Anopheles* species. A number of species of *Anopheles,* including some of the "big shots" in the transmission of malaria, are not uniform species but consist of a number of races, in some instances considered species, which show only very slight morphological differences but which may differ sufficiently in their biological characters to make them very good or very unimportant malaria vectors.

In the northern hemisphere *A. maculipennis* is such a species. It has races which differ in such characters as the color of the eggs and the number of teeth on the maxillae of the adults, and biologically in choice of breeding places, hibernation, and willingness to feed on man. In Europe *A. m. labranchiae,* which breeds in both fresh- and brackish water, and has wedge-shaped black spots on the eggs, is a man-eater and was the most important malaria transmitter in southern parts. *A. m. atroparvus,* which breeds by preference in brackish water, was an important vector in the north and west coasts of Europe. The other European varieties are mainly zoophilic, but may bite man sufficiently to keep malaria going in backward areas where domestic animals are scarce. With the extensive control or even eradication of the more domestic species, as in Italy and Sardinia, the more zoophilic outdoors species may become relatively more important.

In North America, *A. quadrimaculatus* of the southeastern states, and the only important potential malaria vector in that area, is closely related to *maculipennis,* but is regarded as a distinct species. There are four subspecies of *maculipennis: freeborni* in the west, *earlei* in the north, *occidentalis* on the Pacific Coast, and *aztecus* in the central valley of Mexico. Of these only *freeborni* and *aztecus* are (or were) malaria vectors; these two have all dark wings, whereas the others have a lighter coppery spot at the tip of the wing. Some workers consider these to be four distinct species rather than subspecies of *maculipennis.*

Other species of *Anopheles* that have races that differ in food preferences or other characters that influence their importance as malaria vectors are *A. sinensis, maculatus, aquasalis, pseudopunctipennis, gambiae, funestus,* and *punctulatus.* In the United States *pseudopunctipennis* feeds almost entirely on animals and is practically harmless, but in mountainous regions from Mexico to Argentina it is a man-biter and is the principal, in some places the only, malaria vector.

Identification of *Anopheles.* The *Anopheles* mosquitoes fortunately are fairly easy to identify in all stages of their development. The different species vary a great deal in choice of breeding places, habits, and appearance, so that it is necessary in any malarial district to determine which species are malaria carriers, how they may be identified, where they breed, and what their habits are. The majority of the species have mottled or spotted wings, and the arrangement of the markings is usually a good means of identification (Fig. 238).

Figure 237 is a comparative table which shows in a graphic way how *Anopheles* may ordinarily be distinguished from other common mosquitoes, such as *Culex* and *Aedes,* in their different stages. The "floats" on the eggs of *Anopheles* are rarely absent; their size and markings sometimes serve as means of identification of species. Owing to the effects of surface tension the eggs of *Anopheles* tend to assume geometrical patterns on the surface of the water. The larvae, besides the absence of a breathing tube and their horizontal floating position at the surface of the water, have other identifying features such as the rosettelike palmate hairs on some of the segments, which serve to hold the larvae in the characteristic position by surface tension. The species of *Anopheles* larvae are sometimes very difficult to identify, and reliance must be placed on the form, number, and distribution of characteristic hairs. A key to the American species is given on p. 738.

The pupae have short and more flaring breathing tubes than those of *Aedes* or *Culex,* the paddles at the end of the abdomen have an accessory hair in addition to the terminal one, and all but the last abdominal segment have peglike spines on their posterior corners (Fig. 233). The pupae of some of the American *Anopheles* can be identified by coloration (Burgess, 1946). The adult *Anopheles* are usually easily distinguishable by the resting position, with the proboscis, thorax, and abdomen all in a straight line and at an angle to the resting surface, in

Anopheles | *Culex, Aedes, etc.*

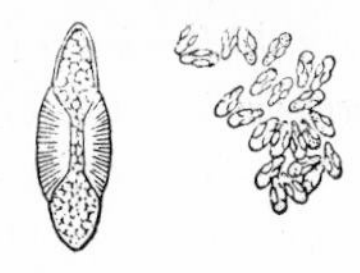

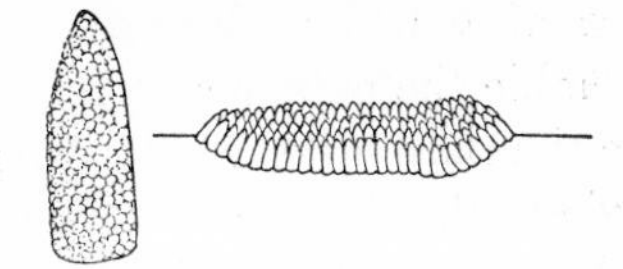

EGGS

Eggs laid singly on surface of water; provided with a "float."

Eggs laid in rafts or egg boats or singly on or near water or where water may accumulate; never provided with a "float."

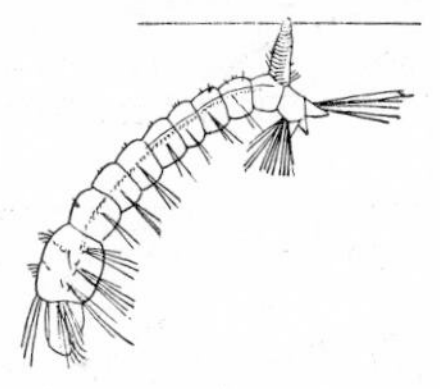

LARVAE

Larvae have no long breathing tube or siphon; rest just under surface of water and lie parallel with it.

Larvae have distinct breathing tube or siphon on eighth segment of abdomen; hang from surface film by this siphon, except in *Mansonia,* which obtains air from aquatic plants.

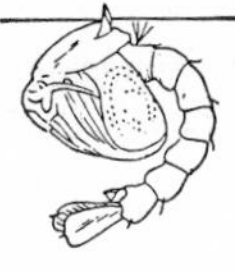

PUPAE

Pupae have short breathing trumpets.

Pupae have breathing trumpets of various lengths.

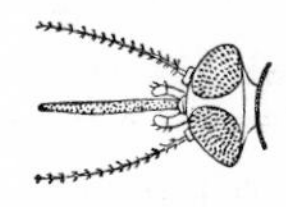

HEADS OF ADULTS

Palpi of both male and female long and jointed, equaling or exceeding the proboscis in both sexes.

Palpi of female always much shorter than proboscis, those of male usually long but sometimes short.

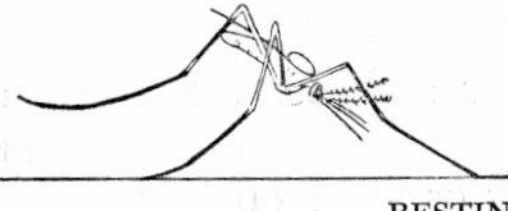

RESTING POSITION OF ADULT

Adult rests with body more or less at angle with surface, the proboscis held in straight line with body.

Adult usually rests with body parallel to surface, though sometimes at an angle. Proboscis not held in straight line with body, giving "humpbacked" appearance.

Fig. 237. Comparison of *Anopheles* and culicine mosquitoes.

contrast to the parallel or drooping abdomen and humpbacked appearance of culicines, but some of them, e.g., *A. culicifacies* of India, resemble the culicines in resting position. *A. barberi* is more culicine-like than are other North American *Anopheles;* it rests at only a slight angle to the surface, whereas *punctipennis* appears almost to stand on its head. Most *Anopheles* have the wings marked with dark or light

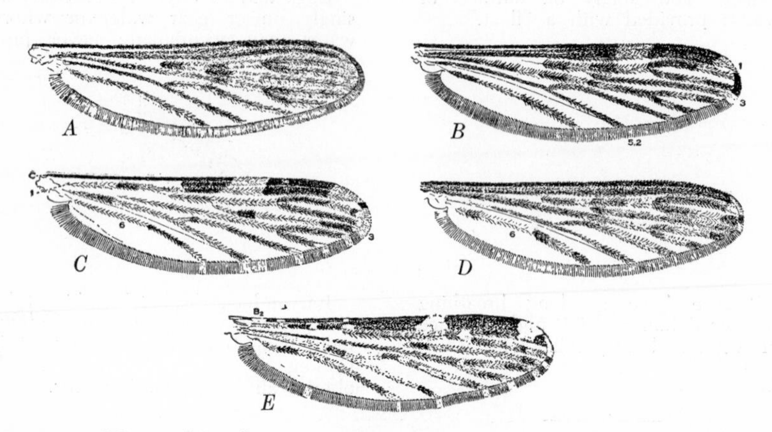

Fig. 238. Wings of North American *Anopheles.* A, *quadrimaculatus;* B, *punctipennis;* C, *pseudopunctipennis;* D, *crucians;* E, *albimanus.* (After Ross and Roberts, *Mosquito Atlas,* 1.)

spots or both, but even this is not constant, since a few culicines have spotted wings and a few *Anopheles*, e.g., *A. atropos, A. barberi,* and *A. walkeri,* have unspotted ones. The long palpus of the female is a character which can always be relied upon.

Following is a key for the identification of the North American species north of Mexico, with comments on their distribution and importance:

1*a*. "White-footed" species with hind tarsi having last 3½ segments all, or nearly all, white; wings marked with both black and white spots. Several species in tropical America, several being very important malaria vectors. Only species reaching Texas and southern Florida, and most important vector in West Indies and Central America **_albimanus_**

1*b*. Dark-legged species without white stripes or areas on legs 2

2*a*. Wings not spotted, or spots indistinct

(1) No white knee spots; inconspicuous rings on palpi; Gulf and Atlantic coasts, breeding in salt water; coloration very dark **_atropos_**

(2) With narrow rings on palpi; eastern United States, breeding in freshwater marshes; white knee spots; not important **_walkeri_**

(3) Small; dark palpi; no white knee spots; breeds in tree holes **barberi**

2*b*. Wings not uniformly colored 3

3*a*. Wings with 2 or 4 dark spots; white knee spots 4

3*b*. Wings with white or yellow spots along costal margin (Fig. 238*C*, *D*) 5

4*a*. Apex of wing uniformly colored, dark; only important malaria carriers in United States — in south, ***quadrimaculatus***; west of Rockies, ***maculipennis freeborni***

4*b*. Apex of wing with coppery spot — in north states, ***m. earlei***; on Pacific slope, ***m. occidentalis***

5*a*. Only 1 white spot on costal margin, at tip of wing (Fig. 238*A*); anal vein with 3 black spots ***crucians***

5*b*. A large yellow spot on outer third of costal margin involving 3 veins, another at tip, and another in basal third; fringe nearly all dark; extends farther north than most species (Fig. 238*C*) ***punctipennis***

5*c*. Four yellowish white spots along front of wing; fringe of wing alternating black and white; southwestern United States to Argentina; not important in the United States, but principal transmitter in mountains from Mexico to Argentina ***pseudopunctipennis***

Habits of *Anopheles*. Most *Anopheles* breed in natural waters such as ponds, swamps, edges of streams, rice fields, and grassy ditches. Our species all breed in standing water, but this is not true everywhere. In Europe and Asia some of the most important species breed in flowing streams, which necessitates entirely different methods for their control. Some species breed in brackish water, some in shaded water, some only in sunlight, and a few in artificial containers around houses. Members of the subgenus *Kerteszia* breed in bracts of leaves of bromeliads growing on forest trees. Some *Anopheles* show much more pronounced preferences than others. *A. quadrimaculatus,* formerly the principal malaria vector in the southeastern states, is one of the least particular species, but optimum conditions for its breeding are afforded by clean, open water with dense aquatic vegetation and abundant floatage. Natural shade restricts the vegetation, thereby decreasing the protection and food of the larvae.

Shade-loving species are *A. darlingi* of tropical America, *A. funestus* of Africa, and *A. umbrosus* of southeast Asia, whereas *A. albimanus* and *aquasalis* of tropical America, *A. gambiae* of Africa, and *A. barbirostris* of Asia breed only in sunlit waters. In Malaya the cutting down of jungle in the flat lands and exposure of sluggish water to sunlight changes a dominance of *umbrosus* to *barbirostris*, with a reduction in malaria; but when ravines in the hills are opened up and the sparkling streams cleared of vegetation and exposed to sunlight, the harmless jungle species disappear and the deadly *A. maculatus* takes their place. Thus in the two localities directly opposite conditions determine the presence or absence of malaria. *A. sundaicus,* an important malaria carrier in southeastern Asia, breeds in strongly brackish or even concentrated sea water, as in the holes of mud lobsters or in sea water pools inside the reefs of coral islands. *A. bellator,* an important vector in Trinidad, breeds in aerial plants in trees planted to shade cocoa groves, and both this species and *A. cruzi* breed in bromeliads in forest trees in southern Brazil, and sometimes even in trees in towns.

The eggs of *Anopheles* are not as resistant to drying as those of *Aedes,* but will survive for as long as 3 weeks on drying mud, and even the larvae can live on moist mud for some time. The eggs hatch only in water and at temperatures above 60°F.; under favorable conditions they hatch in 1 to 2 days. The larvae are surface feeders; they seem to feed on any particles floating on or near the surface which are small enough to swallow. *Anopheles* larvae are not rapid in their development as compared with some mosquitoes; the time required under favorable conditions is 2 to 3 weeks or more. The number of generations a year probably varies greatly with the species and conditions of food and temperature. It has been estimated that *A. quadrimaculatus* has 8 to 10 annual generations in southeastern United States.

Adult *Anopheles* are for the most part twilight feeders; but there are many exceptions. Some species come forth with the first shade of late afternoon, others not until almost dark. A few species, e.g., *A. brasiliensis,* are diurnal; many forest species will bite willingly in the daytime if disturbed. The food preferences of adult females and their important bearing on malaria transmission have already been discussed. An important observation made by Roubaud is that the adults, at least of *A. maculipennis,* fly out into the open and invade other houses or sheds even if there is an abundant food supply where they have been resting after an earlier meal; as a result the *Anopheles* population of any spot is entirely changed in a few days; this flight in the open seems to be indispensable to the life of *A. maculipennis,* and it also has an important bearing on malaria transmission. Mosquitoes do not suck blood daily; usually blood meals are taken at intervals of several days.

Most *Anopheles* are rather sedentary in habit and seldom fly in numbers more than a fraction of a mile from their breeding places, although *A. albimanus* is reported to fly at times as far as 12 miles from Gatun Lake in Panama. Abundant *Anopheles,* nevertheless, usually indicate breeding places within a mile.

Anopheles hibernate either as adults in protected buildings, caves, or tree holes, or as larvae which bury themselves in mud or under debris at the bottom of water in ponds or marshes. The eggs do not live through cold winters.

Malaria-carrying species. More than 200 species of *Anopheles* have been described, and a large percentage of them have been shown to be capable of transmitting malaria experimentally, but only about two dozen of them are important natural vectors of malaria, and some of these only in limited areas. Often the habits of the species, as already shown, are of more importance than the ability to transmit the disease under experimental conditions. The difference in ability of some species of *Anopheles* to nurse one species or strain of malaria more readily than another still further complicates the task of evaluating the roles of different species, for a certain species of *Anopheles* may for this reason be an important transmitter in one place and not in another.

The methods used in determining the importance of particular species are various. The most valuable and reliable criterion is the relative number of individuals with infected salivary glands found in malarial houses. (For methods of dissecting mosquitoes for demonstration of oöcysts and sporozoites, see Barber and Rice, 1936.) Supplementary information is obtained by observations on breeding and feeding habits (aided by precipitin tests), relative abundance and coincidence with outbreaks of the disease, and experimental transmission in the laboratory.

In the United States, until malaria was eradicated as an endemic disease about 1952, *A. quadrimaculatus* (Fig. 239) in the southeast and *A. m. freeborni* in the arid west were the only important vectors. The principal vectors in other parts of the world are as follows: in Mexico, *pseudopunctipennis* on western slopes, *maculipennis aztecus* in the central valley, and *quadrimaculatus* and *albimanus* in humid coastal areas; in Central America and West Indies the white-footed *albimanus;* in Trinidad the bromeliad breeder, *bellator;* in South America, to south latitude 25 in the interior as well as on the coast, the highly domestic shade breeder, *darlingi,* and on the coasts *aquasalis,* breeding in sunlit brackish water, and *albitarsis domesticus;* in southern Brazil, the bromeliad breeders, *bellator* and *cruzi;* in the Andean region, *pseudopunctipennis;* in western Europe certain races of *maculipennis,* especially *labranchiae* and *atroparvus,* which breed in both brackish and fresh water; in the Balkans and Middle East, *superpictus,* breeding in flowing water in hills, *sacharovi* in brackish and freshwater marshes, and *claviger* in marshes, cisterns, and wells; in Africa, *gambiae* and *melas,* breeding in sunlight, and *funestus* in shade, and *pharoensis* in upper Egypt and Sudan; in India numerous species, particularly *culicifacies* (principal vector in Ceylon), *philippinensis, stephensi* (and urban species breeding in man-made containers), *varuna, sundaicus, minimus,* and *fluviatilis;* in southeast Asia the last three of these plus *aconitus, maculatus, latifer,* and *annularis;* in China, Japan, and Korea, *hyrcanus sinensis, minimus,* and *pattoni;* in Borneo, *leucosphyrus;* in the South Pacific and Australian region, *farauti, punc-*

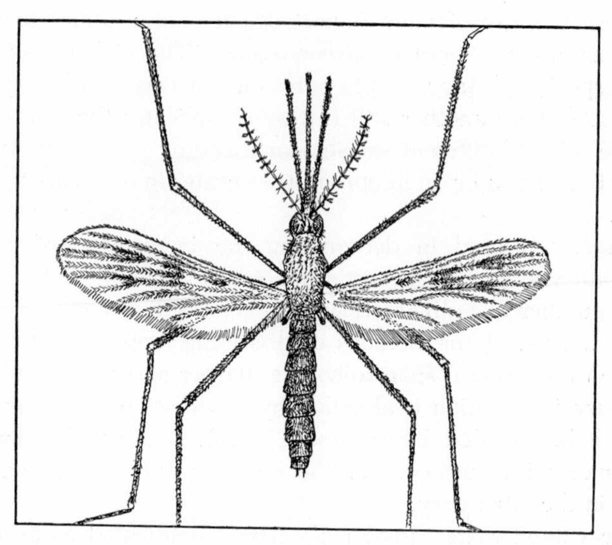

Fig. 239. *Anopheles quadrimaculatus,* in former days the principal malaria transmitter in southeastern United States.

tulatus, and *koliensis* (but only in islands as far east as the New Hebrides; the oceanic ones farther east have no *Anopheles*). These species account for the majority of malaria in the world, but many other species also contribute and may be locally important.

The effect of anti-*Anopheles* campaigns on the prevalence of malaria is discussed in Chapter 8, pp. 190–191.

Mosquitoes and Viruses

I. Yellow Fever

There are very numerous animal viruses that multiply in arthropods, and in most cases are transmitted by their bites. Ticks and mosquitoes harbor especially large assortments. Most, if not all, of these arthropod-borne viruses exist primarily or primitively in wild birds or mammals and are transmitted among these by sylvan or "wild" vectors. Infection of man or domestic animals may be purely accidental and sporadic or, once transferred, these viruses may be spread extensively by the same or other vectors, causing local or widespread epizootics among domestic animals or human epidemics. Sometimes, e.g., yellow fever and dengue, they become entirely divorced, temporarily at least, from their sylvan origins, just as happens in the case of the organisms causing relapsing fever, rickettsial diseases, plague, and others, all of which are primarily zoonoses. Until recently yellow fever was a perfect example of an infection that leads such a "double life," and it still does in Africa.

History and nature of yellow fever. Once a scourge of most of the Western Hemisphere, yellow fever is now confined to forested areas of tropical America and to West and Central Africa. The virus is present in the blood and available to mosquitoes only for 3 to 6 days of incubation prior to the onset of symptoms and during the first 3 days of illness. The disease is ushered in by severe headaches, aches in the bones and fever. The latter subsides in 3 or 4 days, then rises again, accompanied by jaundice and often a black vomit of blood and bile. The mortality in adults is high, but there are many unapparent cases in children in endemic areas, as shown by protective antibodies in the serum. A highly effective protective vaccine is available and has been used on a very large scale in rural South America and in West Africa and for troops in endemic areas.

The discovery of the transmission by *Aedes aegypti* in 1900 by the illustrious work of the American Yellow Fever Commission composed of Reed, Carroll, Lazear, and Agramonte ended what Soper (1937) called the "Dark Age" of yellow fever and began the "Golden Age," during which so much progress was made in the control of the disease that for a few months in 1927 it was thought to have been completely eradicated from the Western Hemisphere. Then came the "Age of Disillusionment," with the dramatic revelation that yellow fever exists in a jungle form, usually silent and unrecognized, over vast areas in tropical America, a situation later found to exist in Africa also. Forest animals serve as sylvan reservoir hosts, particularly spider monkeys, howlers and *Cebus* in America, and all kinds of primates from the tiny bush-baby (galago) to the gorilla in Africa. However, because of the high mortality from yellow fever of most American monkeys, other reservoir hosts are suspected. Opossums and some other animals are known to be susceptible. In tropical America the jungle vectors are mosquitoes, mostly of the genus *Haemagogus,* that live and breed in the forest canopy. Hence, since *Aedes aegypti* is usually absent from small native villages, human cases occur only among people actually working in or near the forests, especially in felling trees, and direct transmission from man to man does not occur. A few other mosquitoes, especially *Aedes leucocelaenus* and species of *Sabethes* play minor roles.

In Africa, unlike America, *Aedes aegypti* is an almost universal inhabitant of small native villages and is a common yellow fever transmitter in them. Among monkeys in forests the infection is transmitted mainly by *A. africanus,* a mosquito that bites at dusk and breeds in rock pools. *A. simpsoni,* a day-biter that breeds in axils of leaves, especially of bananas, transmits the disease from monkeys to man in clearings and plantations.

In the Americas, when persons with yellow fever enter a town or city where *Aedes aegypti* is prevalent, the disease changes from a sporadic to an epidemic form, and then spreads from city to city. This was the only kind of yellow fever known prior to about 1930. In the nineteenth century, even in such places as Philadelphia and Boston, great epidemics broke out late in the year after *A. aegypti,* imported on sailing vessels in which they found abundant breeding places, had become numerous. The toll from yellow fever during the French attempt to build the Panama Canal was appalling.

Before the transmission by *A. aegypti* was discovered and means of control understood, epidemics raged in tropical cities until a high percentage of people were either dead or immune, and in temperate cities until frost stopped the mosquitoes. The last outbreak in the United States was in New Orleans in 1905, when for the first time an epidemic was stopped by intelligent human effort. No urban outbreak of yellow fever has occurred in the Americas since 1933, although small *aegypti*-transmitted outbreaks have followed jungle outbreaks several times. The jungle outbreaks tend to shift from place to place. After 40 years' absence, jungle yellow fever suddenly struck Panama like a bolt from the blue in 1948, and then progressed wavelike through the forests of Central America to Southern Mexico, invading the land of the ancient Mayans as it presumably did periodically in pre-Columbian times, before *Aedes aegypti* entered the picture. In 1954 jungle yellow fever struck Trinidad also.

It commonly requires 10 to 12 days for *Aedes aegypti* to become infective after becoming infected at high temperatures. Transovarial transmission does not occur, but larvae exposed to liquid containing the virus develop into infective adults. In order to keep an epidemic going it is necessary to have a fairly high incidence of *A. aegypti.* Even where there are many nonimmunes an epidemic subsides when the *aegypti* index, i.e., the number of premises on which it is breeding, falls below 5%. It is doubtful whether an epidemic would start if, by inspection and anti-*aegypti* work, the index were held to 2 or 3%, as it can be without too great difficulty.

Because of the prevalence of *Aedes aegypti* in places like India, Malaya, or Australia, where there are enormous nonimmune populations, introduction of yellow fever by modern airplane traffic, either in man or mosquitoes, is a great menace. Southern United States would also be in danger if the *Aedes aegypti* population was permitted to thrive as it did at the beginning of World War II. In India reliance is placed on fumigation or spraying of boats and planes and strict quarantine of individuals coming from endemic areas; in our own country dependence is placed on fumigation, local *aegypti* control, surveillance of exposed persons, and availability of vaccine for whole populations if a case should appear. Control of *A. aegypti* in non-infected countries is important in order to lessen the danger of rapid spread if the disease *should* get in, but the South American countries, which have the jungle disease in their backyards, so to speak, cannot afford to stop short of extermination on a continental scale. Already most of South and Central America, including all of Brazil and Bolivia, have been cleared or is in the final stages of clearance, but less progress has been made in Mexico and the West Indies. There is some question as to whether extermination in the United States should be attempted (see below). In Africa *A. aegypti* has been virtually exterminated from Khartoum, so there is little danger now of yellow fever being carried into Egypt on Nile steamers.

***Haemagogus* spp.** These vectors of jungle yellow fever in South America are closely related to *Aedes* but have brilliantly metallic colors. They lay eggs as do *Aedes,* and the larvae are very similar. In most of the species the males have short palpi like the females. They are day-biting mosquitoes that live primarily in the sunlit tree tops; only a few come down to ground level except where the canopy is more or less open or at the edges of clearings or coffee plantations where the sun gets through. Normally they breed in tree holes or in axils of leaves in the tree tops, but Komp (1952) observed that *H. spegazzinii falco,* the No. 1 transmitter of sylvan yellow fever, can adapt itself to breeding in such places as water-filled hollows in fallen trees, hollow stumps, cut bamboo stems, etc., and *H. splendens* even in old tires and water tanks near plantations. Both *Haemogogus* species and *Aedes leucocelaenus* have been recaptured 4 to 7 miles from points of release, so could disseminate yellow fever from small forest oases surrounded by open country.

All species of *Haemagogus* tested are capable of transmitting yellow fever; in nature their importance as vectors depends on abundance, habitat and habits, particularly as feeders on monkeys and man. From northwestern Brazil to Central America *H. s. falco* (Fig. 240), a beautiful mosquito with metallic blue thorax and violet abdomen, is the most important vector, but as the disease spreads northward into Mexico other species, e.g., *H. equinus* may take over. This species

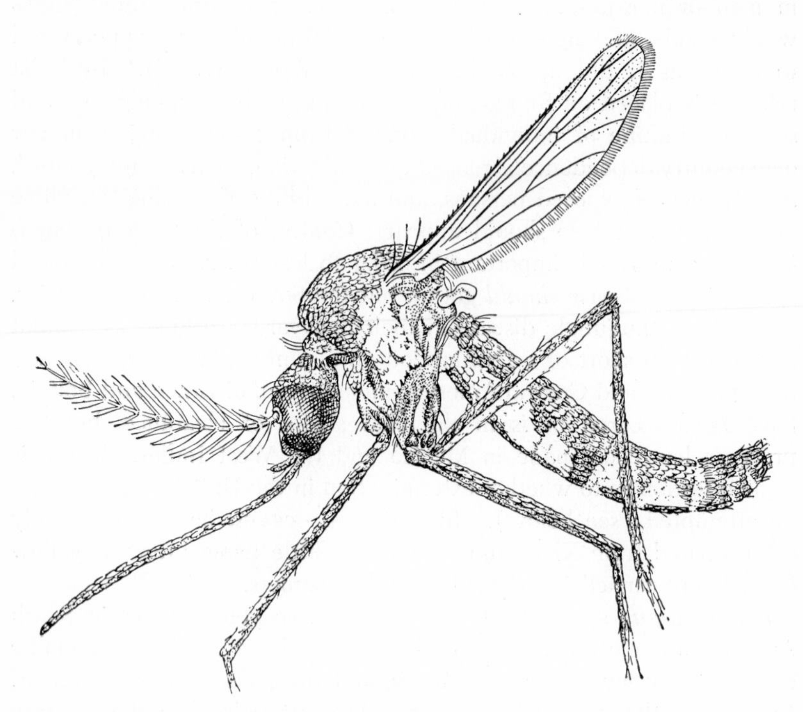

Fig. 240. *Haemagogus spegazzinii falco,* chief vector of yellow fever in northern South America and Central America. (Drawn from a painting by Varela in Kumm, Osorno-Mesa and Boshell-Manrique, *Am. J. Hyg.*, 43, 1946.)

reaches the Lower Rio Grande Valley in Texas but would be negligible as a vector in that dry area.

Other than clearing forest over large areas there is no known method of control of these mosquitoes, and control of jungle yellow fever in man depends on wholesale vaccination of populations in endemic areas.

Aedes aegypti. This mosquito is a member of the subgenus *Stegomyia,* which contains a group of originally tree-hole-breeding mosquitoes, several of which (e.g., *albopictus, scutellaris, polynesiensis,* as well as *aegypti*) have become more or less domestic and have adopted man-made containers as breeding places. *Aedes aegypti* was long known in medical literature as *Stegomyia fasciata.* It is a small black species, conspicuously marked with silvery-white on the legs and abdomen and with a white lyre-shaped design on the thorax (Fig. 241). The female has very short palpi which are white at the the tip. The wings are clear and somewhat iridescent.

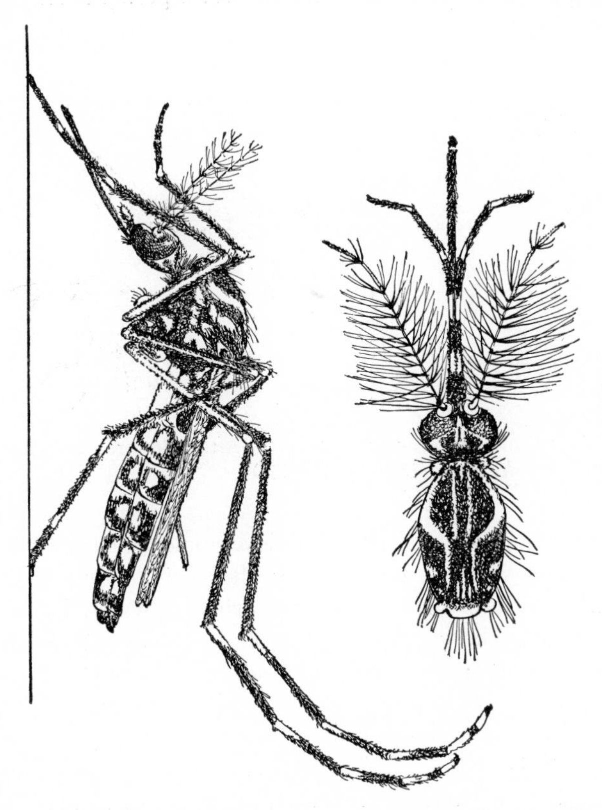

Fig. 241. *Aedes aegypti* ♀, and head of ♂. (After Soper and Wilson, *J. Natl. Malaria Soc.*, 1, 1942.)

Aedes aegypti is a "pet" mosquito, as domestic as a rat or a roach. It is almost never found more than a few hundred feet from human habitations, and feeds readily on human blood. Long familiarity with man has made it an elusive pest. Its stealthy attack from behind or under tables or desks; the suppression of its song; its habit of hiding behind pictures or under furniture; the wariness of its larvae—all these are lessons learned from long and close association with man. It is a diurnal mosquito, biting principally in the morning and late afternoon, with a siesta in the middle of the day, but it will bite at night when hungry.

Aedes aegypti lays its eggs, after a blood meal, on the sides of a container, at or just above the water surface. A number of batches of eggs are laid, 10 to 100 at a time, usually at intervals of 4 or 5 days, until a total of 300 to 750 has been laid over a period averaging about 6 weeks.

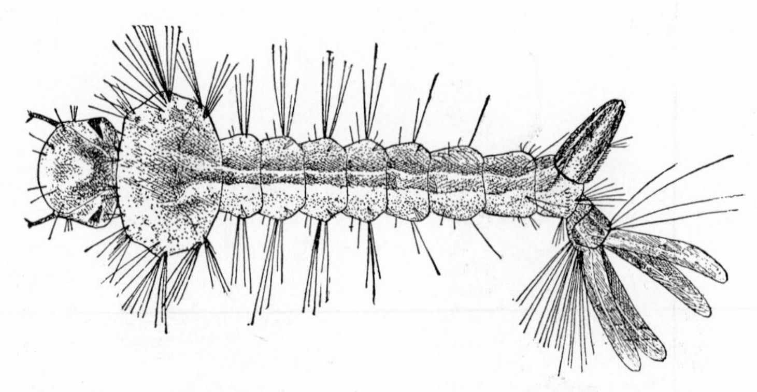

Fig. 242. Larva of *Aedes aegypti*, ×10. (After Howard, Dyar, and Knab, *Carnegie Inst. Wash. Publ.*, 159, 1912–1917.)

The eggs require several days for development of the embryo, and then they hatch within a few minutes after being submerged. Often when containers are filled with fresh drinking water, larvae appear almost at once. The eggs remain viable in the dry state for at least a year, but when wetted, all may not hatch; some may require several wettings. Hatching does not occur readily in perfectly fresh water, but it is favored by presence of bacteria. The larvae (Fig. 242), however, do not thrive in very filthy water such as *Culex quinquefasciatus* delights in, but will tolerate considerable amounts of acid, alkali, or salt.

When disturbed, even by a shadow, the larvae swim to the bottom, which they hug so closely that if a container is dumped, a large proportion of them may remain in a cupful of water that is left behind. Under favorable conditions the time from egg to egg is about 16 days—2 to 3 for hatching, 5 to 6 for the larvae, 1½ to 2 for the pupae (Fig. 243), and 6 to 7 before the adult lays eggs again. The average length of life of the adults is about 60 to 80 days.

Aedes aegypti has more completely forsaken its ancestral breeding places than any other mosquito, only rarely breeding in tree holes or broken bamboo stems but commonly utilizing rain-filled coconut shells around native villages, as well as artificial containers. Inside houses the most important breeding places are drinking-water jars, water-plants, neglected flower vases, unused toilets, and icebox drains; in residential yards they are cisterns, grease traps, tincans, wide-mouthed jars, old tires, animal drinking pans, rain-water barrels, etc., and occasionally sagging roof gutters. In business or industrial areas they breed in barrels or buckets kept for fire protection, basement sumps, elevator pits, trash piles, etc., and in neglected spitoons. Neglected flower containers in cemeteries are a special menace. In places without a piped water supply, barrels or urns containing drinking water, tanks, and wells with wood, brick, or stone sides are important. Other places are the holy-water fonts in churches and the bilges of boats.

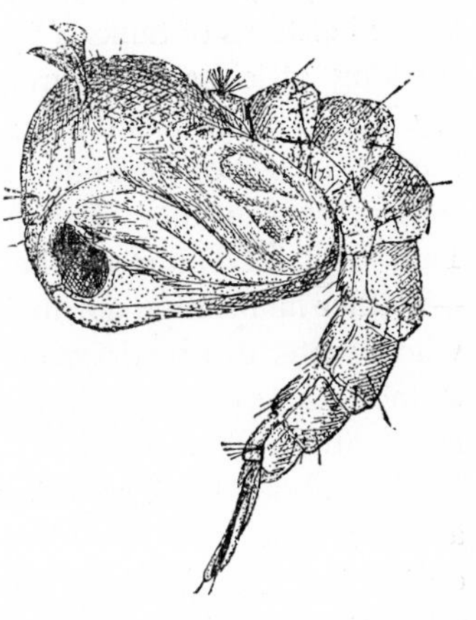

Fig. 243. Pupa of *Aedes aegypti,* ×10. (After Howard, Dyar, and Knab, *Carnegie Inst. Wash, Publ.,* 159, 1912–1917.)

In the southern United States, *A. aegypti* survives the winter in the egg stage in dry containers, some even resisting freezing. Those in wet containers, as Hatchett in 1946 found in Houston, are in a precarious situation since they hatch during warm spells and are then killed later by cold spells. The larvae survive in fire barrels and other protected and more or less permanent receptacles or in large cisterns in which the water does not become too cold. These "mother foci" are also important in seeding secondary, less permanent, containers during the summer. Since this mosquito seldom flies more than a few hundred feet, although a few are often carried by cars, trains, or boats, its presence in numbers indicates a breeding place close at hand.

Aedes aegypti probably originated in Africa, but it has followed man to all parts of the world because of its tendency to "stow away" in boats, trains, airplanes, and automobiles. It is a permanent resident in all tropical and subtropical parts of the world in association with man and is annually carried to places far outside its permanent range. In the United States it lives through the winter in sufficient numbers to become abundant early in the season only on the Gulf Coast, but it survives in numbers sufficient to produce a good population *late* in the year in many interior cities. In the Oriental region, for some unknown reason, it does not thrive so well as other domestic members of the *Stegomyia* group—*scutellaris, pseudoscutellaris* and *polynesiensis* in the South Pacific islands, and *albopictus* in southeast Asia and Honolulu.

Control of Aedes aegypti. Since this mosquito breeds almost exclusively in artificial containers, special control measures apply to it. The principal useful methods are (1) destroying, removing, or turning "bottoms up" unneeded containers exposed to rain; (2) frequently emptying needed containers such as animal drinking pans, flower vases, water plants, and spittoons; (3) covering or treating drinking water in jars, barrels, etc.; (4) sealing, filling, draining, or destroying unused cisterns or tanks; (5) mosquito-proofing used overhead cisterns; stocking attic or underground cisterns, wells, and concrete fish or lily ponds with *Gambusia;* (6) larviciding water in fire-protection buckets or barrels, stranded boats, etc.; (7) puncturing or straightening sagging roof gutters. For drinking water in jars or barrels addition of 1 ml. of 2% DDT in alcohol per 6 quarts is effective for 4 to 5 weeks —neither inebriation nor toxic effects have been observed! For cisterns where water can be drawn off from below and which are not amenable to mosquito-proofing, oil applied to the surface at weekly intervals is most satisfactory.

Since about 1947, spraying with DDT or dieldrin has been used on a large scale in extermination programs; 3 to 5% solutions sprayed on the inside and outside of actual or potential breeding foci, with or without water in them, and on nearby resting places have proved very effective. If thoroughly done one or two applications are adequate in small areas (up to 5000 houses), and four at intervals of 3 months in large cities. Such cities as Georgetown, Barranquilla, and Cartagena have been cleared of *Aedes aegypti* within a year.

II. Dengue

Dengue or "breakbone" fever is another virus disease (see p. 227) transmitted principally by mosquitoes of the subgenus *Stegomyia.* The virus is believed to be related to that of yellow fever but differs strikingly in not attacking the liver and in producing immunity of relatively short duration, sometimes only for a season, though some instances of apparent failure of immunity is undoubtedly due to the existence of at least three different immunological strains. The disease commonly breaks out in explosive epidemics that spread with amazing rapidity. Such an epidemic spread through Texas in 1922; there were estimated to have been 600,000 to 1,000,000 cases in a few months, and 70% of the people in Galveston and Houston were attacked. The disease may occur in all warm parts of the world—north as far as our Gulf states, the Mediterranean countries, and southern China.

The disease starts suddenly with a high fever, flushed face, and severe prostration. Often after a brief let-up there is a return of the fever and a transitory rash. Leucopenia is a marked symptom. The infection is not fatal, but there is a long convalescence. An effective vaccine comparable with that of yellow fever has been developed (Sabin, 1954).

Mosquitoes fed on dengue patients can transmit dengue for 3 days after the onset of the fever. Only *Aedes aegypti* and other members of the subgenus *Stegomyia* are known to be implicated, except a closely related form, *Armigeres obturbans,* in Formosa. In some countries there are denguelike diseases, the transmission of which is uncertain, and it is sometimes difficult to distinguish between true dengue and sandfly fever.

According to the work of Chandler and Rice, *A. aegypti* becomes infected after feeding on patients in the first to fifth days of the disease and can transmit the infection as early as 24 hours after an infective feed, but Siler et al. in the Philippines obtained different results. They found the patient to be infective for the mosquito for only 3 days, and for 6 to 18 hours prior to the onset, and they also failed to transmit the disease in less than 11 days after a mosquito had obtained an infective feed. This incubation period was later shortened to 8 days by Schule. Explanations for these discrepant results are possible transovarial transmission by mosquitoes or use by Chandler and Rice of larvae as a source of experimental mosquitoes which had developed in water contaminated by dead infected adults.

The fact that sandfly fever is transovarially transmitted among sandflies makes it appear probable that this can also occur in the case of dengue and mosquitoes, in spite of some preliminary results to the contrary. The rapid spread of dengue epidemics does not fit in well with a long incubation period in mosquitoes unless hereditary transmission is possible. Once a mosquito becomes infective it appears to remain so for the rest of its life.

Dengue, like yellow fever, is probably a sylvan disease of monkeys, transmitted among them by forest-dwelling mosquitoes of the subgenus *Stegomyia,* such as *Aedes albopictus* and some of the other seventeen members of the *A. scutellaris* group. Human outbreaks presumably begin when some of these mosquitoes become infected from monkeys and then invade human villages where, like *A. aegypti,* they breed in coconut shells and artificial containers. Thence infected human beings carry the disease to other parts of the world where *A. aegypti* takes over as the vector. Monkeys probably serve as interepidemic reservoirs of the disease.

A. albopictus is widely distributed in southeast Asia and is one of the four mosquitoes introduced into the Hawaiian Islands; the others are *Aedes aegypti, Culex quinquefasciatus,* and *Toxorhynchites brevipalpis,* the last being a nonbiting species with cannibalistic larvae introduced in 1950 to help make life miserable for *Aedes albopictus.* In addition to *A. albopictus* other members of the *A. scutellaris* group are found in New Guinea and many other South Pacific Islands; the group includes *A. polynesiensis,* an important vector of nonperiodic filariasis (see p. 478). In most Oriental cities *A. aegypti* is the predominant species of the group, but in Honolulu *albopictus* outnumbers *aegypti* three to one, and was the principal vector in the 1943–1945 outbreak there. An interesting feature of the epidemiology of this outbreak was Usinger's (1944) observation that the incidence of cases is correlated with density of human population rather than density of mosquitoes; where people are crowded a few mosquitoes suffice, but where people are scattered even dense hordes of stay-at-home mosquitoes fail to spread the disease efficiently.

Members of the *scutellaris* group (including *albopictus*) have a white stripe down the middle of the thorax; *albopictus* differs from the majority of the members of the group in having irregular patches of white scales on the sides of the thorax instead of more or less distinct lines. Important members of the group have habits very much like those of *aegypti* except that, in addition to breeding in artificial containers, they also breed in tree holes and, rarely, in leaf axils in jungles far from urban centers and are therefore more difficult to control and probably impossible to exterminate.

For control of *Aedes aegypti* and other container-breeding mosquitoes, see p. 750.

III. Encephalitis and Miscellaneous Viruses

Numerous other mosquito-borne viruses occur in various parts of the world; some, but not all of them, have a predilection for the central nervous system of their hosts and cause encephalitis. There are three of these encephalitis viruses in the United States—western equine (WEE), St. Louis (SLE), and eastern equine (EEE), and in northern South America there is a Venezuelan equine (VEE). Other important mosquito-borne encephalitis viruses are Japanese B in the Far East and Murray Valley in Australia. There are numerous others, besides yellow fever and dengue already discussed, that are *not* primarily neurotropic, e.g., West Nile, Ntaya, Mengo, etc. Some of these viruses have been discovered by isolation from blood of normal or febrile hosts, others only from mosquitoes. At least a dozen different mosquito-borne viruses have been discovered in Africa, probably all infective but not necessarily pathogenic for man, but virologists hit the jackpot in Trinidad where 31 different unknown mosquito-borne viruses were isolated in a single year! Some of the mosquito-borne viruses (e.g., EEE) are transmitted primarily by *Aedes* and *Psorophora,* others (e.g., SLE and Japanese B) by *Culex,* but none by *Anopheles,* although these may harbor the virus outside the alimentary canal (e.g., Murray Valley).

The three encephalitis viruses of the United States are all primarily parasites of birds, but all three can cause serious disease in man, characterized by suddenly developing headache and malaise, fever that reaches a peak about the third day, marked drowsiness and coma (wherefore sometimes called "sleeping sickness,") and often nausea, vomiting, and convulsions. WEE and EEE produce clinical symptoms in horses also, and EEE in birds as well (especially pheasants; SLE produces only unapparent infections in horses).

WEE is found mainly west of the Mississippi and is transmitted primarily by *Culex tarsalis;* sparrows and other small birds are the principal reservoir hosts. *C. tarsalis,* although it feeds largely on birds, also bites man and horses freely, and is therefore both an enzootic and epidemic vector. EEE occurs mainly on the Atlantic and Gulf coasts; it is primarily associated with swamps and transmitted probably most often by *Culiseta melanura* among swamp birds, assisted by other swamp mosquitoes of the genera *Aedes* and *Psorophora,* which are responsible for carrying the infection to horses and man. SLE occurs mainly in the Mississippi Valley and central states; it is transmitted principally by *Culex* mosquitoes—*not Aedes* or *Psorophora.* In California *C. tarsalis* is the main transmitter as it is for WEE, whereas in the Middle West and East human cases are mainly urban or suburban, with members of the *C. pipiens* complex (see p. 754) as carriers to man and chickens; and *C. salinarius* as an enzootic vector to small birds. VEE, in contrast to the United States forms, is primarily a mammalian virus, as are many of the tropical viruses; *A. sollicitans* and other marsh mosquitoes are believed to be important vectors.

Human and equine outbreaks in the United States are rather scattered and often local. WEE is mainly rural; SLE often urban or suburban, and EEE sporadic. There has been much speculation as to how these viruses survive cold winters. What becomes of them in interepidemic periods? When birds were found to be the principal reservoirs, and dermanyssid mites were reported to harbor and even transovarially transmit them, survival in the mites with transmission to nestlings in the spring, and subsequent spread by mosquitoes seemed to provide a perfect explanation; however, subsequent work has failed to support this hypothesis and it is now generally agreed that mites play no significant role. The viruses may sometimes survive in vertebrates, or may be reintroduced each year by migrating birds from permanent foci in the tropics.

Culex tarsalis, the most important vector for both WEE and SLE in the west, may be recognized by its striped legs, band on proboscis, and dark V-shaped spots on the under side of the abdomen. It shows little choosiness as to its breeding places, and breeds in great numbers in irrigation seepage or overflow, rice fields, etc. *Culiseta melanura,* enzootic vector of EEE, is a small dark mosquito with no conspicuous markings.

The *Culex pipens* group includes a number of closely related brown house mosquitoes of world-wide distribution, of which the best known are *pipiens* in temperate climates, *quinquefasciatus* (or *fatigans*) in warm ones, and *molestus* (which lays eggs without a blood meal) in strictly urban areas. These interbreed where their ranges overlap, and are probably best regarded as mere varieties or ecotypes of one species, *C. pipiens.* There is no evidence of differences in their ability to transmit either encephalitis or filariae (p. 755). They are strictly nocturnal and will bite in complete darkness; therefore their activity supplements that of the yellow-fever mosquito, the house mosquito taking the night shift and the yellow-fever species the day shift. They are probably primarily a molester of birds, attacking man and other mammals as a second choice. They breed in almost any standing water but prefer artificial containers and are partial to filthy water. They thrive in cesspools and open sewers. The larvae (Fig. 232) have long breathing tubes and broad heads. Development from egg to adult can probably occur in 5 or 6 days under ideal conditions.

In Colombia and Brazil several strains of viruses in addition to jungle yellow fever have been isolated from forest mosquitoes. In the Far East, from Guam and the Philippines to Japan, Korea, and Manchuria, Japanese B virus causes extensive human outbreaks. It is transmitted primarily by *Culex* mosquitoes (*C. tritaeniorhynchus, C. quinquefasciatus,* and others) although others, e.g., *Aedes albopictus* and *Anopheles hyrcanus,* are also involved. Since the virus has been found in reared mosquitoes, transovarial transmission is suspected (but see p. 751). Japanese B, like VEE, is primarily a mammalian virus, as is that of Kyasanur Forest disease in India (reservoir is monkeys). In Australia the principal vector of Murray Valley virus may be *Culex annulirostris* which resembles *C. tarsalis* in having a banded proboscis. This virus is enzootic in many mammals and birds in tropical Queensland, whence it is carried by migrating birds to southern Australia.

As noted above, mosquito-borne viruses abound in Africa; in one locality in East Africa 297 human sera examined contained antibodies against 8 different viruses, and only 17% were negative. One virus (Bwamba) was present in 44%, and another, Zika, was associated with an outbreak of jaundice. Two of these African viruses which were thought to be rare in man (West Nile and Ntaya) have been shown to be extremely common in Egypt, where over 50% of children show antibodies to West Nile virus by the time they are 5 years old, and 75% of adults show them. Over 45% of adults show antibodies to Ntaya virus. West Nile antibodies are also common in domestic animals and wild and domestic birds, and this virus has also been found in man in Israel.

In Egypt species of *Culex* have been shown to be naturally infected, and *Aedes aegypti* is an experimental vector. Little is known about the pathogenicity of these viruses in infants; in inoculated adults they produce mild symptoms. Another of the African viruses, Rift Valley, is particularly harmful to sheep and cattle, is deadly to small rodents, and sometimes produces denguelike attacks in man. Mosquitoes involved are various species of *Aedes* and *Eretmopodites*. Another virus, Mengo, has been found in four continents but seems nowhere to be prevalent.

Myxomatosis, a virus disease that decimates whole rabbit populations (which may be a blessing in some places and annoying in others), is transmitted primarily by mosquitoes, although fleas and other insects may be involved.

Mosquitoes and Filariasis

Manson's discovery in 1879 that mosquitoes serve as intermediate hosts for filariae marked the beginning of a new era in medical science; it was the first evidence of the development in the bodies of insects of organisms causing human disease. An account of the filarial worms, including the development in mosquitoes, will be found in Chapter 21, pp. 475–485. *Wuchereria bancrofti* and *W. malayi* are the only filarial infections of man known to be transmitted by mosquitoes, though some of the others undergo partial development, up to the "sausage" stage, in these insects. The species of the genus *Dirofilaria* which inhabit the heart and subcutaneous tissues of dogs are also transmitted by mosquitoes.

Transmitters of *Wuchereria bancrofti*. In contrast to the condition existing in malaria and yellow fever, *W. bancrofti* is not limited to one group of mosquitoes for intermediate hosts, though by no means all species of mosquitoes serve equally well as transmitters. Some fail entirely, some allow only partial development to occur, and some allow only a relatively small percentage of the ingested embryos to reach the infective stage; others, on the other hand, are too hospitable and are frequently killed by the heavy infections which develop; apparently the most critical time for the mosquitoes is during the migration of the matured larvae from breast muscles to proboscis.

Although nearly 60 species of mosquitoes, including species of *Anopheles, Culex, Aedes, Psorophora,* and *Mansonia,* are capable of serving as intermediate hosts, only a few species, all closely associated with man and feeding largely on human blood, are common transmitters in nature.

The nonperiodic "pacifica" form of filariasis of the eastern South Pacific islands (see p. 477) is transmitted in most places by *Aedes polynesiensis,* which, along with *A. albopictus,* is one of a number of closely related species constituting a "*scutellaris*" group in the subgenus *Stegomyia,* to which *A. aegypti* (see pp. 727 and 746) also belongs. Locally others of the *scutellaris* group and also *Aedes fijiensis,* and *Culex quinquefasciatus* to a minor degree, may serve as transmitters; all but the last are outdoor mosquitoes. *A. polynesiensis* is a shade-loving, day-biting sylvan mosquito that usually bites outdoors, but may enter houses situated in or near the "bush" to bite but not to stay. It seldom ventures more than 100 yards from where it was bred. The breeding places are tree holes, banana leaves, coconut shells, and other small rain-filled containers, including bottles, cans, etc., in dense shade where available. Obviously such a mosquito is quite unaffected by the residual sprays indoors or by the space sprays outdoors.

The "periodic" form of *Wuchereria bancrofti* in which the microfilariae swarm in the peripheral blood only at night (see p. 477), found elsewhere in the world, is transmitted in cities (except in tropical Africa) primarily by members of the *Culex pipiens* complex (see p. 727), and in small towns and suburbs by man-hunting *Anopheles,* e.g., *A. darlingi* in South America, *A. gambiae, melas,* and *funestus* in Africa, the *A. punctulatus* group in New Guinea and the more western Pacific islands, *A. hyrcanus* in China, and several species in India. *Culex quinquefasciatus* is not a good vector in tropical Africa, suggesting that a different strain either of the filaria or of the mosquito may occur there. It is odd that *Aedes aegypti,* though susceptible, nowhere is an important vector.

Transmission may fail even when good vectors are present in considerable numbers. Humidity and temperature influence the percentage of infected mosquitoes and also successful invasion of the skin when infected mosquitoes bite.

Transmitters of *Wuchereria malayi.* The form of filariasis caused by *Wuchereria malayi,* in contrast to *W. bancrofti* infections in most places, is strictly rural, because it is transmitted mainly by mosquitoes of the genus *Mansonia* assisted in some places by species of *Anopheles,* especially *A. barbirostris. Mansonia* adults commonly enter houses to bite, and sometimes rest there, but not as persistently as does *Culex quinquefasciatus.*

Mosquitoes are the transmitters of the dog heartworm, *Dirofilaria immitis* (see p. 496), and presumably of other species of *Dirofilaria;* successful experimental hosts include species of *Anopheles, Aedes,* and *Culex,* but little is known about which species are important in nature; in the United States, *Anopheles quadrimaculatus* is undoubtedly one, but the unusual frequency and prevalence of these worms in dogs near the coasts suggest the involvement of salt marsh mosquitoes. Probably mosquitoes are intermediate transmitters of some bird filariae also. Mosquitoes of the subgenus *Mansonioides* of the genus *Mansonia* differ from all others in their biology. The eggs of these species are laid in clusters on the under surface of leaves of aquatic plants. The larvae have the breathing tube terminated by a spine (Fig. 236, *11*) with which they pierce the roots of the plants and draw air from them and so never come to the surface. The pupae use their breathing trumpets in a similar manner, and they, too, never rise to the surface until time of emergence.

Most of the species are found in the tropics. Some, including the important *malayi*-carrier in India, *M. annulifera,* are closely associated with the water lettuce, *Pistia stratiotes,* that grows extensively in tropical swamps. This mosquito can be controlled by raking up the plants and killing them with herbicides. Another vector, *M. uniformis,* chooses the water hyacinth, and in parts of China and Indo-China *M. longipalpis* attaches itself to the roots of many plants, including swamp trees; this species is very difficult to control.

Mosquitoes and *Dermatobia*

In many parts of tropical America where the larva of a botfly, *Dermatobia hominis* (see p. 786), infests man and cattle, there has long been a belief among the natives that the maggots which develop under the skin result from mosquito bites. Observations and experiments proved this to be true; mosquitoes normally serve as airplanes for the transportation of *Dermatobia* eggs to a suitable host. Occasionally when a *Dermatobia* is unable to find a mosquito and is under the immediate necessity of depositing eggs, she may oviposit on other captured arthropods, frequently muscoid flies, or even on leaves.

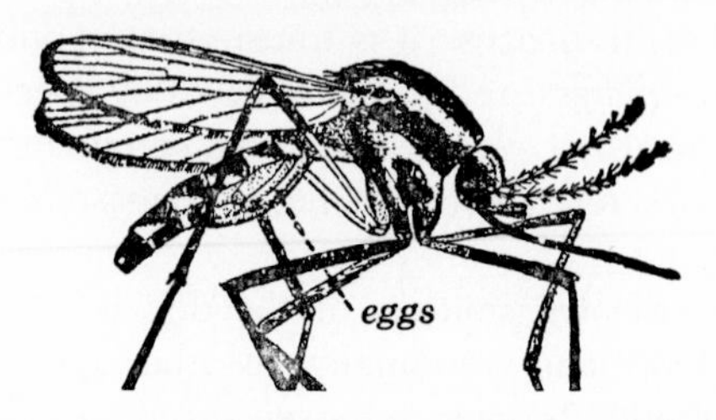

Fig. 244. *Psorophora* (*Grabhamia*) *lutzi*, with eggs of *Dermatobia hominis* attached to abdomen. (After Sambon, *J. Trop. Med. Hyg.*, 25, 1922.)

The mosquitoes involved in nature seem to be, primarily at least, species of *Psorophora*, subgenus *Grabhamia*. In Central America *P. lutzii* alone has been incriminated, but in South America other species are concerned; *P. ferox* is the most frequent vector in Colombia. *P. lutzii* (Fig. 244) is a dark, beautifully colored mosquito, with yellow markings on the thorax and with flashes of metallic violet and sky blue on its thorax and abdomen. It is said by Knab to be one of the most bloodthirsty of American mosquitoes and is found throughout tropical America. The larvae breed in rain puddles, the eggs being laid in dry depressions on the forest floor which will become basins of water after a tropical downpour. The eggs hatch almost with the first drop of rain and mature so rapidly that adult insects may emerge in 4 or 5 days. The larvae feed on vegetable matter and are themselves fed upon by their relatives of the subgenus *Psorophora,* which breed in the same rain pools.

Control and Extermination

Control or protection against mosquitoes and mosquito-borne diseases may be undertaken in the following ways: (1) personal protection from adults, (2) destruction of adults, (3) destruction of larvae, and (4) elimination of breeding places.

Personal protection. This method of dealing with mosquitoes may be indispensable to people operating in mosquito-infested places and for residents in some areas, as in parts of Alaska where control methods are unpracticable. Concerning the use of protective clothing, little need be said; the value of gloves, veils, high boots, leggings, etc., is obvious. As a repellent for either skin or clothing, diethyltoluamide is best, being solidly protective against most mosquitoes for at least 4 hours (see p. 536). In many parts of Alaska protection against mosquitoes outdoors is essential for playing children, etc., and is obtained by more or less stationary aerosol dispensers situated so the insecticidal mist will drift over the areas to be protected. Similar protection is obtained in some cities by mists dispensed from trucks moving up and down the streets, or locally for protection of outdoor gatherings. Indoor protection is obtained by means of screens; their effectiveness is enhanced by spraying them with residual insecticides so that fewer will enter when doors are opened.

Destruction of adults. Immediate destruction of mosquitoes and other insects in houses, barracks, schools, etc., can be obtained by the use of aerosol "bombs" (see p. 535). For continued protection, however, residual sprays with DDT or other chlorinated hydrocarbons in suspensions or emulsions, at the rate of 200 mg. per square foot of the insecticide, are needed. The surfaces sprayed are lethal for 3 to 6 or 8 months to insects resting on them, depending on the nature of the surface, formulation of spray, etc. For details, see p. 534.

Residual spraying of houses has given sensational success against insect annoyance and in control of arthropod-borne diseases wherever the vectors habitually enter houses and rest in them. As noted on p. 750, this method may actually exterminate the domestic *Aedes aegypti*, and it gives such excellent protection against house-frequenting *Anopheles* (e.g., *A. darlingi, A. quadrimaculatus, A. gambiae*) that malaria is quickly reduced to a position of minor importance, or eliminated altogether (see pp. 190–191). On the other hand, residual spraying is of little or no use against the mosquitoes that bite only outdoors, or leave a house immediately after feeding. This includes members of the *Aedes scutellaris* complex, which are important vectors of filariasis and dengue in the Pacific area, and also such *Anopheles* as *A. bellator* and to some extent *A. albimanus* and *A. aquasalis*. Unfortunately insects are adaptive creatures, and we must contemplate three possibilities: (1) that some of the house-frequenting species may develop bite-and-run habits as a result of natural selection and survival of the fleetest, just as they once substituted domesticity for a sylvan life before human houses were converted into lethal traps; (2) that some of the present bite-and-run species, when their house-frequenting competitors are eliminated, may become of greater importance; and (3) that mosquitoes become as resistant to DDT and allied chemicals as flies. Up to the present time only a few species of *Anopheles* have developed any appreciable degree of resistance, and only locally, but several others (*Culex tarsalis* and the salt marsh mosquitoes, *Aedes sollicitans* and *A. taeniorhynchus*) have become fairly resistant in some places.

A novel method of killing adult male mosquitoes is the use of a loudspeaker set up behind an electrified screen to broadcast the sounds made by the female of the species (Kahn and Offenhauser, 1949).

Many birds, especially nighthawks, swifts, and swallows, feed actively on adult mosquitoes. Bats have been exploited as mosquito destroyers, and municipal bat roosts have actually been erected in San Antonio and recommended for other places, but scientific investigation has not substantiated the extravagant claims made for the efficiency of bats as mosquito destroyers; in addition, bats are an important reservoir of rabies. In the tropics wall lizards or geckos and jumping spiders destroy numbers of mosquitoes in dwellings.

Destruction of larvae. Larvicidal measures consist of application of insecticides as sprays, dusts, granules, or pellets (see p. 534), or as solutions in a high-spreading oil (1 quart of 5% solution per acre) for fairly open water; clearing of brush and floating vegetation to permit fish, particularly *Gambusia,* to get access to the larvae; fluctuation of water level; and for *Mansonia,* removal of water lettuce, water hyacinth, and other water vegetation. Fluctuation of water level has been very helpful in preventing breeding in reservoirs. When the level is lowered, most of the larvae are stranded and die; when it is raised, fish can get at larvae that were developing near the lower shore line.

The chlorinated hydrocarbons, at first DDT but now particularly dieldrin (see p. 532), have superseded all of the older larvicides such as oil, Paris green, creosote emulsions, etc. Species of mosquitoes vary somewhat in their susceptibility to particular insecticides, e.g., *C. quinquefasciatus* is considerably more resistant to DDT than most mosquitoes, but highly susceptible to BHC and dieldrin. *Anopheles* larvae are so susceptible to DDT that, if evenly distributed, 1 lb. of DDT would be enough to kill them on 1000 acres; in practice 100 times that amount is used. No harmful effects on fish or other wildlife results from such dosage. In rice fields heavy dusting before flooding greatly retards mosquito breeding.

For culicine mosquitoes, including *Mansonia,* 5% dieldrin granules (2 lb. per acre) have been used to advantage in marshes, ponds, rice fields, etc. Larger pellets (60 grams) of sand and cement (5:1) saturated with insecticides in xylene are very effective in catch basins, crab holes, wells, and even drinking water pots. Dieldrin is the best insecticide for the latter since its solubility (0.05 ppm) is well below the levels considered unsafe.

In treating catch basins, cesspools, sewer inlets, etc., a residual effect good for several months can be obtained by spraying the side walls. For a sewage farm, application of 8 ounces of a 25% emulsion every day or two to the water in the main pipeline did the work of spraying 75 to 100 gallons of diesel oil. By treating a main irrigation ditch with a few gallons of emulsion applied slowly over a period of an hour or so, larvae can be killed for distances of many miles; in one case no larvae were found in 100 miles of irrigation canals.

Natural enemies. Certain kinds of fish are of very great value in control of mosquito larvae in natural waters, lily ponds, etc. The viviparous *Gambusia affinis* (Fig. 245), widely distributed in southeastern United States and extensively introduced elsewhere, is a valuable species because of its hardiness, ability to live in fresh, brackish, or foul water, and rapid multiplication.

Where algae, weeds, and debris are removed to permit free operation of the fish, usually no other control is necessary. Even goldfish may keep lily ponds free if the mosquitoes do not breed more rapidly than the fish can eat them. In salt marshes, various species of killifish (*Fundulus*) are potent factors in destroying mosquito larvae. Great reduction in mosquito output can be obtained by draining marshes in such a way that the fish can get access to most parts of them. If swamps are converted into pools in their deepest parts, fish can sometimes control the mosquito output. *Gambusia* can be used to advantage in cisterns and exposed wells. Many southern cities have hatcheries for these fish to supply them to citizens who have use for them.

The nonbiting *Toxorhynchites* whose tree-hole-breeding larvae are cannibalistic has been introduced into Hawaii and Samoa to prey on the larvae of tree-hole-breeding *Aedes*—with what effect remains to be seen. Many other water inhabitants attack mosquito larvae or eggs, including predaceous insects, bugs, mites, etc., so it is not desirable, when it can be avoided, to kill such life while attempting to kill mosquito larvae. A number of aquatic plants are inimical to some mosquito larvae for one reason or another, *Chara* apparently by producing a high oxygen content of the water, the bladderwort, *Utricularia,* by capturing the larvae in its traplike bladders, and surface-covering plants such as *Lemna* (duckweed) by preventing the larvae from getting access to air.

Elimination of breeding places. The only permanent method of control of most mosquitoes is to eliminate the breeding places entirely wherever possible. The application of this method to container-breeding mosquitoes was discussed on p. 750.

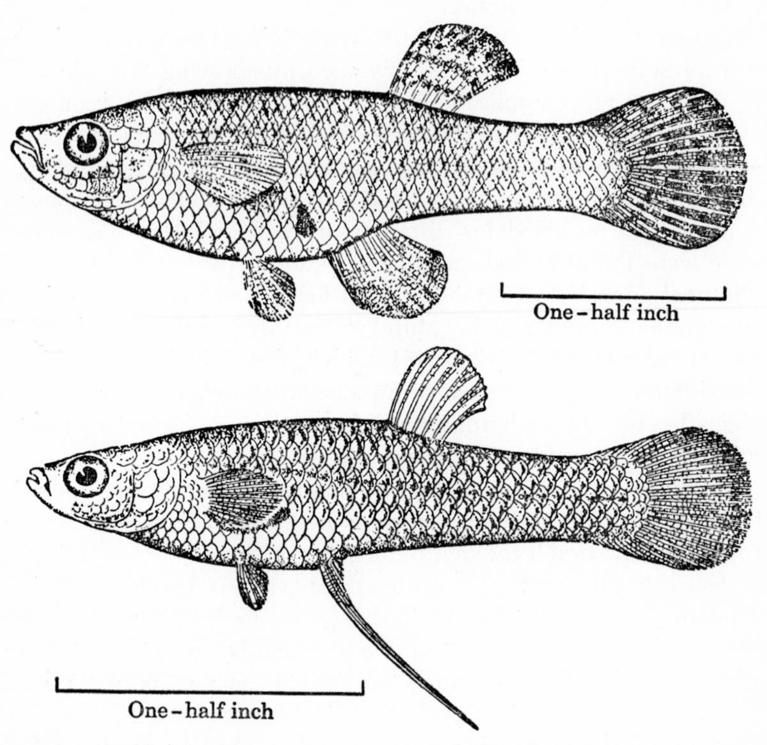

Fig. 245. *Gambusia affinis*, a voracious, mosquito-eating top minnow, useful for stocking ponds, cisterns, wells, etc. (After Jordan and Evermann, *The Fishes of North and Middle America.*)

Drainage is often practicable as a means of eliminating breeding places. Not only must small pools of standing water be eliminated but also the drains themselves must be made unsuitable for breeding. Sometimes other methods of control, such as filling in of depressions or protection of swamps by means of levees, dikes, or tide gates, are more practicable. Drainage ditches with narrow bottoms, the sides of which are kept clean and straight, preferably by cement or board walls, are the best means of draining borrow pits, swampy depressions in streams, or outcrops of seepage water. Seepage water outcrops are the most difficult and often have to be drained by ditches which more or less follow the contours, ultimately connecting with main ditches leading away. Lined ditches cost more to build but are more permanent, more easily kept clean, and cheaper in the long run.

Subsoil drainage by means of tile or pipe is often necessary; in the Malayan hills Watson got wonderful results by thus draining ravines where *A. maculatus* breeds, and excellent results have been obtained from this method in Panama also. In some places effective drainage has been obtained by packing drains or tributaries of ravines with tree trunks and branches and a top covering of grass. Where water is held at the surface by an impervious stratum overlying a pervious one vertical drainage may be successful by drilling holes through which the water can flow down to the deeper pervious strata. Salt marshes may be drained by appropriately placed drains averaging 200 to 300 ft. to the acre, the method successfully used by Headlee in New Jersey, with filling in of parts which cannot be so drained; the falling of the tide carries the water out of the ditches. When the tide is insufficient to do this, engineering projects of diking with tide gates or pumps must be resorted to. In some places dams and automatic siphons, to flush streams in the dry season when pools form in their beds, have been found useful.

Periodic draining of ponds and rice fields results in a great reduction in number of *Anopheles* larvae. In California 10-day intervals between drying have been employed, and in Portugal 16 days. Sometimes provision of shade, sometimes removal of it, eliminates breeding places (see p. 739). For bromeliad-breeding species of *Anopheles* (*A. bellator, A. cruzi*) spraying with 0.5% copper sulfate kills the plants, but in some places hand removal of them is preferable. In either case not *all* the aerial plants need be removed, for some of the smaller species, which are often the most inaccessible, are of little or no importance. Once eliminated there may be no regeneration of the plants for 10 years or more.

REFERENCES

Aitken, T. H. G. 1945. Studies on the anopheline complex of western North America. *Univ. Calif. Publ. Entomol.*, 7, No. 11, 273–364.

American Association for the Advancement of Science. 1941. A symposium on human malaria, with special reference to North America and the Caribbean region, Publ. 15, Section III, Anopheline vectors, 63–130.

Barr, A. R. 1957. The distribution of *Culex p. pipiens* and *C. p. quinquefasciatus* in North America. *Am. J. Trop. Med. Hyg.*, 6: 153–165.

Bates, M. 1949. *The Natural History of Mosquitoes.* Macmillan, New York.

Burgess, R. W. 1946. Pigmentation as a specific character in certain anopheline pupae. *J. Natl. Malaria Soc.*, 5: 189–191.

Carpenter, S. J., and La Casse, W. J. 1955. *Mosquitoes of North America (North of Mexico).* Univ. Calif. Press., Berkeley.

Carpenter, S. J., Middlekanff, W. W., and Chamberlain, R. W. 1946. The mosquitoes of the southern United States east of Oklahoma and Texas. *Am. Midland Naturalist Monogr.* No. 3.

Chamberlain, R. W. 1958. Vector relationships of the arthropod-borne encephalitides in North America. In *Annual Disease and Human Health, Ann. N. Y. Acad. Sci.*, 70, Art. 3: 312–319.

Chandler, A. C. 1945. Factors influencing the uneven distribution of *Aëdes aegypti* in Texas cities. *Am. J. Trop. Med.*, 25: 145–149.

Chandler, A. C. 1956. History of *Aëdes aegypti* control work in Texas. *Mosquito News,* 15: 58–63.

Elliott, R. 1955. Larvicidal control of peridomestic mosquitoes. *Trans. Roy. Soc. Trop. Med. Hyg.*, 49: 528–542.

Elton, N. W. 1952. Public health aspects of the campaign against yellow fever in Central America. *Am. J. Public Health,* 42: 170–174. Progress of the sylvan yellow fever wave in Central America: Nicaragua and Honduras, *ibid.:* 1527–1534.

Frohne, W. C. 1956. The biology of northern mosquitoes. *Public Health Repts.,* 71: 616–621.

Galindo, P., Trapido, H., and Carpenter, S. L. 1950. Observations on diurnal forest mosquitoes in relation to sylvan yellow fever in Panama. *Am. J. Trop. Med.*, 30: 533–574; 1951 (Galindo, Carpenter, and Trapido). Ecological observations on forest mosquitoes of an endemic yellow fever area in Panama. *Am. J. Trop. Med.*, 31: 98–137.

Hackett, L. W. 1934. The present status of our knowledge of the subspecies of of *A. maculipennis. Trans. Roy. Soc. Trop. Med. Hyg.*, 28: 109–128.

Hess, A. D., and Holden, P. 1958. The natural history of the arthropod-borne encephalitides in the United States. In *Animal Disease and Human Health, Ann. N. Y. Acad. Sci.*, 70, Art. 3: 294–311.

Horsfall, W. R. 1955. *Mosquitoes. Their Bionomics and Relation to Disease.* Ronald Press, N. Y.

Howard, L. O., Dyar, H. G., and Knab, F. 1912–1917. The mosquitoes of North and Central America and the West Indies. *Carnegie Inst. Wash. Publ.* 159. Vols. I–V.

Hunter, G. W., Welles, T. H., and Jahnes, W. G., Jr. 1948. An outline for teaching mosquito stomach and salivary gland dissection. *Am. J. Trop. Med.*, 26: 221–228.

Jachowski, L. A., Jr. 1954. Filariasis in American Samoa. V. Bionomics of the principal vector, *Aëdes polynesiensis. Am. J. Hyg.*, 60: 186–203.

Jefferson Medical College. 1955. Yellow Fever. A symposium in commemoration of Carlos Juan Finlay (10 papers by various authors). Philadelphia.

Jenkins, D. W. 1950. Bionomics of *Culex tarsalis* in relation to western equine encephalomyelitis. *Am. J. Trop. Med.*, 30: 909–916.

Kahn, M. C., and Offenhauser, W., Jr. 1949. The first field tests of recorded mosquito sounds used for mosquito destruction. *Am. J. Trop. Med.*, 29: 811–825.

King, W. B., Bradley, G. H., and McNeel, T. E. 1939. The mosquitoes of the southeastern States. *U. S. Dept. Agr. Misc. Publ.* 386.

Komp, W. H. W. 1942. A technique for staining, dissecting and mounting the male terminalia of mosquitoes. *Public Health Repts.*, 57: 1327–1333.

1948. The anopheline vectors of malaria of the world. *Proc. 4th Intern. Congr. Trop. Med. Malaria,* 1: 644–655.

Kumm, H. W., and Cerqueira, N. L. 1951. The *Haemogogus* mosquitoes of Brazil. *Bull. Entomol. Research,* 42: 169–181.

Kumm, H. W., Osorno-Mesa, E., and Boshell-Manrique, J. 1946. Studies on mosquitoes of the genus *Haemagogus* in Colombia. *Am. J. Hyg.*, 43: 13–28.

Macdonald, G., 1956. Theory of the eradication of malaria. *Bull. World Health Organization,* 15: 369–387.

Matheson, R. 1944. *The Mosquitoes of North America.* 2nd ed. Comstock, Ithaca, N. Y.

Mattingly, P. F., et al. 1951. The *Culex pipiens* complex. *Trans. Roy. Entomol. Soc. London,* 102: 331–382.

Penn, G. H. 1947. The larval development and ecology of *Aëdes* (*Stegomyia*) *scutellaris* (Walker, 1859) in New Guinea. *J. Parasitol.*, 33: 43–50.

1949. Pupae of the nearctic anopheline mosquitoes north of Mexico. *J. Natl. Malaria Soc.*, 8: 50–69.

Rosen, L. 1954. Observations on the epidemiology of human filariasis in French Oceania. *Am. J. Hyg.*, 61: 219–248.

Ross, E. S., and Roberts, H. R. 1943. *Mosquito Atlas,* Parts I and II. American Entomological Society, Philadelphia.

Russell, P. F., Rozeboom, L. E., and Stone, A. 1943. *Keys to the Anopheline Mosquitoes of the World.* American Entomological Society, Philadelphia.

Soper, F. L., and Wilson, D. B. 1943. *Anopheles gambiae in Brazil, 1930–1940.* Rockefeller Foundation, New York.

1942. Species eradication. A practical goal of species reduction in the control of mosquito-borne disease. *J. Natl. Malaria Soc.*, 1: 5–24.

South Pacific Commission. 1953. Filariasis in the South Pacific. *Proc. Conf. Specialists,* Papeete, Tahiti, 1951.

Stone, A., Knight, K. L., and Starcke, H. 1960. *Synoptic Catalogue of the Mosquitoes of the World.* Thomas Say Foundation (in press).

Strode, G. K. 1951. *Yellow Fever.* McGraw-Hill, New York.

Taylor, R. M., and Fonseca da Cunha, J. 1946. An epidemiological study of jungle yellow fever in an endemic area in Brazil. Pt. I. Epidemiology of human infections; and Laemmert, H. W., Ferreira, L. de C., and Taylor, R. M. Pt. II. Investigation of vertebrate hosts and arthropod vectors. *Am. J. Trop. Med.*, 26. Suppl., 1–69.

Taylor, R. M., and Theiler, M. 1948. The epidemiology of yellow fever, *Proc. 4th Intern. Congr. Trop. Med. Malaria,* 1, Sect. IV, 506–519.

Thompson, H. V. 1956. Myxomatosis: a survey. *Agriculture,* 63: 51–57.

Trapido, H., and Gallindo, P. 1956. The epidemiology of yellow fever in Middle America. *Exp. Parasitol.*, 5: 285–324.

Trembley, H. L. 1955. Mosquito culture techniques and experimental procedures. *Am. Mosquito Control Assoc. Bull.*, 3: 73 pp.

Usinger, R. L. 1944. Entomological phases of the recent dengue epidemic in Honolulu. *Public Health Repts.*, 59: 423–430.

Whitehead, F. E. 1952. A large scale experiment in rice field mosquito control. *Rept. Ser. Ark. Agr. Exp. Sta.*, 32.

Wilson, C. S. 1951. Control of Alaskan biting insects. *Public Health Repts.*, 66: 911–944.

World Health Organization. 1954. Report on symposium on control of insect vectors of disease. *Tech. Repts. Series.* Geneva.

Chapter 30

DIPTERA
III. FLY MAGGOTS AND MYIASIS

Disgusting as it may seem, man and animals are attacked not only by the numerous adult flies discussed in the last two chapters but also by the maggots or larval stages of many species of flies. Such an infestation by fly maggots is called myiasis. Nearly all cases are caused by larvae of flies of the suborder Cyclorrhapha (see p. 667). Exceptions are a few reported cases of infestation of the skin by a scale insect; of sinuses by the larvae of carpet beetles; and of the rectum by adult dung beetles. The last, and many other unusual cases of beetles or maggots in the feces, stomach, urinary passages, nose, etc., are clearly accidental causes of pseudoparasitism (see Théodoridès, 1948). Mention should also be made of dermatitis produced by products of nonparasitic insects, especially the blisters formed by tiny beetles of the genera *Paederus* (a staphylinid) and *Epicauta* (a cantharid), and the great irritation produced by the poison hairs of a caterpillar, *Megalopyge opercularis,* popularly called an "asp."

The flies most frequently concerned in myiasis belong to two large groups, the Muscoidea (see p. 668) and the botflies. The latter are committed to parasitic life in the larval stage, and live for a very short time, probably not feeding at all, in the adult stage. Their mouth parts are reduced to mere vestiges. Many of the Muscoidea, on the other hand, are a nuisance in the adult stage as bloodsuckers or germ carriers, but some, such as the screwworms and the African tumbu fly, have to be reckoned with as important parasites in the larval stage, and a few, such as the housefly, may cause trouble in *both* stages. Many Muscoidea belonging to the families Calliphoridae and Sarcophagidae, commonly called blowflies and fleshflies, have maggots that feed on dead flesh, and it is not surprising that some of these should have adapted themselves to entering wounds and feeding on living flesh, e.g., the screwworms, or to developing in the foul-smelling soiled wool of sheep and attacking the skin underneath, e.g., the wool maggots.

Identification. Identification of full-grown larvae causing myiasis is usually not difficult so far as the genera are concerned, but accurate determination of species often requires breeding them out. The most important characteristics used for identification are the respiratory openings at the anterior and posterior ends of the abdomen (Fig. 247). The posterior openings consist of two stigmal plates; these are hardened, dark-colored, eyelike spots, in most species surrounded by a sclerotized ring and a buttonlike mark, though in some of the Oestridae the whole plate is chitinized. On the plates the spiracular openings are usually in the form of three slits, which may be straight, bent, or looped. The position and shape of the plates, the development of the ring and button, and the form of the slits are of great value in identification.

First-stage maggots are recognizable as such by the absence of anterior spiracles and posterior spiracular plates, but the genera and species are difficult to identify. Second-stage maggots are also difficult to identify; in the muscoid group they are recognizable as such by the presence of two instead of three spiracular slits on the posterior stigmal plates.

Following are keys to the principal myiasis-producing adult flies and their larvae, including a few forms most likely to be confused with them, but not including adults of those only occasionally found in human feces.

Adults of Myiasis-Producing Flies

I. **Muscoidea (Muscidae, Calliphoridae,** and **Sarcophagidae).** Eyes large (Fig. 248), touching or nearly so in ♂; proboscis well developed.

1*a*. Color metallic blue or green; **Calliphoridae** 2
1*b*. Color gray or yellowish with dark markings 4
1*c*. Color yellowish-brown; **African Calliphoridae** 5
2*a*. Bristles on mesonotum mostly wanting (Fig. 246, 2); blue or greenish blue; face golden or orange-red; palpi short; antennae feathered to tip (Fig. 246, 5) ***Callitroga*** and in Old World, ***Chrysomyia***
2*b*. Bristles well developed on mesonotum (Fig. 246, *1*) 3
3*a*. Small; green or coppery; face silver; squama bare (Fig. 246, 8) ***Phaenicia***
3*b*. Large; blue; face red or golden; squama hairy ***Calliphora***
3*c*. Larger; bluish black; face black ***Phormia***
4*a*. Abdomen yellowish basally, dark at apex, with longitudinal dark stripe; proboscis fleshy ***Musca***

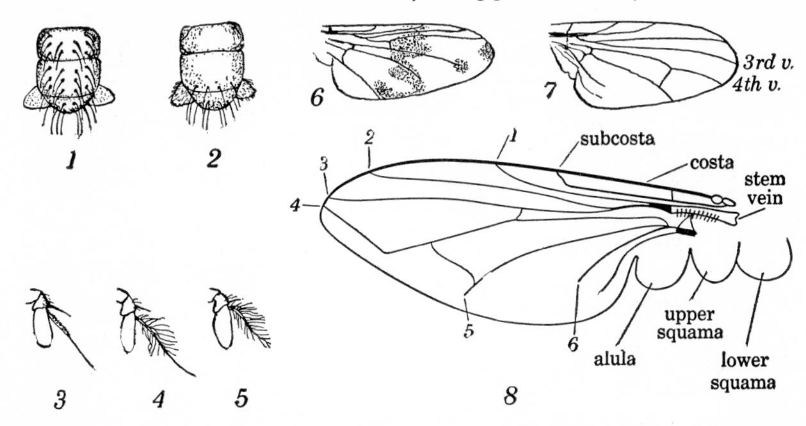

Fig. 246. Details of structure of myiasis-producing flies to illustrate key. *1,* Thorax of a *Lucilia,* dorsal view, showing well-developed bristles in median two rows of mesonotum, and hairless alulae. *2,* Thorax of *Callitroga hominivorax,* dorsal view, showing absence of bristles in median two rows except at hind end of mesonotum, and hairy alulae. *3,* Antenna of *Wohlfartia,* with naked arista. *4,* Antenna of *Sarcophaga,* with arista feathered on basal half. *5,* Antenna of *Callitroga,* with arista feathered to tip. *6,* Wing of *Gasterophilus intestinalis,* showing straight fourth vein not reaching margin, and spotting of wing. *7,* Wing of *Fannia,* showing straight fourth vein. *8,* Wing of a typical muscoid fly, showing veins (with numbers), alula, and squamae.

(1) Similar but more slender; fourth wing vein straight, not curving up towards third (Fig. 246, *7*) ***Fannia***

4*b*. Abdomen checkered gray and black; arista feathered except at tip (Fig. 246, *4*) ***Sarcophaga***

4*c*. Abdomen gray spotted with black; arista bare (Fig. 246, *3*) ***Wohlfartia***

5*a*. Abdominal segments all about equal (Fig. 250) ***Cordylobia***

5*b*. Second abdominal segment of ♂ elongated (Fig. 248); third abdominal segment of ♀ indented ***Auchmeromyia***

II. **Botfly Group (Gasterophilidae, Cuterebridae,** and **Oestridae).** Eyes small, widely separated (Figs. 257, 258); proboscis greatly reduced or absent.

1*a*. Body not markedly hairy; proboscis small, in pit; arista feathered on one side **Cuterebridae**

(1) Body blue, wings brown; tropical American skin maggot of cattle and man (Fig. 255) ***Dermatobia***

(2) Body moderately hairy, black and white; anal vein poorly developed; skin maggots of rodents and cats ***Cuterebra***

1*b*. Body hairy, beelike 2

2*a*. Abdomen elongated; fourth vein of wing straight, not extending to margin of wing (Fig. 246, *6*) (horse bots) **Gasterophilidae**

(1) Abdomen brown, tipped with red (Fig. 258, G) *Gasterophilus haemorrhoidalis*

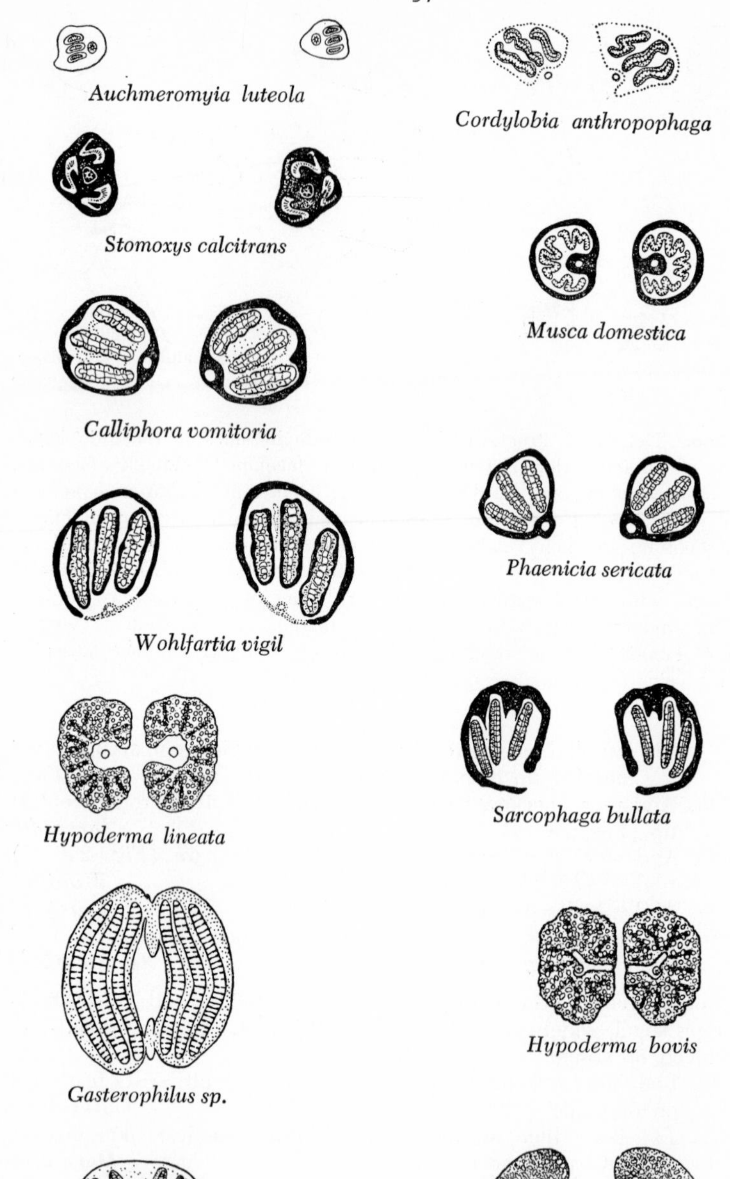

Fig. 247. Stigmal plates and spiracles of various maggots. Note distance apart of stigmal plates, form and position of spiracles, and presence or absence of "button." (Adapted from various authors.)

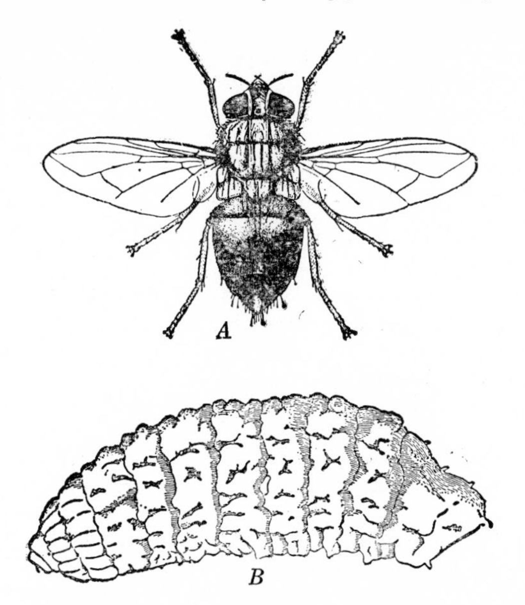

Fig. 248. Congo floor maggot and adult fly, *Auchmeromyia luteola.* *A,* ×3; *B,* ×4. (Adult after Manson-Bahr, *Manson's Tropical Diseases,* Williams and Wilkins; larva after James, *U. S. Dept. Agric. Misc. Publ.* 631, 1947.)

(2) Abdomen light at each end with black band in middle; wings not spotted *Gasterophilus nasalis*
(3) Abdomen brown, dirty white at base; wings spotted (Fig. 246, 6) *Gasterophilus intestinalis*
2*b*. Abdomen short, rounded; fourth vein of wing curved forward at tip, sometimes closing first posterior cell (Fig. 257); mouth parts vestigial; arista bare; **Oestridae** 3
3*a*. Middle part of face narrow; color dirty or grayish ***Oestrus***
3*b*. Middle part of face broad; color mainly blackish ***Hypoderma***
(1) Apex of abdomen orange; thorax not distinctly striped *H. bovis*
(2) Apex of abdomen lemon-yellow; light lines on thorax *H. lineata*

Full-Grown Larvae of Myiasis-Producing Flies

I. Larvae cylindrical, tapering anteriorly (Fig. 249); fairly smooth, without conspicuous colored spines; skin not leathery; stigmal plates separated, well sclerotized, with 3 spiracular slits (**Muscoidea**).
1*a*. Sclerotized ring completely encircles plate; button well developed 2
1*b*. Sclerotized ring incomplete; button region poorly chitinized 5

2*a*. Two mouth hooks; slits straight, or oval and only slightly bent 3
2*b*. One mouth hook; slits S-shaped or in loops 4
3*a*. Button enclosed in ring; slits straight, elongate, directed inward and downward (often nearly horizontal in ***Calliphora***) (Fig. 247) ***Calliphora, Lucilia,*** and ***Phaenicia***
3*b*. Button inside ring; slits oval, may be slightly bent ***Muscina***
4*a*. Slits have several loops; stigmal plates D-shaped, close together (Fig. 247) ***Musca***
4*b*. Slits S-shaped and well separated; stigmal plates separated by nearly twice their diameter (Fig. 247) ***Stomoxys***
5*a*. Stigmata in pits surrounded by fleshy tubercles (Fig. 252); slits vertical, the first one often directed downward and outward **Sarcophagidae**
5*b*. Stigmata not in pits 6
6*a*. Large break in ring; no definite button (Fig. 247); posterior margin of eleventh segment without dorsal spines ***Callitroga***
(1) Main tracheae large, pigmented (Fig. 249) *C. hominivorax*
(2) Main tracheae small, not pigmented (Fig. 249) *Sarcophaga bullata*
6*b*. Ring often nearly complete, but no definite button ***Chrysomyia***
6*c*. A weakly chitinized button present ***Phormia;*** in birds' nests ***Apaulina***

II. Larvae leathery, usually more or less flattened, not tapering from posterior to anterior end; often with conspicuous, colored spines (bots and some maggots).
1*a*. Body with rings of large, dark spines (Fig. 258); stigmal plates in contact, each with 3 bent slits (Fig. 247) (horse bots) ***Gasterophilus***
1*b*. Body entirely covered with black spines; stigmal plates with 3 convoluted spiracles spread laterally, converging medially to lower inner angle; in skin of rodents and cats ***Cuterebra***
1*c*. Body not as above 2
2*a*. Stigmal plates with 3 slits in each 3
2*b*. Stigmal plates solid, with numerous small openings 5
3*a*. Body with last segment retractile; anterior end large; cuticle sparsely studded with dark spines (Fig. 256); stigmal plates close together; slits slightly bent (tropical American skin maggot) (Fig. 247) ***Dermatobia***
3*b*. Stigmal plates well separated 4
4*a*. Plates very poorly developed, less than their own width apart, the slits crooked (Fig. 247); body studded with small yellow spines (African skin maggot) ***Cordylobia***
4*b*. Plates small, very far apart, the slits horizontal (Fig. 247); body without noticeable spines (African bloodsucking maggot) ***Auchmeromyia***
5*a*. Button well inside plate (Fig. 247); body with rows of strong spines on ventral side; in nostrils and other parts of head of sheep ***Oestrus***
5*b*. Button on inner margin of plate; in nostrils of horses ***Rhinoestrus***

5*c*. Button in median indentation of plate (Fig. 247); no conspicuous spines; in skin of cattle (Fig. 257) ***Hypoderma***
(1) Plates kidney-shaped (Fig. 247); no spines on last segment *H. lineata*
(2) Plates deeply indented; no spines on last 2 segments *H. bovis*

III. Larvae of odd types occasionally found in feces.
1. Large, flat, dark-colored, 11-segmented, with distinct head (a member of the soldier-fly family, Stratiomyidae) (Fig. 253*D*) *Hermetia illucens*
2. Cylindrical, with long tail-like process (rat-tailed maggot, member of family Syrphidae) (Fig. 253*C*) *Tubifera* (= *Eristalis*)
3. Small larvae with spiracles on tubercles; acalyptrate flies of families Piophilidae, Drosophilidae, etc. (cheese skippers, fruit flies, etc.) (Fig. 253*B*).
4. Flattened, with fleshy processes (Fig. 253*E*) ***Fannia***

Types of myiasis. Maggots attack their hosts in a number of different ways. The Muscoidea group includes maggots that (1) suck blood; (2) invade wounds and natural cavities (nose, ear, etc.); (3) attack skin under soiled wool, causing "strike" in sheep; (4) live in boils under the skin; and (5) live in or pass through the intestine or urinary tract. The botfly group includes species that (1) live in boil-like lesions in the skin; (2) cause warbles in the skin of cattle; (3) attack the nasal passages, sinuses, or other parts of the head of domestic animals or deer; and (4) live in the stomach or rectum of horses. Each of these will be briefly considered.

MUSCOID MAGGOTS

1. Bloodsucking Maggots

A number of species of flies allied to the blowflies deposit their offspring in the nests of birds, where the maggots attach themselves to the nestlings and suck blood. In northern United States and Canada, species of *Apaulina,* and in the Old World, *Protocalliphora,* have this habit. Hole-nesting passerine birds and hawks suffer most.

The only larva that sucks blood by puncturing the skin of man is the Congo floor maggot, *Auchmeromyia luteola,* found throughout tropical Africa south of the Sahara Desert, wherever there is a "stay-put" population of people who sleep on mats on the floor. Where the people are nomadic or sleep on raised beds, this parasite cannot hope to survive. It is the only known fly that is exclusively parasitic on man. Even where man and animals sleep in the same room, the maggots are found in the sand or dust only where human beings have lain. The adults commonly rest on walls indoors and feed by preference on human feces, though also attracted to fermenting or decaying vegetation. Other species of this genus and the related genus *Choeromyia* live in the burrows of the wart hog and other hairless mammals.

The adult fly (Fig. 248*A*) is a dirty yellowish-brown with the tip of the abdomen rusty black. (See key, p. 768.) The female lays her eggs in batches of about 50 at a time, usually in dust along the edges of sleeping mats; she may have 5 or 6 egg-laying sessions at intervals of a week or so. The eggs hatch within 2 days, and a few hours later the larvae are ready to suck blood, although if the occupant of the sleeping mat happens to be off on a vacation they can live unfed even for several weeks. The larvae (Fig. 248*B*) feed several times between molts, pupate after several weeks, and the adults emerge 10 days or so later. The life cycle probably occupies about 10 weeks, so there could be five generations in a year (Garrett-Jones, 1951). The larvae lie buried in dust under the floor mats in the daytime and come forth every night to pierce the skin with their mouth hooks and suck blood. For most people the bites are not very irritating, and there is no record of their transmitting disease.

2. Myiasis of Wounds and Natural Cavities (Screwworms)

Secondary invaders. A large number of flies belonging to the muscoid group, which normally deposit their larvae in decaying flesh of dead animals, occasionally, probably more or less by accident, deposit their eggs or larvae in neglected wounds or sores when offensive discharges are exuding from them. Included in this group are many blue, green, or coppery-colored species of Calliphoridae belonging to the genera *Calliphora, Phaenicia, Phormia, Callitroga, Chrysomyia,* and others, and gray and black Sarcophagidae of the genera *Sarcophaga* and *Wohlfartia* (see key on p. 768). There are, however, a small number of species which are *commonly* found as secondary invaders of wounds. These include *Sarcophaga bullata, Phormia regina,* several species of *Phaenicia* and *Lucilia,* and one or two other species of *Sarcophaga* (see Fig. 254) in this country, and *Chrysomyia megacephala* and others of this genus in the Old World. Some species deposit their eggs in befouled wool of sheep and later invade the body (see p. 779).

The secondary invaders are not primarily attracted by living tissue but only by decomposed tissue such as would be found in a dead animal. For this reason, and because their excretions have bactericidal properties, some of them were extensively used as a means of removal of dead tissue in cases of osteomyelitis, before the advent of antibiotics. The maggots would, however, attack healthy living tissue when dead tissue was not available, as Stewart demonstrated in 1934 in the case of the supposedly exclusively saprophagous *Phaenicia sericata,* which has been widely used as a "surgical maggot."

Primary invaders (screwworms). Of far greater significance are three species which deposit their eggs on fresh wounds of living animals and feed primarily upon the living tissues. They do not deposit their eggs on the unbroken skin but require only an insignificant wound or scratch, very often a tick bite. Apparently the odor of fresh blood is attractive to them. They are also attracted by the odors emanating from diseased natural cavities of the body and may oviposit in them. Severe infestations in man may lead to a loathsome and horrible death.

As already noted only three species of flies are known *normally* to attack living animals in this manner and to feed on living flesh. These are the American screwworm fly, *Callitroga hominivorax;* the Old World screwworm, *Chrysomyia bezziana* of southern Asia and South Africa; and *Wohlfartia magnifica* of eastern Europe.

CALLITROGA HOMINIVORAX. This highly injurious fly, formerly placed in the genus *Cochliomyia,* and long known as *Callitroga* (or *Cochliomyia*) *americana,* was long confused with a carrion-feeding species, *Sarcophaga bullata* (=*C. macellaria*). The adult flies of these two species are very difficult to distinguish, but the maggots show easily recognizable differences.

C. hominivorax is a true parasite. It only rarely lays its eggs in dead meat, although in the laboratory it will do so if the meat is at body temperature.

Much of the recorded biology of screwworms up to 1933 really applies to S. *bullata,* for *hominivorax* is rarely caught in carcass traps but S. *bullata* commonly is. The parasitic species occurs from the Gulf Coast states to Argentina. In the United States it normally survives the winter south of 30 degrees N. latitude, but in cold winters only in small areas in Arizona, southern Texas, and Florida. It does not survive when the mean daily temperature falls below 49°F. for 3 months or 53°F. for 5 months; adults are killed below 20°, and pupae at 15°F. It has no true hibernation. It survives in wounds or in soil in warm winter weather and spreads gradually during a summer; after a mild winter it may have a good start towards further expansion the next summer. It may also be shipped to distant northern states with infested cattle in the spring, and cause local disturbances until winter comes. Since its pupae are adversely affected by moisture in the soil, it rarely establishes itself where the rainfall exceeds 4 or 5 in. a month.

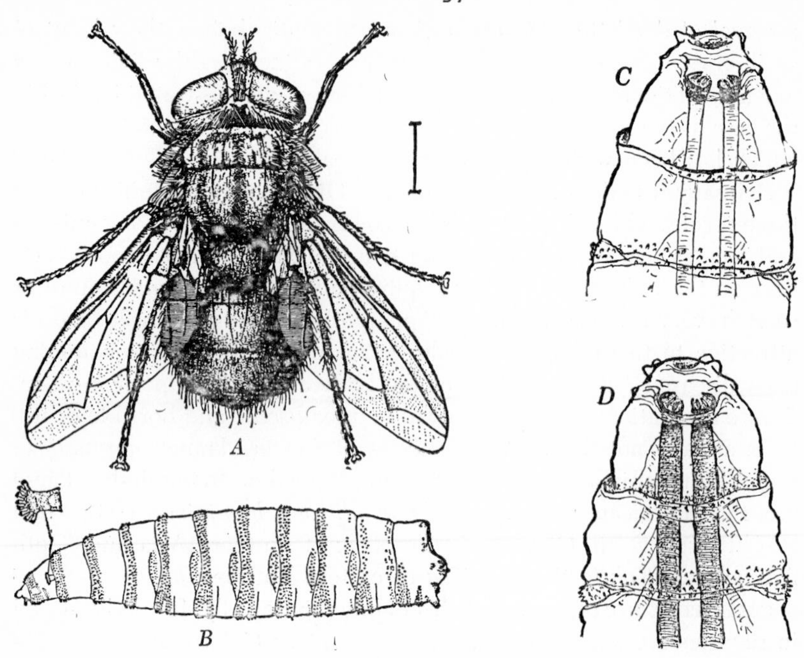

Fig. 249. *A,* Screwworm fly, *Callitroga hominivorax* adult ♀; *B,* full-grown (third stage) larva, anterior spiracles shown separately, enlarged. (Adult after James; larva after Laake, Cushing and Parish, from James, *U. S. Dept. Agric. Misc. Publ.* 631, 1947.) *C,* posterior end of *Sarcophaga bullata* (=*C. macellaria*), ventral view. *D,* same of *C. hominivorax.* Note large, heavily chitinized, dark tracheae of *C. hominivorax* and small, lightly chitinized, uncolored tracheae of *S. bullata.*

The adult screwworm flies (Fig. 249*A*) are large, greenish-blue flies with orange-red faces and eyes (see key, p. 768). The eggs are laid in batches of 150 to over 300, sometimes in more than one wound, and 8 or 10 such batches may be laid at intervals of 4 days. The incubation period is longer than that of *S. bullata* and is seldom less than 12 hours. The larvae (Fig. 249*B*) are whitish with bands of minute spines; they can be distinguished from the larvae of *S. bullata* by the much larger spiracles and large, heavily sclerotized main tracheal tubes (Fig. 249*C*). Eating away at flesh and even bone, they grow to a length of 12 to 15 mm. when mature; they then spontaneously leave the animals, bury themselves in loose earth, and pupate. When infested animals die, the larvae leave within 48 hours and pupate under or within a foot of the carcass in the upper half-inch of soil. In experimental guinea pigs the maggots regularly mature and leave a wound on the fifth or sixth day. The pupal period is

7 to 9 days in summer but may be prolonged to 10 or 12 weeks in winter. *S. bullata* may complete its whole life cycle in 9 or 10 days, but *C. hominivorax* is slower, requiring 18 to 22 days in summer weather.

Screwworms affect cattle, sheep, goats, and hogs most frequently; they are the cause of over 90% of myiasis of wounds. Human cases are rarer, but in 1935 there were over 100 human cases in the southern United States. In man the commonest site of infestation is the nose, whence the sinuses and nasopharynx are invaded, but the mouth, eyes, ears, vagina, and wounds are also attacked. Halitosis seems to be an attraction to screwworm flies as well as a repellent to romance, though the magazine advertisements have neglected to mention it. Sometimes *Dermatobia* lesions, boils, etc., are invaded, though more frequently by saprophagous species.

The damage done may be very extensive and is not infrequently fatal. Reports of 179 cases compiled by Aubertin and Buxton show that 15, or 8%, died. There is usually an abundant discharge of pus, blood, and scraps of tissue, accompanied by intense pain. Often nervous conditions develop, such as delirium, convulsions, visual disturbances, and loss of speech.

Small numbers of larvae are fatal to laboratory animals; guinea pigs usually succumb to more than 3 per 100 grams of body weight. Esslinger (1958) showed that highly toxic substances are produced by the growing larvae in addition to the tissue destruction, which may be serious enough when in the head. Borgstrom (1938) found that the maggots are invariably accompanied by one species of proteolytic but nonpathogenic bacterium (*Proteus chandleri*), which after a day or two is in practically pure culture in the wounds. The wounds usually do not become purulent until after the exit of the larvae but produce a copious sero-sanguinous exudate. Esslinger also showed that animals develop immunity against the toxic effects of the screwworms but not against the maggots themselves.

The damage done by screwworms to domestic animals amounts to millions of dollars in the United States. Wounds made by shears, barbed wire, thorns, ticks, parturition, etc., are commonly invaded, and myiasis of the cloaca of chickens is not infrequent. In our southern coastal areas from August to October about 12% of cattle have infestations; during this season 85% of all screwworm infestations begin in the bites of the Gulf Coast ear tick, *Amblyomma maculatum* (see p. 581). Next in importance are wounds made in shearing. Wounds made by castration, earmarking, and branding are also important. The flies frequently oviposit in the sores made by bots, especially *Dermatobia* in cattle and *Cuterebra* in rabbits (see pp. 786 and 788).

THE OLD WORLD SCREWWORM, *Chrysomyia bezziana.* This fly is widely distributed in Asia and Africa and has habits similar to those of *Callitroga hominivorax,* whereas its close relative, *Chrysomyia megacephala,* is the counterpart of our *Sarcophaga bullata. C. bezziana* is a common cause of human myiasis in India, but in Africa and in the Philippines it confines its attentions largely to animals. The maggots are very destructive and cause horrible, stinking sores.

WOHLFARTIA MAGNIFICA. This fly, a member of the family Sarcophagidae (fleshflies) (see key, p. 768), is found in southeastern Europe, Asiatic Russia, and Asia Minor and has habits similar to those of the screwworms.

The eggs of this fly, as of other sarcophagids, hatch before being deposited. The young larvae are placed directly in the wounds or cavities which the fly chooses for them. According to Portchinsky 150 or more larvae are deposited at a time; one instance is recorded of 70 maggots being extracted from a human eye after about this many had already escaped or been thrown away. The larvae of *Wohlfartia* are larger than those of the calliphorine flies and so are capable of even greater damage.

Treatment and control of screwworms. When discovered, the larvae should be removed as speedily as possible, especially if in the head. In animals the wounds should be treated with a smear developed by the U. S. Bureau of Entomology called EQ 335, containing 3% lindane and 35% pine oil, made up with mineral oil, emulsifier, and thickener (see Bruce, 1952). This is worked into the wound with a brush, and repeated in 5 to 7 days until it is healed. Substances like creosote, coal tars, etc., aggravate the wounds and retard healing.

In man removal of the maggots as soon as possible is indicated. Application of 5% chloroform in a light vegetable oil or in milk is helpful if applied by douching for 30 minutes, or application of saturated dressings to wounds from which the maggots are not easily removed. Even salt water is helpful if no better wash is available. Infestations of the nose, sinuses, ear, etc., if not attended to promptly, may require surgery.

Control consists in treating infested wounds promptly and in taking all precautions possible against wounds, especially in the fly season. Spread of the infestation would be greatly curtailed if there were enforced examination and treatment of all animals before shipment from an infested area to a distant part of the country.

A novel method proposed by Bushland and Hopkins (1951), and

successfully used in Curaçao in 1954, is the mass liberation of laboratory-reared male flies after sterilization by irradiation. A female fly mates only once, and if with a sterilized male none of her eggs will hatch. A male, on the other hand, may mate as many as eleven times if virgin females are available, so he may have a considerable effect on the next generation. Preliminary tests indicate that if there are five to ten times as many sterile as normal males, there is very little reproduction. Since the fly survives the winter in rather small numbers and in limited areas, extermination in our southeastern states might be possible by the mass liberation of treated males over two winters and a summer. In Texas it would be more difficult because of the vast areas involved, and because of reintroduction from Mexico, but Knipling and collaborators have recently reported successful results in Florida.

3. Wool Maggots Causing "Strike"

Sheep suffer from attacks by maggots which develop from eggs laid by carrion-feeding flies in damp wool soiled by feces or urine. Bacterial action produces ammonia, causes dermatitis, and attracts the flies. Such an infestation is called a "strike." The maggots eat into the flesh and often cause the death of sheep. Wool maggots are said to cause as much loss of sheep in parts of Australia as all other factors combined. Breeding in wool is a recently developed habit on the part of the flies—a result of changing conditions making for more blowflies, more vulnerable types of sheep, and perhaps less natural food for the flies. *Phaenicia* (or *Lucilia*) *cuprina* causes 96% of the wool maggot trouble in Australia and is important, along with certain species of *Chrysomyia,* in South Africa. In New Zealand the viviparous *Calliphora stygia* is the chief cause of strike early in the year, and *Phaenicia sericata* in the fall; the latter species is the main one concerned in Europe also. In southwestern United States, where sheep and goats frequently suffer, maggots of the blue-black *Phormia regina* (see p. 772) are commonest in spring and fall, those of *Sarcophaga bullata* in summer, and S. *bullata* has the bad habit of leaving infested wounds to attack soiled fleece.

Good control is obtained by dipping, power-spraying, or preferably high-power "jetting" of the crotch region of ewes and heads of rams with chlorinated hydrocarbons. Lindane, dieldrin, and aldrin are superior to the others because they diffuse along the wool fibers as the latter grow and give protection for 6 months; the DDT group do not do this, and chlordane does to a less degree. Prophylactic "jetting" of head, back, and crotch with diazinon or dieldrin gives protection for several months, and acts as a trap for the flies during the rainy season. Surgical removal of loose skin has been used as a prophylactic measure with some success.

4. Muscoid Skin Maggots

Two genera of muscoid flies, *Cordylobia* and *Wohlfartia,* pierce the skin of animals and develop in boil-like lesions in the skin after the manner of certain bots (*Dermatobia, Cuterebra,* and *Hypoderma*) which are discussed on pp. 786 and 789.

African skin maggots (*Cordylobia anthropophaga*). This yellowish-brown fly (Fig. 250), called the tumbu fly, is related to the floor maggot (see p. 774) and, like it, is found throughout tropical Africa. Although rodents are probably the primary hosts, a large number of tender-skinned wild and domesticated animals, especially dogs, are attacked, and man is a frequent victim.

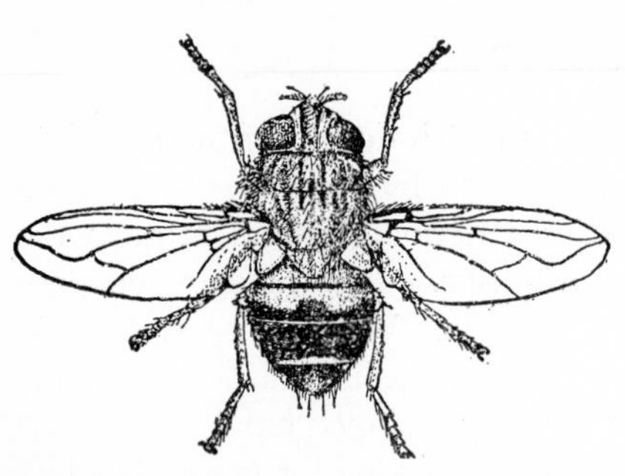

Fig. 250. Adult ♀ of African skin maggot, *Cordylobia anthropophaga.* ×3. (After Castellani and Chalmers, *A Manual of Tropical Medicine,* 1920.)

According to Blacklock and Thompson (1923) the eggs are laid by preference in dry sand and occasionally in cloth if either has been contaminated by excreta or has body odors, so clothing left exposed to flies may become dangerous. The eggs hatch in about 4 days. Upon stimulation by heat or touch the young larva becomes alert and active, attaches itself to skin, crawls to the nearest wrinkle or crevice, tears a hole with its mouth hooks, and within a minute or two has buried itself under the surface if the skin is not too tough. First attacks are painless, but there are marked reactions to subsequent attacks. The three larval stages (Fig. 251) are passed through in 8 days or more; the mature larvae then leave the tumor and pupate in the ground, the adult flies emerging after 8 or 10 days under favorable conditions.

Blacklock and Gordon (1927) made some interesting observations on immunity to this infestation, showing that larvae are unable to develop in previously infested skin. This the senior writer has interpreted as a specific reaction of the skin tissue which makes it unavailable as food for the larvae. The immunity is local and temporary in nature, gradually spreading in the skin, and is retained even when immune skin is grafted into another animal.

The boil-like lesions are often considerably excavated, apparently by a histolytic action of the larvae, and heavy infestations in animals may even cause death.

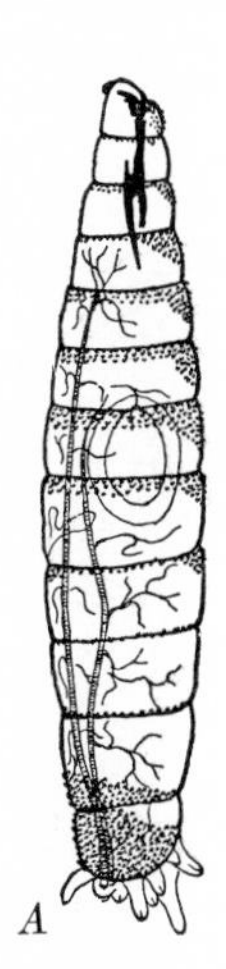

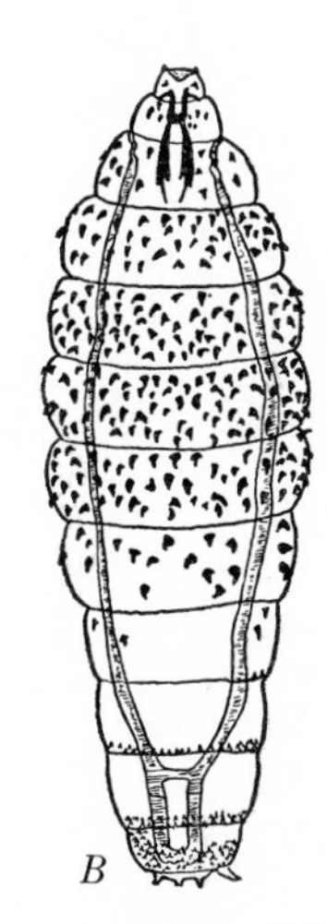

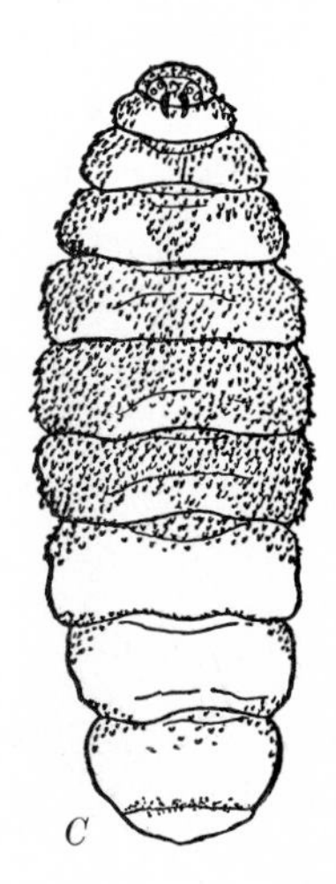

Fig. 251. *A, B,* and *C,* first-, second-, and third-stage larvae of *Cordylobia anthropophaga; A,* ×60; *B,* ×15; *C,* ×4. (After Blacklock and Thompson, *Ann. Trop. Med. Parasitol.,* 17, 1923.)

Large maggots can be removed with forceps, but smaller ones are best removed by applications of liquid paraffin. The larvae back out into the paraffin searching for air and by addition of more drops can usually be induced to emerge far enough to be captured or squeezed out.

***Wohlfartia vigil* and *W. opaca*.** Most of the species of *Wohlfartia,* like those of *Sarcophaga* (both members of the Sarcophagidae—see keys, pp. 768 and 771), are not primary myiasis producers; but, as we have already seen, *W. magnifica* is an Old World screwworm, and two American species, *W. vigil* and *W. opaca,* are invaders of healthy skin.

Walker in 1920 called attention to a number of cases of infestation of the otherwise healthy skin of young children by the maggots of *Wohlfartia vigil* (Fig. 252) in Toronto, Canada. Later, Ford (1936) reported additional human cases and many in small animals. In very small animals 5 to 20 larvae are said to cause death within 10 days. Young mink are frequently killed by them on farms in the upper Mississippi valley. Ford observed that the female flies habitually deposit their larvae on the skin of animals, especially young and tender ones, when available; the larvae are unable to penetrate adult human skin.

Tender-skinned babies sleeping outdoors unscreened are liable to nasty infestations. Larval development is rapid, sometimes requiring only 5 days in hot weather but usually occupying 6 to 9 days. The

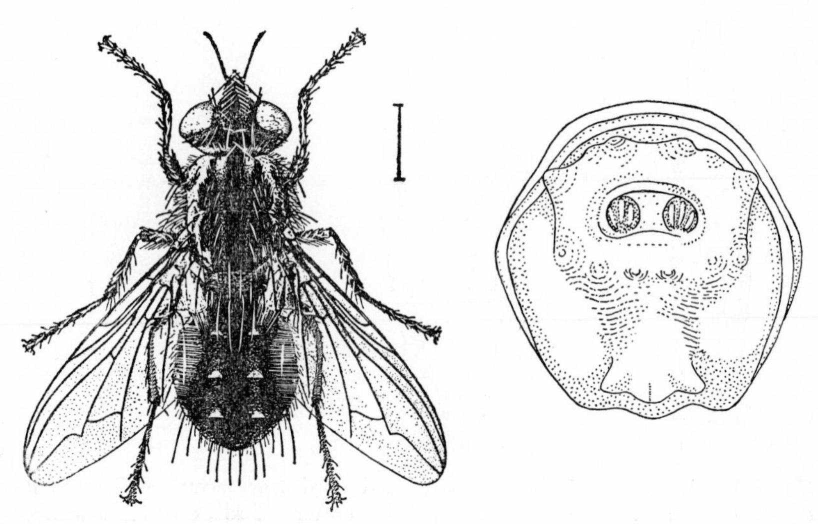

Fig. 252. *Left, Wohlfartia vigil,* adult ♀. (After James, *U. S. Dept. Agric. Misc. Publ.* 631, 1947.) *Right,* posterior end of third-stage larva of same. (Adapted from Walker, *J. Parasitol.*, 7, 1920.)

majority of human cases have been reported from the Toronto region, but scattered cases have been noted in various localities in the west where *W. vigil* is replaced by *W. opaca.* This species, called the fox maggot, is a common and important parasite of foxes and mink in the west and causes losses running to thousands of dollars on farms where these animals are raised. Rarely do cases occur in dogs.

5. Myiasis of the Intestine and Urinary Tract

Intestinal myiasis. Many species of fly maggots may accidentally be taken into the intestine of man. To quote from Banks, "When we consider that these dipterous larvae occur in decaying fruits and vegetables and in fresh and cooked meats; that the blowfly, for example, will deposit on meats in a pantry; that other maggots occur in cheese, oleomargarine, etc., and that pies and puddings in restaurants are accessible and suitable to them, it can readily be seen that a great number of maggots must be swallowed by persons each year, and mostly without any serious consequences." The reason for the lack of serious consequences is the fact that most maggots are killed in the stomach and thus fail to establish themselves. This is rather surprising since the maggots are unusually resistant to many chemicals that would quickly destroy other animals. Causey in 1938 fed larvae of a number of species to dogs and cats and found the larvae to be killed or immobilized in the stomach within 3 hours; none of them passed through the alimentary canal alive. In another experiment (Kenney, 1945) fifty human volunteers were fed living maggots of *Musca domestica, Calliphora,* and *Sarcophaga* under conditions planned to avoid destruction in the stomach. Fifty per cent had gastrointestinal disturbances—nausea, vomiting, cramps, and diarrhea—but the symptoms disappeared in 48 hours after elimination of the larvae, only a few of which were recovered alive after being vomited or passed in the feces.

Possibly intestinal myiasis is associated with low hydrochloric acid in the stomach or with particular conditions favoring rapid passage into the intestine. It is also possible that sometimes the flies get access to the intestine via the anus rather than the mouth, for some of the flies involved, particularly *Sarcophaga* and two species of *Fannia* (see p. 768), normally deposit their eggs in feces and decaying organic matter, and their eggs or larvae would rarely be found in edible food.

Excluding the species of *Gasterophilus,* which are true parasites of the alimentary canal (see p. 792), all the fly larvae recorded as causing intestinal myiasis are accidental parasites and probably in most cases pseudoparasites, actually no more parasitic than a swallowed goldfish. Some workers doubt that any of the numerous species found in human feces stop to nourish themselves, much less multiply, en route, but the evidence is against this extreme view, for there are well-authenticated cases of digestive disturbances occasioned by them. Even if they do not attack the mucous membranes they may cause nausea and abdominal discomfort by their movements.

Occasional remarkable cases are recorded of long-standing infections, even when opportunities for reinfection do not appear to exist. There are reports of living larvae of fleshflies or their allies persisting and causing symptoms for periods of weeks or even months, in one case (Herms and Gilbert, 1933) for several years. In most of these long-standing cases the larvae concerned are those of *Calliphora, Phaenicia,* or *Sarcophaga* or small acalyptrate flies of the genus *Apheochaeta* (see p. 667) which might be expected to lay their eggs on meat, cheese, or other foods. Micks and McKibben in 1956 reported a case in which the larvae *and pupae* of another small fly, *Leptocara venalicia,* related to the cheese skipper, were found at intervals over a period of 5 months. The presence of the pupae indicates that the larvae were capable of partial development, as well as survival, in the intestine.

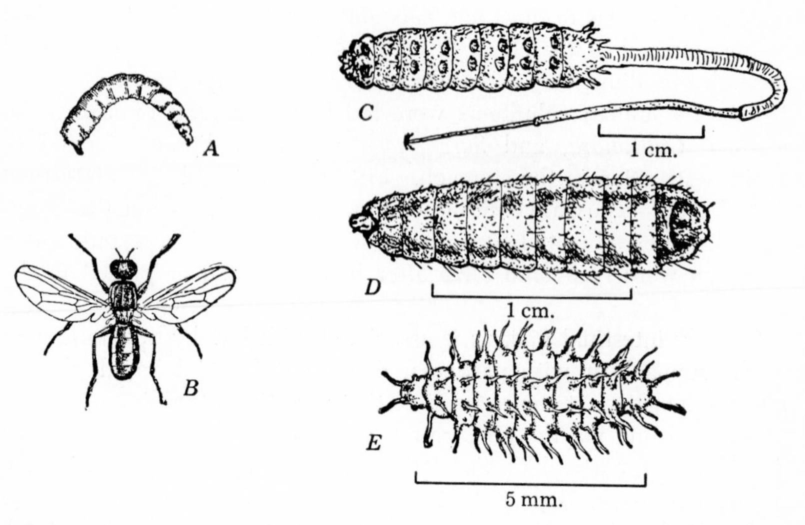

Fig. 253. Maggots occasionally found in human feces. *A* and *B,* cheese skipper and adult, *Piophila casei,* ×3. *C,* rat-tailed maggot, *Tubifera tenax. D, Hermetia illucens. E,* larva of lesser housefly, *Fannia canicularis.* (Adapted from various authors.)

In some cases, however, it is a problem for a psychiatrist rather than an entomologist. The senior writer was once informed by a woman that she had been passing worms in her stool for over a year and suffering gastrointestinal disturbances from them, and she brought a stool swarming with larvae of *Aphiochaeta* to prove it. She had been fascinated by seeing the larvae develop in stools saved in covered receptacles. When, however, a flytight jar was supplied for additional specimens, no more larvae were found.

The commonest fly maggots found in human feces are small species, about 5 mm. long, that breed in dead vegetable or animal matter, including *Piophila casei,* the cheese skipper (Fig. 253*A* and *B*); *Drosophila,* the fruit fly, famous in genetics; *Aphiochaeta; Sepsis;* etc. The larvae of most of these have the posterior spiracles on tubercles but otherwise resemble miniature housefly larvae. Some, like the cheese skipper, can flick themselves about; the presence of this species is sometimes considered a mark of particularly good cheese. Other maggots occasionally found and connected with digestive disorder are two species of *Fannia* (Fig. 253*E*); *Musca domestica;* several species of fleshflies (*Sarcophaga*) (Fig. 254); rat-tailed maggots (*Tubifera* (=*Eristalis*)) (Fig. 253*C*); and a soldier fly, *Hermetia illucens* (Fig. 253*D*). The characteristics of these will be found on pp. 768 and 771.

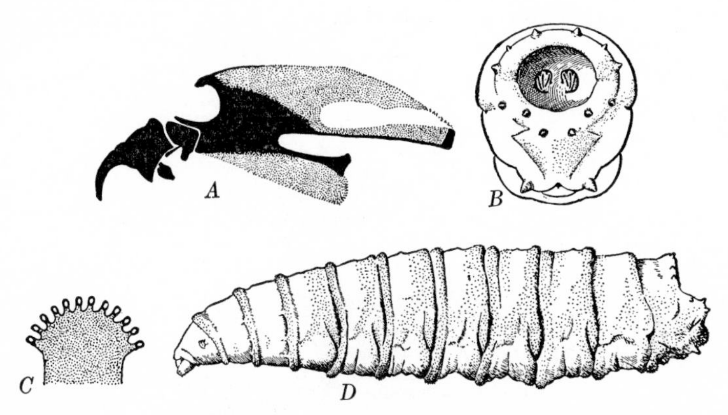

Fig. 254. Larva of *Sarcophaga crassipalpis,* a secondary wound invader. *A,* larva; *B,* posterior end of same showing stigmal plates in pit or recess; *C,* mouth hook and cephalo-pharyngeal skeleton characteristic of muscoid larvae; *D,* anterior spiracle. (*A* and *B* after James. *C* and *D* after C. N. Smith from James, *U. S. Dept. Agric. Misc. Publ.* 631, 1947.)

Although undoubtedly harmless in most cases, these fly maggots sometimes damage the intestinal mucosa and cause loss of appetite, diarrhea, vomiting, colicky pains, headache, vertigo, etc.

Fly maggots can usually be expelled readily by means of the purges and various anthelmintics used for intestinal worms. Prevention, of course, consists principally in being careful of what is eaten, especially in regard to such foods as raw vegetables, cheese, and partly decayed fruits and meats exposed to flies.

Myiasis of urinary passages. Myiasis of the urinary passages, both urethra and bladder, is a rare but occasional occurrence. The flies implicated are usually the lesser housefly, *Fannia canicularis,* and the closely allied latrine fly, *F. scalaris.* The senior writer in 1941 reported a case in which *Phaenicia* larvae were recovered. In most cases infection occurs from eggs laid near the external opening of the urethra, the larvae working their way up into this tube and even into the bladder; apparently they need very little oxygen. One case of infection of a boy's bladder by the larvae of a *Psychoda* was recorded by Patton; he thinks that the larvae burrowed through from the rectum to the bladder. Hoeppli and Watt found that larvae of *Chrysomyia megacephala* when placed in the urinary bladder taken from a freshly killed pig and filled with human urine, if fed daily, would live for 9 days, whereas *Phaenicia sericata* lived only 3 days. Contamination is favored by sleeping without covers in hot weather, giving flies free access to the anal and genital region.

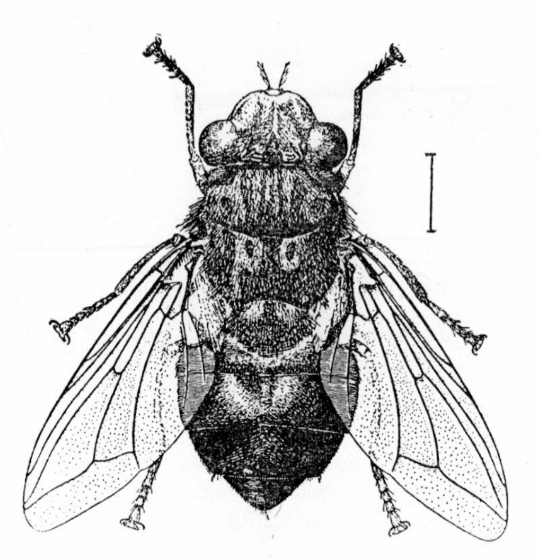

Fig. 255. *Dermatobia hominis* adult ♀. (After James, *U. S. Dept. Agric. Misc. Publ.* 631, 1947.)

BOTFLIES

1. Skin Bots (Cuterebridae)

The flies of this family are robust hairy flies distinguished from other bots (Oestridae and Hypodermatidae) by having a deep groove under the head containing a reduced proboscis.

Dermatobia hominis. This big, blue, brown-winged fly (Fig. 255) is found from Mexico to northern Argentina. Its larvae develop in many animals, including cattle, dogs, hogs, goats, turkeys, and, more rarely, horses, mules, or man. The maggots of this fly, called "berne" in Brazil and "torsalo" in Central America, are undoubtedly the most serious and damaging parasites of cattle in Central and South America.

They retard growth, lower meat and milk production, cause anemia and digestive disturbances, and riddle hides until they become worthless. Human infestations are contracted chiefly in low forest regions and seldom in houses. Young children exposed outdoors may be severely affected, and occasionally are killed by them, as are cattle. Many calves are killed by secondary infestation by screwworms (see p. 776).

The adult fly is about the size of a large blowfly, with the legs and face yellowish, the thorax bluish-black with a grayish bloom, the abdomen a beautiful metallic violet blue, and the wings brown.

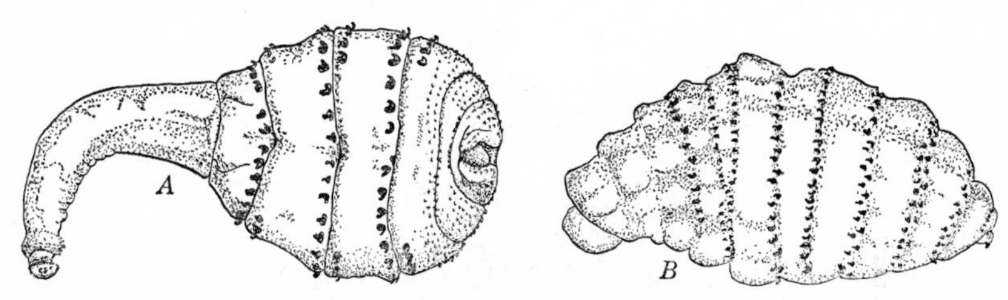

Fig. 256. *Dermatobia hominis: A,* first-stage larva; *B,* third-stage larva. (*A,* adapted from Blanchard, from Neveu-Lemaire, *Traité de zoologie médicale et vétérinaire.* II. *Entomologie,* Vigot Frères. *B,* original.)

The method by which these flies give their offspring a start in life is unique. When ready to oviposit, the female captures an insect, usually a large mosquito of the genus *Psorophora* but occasionally various other Diptera or even ticks, and glues her eggs by means of an adhesive, quick-drying cement to the underside of the abdomen of her captive (see p. 758 and Fig. 244). A total of 200 eggs may be laid by one female fly, 8 or 10 to several dozen on individual mosquitoes. The eggs require several days' incubation before they are ready to hatch.

When mosquitoes burdened with ripe eggs alight upon the skin of warm-blooded animals the maggots emerge, penetrate the skin of the host, and begin their development. If the young larva does not have time to emerge while its mosquito transporter is biting it is said to draw back into the egg shell and await another opportunity.

The larvae mature in the host's skin in 5 to 10 weeks. They ultimately reach a length of 18 to 24 mm. (Fig. 256). The anterior end of the larva is broad and is provided with double rows of thorn-shaped spines; the posterior portion is slender, smooth, and retractile. As the larva develops, a boil-like cyst forms about it, opening to the surface of the skin by a little pore which is plugged by the posterior end of the maggot and is used for obtaining air. At intervals these warblelike boils cause excruciating pain. When mature, the larvae voluntarily leave their host and fall to the ground to pupate. They transform into the adult form in the course of several weeks; the entire life cycle requires 3 to 4 months.

After the larvae have evacuated their cysts or have been removed, the wounds sometimes develop serious or even fatal infections, or are invaded by screwworms. To remove the maggots, frequently tobacco juice or tobacco ashes are applied to the infested spots, thus killing the worms and making their extraction easy. Another method used by natives in some parts of South America is to tie a piece of fat tightly over the entrance to the boil. The larva, deprived of air, works its way out into the fat, being thus induced to extract itself. A much more satisfactory method is to enlarge the entrance to the cyst with a sharp clean knife and remove the worm with a forceps. Antiseptic treatment of the wound obviates danger of subsequent infection. The wound heals quickly but leaves a scar. Treatment of *Dermatobia* lesions in animals with smear EQ 335 (see p. 778) or 4% lindane in lubricating grease and used engine oil is very effective.

Good control is obtained by spraying animals with chlorinated hydrocarbons twice monthly, or by means of arsenic dips. On one farm in Nicaragua about 850 *Dermatobia* larvae were extracted per animal during 9 months preceding treatment; 9 months *after* treatment there was less than 1 per animal and no *Boophilus* ticks (Laake, 1953). Hypodermic injection of 10 mg. per kilogram of lindane at intervals of 20 days, or addition of that amount to food at those intervals, reduced the infestation per animal very markedly (de Toledo and Sauer, 1950). Injection of certain organic phosphorus compounds at this rate at about 10 mg. per kilogram, or their oral administration at about 100 mg. per kilogram, which was found effective for *Hypoderma* and other ectoparasites in cattle, should be worth a trial.

***Cuterebra* spp.** This genus contains a number of species of large beelike flies, the larvae of which develop individually in the skin of rodents and rabbits, fairly frequently in cats, and rarely in dogs. The full-grown larvae are large robust maggots, sometimes over an inch long; they are easily recognized by their complete covering of black spines which gives them a jet black color. The younger instars have only rings of spines. Usually an animal harbors only one or a few maggots, but occasionally there are more.

According to Dalmat (1942) the flies lay very large numbers of eggs, depositing them in the burrows or habitats of the host, where they hatch intermittently. The larvae attach themselves to a host when the opportunity comes. They live in the host about a month and have a long pupal period in the soil; probably there is usually only one brood a year.

The maggots seem to be definitely injurious to their hosts, sometimes causing parasitic castration. *Cuterebra* lesions in cats and dogs are remarkably dirty and persistent. Most screwworm infections in rabbits develop in *Cuterebra* sores. Infested squirrels are considered inedible by hunters and are thrown away.

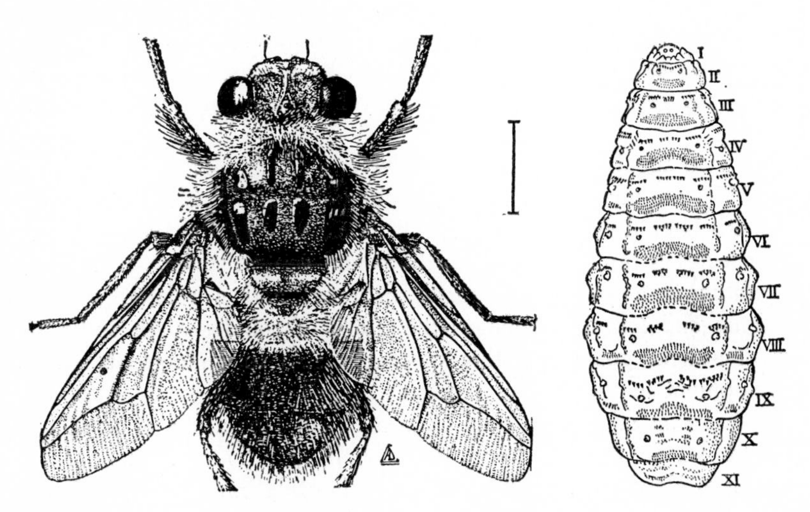

Fig. 257. *Left,* common cattle grub or warble fly, *Hypoderma lineata,* adult ♀ (after James, *U. S. Dept. Agric. Misc. Publ.* 631, 1947). *Right,* third-stage larva of same (after Cameron, *Trans. Highland and Agric. Soc. Scotland,* Ser. 5, 49, 1937, from James).

2. Bots Causing Warbles in Cattle (*Hypoderma* spp.)

Hypoderma. The warble or heel flies, *H. bovis* and *H. lineata* (see key, p. 769, and Fig. 257), are hairy, black and yellow flies which lay their eggs on the hairs of the lower part of the legs or flanks of cattle, causing them great annoyance. The annoyance is purely instinctive, for the flies do not bite or sting, yet the animals act terror-stricken. One fly may deposit 100 or more eggs on one animal. The eggs hatch in a few days; the spiny larvae, 1 mm. long, burrow into the skin, and then for several months they ramble about among the viscera in the abdomen and thorax. Those of *H. lineata* commonly invade the walls of the esophagus and only occasionally enter the spinal canal, whereas for *H. bovis* the reverse is true. During this migratory phase the larvae are glassy smooth and grow to a length of 12 mm.

Toward the end of the winter they begin to appear in the skin of the back, where they form little cystlike lumps or warbles. The larvae (Fig. 257, *right*) develop in the warbles for about 1 to 3 months, molt twice, and become opaque, warty maggots which make a little breathing hole in the skin, into which they thrust the posterior end with its spiracles. The warbles usually appear in early December in the South, in mid-February in the North. The grubs usually emerge in spring or early summer, fall to the ground, and pupate. The pupae are very resistant to cold but are killed by excessive moisture. The adults emerge in 2 to 7 weeks. They have a flight range of about 3 miles.

The two species in cattle, *H. bovis* and *H. lineata* (see keys and Fig. 247), are much alike in their biology, but *H. bovis* appears about a month later than *H. lineata* in all its stages. In the United States this species is limited to the northern and central states. In Europe *H. diana* is a common parasite of deer, and a related species, *Oedemagena tarandi,* damages the hides of reindeer. In India *H. crossii* attacks goats, but this species is said to undergo its entire development in the skin of the back.

Hypoderma larvae are occasional accidental parasites of man, but being in an abnormal host they do not behave in a normal manner but wander aimlessly in the skin, causing "migrating lumps." *Hypoderma* larvae are the commonest species that invade the interior of human eyes; possibly the flies lay their eggs on the eyelashes or brows; usually the cases come to the attention of ophthalmologists in winter or early spring, long after the fly season (Krümmel and Brauns, 1956). *Hypoderma* has been reported to cause myiasis of the human eye in Norway and Russia.

Warbles cause losses amounting to millions of dollars by irritation to cattle caused by the larvae, annoyance caused by the flies, and damage to hides.

Since there is no wild animal reservoir for the cattle grubs, community effort has brought about great reduction of them. By continued effort and broadening of control areas, extermination would be possible, and then *Hypoderma,* like *Boophilus,* would be only a memory in this country.

For external treatment derris or cubé powder containing 5% rotenone is rubbed into the warbles, or the backs of the animals are sprayed with a rotenone emulsion (12 ounce 5% powder per gallon, plus a detergent and Toluene). The applications are made 30 days before the time for the warbles to appear, and then every 30 days during the season. Such treatments, however, fail to prevent injury to the hides. Oral administration of insecticides in capsules or in food, or subcutaneous injections, destroy the grubs during their internal migration. Certain of the organic phosphorus compounds in capsules at about 100 mg. per kilogram given orally twice, 3 months apart, kill the larvae before they injure the hide but does not kill them after establishment in the warbles. Such treatments are also effective against most ectoparasites and also *Dermatobia*, but the possibility of toxic effects has not yet been entirely eliminated. Addition of dieldrin or lindane to cattle feed at 100 ppm, or of phenothiazine in salt mixtures are also effective; according to Russian reports, subcutaneous injections of 10% aqueous suspensions of phosphorus compounds at about 10 mg. per kilogram at appropriate times 30 days apart also kill the larvae before the warbles develop; the injections are said to be harmless, and not to appear in milk.

3. Head Bots (Oestridae and Cuterebridae)

The family Oestridae contains robust flies with a hairy "pile" of black, yellow, or gray. These, while on the wing, deposit their newly hatched larvae (or eggs in some species) in or on the nostrils of sheep, goats, deer, and camels, or, rarely, in man. The larvae are large grubs an inch or more long. The important species are *Oestrus ovis*, a world-wide parasite of sheep and goats, *Rhinoestrus purpureus* of horses in the Old World, *Cephalopina titillator* of camels in North Africa and Asia, and *Cephenemyia* spp. of members of the deer family in North America and Europe.

Oestrus ovis. This grayish-brown fly, imported from Europe, is a pest in southwestern United States, where over 95% of sheep and goats are infected. Sheep have an instinctive fear of the flies; they become nervous and collect in groups with their noses close to the ground, and stop feeding, with great loss of weight.

The flies deposit their larvae in the nares, but after 2 or 3 weeks the larvae invade the frontal sinuses and sometimes other parts of the head. In 2 or 3 more weeks in summer they are full-grown inch-long grubs which work their way back into the nostrils, drop out, and pupate in the ground to emerge 4 to 6 weeks later as adult flies. Other half-grown larvae enter the sinuses vacated by the grown grubs, and so the process continues as long as warm weather lasts. The grown larvae do not survive freezing, but nature has provided a clever protection against that hazard. When cold air enters the nostrils, the young larvae become dormant and remain so until the next spring, when they migrate into the sinuses and finish out the life cycle. The full-grown third-stage larvae are recognizable by their characteristic spiracles (Fig. 247). The larvae cause a profuse discharge of mucus (snotty nose) which causes considerable distress in old or weak sheep, but little in healthy ones.

This fly sometimes deposits its eggs on the eyes, nostrils, and lips of shepherds whose breath smells of fresh sheep or goat cheese or curds. These grubs, unlike *Hypoderma,* are rarely found inside human eyes but on the surface of the conjunctiva where the young larvae, 1 mm. long, are deposited. The pain compels prompt treatment.

Effective treatment is possible only against the young larvae still in the nares, and then only in cold climates where the dormant larvae can be killed in late fall or winter. Recommended treatments are irrigation of the nares with 3% lysol, with the help of special equipment (Cobbett, 1956), or spraying the nostrils with 10 ml. of emulsified ether extract of male fern (see Lindquist and Knipling, 1957).

Other head bots. The larvae of the purplish-hued *Rhinoestrus purpureus* are frequent parasites of the heads of horses in Europe, Siberia, and North Africa. Like the sheep bot, this fly sometimes darts at human beings and deposits its larvae in the eyes, where they may cause serious damage if not promptly removed. One oestrid, *Booponus intonosus,* the hoof maggot of Celebes and the Philippines, lays its eggs on the feet of cattle and the larvae develop there. *Cephalopina titillator* causes great discomfort to camels; it lives in the nostrils and nasopharynx for 10 or 11 months.

The deer head bots, *Cephenemyia* spp. (family Cuterebridae), are well known to hunters. Nearly all deer are infested by them, but they seem usually to do little damage, though sneezing fits are sometimes observed as are occasional cases of "craziness," possibly due to rare penetration of the parasites into the brain. The adults of *Cephenemyia* are among the swiftest flying insects known; they are said to get up a speed of 800 miles an hour! Females are rarely seen. The males of our western *C. jellisoni* rest on sun-warmed rocks on inaccessible mountaintops. Bagging one is more of a feat for a hunter than getting a mountain goat or a condor; when found in their remote retreats they have to be shot with .22-caliber dust shells!

4. Horse Bots (*Gasterophilus* spp.)

The genus *Gasterophilus,* constituting a separate family Gasterophilidae (see key, p. 769), contains flies, the larvae of which develop

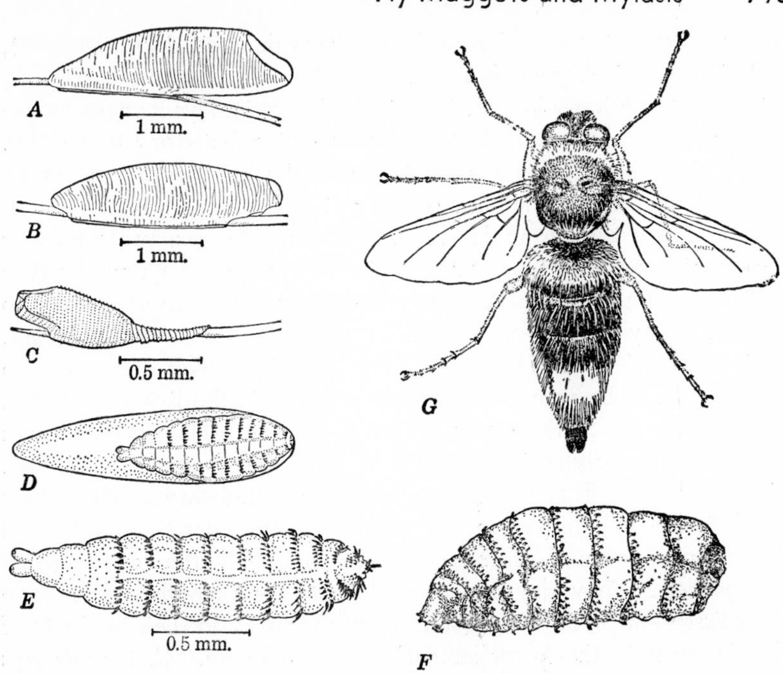

Fig. 258. *Gasterophilus.* *A*, egg of *G. intestinalis;* *B*, egg of *G. nasalis;* *C*, egg of *G. haemorrhoidalis;* *D*, egg membrane enclosing unhatched larva of *G. intestinalis;* *E*, first-stage larva of same; *F*, third-stage larva of same; *G*, *G. haemorrhoidalis.* (*A–E* and *G* adapted from Hadwen and Cameron, *Bull. Ent. Research,* 9, 1918.)

in the stomach or rectum of horses. The adults are hairy, beelike flies, clothed in dark brown or black with yellow markings, and one species has an orange-red tip to the abdomen (Fig. 258, *G*). The wings of the commonest species, *G. intestinalis,* have smoky markings (Fig. 246, *6*). The abdomen in the females is elongated.

Each of the species has somewhat different habits, but all cause an amazing amount of annoyance while laying eggs, the horses becoming excited and often frantic; on warm days a horse may be so worried fighting botflies (gadflies) that he cannot graze at all. The adults usually live for only about 3 to 10 days, but during that time a female lays several hundred eggs which she attaches to hairs on parts of the horse which vary with the insect species. *G. percorum* is an exception (see below). The first-stage larvae (Fig. 258, *E*) have rings of black spines; third-stage larvae (Fig. 258, *F*) are heavy-bodied and rather squarish posteriorly; all but *inermis* have heavy spines on some of the segments—one row in *nasalis,* two in the others. In *G. intestinalis* the spines of the first row are larger than those of the second (Fig. 258, *F*), and vice versa in *haemorrhoidalis.*

The common horse bot, *G. intestinalis* (=*equi*), lays its eggs on the hairs of the forelegs of the horse, where they incubate for 1 to 2 weeks. When ripe, the warmth and moisture of the animal's tongue when licking cause them to hatch and adhere to the tongue. In the mouth they excavate tunnels under the mucous membranes, principally of the tongue, and after 3 or 4 weeks migrate to the stomach, where they live in the lumen, often in large colonies, until mature and ready to pupate. They then release their hold and are passed in the droppings.

G. haemorrhoidalis (nose fly) (Fig. 258, *G*) strikes at the lips to lay its eggs. The young larvae burrow about in the lips and tongue, later developing in the stomach and duodenum. This species leaves these parts in early spring and finishes its development in the rectum. *G. nasalis* (chin fly) lays its eggs on the chin and throat, where they hatch unaided. The larvae crawl to the lips to enter the mouth, and invade spaces around and between the teeth, below the gums, causing pus pockets. They may live in this locality for a month and undergo their first molt before they continue on their way to their final site of development in the lower stomach and duodenum. *G. inermis* lays its eggs on the cheeks, where it causes a dermatitis; the larvae burrow through the tissues to the mouth and then go to the rectum. *G. pecorum* lays its eggs on the hoofs sometimes, but usually on food or in pastures. When the eggs are ingested they hatch in the mouth, burrow into the mucosa, make their way to the esophagus and stomach, and eventually go to the rectum and reattach before finally leaving the body.

A few horse bots do very little damage, and some farmers think that a horse just naturally ought to have a few, but when numerous the bots cause gastrointestinal disturbances. The worms in the stomach are persuaded to let go by giving carbon bisulfide in gelatin capsules at the rate of 1.5 drams per 250 lb. of horse. Toluene given by stomach tube (10 ml. per 100 lb.), being tested against *Ascaris* in horses, was found by Todd et al. to be very effective against *G. intestinalis* but less so against *G. nasalis.*

Gasterophilus occasionally penetrates into man, but instead of behaving in an orthodox manner, the larva wanders about under the skin and is called a "larva migrans." *G. intestinalis* is the most common species concerned in this.

REFERENCES

Aubertin, D., and Buxton, P. A. 1934. *Cochliomyia* and myiasis in tropical America. *Ann. Trop. Med. Parasitol.,* 28: 245–254.

Bennett, G. F. 1955. Studies on *Cuterebra emasculator* Fitch, 1856, and a discussion of the status of the genus *Cephenemyia. Can. J. Zool.,* 33: 75–98.

Bishopp, F. C. 1941. The horse bots and their control. *U. S. Dept. Agr. Farmers' Bull.* 1503.

Bishopp, C., Laake, E. W., Brundrett, H. M., and Wells, R. W. 1927. The cattle grubs or ox warbles, their biologies and suggestions for control. *U. S. Dept. Agr. Bull.* 1369.

Blacklock, D. B., and Gordon, R. M. 1927–1930. The experimental production of immunity against metazoan parasites and an investigation of its nature. *Ann. Trop. Med. Parasitol.,* 21: 181–224; 24: 5–54.

Blacklock, D. B., and Thompson, M. G. 1923. A study of the Tumbu fly, *Cordylobia anthropophaga,* in Sierra Leone. *Ann. Trop. Med. Parasitol.,* 17: 443–510.

Borgstrom, F. 1938. Experimental *Cochliomyia americana* infestations. *Am. J. Trop. Med.,* 18: 395–411.

Bushland, R. C., and Hopkins, D. E. 1951. Experiments with screwworm flies sterilized by X-rays. *J. Econ. Entomol.,* 44: 725–731.

Cobbett, N. G. 1956. Head grubs of sheep. In *Annual Diseases, Yearbook of Agriculture,* pp. 407–411. Government Printing Office, Washington.

Dalmat, H. T. 1943. A contribution to the knowledge of the rodent warble flies. *J. Parasitol.,* 29: 311–318.

Dove, W. E. 1937. Myiasis of man. *J. Econ. Entomol.,* 30: 29–39.

Esslinger, J. H. 1958. Host-parasite relations of the screw-worm *Callitroga hominovorax.* Ph.D. Thesis, Rice Institute.

Ford, N. 1936. Further observations on the behavior of *Wohlfartia vigil,* with notes on the collecting and rearing of the flies. *J. Parasitol.,* 22: 309–328.

Garret-Jones, C. 1951. The Congo floor maggot, *Aucheromyia luteola* (F.) in a laboratory culture. *Bull. Entomol. Research,* 41: 679–708.

Gassner, F. X., and James, M. T. 1948. The biology and control of the fox maggot, *Wohlfartia opaca* (Coq.). *J. Parasitol.,* 34: 44–50.

Hadwen, S., and Cameron, A. E. 1918. A contribution to the knowledge of the botflies, *Gasterophilus intestinalis* DeG., *G. haemorrhoidalis* L., and *G. nasalis* L. *Bull. Entomol. Research,* 9: 91–106.

Hall, D. G. 1948. *The Blowflies of North America.* Thomas Say Foundation, Baltimore.

Herms, W. B., and Gilbert, Q. O. 1933. An obstinate case of intestinal myiasis. *Ann. Internal Med.,* 6: 941–945.

James, M. T. 1947. The flies that cause myiasis in man. *U. S. Dept. Agr., Misc. Publ.,* 631.

Kenney, M. 1945. Experimental intestinal myiasis in man. *Proc. Soc. Exp. Biol. Med.,* 60: 235–237.

Knipling, E. F., and Rainwater, H. T. 1937. Species and incidence of Diptera concerned in wound myiasis. *J. Parasitol.,* 23: 451–455.

Krümmel, H., and Brauns, A. 1956. *Myiasis des anges.* Duncker and Humblot, Berlin (*or Zeitsch. f. angewandte Zool.,* Heft 2, 1956).

Laake, E. W. 1953. Torsalo and tick control with toxaphene in Central America. *J. Econ. Entomol.,* 46: 454–458.

Laake, E. W., Cushing, E. C., and Parish, H. E. 1936. Biology of the primary screwworm fly, *Cochliomyia americana,* and a comparison of its stages with those of *C. macellaria. U. S. Dept. Agr. Tech. Bull.* 500.

Micks, D. W., and McKibbin, J. W. 1956. Report of a case of human intestinal myiasis caused by *Leptocera venalicia. Am. J. Trop. Med. Hyg.,* 5: 929–932.

Miller, M. F., and Lockhart, J. A. 1950. Hypodermal myiasis caused by larvae of the ox-warble (*Hypoderma bovis*). *Can. Med. Assoc. J.,* 62: 592–594.

Paramonow, S. J. 1949. Bestimmungstabelle sämtlicher Entwicklungstadien der Magendasseln (Fam. Gasterophilidae; Dipt.) *Z. Parasitenk.,* 14: 27–37.

Parman, D. C. 1945. Effect of weather on *Cochliomyia americana* and a review of methods and economic applications of the study. *J. Econ. Entomol.,* 38: 66–76.

Pfadt, R. E. 1947. Effects of temperature and humidity on larval and pupal stages of the common cattle grub. *J. Econ. Entomol.,* 40: 293–300.

Roberts, I. H., and Lindquist, A. W. 1956. Cattle grubs. In *Animal Diseases, Yearbook of Agriculture,* pp. 300–306. Govt. Printing Office, Washington.

Sambon, L. W. 1922. Tropical and subtropical diseases (*Dermatobia*). *J. Trop. Med. Hyg.,* 25: 170–185.

Scharff, D. K. 1950. Cattle grubs—their biologies, their distribution and experiments on their control. *Bull. Montana Agr. Exptl. Sta.* No. 47.

Théodoridès, J. 1948. Les coléoptères parasites accidentels de l'homme. *Ann. parasitol. humaine et comparée,* 23: 348–363.

de Toledo, A. A., and Sauer, H. F. G. 1950. Efeito de alguns inseticidas clorados sôbre or berne. *Biologico,* 16: 25–34.

SOURCES OF INFORMATION

The following periodicals are those in which a very considerable part of the literature on parasitology can be found, either in original or abstract form, and which are therefore desirable in libraries where parasitological study or research is being carried on. The starred periodicals are important for their abstracts or references. Important general books on parasitology are listed at the end of Chapter 1.

United States and Canada

American Journal of Hygiene, Baltimore, 1921–
American Journal of Public Health, New York, 1911–
American Journal of Tropical Medicine, Baltimore, 1921–1951.
American Journal of Tropical Medicine and Hygiene, Baltimore, 1952–
American Journal of Veterinary Research, Chicago, 1940–
American Midland Naturalist, Notre Dame, Indiana, 1909–
**Biological Abstracts,* Philadelphia, 1926–
Canadian Journal of Research (Sect. C–D, from Vol. 13), Ottawa, 1929–
Cornell Veterinarian, Ithaca, N. Y., 1911–
Experimental Parasitology, New York, 1951–
**Index Medicus,* Washington, 1879–
**Journal of the American Medical Association,* Chicago, 1883–
Journal of the American Veterinary Medicine Association, New York, 1877–
Journal of Economic Entomology, Menasha, Wisconsin, 1908–
Journal of the National Malaria Society, Tallahassee, 1942–1951.
Journal of Parasitology, Lancaster, Pa., 1914–
North American Veterinarian, Evanston, Ill., 1920–
Proceedings of the Society for Experimental Biology and Medicine, N. Y., 1903–
Proceedings of the Helminthological Society of Washington, Washington, 1934–
Public Health Reports, Washington, 1878–
**Quarterly Cumulative Index Medicus,* Chicago, 1927–
Rockefeller Foundation, International Health Division, *Annual Reports,* N. Y., 1913–
Rockefeller Institute for Medical Research, *Monographs,* New York, 1910–
Transactions of the American Microscopical Society, Columbus, Ohio, 1892–
U. S. Bureau of Animal Industry, *Bulletin* No. 1–167, Washington, 1893–1914.
U. S. Bureau of Entomology, *Bulletin* No. 1–33, new series No. 1–127, Washington, 1883–1916.
U. S. Department of Agriculture, Farmer's Bulletins, Circulars, Leaflets, Technical Bulletins, and Yearbooks of Agriculture, Washington.
U. S. Hygienic Laboratory, *Bulletins* (continued as *National Institutes of Health Bulletins*), Washington, 1900–
Zoonoses Research, Lyceum Press, New York, 1960–

Foreign

Acta Medica Scandinavica, Stockholm, 1919–
Acta Tropica, Basel, 1944–

Anales del Instituto de Biologia, Mexico City, 1930–
Annales de la société belge de médecine tropicale, Antwerp, 1920–
Annales de parasitologie humaine et comparée, Paris, 1922–
Annals of Tropical Medicine and Parasitology, Liverpool, 1907–
Archiv für Schiffs- und Tropen-Hygiene, Leipzig, 1897–1945.
Archives de parasitologie, Paris, 1898–1919.
Boletin de la oficina sanitaria panamericana, Washington, 1920–
Brasil-Medico, Rio de Janeiro, 1887–
British Medical Journal, London, 1857–
Bulletin de la société de pathologie exotique, Paris, 1908–
Bulletin of Entomological Research, London, 1910–
China Medical Journal, Shanghai, 1887–1931, continued as *Chinese Medical Journal,* 1932–1940.
Comptes rendus de la société de biologie, Paris, 1849–
**Excerpta medica,* Sect. IV, Medical Microbiology and Hygiene. Amsterdam, 1948–
Geneeskundig Tijdschrift voor Nederlandsch-Indie, Batavia, 1852–1941.
**Helminthological Abstracts,* St. Albans, England, 1932–
Indian Journal of Medical Research, Calcutta, 1913–
Indian Medical Research Memoirs, Calcutta, 1924–
Journal of Helminthology, London, 1923–
Journal of the London School of Tropical Medicine, London, 1911–1913.
Journal of Tropical Medicine and Hygiene, London, 1898–
Lancet, London, 1823–
Medical Journal of Australia, Sydney, 1914–
Memoirs, Liverpool School of Tropical Medicine, 1901–1906, 1924–
Memoirs do instituto Oswaldo Cruz, Rio de Janeiro, 1909–
Parasitology, Cambridge, 1908–
Philippine Journal of Science, Manila, 1906–
Publicaciones, Instituto de parasitologia y enfermedades parasitarias, Universidad nacional, Buenos Aires, 1938–
Publicaciones, Mision de estudos de patologia regional argentina. Jujuy, 1930–
Puerto Rico Journal of Public Health and Tropical Medicine, San Juan, 1925–
**Review of Applied Entomology,* Series B (Medical and Veterinary), London, 1913–
Revista brasileira de malariologia e doenças tropicais, Rio de Janeiro, 1949–
Revista de medicina tropical y parasitologia, clinica y laboratorio, Havana, 1935–
Revista de medicina veterinaria y parasitologia, Caracas, 1944–
Revista iberica de parasitologia, Madrid, 1940–
Revista Kuba de medicina tropical y parasitologia, Havana, 1945–
Revista serviço especial de saúde pública, Rio de Janeiro, 1947–
Revue pratique des maladies des pays chauds, Paris, 1922–1940.
Rivista de parassitologia, Rome, 1937–
Transactions of the Royal Society of Tropical Medicine and Hygiene, London, 1907–
**Tropical Diseases Bulletin,* London, 1913–
**Tropical Veterinary Bulletin,* London, 1912–1930, succeeded by *Veterinary Bulletin, Weymouth,* 1931–
World Health Organization, *Monograph Series, Technical Reports Series, Reports of Expert Committees, Bulletin,* and *Chronicle,* Geneva.

Zeitschrift für Parasitenkunde, Berlin, 1928–
(Vols. 1 to 6 as *Zeitschrift für Wissenschaftliche Biologie,* Abt. F.)
Zeitschrift für Tropenmedizin und Parasitologie, Stuttgart, 1950–
**Zentralblatt für Bakteriologie und Parasitologie,* I Abteilung, Original und Referat, Jena, Germany, 1887– (Referat contains references and reviews of many articles dealing with infectious diseases.)

INDEX